MW01633444

INTRODUCTION TO MARCUS THEORY OF ELECTRON TRANSFER REACTIONS

INTRODUCTION TO MARCUS THEORY OF ELECTRON TRANSFER REACTIONS

Francesco Di Giacomo
Sapienza University of Rome

World Scientific

NEW JERSEY • LONDON • SINGAPORE • BEIJING • SHANGHAI • HONG KONG • TAIPEI • CHENNAI • TOKYO

Published by

World Scientific Publishing Co. Pte. Ltd.
5 Toh Tuck Link, Singapore 596224
USA office: 27 Warren Street, Suite 401-402, Hackensack, NJ 07601
UK office: 57 Shelton Street, Covent Garden, London WC2H 9HE

British Library Cataloguing-in-Publication Data
A catalogue record for this book is available from the British Library.

INTRODUCTION TO MARCUS THEORY OF ELECTRON TRANSFER REACTIONS

Copyright © 2020 by World Scientific Publishing Co. Pte. Ltd.

All rights reserved. This book, or parts thereof, may not be reproduced in any form or by any means, electronic or mechanical, including photocopying, recording or any information storage and retrieval system now known or to be invented, without written permission from the publisher.

For photocopying of material in this volume, please pay a copying fee through the Copyright Clearance Center, Inc., 222 Rosewood Drive, Danvers, MA 01923, USA. In this case permission to photocopy is not required from the publisher.

ISBN 978-981-120-846-1 (hardcover)
ISBN 978-981-120-847-8 (ebook for institutions)
ISBN 978-981-120-848-5 (ebook for individuals)

For any available supplementary material, please visit
https://www.worldscientific.com/worldscibooks/10.1142/11503#t=suppl

Typeset by Stallion Press
Email: enquiries@stallionpress.com

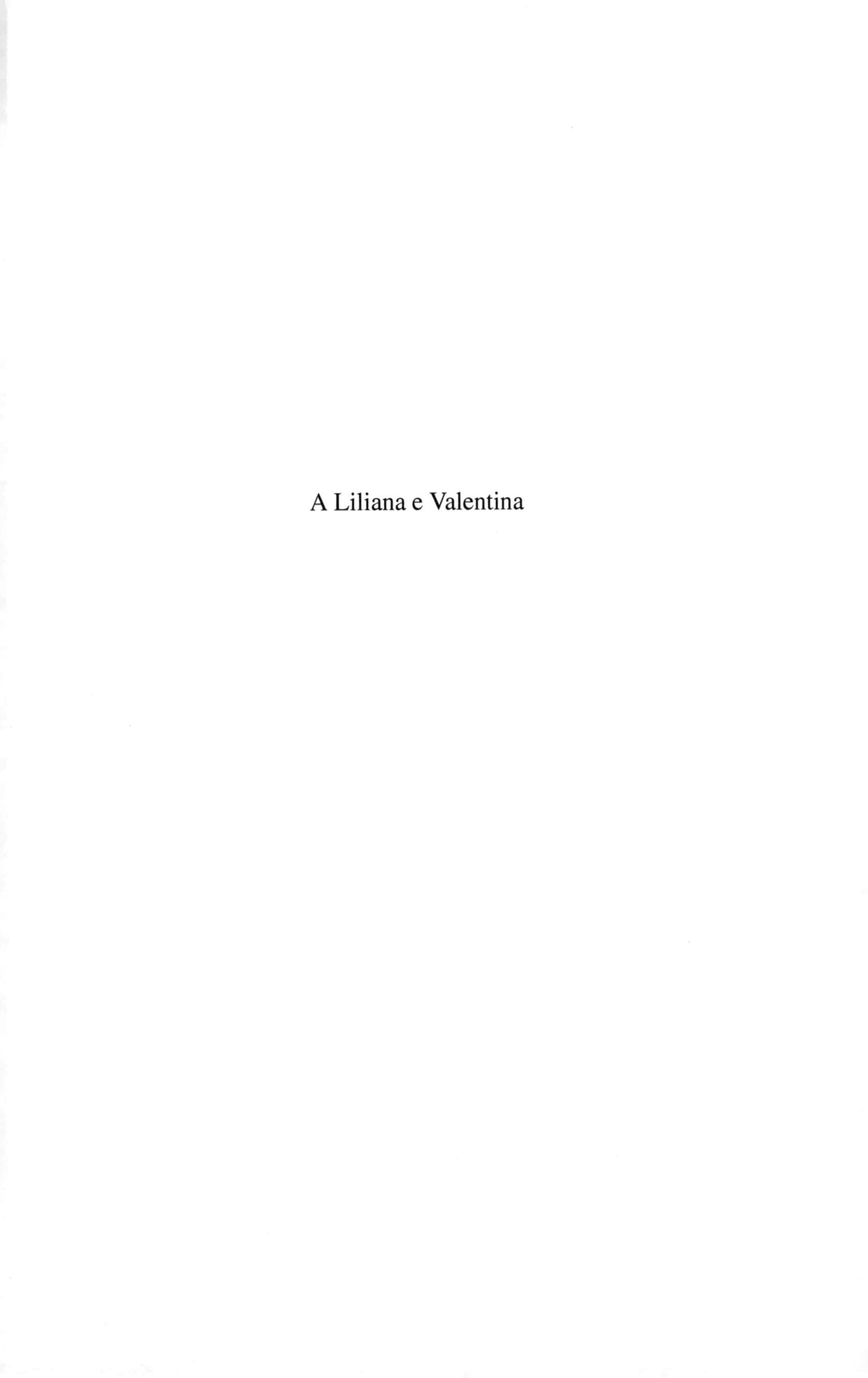

A Liliana e Valentina

Foreword

The author of this volume, Professor Francesco Di Giacomo, a member of the engineering faculty of the University of Rome, was also, some 40+ years ago, a visitor in my group, at that time at the University of Illinois. More recently he set himself the daunting task of learning about the theory of electron transfer reactions, with a background that included some research on the quantum mechanics of potential energy surfaces but none on key aspects of electron transfer theory—the dynamical features of quantum and classical theory, the statistical mechanics of chemical reactions, and features such as nonequilibrium dielectric polarization. They form the core of the theory and its applications. To this end he supplemented his study of my electron transfer articles by many hour-long question and answer sessions that we held, largely via Skype. Francesco comes with a delightful background rich in cultural and scientific history, especially of the ancient Greek and Roman periods. It added another side to our many discussions, an aspect that does not come out in this book but is seen in articles he has published elsewhere. Our discussions on the electron transfer articles involved many fine points of the theory and of its history, points that are described in this book. With his background Professor Di Giacomo's scientific questions have a fresh and novel point of view that was both interesting and challenging.

It adds to the novelty of the book. Hopefully the readers will find it equally so, in their study of the theory, of its history and in exploring the many distinct concepts that electron transfer processes present.

Rudolph A. Marcus
California Institute of Technology, March 2019

Preface

To learn about the theory of electron transfer reactions, I have followed, with David Mermin, "the time honored method of writing a book," systematically interviewing the most famous scientist in the field, Professor Rudolph A. Marcus for whom I had the honor of being Research Associate at the University of Illinois at Urbana-Champaign. Most of the interviews took place through Skype connections between Rome and Pasadena, but some occurred at Caltech, and the first one happened on board of a Southwestern Airlines flight between Oakland and Burbank airports, while we were coming back to Pasadena from San Francisco. There, in October 2006, a Special Symposium of the American Chemical Society had been held to celebrate the 50th year from the publication of Marcus' first paper on the theory, back in 1956.

I have written this book mainly to teach myself and I believe it may be useful for other beginners. The book aims at introducing the Marcus Theory of Electron Transfer Reactions from reading of Marcus' papers. Only excerpts from each paper are cited which are important at first reading. The layout is, when possible, like that of lecture notes, more "open" and reader friendly than the layout of the original paper. In the text simple notes of the author are presented which may be of help to the student or the beginner. At the end of each paper are Marcus' Notes with his comments and remarks, an accompaniment interesting to students and scholars in the field of electron transfer reactions. Finally, during the transcription of excerpts from the original papers, typos and small errata have been corrected.

In this book, 57 papers on the Marcus theory of electron transfer are considered, spanning the 30 years from 1956 to 1986. After an introductory chapter by the author, Marcus' papers are described, either closely following their content or summarizing and abridging the original text. At the end of the first eight chapters, there are lists of references, in the subsequent three chapters only references to Marcus' papers are reported in the form **MN** where **M** stands for Marcus and **N** is the order number in the official list of Marcus' publications. The last two chapters report oral interviews with Marcus with notes on his papers.

I shall consider this work successful if it will help the reader to better appreciate the Marcus theory of electron transfer from its primary source, Marcus' papers. A following book will contain oral interviews with Marcus with notes on his papers ranging from 1987 onward, up to the most recent ones.

F. Di Giacomo
Roma 2019

Scheme of the Book, Figures, Notes, and Abbreviations

The book has the format of a textbook but is not a textbook. It is a collection of excerpts from many of Marcus' papers with explanatory notes from the author and, much more importantly, with notes by Marcus on his papers.

The book may of course be read from cover to cover as a textbook, but it is more in general intended for readers interested in particular papers of Marcus to get an introduction to them and to read the related Marcus' notes. It is an anthology of Marcus' papers. Passages from the papers are reported in the book as they appear in the original publication. It so happens then that a topic may be treated in different papers in a similar way so that in the book there are sections which may be partly repetitive because Marcus normally treats the same topic in slightly different ways in his papers, moreover the sections are reported with figures from the original papers because text, figures, and Marcus's notes are closely connected. The ancient Roman dictum "Repetita iuvant," it helps to repeat, may have some truth in it, particularly for beginners.

The Figures from Marcus' papers appear with their original numbers and legends. The numbers of Figures from the author are followed by an asterisk, like, say, Fig. 1*. They have no legends, their meaning being explained in the text.

Excerpts from Marcus' papers are reported in quotes. Explanatory notes and remarks from the author are designated by NOTE, Notice or author's comment. Notes appearing in the papers are denoted by Note followed by the number of the note in the paper. Marcus' **NOTES** are reported at the end of the chapters or of the papers. Marcus is often indicated by M. or **M**. The symbol of the splitting at the avoided crossing region may be S_c, $2\varepsilon_{12}$, V_{12} or $2\Delta\varepsilon$. The author's questions in the **NOTES** may be numbered like (1), (2)... and each question may be divided in subquestions (i), (ii), and so on.

Abbreviations

AT atom transfer
EE electron exchange reaction
ET electron transfer
FC Franck Condon factor
PEC Potential energy curve
PES potential energy surface
PT proton transfer
TS transition state
TST transition state theory

Acknowledgments

This book would have never been written without input and help by Professor Marcus. Professor Jay Winkler from Caltech is gratefully acknowledged for his words of appreciation and encouragement after having read the Preface and the first Chapter. Doctor Marshall Newton of Brookhaven National Laboratories has been very helpful with his comments, review, and his kind words of appreciation for this work. Very special thanks are deserved to Professor Robert J. Cave of Harvey Mudd College for his most helpful very detailed review, his comments, and friendly advice. The line drawings of my figures have been beautifully drawn by Architect Andrea Cataldi.

Contents

List of 42 Marcus' Papers Closely Followed, Summarized or Abridged in Chapters 2–11 and of 10 Papers Considered in the Oral Interviews of Chapters 12 and 13

Chapter 1

Introduction

Chapter 2

M16. R. A. Marcus, On the Theory of Oxidation-Reduction Reactions Involving Electron Transfer. I. *J. Chem. Phys.* **24**, 966–978, (1956).

M17. R. A. Marcus, Electrostatic Free Energy and Other Properties of States Having Nonequilibrium Polarization. I. *J. Chem. Phys.* **24**, 979–989, (1956).

Chapter 3

M19. R. A. Marcus, On the Theory of Oxidation-Reduction Reactions Involving Electron Transfer. II. Applications to Data on the Rates of Isotopic Exchange Reactions. *J. Chem. Phys.* **26**, 867, (1957).

M20. R. A. Marcus, On the Theory of Oxidation-Reduction Reactions Involving Electron Transfer. III. Application to Data on the Rates of Organic Redox Reactions. *J. Chem. Phys.* **26**, 872, (1957).

M21. R. A. Marcus, On the Theory of Oxidation-Reduction Reactions and of Related Processes. *Trans. N.Y. Acad. Sci.* **19**, 423, (1957).

M28. R. A. Marcus, On the Theory of Electrochemical and Chemical Electron Transfer Processes. *Can. J. Chem.* **37**, 155, (1959).

M33. R. A. Marcus, A Theory of Electron Transfer Processes at Electrodes, in *Transactions of the Symposium on Electrode Processes*, p. 239, E. Yeager, Editor, Wiley, New York (1961).

M41. R. A. Marcus, On the Theory of Oxidation-Reduction Reactions Involving Electron Transfer. V. Comparison and Properties of Electrochemical and Chemical Rate Constants. *J. Phys. Chem.* **67**, 853, 2889, (1963).

M51. R. A. Marcus, Electron Transfer at Electrodes, in *Encyclopaedia of Electrochemistry*, p. 529, C. A. Hampel, Editor, Reinhold, New York (1964).

Chapter 4

M129. R. A. Marcus, On the Theory of Overvoltage for Electrode Processes Possessing Electron Transfer Mechanism. I. (ONR, Report No. 12, dated 1957), in *Special Topics in Electrochemistry*, p. 181, P. A. Rock, Editor. Elsevier, New York (1977).

M130. R. A. Marcus, Electrostatic Free Energy and Other Properties of States Having Nonequilibrium Polarization. II. Electrode Systems. (ONR Report No. 11, dated 1957), in *Special Topics In Electrochemistry*, p. 210, P. A. Rock, Editor, Elsevier, New York (1977).

Chapter 5

M30. R. A. Marcus, Theory of Oxidation-Reduction Reactions Involving Electron Transfer. IV. A Statistical Mechanical Basis for Treating Contributions from Solvent, Ligands, and Inert Salt. *Discuss. Faraday Soc.* **29**, 21, (1960).

M42. R. A. Marcus, Interactions in Polar Media. I. Interparticle Interaction Energy. *J. Chem. Phys.* **38**, 1335 (1963).

M43. R. A. Marcus, Free Energy of Nonequilibrium Polarization Systems. II. Homogeneous and Electrode Systems. *J. Chem. Phys.* **38**, 1858, (1963).

M44. R. A. Marcus, Interactions in Polar Media. II. Continua. *J. Chem. Phys.* **39**, 460, (1963).

M45. R. A. Marcus, Free Energy of Nonequilibrium Polarization Systems. III. Statistical Mechanics Mechanics of Homogeneous and Electrode Systems. *J. Chem. Phys.* **39**, 1734, (1963).

M129. R. A. Marcus, On the Theory of Overvoltage for Electrode Processes Possessing Electron Transfer Mechanism. I. (ONR, Report No. 12, dated 1957), in *Special Topics in Electrochemistry*, P. A. Rock, Editor. Elsevier, New York (1977).

M16. R. A. Marcus, On the Theory of Oxidation-Reduction Reactions Involving Electron Transfer. *J. Chem. Phys.* **24**, 966, (1956).

Chapter 6

M53. R. A. Marcus, On the Theory of Electron-Transfer Reactions. VI. Unified Treatment of Homogeneous and Electrode Reactions. *J. Chem. Phys.* **43**, 679, (1965).

Chapter 7

M55. R. A. Marcus, Theoretical Study of Electron Transfer Reactions of Solvated Electrons. *Adv. Chem. Ser.* **50**, 138, (1965).

M61. R. A. Marcus, Theory of Electron Transfer Reactions and Related Phenomena. *Proceedings of Symposium on Exchange Reactions* (Atomic Energy Agency 1965) p. 1.

M57. R. A. Marcus, On the Theory of Chemiluminescent Electron-Transfer Reactions. *J. Chem. Phys.* **43**, 2654, (1965). Erratum: **52**, 1018, (1970).

M60. R.A. Marcus, Theory of Electron-Transfer Reaction Rates of Solvated Electrons. *J. Chem. Phys.* **43**, 3477, (1965).

Chapter 8

M69. R. A. Marcus, Theoretical Relations Among Rate Constants, Barriers, and Brønsted Slopes of Chemical Reactions. *J. Phys. Chem.* **72**, 891, (1968).

M71. R. A. Marcus, Electron Transfer at Electrodes and in Solution. *Electrochimica Acta*, **13**, 995, (1968).

M77. R. A. Marcus, General Introduction to Faraday Society Discussions on Electrode Reactions of Organic Compounds. *Discuss. Faraday Soc.* **45**, 7, (1968).

Chapter 9

M49. R. A. Marcus, Generalization of the Activated Complex Theory of Reaction Rates. I. Quantum Mechanical Treatment. *J. Chem. Phys.* **41**, 2614 (1964).

M50. R. A. Marcus, Generalization of the Activated Complex Theory of Reaction Rates. II. Classical Mechanical Treatment. *J. Chem. Phys.* **41**, 2624 (1964).

M74. R. A. Marcus, Electron Transfer Reactions, in *Chemische Elementarprozesse*, p. 348, H. Hartmann, Editor. Springer-Verlag, New York (1968).

M105. R. A. Marcus, Activated-Complex Theory: Current Status, Extensions and Applications. In *Techniques of Chemistry*, Vol. 6, Part 1, Investigations of Rates and Mechanisms of Reactions, E. S. Lewis, ed. (Wiley, New York, 1974) Chapter 2.

M118. R. A. Marcus, Electron Transfer in Homogeneous and Heterogeneous Systems. *Phys. Chem. Sci. Res. Rep.* **1**, 477, (1975).

M119. R. A. Marcus, Energetic and Dynamical Aspects of Proton Transfer Reactions in Solution. *Faraday Symposia Chem. Soc.* **10**, 60, (1975).

M134. R. A. Marcus, Theories of Electrode Kinetics. *Physicochem. Hydrodyn.* **1**, 473, (1977).

Chapter 10

M110. R. A. Marcus, N. Sutin, Electron-Transfer Reactions with Unusual Activation Parameters. A Treatment of Reaction Accompanied by Large Entropy Decreases. *Inorg. Chem.* **14**, 213, (1975).

M139. R. A. Marcus, Electron and Nuclear Tunneling in Chemical and Biological Sytems. In *Tunneling in Biological Systems*, B. Chance, D. C. Devault, H. Frauenfelder, R. A. Marcus, J. R. Schrieffer, and N. Sutin, eds. (Academic Press, New York, 1979) p. 109.

M140. R. A. Marcus, Electron Transfer and Tunneling in Chemical and Biological Systems. *Life Sci. Res. Rep.* **12**, 15, (1979).

M141. Joussot-Dubien, A. C. Albrecht, H. Gerischer, R. S. Knox, R. A. Marcus, M. Schott, A. Weller, Mechanisms of Charge Separation and Subsequent Processes Group Report. *Life Sci. Res. Rep.* **12**, 129 (1979).

M145. R. Haberkorn, M. E. Michel-Beyerle, and R. A. Marcus, On Spin Exchange and Electron-Transfer Rates in Bacterial Photosynthesis. *Proc. Natl. Acad. Sci.* **76**, 4185 (1979).

M146. Tunneling in Biological Systems (Book). B. Chance, D. C. DeVault, H. Frauenfelder, R. A. Marcus, J. R. Schriffer, and N, Sutin, eds. (Academic Press, New York, 1979).

M154. R. A. Marcus, Similarities and Differences Between Electron and ProtonTransfers at Electrodes and in Solution. Theory of a Hydrogen Evolution Reaction. In *Proc. Third Symposium Electrode Processes*, 1979. S. Bruckenstein, J. D. E. McIntyre, B. Miller, E. Yeager, eds. (Electrochemical Society, Princeton, 1980) p. 1.

M156. R. A. Marcus, On Quantum, Classical and Semiclassical Calculations of Electron Transfer Rates. In *Oxidases and Related Redox Systems*, T. E. King, M. Morrison, and H. S. Mason, eds. (Pergamon, New York, 1982) p. 3.

M163. P. Siders, R. A. Marcus, Quantum Effects in Electron Transfer Reactions. *J. Am. Chem. Soc.* **103**, 741, (1981).

Chapter 11

M30a. R. A. Marcus, Discussion Comment on Mixed Reaction-Diffusion Controlled Rates. *Discuss. Faraday Soc.* **29**, 129 (1960).

M164. P. Siders, R. A. Marcus, Quantum Effects for Electron Transfer Reactions in the "Inverted Region." *J. Am. Chem. Soc.* **103**, 748, (1981).

M170. R. A. Marcus, On the Frequency Factor in Electron Transfer Reactions and its Role in the Highly Exothermic Regime. *Int. J. Chem. Kinetics* **13**, 865 (1981).

M173. R. A. Marcus and P. Siders, Further Developments in Electron Transfer. *ACS Symp. Ser.* **198**, 235, (1982).

M174. R. A. Marcus and P. Siders, Theory of Highly Exothermic Electron Transfer Reactions. *J. Phys. Chem.* **86**, 622, (1982).

M177. R. A. Marcus, Electron, Proton and Related Transfers. *Faraday Disc. Chem. Soc.* **74**, 7, (1982).

M194. R. A. Marcus, Nonadiabatic Processes Involving Quantum-Like and Classical-Like Coordinates with Applications to Nonadiabatic Electron Transfers. *J. Chem. Phys.* **81**, 4494, (1984).

Chapter 12

M126. E. Weissman, G. Worry and R. A. Marcus, A Study of Entropic and Electrolyte Effects in Electron Transfer Reactions. *J. Electroanal. Chem.* **82**, 9, (1977).

M201. P. Siders, R, J. Cave and R. A. Marcus, A Model for Orientation Effects in Electron-Transfer Reactions. *J. Chem. Phys.* **81**, 5613, (1964).

M204. R. A. Marcus and N. Sutin, Electron Transfers in Chemistry and Biology. *Biochim. Biophys. Acta.* **811**, 265, (1985).

M206. R. A. Marcus and N. Sutin, Application of Electron-Transfer Theory to Several Systems of Biological Interest. *Antennas and Reaction Centers of Photosynthetic Bacteria*, 226, (1985).

M207. R. A. Marcus and N. Sutin, The Relation Between the Barriers for Thermal and Optical Electron Transfer Reactions in Solution. *Comments Inorg. Chem.* **5**, 119, (1986).

M209. R. J. Cave, S. J, Klippenstein and R. A. Marcus, A Semiclassical Model for Orientation Effects in Electron Transfer Reactions. **84**, 3089, (1986).

M210. R. J. Cave, S. J, Klippenstein and R. A. Marcus, Mutual Orientation Effects on Electron Transfer Between Porphyrins. **90**, 1436, (1986).

Chapter 13

M211. H Sumi and R. A. Marcus, Dinamical Effects in Electron Transfer Reactions. *J. Chem. Phys.* **84**, 4894, (1986).

M213. H. Sumi and R. A. Marcus, Dielectric Relaxation and Intramolecular Electron Transfers. *J. Chem. Phys.* **84**, 4272, (1986).

M214. R. A. Marcus and H. Sumi, Solvent Dynamics and Vibrational Effects in Electron Transfer Reactions. *J. Electroanal. Chem.* **204**, 59, (1986).

List of Marcus' Papers Considered in the Chapters of the Book

The papers are numbered as in the official list of Marcus' publications.

16. R. A. Marcus, On the Theory of Oxidation-Reduction Reactions Involving Electron Transfer. *J. Chem. Phys.* **24**, 966, (1956) I. **M16**, Ch. 2
17. R. A. Marcus, Electrostatic Free Energy and Other Properties of States Having Nonequilibrium Polarization. *J. Chem. Phys.* **24**, 979, (1956) I. **M17**, Ch. 2
19. R. A. Marcus, On the Theory of Oxidation-Reduction Reactions Involving Electron Transfer. II. Applications to Data on the Rates of Isotopic Exchange Reactions. *J. Chem. Phys.* **26**, 867, (1957) **M19**, Ch. 3
20. R. A. Marcus, On the Theory of Oxidation-Reduction Reactions Involving Electron Transfer. III. Applications to Data on the Rates of Organic Redox Reactions. *J. Chem. Phys.* **26**, 872, (1957) **M20**, Ch. 3
21. R. A. Marcus, On the Theory of Oxidation-Reduction Reactions and of Related Processes. *Trans. N. Y. Acad. Sci.* **19**, 423, (1957) **M21**, Ch. 3
28. R. A. Marcus, On the Theory of Electrochemical and Chemical Electron Transfer Processes. *Can. J. Chem.* **37**, 155, (1959) **M28**, Ch. 3

33. R. A. Marcus, A Theory of Electron Transfer Processes at Electrodes. In *Transactions of the Symposium on Electrode Processes*, E. Yeager, ed. (Wiley, New York, 1961) p. 239. **M33**, Ch. 3
41. R. A. Marcus, On the Theory of Oxidation-Reduction Reactions Involving Electron Transfer. V. Comparison and Properties of Electrochemical and Chemical Rate Constants. *J. Phys. Chem.* **67**, 853, 2889, (1963) **M41**, Ch. 3
51. R. A. Marcus, Electron Transfers at Electrodes. In *Encyclopedia of Electrochemistry*, C. A. Hampel, ed. (Reinhold, New York, 1964) p. 529. **M51**, Ch. 3
128. R. A. Marcus, Theory and Applications of Electron Transfers at Electrodes and in Solution. In *Special Topics in Electrochemistry*, P. A. Rock, ed. (Elsevier, New York, 1977) p. 161. **M128**, Ch. 4
129. R. A. Marcus, On the Theory of Overvoltage for Electrode Processes Possessing Electron Transfer Mechanism. I. (ONR Report No. 12, dated 1957). In *Special Topics in Electrochemistry*, P. A. Rock, ed. (Elsevier, New York, 1977) p. 181. **M129**, Ch. 4
130. R. A. Marcus, Electrostatic Free Energy and Other Properties of States Having Nonequilibrium Polarization. II. Electrode Systems. (ONR Report No. 11, dated 1957). In *Special Topics in Electrochemistry*, P. A. Rock, ed. (Elsevier, New York, 1977) p. 210. **M130**, Ch. 4
30. R. A. Marcus, Theory of Oxidation-Reduction Reactions Involving Electron Transfer. 4. A Statistical-Mechanical Basis for Treating Contributions from Solvent, Ligands, and Inert Salt. *Discussions Faraday Soc.* **29**, 21, (1960) **M30**, Ch. 5
42. R. A. Marcus, Interactions in Polar Media. I. Interparticle Interaction Energy. *J. Chem. Phys.* **38**, 1335, (1963) **M42**, Ch. 5

43. R. A. Marcus, Free Energy of Nonequilibrium Polarization Systems. II. Homogeneous and Electrode Systems. *J. Chem. Phys.* **38**, 1858, (1963) **M43**, Ch. 5
44. R. A. Marcus, Interactions in Polar Media. II. Continua. *J. Chem. Phys.* **39**, 460, (1963) **M44**, Ch. 5
45. R. A. Marcus, Free Energy of Nonequilibrium Polarization Systems. III. Statistical Mechanics of Homogeneous and Electrode Systems. *J. Chem. Phys.* **39**, 1734, (1963) **M45**, Ch. 5
53. R. A. Marcus, On the Theory of Electron-Transfer Reactions. VI. Unified Treatment of Homogeneous and Electrode Reactions. *J. Chem. Phys.* **43**, 679 (1965) **M53**, Ch. 6
55. R. A. Marcus, Theoretical Study of Electron Transfer Reactions of Solvated Electrons. *Advan. Chem. Ser.* **50**, 138 (1965) **M55**, Ch. 7
57. R. A. Marcus, On the Theory of Chemiluminescent Electron-Transfer Reactions. *J. Chem. Phys.* **43**, 2654 (1965) **M57**, Ch. 7. Erratum: **52**, 2803 (1970)
60. R. A. Marcus, Theory of Electron-Transfer Reaction Rates of Solvated Electrons. *J. Chem. Phys.* **43**, 3477 (1965) **M60**, Ch. 7
61. R. A. Marcus, Theory of Electron-Transfer Reactions and of Related Phenomena. *Proc. Symp. Exchange Reactions* (Atomic Energy Agency 1965) p. 1. **M61**, Ch. 7
69. R. A. Marcus, Theoretical Relations Among Rate Constants, Barriers, and Brønsted Slopes of Chemical Reactions. *J. Phys. Chem.* **72**, 891 (1968) **M69**, Ch. 8
71. R. A. Marcus, Electron Transfer at Electrodes and in Solution: Comparison of Theory and Experiment. *Electrochim. Acta* **13**, 995 (1968) **M71**, Ch. 8
77. R. A. Marcus, General Introduction to Faraday Society Discussions on Electrode Reactions of Organic Compounds. *Disc. Faraday Soc.* **45**, 7 (1968) **M77**, Ch. 8

49. R. A. Marcus, Generalization of the Activated Complex Theory of Reaction Rates. I. Quantum Mechanical Treatment. *J. Chem. Phys.* **41**, 2614 (1964) **M49**, Ch. 9
50. R. A. Marcus, Generalization of the Activated Complex Theory of Reaction Rates. II. Classical Mechanical Treatment. *J. Chem. Phys.* **41**, 2624, (1964) **M50**, Ch. 9
74. R. A. Marcus, Electron Transfer Reactions. In *Chemische Elementarprozesse*, H. Hartmann, ed. (Springer-Verlag, New York, 1968) p. 348. **M74**, Ch. 9
105. R. A. Marcus, Activated-Complex Theory: Current Status, Extensions and Applications. In *Techniques of Chemistry*, Vol. 6, Part 1, Investigations of Rates and Mechanisms of Reactions, E. S. Lewis, ed. (Wiley, New York, 1974) Chapter 2 **M105**, Ch. 9
118. R. A. Marcus, Electron Transfer in Homogeneous and Heterogeneous Systems. *Phys. Chem. Sci. Res. Rep.* **1**, 477, (1975) **M118**, Ch. 9
119. R. A. Marcus, Energetic and Dynamical Aspects of Proton Transfer Reactions in Solution. *Faraday Symposia Chem. Soc.* **10**, 60, (1975) **M119**, Ch. 9
134. R. A. Marcus, Theories of Electrode Kinetics. *Physicochem. Hydrodyn.* (Pap. Conf.) **1**, 473, (1977) **M134**, Ch. 9
110. R. A. Marcus and N. Sutin, Electron-Transfer Reactions with Unusual Activation Parameters. A Treatment of Reactions Accompanied by Large Entropy Decreases. *Inorg. Chem.* **14**, 213, (1975) **M110**, Ch. 10
139. R. A. Marcus, Electron and Nuclear Tunneling in Chemical and Biological Systems. In *Tunneling in Biological Systems*, B. Chance, D. C. Devault, H. Frauenfelder, R. A. Marcus, J. R. Schrieffer, and N. Sutin, eds. (Academic Press, New York, 1979) p. 109. **M139**, Ch. 10

140. R. A. Marcus, Electron Transfer and Tunneling in Chemical and Biological Systems. *Life Sci. Res. Rep.* **12**, 15, (1979) **M140**, Ch. 10
141. J. Joussot-Dubien, A. C. Albrecht, H. Gerischer, R. S. Knox, R. A. Marcus, M. Schott, A. Weller, Mechanisms of Charge Separation and Subsequent Processes Group Report. *Life Sci. Res. Rep.* **12**, 129, (1979) **M141**, Ch. 10
145. R. Haberkorn, M. E. Michel-Beyerle, and R. A. Marcus, On Spin-Exchange and Electron-Transfer Rates in Bacterial Photosynthesis. *Proc. Natl. Acad. Sci.* **76**, 4185, (1979) **M145**, Ch. 10
146. Tunneling in Biological Systems (Book). B. Chance, D. C. DeVault, H. Frauenfelder, R. A. Marcus, J. R. Schrieffer, and N. Sutin, eds. (Academic Press, New York, 1979) **M146**, Ch. 10
154. R. A. Marcus, Similarities and Differences Between Electron and Proton Transfers at Electrodes and in Solution. Theory of a Hydrogen Evolution Reaction. *In Proc. Third Symposium Electrode Processes, 1979*. S. Bruckenstein, J. D. E. McIntyre, B. Miller, E. Yeager, eds. (Electrochemical Society, Princeton, 1980) p. 1. **M154**, Ch. 10
156. R. A. Marcus, On Quantum, Classical and Semiclassical Calculations of Electron Transfer Rates. In *Oxidases and Related Redox Systems,* T. E. King, M. Morrison, and H. S. Mason, eds. (Pergamon, New York, 1982) p. 3. **M156**, Ch. 10
163. P. Siders and R. A. Marcus, Quantum Effects in Electron-Transfer Reactions. *J. Am. Chem. Soc.* **103**, 741, (1981) **M163**, Ch. 10
30a. R. A. Marcus, Discussion Comment on Mixed Reaction-Diffusion Controlled Rates. *Discussions Faraday Soc.* **29**, 129, (1960) **M30a**, Ch. 11
164. P. Siders and R. A. Marcus, Quantum Effects for Electron-Transfer Reactions in the Inverted Region. *J. Am. Chem. Soc.***103**, 748, (1981) **M164**, Ch. 11

170. R. A. Marcus, On the Frequency Factor in Electron Transfer Reactions and Its Role in the Highly Exothermic Regime. *Intl. J. Chem. Kinetics* **13**, 865, (1981) **M170**, Ch. 11
173. R. A. Marcus and P. Siders, Further Developments in Electron Transfer. *ACS Symp. Ser.* **198**, 235, (1982) **M173**, Ch. 11
174. R. A. Marcus and P. Siders, Theory of Highly Exothermic Electron Transfer Reactions. *J. Phys. Chem.* **86**, 622, (1982) **M174**, Ch. 11
177. R. A. Marcus, Electron, Proton and Related Transfers. *Faraday Disc. Chem. Soc.* **74**, 7, (1982) **M177**, Ch. 11
194. R. A. Marcus, Nonadiabatic Processes Involving Quantum-Like and Classical-Like Coordinates with Applications to Nonadiabatic Electron Transfers. *J. Chem. Phys.* **81**, 4494, (1984) **M194**, Ch. 11
126. E. Waisman, G. Worry, and R. A. Marcus, A Study of the Entropic and Electrolyte Effects in Electron Transfer Reactions. *J. Electroanal. Chem.* **82**, 9, (1977) **M126**, Ch. 12
201. P. Siders, R. J. Cave and R. A. Marcus, A Model for Orientation Effects in Electron-Transfer Reactions. *J. Chem. Phys.* **81**, 5613, (1984) **M201**, Ch. 12
204. R. A. Marcus and N. Sutin, Electron Transfers in Chemistry and Biology. *Biochim. Biophys. Acta* **811**, 265, (1985) **M204**, Ch. 12
207. R. A. Marcus and N. Sutin, The Relation Between the Barriers for Thermal and Optical Electron Transfer Reactions in Solution. *Comments Inorg. Chem.* **5**, 119, (1986) **M207**, Ch. 12
209. R. J. Cave, S. J. Klippenstein, and R. A. Marcus, A Semiclassical Model for Orientation Effects In Electron Transfer Reactions. *J. Chem. Phys.* **84**, 3089, (1986) **M209**, Ch. 12
210. R. J. Cave, P. Siders, and R. A. Marcus, Mutual Orientation Effects on Electron Transfer Between Porphyrins. *J. Phys. Chem.* **90**, 1436, (1986) **M210**, Ch. 12

211. H. Sumi and R. A. Marcus, Dynamical Effects in Electron Transfer Reactions. *J. Chem. Phys.* **84**, 4894, (1986) **M211**, Ch. 13
213. H. Sumi and R. A. Marcus, Dielectric Relaxation and Intramolecular Electron Transfers. *J. Chem. Phys.* **84**, 4272, (1986) **M213**, Ch. 13
214. R. A. Marcus and H. Sumi, Solvent Dynamics and Vibrational Effects in Electron Transfer Reactions. *J. Electroanal. Chem.* **204**, 59, (1986). **M214**, Ch. 13

CHAPTER 1

Introduction

Charge transfer is present in the phenomenon of contact electrification at the very beginning of the history of electricity, with Thales of Miletus, as well as at the dawn of electricity as science, with William Gilbert. Triboelectricity, the contact electrification by tribocharging, was first observed, as far as we know, by Thales of Miletus when he rubbed amber with wool, the Greek word for friction being $\tau\rho\iota'\beta o\varsigma$ and that for amber $\eta'\lambda\varepsilon\kappa\tau\rho o\nu$ (or $\eta'\lambda\varepsilon\kappa\tau\rho o\varsigma$). Tribocharging is due, we today know [1], to ions or electrons transferring from one to the other of two surfaces in contact. In a sense, we can consider Thales' tribocharging also as the first experiment in chemical kinetics because, as Marcus remarked, that rubbing is overcoming the activation energy barrier for transfer of an ion or of an electron.

We have an example of charge transfer that people are familiar with in the form of proton transfer, in acid–base reactions such as

$$H_3O^+ + OH^- \rightarrow H_2O + H_2O \tag{1.1}$$

If we now focus on electron transfer, then probably the first one we meet in Physics is the transfer of the electron in a hydrogen atom from an atomic orbital in the ground state to another [2] in an excited state

$$H(1s) + h\nu \rightarrow H(2p) \tag{1.2}$$

a process we might call photon induced *intra-atomic* ET.

The simplest molecule of Chemistry is, as we all know, H_2^+. And the simplest ET reaction [3] is

$$H + H^+ \rightarrow H^+ + H \tag{1.3}$$

More in general we have gas phase interatomic ET reactions like

$$A + B^+ \rightarrow A^+ + B \tag{1.4}$$

The simplest ET reaction between neutral atoms giving a cation and an anion [4, p. 26] is

$$H + H \rightarrow H^+ + H^- \tag{1.5}$$

In the last two reactions, the reactants and the products are different. Reaction (1.3) in which reactants and products are the same is the simplest example of an *electron exchange* reaction or *self-exchange* reaction.

A much more complicated kind of ET occurs when an ion dissolved in a polar solvent absorbs light. In this case, as in Eq. (1.2), the electron jumps to an excited state almost instantaneously, but a considerable longer time is required for the slower moving solvent dipoles to readjust to the new electronic configuration, electron density and associated electric field of the ion. Thus, for some time after the absorption act, the overall electrical polarization of the solvent medium in the neighborhood of the ion will not be in *electrostatic equilibrium* with the ion's electric field, that is, the solvent polarization will not be the one dictated by the ion's new charge distribution. This concept served as a basis of a theory of the absorption spectrum of various halide ions in solution [5].

Even more complicated ET's occur in the usual oxidation–reduction reactions. For instance, in:

$$MnO_4^- + 5Fe^{+2} + 8H_3O^+ \rightarrow Mn^{+2} + 5Fe^{+3} + 12H_2O \tag{1.6}$$

five electrons are altogether exchanged between MnO_4^- and Fe^{+2}, in *successive elementary reaction steps*, with an extensive rearrangement of chemical bonds among the atoms and of solvent molecules around the ions. In this case, the ET is *nonradiative*, that is, it doesn't happen because of absorption or emission of light, but it is thermal: it happens because of suitable *thermal fluctuations* in nuclear configurations of reactants and solvation molecules. This last statement may sound obscure to the uninitiated but it will be made clear in the following.

The simplest oxidation–reduction reactions in solution are those in which no bonds are broken or formed when the electron is transferred between reagents. Consider one such reaction:

$$^{*}\mathrm{Fe}_{\mathrm{aq}}^{+3} + \mathrm{Fe}_{\mathrm{aq}}^{+2} \rightarrow {}^{*}\mathrm{Fe}_{\mathrm{aq}}^{+2} + \mathrm{Fe}_{\mathrm{aq}}^{+3} \tag{1.7}$$

The symbol "aq" means that the ions in water solution are solvated, the number of water dipoles and/or their orientations around the ions being clearly dependent on the ionic charges. In this electron exchange or "self-exchange" reaction, two isotopes of iron are used—one of them, Fe*, radioactive—to follow the ET between the ions because without the use of isotopes the reagents and products systems look the same. It was "in this small corner of inorganic chemistry" (**M.**), that of *isotopic exchange reactions*, where from the story of ET in polar solvents began. In reactions such as these, in which reactants and products are the same, the standard free energy difference between final and initial states is zero and the thermodynamic control on the reaction is missing. They are very interesting because it is in such reactions that those "intrinsic factors" which control their chemical kinetics, that is, the structure of the transition state (TS) and the nature of the reaction coordinate, come to the fore.

As for reactions in which reactants and products are the same, we may also remember the simplest of all bimolecular ones, the hydrogen atom exchange reaction:

$$\mathrm{H}_{\alpha} - \mathrm{H}_{\beta} + \mathrm{H}_{\gamma} \rightarrow \mathrm{H}_{\alpha} + \mathrm{H}_{\beta} - \mathrm{H}_{\gamma} \tag{1.8}$$

the simplest reaction of Chemistry where a bond breaks while another bond forms [6, 7]. Here reactants and products are the same although a reaction really happens because a covalent bond breaks while another one forms. The Greek subscripts correspond here to the isotopic labels in (1.7).

If we now turn to Electrochemistry looking for examples of elementary reactions in which an electron is transferred from the electrode to the ion without formation or rupture of chemical bonds, we could consider for instance the reduction of permanganate to manganate [8]:

$$MnO_4^- + M(e) \rightarrow MnO_4^{-2} + M \tag{1.9}$$

or of, say, protonated nitrogen oxide to NOH [9]:

$$NOH^+ + M(e) \rightarrow NOH + M \tag{1.10}$$

where M is the metal electrode.

Today, among the topics included in the field of ET, are: inorganic, organic and biological ET, charge transfer spectra, ET at interfaces other than metal electrode–liquid, like semiconductor–liquid, modified electrode–liquid, polymer–liquid and liquid–liquid, ET at colloids and micelles and many others, see Fig. 1, p. 51 in [10]. Among them a recent one is ET in quantum dots [11].

Electron transfer processes are also implicitly present when the cohesive energy is defined for ionic or metallic crystals in solid state physics [12].

1.1. Description of Electron Transfer Reactions with Potential Energy Curves

Adiabatic and Nonadiabatic Processes

Before delving into the treatment of bimolecular ET reactions in polar solvents, we first consider—as a warm up and to introduce some fundamental concepts—the much simpler case of a diatomic ET in vacuo. Let us consider the reaction:

$$Na + Cl \rightarrow Na^+ + Cl^- \tag{1.11}$$

Without in general being aware of it, we pass by this reaction when studying the formation of the ionic bond in alkali halides. Let us then consider the process of formation of the ionic bond in the molecule of NaCl starting with an atom of Na far apart from one of Cl and letting the two atoms slowly (adiabatically) approach each other to form a chemical bond. "Slow collisions of atoms and molecules (neutral or charged) are defined as collisions in which the velocity υ of relative motions of colliding particle is substantially lower than the velocity of valence electrons υ_e,

$$\upsilon/\upsilon_e \ll 1$$

υ_e is estimated as $\upsilon_e \approx 1\text{au} \approx 10^8\,\text{cm/s}$" [13].

A slow electronically adiabatic diatomic collision is a collision such that "the electron cloud is able to adjust its state instantaneously to the nuclear framework" and, on the other hand, "the nuclear motion will not be influenced by the momentary spatial arrangement of the single electrons but only by the mean force field (averaged over many periods of the motion) of the whole electron cloud" [14, p. 7]. To discuss the energetics of such a collision, it is necessary to represent the total electronic energy of the Na + Cl system as a function of the internuclear distance. By total electronic energy, we mean the sum of the electronic energy of the electrons plus the repulsive potential energy of the nuclei [15, p. 3]. The curve representing this function is called adiabatic potential energy curve (PEC), or adiabatic molecular interaction potential, because the total electronic energy plays the role of potential energy governing the motion of atoms, that is, of the nuclei and of the attached electron clouds, when discussing the dynamics of the process, see, for example, Refs. [14, 15]. Such a curve is schematically shown by the lowest solid line in Fig. 1*, [cf. 16, Fig. 14.3, p. 537], we see that it is almost a straight line parallel to the axis of the internuclear R distances until, at $R_c \sim 10.15$ Å, it suddenly turns down changing into the segment of hyperbola which goes down in energy until it changes into the segment of a parabola around the minimum. Beyond that, the curve goes up steeply to higher energies. Let's now explain this behavior. Starting with atoms infinitely far apart and with zero kinetic energy, that is, in their asymptotic state,

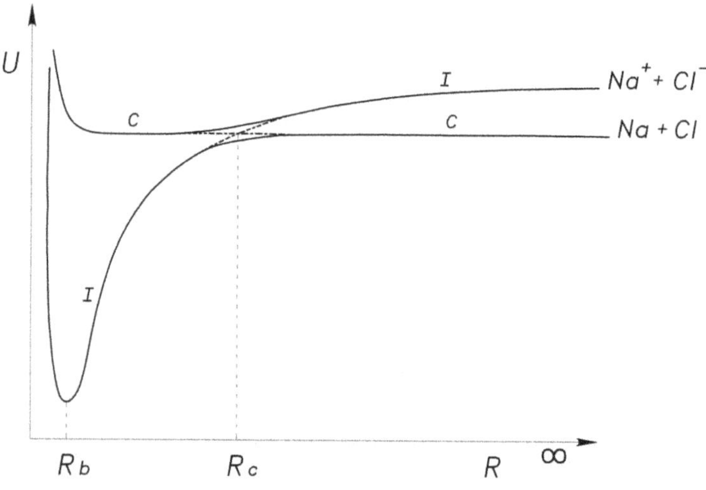

Fig. 1*. (Adapted from Ref. [22]).

the potential energy of the system decreases very slightly, remaining almost constant, because between neutral atoms there is only a small Van der Waals attraction. At internuclear distances in the neighborhood of R_c it so happens that an electron transfer may occur and $Na + Cl$ changes then into $Na^+ + Cl^-$. The two ions will attract each other with an electronic potential energy equal to:

$$U = -\frac{1}{4\pi\varepsilon_0}\frac{e^2}{R} \tag{1.12}$$

This function represents a hyperbola. We mark the segment of the curve corresponding to $R > R_c$ with a C (from "covalent") and the segment for $R < R_c$ with an I (from "ionic"). When the atoms are close enough, a Pauli repulsion (exchange repulsion) sets in between the atomic cores and the combination of Pauli repulsion, of nuclear repulsion and of nuclear-electronic attraction gives the segment of parabola around the minimum. At minimum energy, the molecule Na^+Cl^- is formed, whose bonding distance we denote with R_b. Going to shorter R, the Pauli repulsion—to which one should also add the internuclear repulsion—overwhelms the attraction and the curve shows a steeply rising repulsive branch. At the minimum potential energy, the system reaches its maximum stability and the parabola at the bottom of the curve describes the low energy vibrational motion of the Na^+Cl^- molecule like that of a harmonic oscillator. As a whole we can then symbolically describe in the following way the process of formation of the ionic bond:

$$\begin{aligned}(\mathrm{Na}+\mathrm{Cl})_{R=\infty} &\rightarrow (\mathrm{Na}+\mathrm{Cl} \rightarrow \mathrm{Na}^+ + \mathrm{Cl}^-)_{R\sim Rc} \\ &\rightarrow (\mathrm{Na}^+\mathrm{Cl}^-)_{R=Rb} \end{aligned} \tag{1.13}$$

Notice that the systems above are considered each with a total energy equal to the potential energy at the various values of R. Na and Cl or Na^+ and Cl^- are fixed at the various distances R. The symbols are not describing a scattering process in which kinetic energy would be present.

The description of the process using only the lower PEC in the figure is unsatisfactory because we cannot explain why the nature of the two branches of the lower solid curve changes right around $R \sim R_c$, that is, why R_c has that particular value and why just around there an ET becomes possible. Moreover, a question comes naturally to mind: how would one describe the formation of NaCl starting not with atoms but rather with the ions Na^+ and Cl^-? To describe this process and to answer the above two questions, we need to consider also another PEC, the upper solid line in Fig. 1*.

At the right panel side of the upper curve, we see the system of the ions in their asymptotic state. Their energy differs from that of the atoms by the difference of the ionization potential of Na, I_P (Na), and the electron affinity of Cl, E_A (Cl). These energies correspond to the processes:

$$\mathrm{Na} \rightarrow \mathrm{Na}^+ + \mathrm{e} \quad I_P; \quad \mathrm{Cl}^- \rightarrow \mathrm{Cl} + \mathrm{e} \quad E_A$$

The energy associated with the process $(\mathrm{Na} + \mathrm{Cl})_{R=\infty} \rightarrow (\mathrm{Na}^+ + \mathrm{Cl}^-)_{R=\infty}$ is $I_P - E_A$ and the ET is radiative at large distances between the atoms because a photon is needed to ionize the

Na atom. The new PEC correlated to $(Na^+ + Cl^-)_{R=\infty}$ begins with a hyperbola of equation:

$$U = -\frac{1}{4\pi\varepsilon_0}\frac{e^2}{R} + I_P - E_A \tag{1.14}$$

using as zero of energy the energy of the asymptote of the lower solid curve. The upper curve decreases to reach R_c, in whose neighborhood an ET may happen:

$$Na^+ + Cl^- \rightarrow Na + Cl \tag{1.15}$$

and the curve flattens because the strong Coulombic attraction has disappeared due to the formation of neutral atoms. Going to shorter distances, the Pauli and nuclear repulsions set in, and the curve goes up steeply. Even in this case a letter "I" marks the branch of the curve corresponding to the ions and a "C" the one corresponding to the atoms.

Now we can understand why the ET reaction (1.11) occurs right in the neighborhood of R_c. It so happens because at that interatomic distance in the *narrow region of the avoided crossing* shown in the figure the systems $Na^+ + Cl^-$ and $Na + Cl$ may have *the same energy* because there the large amount of energy required to transfer an electron from Na to Cl forming Na^+ and Cl^- is completely counterbalanced by the mutual Coulomb energy of the ions [4, p. 73] and it is therefore possible to go from the atoms to the ions and vice versa without the help of photons making up for the energy difference between reagents and products. We have here an example of a *resonant* ET. In reaction (1.7), we have an example of a *thermal resonant* ET because the resonant condition is reached thanks to an appropriate thermal fluctuation of the solvent.

The second question—about the possibility of forming the molecule starting from the ions—has thus far received a negative answer: if we move only on the higher curve, we cannot directly go from NaCl in its electronic ground state. The two curves we have been considering until now are "adiabatic" curves, the meaning of the term "adiabatic" being explained shortly below. They can be obtained by accurate *ab initio* quantum mechanical calculations, in which the total electronic energy at each point of the curve is computed at the fixed nuclear distance given by the abscissa of the point and each curve has either a "C" character before the avoided crossing and an "I" character after the covalent–ionic crossing or vice versa. Looking at the two curves, we see that in the neighborhood of R_c they come very close to each other but they don't cross forming, in a very small range of the R coordinate around R_c, an "*avoided crossing pattern*" or "avoided crossing seam" limited by two short branches of hyperbola showing a narrow energy splitting between them. The reason for this is explained, for instance, in [14, pp. 123 ff.]. We want only mention here that the splitting is due to the "Pauling resonance" [17, p. 204], see "the prototype case of H_2^+" [17, p. 204] in Ref. [4, p. 14 ff.] (and see Appendix) between the ionic and covalent forms of NaCl, usually symbolized $\{Na+Cl \leftrightarrow Na^+ + Cl^-\}$, which happen to be in a "perfect" resonance at R_c because at that R the two forms have the same energy. Joining now the two "I" branches with a dashed line along a hyperbola's asymptote, we have a new curve which is called the ionic "diabatic" curve and which away from the avoided crossing seam "will merge and be essentially the same" [14, p. 153] with two branches of the adiabatic curves, see Fig. 1* [cf. 16, Figs. 14.2 and 14.3, p. 537]. Likewise, the two C branches and a dashed line connecting them across the avoided crossing region represent the "diabatic" covalent curve. The two C branches merge out of the avoided crossing seam and become essentially the same with two branches of the adiabatic curves. If we now imagine starting with $(Na^+ + Cl^-)_{R=\infty}$ and follow the ionic diabatic curve, jumping from the above adiabatic curve to the one below in the avoided crossing region, is then possible to go from the ions directly to the molecule, with its mainly ionic bond, without the need of an intermediate ET. Processes like the first two we considered, in which the system follows a single adiabatic PEC, are called "adiabatic." The word "adiabatic" is almost a transliteration from the ancient Greek "αδιάβατος" meaning "nonpassable," "noncrossable." Those in which the system follows first a branch of an adiabatic curve, jumps then on another adiabatic curve across an avoided crossing and subsequently follows the second curve, are called "nonadiabatic." The nonadiabatic processes follow a diabatic curve. "Diabatic" comes from "διαβαίνω," "I pass." In other words, the ET process happens when the system follows an adiabatic curve, and it doesn't if it follows a diabatic curve. If the system follows a branch of one of the two adiabatic curve and then, in the interaction region of the avoided crossing, jumps to the other adiabatic curve, we have a nonadiabatic process, no ET has happened, it is as if the process had happened along a single diabatic curve.

The question as to when we have nonadiabatic processes and when adiabatic ones is answered by the celebrated Landau–Zener–Stueckelberg–Majorana formula for nonadiabatic transition probabilities:

$$P = \exp\left[-\frac{\pi S_c^2}{2\hbar \upsilon_c |\Delta F|}\right] \tag{1.16}$$

The formula is more usually designated simply as the Landau–Zener formula. For a discussion see Refs. [14, 16, 18, 19]. A short summary of the Landau–Zener theory is given in Ref. [18]. In Eq. (1.16), $S_c \equiv S(R_c)$ is the splitting between the adiabatic curves at the avoided crossing, equal to twice the resonance interaction between the resonant structures NaCl and Na^+Cl^- when Na and Cl (and likewise Na^+ and Cl^-) are at a distance R_c (see Appendix). $\Delta F = F_{ion} - F_{cov}$ where F_{ion} and F_{cov} are the slopes of the ionic and covalent diabatic curves at R_c and $\boldsymbol{\upsilon}_c$ is the velocity with which the system passes across the avoided crossing region. Looking at Eq. (1.16) we see that "in those region of the nuclear configuration space where adiabatic potential surfaces are close together or intersect, where electronic wave-functions are changing very rapidly for varying nuclear coordinates, where nuclei are moving with high velocity, non-adiabatic transitions become probable" [14, p. 22], and so if the system passes through the avoided crossing with high velocity, it jumps through the splitting with a nonadiabatic transition probability close to 1 and no ET happens because the electronic structure of Na + Cl doesn't have time to change to that of $Na^+ + Cl^-$ and the system follows the diabatic path of the covalent PEC, that is, for collisions velocities high enough the diabatic terms can be interpreted as potential surfaces which govern the motion of the nuclei [14, p. 153]. On the other hand, when Na and Cl approach each other with a vanishingly small velocity, the probability P of a nonadiabatic jump from the lower to the upper curve in the avoided crossing region is almost zero and so the probability $1 - P$ of staying on the lower curve is almost 1. This probability is the probability of the adiabatic ET reaction $Na + Cl \rightarrow Na^+ + Cl^-$.

The term "diabatic" has been introduced in the physical literature only since 1963 by Lichten [20]. Kauzmann [16] calls the adiabatic and diabatic curves "slow" and "fast" curves. A recent description of diabatic curves is given in Ref. [21].

The above adiabatic curves do not cross because of the Wigner–Witmer noncrossing rule [13, p. 563]. Similar curves are also reported on p. 77 of Ref. [4]. There Pauling used valence bond calculations and the curves cross because of the approximate calculations. This is why the avoided crossing is also designated as pseudocrossing, that is, false or spurious crossing. Pauling's valence bond energies are diabatic. The curves are also reported on p. 372 of Ref. [22]. Note that noncrossing rule is only true for diatomics. Conical intersections are the rule for polyatomics.

One *very* important point needs here to be emphasized. We observe that the crossing point distance $R_c \sim 10.15$ Å is much larger than the sum of covalent radii of the atoms in NaCl ($\sim$2.53 Å) or of the ionic radii in Na^+Cl^- ($\sim$2.45 Å) [15]. This means that the charge transfer inducing interaction, measured by the small energy splitting between the curves around R_c, and proportional to the overlap of the orbitals of the atomic wave functions, is a *weak* electronic interaction (see Appendix) if compared to the strong chemical interaction typical of the covalent bonds, where there is a much greater orbitals' overlap. A. C. Wahl et al. [23] in a calculation reported in Refs. [15, 24] showed that when the lithium and fluorine atoms approach, to form the lithium fluoride molecule, an ET occurs at $R_c \sim 7.35$ Å and the lithium cation and fluorine anions form. In this case, that distance is about four to five times larger the sum of the atomic or ionic radii!

I want to end up this section citing the fascinating study by A. Zewail of the reaction, analogous to Eq. (1.11)

$$\mathrm{NaI} \rightarrow \mathrm{NaI}^* \rightarrow \mathrm{Na} + \mathrm{I} \quad \text{or} \quad \mathrm{Na}^+ + \mathrm{I}^- \tag{1.17}$$

In Ref. [25, p. 264], one finds that curves similar to the ones discussed above have been used to follow processes (1.17) studied by femtoseconds laser spectroscopy.

1.2. Molecular Polarization, Reactants Model

In the beginning... there were experimental kinetics studies. "It was found that isotopic exchange between ions differing only in their valency are generally slow if single cations are involved and fast if the ions are relatively large, such as complex ions" [26].

W. Libby [27] surmised that this behavior was dependent on the orientation of the solvent dipoles around the ions. Immediately after an ET event, the charges on the ions change and the solvent dipoles around them are not anymore in electrostatic equilibrium with the charges (vide infra). This means that the new state, not being in stable equilibrium, is of higher energy than the original and this fact could explain the slow reaction velocity due to the high activation energy barrier for the small ions' reactions, being more highly solvated than the big ones. The insight of Libby, that the barrier depended on the nonequilibrium polarization of the solvation molecules, was correct.

Moreover, he was right in thinking that the quantitative explanation should have been found in applying the Franck–Condon (FC) principle. But he was wrong in the way he applied it [28].

At this point, a brief reminder on molecular polarization is in order.

An electric field—that of an ion in particular—induces in a molecule electronic, atomic and orientational polarization. The electronic polarization is due to the shift of electrons relative to the nuclei, atomic polarization means that atoms are displaced relative to one another [29], with consequent variation of interatomic distances, bond lengths, and angles. The orientational polarization is due to the orienting effect on the molecular dipole by the directing electric field. Marcus uses the symbol $\mathbf{P}_e$ for the electronic polarization that he designated as of "E type," while the symbol $\mathbf{P}_u$ is collectively used for atomic and orientation polarization, which is of "U type."

The relaxation times are of the order of 10^{-15} sec for electronic polarization, 10^{-13} sec for the atomic and 10^{-11} sec or slower for orientational polarization [26].

The potential energy of orientation between an ion and a dipole [30] is given by:

$$U = -\mu\, E \cos\theta \quad (U = -\mathbf{p} \cdot \boldsymbol{E}) \tag{1.18}$$

where θ is the angle between the direction of the electric field $\boldsymbol{E}$ of the ion and that of the molecular dipole p. The ion–dipole system is in *electrical equilibrium* when $\theta = 0$, that is, when the dipole is lined up with the field. In this case, the direction of the dipole is *the one dictated by the electric field and U is at its minimum value.*

When a molecule is in a medium, the orienting effect of the electric field is counteracted by the thermal agitation of the molecules, by their mutual interactions and orientation correlation of neighboring molecules [29] so that at temperature T the *average* equilibrium θ of the solvation molecules in the ion's field may be different from zero. Moreover—and this fact is of paramount importance for the ET theory—the instantaneous polarization continuously fluctuates around its equilibrium value because of thermal agitation, a situation reminiscent of the vibrations of a harmonic oscillator around its equilibrium geometry.

In the first model used by Marcus for the reacting ions, they are supposed to be spheres of radii a_1 and a_2. The spheres are rigid, formed by the bare ions and *possibly* by a spherical region of saturated dielectric made up by solvent molecules fully oriented in the ions' fields. *Outside these saturated spheres* are the molecules whose $\mathbf{P}_u$ polarization is determined, as described above, by the counteracting ordering and disordering effects of electric fields and thermal agitation.

Initially, under the influence of Born's description of the charging of ions in solution, M. considered the spheres as conducting. This restrictive hypothesis was dropped later on.

1.3. Solvation Molecules' Contribution to the Barrier in ET Reactions

In order to understand—on a qualitative level—the barrier to ET reactions due to the orientation of the solvent molecules around the ions, let us first consider the barrier to reaction for the most simple three-center atom transfer reaction involving a linear activated complex, that is the H exchange reaction:

$$\mathrm{H}_\alpha{-}\mathrm{H}_\beta + \mathrm{H}_\gamma \;\rightarrow\; \mathrm{H}_\alpha + \mathrm{H}_\beta{-}H_\gamma \tag{1.19}$$

The energy barrier to reaction is due to the fact that an activation energy is necessary to go from the reagents to the products because a chemical bond between two hydrogen atoms is to be broken *while* another one forms. If we imagine, for simplicity, that the reaction happens on a line, we may represent schematically the process as:

$$\underbrace{\mathrm{H}_\alpha - \mathrm{H}_\beta + \mathrm{H}_\gamma}_{\mathrm{R}} \rightarrow \mathrm{H}_\alpha - \mathrm{H}_\beta \cdots\mathrm{H}_\gamma \rightarrow \underbrace{\mathrm{H}_\alpha \cdots \mathrm{H}_\beta \cdots \mathrm{H}_\gamma}_{\ddagger} \rightarrow \mathrm{H}_\alpha \cdots \mathrm{H}_\beta - \mathrm{H}_\gamma \rightarrow \underbrace{\mathrm{H}_\alpha + \mathrm{H}_\beta - \mathrm{H}_\gamma}_{\mathrm{P}}$$

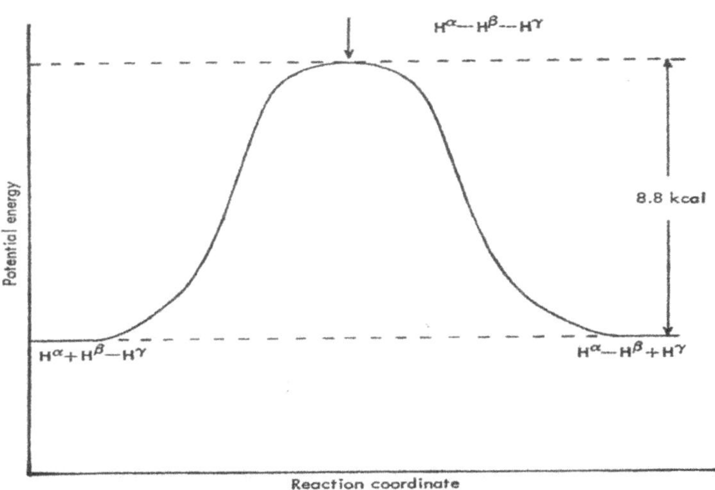

Fig. 2*. (Adapted from Ref. [31]).

where R stands for the reagents, P for the products and the double dagger symbol "‡" for the transition state at the *nuclear* configuration *intermediate* between that for R and P.

The energy barrier to reaction is represented in the simplest possible way as in Fig. 2* [31–33]. The potential energy at the maximum along the reaction coordinate corresponds to the nuclear configuration of the TS. Note though that, contrary to the gradual process in Eq. (1.19), the jump of the electron in the case of ET is an abrupt process.

Imagine now having a system of two ions of charges +3 and +2, for example, of iron, in a polar solvent, and that an ET reaction happens between them. The distribution of orientations of solvation dipoles around ions with different charge is different because the directing electric field is greater for more highly charged ions and so the average angle θ is smaller for the dipoles around the ion with charge +3 than for those around the one of charge +2. After the charge exchange reaction, the *equilibrium* orientational distribution of the dipoles around the ions is also switched. This is described in the ET literature by means of figures where the authors either pictorially

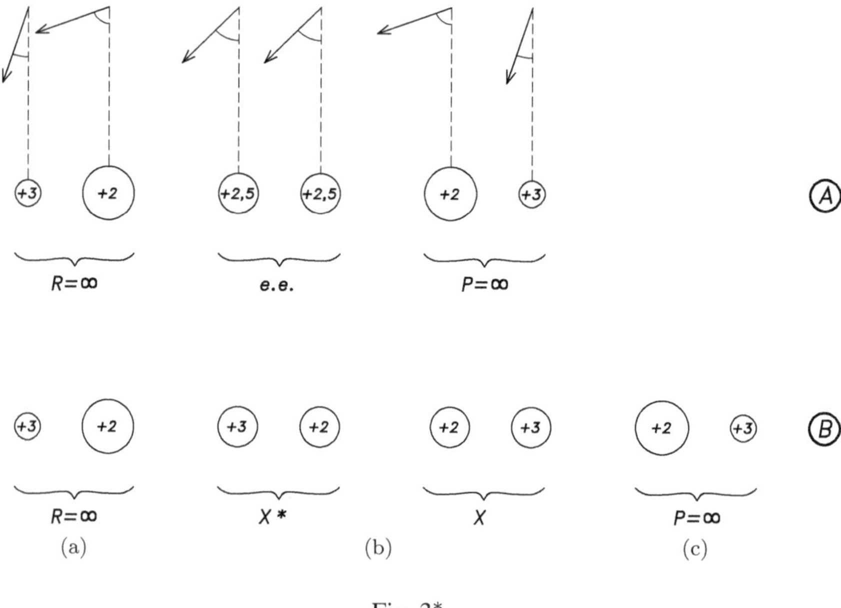

Fig. 3*.

show the different orientation of the solvent molecules around the ions [34] or more simply represent the dipoles by swarms of small ellipses [35] or arrows [36–38]. The number of arrows varies from many to three used by Krishtalik and two by Kuznetsov. It is possible to give the reader the sense of the shift in dipole orientation using a single arrow, as suggested by Eq. (1.18), to represent the average dipoles distribution.

In the left side of Fig. 3* panel A, small circles represent two ions with charges +3 and +2 and, above them, are arrows representing, for simplicity, only the orientational part of $\mathbf{P}_u$ (in order to represent the atomic part of it one should use arrows of different length), all at an infinite separation distance of the two ions, for simplicity. The angles between field and average dipole directions are only meant to be illustratively effective.

The left side of the figure represents the reagents system R. After ET, the system relaxes to the new equilibrium system P at the right side of the figure. As in the case of the H atom transfer reaction, *the nuclear configuration in the TS will be intermediate between those of R and P*. In the words of Marcus the TS: "... can be reached by any suitable fluctuation of atomic coordinates to produce some atomic configuration which is usually a *compromise* between the stabler ones of the redox orbitals.... Fluctuations of this nature involve simultaneous changes in orientation, position and atomic polarization of the solvent molecules, in internuclear distances in the coordination shell, in relative motion of the reactants and in configuration of the ionic atmosphere." [39, p. 22] Such a configuration is represented in the middle of the figure by arrows with orientations intermediate between those in R and P, and is indicated by e.e., for "equivalent equilibrium" [39, p. 25] $\mathbf{P}_u$ polarization. Such polarization would be the one in equilibrium with *fictitious* charges $+2.5$ on each ion. But the system so designated cannot be the *real* TS for the following reasons. First of all the fractional $+2.5$ charges are obviously only hypothetical. Moreover, the transition state is a nonequilibrium state whereas in the e.e. system there is equilibrium between charges and polarization, like in the R and P systems. Finally, and most importantly, the $\mathbf{P}_e$ polarization is *always in equilibrium with the charges* and so we have here, in a misleading depiction, a fictitious electronic polarization in equilibrium with fictitious charges. This hypothetical intermediate system is nonetheless of great importance in ET theory because it has the *correct* $\mathbf{P}_u$ polarization equivalent to that in the transition state and, moreover, *it suggests a way to find it*: $\mathbf{P}_u$ is *that* polarization which is in equilibrium with a hypothetical intermediate charge and can so be found through a suitable charging process of the ions [40].

In Fig. 3* panel B, the *real* ET process is schematically described. Between the R and P systems there is, *just before* the TS, the nonequilibrium state X^* and, *just after* the TS, there is the nonequilibrium state X, the one in which X^* transforms immediately after the ET (vide infra). At the TS, the wave function of the system is a linear combination of those of X^* and of X. In the following, the R, P, X^*, and X symbols will represent, as they do in Marcus' first ET paper, thermodynamic states each one made up, as always in thermodynamics, of very many complexions or microstates and the X^* in the figure is really meant to be representative of one of the x^* microstates making up the X^*. Let us now consider two microstates x^* and x, one having the electronic structure and the charges of the reagents, belonging to the state X^*, the other having the electronic structure and the charges of the products and belonging to the successor state X. $\mathbf{P}_u$ refers to the polarization of the thermodynamic states. Each of the microstates x^* contributes with its own $\mathbf{P}_u(x^*)$ to the $\mathbf{P}_u$ polarization.[(1)] In x^*, the polarization is $\mathbf{P}_u(x^*)$ and the ions' charges are those of the reagents. In x, $\mathbf{P}_u(x)$ is equal to that of x^* but the charges equal to those of the products. Both x^* and x are unstable nonequilibrium systems because $\mathbf{P}_u$ in them is not in equilibrium with the ionic charges. The TS has equal total energies, including $\mathbf{P}_e$ for either electron localization.

We are now in the position of correctly describing the thermal ET process in solution. The process begins with a suitable thermal fluctuation of the nuclear coordinates bringing a microscopic system belonging to R to a system x^* belonging to X^*.[(2)] Such fluctuations are of orientations of solvent molecules and of their bond lengths and angles. At the hypersurface representing the TS, after an electron transfer involving the *coupled motion of nuclei and electrons*, the successor state x forms, belonging to X, which has the *same nuclear configuration* of x^* but the electronic configuration of the products. The electron transfer probability in this dynamical process is given by the Landau–Zener formula.[(3)] The system x belonging to X, finally relaxes to a microscopic system belonging to P. Of course, each of the above three steps must run in the direction of the products for a successful ET finally to happen.

Marcus has taught us how to apply the FC principle to thermal ET processes. The states FC connected are here states x^* and x of equal nuclear configurations *and of equal energy* contrary to the usual situation in which the FC principle is applied in spectroscopy to *vertical* transitions between states of equal nuclear configurations but of different energy, like the ones shown, for the NaI system, in

Fig. 6.44, p. 356 of Ref. [41]. Marcus applied the FC principle to an energetically *horizontal* transition between states of equal energy.

We want to emphasize at this point that although the nonequilibrium $\mathbf{P}_u$ contribution of the solvation molecules to the reaction barrier is important and was the first to be studied in the development of the Marcus theory, an important contribution is also that of the vibrational motions of the reactants and of the configurations of the ionic atmospheres. The relative importance of the different contributions varies for different reactions. We shall take up later these further contributions to the reaction barrier. In Fig. 3* are represented ions with charges +3 and +2 participating in an isotopic exchange reaction. The ions are supposed to be surrounded by polarized solvent molecules and ionic atmospheres. They are represented by circles with arbitrary different radii intended to simply schematically summarize their different atomic configurations in order to represent the processes illustrated in the figures.

1.4. Electronic Configuration of the Activated Complex

On pp. 967 and 968 of Ref. [26], M. gives a crystal clear description of the electronic configuration of the activated complex. Two remarks are in order. First, M. uses there the older Eyring's terminology "activated complex" instead of the modern "transition state" which he adopted successively.(4) Secondly, he imagines that the activated complex is made up of two electronic forms in equilibrium with each other, that is, $X^* \leftrightharpoons X$, where X^* is the activated complex with the electronic structure of the reactants and X is the one with the electronic structure of the products. Such formulation represents a good approximation but it is "static" and has been superseded by a "dynamical" one in which "X^* and X are two *electronic participants in the TS. Just before* the TS there is the X^* form, *just after* the TS there is the X form, in between there is a combination of the two, you may call in resonance combination. In other words, if a wave function ψ_1 refers to X^*, a wave function ψ_2 refers to X, at the TS we have $\psi_1 + \psi_2$" (**M**, personal communication). The initial state X^* goes to the successor state X by a Landau–Zener dynamical process, as discussed in the Appendix.

In the following, I shall briefly mention the main characteristic properties of the activated complex for ET reactions:

(1) In the usual chemical reactions, there are transfers and/or rearrangement of atoms between the reactants. This happens because of *strong interactions* between atoms as a consequence of considerable spatial overlaps of the atomic and molecular orbitals of the reactants. In the case of ET reactions, there is a *slight electronic interaction*[(5)] which is sufficient to electronically couple the reacting molecules and permit the ET to occur. We have seen above the examples of ET between atoms of Na and Cl or of Li and F atoms happening at interatomic distances such that the interaction between atomic partners is very weak.

(2) In the ET process, the electronic configuration changes, in a successful collision, from the one characteristic of the reactants to that of the products. The process is one of *abrupt* transfer of an electron, not that of a gradual transfer of electron density from one reactant to the other as, for an example, in the case of the reaction $H^+ + H_2O \rightarrow H_3O^+$ [42]. The collision process is a dynamical process in which the motion of nuclei (better, of atomic cores) and of electrons (valence or outer electrons) is largely decoupled before reaching the crossing region and the motion is governed by a single PES and the single associated wave function but it is *coupled in the crossing region* where it is governed by *both* PESs and by a combination of *both* associated wave functions, vide infra.

(3) Because of the slight electronic interaction, the energy splitting between the adiabatic potential energy surfaces is small and M. normally approximates the adiabatic surfaces with the diabatic ones, like in this first treatment. But in the diabatic case of zero

splitting, there would be no overlap of the electronic orbitals, so the M. approximation is "the better the less the overlap." He needs the splitting small, or the energy of the TS would be wrong.

(4) Let x^* be one of the microstates belonging to the state X^* and x a microstate of the state X. The wave function of the electrons in the microstates not only describe the reacting particles but they take account also of the solvent molecules.[6] The energies of the two states are the same, $E_{x^*} = E_x$, for every x^* system in X^*, there is an x system in X which has the same energy as x^*, the X^* and X, which are made up of x^*'s and x's, have the same energy[7] and the wave function of the system at the TS is a linear combination of the wave functions of the two states, that is, $\phi_{x^*} + c\phi_x$, like in the case of a wave function describing a quantum mechanical resonant structure built up of Lewis resonant structures. Notice that the frequency of hopping from x^* to x and back and forth is slow because of the weak electronic interaction but the time for the single jump is very short (see Appendix).

(5) The average configuration is the same in the two states X^* and X and the energy must be the same for the two states. But the charges in X^* and X are different, so that the state of the solvent must be one of nonequilibrium. **M.** so rephrases the foregoing discussion in terms of the FC principle: "When one electron configuration is formed from the other by an electronic transition, the electronic motion is so rapid that the solvent molecules do not have time to move during the electronic jump."

1.5. Reaction Scheme

An important consequence of the small orbitals overlap in the activated complex is that the ET process may be slow, so determining the rate of the overall process to which the ET step belongs.

After having started from the reactants, $A + B$, say, once the nonequilibrium state x^* is reached, isoenergetic with the x state of the products, there will be a certain probability of the *electronic* transition

$x^* \to x$. This transition is discussed in the Appendix. There is also the possibility that the state x^* will reform the reactants "by disorganization of some of the oriented solvent molecules." The state x can either reform x^* by an electronic transition $[x \to x^*]$ or, alternatively, the products in this state can merely move apart, say. The detailed reaction scheme for bimolecular ET reactions will be dealt with in the next chapter.

1.6. Potential Energy Hypersurfaces and Schematic Diagrams for ET Reactions—A Summary in Marcus' Words

"To treat rates of reactions in general , regardless of whether they involve transfers of electrons, atoms, or protons, bond scission, or molecular isomerization, it is useful to plot potential energy curves. The potential energy U is a function of the positions of all the atoms in the system. Thereby, U depends, for example, on all the bond lengths and angles, on orientations of reacting molecules, and on distances and orientations of molecules in the surrounding environment. Because there are so many position coordinates involved, only a profile of U versus some general coordinate can be plotted, which has as components all of the coordinates above. Such a plot is useful for pictorial purposes, although the actual calculations themselves involve all instead of one general coordinate.

The position of each atom in the entire system is subject to thermal fluctuations and the reactive system thereby wanders over the curve R (really surface R) in Fig. 4*. No reaction occurs until the system reaches the coordinates at the intersection of the R and P surfaces. At that intersection, the system can go from the reactants' surface R to the products' surface P when there is a coupling between the orbitals of the two reactants. The extent of coupling of two electronic orbitals (one on the reactant, occupied by the electron to be transferred, and an orbital on the other reactant, waiting to be occupied by

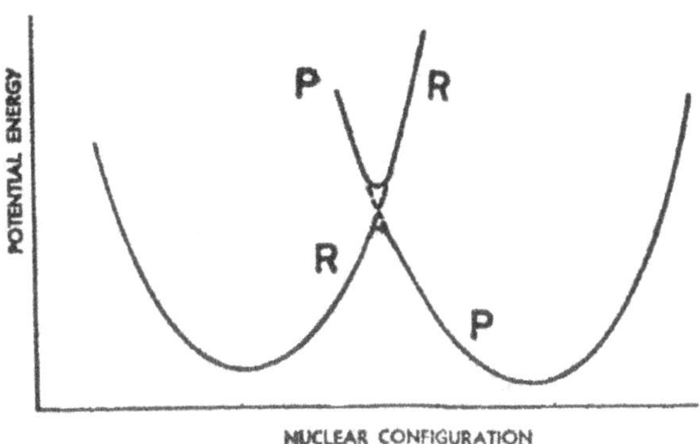

Fig. 4*.

the transferred electron) is reflected in the splitting 2ϵ of the intersecting R and P curves, as in Fig. 4*.

The probability of reaching the intersection can be calculated by statistical mechanics or by some related formalism. The probability of the system's crossing from the R to the P curve can be calculated by quantum mechanics with a velocity-weighted Landau–Zener transition probability κ. The weaker the electronic coupling of electronic orbitals of the reactants with each other, the smaller is this κ. The probability of transition κ at the intersection increased with increasing ϵ" (from **M140**, pp. 15–16).

It has a maximum value of unity at strong enough coupling.

Note that the figure refers to a thermoneutral reaction, like that of isotopic exchange reactions, and that the ET is supposed to happen at temperatures high enough that nuclear tunneling is neglected. Such a point will be dealt with later in the book.

In the case of the gas phase ET reaction between Na^+ and Cl^-, the dynamics of the process was described using two-dimensional PECs of the kind $U = U(R)$ where the abscissa is the internuclear distance and the ordinate is the total electronic energy. The two-dimensional space of a page is appropriate to represent such curves. In the case of the hydrogen transfer reaction along a line, the potential energy is represented by a two-dimensional potential energy surface (PES) in the three-dimensional space of U and of the internuclear distances $R(\mathrm{H}_\alpha - H_\beta)$ and $R(\mathrm{H}_\beta - \mathrm{H}_\gamma)$. The famous PES for such reaction is described, for instance, in Refs. [14, 31–33, 43].

If one considers a nonlinear three-atomic system like, for instance, the $[\mathrm{HO}_2]^+$ system, and the gas phase ET reaction between H^+ and O_2:

$$\mathrm{H}^+ + \mathrm{O}_2 \rightarrow \mathrm{H} + \mathrm{O}_2^+$$

three coordinates are necessary to describe the relative positions of the proton and the oxygen atoms in the plane of the nuclei [44]. In these cases, we have then potential energy hypersurfaces (PESs) in *four* dimensions, one for U and three for the geometrical coordinates, which obviously cannot be visualized as a whole in three-dimensional space. Holding two coordinates fixed, it is possible to compute PECs which are cuts or cross sections, as they are called, of the PES.

When we pass to a system of two "central ions" [26], each one surrounded by solvent molecules and, more in general, even by ionic atmospheres, the N nuclear coordinates necessary to describe the spatial configurations number in the order of thousands [10] if one considers only the reactants and the solvent molecules and the ions closer to the reactants, while the total number of coordinates necessary to describe the whole macroscopic system is of the order of 10^{23}. The generalized coordinates [45] to describe the configurations space of the system are the distance R between the central ions, the vibrational coordinates, the angles describing molecular orientations and intermolecular distances, all of them *parametrically* dependent on R, that is, the potential energy hypersurface of the reactants R is $U = U(q(R))$ where $q(R) = \{q_1 = R, q_2(R), \ldots, q_N(R)\}$. Because of the high number of nuclear coordinates, the potential energy hypersurface U could only in principle be represented in the same way as for the simple systems considered above.

A schematic potential energy diagram used by Marcus to describe the ET process is reported in Fig. 4*. It is very often found in Marcus' papers, for example, in Refs. [10, 28, 32, 46]. This figure is really almost a logo of the Marcus theory of electron transfer. Such potential energy "profiles" are not quantitative nor qualitative PECs obtained from the N-dimensional $U_{\mathrm{Reactants}}$ and U_{Products} hypersurfaces cutting them as described above. They are just "profiles of the actual potential energy surfaces plotted *along the reaction coordinate q*," schematic potential energy diagrams [31, 47] which graphically summarize the information described below.

The R and P surfaces are sketched as two potential wells in parabola-like forms to represent "some sort of vibrational-like motion" [32, 33] where the abscissa is that of "a collective mode of the donor, acceptor and solvent" [48]. They are similar in shape to the unidimensional symmetric bistable PECs used to describe isomerization processes, such as the ammonia inversion [49, 50], or the inversion of its isoelectronic ion H_3O^+ represented in Fig. 3 of Ref. [51]. In Fig. 5*(a), the "profiles" of PESs U_{R} and U_{P} of reactants and products cross at a point $\{R, q_2(R), \ldots, q_N(R)\}$ in N-dimensional space where $U_{\mathrm{R}}(R, q_2(R), \ldots) = U_{\mathrm{P}}(R, q_2(R), \ldots)$. The systems R and P have the same q configuration and the same energy *but* the distance R between the central ions is not small enough to allow for an electronic interaction between them, so that there is no splitting of the PESs and no ET.

When the central ions are at a distance $R = R_c$ at which they electronically interact, the PESs cross at the point $\{R_c, q_2(R_c), \ldots, q_N(R_c)\}$ so that:

$$U_{\mathrm{R}}(R_c, q_2(R_c), \ldots, q_N(R_c)) = U_{\mathrm{P}}(R_c, q_2(R_c), \ldots, q_N(R_c)) \tag{1.20}$$

the ET act is possible and the curves in Fig. 5*(a) change into those of Fig. 5*(b). The energy splitting appears and we shall have a possible adiabatic ET reaction. In this figure, the potential minima of the R and P profiles are equal, so that such curves are suitable to describe, for instance, isotopic exchange reactions in which reactants and products are the same.

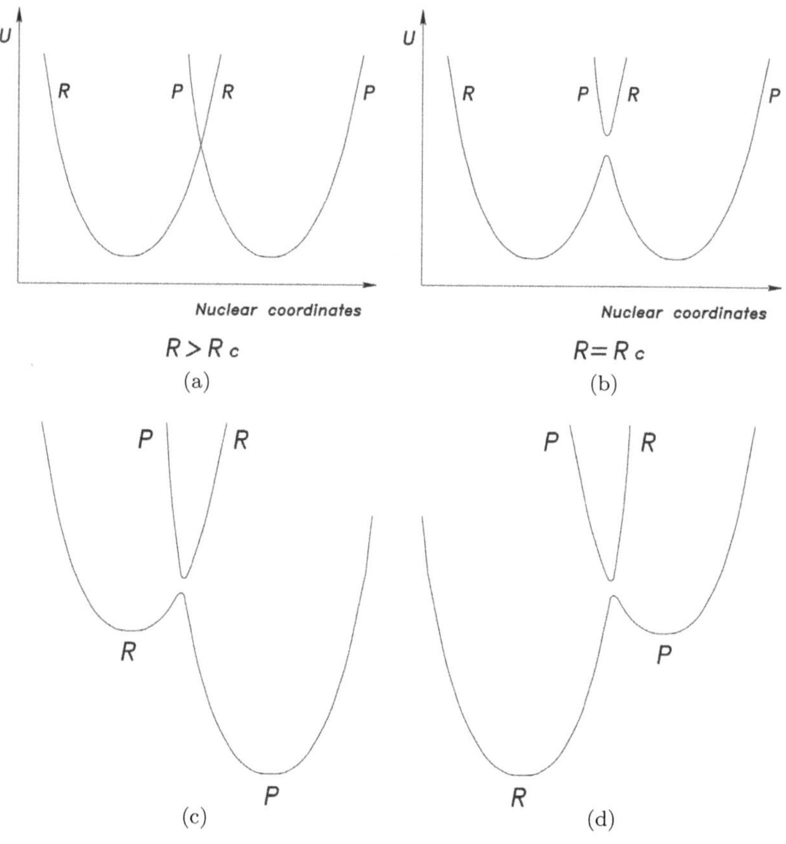

Fig. 5*.

In Fig. 5*(c), a more general profile considers the possibility that reactants and products are different and that the products are more stable than the reactants. In Fig. 5*(d), the reactants are instead more stable than the products.

In a following development of the theory Marcus introduced a *generalized reaction coordinate* (the energy difference of the two energy surfaces at each point) and the problem of the multidimensional PESs was reduced, by a statistical mechanical averaging, to a discussion in terms of *free energy curves.*

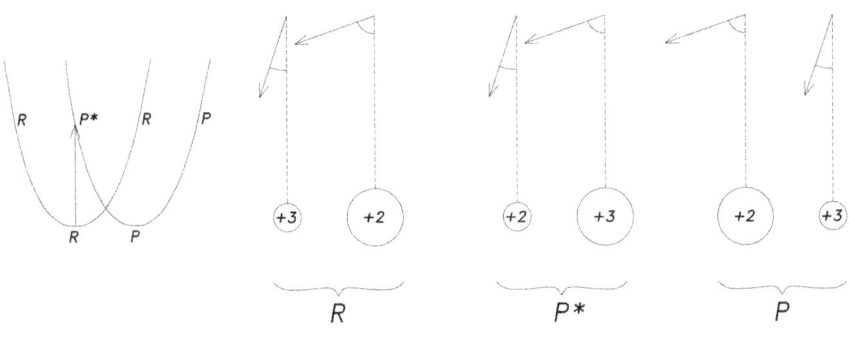

Fig. 6*.

Using the energy diagrams with the above schematic orientational polarization diagrams, it is possible to show very clearly what was wrong in Libby's description of thermal ET process.

In Fig. 6*, an arrow shows the vertical excitation of the R system to an excited products state, P* say, wherefrom the system relaxes to the equilibrium state corresponding to the minimum of the P diagram. The vertical excitation corresponds to an *optical* ET in which the R and P* states have the same nuclear configuration—as expected from the *usual* way of applying the FC principle *to spectroscopic transitions*—but while in R the solvent polarization is in equilibrium with the electric field, in P* it is not. The P* system then thermally relaxes to P. The polarization diagrams corresponding to R, P*, and P are shown in the Figure. The P* system has two characters in common with the correct activated complex for thermal ET. The P* state is in fact of higher energy than the reactants state and is in a nonequilibrium polarization state. But instead of having a nuclear configuration *intermediate* between that of reactants and products, as one expects in a TS, it has the same configuration as that of reactants. And in the correct theory not only the excited system relaxes to the equilibrium state with a thermal fluctuation but an *a priori suitable thermal fluctuation* is necessary for the reactants system to reach the TS nuclear configuration region.

Appendix

In the crossing point of the diabatic curves, the system is degenerate because the two different electronic configurations have the same energy and so a resonance energy appears that separates the two diabatic curves. If ϕ_1 and ϕ_2 denote the electronic wave functions of the "pure" covalent and ionic valence bond structures, then the diabatic energy V_1 of structure ϕ_1 at the internuclear distance R is given by $V_1(R) = \int \phi_1 H_{el} \phi_1 d\boldsymbol{x}$, where H_{el} is the electronic Hamiltonian of the system depending parametrically on the internuclear distance R, the diabatic energy V_2 of structure ϕ_2 is given by $V_2(R) = \int \phi_2 H_{el} \phi_2 d\boldsymbol{x}$. The interaction energy of the two structures is given by the resonance (or exchange) integral $V_{12}(R) = \int \phi_1 H_{el} \phi_2 d\boldsymbol{x} = \int \phi_2 H_{el} \phi_1 d\boldsymbol{x}$, a perturbation which, in the Landau–Zener theory is assumed to be constant and equal to V_{12} in the narrow avoided crossing (or pseudocrossing) region [14, p. 151]. The energies $W_\pm$ of the adiabatic curves are then given by:

$$W_\pm(R) = \frac{1}{2}[V_1(R) + V_2(R)] \pm \frac{1}{2}\{[V_1(R) + V_2(R)]^2 + 4V_{12}^2\}^{\frac{1}{2}} \tag{A.1}$$

where $2V_{12}$ is the separation of the levels due to as shown in Fig. 5.7 on p. 150 of Ref. [14] or on Fig. 8.9(b) on p. 173 of Ref. [52] and in the interaction region of the avoided crossing the wave functions representing the two states system become linear combinations of the resonant forms, that is,

$$\psi_+ = \frac{1}{\sqrt{2}}(\phi_1 + \phi_2) \quad \text{and} \quad \psi_- = \frac{1}{\sqrt{2}}(\phi_1 - \phi_2) \tag{A.2}$$

Using perturbation theory involving time, it is possible to show that the probabilities P_1 and P_2 of finding the system in state 1 and 2, respectively, at time t are:

$$P_1 = \cos^2\left(\frac{2\pi V_{12}}{h}t\right) \quad \text{and} \quad P_2 = \sin^2\left(\frac{2\pi V_{12}}{h}t\right) \tag{A.3}$$

that is, "We see that these probabilities vary harmonically between the values 0 and 1. The period of a cycle (from $P_1 = 1$ to 0 and back to 1 again) is seen to be $h/2V_{12}$ and the frequency $2V_{12}/h$, this being ... just $1/h$ times the separation of the levels due to the perturbation" [53, p. 323], see also Ref. [16, p. 534 ff.]. The situation is then the following: if the system were to remain static in the pseudocrossing region, there would be no final electron transfer: the electron would simply jump back and forth between the two resonant structures 1 and 2. But the system *moves* across the pseudocrossing region with some velocity, the electronic and nuclear motions are coupled in the avoided crossing region, *both* PECs govern the dynamics there through their splitting and through the difference of their slopes and following the Landau–Zener theory there may be two possible outcomes. If the system will move very rapidly across the interaction region, the electronic cloud will not have time to change from structure 1 to structure 2, no ET will happen, the system will simply go along a diabatic curve. But if the system will pass through the pseudocrossing at a velocity such that the electron will jump from 1 to 2 but it will not have time to go back, then we shall have an electron transfer. As a matter of fact in the so-called Massey parameter $(r_c) = \frac{2\pi V_{12} l}{\hbar u}$, where $l(R_c)$, the linear dimension of the pseudocrossing region, is "a characteristic length over which the corresponding electronic wave-functions substantially change and u some average (classical) velocity of the nuclei at point R_c" [14, p. 21, 22], we see that we have the ratio of the passage time of the nuclei through the interaction region to the transfer time h/V_{12}.

"We may state the general rule that exchange (resonance) integrals will tend to be large only if the orbitals concerned overlap effectively" [16, p. 298]. Which means that the smaller the distance R between the colliding partners the greater the orbitals overlap, the greater the exchange (resonance) integrals, that is, the greater V_{12} and the greater the splitting $2V_{12}$ between the adiabatic curves at the pseudocrossing. An interesting example is presented by the PECs for the system HF and H^+F^- reported on Fig. 3.4, p. 74 in Ref. [4]. There we see that the crossing of the ionic and covalent curves happen at somewhat less than 1 Å and the splitting is then very large. On the other hand, the crossing of the curves for NaCl and Na^+Cl^- happens at about 10.15 Å and the splitting is very small, as reported on p. 536 with Fig. 14.3 on p. 537 of Ref. [16]. The first case is an example of Pauling's resonance between Lewis structures, while the second is an example of the weak interactions considered in Marcus' theory of electron transfer. Another example is that, previously cited in the text, of the LiF system. In the case of Pauling's resonance, the frequency with which the electron jumps between the two structures is "the frequency of resonance among structures ... is very large, of the order of magnitude of electronic frequencies in general ..." [4, p. 186] so high that there is no chemical equilibrium between the two electronic tautomers and the resonance between the Lewis structures is sometimes indicated with a double-pointed arrow, a symbol suggested by Fritz Arndt and Bern Eistert to indicate resonance [4, p. 187]. We then have $H - F \leftrightarrow H^+F^-$ but in the case of electron transfer *reaction* supposed to happen at fixed internuclear distance there is a real chemical equilibrium $NaCl \rightleftarrows Na^+Cl^-$ between the electronic tautomers which exchange the electron at a lower frequency than that in the case of the usual Pauling resonance.

In the first paper on electron transfer M. considered the states X^* and X to be two forms of the activated complex, a *static* formulation then like the one considered above, and abandoned since 1964 in favor of the *dynamic* Landau–Zener treatment of ET probability.

In the first paper M. considered two methods of calculating the splitting $2V_{12}$, in one of them making use, as above, of time-dependent perturbation theory and in the other, already used by previous authors [54, 55] considering the electron tunneling of the transferred electron, as explained below.

Once the Na + Cl colliding system has reached the pseudocrossing region, in order to transfer from Na to Cl the electron must pass through an *electronic* potential energy barrier due to the attraction exerted on the leaving electron by the Na^+ ion left behind.

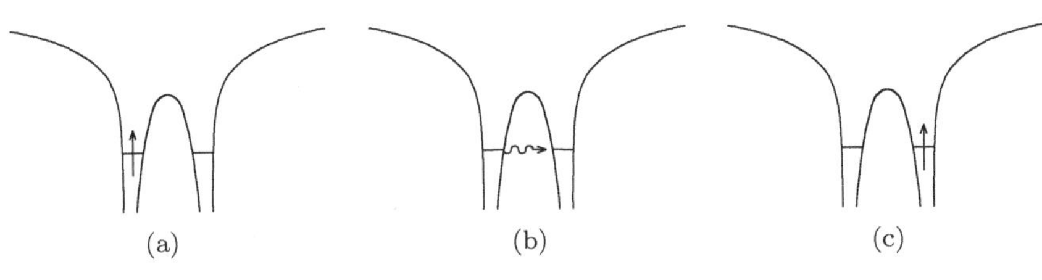

Fig. 7*.

The penetration of the barrier is classically impossible but "It is possible in quantum mechanics to sneak quickly across a region which is illegal energetically" [56, 8.12]. So that the electron tunnels through the potential barrier separating them: In Fig. 7*(b), I have used a wiggly arrow to symbolically represent the electron "worming its way"—in the words of Feynman [56, 10.2]—through the barrier.[(8)] In Fig. 7*(a) and 7*(c), the potential wells are represented, as in Ref. [24], Fig. 6.4, p. 153 for the symmetric electronic potential energy barrier to be penetrated for electron transfer reaction (1.3) when H and H^+ are at a large enough distance, in Ref. [57], Fig. 3.6, p. 58 for electron tunneling between two metal nuclei, and in Ref. [36], Fig. 3.2, p. 91. In Fig. 7*(a), the electron symbolically represented by an arrow is on the first iron ion, in Fig. 7*(c) on the second.

The resonance splitting is one half of the energy splitting produced by electron "tunneling among equivalent Lewis structures" [17, p. 204] (small R's) and more in general among resonant structures even at larger R's, like in the case of Na – Cl and Na^+Cl^- at the curves crossing. The tunneling across the potential barrier through which the electron leaving Na – Cl is to pass to reach the state Na^+Cl^- removes the energy degeneracy at the crossing because the separated resonant states combine linearly as in Eq. (A.2) to describe the system present in two potential wells with probabilities changing with time as in Eq. (A.3) and the two states described by wave functions ψ_+ and ψ_-, with which the time evolution appears, have their energies separated by the above energy splitting. A modern, detailed and very clear description of the tunneling-generated energy splitting can be found, for the analogous two states problem with tunneling of ammonia inversion, in Ref. [50, p. 455 ff.], but there the abscissa coordinate is measuring the distance of the plane containing the three hydrogen atoms from the N atom in NH_3, while here the abscissa is the coordinate of the transferring electron.

A short account of Marcus theory of electron transfer is given in Ref. [58] and a discussion of it in the light of the history of science is given in Ref. [59].

NOTES

1. **M**: "The different microscopic states contribute differently towards $\mathbf{P_u}$, they are microscopic, $\mathbf{P_u}$ is macroscopic. The dielectric polarization involves a number of microscopic states the order of 10^{23}, a huge number of states. $\mathbf{P}_u$ refers to X^*, not to x^*. Every microscopic state has its own contribution towards $\mathbf{P}_u$ but each contribution is different, it depends on the microscopic state, just like any microscopic property."
2. **M**: "You get the fluctuation [of the reactants] *at the same* electronic configuration that you start with, *exactly*. You have a whole new distribution and then, provided that the electronic configuration and nuclear configuration is isoenergetic with the other one for the products, then you (may) get a transfer. If it isn't, you won't get a transfer."
3. **M**: "The whole process involves going to the crossing, going pass it. In LZ you have two electronic states and one nuclear coordinate. It's the same nuclear coordinate for both, *perpendicular* to the TS hypersurface. When you treat the system with LZ, you don't just treat the system at the crossing of the diabatic states, you treat the whole dynamics before and afterwards too. That of Landau and Zener is a semi-classical approximation to handle a full quantum mechanical nuclear and electronic problem."

When the electron goes from one x^* state to an x state on the intersection surface, one has two *electronic* wells and the electron tunnels going from one to the other.

M: "That just corresponds to what is the probability for going from one of the surfaces to the other in the intersection region, a Landau-Zener type situation. Those wells are not explicitly seen on the potential energy surfaces because they are *electronic* wells, while the potential energy surfaces are plots of energy versus *nuclear* configurations, not a plot of potential energy versus *electronic* configuration. Note that when you are on the intersection surface the bottoms of the two *electronic* surfaces are of the same height and so you can transfer the electron without changing energy. The molecular energies are equal for every state on that intersection surface, and of course you have a thermal distribution of points on that intersection surface."

NOTE: This is very well shown in Fig. 3.2, p. 91 of Ref. [36], where the nuclear coordinate is shown in the left panel and the electronic coordinate in the right one.

4. **M**: "Following Eyring I used the term 'activated complex' for the longest time until I finally bent over to the Polanyi' term 'transition state,' which is what everybody uses nowadays. Probably neither terminology is great because the 'state' doesn't exist, it is a particular arrangement in the 10^{23}-dimensional space, it is a hypersurface, so that's not fictitious but typically you don't observe it."
5. **M**. considered, in his first formulation of the theory, a double form for the activated complex that was later abandoned:

   M: "Pauling was interested in strongly interacting systems and on how the energy of the strongly interacting system compared with that in one resonant form or another. Here I am interested in weakly interacting systems. I don't have a strongly interacting system, so formally *in the activated complex electrons can hop from one form of the activated complex to the other*, and that would be a legitimate description, but in Pauling's case you don't think of one resonant structure hopping to another, back and forth, it is not a real equilibrium."

M: "An activated complex in the ET theory consists really of two electronic structures. The activated complex in this case is *an unusual activated complex* because usually an activated complex with a single electronic structure was used when one was thinking of the strong interactions systems, but this ET system is a weakly interacting system. So, this is a key difference and it is probably the reason why this type of treatment was missed in earlier work."

NOTE: As a matter of fact, as discussed in the Appendix, the two forms of the activated complex are in Marcus' case symbolically connected by two arrows, $X^* \rightleftarrows X$, while in Pauling's case one often uses the Arndt–Eistert double-pointed arrow: $\leftrightarrow$ [4, p. 187].

6. **M**: "The reactants are really imbedded in the solvent in the fluctuated state. You never speak of the reacting particles alone, you don't separate them from the other [surrounding particles]. You are talking about what you may call dressed particles, reacting particles with solvent around, influencing them, although it is true that if you are talking of electronic interaction, that's only weakly influenced by the interaction of the solvent molecules surrounding the reactants except when the solvent molecules come between the reactants, that of course strongly influences the electronic interaction."
7. NOTE: **M**. refers below to his **M30** paper of 1960 to make clear the relations of the microscopic states x^*'s to the thermodynamic state X^*. There the potential energy surface $\mathcal{E}_k$ of the reactants crosses that $\mathcal{E}_k^p$ of the products at the surface representing the transition state TS.

   M: "In the 1960 paper you have the intersection of two surfaces [order of] 10^{23}-dimensional. On the intersection, which is one dimension less than 10^{23}, i.e. 10^{23}–1, on the intersection of those surfaces then, you can see that if you regard one member of the pair of the surfaces to be that of one electronic configuration and the other member of the pair to be of the other electronic configuration, they share the same nuclear configurations, they share *the individual distribution of microscopic energies*, they share the same average energy, and they share *the same distribution of microscopic states*, so they share the same entropy. *It is that intersection which is the TS*. One of the members of the intersection, one of the surfaces, consists of all the microscopic configurations x^* on the intersection, thermally distributed. That is the thermodynamic system X^*. X^* is a thermodynamic ensemble and so it consists of an enormous number of microstates x^*."

NOTE: At an "infinite" number of places on the intersection surface, there may be a microscopic reacting system $\boldsymbol{x}^*$ with the same nuclear configuration and the same energy of a microscopic products system x where the electron transfer reaction may happen.

M: "The activated complex or the transition state is of 10^{23}–1 configurations, if the dimension of the space is 10^{23}, it's a hypersurface one dimension less than the entire phase space. That's a transition state."

8. **M**: "*The electron tunneling . . . that's a different way of describing the overlap of electronic wave functions if they don't overlap well, if you have a weak overlap of wave functions.* An alternative description is that to go from one to the other you tunnel and there is a quantitative relation of the two, you can take the overlap of the wave functions, you can use *semi-classical theory* and get a tunneling probability out of it, so there is a whole theory associated with that."

NOTE: **M.** comments below on Pauling's description of the oscillation back and forth between two configurations in relation with the configuration of the system at the TS and that immediately after the electron transfer:

M: "That description is OK for a static system, but your system isn't static, so you don't have that kind of oscillation. If you calculate that way the probability of going from one to the other, you get that right ballpark, but conceptually that theory is not right because the nuclei are moving, so you don't have that oscillation back and forth, you would have it if the nuclei were static and if you were exactly at the crossing point of those crossing energy curves, but that's not what you have, so you get a result that is on the right ballpark, but it's more conceptual than actually describing the process as it occurs when you take into account the combined electron-nuclear motion. But if you had a static picture ... that $2H'_{AB}/h$ gives you sort of the frequency of tunneling. One would use that way if one tried to be moderately modern. You don't do that way because you don't have static nuclei, nevertheless you get some result out of it, which is right ballpark, but you have to take into account that [in the case of Pauling's treatment] you are dealing with static nuclei which don't exist. This is a question almost of beauty of theory, I mean not that the theory which is unbeautiful gives you answers wrong by any magnitude, it's just that's not really the most *complete* picture. Pauling gives you a picture of what it would be statically and in a ballpark, but the system doesn't oscillate back and forth, so physically that part isn't right [for ET] but the oscillation is related to the probability of tunneling. You won't use that equilibrium description, that wouldn't be a good quantum mechanical description, you wouldn't say you have a static system and that there is an equilibrium with the other electronic structure . . . there is no equilibrium there, you would treat the transition. Once you reach the TS you have one electronic configuration and then there is a certain probability that you form the other at the same nuclear configuration, but there is not a back and forth going. Look at how LZ treats it. You go through identical nuclear configurations and you solve the Schrödinger equation that is appropriate for that whole process. In other words, you don't sit still at the intersection region, it's quite un-dynamical, it's quite un-quantum-mechanical. In Pauling's case the interaction is large and you don't start with one configuration and go over to the other in his description."

References

1. L. S. McCarty, G. M. Whitesides, *Angew. Chem. Int. Ed.* **47**, 2188–2207, (2008).
2. G. Herzberg, *Atomic Spectra and Atomic Structure*, p. 26, Dover Publications (1945).
3. N. F. Mott, H. S. W. Massey, *The Theory of Atomic Collisions*, p. 619, Oxford University Press (1965).
4. L. Pauling, *The Nature of the Chemical Bond*, Cornell University Press (1960).
5. R. Platzman, J. Franck, *L. Farkas Memorial Volume*, Oth Cooperative Printing Press, Haifa (1952).
6. The reaction was cited by Martin Karplus in his 2013 Nobel Lecture. A review on it can be found in Ref. [7].
7. D. G. Truhlar, R. E. Wyatt, "$H + H_2$: Potential Energy Surfaces and Elastic and Inelastic Scattering," in *Advances in Chemical Physics*, Vol. 36, pp. 141–204, I. Prigogine, S. A. Rice, Editors, John Wiley & Sons, New York (1977).
8. R. A. Marcus, in *Transactions of the Symposium on Electrode Processes*, pp. 239–248, E. Yeager, Editor, Wiley, New York (1961).
9. F. Di Giacomo, F. Rallo, *Gazz. Chim. Ital.* **101**, 581–590, (1971).
10. R. A. Marcus, P. Siddarth, in *Photoprocesses in Transition Metal Complexes, Biosystems and Other Molecules. Experiment and Theory*, pp. 49–88, Kluwer Academic Publishers (1992).
11. Extension of the Diffusion Controlled Electron Transfer Theory for Intermittent Fluorescence of Quantum Dots: Inclusion of Biexcitons and the Difference of "On" and "Off" Time Distributions Z. Zhu, R. A. Marcus, *Phys. Chem. Chem. Phys.* **16**, 25694–25700, (2014). doi:10.1039/C4CP01274G.
12. C. Kittel, *Introduction to Solid State Physics*, p. 49, 8th Edition, John Wiley & Sons, Inc. (2005).
13. E. E. Nikitin, "Adiabatic and Diabatic Collision Processes at Low Energies," in *Atomic, Molecular and Optical Physics Handbook*, p. 561, G. W. F. Drake, Editor, AIP Press (1996).
14. E. E. Nikitin, L. Zülicke, *Selected Topics of the Theory of Chemical Elementary Processes*, Springer-Verlag Berlin Heidelberg, New York (1978).
15. R. McWeeny, *Coulson's Valence*, 3rd Edition, Oxford University Press (1979).
16. W. Kauzmann, *Quantum Chemistry*, Academic Press (1957).

17. D. R. Herschbach, "Chemical Reaction Dynamics and Electronic Structure," in *The Chemical Bond, Structure and Dynamics*, pp. 175–222, A. Zewail, Editor, Academic Press Inc., Boston, San Diego, New York (1992).
18. F. Di Giacomo, E. E. Nikitin, *Phys. Usp.* **48**(5), 515–517, (2005).
19. J. Lacki, "Stueckelberg and Molecular Physics," in *E. C. G. Stueckelberg, An Unconventional Figure of Twentieth Century Physics, Selective Scientific Papers with Commentaries*, p. 13, J. Lacki, H. Ruegg, G. Wanders, Editors, Birkhäuser (2009).
20. W. Lichten, *Phys. Rev.* **131**, 229, (1963).
21. H. Köppel, "Diabatic Representations: Methods for the Construction of Diabatic Electronic States," in *Conical Intersections, Electronic Structure, Dynamics and Spectroscopy*, World Scientific (2004).
22. G. Herzberg, *Molecular Spectra and Molecular Structure I. Spectra of Diatomic Molecules*, 2nd Edition, Fig. 170 (b), p. 372, Van Nostrand Reinhold Company (1950).
23. A.C. Wahl, et al., *Int. J. Quantum Chem. Symp.* **3** (Pt. 2), 479, (1970).
24. R. S. Berry, S. A. Rice, J. Ross, *Physical Chemistry*, 2nd Edition, Oxford University (2000).
25. A. Zewail, R. Bernstein, "Real Time Laser Femtochemistry: Viewing the Transition from Reagents to Products," in *The Chemical Bond Structure and Dynamics*, p. 223 ff., A. Zewail, Editor, Academic Press, Inc. (1992).
26. R. A. Marcus, *J. Chem. Phys.* **24**, 966–978, (1956).
27. W. Libby, *J. Phys. Chem.* **56**, 863, (1952).
28. R. A. Marcus, Nobel Lecture, *Angew. Chem.* **32**, 1111–1121, (1993).
29. C. J. F. Böttcher, *Theory of Electric Polarization*, Elsevier Scientific Publishing Company, Amsterdam, London, New York (1973).
30. J. O' M. Bockris, A. K. N. Reddy, *Modern Electrochemistry 1*, 2nd Edition, Plenum Press, New York and London (1998).
31. K. J. Laidler, *Chemical Kinetics*, 2nd Edition, McGraw-Hill (1965).
32. R. A. Marcus, in *Lecture Notes, International Summer School on the Quantum Mechanical Aspects of Electrochemistry*, pp. 1–75, P. Kirkov, Editor, Ohrid, Yugoslavia (1971).
33. R. A. Marcus, "Investigation of Rates and Mechanisms of Reactions," in *Techniques of Chemistry*, Vol. 6, Part 1, pp. 13–46, E. S. Lewis, Editor, Wiley, New York (1974).
34. S. S. Shaik, H. B. Schlegel, S. Wolfe, *Theoretical Aspects of Physical and Organic Chemistry The S_N2 Mechanism*, John Wiley & Sons, New York, Chichester, Brisbane (1992).
35. R. J. D. Miller et al., *Surface Electron Transfer Processes*, VCH Publishers, Inc., New York (1995).
36. L. I. Krishtalik, *Charge Transfer Reactions in Electrochemical and Chemical Processes*, Consultants Bureau, New York and London (1986).

37. D. F. Calef, in *Photoinduced Electron Transfer, Part A, Conceptual Basis*, pp. 362–390, M. A. Fox, M. Chanon, Editors, Elsevier, Amsterdam, Oxford, New York (1988).
38. A. M. Kuznetsov, *Charge Transfer in Physics, Chemistry and Biology. Physical Mechanisms of Elementary Processes and an Introduction to the Theory*, Gordon and Breach Publishers, Australia, Austria, China (1995).
39. R. A. Marcus, *Faraday Soc. Discuss.* **29**, 21–31, (1960).
40. R. A. Marcus, *Ann. Rev. Phys. Chem.* **15**, 155–196, (1964).
41. R. D. Levine, R. B. Bernstein, *Molecular Reaction Dynamics and Chemical Reactivity*, Oxford University Press (1987).
42. F. Di Giacomo, F. A. Gianturco, E. E. Nikitin, F. Schneider, *J. Phys. Chem. A* **103**, 7116–7126, (1999).
43. W. J. Moore, *Physical Chemistry*, 4th Edition, Prentice Hall, Englewood Cliffs, NJ (1972).
44. F.Schneider, L. Zülicke, F. Di Giacomo, F. A. Gianturco, I. Paidarovà, R. Polàk, *Chem. Phys.* **128**, 311–320, (1988).
45. H. Goldstein, C. Poole, J. Safko, *Classical Mechanics*, 3rd Edition, Addison Wesley, San Francisco, Boston, New York (2002).
46. R. A. Marcus, *J. Chem. Phys.* **43**, 679–701, (1965).
47. W. C. Gardiner, Jr., *Rates and Mechanism of Chemical Reactions*, W.A. Benjamin, Inc., New York, Amsterdam (1969).
48. P. Atkins, J. De Paula, *Atkins' Physical Chemistry*, p. 896, 8th Edition, Oxford University Press (2006).
49. B. H. Bransden, C. J. Joachain, *Quantum Mechanics*, 2nd Edition, Prentice Hall (2000).
50. C. Cohen Tannoudji, B. Diu, F. Laloe, *Quantum Mechanics*, Vol. 1, Wiley (1977).
51. F. Di Giacomo, F. A. Gianturco, F. Raganelli, F. Schneider, *J. Chem. Phys.* **101**, 3952–3961, (1994).
52. M. S. Child, *Molecular Collision Theory*, Academic Press (1974).
53. L. Pauling, E. Bright Wilson, *Introduction to Quantum Mechanics*, McGraw-Hill Book Company (1935).
54. R. J. Marcus, B. J. Zwolinski, H. Eyring, *J. Phys. Chem.* **48**, 432, (1954).
55. J. Weiss, *Proc. Roy. Soc. (London).* **A222**, 128, (1954).
56. R. P. Feynman, R. B. Leighton, M. Sands, *The Feynman Lectures on Physics. Quantum Mechanics*, Vol. III, Addison–Wesley, Reading, Massachusetts (1965).
57. J. Millman, C. C. Halkias, *Electronic Devices and Circuits*, McGraw-Hill Book Company, Kogakusha Company Ltd, (1967).
58. F. Di Giacomo, *J. Chem. Ed.* **92**(3), 476–481, (2015).
59. F. Di Giacomo, *Found Chem.* **17**, 67–78, (2015).

CHAPTER 2

Foundations of the Theory, Its First Formulation

In this chapter I have summarized the two Marcus' landmark papers of 1956 – Refs. [1] (Part I of the theory) and [2], which laid the foundation of the whole theory. References have been given in keeping with the educational character of this work. The most important results have been emphasized and mathematical proofs have been given in separate starred sections following the results, so they can be omitted by the noninterested reader. In Chapters 5 and 6, the more general and more abstract formulations of the theory in terms of potential energy surfaces and statistical mechanics is dealt with, but the present earlier treatment in terms of the dielectric continuum theory represents an easier introduction to Marcus theory, with its detailed explanation of all the different steps involved in the ET process, and by itself allows an understanding of the applications given by M. in Part II and in Part III. Similarly, one deals with ideal gases before studying the real ones.

2.1. The Reaction Scheme for Bimolecular ET Reactions

The usual way of calculating a reaction rate is that of first determining the free energy of activation and of then introducing it in the absolute reaction rate theory formula [3–7] for the rate constant. But the ET overall reaction scheme corresponds in general to a sequence of steps, some of which may be slow and so must be considered in *calculating the overall rate constant* to be compared with *the observed rate constant* of the reaction sequence. The reaction scheme considered by M. for the bimolecular ET reactions is the following:

$$A + B \underset{k_{-1}}{\overset{k_1}{\rightleftarrows}} X^* \tag{2.1}$$

$$X^* \underset{k_{-2}}{\overset{k_2}{\rightleftarrows}} X \tag{2.2}$$

$$X \xrightarrow{k_3} \text{Products} \tag{2.3}$$

The reverse step of Eq. (2.3) is not considered because we are interested in calculating the rate constant of the overall forward reaction and, moreover, the concentration of the products, starting from a very dilute solution of reactants, would be very low.

The process relative to the constant k_{-1} corresponds to the probability of "a disorganizing motion of the solvent, destroying the polarization appropriate to the intermediate state" so leading to deactivation. Likewise, X can go back to X^* and so a rate constant k_{-2} must be considered. Because of the small electronic interaction that was supposed to exist between reactants, the $X^* \rightarrow X$ process can be slow and is the reason why *two* electronic structures of the TS were considered.

Note that A and B are not necessarily the actual compounds introduced in the reaction system, they may rather be "active entities formed from them" (M.).

The overall rate constant of the reaction sequence is $k_{bi}c_ac_b$ where the c's denote concentrations and k_{bi} is the observed bimolecular rate constant. According to Eq. (2.3), the rate is also given by k_3c_x, so that:

$$k_{bi}c_ac_b = k_3c_x \tag{2.4}$$

The *steady-state equations* [3–7] for the concentrations of X^* and X are given by Eqs. (2.5) and (2.6):

$$\frac{dc_x^*}{dt} = 0 = k_1 c_a c_b - (k_{-1} + k_2)c_x^* + k_{-2}c_x \tag{2.5}$$

$$\frac{dc_x}{dt} = 0 = k_2 c_x^* - (k_{-2} + k_3)c_x \tag{2.6}$$

The meaning of Eq. (2.5) (standard chemical kinetics...) is the following: if during the reaction the concentration of X^* remains constant, this means that the effect of the two reactions leading to *formation* of X^*, with velocities $k_1 c_a c_b$ and $k_{-2}c_x$ is equal to the effect of the reactions leading to *depletion* of it with velocities $k_2 c_x^*$ and $k_{-1}c_x^*$. The same argument is valid for Eq. (2.6). The two equations allow for the determination of those values of c_x^* and c_x for which the conditions $dc_x^*/dt = 0$ and $dc_x/dt = 0$ are valid.

Introducing the value obtained for c_x in Eq. (2.4), Eq. (2.7) is obtained [1]:

$$k_{bi} = k_1/[1 + (1 + k_{-2}/k_3)k_{-1}/k_2] \tag{2.7}$$

M. has shown that when the probability of forming X from X^* is $\geq$ than the probability of X^* reforming A and B, then

$$k_{bi} \sim k_1 \tag{2.8}$$

If the above condition is not valid, the more complex Eq. (2.7) is to be used. A more detailed analysis, given in Marcus' later papers [8–10] uses a nonadiabatic/adiabatic transition probability, replacing the factor $1/[1 + (1 + k_{-2}/k_3)k_{-1}/k_2]$.[(1)]

In an impressive tour de force, M. estimated in his landmark paper [1] all of the rate constants appearing in Eq. (2.7). The description of their calculation is the main purpose of this chapter.

2.2. A Model for Reactants, the Medium, and Their Interactions

We begin with a *model for reactants* that is the simplest and the first one used in the theory.

M. assumes that each reactant may be treated as a sphere which, if the reactant is an ion, may be surrounded by a concentric spherical shell of *saturated dielectric*, that is, of dielectric completely oriented in the ion's electric field, with which is in equilibrium. Outside this spherical shell the solvent medium is supposed to be dielectrically unsaturated, in the 1956 paper [1]. The ion plus the rigid saturated dielectric region is treated as a *conducting* sphere of radius a,[(2)] and in a more general manner in later papers. Let us now have two reacting ions of radii a_1 and a_2. In the first treatment of M., the radii are supposed to be of a fixed value before, during, and after ET. Each reactant is made up also by all of the surrounding solvent molecules (in theory, considering the long-range nature of the electrostatic interactions). The solution is supposed to be dilute enough that other couples of reactants do not interact with the pair we are considering. Moreover, in this first formulation of the theory no ionic atmosphere is considered.

Let us now consider *the free energy of the system* made up of two ions with charges q_1 and q_2 at distance R from each other. There are *four* distinct contributions to it:

(i) The free energy of interaction of all atoms within the first sphere with each other and with the central ionic charge
(ii) The same for the second sphere
(iii) The free energy of interaction of all molecules *outside of the two spheres* with each other and with the charges of the spheres
(iv) The interaction of the two ionic spheres with each other

If we assume that the atoms within the spheres do not change their average position during the mutual approach of the ions, the first two contributions to the formation of the TS are independent of the interionic distance R and therefore do not contribute to the free energy of its formation from the reactants.

In order to calculate contributions (iii) and (iv), it is necessary to consider the properties of the dielectric *outside the spheres.*

2.3. Electrostatic Characteristics of the Transition State

State X^* is considered as a *macroscopic system*, made up by x^* microstates, in which the dielectric medium surrounding the spheres is a *continuum* characterized by a definite value of the *macroscopic* polarization at each point of the system. In order to calculate contribution (iii), we need to determine the *polarization function* in the volume outside that occupied by the spheres. The electronic polarization for X^* will be denoted as $\mathbf{P}_e^*(\mathbf{r})$ and the atomic plus orientation as $\mathbf{P}_u^*(\mathbf{r})$ where $\mathbf{r}$ is the position vector with respect to an arbitrarily chosen origin. The total polarization function is $\mathbf{P}^*(\mathbf{r})$, that is:

$$\mathbf{P}^*(\mathbf{r}) = \mathbf{P}_e^*(\mathbf{r}) + \mathbf{P}_u^*(\mathbf{r}) \tag{2.9}$$

For state X, the corresponding polarization functions are $\mathbf{P}_e(\mathbf{r})$, $\mathbf{P}_u(\mathbf{r})$, and $\mathbf{P}(\mathbf{r})$. These functions have the following properties:

(i) As previously seen, the X^* and X states have the same atomic and orientation polarization functions, that is, $\mathbf{P}_u^*(\mathbf{r}) = \mathbf{P}_u(\mathbf{r})$

(ii) $\mathbf{P}_e^*(\mathbf{r})$ and $\mathbf{P}_e(\mathbf{r})$[(3)] are *always* in equilibrium with the respective electric field strengths, that is, they are determined by the fields:

$$\mathbf{P}_e^*(\mathbf{r}) = \alpha_e \mathbf{E}^*(\mathbf{r}) \quad \text{and} \quad \mathbf{P}_e(\mathbf{r}) = \alpha_e \mathbf{E}(\mathbf{r}) \tag{2.10}$$

where α_e is the polarizability associated with the E-type polarization.

(iii) $\mathbf{P}_u(\mathbf{r})$ is unrelated to the electric field strengths in the *nonequilibrium states.*[(4)]

The meaning of the *macroscopic* polarization functions in relation to the *microscopic structure* of the dielectric media will be now defined following Ref. [11] where it is very clearly explained.

The classical electrostatics definition of **P** is

$$\text{Average } \mathbf{P}(\mathbf{r}) = \langle \mathbf{P}(\mathbf{r}) \rangle$$
$$\equiv \text{dipole moment for } \textit{unit volume} \text{ at position } \mathbf{r}$$

The macroscopic $\mathbf{P}(\mathbf{r})$ is the result of very many microscopic dipoles pointing along the direction of the polarizing electric field. What does it mean to represent *discrete* molecular dipoles by a *continuous macroscopic density function* $\mathbf{P}(\mathbf{r})$? The microscopic electric field inside matter varies abruptly from point to point and also so does in time: the field can be, for instance, very great near an electron at a certain instant and it can be completely different an instant later as the atoms move about because of thermal agitation. One is unable to calculate this wildly varying microscopic field which is also in general of no interest. The problem would be similar to that of keeping track of the instantaneous microscopic density at a point in a liquid or gas. What one does is to calculate the *macroscopic* field inside the matter defined as the *average field* in a region *large enough* to contain *thousands of atoms*, say, so as to allow a *smoothing out* of the uninteresting microscopic fluctuations and be describable by a $\mathbf{P}(\mathbf{r})$. The region must also be small enough to follow the large-scale variations of the field. This is what is meant by macroscopic average field inside matter. So this is the meaning of the $\mathbf{E}(\mathbf{r})$ and $\mathbf{P}(\mathbf{r})$ fields we are considering.

We now want to determine the *contribution to the electric field caused by the polarization itself.*

We know from electrostatics that the individual microscopic dipole **p** exerts an electric potential:

$$\psi_{\text{dip}}(\boldsymbol{r}) = \mathbf{p} \cdot \hat{\boldsymbol{r}}/r^2$$

What is now the potential caused by the polarization $\mathbf{P}(\mathbf{r})$? In each tiny volume element dV we have

$$\mathbf{p} = \mathbf{P}dV$$

so that the potential exerted in **r** by the polarization **P** will be

$$\psi = \int \frac{\mathrm{P} \cdot \hat{\boldsymbol{r}}}{r^2} dV$$

Observing that [11, p. 15]:

$$\nabla \left(\frac{1}{r} \right) = -\frac{\hat{\boldsymbol{r}}}{r^2}$$

we finally have:

$$\psi = -\int \mathbf{P} \nabla \left(\frac{1}{r} \right) dV$$

The *complete potential* when dielectrics are present [12, p. 281] is in general the sum of a contribution from volume charges with volume density $\rho(\mathbf{r})$, of surface charges with surface density $\sigma(\mathbf{r})$ plus the contribution earlier considered of the polarized volume elements *dV*. On the overall we have [1]:

$$\psi(\mathbf{r}') = \int \frac{\rho(\mathbf{r}) dV}{|\mathbf{r} - \mathbf{r}'|} + \int \frac{\sigma(r) dS}{|\mathbf{r} - \mathbf{r}'|} + \int \mathbf{P}(\mathbf{r}) \cdot \nabla_{\mathbf{r}} \frac{1}{|\mathbf{r} - \mathbf{r}'|} dV^{(5)} \tag{2.11a}$$

Note that the integrals run over the entire volume of the dielectric and overall surfaces present.

The problem is now that of determining the free energy of a system whose "solvent configuration" is not in equilibrium with the ionic charges of the TS and so cannot be determined by standard electrostatics, just knowing the ionic charge distribution, the dielectric constant and the overall radius of the TS considered as a sphere whose charge is the sum of the charges of the reactants, as is usually done when the polarization is in equilibrium with the charges [5]. Ref. [2] was devoted to finding an expression for this free energy in terms of equilibrium and nonequilibrium electrostatic macroscopic functions for a two spheres system.

For the description of systems with nonequilibrium electrical polarization, M. uses *three* vectors [2], the electric field strength **E**, the polarization **P**, and the electric field strength $\mathbf{E}_c$ which the charge distribution would exert "if it were in a vacuum rather than in a polarized medium" [2]. The electric field strength in the absence of polarized medium is the negative gradient of the potential due to the polarized spheres and so

$$\mathbf{E}_c(\mathbf{r}') = -\nabla_{\mathbf{r}'}\left\{\int \frac{\rho(\mathbf{r})dV}{|\mathbf{r}-\mathbf{r}'|} + \int \frac{\sigma(\mathrm{r})dS}{|\mathbf{r}-\mathbf{r}'|}\right\}$$

Eq. (2.11a) can be expressed in the form:

$$\psi(\mathbf{r}') = \int (\mathbf{P} - \mathbf{E}_c/4\pi)\cdot\nabla\frac{1}{|\mathbf{r}-\mathbf{r}'|}dV \tag{2.11b}$$

which is valid for any system, equilibrium or not. Only in an equilibrium system can **P** be expressed in terms of **E**.

$$\mathbf{P}(\mathbf{r}') = \alpha\mathbf{E} = -\alpha\nabla\psi \tag{2.12}$$

where α is the total polarizability of the medium and considering that $\mathbf{E} = -\nabla\psi$. The values of $\mathbf{E}_c$, $\mathbf{P}_e$, $\mathbf{P}_u$, **P**, and **E** which obtain in the intermediate state X^* will be designated by an asterisk, while those characteristic of state X will bear no asterisk. Since the U-type polarization is the same in both states, $\mathbf{P}_u^*$ equals $\mathbf{P}_u$.

The electrostatic free energy of any state is generally defined as the reversible work to charge up that state [1, 13, 14]. The electrostatic free energy of the nonequilibrium systems X^* and X will be derived in the next starred paragraph.

We give here the final results of the derivation:

$$F(X^*) = \frac{1}{2}\int\left\{\frac{\mathbf{E}_c^{*2}}{4\pi} - \mathbf{P}^*\cdot\mathbf{E}_c^* + \mathbf{P}_u\cdot\left(\frac{\mathbf{P}_u}{\alpha_u} - \mathbf{E}^*\right)\right\}dV^{(6)} \tag{2.13}$$

$$F(X) = \frac{1}{2}\int\left\{\frac{\mathbf{E}_c^{2}}{4\pi} - \mathbf{P}\cdot\mathbf{E}_c + \mathbf{P}_u\cdot\left(\frac{\mathbf{P}_u}{\alpha_u} - \mathbf{E}\right)\right\}dV \tag{2.14}$$

Where the dot denotes the dot product of two vectors and where α_u is the polarizability of the U-type polarization. α_u can be expressed [2] in terms of the static dielectric constant D_s *and* the optical constant

D_{op}. D_{op} is the square of the refractive index in the visible region of the spectrum [1, 12]:

$$4\pi\alpha_u = D_s - D_{op} \tag{2.15}$$

If the solvent is water, $D_s = 78.5$ and $D_{op} \sim 1.8$ at 25°C.

"The electrostatic contribution ΔF^* to the free energy of formation of the intermediate state X^* from the reactants in the dielectric medium" [1] is found by subtracting from F^* the reversible work $W_{iso}^{s^*}$ required to charge them up when they are isolated, that is, far apart *in the dielectric medium*:

$$\Delta F^* = F^* - W_{iso}^{s^*} \tag{2.16}$$

Similarly for X:

$$\Delta F = F - W_{iso}^{s} \tag{2.17}$$

The reversible work required to charge up a conducting sphere of radius a in vacuum with charge e is $e^2/2a$ and $e^2/2aD_s$ in the dielectric medium [15, p. 204 ff.].

In Fig. 1*, the uncharged system, the charged one with reactants at infinite distance from each other and the charged system X^* with

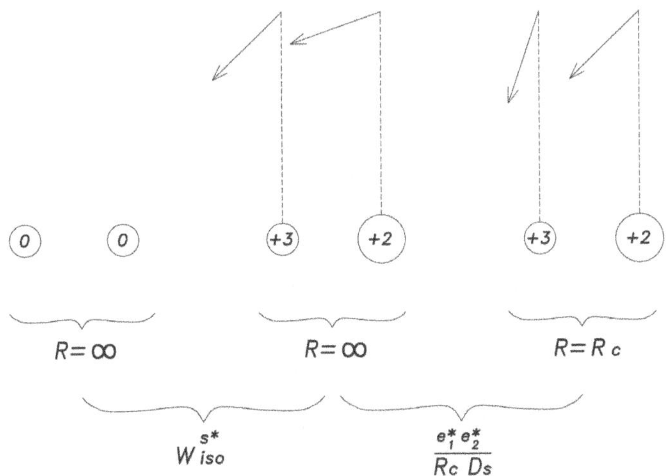

Fig. 1*.

reactants at distance R, are represented using schematic orientational polarization diagrams.

2.4. *The Electrostatic Free Energy for the Nonequilibrium Systems X^* and X

Derivation of the Formulas

In [2], there are two derivations of Eq. (2.14). For Marcus' famous expression of the nonequilibrium polarization expression by a reversible two steps charging process see Chapter 3; a second one, more intuitive, is reported here.

We have just seen how much free energy is stored in a charged sphere. Likewise, the free energy stored in an *induced* dipole $\mathbf{p}$ is given by:

$$F_i = \mathbf{p}^2/2\alpha_o \tag{2.18}$$

where α_0 is the electronic polarizability. It is the energy necessary to *create* the induced dipole in a field $\mathbf{E}$. The free energy of interaction of the dipole with the field is [12, p. 285]:

$$-\mathbf{p}\cdot\mathbf{E} \tag{2.19}$$

If we now imagine turning off the field $\mathbf{E}$ but of *holding fixed* the dipole moment, the energy of interaction (2.19) would become zero, but the energy (2.18) stored in the dipole would remain there.

We are interested in a system made up of a distribution of charges in a polarized medium with polarizabilities $\mathbf{P}_u(\mathbf{r})$ and $\mathbf{P}_e(\mathbf{r})$. The free energy of the whole system is the sum of the free energies of each volume element, considered isolated, and the free energy of interaction among different volume elements. It follows from Eq. (2.18) that the free energy stored in each volume element is

$$\mathbf{P}_u\cdot\mathbf{P}_u/2\alpha_u + \mathbf{P}_e\cdot\mathbf{P}_e/2\alpha_e$$

summing up the two contributions to the total polarization. This follows from Eq. (2.18) successively substituting in it $\mathbf{p} = \mathbf{P}_u dV$ and

$\mathbf{p} = \mathbf{P}_e dV$, $\alpha_0 = \alpha_u dV$ and $\alpha_0 = \alpha_e dV$. We have $\alpha_0 = \mathbf{p}/\mathbf{E} = \mathbf{P}_e dV/\mathbf{E} = (\mathbf{P}_e/\mathbf{E})dV = \alpha_e dV$. The same for α_u. We determine now the contributions arising from the various interactions among different volume elements. To this aim, we consider the contribution to the whole field strength $\mathbf{E}$ given by the fields: $\mathbf{E}_c$, arising from the charges in vacuum, $\mathbf{E}_u$, arising from the U-type polarization and $\mathbf{E}_e$ from the E-type, that is,

$$\mathbf{E} = \mathbf{E}_c + \mathbf{E}_u + \mathbf{E}_e \tag{2.20}$$

We see from Eq. (2.19) that the free energy of interaction of the entire field with the dipole of E-type polarization in a volume dV is:

$$-\mathbf{P}_e dV(\mathbf{E}_c + \mathbf{E}_u + \mathbf{E}_e)$$

Similarly, for the U-type polarization in the same volume dV we have the interaction energy:

$$-\mathbf{P}_u dV(\mathbf{E}_c + \mathbf{E}_u + \mathbf{E}_e)$$

Finally, one should consider the free energy of interaction among charges. This term is the same as that among charges in vacuum because the terms describing the interactions among charges and polarizations have already been considered. We have, for unit volume [12, p. 288]:

$$\mathbf{E}_c \cdot \mathbf{E}_c/8\pi$$

The expression for the electrostatic free energy F of the system will be the sum of the earlier terms integrated over the whole volume of the system. In performing the integration, some terms must be divided by two in order not to count the interactions twice. Let us consider, for example, the integral

$$-\int \mathbf{P}_e \cdot \mathbf{E}_e dV$$

The differential term $\mathbf{P}_e(\mathbf{r}) \cdot \mathbf{E}_e(\mathbf{r})dV$ is the interaction between the dipole $\mathbf{P}_e(\mathbf{r})dV$ and the electric field $\mathbf{E}_e(\mathbf{r})$ which is the resultant in

$\mathbf{r}$ of the electric field strengths due to all the electronically polarized volume elements $dV' \neq dV$. At the same time, the dipole $\mathbf{P}_e(\mathbf{r})dV$ will participate in the electric field $\mathbf{E}_e$on all $dV' \neq dV$. This means that the interaction between two dipole elements, $\mathbf{P}_e(\mathbf{r}_1)dV_1$ and $\mathbf{P}_e(\mathbf{r}_2)dV_2$ say, is counted twice in the integral, which is then to be divided by two in order to correctly take into account the interactions among different electronically polarized volume elements.

The terms which are to be divided by a factor of two are those in the integral:

$$-\int (\mathbf{P}_e \cdot \mathbf{E}_e + \mathbf{P}_u \cdot \mathbf{E}_u + \mathbf{P}_e \cdot \mathbf{E}_u + \mathbf{P}_u \cdot \mathbf{E}_e)dV$$

Marcus thus obtained:

$$F = \int \left\{ \frac{\mathbf{E}_c^2}{8\pi} + \frac{\mathbf{P}_u^2}{2\alpha_u} + \frac{\mathbf{P}_e^2}{2\alpha_e} - \mathbf{P}_e \cdot \left(\mathbf{E}_c + \frac{\mathbf{E}_u}{2} + \frac{\mathbf{E}_e}{2} \right) - \mathbf{P}_u \cdot \left(\mathbf{E}_c + \frac{\mathbf{E}_u}{2} + \frac{\mathbf{E}_e}{2} \right) \right\} dV \quad (2.21)$$

Using Eq. (2.20) and the relations $\mathbf{P}_e = \alpha_e \mathbf{E}$, $\mathbf{P} = \mathbf{P}_e + \mathbf{P}_u$ into Eq. (2.21), he finally obtained:

$$F = \int \left\{ \frac{\mathbf{E}_c^2}{8\pi} + \frac{\mathbf{P}_u^2}{2\alpha_u} - \frac{\mathbf{P} \cdot \mathbf{E}_c}{2} - \frac{\mathbf{P}_u \cdot \mathbf{E}}{2} \right\} dV \quad (2.22)$$

which is identical with Eq. (2.14).

2.5. Restraints Imposed on the Forms X^* and X of the Transition State

In the framework of the 1956 theory, we have seen that the two forms X^* and X of the TS have the following properties:

(i) Same atomic configuration
(ii) Same total electronic energy
(iii) Same U-type polarization, that is, $\mathbf{P}_u^* = \mathbf{P}_u$
Moreover, the transition $X^* \to X$ is of a Franck–Condon type [16] and so

(iv) The momentum distribution of the atoms in X^* and X is the same

Properties (i) and (iv) say that the configurational and the thermal entropy [17] of X^* and X are the same so that the only possible entropy difference between them can arise from a possible difference in electronic degeneracy between products and reactants, denoted by M. as ΔS_e. Designating with Ω^* the product of the electronic degeneracies of the reactants and with Ω the same for the products, ΔS_e is given by [1, 18]:

$$\Delta S_e = k \ln \Omega / \Omega^* \tag{2.23}$$

ΔS_e is usually equal or close to zero [1].

Since $F^* = U^* - TS^*$, $F = U - TS$ and $U^* = U$ for X^* and X, we have then $F - F^* = -T\Delta S_e$.

In his first treatment of the kinetics of ET reactions, M. considers a further restriction on the reactants: all atoms within the sphere of radius a "maintain their relative positions throughout the reaction" so that consequently the radii a_1 and a_2 remain fixed throughout the reaction, as seen earlier.

Because of all of the earlier restrictions, the overall standard free energy of formation of the products from the reactants ΔF^0 can be written as sum of three terms:

(i) $F^* - W_{iso}^{s^*}$, free energy of formation of state X^* from the reactants
(ii) $-T\Delta S_e$, free energy of formation of X state from X^*
(iii) $-(F - W_{iso}^{s})$, free energy of formation of the products from X state:

$$\Delta F^0 = (F^* - W_{iso}^{s^*}) - T\Delta S_e - (F - W_{iso}^{s}) \tag{2.24}$$

This equation "summarizes the restraints imposed upon the two intermediate states."

2.6. Minimization of the Free Energy Subject to the Free Energy Restriction (2.24)

The Formula for ΔF^*

There is an infinite number of pairs of intermediate states X^* and X which could satisfy the free energy restriction (2.24). All of the pairs have the same charge distribution but the **r**-dependent $\mathbf{P}_u(\mathbf{r})$ is different for different pairs. The problem is now of determining *the* $\mathbf{P}_u(\mathbf{r})$ of *that* pair of "groups of atomic configurations" that "contribute to the macroscopic or 'thermodynamic' properties of the transition state TS" [19]. Such a pair, subject to the free energy restriction (2.24), is characterized by *the minimum free energy* F^*, because we are looking for *that* thermodynamic state which has *the maximal probability of formation from the reactants.*(7) In terms of atomic configurations, one can say that many of them can satisfy the energy restriction, but only a group of atomic configurations contribute to the macroscopic or "thermodynamic" properties of the TS, namely those which minimize its free energy of formation from the reactants [19]. I give here the final results for $\mathbf{P}_u$ and F^*. They allow the calculation of ΔF^*, which is one of the major results of the Marcus theory.

The work $W_{iso}^{s^*}$ required to charge up two conducting spheres of charges e_1^* and e_2^* and of radii a_1 and a_2 in the dielectric medium is given by(8):

$$W_{iso}^{s^*} = \frac{e_1^{*2}}{2a_1 D_s} + \frac{e_2^{*2}}{2a_2 D_s} \tag{2.25a}$$

Similarly for W_{iso}^{s}:

$$W_{iso}^{s} = \frac{e_1^{2}}{2a_1 D_s} + \frac{e_2^{2}}{2a_2 D_s} \tag{2.25b}$$

Marcus obtained for *the* $\mathbf{P}_u(\mathbf{r})$ and *the* F^*:

$$\mathbf{P}_u = \alpha_u\{\mathbf{E}^* + (\mathbf{E}^* - \mathbf{E})m\} \tag{2.26}$$

$$F^* = \frac{1}{8\pi}\int\left\{\frac{\mathbf{E}_c^{*2}}{D_s} + m^2\left(\mathbf{E}_c^* - \mathbf{E}_c\right)^2\left(\frac{1}{D_{op}} - \frac{1}{D_s}\right)\right\}dV$$

Using the preceding results Marcus obtained the famous formula for ΔF^*:

$$\Delta F^* = F^* - W_{iso}^{s^*} = \frac{e_1^* e_2^*}{RD_s} + m^2(\Delta e)^2 \left(\frac{1}{2a_1} + \frac{1}{2a_2} - \frac{1}{R} \right) \left(\frac{1}{D_{op}} - \frac{1}{D_s} \right) \quad (2.27)$$

where we have used the conservation of charge relation $e_1^* - e_1 = -(e_2^* - e_2) = \Delta e$ and R is the mean activation distance in the activated complex.

In order to properly appreciate the earlier astonishing final result, the reader should consider that it was obtained by *three* very clever steps. In the first one, M. discovered the beautiful general formulas (2.13) and (2.14) (appearing in the demonstration earlier in the form of Eq. (2.22)) for the electrostatic free energies F^* and F. He then devised the minimization procedure to get the formulas for the free energies of the activated complexes. Finally, starting from an expression for F^* depending on $\mathbf{E}^*$, $\mathbf{P}_u$, and $\mathbf{E}_c$ he ended up—as if by magic—with a formula of great simplicity and elegance and of immediate applicability depending on the charges (e_1^*, e_2^*, Δe), the geometry of the system, (a_1, a_2, R), the dielectric properties of the medium (D_{op} and D_s), and the thermodynamic quantities ΔF^0 and $T\Delta S_e$ which appear through the Lagrangian multiplier m obtained from the formula:

$$-(2m+1)(\Delta e)^2 \left(\frac{1}{2a_1} + \frac{1}{2a_2} - \frac{1}{R} \right) \left(\frac{1}{D_{op}} - \frac{1}{D_s} \right) = \Delta F^0 + T\Delta S_e + \frac{e_1 e_2 - e_1^* e_2^*}{D_S R} \quad (2.28)$$

Let us use it to calculate the value of m for the electron exchange reaction:

$$^*Fe^{+3} + Fe^{+2} \rightarrow {}^*Fe^{+2} + Fe^{+3}$$

We see that for this case each term on the RHS of Eq. (2.28) is zero. On the LHS, the terms in parenthesis are different from zero,

$\Delta e = 1$ and so Eq. (2.28) can only be satisfied if $2m + 1 = 0$, that is, if $m = -\frac{1}{2}$.

To understand what this means, let us consider again formula (2.26) for the $\mathbf{P}_u$ which minimizes F^*.

We see that if $m = -\frac{1}{2}$

$$\mathbf{P}_u = \alpha_u \frac{\mathbf{E}^* + \mathbf{E}}{2}$$

that is, the value of $\mathbf{P}_u$ which minimizes F^* is the one that would be in equilibrium with a hypothetical electric field equal to the arithmetic average of the fields $\mathbf{E}^*$ and $\mathbf{E}$. And is consistent with the intuitive idea that the $\mathbf{P}_u$ of the TS would be in equilibrium with fictitious charges on the ions which would be averages between initial and final charges!

*Proof of Eqs. (2.26), (2.27), and (2.28)

At the minimum of F^*, we shall have $\delta F^* = 0$ (M. uses here the Calculus of Variations, elements of which can be found in books of Advanced Calculus or of Mathematical Methods of Physics and Chemistry, for example in Refs. [20–24] and, more extensively, in Ref. [25]).

M. expresses the variation δF^* as function of the variations $\delta \mathbf{P}_u$ of the function $\mathbf{P}_u$.

The charges are fixed, so $\mathbf{E}_c^*$ is fixed and $\delta \mathbf{E}_c^* = 0$. Computing δF^* from Eq. (2.13) one obtains:

$$\delta F^* = \frac{1}{2} \int \left\{ -\delta \mathbf{P}^* \cdot \mathbf{E}_c^* + \frac{2\mathbf{P}_u}{\alpha_u} \cdot \delta \mathbf{P}_u - \mathbf{P}_u \cdot \delta \mathbf{E}^* - \mathbf{E}^* \cdot \delta \mathbf{P}_u \right\} dV$$

We have seen that $\mathbf{P}^*(\mathbf{r}) = \mathbf{P}_u(\mathbf{r}) + \alpha_e \mathbf{E}_e^*$, so that:

$$\delta F^* = \frac{1}{2} \int \left\{ \left(-\mathbf{E}_c^* + \frac{2\mathbf{P}_u}{\alpha_u} - \mathbf{E}^* \right) \cdot \delta \mathbf{P}_u - \left(\alpha_e \mathbf{E}_c^* + \mathbf{P}_u \right) \cdot \delta \mathbf{E}^* \right\} dV \tag{2.29}$$

δF^* depends on the variations of $\mathbf{P}_u$ and $\mathbf{E}^*$. But we know that $\mathbf{P}_u$ and $\mathbf{E}^*$ *are not independent* because $\mathbf{E}^*$ is the negative gradient of a potential dependent on the charges *but also* on the polarization $\mathbf{P}_u$, as we have seen in Eq. (2.11b). In [2] it is demonstrated that Eq. (2.29) can be expressed as:

$$\delta F^* = \int \left(\frac{\mathbf{P}_u}{\alpha_u} - \mathbf{E}^* \right) \cdot \delta \mathbf{P}_u(\mathbf{r}) dV \tag{2.30a}$$

Putting $\delta F^* = 0$ we have:

$$\int \left(\frac{\mathbf{P}_u}{\alpha_u} - \mathbf{E}^* \right) \cdot \delta \mathbf{P}_u(\mathbf{r}) dV = 0 \tag{2.31a}$$

Note that if there were no constraints on $\delta \mathbf{P}_u(\mathbf{r})$, i.e. if its arbitrary variations would not affect $\mathbf{E}^*$, Eq. (2.24) would be satisfied for arbitrary variations $\delta \mathbf{P}_u(\mathbf{r})$ only if

$$\frac{\mathbf{P}_u}{\alpha_u} - \mathbf{E}^* = 0$$

that is, if

$$\mathbf{P}_u = \alpha_u \mathbf{E}^{*(\mathbf{9})}$$

which is the usual electrostatic relation between $\mathbf{P}_u$ and $\mathbf{E}^*$ for systems with equilibrium U-type polarization. In our nonequilibrium system, the variations $\delta \mathbf{P}_u$ must satisfy the equation of constraint (2.24) where from we see that:

$$\delta F^* - \delta F = 0^{(\mathbf{10})}$$

The equation for the variation δF is (see Eq. (2.30a)):

$$\delta F = \int \left(\frac{\mathbf{P}_u}{\alpha_u} - \mathbf{E} \right) \cdot \delta \mathbf{P}_u(\mathbf{r}) dV \tag{2.30b}$$

Subtracting Eqs. (2.30b) from (2.30a) one has:

$$\delta F^* - \delta F = \int (\mathbf{E} - \mathbf{E}^*) \cdot \delta \mathbf{P}_u dV = 0 \tag{2.31b}$$

Equations (2.30a) and (2.31b) are to be satisfied simultaneously. Using the method of Lagrangian multipliers (see, e.g., Refs. [20–25]), the second condition multiplied by a Lagrangian multiplier m is added to the first one to have:

$$\int \left\{ \frac{\mathbf{P}_u}{\alpha_u} - \mathbf{E}^* + (\mathbf{E} - \mathbf{E}^*)m \right\} \cdot \delta \mathbf{P}_u dV = 0$$

This is an identity valid for every arbitrary variation of $\mathbf{P}_u$ in each volume element. It can be equal to zero only if the term in braces is equal to zero and this condition gives:

$$\mathbf{P}_u = \alpha_u \{\mathbf{E}^* + (\mathbf{E}^* - \mathbf{E})m\} \tag{2.26}$$

So this is the $\mathbf{P}_u$ we were looking for. The first term is the usual electrostatic equilibrium relation between $\mathbf{P}_u$ and $\mathbf{E}^*$. A physical interpretation of Eq. (2.26) was given in the preceding section.

M. expresses now $\mathbf{E}^*$ and $\mathbf{E}$ in terms of $\mathbf{E}_c^*$ and $\mathbf{E}_c$ which can be easily calculated from the known charge distribution. The formulas for $\mathbf{E}^*$ and $\mathbf{E}$ are [1]:

$$\mathbf{E}^* = \frac{\mathbf{E}_c^*}{D_s} - m(\mathbf{E}_c^* - \mathbf{E_c}) \left(\frac{1}{D_{op}} - \frac{1}{D_s} \right)$$

$$\mathbf{E}^* - \mathbf{E} = (\mathbf{E}_c^* - \mathbf{E}_c)/D_{op}$$

With the aid of these equations, Eq. (2.26) for $\mathbf{P}_u$ becomes:

$$\mathbf{P}_u(\mathbf{r}) = \alpha_u \left\{ \frac{\mathbf{E}_c^*}{D_s} - m(\mathbf{E}_c^* - \mathbf{E}_c) \left(\frac{1}{D_{op}} - \frac{1}{D_s} \right) + \frac{m(\mathbf{E}_c^* - \mathbf{E}_c)}{D_{op}} \right\}$$

Introducing the equations for $\mathbf{E}^*$, $\mathbf{E}$, and $\mathbf{P}_u$ into Eq. (2.13) for the electrostatic free energy of state X^*, we obtain:

$$F^* = \frac{1}{8\pi} \int \left\{ \frac{\mathbf{E}_c^{*2}}{D_s} + m^2 (\mathbf{E}_c^* - \mathbf{E}_c)^2 \left(\frac{1}{D_{op}} - \frac{1}{D_s} \right) \right\} dV \tag{2.32}$$

We also have:

$$F^* - F = \frac{1}{8\pi} \int \left\{ \frac{\mathbf{E}_c^{*2} - \mathbf{E}_c^2}{D_s} - (2m+1)(\mathbf{E}_c^* - \mathbf{E}_c)^2 \left(\frac{1}{D_{op}} - \frac{1}{D_s} \right) \right\} dV = \Delta F^0 + T\Delta S_e + W_{iso}^{s^*} - W_{iso}^s \tag{2.33}$$

where from m can be calculated.

We designate now the charges of the reactants 1 and 2 in state X^* by e_1^* and e_2^*. The corresponding charges in X will be e_1 and e_2. The radii a_1 and a_2 are assumed not to change essentially during the course of the reaction.

The vector $\mathbf{E}_c^*$ is the negative gradient of the potential which the reacting ions in state X^* would exert if they were in a vacuum rather than in a polarized medium. The potential exerted at point $\mathbf{r}$ by two ions of charges e_1 and e_2 is:

$$\psi(\mathbf{r}) = \frac{e_1}{r_1} + \frac{e_2}{r_2} \text{ for } \mathbf{r} \text{ outside the ions or on their surfaces, i.e. } r_1 \geq a_1 \text{ and } r_2 \geq a_2$$

$$\psi(\mathbf{r}) = \text{constant for } \mathbf{r} \text{ inside one of the ions, i.e. for } r_1 < a_1 \text{ or } r_1 < a_2$$

where r_1 and r_2 are the distances of the field point r from the centers of the ions. The vector $\mathbf{E_c}$ is $-\nabla\psi$, ψ being given by the earlier equations. We thus have:

$$\begin{aligned} \mathbf{E}_c &= -e_1\nabla\frac{1}{r_1} - e_2\nabla\frac{1}{r_2} \quad && r_1 \geq a_1 \quad \text{and} \quad r_2 \geq a_2 \\ \mathbf{E}_c &= 0 && r_1 < a_1 \quad \text{or} \quad r_1 < a_2 \end{aligned} \tag{2.34}$$

The expressions for $\mathbf{E}_c$ and $\mathbf{E}_c^*$ are introduced into Eq. (2.32) and into Eq. (2.33) and the integrations are performed. The following integrals are used for this purpose:

$$\int \nabla\frac{1}{r_1}\cdot\nabla\frac{1}{r_2}dV = \frac{4\pi}{R}$$
$$\int \nabla\frac{1}{r_i}\cdot\nabla\frac{1}{r_i}dV = \frac{4\pi}{a_i} \tag{2.35}$$

where i can be 1 or 2, R is the distance between the centers of the ions and the integration volume excludes the volume physically occupied by the two ionic spheres, that is, $r_1 \geq a_1$ and $r_2 \geq a_2$ simultaneously.

With the aid of Eqs. (2.24), (2.32), (2.34), (2.35), (2.25), M. obtained Eq. (2.27) and with the aid of Eqs. (2.33), (2.34), (2.35), (2.25) he obtained Eq. (2.28).

2.7. Rate Constants of the Elementary Steps

(a) *Estimation of* k_{-1} *and* k_3

The rate constants k_{-1} and k_3 are associated with the unimolecular reactions of dissociation of X^* to reform the reactants and of X to form the products, respectively. M. considers a model for X^* in which the collision complex in solution is made up of the two reactants contained in a solvent cage [26, pp. 156, 403–404]. The reactants are supposed to vibrate in the solvent cage striking the walls about 10^{13} times a second. The chance of one of them escaping from the cage is α per collision with the walls of the cage, with $\alpha < 1$. In a first mode of dissociation of X^*, this reaction happens every time one of the reactants escapes from the cage, and so the unimolecular rate constant for this mode of X^* decomposition is $10^{13}\alpha$ sec^{-1}.

A second mode of decomposition is "a disorganizing motion of the solvent destroying the polarization appropriate to the intermediate state." We have seen that the relaxation times for orientation and atomic polarizations are from 10^{-11} to 10^{-13} seconds. The atomic polarization is an appreciable fraction of the total U-polarization $\mathbf{P}_u(\mathbf{r})$ so that if it reverts to an unsuitable value, this is enough to destroy the X^* state. The unimolecular rate constant for this mode of decomposition would then be 10^{-13} $\sec^{-1}$. This is no less than the value for the solvent escape mechanism and therefore it is to be considered a *prevalent* mode of decomposition. What has been said for X^* is also valid for X, so that:

$$k_{-1} = k_3 = 10^{13}\ \sec^{-1}$$

(b) *Estimation of* k_1

The equilibrium constant of reaction (2.1) is k_1/k_{-1}. M. calculates the equilibrium constant using the recipe given by Statistical Mechanics, using in this case elementary collision theory and partition functions (see, e.g., Ref. [27], Ref. [17, p. 382], and Ref. [7, p. 93]) and, knowing k_{-1} from the previous section, obtains for k_1 the formula:

$$k_1 = Z \exp(-\Delta F^*/kT) \tag{2.36}$$

where Z is the collision number in the gas phase, see Ref. [7, p. 93, 130], approximately the number of collisions occurring between two neutral species in unit volume in unit time at the mean separation distance in the TS [9]. Z is about 10^{11} liter mole^{-1} $\sec^{-1}$, which is adequate also for the majority of reactions in solution, see Ref. [7, p. 130]. Marcus will later correct this number, using the more precise value 10^{12} he obtained in Ref. [28].

Proof of Eq. (2.36)*

Each of the two reactants A and B has three translational degrees of freedom. In the intermediate state X^*, they become three translational degrees of freedom of the center of gravity of the two reactants, two rotational degrees of freedom about this center and one degree of freedom of the reactants vibrating with respect to each other in the solvent cage. The partition function of the three translational degrees of freedom of reactant 1, a rigid sphere A say, in the gas phase would be

$$(2\pi m_1 kT/h^2)^{1/2}$$

The corresponding factors for reactant 2 (B), and those for the state X^*, are obtained from the preceding formula replacing m_1 by m_2 and

$(m_1 + m_2)$ respectively. The rotational partition function is [27, p. 32]:

$$8\pi^2 \mu R^2 kT / h^2$$

where μ is the reduced mass:

$$\mu = \frac{m_1 m_2}{m_1 + m_2}$$

and R is the distance between the centers of gravity of the reactants. The vibrational partition function for motion within the cage is equal to one, within a factor of three.

Introducing the earlier results into a statistical mechanical expression for the equilibrium constant of reaction (2.1) [27, 7, p. 93], one then gets:

$$\frac{k_1}{k_{-1}} \cong \frac{(2\pi(m_1 + m_2)kT/h^2)^{\frac{3}{2}}(8\pi^2 \mu R^2 kT/h^2)\exp(-\Delta F^*/kT)}{(2\pi m_1 kT/h^2)^{\frac{3}{2}}(2\pi m_2 kT/h^2)^{\frac{3}{2}}}$$

$$= \frac{h}{kT}(8\pi kT/\mu)^{\frac{3}{2}}$$

It now so happens that k_{-1} has the value of 10^{13} sec^{-1}, which is also the value of kT/h. Substituting then k_{-1} with kT/h one gets, after cancellations,

$$k_1 = (8\pi kT/\mu)^{\frac{1}{2}} R^2 \exp(-\Delta F^*/kT)$$

This expression is Eq. (2.36) with Z the collision number in solution taken as approximately equal to that in the gas phase [7, p. 130]. ΔF^*, the free energy of formation of the intermediate state X^* from the reactants

$$A + B \underset{k_{-1}}{\overset{k_1}{\rightleftarrows}} X^* \qquad (2.1)$$

is the free energy for the reaction $A + B \rightarrow$ products.

(c) *Estimation of* k_2 *and* k_{-2}

The free energy difference between states X^* and X is $-T\Delta S_e$ where ΔS_e is given by Eq. (2.23).

The equilibrium constant for the interconversion reaction of X^* and X, Eq. (2.2), is then:

$$k_2/k_{-2} = \exp(\Delta S/k) = \Omega/\Omega^*$$

This ratio will in general be approximately or exactly equal to 1.

In order to estimate k_2 and k_{-2}, some model is to be assumed for the electronic jump process. Such process was treated as an electronic tunneling process in Ref. [29].[(11)] The probability κ_e of an electron tunneling through a barrier from one reactant to the other was estimated to depend exponentially on the tunneling distance r_{ab} between A and B in X^*, that is:

$$\kappa_e = \exp(-\beta r_{ab}) \tag{2.37a}$$

For the ferrous-ferric isotopic exchange reaction in water a value of $\beta = 1.23\,\text{Å}^{-1}$ was estimated. The tunneling distance between the ions is taken by M. as about twice the diameter of a water molecule, that is, 5.5 Å, because each ion is supposed to be strongly bound to an innermost layer of water molecules. M. calculated $\kappa_e = 10^{-3}$ for this distance.

For k_2 we have the following:

$$k_2 = \kappa_e \times \text{number of times per second that the electron strikes the electronic potential barrier} \tag{2.37b}$$

The above number of times per second is given by the frequency of motion of the valence electron in the ground state of the ferrous ion, which is of the order of the frequency of excitation of this electron to the next higher principal quantum number (for a brief discussion of the correlation between classical frequency of revolution in electron orbits and quantum frequency of energy emission see, e.g., Ref. [30]). From data on the energy levels of the ferrous ion [31] M. estimated the frequency to be $2 \times 10^{15}\ \text{sec}^{-1}$, so that:

$$k_2 \cong 2 \times 10^{12}\ \text{sec}^{-1}$$

which is of the same order of magnitude as k_{-1} considering the approximations which have been made.

2.8. Validity of the Assumed Small-Overlap Transition State

In the Marcus' theory of ET reactions, states X^* and X are supposed to be isoenergetic with weak electronic interaction of the reacting particles, that is, with a small overlap of their electronic orbitals.

A fundamental limitation to the statement of equal energy for X^* and X comes from the energy–time uncertainty relation (a very nice discussion of it can be found in Ref. [32]. See also, e.g., Refs. [33, 34]). The energy–time uncertainty relation relates the lifetime τ of a state to the uncertainty $\delta\varepsilon$ of its energy:

$$\delta\varepsilon \cdot \tau \geq \hbar \tag{2.38}$$

which means that the energy of state X^* is broadened by an amount $\delta\varepsilon$ and the same is true for state X. The energies of the two states can then be equal only within an energy interval of $2\delta\varepsilon$ prescribed by the above uncertainty principle. The energy broadening $2\delta\varepsilon$ is then like a *minimal splitting* of PESs in the neighborhood of the nuclear configurations space region where the ET process may happen. Such a splitting is related, as we have seen, to the orbitals' overlap in the activated complex and "*The greater the overlap the shorter will be the lifetimes of* X^* *and* X." But $2\delta\varepsilon$ is related to τ in Eq. (2.38) and τ depends, on its turn, on the rate constants k_2, k_{-2}, k_3, and k_{-1} making up the factor $1/[1+(1+k_{-2}/k_3)k_{-1}/k_2]$ in Eq. (2.7). Their knowledge allows then—through knowledge of the lifetime of the activated complex—an estimate of the splitting $2\delta\varepsilon$ and so of the extent of the overlap in the complex.

From Eq. (2.38), we see then that overlaps and lifetimes are inversely proportional. Let us consider the lifetime of X^* which can disappear through the reactions:

$$X^* \xrightarrow{k_{-1}} A + B$$

$$X^* \xrightarrow{k_2} X$$

Its lifetime is then equal to $1/(k_{-1} + k_2)$ and is essentially the same as that of state X. For $\tau \cong 10^{-13}$ sec $\delta\varepsilon \cong 0.075$ kcal mole^{-1} and $2\delta\varepsilon \cong 0.15$ kcal mole^{-1}. In Eq. (2.24):

$$\Delta F^0 = (F^* - W_{iso}^{s^*}) - T\Delta S_e - (F - W_{iso}^{s})$$

which was derived supposing exactly equal energies for X^* and X, we must replace ΔF^0 with $\Delta F^0 \pm 0.15$ kcal mole^{-1}, a negligible correction with negligible effects on the calculated ΔF^*.

Even if, as a result of large values of k_2 and k_{-2}, the lifetimes were as small as 10^{-14} sec, $2\delta\varepsilon$ would be only 1.5 kcal mole^{-1}, with a relatively small effect on ΔF^*.

A large-overlap activated complex would be characterized by a switching of electronic structures from X^* to X of the order of the electronic frequency in molecules. These frequencies are of about 10^{15} sec^{-1}. The lifetime τ would be about a femtosecond, 10^{-15} sec and $2\delta\varepsilon$ would have the value of 15 kcal mole^{-1}, that is, 0.65 eV, a large value. In Eq. (2.24), ΔF^0 should be replaced by $\Delta F^0 \pm 15$ kcal mole^{-1}, and the restraint involved in the equation would not be very strong anymore.

As a result of the earlier discussion, we have now a quantitative measure to distinguish a small-overlap activated complex from a large overlap one. If the lifetimes of the intermediate states X^* and X are greater than 10^{-14} sec, we have a small-overlap activated complex. "On the basis of the calculations for k_3 and k_2 given previously we infer that a small-overlap activated complex complex may well prevail for many ET reactions." (M.)

NOTE: This first formulation of the theory was later replaced by a later derivation. In this first paper, **M16**, there is a quasiequilibrium between X^* and X with rate constants in the opposite directions of the reaction. In the later derivation in **M53** of 1965, Marcus has a more modern formalism which applies the Transition State Theory in the spirit of Wigner in his paper in the Transactions of the Faraday Society of 1938 and of Eyring, Walter, and Kimball in their book

"Quantum Chemistry" of 1944. There is there a ballistic treatment and a nonadiabatic formalism. The $\boldsymbol{k_{-1}}$ or the equivalent for the electron going through is the rate constant for the fastest velocity at which an ET could occur, with the ET happening in 100 femtoseconds according to the more recent experiments.

2.9. The Interionic Distance *R*

The interionic distance R of the reactants in the intermediate states X^* and X affects the overall reaction rate because it appears in (i) the tunneling probability κ_e, Eq. (2.37), and (ii) in ΔF^*, Eq. (2.27).

(i) κ_e determines the rate constant k_2 in (2.37b). We saw that when $k_2 \geq 10^{13}$ sec^{-1}, the overall reaction rate is independent of k_2. But κ_e depends exponentially on R (r_{ab} in Eq. (2.37)) and, for large R, k_2 will become small and will then affect the overall reaction rate. Because of the exponential decrease of the electron jump probability with R, the reaction rate k_2 will be maximal for minimum R.

(ii) R appears in ΔF^*, Eq. (2.27), in the first term, the one of the Coulombic interaction between ions, and in the second, related to the barrier due to difference in solvation of the ions. For an electron exchange between positive ions, we see that a decrease of R increases the contribute to ΔF^* in the first term and decreases it in the second. If the second term in Eq. (2.27) is greater than the first, so that the major barrier to reaction lies in the difference of solvation about reactants, the reaction will occur most readily for the smallest possible R. As a matter of fact, until now, when the ions come close to each other, we have only considered the overlap of their electronic orbitals. But another kind of overlap should be considered: that of their solvation atmospheres, because, according to M., when the solvation atmospheres overlap they become more similar to each other so as to lower the solvation barrier to reaction and to favor the ET process.

In summary, while in theory the most appropriate value of R is to be found maximizing the overall rate constant with respect to R, limitations in the quantitative knowledge of the electronic jump process suggested to Marcus in 1956 to consider as the most suitable value for R its minimum value, that is, the sum of the radii of the separated reactants:

$$R = a_1 + a_2$$

2.10. The Radius *a*

The radii a_1 and a_2 of the reacting ions enter the expression for rate constant k_1 through ΔF^*, Eq. (2.27). M. considers in its first paper the radii for simple cations and anions and for complex cations and anions.

In the case of simple cations, the radius is taken as the sum of the crystallographic radius plus the diameter of a solvent molecule, water in particular, since the cations have a fairly tightly bound hydration layer. The recipe follows the previously assumed model of a reactant considered as a sphere formed by the ion inside a spherical shell made up by an innermost layer of dielectrically saturated solvent molecules, outside of which the solvent is assumed to be dielectrically unsaturated.

In the case of simple anions, this recipe needs to be modified because the solvent in the first layer appears in this case not to be completely saturated. After this warning, M. did not give a recipe for this case in Ref. [13]. The topic of *effective polarizing radius* will be introduced in the "Effective Radii" and "Ionic Radii" sections of Chapter 3.

In the case of complex ions such as MnO_4^- or $Fe(CN)_6^{-3}$, the degree of dielectric saturation in the first solvation layer is much less than for monoatomic ions because the orienting effect of the ions' electric field varies roughly as the inverse square of the distance from the center of the ion. In this case then the radius of the ion is to be considered as only that of the *naked* ion.

NOTES

1. M. will give in his 1965 paper an expression for k_{bi} given by

$$k_{bi} = \kappa \varrho Z_{bi} \exp(-w^r/kT) \exp[-\Delta F^*(R)/kT].$$

   M: "The 1965 expression is derived from statistical mechanics, the other,

$$k_{bi} = k_1/[1 + (1 + k_{-2}/k_3)k_{-1}/k_2]$$

   is an *approximate macroscopic expression*. In the 1965 expression there are all sorts of microscopic details that are not present in the 1956 paper."

2. M's correction of the statement on p. 970 of [1], "We shall treat an ion plus its rigid, saturated dielectric region as a conducting sphere of radius a.":

   M: "I think it is unfortunate that I called it a conducting sphere... in the later papers of course I didn't have the system as a conducting sphere."

   NOTE: M. believes that he used that model in the first two papers because he was under the influence of Born's work on the charging of ions, where they were considered as conducting spheres.

3. **M**: "Both α_e and α_u depend on temperature. α_e is really related to an optical dielectric constant, so it depends very little on temperature, but it depends a little. α_u depends upon a difference of dielectric constants, one of which is the static dielectric constant and so it definitely depends on temperature."

4. **M**: "The polarization in an equilibrium system is always proportional to the electric field, so that means that as soon as you specify the polarization in an equilibrium system, you specify the electric field everyplace. In a non-equilibrium system that is not true."

5. **Q**: The macroscopic system is formed by the couple of the reactants plus all the other solvent molecules and ions surrounding them, so in theory the potential is exercised by all of the 10^{23} molecules and ions which make up the macroscopic system. But in practice how many are the solvent molecules and ions interacting with the couple of electron exchanging reactants?

   M: "I'd say, maybe a few thousands, because Coulomb forces are long range, and of course it depends on how dilute the solution is. But it is true that usually you have solvent there which dampens the Coulomb forces, so the distance of the solvent molecules and ions affecting the reacting pair doesn't go beyond a few Debye lengths, and you can call *mesoscopic* the system that would cover most of the relevant part of the system. The main advantage of considering it would be for numerical calculations, you wouldn't have to use 10^{23} coordinates."

6. **Q**: In the **M16** paper, you say nothing about the nature of the *reaction coordinate* but looking at the nonequilibrium free energy formula $F = \int \left\{ \frac{\mathbf{E}_c^2}{8\pi} + \frac{\mathbf{P}_u^2}{2\alpha_u} - \frac{\mathbf{P}\cdot\mathbf{E}_c}{2} - \frac{\mathbf{P}_u\cdot\mathbf{E}}{2} \right\} dV$ one sees that at constant E_c the functional F depends on the function $\mathbf{P}_u$, that is, apparently you tacitly consider the $\mathbf{P}_u$ polarization as the ET reaction coordinate.

   M: "In the 1956 paper the reaction coordinate is really $\mathbf{P}_u(m)$. m is really the reaction coordinate. You see, P_u varies all over the place, every point has a different value of P_u, but those P_u's are connected in a way in a form of reaction coordinate with that m, the Lagrangian multiplier, so really if one would ask what really was the reaction coordinate in the 1956 paper, a single coordinate, you could say that m was the coordinate. $m = 0$ when the function P_u is everywhere appropriate to the reactants, if it's $m = -1$ then the function everyplace is appropriate to the products, but in the TS has a value given by that $(2m + 1)\lambda$

formula. . . so m is really the reaction coordinate in the first paper, although I didn't say it explicitly, and it describes how P_u changes along the reaction coordinate everyplace. . . m can serve as a reaction coordinate, in the 1956 paper, but taking on for the actual system the value given by $-(2m + 1)\lambda = \ddot{A}G^0$."

Q: But what exactly is a reaction coordinate?

M: "Every coordinate that leads from reactants to products, and of course that means that it could be some terrible reaction coordinate and the whole idea is to try to find the best within some framework of reaction coordinates. Nothing unique. I mean, there are different reaction coordinates, then better ones, then better ones. . . and getting the best one is a big question and one may not be able to do it. So one does the best one can. One should speak of the best possible reaction coordinate that the writer can think of."

Q: My ideas are much more clear now.

M: So are mine.

Q: Through these discussions things become more clear.

M: Yes, and that's the beauty of discussions.

7. **M**. "If you are thinking of each pair X^* and X as occupying some section of the 10^{23}–1 dimensional hypersurface that constitutes the [whole] TS and if you are assigning to each pair, to each member of the pair a free energy, then the one that has the minimum free energy, that would be the TS [considered in the paper]. You don't have to consider the full thing, you can consider just an important part of it. In other words, it is wise to consider the subset of the hypersurface which is associated with the minimum free energy. The different parts of the hypersurface correspond to different thermodynamic systems and you pick the one that has the minimum free energy. . . you think of that 10^{23}–1 dimensional hypersurface effectively broken into different systems. . . each polarization function being specific of each system."

8. **M**: "The 1956 theory neglects interactions of other pairs of reactants on this particular pair, it neglects the effects of other reactants, it focuses only on this pair... it treats the different pairs as independent of each other. A later paper treats the effects of other ions, of the ionic atmosphere, on the particular pair."

9. **M**. used the variational condition $\delta F = \frac{1}{2}\int\left(\frac{2\mathbf{P}_u}{\alpha_u} - 2\mathbf{E}\right) \cdot \delta\mathbf{P}_u dV = 0$ to demonstrate that $\mathbf{P}_u = \alpha_u\mathbf{E}$. He did so because he wanted to be sure that there were no other constraints that would give the earlier standard equilibrium result $\mathbf{P}_u = \alpha_u\mathbf{E}$, that is, he wanted to be sure that if he didn't impose other conditions he would get the usual expected result.

10. **M**. points out that when one considers the condition $F^* = F$, that means that one is at the intersection of the potential energy surfaces and one should consider that F^* is a function of $\mathbf{P}_u$ and that F^* equals F as function of $\mathbf{P}_u$ and that when one considers δF one means δF obtained by varying $\delta\mathbf{P}_u$. It is from the condition $F^* = F$ that $\delta F^* - \delta F = 0$ follows. If one plots the free energies of reactants and products as functions of $\mathbf{P}_u$, one has $\delta F^* = 0$ at the bottom of the reactants free energy well, $\delta F = 0$ at the bottom of the free energy well of the products with different values of $\mathbf{P}_u$ corresponding to those two minima. One has $\delta F^* - \delta F = 0$ *at the TS* at the same common value of $\mathbf{P}_u$.

11. **M**. comments on the mistake that was made by the authors in Ref. [27] in the calculation of tunneling probability:

    M: "They calculated the probability of tunneling, but that's tunneling of the electron, what they should have done is multiply that tunneling by the number of times the electron is hitting that tunneling barrier, in this classical picture you have sort of classical frequency, but that's 10^{13} per second, so [overlooking that] they made a big error."

References

1. R. A. Marcus, *J. Chem. Phys.* **24**, 966–978, (1956).
2. R. A. Marcus, *J. Chem. Phys.* **24**, 979–989, (1956).
3. K. J. Laidler, *Chemical Kinetics*, 3rd Edition, Harper Collins Publishers (1987).
4. W. C. Gardiner, Jr., *Rates and Mechanism of Chemical Reactions*, W.A. Benjamin, Inc., New York, Amsterdam (1969).
5. S. Glasstone, K. J. Laidler, H. Eyring, *The Theory of Rate Processes*, McGraw-Hill, New York and London (1941).
6. R. E. Weston, Jr., H. A. Schwarz, *Chemical Kinetics*, Prentice-Hall, Inc., Englewood Cliffs, New Jersey (1972).
7. A. A. Frost, R. G. Pearson, *Kinetics and Mechanism*, 2nd Edition, John Wiley & Sons, Inc., New York, London, Sydney (1961).
8. R. A. Marcus, *Discuss. Faraday Soc.* **29**, 21–31, (1960).
9. R. A. Marcus, *Ann. Rev. Phys. Chem.* **15**, 155–196, (1964).
10. R. A. Marcus, *J. Chem. Phys.* **43**, 679–701, (1965).
11. D. J. Griffiths, *Introduction to Electrodynamics*, 3rd Edition, Prentice Hall International Inc. (1999).
12. G. Joos, *Theoretical Physics*, 3rd Edition, Dover Publications, Inc., New York (1986).
13. W. K. H. Panofsky, M. Phillips, *Classical Electricity and Magnetism*, 2nd Edition, Addison-Wesley, Reading, MA (1962).
14. C. Kittel, *Thermal Physics*, John Wiley & Sons, Inc., New York, London, Sidney, Toronto (1969).
15. J. O' M. Bockris, A. K. N. Reddy, *Modern Electrochemistry 1*, 2nd Edition, Plenum Press, New York and London (1998).
16. G. Herzberg, *Molecular Spectra and Molecular Structure, I. Spectra of Diatomic Molecules*, 2nd Edition, Van Nostrand Reinhold Company, New York, Cincinnati, Toronto (1950).
17. K. Denbigh, *The Principles of Chemical Equilibrium*, 4th Edition, Cambridge University Press (1981).
18. E. A. Guggenheim, *Thermodynamics*, 6th Edition, North-Holland Publishing Company, Amsterdam, New York, Oxford (1977).
19. R. A. Marcus, *Can. J. Chem.* **37**, 155, (1959).
20. W. Kaplan, *Advanced Calculus*, 2nd Edition, Addison-Wesley Publishing Company, Reading, Massachusetts, Menlo Park, California, London, Don Mills, Ontario (1973).
21. H. Margenau, G. M. Murphy, *The Mathematics of Physics and Chemistry*, 2nd Edition, D. Van Nostrand Company, Inc., Princeton, NJ (1956).
22. J. Mathews, R. L. Walker, *Mathematical Methods of Physics*, 2nd Edition, W.A. Benjamin Inc., New York (1970).

23. M. Boas, *Mathematical Methods in the Physical Sciences*, 3rd Edition, Wiley, New York (2005).
24. G. Arfken, H.-J. Weber, *Mathematical Methods for Physicists*, 6th Edition, Elsevier Academic Press (2005).
25. F. B. Hildebrand, *Methods of Applied Mathematics*, Dover Publications, Inc., New York (1992).
26. J. I. Steinfeld, J. S. Francisco, W. L. Hase, *Chemical Kinetics and Dynamics*, Prentice Hall (1989).
27. B. Widom, *Statistical Mechanics, A Concise Introduction for Chemists*, Cambridge University Press (2002).
28. R. A. Marcus, *Int. J. Chem. Kinet* **13**, 365, (1981).
29. R. J. Marcus, B. J. Zwolinski, H. Eyring, *J. Phys. Chem.* **58**, 432, (1954).
30. H. E. White, *Introduction to Atomic Spectra*, McGraw-Hill Book Company Inc., New York, St. Louis, San Francisco (1934).
31. C. E. Moore, *Atomic Energy Levels* (National Bureau of Standards, 1952), circular *467*, Vol. II.
32. F. Mandl, *Quantum Mechanics*, John Wiley & Sons, Chichester, New York, Brisbane (1992).
33. B. H. Bransden, C. J. Joachain, *Quantum Mechanics*, p. 75, 2nd Edition, Prentice Hall (2000).
34. J. J. Sakurai, *Modern Quantum Mechanics*, The Benjamin/Cummings Publishing Company Inc., Menlo Park California, Reading, Massachusetts (1985).

CHAPTER 3

Extensions, Electron Transfer at Electrodes, Applications

After the fundamental results of the wonder year 1955 (published ten months later in 1956), M. developed his theory and applied it to a variety of chemical and electrochemical systems. This chapter is devoted to the description of results mainly reported in Refs. [1–7], which span the years 1956–1964. I have also freely resorted to the reviews in Refs. [8–13], altogether a real "Guide of the Perplexed" [14]. Among the early applications are comparisons of calculated and experimental data, examples of how to use the theory to investigate reaction mechanisms and of how to *systematize and correlate* experimental data. These last characteristics have been typical of chemical theories like, say, the theory of the periodic system of elements or that of electrode potentials.

3.1. Theoretical Equations

In Chapter 1, the Landau–Zener–Stueckelberg–Majorana Equation was introduced, giving the probability P of a nonadiabatic transition in the narrow avoided crossing region.

If a transition from the R curve on the left of the col in Fig. 1.4a to the R on the right occurs, the ET *does not* occur because the system remains on the reagents R set of configurations. The probability that the system remains on the lower potential energy surface (PES) passing, through the col of the surface, to the products P set of nuclear configurations is $\gamma = 1 - P$. Marcus denotes with κ the nuclear velocity weighted average of γ [11, p. 181]. The formula for k_{rate} is written by M. as:

$$k_{\text{rate}} = \kappa\rho Z e^{-\Delta F^*/kT} \tag{3.1}$$

The same equation can obviously be written as

$$k_{\text{rate}} = \kappa\rho Z e^{-\Delta F^*/RT}$$

where ΔF^* refers now to a mole of ET reactions.

The M. theory in its early formulation is adiabatic, that is, κ is supposed equal to 1. Note that even if κ were 0.01 the value of ΔF^*_{calc} would only increase by $kT \ln 100$, that is, 2.5 kcal mol^{-1} at 0°C. The symbol ρ is "the ratio of the root mean square fluctuations in separation distance in the activated complex to the root mean square fluctuations of a coordinate for motion away from the intersection surface... and should have *a value of the order of unity*." [11, p. 178]

M. introduces in Ref. [1] the symbol λ:

$$\lambda = \left(\frac{1}{2a_1} + \frac{1}{2a_2} - \frac{1}{R}\right)\left(\frac{1}{D_{op}} - \frac{1}{D_s}\right)(\Delta e)^2 \tag{3.2}$$

in terms of which

$$2m + 1 = -\left[\Delta F^0 - \left(e_1 e_2 - e_1^* e_2^*\right)/D_s R\right]/\lambda \tag{3.3}$$

$$\text{and} \quad \Delta F^* = m^2\lambda + e_1^* e_2^*/D_s R \tag{3.4}$$

which shows very neatly the *two components* of the activation free energy:

(i) the reorganization free energy $m^2\lambda$ from an equilibrium state at distance of approach R and (ii) the Coulombic interaction $e_1^* e_2^*/D_s R$ at R, with the usual meaning of the symbols. R is the average distance between the centers of the reactants in the activated complex [6]. Note that in the following the symbol R is also used for the ideal gas constant.

The energy of activation of a bimolecular reaction is defined in chemical kinetics as:

$$E_a = -\partial \ln k_{bi}/\partial(1/T) \tag{3.5}$$

ΔS^* is expressed in thermodynamics in terms of ΔF^* and T as:

$$\Delta S^* = -\partial \Delta F^*/\partial T \tag{3.6}$$

Introducing Eq. (3.1; with $\kappa \cong 1$) into the defining equation of E_a and neglecting the small dependence of Z on T one gets (using the symbol k_{bi} for k_{rate} and R instead of k):

$$k_{bi} = Z \exp(\Delta S^*/R) \exp(-E_a/RT) \tag{3.7}$$

The terms ΔS^* and ΔF^* are not the same as the usual free energy and entropy of activation $\Delta F^\ddagger$ and $\Delta S^\ddagger$ as in, for example, Ref. [15, p. 195 ff.]. There we find $\Delta F^\ddagger$ and $\Delta S^\ddagger$ defined by the Eyring relation:

$$\begin{aligned} k_{bi} &= (kT/h) \exp(-\Delta F^\ddagger/RT) \\ &= e^2 (kT/h) \exp(\Delta S^\ddagger/R) \exp(-E_a/RT) \end{aligned} \tag{3.8}$$

At room temperature, the term kT/h equals 10^{13} sec^{-1} while Z, the collision number in solution, is about equal to 10^{11} liter mole^{-1} sec^{-1} [16]. M. will later use the more precise value 10^{12} [16a].

Marcus calls ΔS^* and ΔF^* excess entropy and excess free energy of activation. The excess for ΔF^* is the free energy of formation of the intermediate state from the reactants in excess of what it would be if the state were simply the usual transient collision complex at the thermally averaged geometry of two neutral, nonreactive (i.e., not forming or breaking chemical bonds) molecules in a solution. ΔS^* and $\Delta S^\ddagger$, ΔF^* and $\Delta F^\ddagger$ are related in Note 13 [1, p. 868, **M19**, Part II of the Theory].

Equations (3.1) and (3.7) express the rate constant of the redox step in terms of ΔF^* and ΔS^*, and Eqs. (3.4) and (3.6) relate ΔF^*

and ΔS^* to ΔF^0, the standard free energy change of the redox step, and to the *polarizing radii* (vide infra) of the reactants a_1 and a_2. Defining now ΔF_R^0 and ΔS_R^0 as:

$$\Delta F_R^0 \equiv \Delta F^0 + \frac{e_1 e_2 - e_1^* e_2^*}{D_s R} \quad \Delta S_R^0 \equiv -\partial \Delta F_R^0 / \partial T$$

from Eqs. (3.4) and (3.6) we get:

$$\Delta S^* = \frac{1}{2}\Delta S_R^0 \left(1 + \frac{\Delta F_R^0}{\lambda}\right) - \frac{e_1^* e_2^*}{R}\frac{d}{dT}\left(\frac{1}{D_s}\right) - \frac{1}{4}\left[1 - \left(\frac{\Delta F_R^0}{\lambda}\right)^2\right]\frac{d\lambda}{dT} \tag{3.9}$$

In Eq. (3.9), the last term is generally very small and we then have as an approximate equation for ΔS^*:

$$\Delta S^* \cong \frac{1}{2}\Delta S_R^0 \left(1 + \frac{\Delta F_R^0}{\lambda}\right) - \frac{e_1^* e_2^*}{R}\frac{d}{dT}\left(\frac{1}{D_s}\right) \tag{3.10}$$

3.2. Frequency Factors, Activation Energy, and Transition Probability Factor κ

Equation (3.1) is also written [4, p. 156] as:

$$k = A' \exp(-\Delta F^*/RT) \tag{3.11}$$

and the usual Arrhenius expression for k is:

$$k = A \exp(-E_a/RT)$$

The preexponential coefficient A is generally known as the *frequency factor*. The name derives from the interpretation of the expression in terms of the energy barrier to reaction: the exponential states the probability of surmounting the barrier E_a while A is related to the frequency of attempts [17, p. 87]. Writing E_a as $E_a = \Delta F^* + T\Delta S^*$ we see that:

$$A = A' \exp(-\Delta S^*/R) = \kappa Z \exp(-\Delta S^*/R)^{(1)} \tag{3.12}$$

The calculated and experimental frequency factors provide the most direct measure of the probability factors κ [4, p. 160]. M. found a good agreement between experimental and calculated frequency factors for the $MnO_4^{-1,-2}$ and for the $Fe(CN)_6^{-3,-4}$ systems, supporting — at least for these systems — a probability of adiabatic reaction of the order of one (vide infra).

3.3. The Dependence of ΔF^* on ΔF^0 for a Series of Compounds at Constant λ

From Eqs. (3.2) and (3.4), a simple dependence of ΔF^* on ΔF^0 can be deduced for the reaction of a reagent with a series of compounds *chemically similar to each other* in the sense that *the charged part of the molecule remains the same throughout the series* and the molecules differ only in some substituent.

Two such series will be discussed later. The *polarizing radius* is that of the *charged part* of the molecule which causes the electrical polarization of the medium. For each member of a series, it will be taken to be the same, since each has the *same charged group* with the same radius which is then a constant for all members of the series. It will therefore be seen from Eq. (3.2) that λ *is the same for each member of the series.* Accordingly, we are interested in the dependence of ΔF^* on ΔF^0 *at a given* λ, that is, in $(\partial \Delta F^*/\partial \Delta F^0)_\lambda$. We find, from Eqs. (3.3) and (3.4) that:

$$\left(\frac{\partial \Delta F^*}{\partial \Delta F^0}\right)_\lambda = \frac{1}{2}\left(1 + \frac{\Delta F_R^0}{\lambda}\right) \tag{3.13}$$

This predicts that when $\Delta F_R^0/\lambda$ is very small, a plot of ΔF^* *versus* ΔF^0 should be a straight line with a slope of $\frac{1}{2}$. When $\Delta F_R^0/\lambda$ is large, a plot of ΔF^* *versus* ΔF^0 will be approximately linear for small changes in ΔF^0; the slope will be about unity if $\Delta F_R^0/\lambda$ is as large as unity.

3.4. Theoretical Equations for Isotopic Exchange Reactions

For such reactions, the expression for ΔF^* simplifies because $e_1^* = e_2$, $e_2^* = e_1$, and $m = -\frac{1}{2}$.

Moreover, "since, in the present case, reacting particle 1 as a product is the same as 2 as a reactant, and conversely, we have $a_1 = a_2$ and will denote these by a. Since R is taken to equal $a_1 + a_2$ it therefore equals $2a$." The expression for ΔF^* simplifies then to:

$$\Delta F^* = \left\{ \frac{e_1^* e_2^*}{D_S} + \frac{(\Delta e)^2}{4} \left(\frac{1}{D_{op}} - \frac{1}{D_S} \right) \right\} \frac{1}{2a} \tag{3.14}$$

We immediately see that the reorganization term increases as a decreases and Δe increases, as expected from the discussion in Chapter 2.

As D_{op} is approximately independent of temperature, ΔS^* is obtained from Eqs. (3.4) and (3.6):

$$\Delta S^* = -\frac{1}{2a}[e_1^* e_2^* - (\Delta e)^2/4]\frac{\partial\,(1/D_s)}{\partial T} \tag{3.15}$$

When only one electron is transferred, the term $(\Delta e)^2/4$ will generally be appreciably less than $e_1^* e_2^*$ and we would then have:

$$\Delta S^* \cong -\frac{e_1^* e_2^*}{2a}\frac{\partial(1/D_S)}{\partial T} \tag{3.16}$$

The first term of Eq. (3.15) is the entropy change which occurs when two ions approach each other under electrostatic equilibrium conditions. The lowering of ΔS^* for ions of the same sign *when they get close to each other*, that is, the *entropy change which results when two ions are brought together* is due to the enhanced local electric field causing an ordering of the solvent molecules, that is, when e_1^* and e_2^* are like charges, $\Delta S^* < 0$. This is demonstrated in Section 3.5

because it may well not be immediately evident. In the opinion of J. Randles, later seen to be incorrect, for instance, in a discussion in Ref. [5, p. 246]: "with like charges the solvent molecules do not know which way to turn: I would expect that this would represent an increase in entropy." A correct interpretation is given there by Marcus' answer to Randles' remark.

The second term in the equation is the entropy change resulting from the reorganization of the solvent molecules which accompanies the formation of the nonequilibrium state from the equilibrium one at distance R. The smallness of the second term relative to the first is due to a cancellation effect, because accompanying the solvent reorganization there is a decrease of entropy of solvation around one ion which approximately cancels the increase around the other: the solvent near the more highly charged reacting ion becomes less oriented while it becomes more oriented near the less charged ion.

3.5. *Demonstration That the Solvation Entropy of Ions of Like Sign Decreases with Their Distance [5, p. 246]

The free energy of Coulombic repulsion between two ions of like sign and charge is $\Delta F = q^2/D_s R$.

Moreover, the free energy difference between two states of a system at the same temperature is $\Delta F = \Delta U - T\Delta S$. If we now show that for our system the ratio $-T\Delta S/\Delta F \cong 1$, we will have demonstrated that $\Delta S^* < 0$ and that ΔS depends on R as ΔF does.

Recalling that $\Delta S = -(\partial \Delta F/\partial T)$ we have:

$$-\frac{T\Delta S}{\Delta F} = -\left[\frac{T}{\Delta F}\left(-\frac{\partial \Delta F}{\partial T}\right)\right] = \frac{T}{\Delta F}\frac{\partial\left(q^2/Dr\right)}{\partial T}$$

$$= \frac{T}{\frac{q^2}{Dr}}\left[\frac{q^2}{r}\left(\frac{\partial D^{-1}}{\partial T}\right)\right] = \frac{T}{D^{-1}}\left(\frac{\partial D^{-1}}{\partial T}\right)$$

$$= \frac{T}{D^{-1}}\left(-D^{-2}\frac{\partial D}{\partial T}\right) = -T\frac{\partial \ln D}{\partial T}$$

and for water at room temperature we have $-T\frac{\partial \ln D}{\partial T} = 1.1$.

Marcus' comment: "D decreases with increasing T so the RHS of your equation above is positive, for water at room temperature it is 1.1. This makes physical sense, ΔS will get smaller with increasing T (less ordered) and ΔF will get larger with increasing T (smaller D), so the net result for the LHS of eq 1 is positive."

3.6. The Work Terms w and w^p (w^* and w, w^r, and w^p)

The work term is "the free energy change when the reactants are brought together to the separation distance R" [18, p. 690], it is the work required to bring the reactants from infinity to their separation distance R. It was initially taken as e_1e_2/D_sR, which is the Coulombic work to bring the reactants together at the distance R at infinite dilution. M. subsequently considered the possibility of polar (e.g., electrostatic) and nonpolar contributions to the work, and a more realistic situation in which the reactants were in a solution with some electrolyte concentration. For a discussion of the most appropriate value of R, see Section 2.9 of Chapter 2. One should also consider that "When R becomes large κ tends to zero and when R is small the van der Waals' repulsion makes $F^*(R)$ large" [18, p. 685]. The symbol w was then used for the earlier work term "*in the prevailing medium*," and the symbol $-w^p$ was used for the work required to separate the products from R to infinity in that same medium.

Using the new symbols, Eq. (3.4) is written as:

$$\Delta F^* = w + m^2\lambda \tag{3.17}$$

and Eq. (3.3) as:

$$-(2m+1)\lambda = \Delta F^0 + w^p - w \tag{3.18}$$

with m equal to:

$$m = -\left(\frac{1}{2} + \frac{\Delta F^0 + w^p - w}{2\lambda}\right) \tag{3.19}$$

Eq. (3.14) for ΔF^* for isotopic exchange reactions can be written as:

$$\Delta F^* = w + \lambda/4 \tag{3.20}$$

and Eq. (3.15) becomes:

$$\Delta S^* = -\frac{\partial w}{\partial T} - \frac{1}{4}\frac{\partial \lambda}{\partial T} \tag{3.21}$$

I have used here the symbols w and w^p instead than the symbols w^* and w used earlier. A certain confusion may arise because in terms of the older symbolism Eq. (3.15) is written as:

$$\Delta F^* = w^* + m^2\lambda$$

Moreover, w^* is also indicated by Marcus as w^r. One finds then the following couples of symbols for the work terms in Marcus' papers:

$$w^* \quad w, \qquad w \quad w^p, \qquad w^r \quad w^p$$

In the following only the symbols w^r and w^p will be used. M. uses also sometimes the symbol F and sometimes the symbol G for the free energy.

3.7. Inner and Outer Contributions to λ

In the first formulation of the theory (1956), the ion was treated as a sphere inside of which no changes in interatomic distances occurred during the reaction. This assumption was later eliminated when the theory was extended to include the effect of changes Δq_j^0 in bond distances and bond angles in the inner coordination shell of each reactant [6, p. 853]. In the earlier equations, λ will then be in general equal to:

$$\lambda = \lambda_i + \lambda_o \tag{3.22}$$

where λ_o (λ outer) is what we have until now indicated with λ, that is:

$$\lambda_o = \left(\frac{1}{2a_1} + \frac{1}{2a_2} - \frac{1}{R}\right)\left(\frac{1}{D_{op}} - \frac{1}{D_s}\right)(ne)^2 \tag{3.23}$$

and λ_i (λ inner) is given by:

$$\lambda_i = \sum_j \frac{k_j k_j^p}{k_j + k_j^p}(\Delta q_j^o)^2 \tag{3.24}$$

where k_j and k_j^p denote the force constants of the jth vibrational coordinate in a species involved in the reaction when that species is a reactant and when it is a product, respectively. The summation is over both reactants in the homogeneous case and over the one reactant in the electrode case (vide infra). When $(\Delta F^0 + w^p - w^r)/\lambda$ is small, say 1/4, then:

$$\Delta F^* \cong \frac{w^r + w^p}{2} + \frac{\lambda}{4} + \frac{\Delta F^0}{2} \tag{3.25}$$

ΔF^* is then linear in ΔF^0 with a slope 0.5. When w^r and w^p are small and when $\Delta F^0 = 0$ we have $\lambda = 4(\Delta F^*)_0$ (the subscript indicates of $\Delta F^0 = 0$). Thus, the earlier condition for linearity in ΔF^0 can be written as

$$|\Delta F^0/(\Delta F^*)_0| \leq 1 \tag{3.26}$$

a condition often fulfilled in practice. More generally, the *instantaneous* slope of a plot of ΔF^* *versus* ΔF^0 is, according to Eqs. (3.17) and (3.19), $\frac{1}{2}[1 + \Delta F^0/4(\Delta F^*)_0]$ when the work terms are small.

3.8. An Outline of the Theory of Electron Transfer Processes at Electrodes—Introduction

The Marcus theory of the rates of electron transfer reactions at electrodes follows in the steps of the ET theory in solution because chemical and electrochemical redox reactions have many characteristics in common from the point of view of experimental results and theory.

I shall follow here mainly an abridged very clear qualitative description of the theory given by M. in Ref. [5]. The theory had been originally developed in two ONR Reports [19, 20] later published in Ref. [13]. The original theory of 1956 appears here already extended considering not only the effect of the nonequilibrium solvent polarization in inducing ET processes but also the effect of *molecular vibrations* and of *ionic atmospheres*.

The electrochemical processes are characterized by various levels of complexity. In the simpler case, they consist of only one elementary reaction, the redox step. But the process is usually more complicated because the elementary step may be preceded or followed by various chemical reactions and equilibria involving the electrochemically active species. It is then essential to analyze the experimental data in sufficient detail so that the electric current is known as a function of the overpotential *of the redox step*, and not simply of the more usual overpotential of the overall reaction sequence. "When such an analysis has been performed for a complicated electrode reaction, attention can then be focused on *the actual redox step* itself," that is, the one dealt with in the M. theory.

The redox step can be visualized in the way usual in reaction rate theory: the atomic motions in the electrochemical systems happen on PESs which are functions of the coordinates of *all* atoms in the system, which means that to the usual atomic coordinates of solvent and solute we should add the coordinates of the atoms making up:

(i) The electrical double layer at the electrode/solution interface [17, 21–23]
(ii) The electrode itself

By a suitable fluctuation, that is, by a suitable concerted motion of the atoms, the system moves from a region of the many dimensional PES "where the electrochemically active species exists in one valence state to a region where it exists in the other valence state, with the electrode having undergone a corresponding change."

Figure 1 from Ref. [5] represents the simplest schematic potential energy diagram for the electrochemical process;

$$A_{\mathrm{ox}} + ne^{-}\ (\mathrm{metal}) \rightarrow A_{\mathrm{red}} + (\mathrm{metal}) \qquad (3.27)$$

The diagram is very similar to those already seen in Chapter 1 and the electrode appears here as a "giant central ion" [21, p. 774] but with at least two important differences from those of the usual ions. First, the electrode has numerous energy levels and then connecting

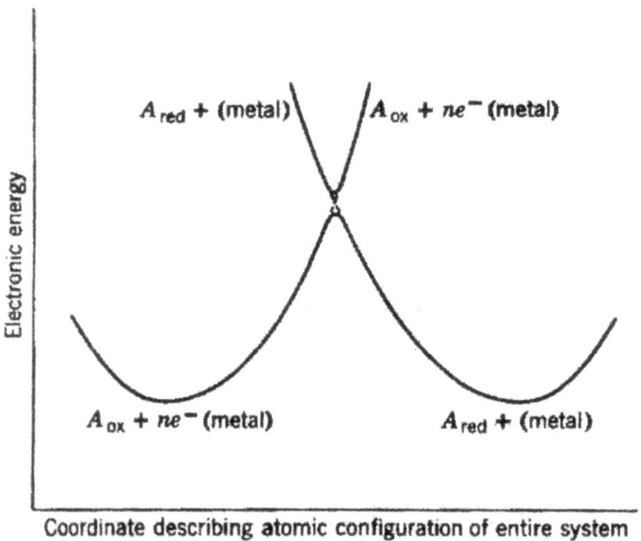

Fig. 1. Cross section of two intersecting electronic energy surfaces in N-dimensional atomic configuration space. Electrochemical process: A_{ox} + ne^- (metal) $\rightarrow$ A_{red} + (metal). The intersecting dashed lines indicate zero overlap of the electronic orbitals of A and metal (From Ref. [5]).

it to an external potential source it is possible to control its potential, charge, and energy levels. If the curves in Fig. 1 refer to a situation of electrodic equilibrium potential E_{eq}, changing the external potential making the electrode more positive the curve to the left is lowered and is instead shifted to higher energies setting the potential to a level more negative [17, 21, 22, 24, pp. 148, 149]. This gives the possibility of changing the activation energies for the cathodic and anodic currents at the electrode, as described in Refs. [17, 21, 22, 24].

The qualitative description of an electrodic ET using Fig. 1 from Ref. [5] is very simplified because the potential energy diagram represents only a mean of the electronic energy levels of the electrode taking part in the ET process. For a metal piece of finite size, a more realistic representation is given in Fig. 2 from Ref. [11] — another Marcus' logo. Each surface there is "a many-electron energy level of the entire reacting system and is a function of the nuclear coordinates." The different R and P surfaces differ in the distribution of the electrons among the one-electron quantum states in the metal piece. As one sees there are many intersections among the different R and P surfaces and consequently the calculation of the velocity averaged adiabatic transition probability κ cannot be dealt with by a simple straightforward application of the Landau–Zener equation.

The use of a single surface of the PESs in Ref. [5] makes sense because the electron from the metal comes from levels in a neighborhood of width kT around the Fermi level E_F [11].

For our purposes, it is possible to classify the electrode reactions into at least three classes.

To a first class belong electrode reactions which happen with chemical bonds broken or formed, as, for instance, in the famous reaction:

$$H_3O^+ + M(+\text{electron}) \rightarrow H_2O + H{-}M \tag{3.28}$$

where M is the electrode. To this class of bond rupture atom transfer type belong also the electrode processes involving deposition of metal cations [17].

In a second class are reactions in which no chemical bonds are broken or formed, such as:

$$MnO_4^- + M(+\text{electron}) \rightarrow MnO_4^{-2} + M \tag{3.29}$$

$$Fe(CN)_6^{-3} + M(+\text{electron}) \rightarrow Fe(CN)_6^{-4} + M \tag{3.30}$$

The ions in the above reactions are "tightly knit" (M.), that is, the interatomic distances in the oxidized and reduced forms are about the same (vide infra). In these reactions, the changes of atomic configurations consist primarily of a large reorientation of the solvent molecules about the species. Such a reorientation, which arises from the change in ionic charge, naturally occurs in reaction (3.28) as well, but is of particular importance in the cases (3.29) and (3.30) in determining the reaction kinetics. To this class also belong reactions in which bonds are considerably stretched [25] but not broken, such as the Co–N bond in:

$$Co(NH_3)_6^{+3} + M(+electron) \rightarrow Co(NH)_6^{+2} + M \qquad (3.31)$$

To a third class might belong electrode reactions in which bonds between ions and electrode surface are formed to facilitate ET and then broken. Such a class of "*bridged activated complexes*" occurs in certain electron transfers in solution where an anion may form a bridging species between two cations [26]. One may consider a reaction in which the ET between cation and electrode happens through a bridging anion adsorbed on the electrode.

From the point of view of Marcus theory, it is then convenient to classify the ET reactions according to whether:

(a) Chemical bonds are broken or formed during ET.
(b) No bond rupture or formation occurs, but merely an ET.
(c) Bonds are formed to facilitate ET and then broken.

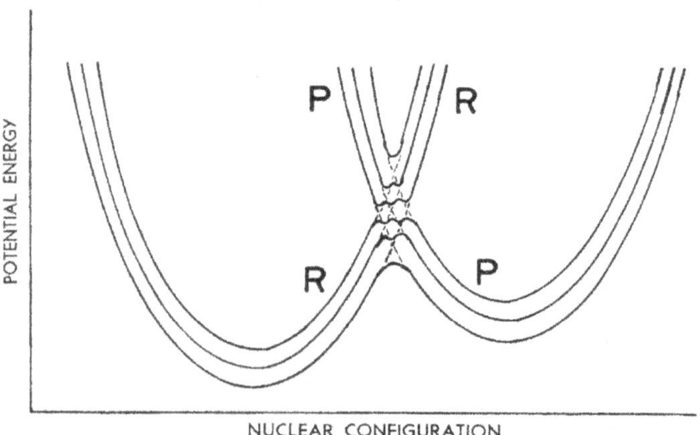

Fig. 2. Same plot as in Fig. 1 but for an electrode reaction. The finite spacing between levels, reflecting the finite size of the electrode, is enormously exaggerated. Only three of the numerous electronic energy levels of this system are indicated. The splitting differs from level to level, and the spacing decreases as the size of the metal increases. (From Ref. [11]).

The Marcus theory of ET at electrodes deals with (b) processes. In these reactions, the reactant approaches the electrode close enough to effect those electronic interactions between its orbitals and those of the electrode which are necessary to induce ET. After ET, the product recedes from the electrode.

3.9. The Theory

In the electrochemical case, the rate of the reaction is measured by the electric current passing through the electrode and that equals the total probability of reaching the saddle point connecting the valley of the reagents and that of the products multiplied by an appropriate frequency factor for passage through and by the concentration of active ions in the solution. "This probability is calculated by means of equilibrium statistical mechanics and a knowledge of the potential energy surface in the valley of the reactants and at the passes (saddle points). The assumption normally made for this purpose is that the reaction hardly disturbs the equilibrium between systems in these two regions." [17]

In describing the PESs of Fig. 1, "let us first examine the hypothetical case of zero electronic interaction." There are then two electronic states to consider, "each having its own many-dimensional PES. In the first of these states, the ion is in an oxidized form A_{ox} and the electrode has the electronic charge distribution appropriate to the metal-solution potential difference. In the second state, the ion is in the reduced form A_{red}, the electrode has lost to it n electrons and again has a charge distribution appropriate to the same potential difference.

The valley of one surface is centered at quite different atomic coordinates from that of the other surface. The difference in ionic charge between reactants and products results in differences in:

(i) Average stable orientation of the solvent molecules outside the coordination shell
(ii) Ionic atmospheres about the ion and electrode
(iii) Interatomic distances inside the inner shells

If these surfaces are plotted versus N atomic coordinates of the entire system, they will intersect on some $(N - 1)$-dimensional surface. At the intersection surface, the atomic coordinates have values which represent a *compromise configuration* of *solvent molecules*, *atmospheric ions*, and *coordination shell*. The compromise is between the stable atomic configurations of the reactants and those of the products. Because of the assumed absence of electronic interactions between ion and electrode, there will be no electron transfer if a system moving on one diabatic PES reaches the intersection region. The system merely stays on the surface appropriate to its electronic configuration. During the *reverse fluctuation*, the system again passes through this intersection region, always staying on the initial PES, and reverts to the stable atomic configuration of the latter.

Let us consider next the case of a weak electronic interaction between ion and electrode. Such a *weak interaction* hardly affects the two PESs, but it *removes the degeneracy* at the intersection, as indicated in Fig. 1, so that we no longer have an intersection and the system is now described by adiabatic PESs. The quantum mechanical reason for this is briefly the following. If ψ_I denotes the electronic wave function for one of the two electronic states in the hypothetical zero-interaction case of Fig. 5*(a) of Chapter 1 and if ψ_{II} denotes the other, ψ_I and ψ_{II} are functions of the atomic coordinates. They describe not only the motion of electrons of the central ion but also those in the molecules outside this ion. They describe for instance the electronic polarization $\mathbf{P}_e$ of the solvent by a central ion in each of the two valence states. Because of the ionic charge difference in the two states, this electronic polarization is quite different. The weak interaction may be shown from quantum mechanics to produce two new wave functions which in the quantum mechanical adiabatic case, the one now considered by M., are essentially $\psi_I + \psi_{II}$ and $\psi_I - \psi_{II}$ for atomic coordinates corresponding to the intersection surface. Because of the weak interaction, these wave functions have energies *slightly different* from those of ψ_I and ψ_{II}, as indicated in Fig. 1.

The wave functions $\psi_I \pm \psi_{II}$ describe the state better than ψ_I or ψ_{II} alone, one of them will have an energy lower than either ψ_I or ψ_{II}, lower by the *resonance energy* in the usual manner familiar to chemists, as indicated in the figure.

The behavior of the system on passing through the avoided crossing region of the PES will depend on the magnitude of this resonance energy. If it is large enough, about 1 kcal mole^{-1}, $\cong$0.043 eV, that is, of the order of 1/100 of the energy of a normal chemical bond, a system passing through this region and initially residing on the lower surface, on the left side of Fig. 1, will remain throughout on the lowest surface and end up on the right side of the lowest surface.

If we consider next the case of an extremely small resonance energy, that is, of an extremely small electronic interaction between ion and electrode, we see that "On passing through the intersection region, the system would tend to stay on the same surface that it stays on in the case of zero interaction. That is, if the system is started on the lower surface in the left of Fig. 1, the atomic motion would tend to carry it to the upper surface on the right and conversely during the return fluctuation. We see that when the interaction is extremely weak, the system 'jumps' from the lower surface to the upper one at the intersection. Actually, *in the intersection region* the concept of PESs and of the Born–Oppenheimer approximation on which it is based breaks down to this extent when the interaction is so extremely weak. During a passage through this region there is, however, a certain probability that the system will not 'jump' and hence a certain probability that the system will end up on the lower surface on the right in Fig. 1, and so have effected an electron transfer." [5]

M. calls electron transfer mechanisms involving *weak interactions* and *extremely weak interactions* the quantum mechanically *adiabatic* and *nonadiabatic* mechanisms respectively, according to customary terminology. At the time of writing of Ref. [5], the few experimental data on preexponential factors tended to favor the adiabatic mechanism.

I shall now report the results of the Marcus treatment of ET processes at electrodes. The demonstration of how they were obtained, in Refs. [19] and [20], will be outlined in Chapter 4.

In order to understand these results, it is necessary to anticipate that M. uses the method of electrostatic images to determine the ion-electrode potential and so the electron released by, say, the reactant to the electrode is *formally* released by the reactant to its *electrostatic image*. The use of images permits the condition of zero electric field at a metal electrode to be satisfied. If the ion lies in solution at a distance $R/2$ from the electrode surface, its image lies inside the electrode also at a distance $R/2$ from the surface. Consequently, the distance between the ion and its image will be R but the electrostatic image is just a *fictitious ion* and so the energy λ for the process will be one half of the value for the homogeneous case where *two real ions* exchange an electron.

The electrostatic method of images is described, for example, in Refs. [27–30].

The current density i as a function of the activation overpotential η_a is:

$$i = neFA' \exp(-\Delta F^*/kT) \tag{3.32}$$

where

$$\Delta F^* = w^r + m^2\lambda/2 \tag{3.33}$$

$$-(2m+1)\lambda/2 = -ne\eta_a + w^p - w^r \tag{3.34}$$

$$\lambda = m^2e^2[(1/a) - (1/R)][(1/D_{op}) - (1/D_s)] \tag{3.35}$$

In these equations, w^r and w^p are the works required to bring the reactant and the product to the electrode, respectively, m and λ are defined by Eqs. (3.34) and (3.35), n is the number of electrons transferred, e is the unit electronic charge, F is the Faraday, a is the ionic radius, R is twice the distance between the electrode and the center of the ion, D_{op} and D_s have the usual meaning, A' is of the order of

1×10^4 to 5×10^4 cm sec^{-2}. The activation overpotential is:

$$\eta_a = -nF(E - E^{0\prime}) \qquad (3.36)$$

where E is the electric potential, $E^{0\prime}$ is the value the latter would have under equilibrium condition if the reactant and product concentrations were equal (thereby $E^{0\prime}$ is the *formal standard potential in the prevailing medium*, that is, at given salt concentration [11, 17]).

If the activation overpotential is small relative to the free energy barrier prevailing at zero overpotential, on expanding m^2 in Eq. (3.33) about its value at $\eta_a = 0$ one obtains a linear dependence of ΔF^* on η_a:

$$\Delta F^* \cong \frac{w^p + w^r}{2} + \frac{\lambda}{8} + \frac{(w^p - w^r)^2}{2\lambda} - \frac{ne\eta_a}{2}\left[1 + \frac{2(w^p - w^r)}{\lambda}\right] \qquad (3.37)$$

One sees from this equation that for a salt concentration sufficiently large for w^p and w^r to be small, the *transfer coefficient*, $\partial \ln k/\partial(ne\eta_a)$, the slope of the Tafel plot [17, 21, 23], should be equal to 0.5.

The reason why, at sufficiently high salt concentration, the electric work necessary to bring the ion to the electrode or away from it is close to zero, is that in the presence of a sufficiently high salt concentration the transference number of the electroactive ion decreases practically to zero and the electrical resistance of the solution and thus the potential gradient throughout is made small. In such conditions, the electroactive ion plays only a negligible part in the electrolyte conduction and its motion is affected by the diffusive force alone, see, for example, Refs. [17, 31–33].

As in the case of homogeneous ET, the Franck–Condon principle plays here an essential role as it was first recognized by J. Randles [4]. Even now a nonradiative ET becomes possible with little or no change of electronic configurations and momenta only at the intersection region of PESs. The electron can be released from the electrode to the oxidized form A_{ox} of the ionic or neutral electroactive molecule only if a suitable previous thermal fluctuation of polarized solvent molecules, of ionic atmospheres and of inner shell vibrations have brought A_{ox} in an X* configuration of energy equal to that of an X configuration of A_{red} [22].

3.10. The Vibrational Motion within the Reactants

After having extended the use of PESs to the description of ET reactions at electrodes, I shall now show how potential energy curves (PECs) can describe the vibrations of the electroactive species inside the inner shell which, in the first formulation of the theory, was approximated as a rigid sphere. The description is here very simple and follows a formulation to be found in Ref. [13].

A simple ET reactions at an electrode can be written as

$$A_1(\mathrm{ox}) + \mathrm{M}(ne) \rightarrow A_1(\mathrm{red}) + \mathrm{M} \tag{3.38a}$$

while a simple ET between species A_1 and A_2 in solution is represented as

$$A_1(\mathrm{ox}) + A_2(\mathrm{red}) \rightarrow A_1(\mathrm{red}) + A_2(\mathrm{ox}) \tag{3.38b}$$

with by now evident meaning of the symbols. In Fig. 1 of Ref. [13] curve R represents the PEC that describes how the potential energy of the reactant system, $A_1(\mathrm{ox}) + \mathrm{M}(ne)$, varies as function of a single vibrational coordinate q varied in $A_1(\mathrm{ox})$. The P curve likewise represents the products system $A_1(\mathrm{red}) + \mathrm{M}$, when a single vibrational coordinate is varied in $A_1(\mathrm{red})$, all the other coordinates remaining fixed. Considering the simple but realistic case of harmonic vibrations, the figure represents the potential energies of R and P as two parabolas crossing at a point. Several facts are noted:

(i) The minima of the two curves occur at different values of q, because the equilibrium bond length in A_{ox} is different (and usually shorter in the case of transition metal ions and a metal–ligand bond) from that in A_{red}.

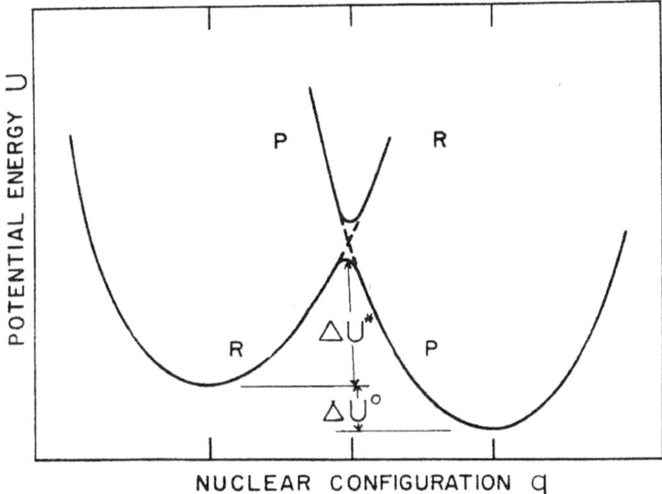

Fig. 3. A plot of the potential energy U of the system consisting of reactants plus solvent (R), along some coordinate q, and of the system consisting of products plus solvent (P), holding all other coordinates fixed, for reaction [3] or [4]. (From Ref. [13]).

(ii) The relative height of the two minima ΔU° depends on the electrostatic potential, the P curve being lowered vertically relative to the R curve by making the electrode more negative, that is, by decreasing $e(\varphi_M - \varphi_S)$, where the φ's represent the Galvani potentials [21, 24] of the metal and solution.

(iii) For a given ΔU°, the barrier height ΔU^* is lower when the difference Δq° of equilibrium bond lengths is smaller.

(iv) When the P curve is lowered relative to the R, the height ΔU^* of the barrier between the minimum of the R curve and the intersection of the R and P curves, is reduced. This is not true in the "inverted region," see Chapter 5.

The two curves have properties which are by now familiar. When the reactant is far from the electrode, the curves in Fig. 3 merely cross in the intersection region (dotted lines there, see also Fig. 5*(a)). When the reactant is close to the electrode, the electronic interaction of the orbitals of the reactant and the electrode perturbs the dotted line curves in Fig. 3 in the intersection region.

At the intersection, the unperturbed electronic R and P quantum states are degenerate, and the degeneracy is broken by the interaction. The new curves are the solid curves. The energy at the maximum of the lower solid curve is less than that at the intersection by an electronic interaction energy ε_{12}, the resonance energy. In this case we see that the *fluctuation* needed for ET to occur is that *of the vibrational coordinate* q which may induce the ET, of course, only if the reactant—reactant distance is sufficiently small for the electronic overlap to occur. Here the estimate of the barrier height in Fig. 3 is very simple. It will be assumed that the resonance energy ε_{12} is small enough that the height of the crossing point is approximately that of the maximum of the lower solid curve in the figure. We let the potential energy of the R and P curves to be approximated by [34, p. 7]:

$$U^r(q) = \frac{1}{2}k(q - q_0^r)^2 \tag{3.39}$$

$$U^p(q) = \frac{1}{2}k(q - q_0^p)^2 + \Delta U^0 \tag{3.40}$$

Where q_0^r and q_0^p are the equilibrium bond lengths for $A(\text{ox})$ and $A(\text{red})$, respectively. ΔU^0 is a linear function of $ne(\varphi_M - \varphi_S)$. At the intersection of the dotted curves we have:

$$U^r = U^p$$

and $q = q^\ddagger$. From the earlier condition one gets:

$$q^\ddagger = \frac{q_0^p - q_0^r}{2} + \frac{\Delta U^0}{k\Delta q^0} \tag{3.41}$$

where $\Delta q^0 = q_0^p - q_0^r$.

Introducing $q^\ddagger$ into Eq. (3.39) one obtains:

$$\Delta U^* = (\lambda_i/4)[1 + (\Delta U_0/\lambda_i)]^2 \tag{3.42}$$

where

$$\lambda_i = \frac{1}{2}k(\Delta q_0)^2. \tag{3.43}$$

And we see that $\lambda_i/4$ is the barrier height when $\Delta U^0 = 0$.

Figure 3 suffices to describe also the vibrational motion within the reactants in the ET reaction in solution:

$$A_1(\text{ox}) + A_2(\text{red}) \rightarrow A_1(\text{red}) + A_2(\text{ox}) \tag{3.44}$$

with completely analogous results.

The rate constant k_r is given by the collision frequency Z (per unit area of electrode and unit time in the electrode case, per unit concentration and unit time in the homogeneous case) multiplied by the Boltzmann factor $\exp(-\Delta U^*/RT)$ and by a Boltzmann factor $\exp(-w^r/RT)$, where w^r is the work term, if any, required to bring the reactants together to some separation distance R, and finally by κ:

$$k_r = \kappa Z \exp[-(w^r + \Delta U^*/RT)] \tag{3.45}$$

When such work terms occur, the ΔU° in Eq. (3.40) is replaced by ΔU_R^0, that is, the ΔU^0 at separation distance R. Thus Eq. (3.42) is replaced by:

$$\Delta U^* = (\lambda_i/4)[1 + (\Delta U_R^0/\lambda_i)]^2 \tag{3.46a}$$

$$\Delta U_R^0 = \Delta U^0 + w^p - w^r, \tag{3.47a}$$

where ΔU^0 is the ΔU at infinite separation and w^p is the work to bring the products together to the separation distance R.

The earlier equations refer to the contribution given by λ_i to the total $\lambda = \lambda_i + \lambda_o$. The equations found here in a very simple way considering a cross section of the PES where only simple harmonic forces are involved are the same as those found using the nonequilibrium dielectric polarization theory for the complicated atomic motions and interactions of the solvent, that is,

$$\Delta F^* = (\lambda/4)[1 + (\Delta F_R^0/\lambda)] \tag{3.46b}$$

$$\Delta F_R^0 = \Delta F^0 + w^p - w^r. \tag{3.47b}$$

This “method of the intersecting parabolas” is used, for instance, in Ref. [35] in a simple description of Marcus theory.

The above calculation of ΔU^* is *classical* and indeed classical mechanics is commonly used nowadays to treat reactive collisions but for some problems a quantum mechanical treatment is needed as, for example, for treating a proton vibration in the equation:

$$H_3O^+ + M(e) \rightarrow H_2O + H{-}M$$

A quantum treatment is given in Refs. [36, 37].

The results of Eqs. (3.45–3.47a) may be extended to all vibrations of the reactant(s) in reactions 3.38 or 3.44. The PECs of Fig. 3 are replaced by PESs plotted as a function of all the q's in the system rather than just one. The coordinate q in the figure represents then in this case some *path in the N-dimensional q-space* and the R and P curves are *profiles* of the actual PESs plotted along that path. λ_i becomes a sum of terms of the type in Eq. (3.43), summed over all vibrations, that is,

$$\lambda_i = -\frac{1}{2}\sum_i k_i(\Delta q_i^0)^2 \tag{3.48}$$

k_i is related to the force constants of a bond, k_i^r and k_i^p, in the oxidized and reduced form for Eqs. (3.38) or (3.44). A more sophisticated treatment of λ_i is given in Ref. [18], leading to k_i being a certain average of k_i^r and k_i^p. More on this in Chapter 6.

“There remain the solvent fluctuations outside of the inner coordination shell of the reactant in Eq. (3.38a) or reactants in Eq. (3.38b). Here, the potential energy functions do not depend on the solvent coordinates (orientations, translations) in the simple quadratic fashion in Eqs. (3.39) and (3.40) of Fig. 1. The treatment of the solvent coordinates is correspondingly more complicated. However, one feature is immediately clear: Just as a thermal fluctuation of vibrational coordinates was needed to reach the intersection region in Fig. 1, a suitable thermal fluctuation of solvent orientation coordinates or reactant's vibrations also permits the system to reach the $N - 1$ dimensional hypersurface (the intersection region). A statistical mechanical treatment of the free energy associated with these fluctuations is given in Ref. [18]. Dielectric continuum theory also permits an estimate of the latter to be made." [13, p. 166]

In the very instructive review of Ref. [13], M. gives a simple and elegant derivation of the free energy change needed to reach the intersection region by the fluctuations of solvent dielectric polarization.

I shall report it now almost verbatim with explanatory notes.

3.11. *The Nonequilibrium Polarization Expression by a Two Steps Charging Process—A Simplified Derivation

Let us consider the homogeneous reaction system in Eq. (3.44) and the modification for the electrode case in Eq. (3.38a). The charges of the reactants are denoted by e_i and their radii by a_i for reactants 1 and 2 ($i = 1, 2$). A superscript p to the e_i denotes the charges of the products. Let D_s and D_{op} be the static and optical dielectric constants of the solvent. The separation distance of the centers of the reactants is denoted by R. Marcus describes here in a simplified way his famous method [38] of calculating the nonequilibrium dielectric polarization of the medium. He produced the state of nonequilibrium polarization $\mathbf{P} = \mathbf{P}_u + \mathbf{P}_e$ by a *two stages reversible charging process*. "Since each step is reversible, the free energy of formation of this non-equilibrium system, i.e. *the free energy of this polarization fluctuation*, can be calculated in a relatively straightforward manner."

The two-step charging process at a given separation distance R is the following:

(i) The charge of each reactant i is changed from e_i to e_i' where e_i' is so chosen as to produce the desired orientational–vibrational dielectric polarization $\mathbf{P}_u$

(ii) The charge of each particle i is changed back from e_i' to e_i *holding the* earlier *orientational–vibrational dielectric polarization* $\mathbf{P}_u$ *fixed.* After step (ii), the desired electronic polarization $\mathbf{P}_e$ is also obtained because $\mathbf{P}_e$ is now in equilibrium with the electrostatic potential, that is, it is determined by $\psi(\mathbf{r})$ which is, on its turn, determined by the ionic charges and by $\mathbf{P}_u$ (see Eq. 2.11b). Note then that, once back to the initial charge e_i this charge is not alone in determining $\mathbf{P}_e$ because $\mathbf{P}_e$ is also determined by $\mathbf{P}_u$.

In the detailed following calculations, the electrostatic potential in the solvent medium at every point $\mathbf{r}$ is denoted by $\psi(\mathbf{r})$.

Step 1

At any stage ν of the charging process, the values of e_i and $\psi(\mathbf{r})$ are denoted by e_i^ν and $\psi^\nu(\mathbf{r})$.

They are given by:

$$e_i^\nu = e_i + \nu(e_i' - e_i) \tag{3.49a}$$

$$\psi^\nu(\mathbf{r}) = \frac{e_1^\nu}{D_s r_1} + \frac{e_2^\nu}{D_s r_2} \tag{3.50}$$

r_i is the distance from the field point r to the center of ion i. ν varies between 0 at the beginning of the charging process an 1 at the end, ψ^ν(r) can then be written as in Eq. (3.50). The potential at the surface of ion 1, due to the medium and to ion 2, is obtained replacing r_1 by a_1 in Eq. (3.50). Φ_1^ν, the potential there minus the self-potential, is obtained by subtracting e_1^ν/a_1 (typo in the original) from Eq. (3.50)

$$\Phi_1^\nu = \frac{e_2^\nu}{D_s r_2} + \frac{e_1^\nu}{a_1}\left(\frac{1}{D_s} - 1\right) \tag{3.51}$$

For the potential at the surface of a spherical conducting ion of charge e in a solvent medium of static dielectric constant D_s, compare, the Born equation in Ref. [21].

The average of Φ_i^ν over the surface of ion 1 is denoted by $\bar{\Phi}_i^\nu$ and is found to be:

$$\bar{\Phi}_1^\nu = \frac{e_2^\nu}{D_s R} + \frac{e_1^\nu}{a_1}\left(\frac{1}{D_s} - 1\right) \tag{3.52}$$

where R is the distance between the centers of the two ions.

The average leading from Eqs. (3.51) to (3.52) is the electrostatic result that the average value of a $1/r_2$ potential from a uniform distribution over a sphere is $1/R$ [39, 40].

On multiplying Φ_i^ν by an increment of charge $d(e_1^\nu)$ where:

$$d(e_1^\nu) = (e_1' - e_1),$$

integrating over ν from 0 to 1, performing the same integration for ion 2 and summing both terms, we obtain the work term W_I required in charging step I:

$$W_1 = \int_0^1 \bar{\Phi}_1^\nu (e_1' - e_1) d\nu + \int_0^1 \bar{\Phi}_2^\nu (e_2' - e_2) d\nu \tag{3.53}$$

Eqs. (3.52, 3.53) yield:

$$W_1 = \frac{e_1 \Delta e_2 - e_2' \Delta e_1}{D_s R} + \left[e_1 \Delta e_1 + \frac{1}{2}(\Delta e_1)^2\right]\left(\frac{1}{D_s} - 1\right)\frac{1}{a_1} + \left[e_2 \Delta e_2 + \frac{1}{2}(\Delta e_2)^2\right]\left(\frac{1}{D_s} - 1\right)\frac{1}{a_2} \tag{3.54}$$

where

$$\Delta e_i = e_i' \tag{3.55}$$

When the initial charges e_1 and e_2 are both zero, the $1/R$ term becomes the usual Coulomb repulsion $e_2' e_1' / D_s R$, the $1/a_1$ term becomes the well-known Born charging term for ion 1, $(e_1'^2/2a_1)[(1/D_s) - 1]$, and the $1/a_2$ term the Born charging term for ion 2.

Step 2

The charges are given by Eq. (3.49b), where ν goes from 0 to 1.

$$e_i^{\nu} = e_i' + \nu(e_i - e_i') \tag{3.49b}$$

For $\nu = 0$, $e_i^{\nu} = e_i'$ (starting point of step 2), for $\nu = 1$, $e_i^{\nu} = e_i$ (the initial charge *to which one goes back in step* 2).

Let $\psi_I(\mathbf{r})$ denotes the potential at the end of step 1 and $\psi^{\nu}(\mathbf{r})$ the potential at any state ν of step 2. The change of potential during step 2 is, for any ν, $\psi^{\nu}(\mathbf{r}) - \psi_I(\mathbf{r})$. Since the medium responds now to the change of charge $\delta\varepsilon_i^{\nu}$ only via the optical dielectric constant D_{op} (remember that $\mathbf{P}_u$ is supposed fixed while $\mathbf{P}_e$ is following the charge) during step 2 we have:

$$\delta\psi^{\nu}(\mathbf{r}) = \frac{\delta e_1^{\nu}}{D_{op}r_1} + \frac{\delta e_2^{\nu}}{D_{op}r_2} \tag{3.56}$$

compare Eq. (3.50). Writing $\delta e_i^{\nu} = \nu(e_i - e_i')$, and $\delta\psi^{\nu} = \psi^{\nu} - \psi_I$ we have:

$$\psi^{\nu}(\mathrm{r}) = \psi_I(\mathrm{r}) + \frac{\nu(e_1 - e_1')}{D_{op}\mathrm{r}_1} + \frac{\nu(e_2 - e_2')}{D_{op}r_2} \tag{3.57}$$

Where

$$\psi_I(\mathrm{r}) = \frac{e_i'}{D_s r_1} + \frac{e_2'}{D_s r_2}. \tag{3.58}$$

Keeping now in mind the process that brought from Eqs. 3.50 to 3.52, we see that "$\bar{\Phi}_1^{\nu}$, the average potential on the surface of ion 1 minus the self-potential, is obtained by subtracting e_1'/r_1 from Eq. (3.58) and $\nu(e_1 - e_1')/r_1$ from the last term but one in Eq. (3.57), then replacing r_1 in those equations by a_1 and averaging the $1/r_2$ in those equations over the surface of ion 1, thereby replacing $1/r_2$ by $1/R$" we thus have:

$$\overline{\Phi_1^{\nu}} = \frac{e_1'}{a_1}\left(\frac{1}{D_s} - 1\right) + \frac{e_2'}{D_s R} + \frac{\nu\left(e_1 - e_1'\right)}{a_1}\left(\frac{1}{D_{op}} - 1\right) + \frac{\nu\left(e_2 - e_2'\right)}{D_{op}R} \tag{3.59}$$

The charging work of both ions done during this step is W_{II}

$$W_{II} = \int_{\nu=0}^{1} \bar{\Phi}_1^{\nu} de_1^{\nu} + \int_{\nu=0}^{1} \bar{\Phi}_2^{\nu} de_2^{\nu}$$

$$\Rightarrow W_{II} = \left(\frac{\Delta e_1 \Delta e_2}{D_{op}} - \frac{e_2' \Delta e_1 + e_1' \Delta e_2}{D_s} \right) \frac{1}{R}$$
$$+ \left[\frac{1}{2} (\Delta e_1)^2 \left(\frac{1}{D_{op}} - 1 \right) - e_1' \Delta e_1 \left(\frac{1}{D_s} - 1 \right) \right] \frac{1}{a_1}$$
$$+ \left[\frac{1}{2} (\Delta e_2)^2 \left(\frac{1}{D_{op}} - 1 \right) - e_2' \Delta e_2 \left(\frac{1}{D_s} - 1 \right) \right] \frac{1}{a_2}. \tag{3.60}$$

The total work ΔG^r done is the sum of W_I and W_{II} and is *the free energy of this fluctuation*. It is:

$$\Delta G^r = W_I + W_{II} = \left(\frac{\Delta e_1^2}{2a_1} + \frac{\Delta e_2^2}{2a_2} + \frac{\Delta e_1 \Delta e_2}{R} \right) \left(\frac{1}{D_{op}} - \frac{1}{D_s} \right) \tag{3.61}$$

with Δe_i given by Eq. (3.55). "If $\Delta G_R^{0\prime}$ denotes the free energy change for unit concentration of reactants at a separation distance R, it is related to the same quantity at infinite separation $\Delta G^{o\prime}$ by an equation similar to Eq. (3.47a)," namely:

$$\Delta G_R^{o\prime} = \Delta G^{0\prime} + w^p - w^r \tag{3.62}$$

Reactants and products have the same *distribution of configurations* (the same set of values of $q^{\ddagger}$) on the intersection hypersurface of Fig. 3, and the same potential energy, $U^r = U^p$, averaged over a distribution of such configurations. Being the distribution of *configurations* and *momenta*, the same for reactants and products on the intersection hypersurface, the entropy is also the same and consequently the free energy of the reactants is there the same as that for the products.

We can write:

$$\Delta G^r - \Delta G^p = \Delta G_R^{0\prime} \tag{3.63}$$

where ΔG^r is given by Eq. (3.61) and ΔG^p has the same expression with e_i replaced by e_i^p.

In order to find $\boldsymbol{e}_1'$ and $\boldsymbol{e}_2'$, that is, the *fictitious* charges of the activated complex that would be in electrostatic equilibrium with the correct *and real* nonequilibrium $\mathbf{P}_u$, M. minimizes ΔG^r in Eq. (3.61) subject to the constraint imposed by Eq. (3.63):

$$\delta \Delta G^r = \left(\partial \Delta G^r / \partial e_1'\right) \delta e_1' + \left(\partial \Delta G^r / \partial e_2'\right) \delta e_2' \tag{3.64}$$

$$\delta \Delta G^r - \delta \Delta G^p = 0 \tag{3.65}$$

Following the recipe of the calculus of variation, one multiplies the second equation by a Lagrange multiplier m and adds the equation so multiplied to the first, introducing then expressions such as Eq. (3.64) for δG^r and δG^p into Eq. (3.65). Setting finally the coefficients of $\delta e_1'$ and $\delta e_2'$ equal to zero, one finds:

$$e_i' = e_i + m(e_i - e_i^p) \quad (i = 1, 2) \tag{3.66}$$

Introducing this result into Eqs. (3.54) and (3.56), one obtains:

$$\Delta G^r = m^2 \lambda_0 \tag{3.67}$$

and

$$-(2m + 1)\lambda_0 = \Delta G_R^{0\prime} \tag{3.68}$$

where

$$\lambda_0 = (\Delta e)^2 \left(\frac{1}{2a_1} + \frac{1}{2a_2} - \frac{1}{R}\right)\left(\frac{1}{D_{op}} - \frac{1}{D_s}\right) \tag{3.69}$$

and $\Delta e = e_i^p - e_i$, the charge transferred. Solving for m from Eq. (3.68), we have:

$$m = -\frac{1}{2} - \frac{\Delta G_R^{0\prime}}{2\lambda_0} \tag{3.70}$$

The free energy barrier to reaction ΔG^* consists of *two terms*: the work term w^r to bring the reactants together and ΔG^r. Introducing m in Eq. (3.67), one obtains:

$$\Delta G^* = w^r + (\lambda_0/4)[1 + (\Delta G_R^{0\prime})/\lambda_0]^2 \tag{3.71}$$

Turning now to the electrode case, the electrostatic potential in step 1 is given by Eq. (3.50) with e_2 replaced by the image charge of reactant 1 inside the electrode, that is, $e_2 \to -e_1$. "The image charge ensures that the potential given by Eq. (3.50) is constant on the surface of the electrode where $r_1 = r_2$." The derivation parallels the one above, where R denotes now the distance from ion 1 to its image, namely twice the distance to the electrode. W_I and W_{II} are computed considering that in the electrode case, only ion 1 needs to be charged, the image being only a fictitious ion (cf. Ref. [27], p. 124). Expressions similar to Eqs. (3.67)–(3.71) are ultimately found, but now we have:

$$-(2m + 1)\lambda_0 = -nF(E - E^{o\prime}) + w^p - w^r \tag{3.72}$$

and

$$\lambda_0 = \frac{1}{2}\left(\frac{1}{a_1} - \frac{1}{R}\right)\left(\frac{1}{D_{op}} - \frac{1}{D_s}\right) \tag{3.73}$$

that is, one obtains for the electrode case:

$$\Delta G^* = w^r + (\lambda_0/4)\,[1 + \{-nF(E - E^{0\prime}) + w^p - w^r\}/\lambda_0]^2 \tag{3.74}$$

where F is the Faraday.

This type of simplified derivation of the earlier results was given earlier in Ref. [10] and made more readily available in a review which is also an excellent review of the theoretical literature up to its time [41]. In order to consider *simultaneously* these fluctuations in solvent polarization and those in reactants' vibrations, one could do so: add to the right side of Eq. (3.61) the term $\frac{1}{2}\sum_i k_i(q_i - q_i^{0r})^2$ and add $\frac{1}{2}\sum_i k_i(q_i - q_i^{0p})^2$ to the corresponding expression for ΔG^p. One proceed then as before using Eqs. (3.64), (3.65) but now Eq. (3.64) contains an extra term $\sum_i k_i(q_i - q_i^{0r})\delta q_i$ (because

$$\partial\left(\frac{1}{2}\sum_i(q_i - q_i^{0r})^2\right)\Big/\partial q_i = \sum_i k_i(q_i - q_i^{0r})$$

and $\delta\Delta G^p$ contains an analogous extra term. One finds, in addition to Eq. (3.66), the result that:

$$q_i^{\ddagger} = q_i^{0r} + m(q_i^{0r} - q_i^{0p}) \tag{3.75}$$

When this $q_i^{\ddagger}$ and the e_i', again given by Eq. (3.66), are introduced into the expression for ΔG^r, Eq. (3.67) follows again but with λ_0 replaced by $\lambda_0 + \lambda_i$, λ_0 being given by Eq. (3.69) and λ_i by Eq. (3.48), with similar remarks for the electrode reaction case.

3.12. Additive Properties of the Reorganization Term λ

For the reorganization term λ, there is an additivity property which is important to establish for subsequent correlations. Moreover, the λ's for homogeneous and electrochemical systems are also related with each other.

Let us consider the electron exchange reaction:

$$A(\mathrm{ox}) + A(\mathrm{red}) \rightarrow A(\mathrm{red}) + A(\mathrm{ox}) \tag{3.76}$$

and let us indicate its λ_0 as λ_0^{aa}:

$$\lambda_0^{aa} = \left(\frac{1}{2a_1} + \frac{1}{2a_2} - \frac{1}{R}\right)\left(\frac{1}{D_{op}} - \frac{1}{D_s}\right)$$

where a_1 and a_2 are the radii of the oxidized and of the reduced form respectively.

Defining λ_0^{a1} as $\lambda_0^{a1} = \frac{1}{2a_1}\left(\frac{1}{D_{op}} - \frac{1}{D_s}\right)$ and likewise $\lambda_0^{a2} = \frac{1}{2a_2}\left(\frac{1}{D_{op}} - \frac{1}{D_s}\right)$, $\lambda_0^R = \frac{1}{R}\left(\frac{1}{D_{op}} - \frac{1}{D_s}\right)$, we have that

$$\lambda_0^{aa} = \lambda_0^{a1} + \lambda_0^{a2} - \lambda_0^R \tag{3.77}$$

This equation is more simply written as:

$$\lambda_{\text{soln}} = \lambda_1 + \lambda_2 - \lambda_r \tag{3.78}$$

If the distance R between the ions is very large, so large that the force field from one reactant does not influence the other, the fluctuations around each reactant are independent [18] and one can write:

$$\lambda_0^{aa} = \lambda_0^{a1} + \lambda_0^{a2} \quad (R \text{ large})$$

λ_0^{a1} and λ_0^{a2} will be simply indicated in the following as λ_1 and λ_2.

If, moreover, $a_1 \cong a_2$ and we indicate with a their mean, we have:

$$\lambda_0^{aa} \cong \lambda_0^{a} \equiv \frac{1}{a}\left(\frac{1}{D_{op}} - \frac{1}{D_s}\right)$$

Considering now the reaction:

$$B(\text{ox}) + B(\text{red}) \rightarrow B(\text{red}) + B(\text{ox}) \tag{3.79}$$

with parallel reasoning and symbols we get:

$$\lambda_0^{bb} \cong \lambda_0^{b} \equiv \frac{1}{b}\left(\frac{1}{D_{op}} - \frac{1}{D_s}\right)$$

Considering finally the *cross-reaction* of the earlier electron exchange reactions:

$$A(\text{ox}) + B(\text{red}) \rightarrow A(\text{red}) + B(\text{ox}) \tag{3.80}$$

always in the hypotheses $a_1 \cong a_2 = a$, $b_1 \cong b_2 = b$, we have:

For R large:

$$\lambda_0^{ab} = \left(\frac{1}{2a} + \frac{1}{2b}\right)\left(\frac{1}{D_{op}} - \frac{1}{D_s}\right)$$

so that finally a cross relation follows among the λ's of reactions (3.76), (3.79), and (3.80):

$$\lambda_0^{ab} = \frac{\lambda_0^a + \lambda_0^b}{2} \tag{3.81}$$

At finite distance, we have the approximate relation

$$\lambda_0^{ab} = \left(\frac{1}{2a} + \frac{1}{2b} - \frac{1}{R}\right)\left(\frac{1}{D_{op}} - \frac{1}{D_s}\right)$$

with $R = a + b$.

For the electrochemical reaction:

$$A(\text{ox}) + \text{M}(ne) \rightarrow A(\text{red}) + \text{M}$$

λ is given by Eq. (3.35) so that:

$$\lambda_{el} = \lambda_0^a - \frac{\lambda_0^R}{2} \tag{3.82}$$

More simply:

$$\lambda_{el} = \lambda_1 - \frac{\lambda_r}{2} \tag{3.83}$$

and we see that for large R:

$$\lambda_{ex} \cong 2\lambda_{el} \tag{3.84}$$

where λ_{ex} and λ_{el} are the λ's for the exchange and electrode reactions, respectively.

The equation agrees with the fact that "in the electrochemical case there is only a contribution" to λ "from one ion" [18].

3.13. A Synopsis of Equations for Chemical and Electrochemical ET Reactions

$$\text{Solution} \quad \Delta F^* = w^r + m^2\lambda \tag{3.85}$$

$$-(2m+1)\lambda = \Delta F^0 + w^p - w^r \tag{3.86}$$

$$\text{Electrode} \quad \Delta F^* = w^r + m^2\lambda/2 \tag{3.87}$$

$$-(2m+1)\lambda/2 = -ne\eta_a + w^p - w^r \tag{3.88}$$

3.14. Correlation between the Rate Constants of Two Electron Exchange Reactions and That of Their Cross-Reaction

It follows from Eqs. (3.1, 3.78, 3.25) that when condition (3.26) is fulfilled, the forward rate constant k_{12} of the cross-reaction (3.80) is correlated to the rate constants k_{11} and k_{22} of the isotopic exchange reactions (3.76) and (3.79).

The cross relation is:

$$k_{12} \cong \sqrt{k_{11}k_{22}K_{12}}\exp\{-(w_{12}^r + w_{12}^p - w_{11}^r - w_{22}^r)/2RT\} \tag{3.89}$$

w_{11}^r and w_{22}^r are the work terms for the isotopic exchange reactions, K_{12}, w_{12}^r, w_{12}^p denote the equilibrium constant and the work terms of Eq. (3.80).

When $w_{12}^r = w_{12}^p = w_{11}^r = w_{22}^p$ like in the case of charge interchange among aquo ions, the formula simplifies to:

$$k_{12} \cong \sqrt{k_{11}k_{22}K_{12}} \tag{3.90}$$

For a somewhat more accurate comparison, k_{12} may be estimated from k_{11} and k_{22} using the complete Eqs. (3.1, 3.17, 3.19) and noting that $\lambda_{12} = (\lambda_{11} + \lambda_{22})/2$. When the work terms are negligible, we have:

$$k_{12} = \sqrt{k_{11}k_{22}K_{12}f} \tag{3.91}$$

where $\ln f = (\ln K_{12})^2/4\ln(k_{11}k_{22}/Z^2)$ [6, p. 856].

The cross relation was studied very extensively by N. Sutin and coworkers [42, 43].

In the words of Henry Taube: "The Marcus correlation is a powerful one, leading as it does to a calculation in most instances of a specific rate to an order of magnitude or so" [44]. It is "a touchstone of normal behavior" for outer sphere activated complexes [45].

3.15. Correlation between Isotopic Electron Exchange Rate and Corresponding Electrochemical Rate Constants

"For an isotopic exchange reaction between ions differing only in valence state, $\Delta F^0 = 0$, $w^r = w^p$, and hence $m = -0.5$ in Eq. (3.19). In the 'exchange current' of the corresponding electrochemical system $\eta_a = 0$ by definition, and $m = -0.5$, if the work term $w^r - w^p$ is small. The λ_1's and λ_2's in Eq. (3.78) are all equal" [6, p. 854]. Considering Eqs. (3.78) and (3.83) "It then follows that:

$$\lambda_{ex} \leq 2\lambda_{el} \tag{3.92}$$

(= or < according as the reactant can or cannot penetrate the solvent layer adjacent to the electrode). From a physical viewpoint, the factor of two enters in the exchange system because two ions and their solvation shells are undergoing rearrangement in forming the activated complex, while in the electrochemical system there is but one such a particle."

*Demonstration by Marcus: "In solution and if $a_1 = a_2$

$$\lambda_o^{ex} = (ne)^2 \left(\frac{1}{2a} + \frac{1}{2a} - \frac{1}{r}\right)\left(\frac{1}{D_{op}} - \frac{1}{D_s}\right)$$

When $r = 2a$

$$\lambda_o^{ex} = (ne)^2 \left(\frac{1}{2a}\right)\left(\frac{1}{D_{op}} - \frac{1}{D_s}\right) \tag{1}$$

For the electrode

$$\lambda_o^{el} \equiv (ne)^2 \left(\frac{1}{2a} - \frac{1}{2R}\right)\left(\frac{1}{D_{op}} - \frac{1}{D_s}\right)$$

Where R = distance from the ion to its charge image. So

$$R \geq 2a.$$

(equal $2a$ when the ion is in contact with the surface of the electrode). For the case that $R = 2a$ we have

$$\lambda_o^{el} = (ne)^2 \left(\frac{1}{4a}\right)\left(\frac{1}{D_{op}} - \frac{1}{D_s}\right) \tag{2}$$

which is 1/2 of the value in Eq. (1)."

Marcus' comment: "Let's pick the simplest mode, namely that you have that dielectric continuum, there are no subtleties about saturation, and things like that. Now, if the ion can get up and touch the electrode, then the equality sign would prevail, just looking at the equations. If the ion can't get that close, then that means that it doesn't interact much with its image charge, and its image charge would cancel some of the field, and that's all part of this inequality here, so the image charge is cancelling less if the ion can't come up close, because of some absorbed layer, and then in that case, because of the image effect, for R bigger than $2a$ the $\boldsymbol{\lambda_{el}}$ is bigger."

"It thus follows that:

$$\Delta F_{ex}^* \leq \Delta F_{el}^* \tag{3.93}$$

when w^r and w^p are small in both the *ex* and *el* experiments. From Eq. (3.1), we then expect that:

$$\sqrt{k_{ex}/10^{11}} \geq k_{el}/10^4 \tag{3.94}$$

where k_{ex} and k_{el} are in units of liter mole sec^{-1} and cm sec^{-1}, respectively. Another factor tending to favor the '>' sign is the existence, if any, of inactive sites due, say, to any strongly absorbed foreign particles. More recently it has been concluded theoretically that under conditions neither the earlier deduction of this $\sqrt{k_{ex}}/k_{el}$ relation nor that of the 0.5 slopes of the ΔF^* plots should be affected if one or both of the reactants form relatively weak complexes with other ions. The ΔF^*'s are then corresponding to the *pseudo-rate constants*, 'constants' which depend on the concentrations of these other ions." [6, pp. 854–855].

3.16. Chemical and Electrochemical Transfer Coefficients

When the work terms can be made small by using high electrolyte concentration, or when they are essentially constant, one may draw from Eq. (3.25) the following variations in the plots of ΔF^* *versus* ΔF^0 and of ΔF^* *versus* $-nF\eta_{\mathrm{a}}$:

1. In the oxidation—reduction reaction of *a given reagent* with a series of *related compounds* such that the reactions' ΔF^0 is essentially the only parameter varied, a plot of ΔF^* *versus* ΔF^0 and hence of $\log k$ *versus* $\log K$ should be linear with a slope of 0.5 for ΔF^0's satisfying Eq. (3.26).
2. In the electrochemical case, the corresponding plot of ΔF^* *versus* $-nF\eta_a$ (or of $-RT/nF \ln k$ *versus* electrode potential), the *electrochemical transfer coefficient*, should also be linear with a slope of 0.5.

By analogy, M. calls the slope of the ΔF^* *versus* ΔF^0 plot in case (1) the "*chemical transfer coefficient*" of the reaction [6].

3.17. Taube's Definition of Inner-Sphere and Outer-Sphere Electron Transfer Reactions

An important classification of ET reactions due to H. Taube, is that of *inner-sphere* and *outer-sphere* ET reactions: "The distinction between inner-sphere and outer-sphere activated complex... is fundamentally between reactions in which electron transfer takes place from one primary bond system to another (outer-sphere mechanism), and those in which electron transfer takes place within a single primary bond system (inner-sphere mechanism)," from Ref. [45, p. 28].

Using the former nomenclature, we can say that the Marcus theory generally applies to ET reactions with outer-sphere activated complex mechanism. This kind of mechanism operates, for example, for redox reactions both self-exchange and cross-reactions among the complexes in Table 1 of Ref. [44]. Many reactions to which they apply are reported in Ref. [24].

3.18. Atom versus Electron Transfer

"Isotopic exchange reactions may occur by the alternative mechanisms of atom transfer and of electron transfer." [1, p. 868]. Let us consider for instance the reaction:

$$Cr^{+2} + CrCl^{+2} \rightarrow Cr^{+3} + CrCl^{+1}$$

In the case of an electron transfer mechanism, an electron would jump from Cr^{+2} to $CrCl^{+2}$ and we would directly have the products. In the case of an atom transfer mechanism, a Cl^{-} would transfer from $CrCl^{+2}$ to Cr^{+2}. The Cr^{+2}–$CrCl^{+2}$ isotopic exchange reaction has been shown to possess an atom transfer mechanism [46]. "On the other hand reasonable evidence was obtained for a bridge-activated complex electron transfer mechanism for the reduction of $IrCl_6^{-3}$ and of $Fe(CN)_6^{-4}$ by Cr^{+2}" [47]. The Marcus theory does not in general apply to atom or group transfer reactions or to bridged complex electron transfer reactions.

3.19. Three Groups of ET Reactants

M. classifies the reactants "somewhat loosely" in three groups [4, p. 156].

Class I consists of species for which the interatomic distances in the coordination shells of the oxidized and the reduced form are essentially the same.

Class II consists of species in which the bonds of one form are slightly stretched or compressed compared with the other form.

Class III consists of those in which large stretchings or compressions appear.

Examples of Class I: MnO_4^{-} – MnO_4^{-2} and probably $Fe(CN)_6^{-3}$ – $Fe(CN)_6^{-4}$.

Examples of Class II: many hydrated metal cations such as $Fe(H_2O)_6^{+3}$ – $Fe(H_2O)_6^{+2}$.

Examples of Class III: $Co(NH_3)_6^{+3} - Co(NH_3)_6^{+2}$ and $Cr(H_2O)_6^{+3} - Cr(H_2O)_6^{+2}$" [4, p. 156].

The theory of Marcus in its first formulation applies in general to the "tightly knit covalently bound ions" reactants [3, p. 426] of Class I. The extension of the theory to the consideration of harmonic vibrations in the reactants allows its application to Class II reactants. The theoretical expressions for ΔF^* and $\Delta S^*$of Class III reactants would contain additional terms beyond those considered in the present form of the theory [4, p. 158].

3.20. Interpretation of the Equations

The former set of equations suggests experimental studies on the dependence of the rate constants, for both chemical and electrochemical ET, from parameters such as:

(i) Standard free energy change or activation overpotential
(ii) Ionic structure and ligand field
(iii) Frequency factor and activation energy
(iv) Added salts
(v) Solvent medium
(vi) Electrode material and surface contamination

It also suggests experiments on:

— The extent of parallelism between both rate constants for a series of reactants
— Correlation between rate constants and between chemical and electrochemical transfer coefficients [4, p. 157].

There is a close relation between the theoretical equations of the chemical and electrochemical processes. In each case, ΔF^* *decreases* with increasing a, decreasing ne, decreasing R, increasing D_{op}, decreasing D_s (the latter's effect is perhaps more important on

w^r and w^p), and increasingly negative ΔF^0 or $-ne\eta_a$. The physical interpretation of this behavior is simple and related to the ease with which the energy restriction can be satisfied [1, 2]. "For example, ion–solvent interactions decrease with increasing ionic radius a, so that there would be a smaller energy difference between the two hypothetical charge distributions in the original atomic configuration of the reactants and therefore a smaller reorganization energy would be needed to equalize the two energies" [4, p. 158].

"Similarly, the smaller the charge transferred ne, or the smaller the distance between the reactants (or between the ion and its electrostatic image in the electrode), the less the medium can discriminate energywise" between initial and final charge distributions "and therefore the lesser will be the reorganization needed for energy equalization. A larger optical dielectric constant, D_{op}, serves to partially neutralize the electrostatic fields of the charges and therefore to reduce their energy difference for a given atomic configuration. The more D_s approaches D_{op}, the smaller the dipolar contribution to the electrical polarization of the medium and the less the necessary rearrangement of the atoms (at least for zero ΔF^0 or zero η_a)." [4, p. 158]

From Eqs. (3.18) and (3.34), we see that the *effective driving force* of the reaction is not just given by the free energy difference (ΔF^0 or $-ne\eta_a$) between reactants and products when the reacting species are far apart but rather when they are in the positions they occupy in the activated complex. By lowering the free energy of the final state of the system relative to that of the initial state, these terms reduce the amount of reorganization of atomic configuration of the initial state necessary to satisfy the energy restriction [1].

We have seen that λ in Eqs. (3.17) and (3.18) has become $\lambda/2$ in Eqs. (3.33) and (3.34) for the electrode system. "This difference can be attributed to the absence of one half of the dielectric medium in the electrode case. (Effectively, the charge is transferred from the ion to its electrostatic image, and the metal surface bisects the line drawn between the two)." [4, p. 158]

3.21. Interpretation of the Data

A discussion of data follows on experiments reported in Refs. [4] and [6], from the point of view of the various factors influencing the reaction rate.

3.21.1. *"Standard" Free Energy Change or Activation Overpotentials*

The influence of this only factor alone on the rate constant is more easily investigated for the electrode processes. As a matter of fact, it is obviously much easier to vary the potential of the working electrode at which a substance reacts, than studying the redox reactions of the substance with a variety of reactants.

"When w^p and w^r are sufficiently small, it can be deduced from Eqs. (3.11), (3.33), and (3.34) that the transfer coefficient is 0.5, a value found in a number of simple electron transfer processes [48–50]." "The quantitative effect of ΔF^0 on the homogeneous bimolecular rate constant has been measured experimentally for some nonspherical reactants [51, 52]. While ion–solvent interactions in these systems (oxidation of hydroquinone-like compounds) were not as simple as assumed in Eq. (3.17) and an atom transfer mechanism could not be ruled out, reasonable agreement with the experimental rate constant was found [2]." [4, p. 158]

The detailed theoretical description from Ref. [2] is reported in a following section [4, p. 159].

3.21.2. *Ionic Structure and Ligand Field Effects*

In order to study *the effect of ion size* on the rates of homogeneous reaction, it is most convenient to study isotopic exchange reactions because there the reactants merely exchange their charges, ΔF^0 and $w^r - w^p$ are both zero, and m of Eq. (3.85) is simply $-\frac{1}{2}$. Wahl and coworkers have extensively investigated the factor of ion size [53, 54]. The rates of large ions, probably of Class I, were found, as expected, to be much larger than those of the smaller ions, $MnO_4^{-1,-2}$

and $Fe(CN)_6^{-3,-4}$. It is interesting that "the latter two systems had comparable rates in spite of the greater Coulombic repulsion in the second case, perhaps because of larger size of the iron cyanide ion (4.5 vs. 2.9 Å). Rather good agreement was obtained between experimental and calculated rate constants using the theoretical equations, taking into consideration the absence of adjustable parameters [55]."

Marcus theory incorporates the influence of ligand field effects on the rate constants of oxidation–reduction reactions; they influence, in particular, k_i, Δq_i^0, and ΔF^0. Accordingly, the discussions of ligand field effects on kinetic problems is then converted into a discussion of the problem of estimating k_i, Δq_i^0, and ΔF^0 [6, p. 67].

"The cobalt–nitrogen bonded complexes probably are members of Class III [25, 56] and the rates are therefore extremely slow. The marked increase in rate [56, 57] in the sequence $Co(NH_3)_6^{+3,+2}$, $Co(en)_3^{+3,+2}$, $Co(phen)_3^{+3,+2}$ may be partly a reflection of increasing ion size, and partly a reflection of ligand field [58] effects.

The ferrous–ferric system is probably of class II, and its rate [59] is correspondingly much less than that of the MnO_4^- or $Fe(CN)_6^{-3}$ systems. Nevertheless, its rate greatly exceeds that [60] of $Cr(H_2O)_6^{+3,+2}$, a probable member of Class III [56]. Reasonable agreement between calculated and experimental between calculated and experimental results for the $Fe(H_2O)_6^{+3,+2}$ was found by adapting the theory to Class II reactants [3]. Thus far [1958...] there is no experimental evidence which definitely establishes either an electron or an atom transfer mechanism for these homogeneous isotopic exchange reactions of hydrated cations. *The general parallelism of various rates* in the two processes discussed in the following suggests, however, *a common electron transfer mechanism*." [4, p. 159]

3.21.3. *Parallelism between Rates in Solution and at the Electrodes*

The theoretical equations suggest a close parallelism between the rates of the two processes, when similar mechanisms are operative, in particular when ΔF^0 and η_a are zero and the work terms are small, a situation typical of isotopic exchange reactions. "The parallelism becomes very close, therefore, when the rates of isotopic exchange reactions are compared with the corresponding electrochemical exchange currents. The chemical and electrochemical electron transfer rates of the systems $Cr(H_2O)_6^{+3,+2}$, $Fe(H_2O)_6^{+3,+2}$, and $Fe(CN)_6^{-3,-4}$ [54, 56, 59, 61] *both* increase in the order given as do those of the systems $Co(NH_3)_6^{+3,+2}$ and $Co(en)_3^{+3,+2}$ [55, 57, 62]. The absolute rates of various electrochemical processes [60] are in the general range expected from the values of the chemical rate constants, but further work is desirable, both theoretically on Class II and Class III systems as well as experimentally on salt effects and hydrolysis effects." [4, p. 159]

3.21.3.1. *Frequency Factor and Activation Energy*

For simple isotopic exchange reactions: "ΔF^* equals $w^r + \lambda/4$. Correspondingly, ΔS^* *is the sum of two terms.* The first, $-\partial w^r/\partial T$, is the usual entropy change which results when two ions are brought together. The second, $-(d\lambda/dT)/4$, is the entropy of formation of the 'nonequilibrium' atomic configurations from the equilibrium ones at the same r in order to satisfy the energy restriction. Its value may be computed from Eq. (3.35). It proves to be *very small for reactants of Class I.*

Applications of the theoretical equations to Class I reactants by the writer" (M.) "led to good agreement between experimental and calculated frequency factors for the $MnO_4^{-1,-2}$ system and (within a factor of 50) for the $Fe(CN)_6^{-3,-4}$ system, the latter system having *an extremely small frequency factor.*"

The last frequency factor was measured by A. C. Wahl. "The data were extrapolated to infinite dilution, taking cognizance of the comparative insensitivity of the theoretical activation energy to salt effects." [4, pp. 159–160]

Recalling Eq. (3.12) $A = A' \exp(-\Delta S^*/R) = \kappa Z \exp(-\Delta S^*/R)$ relating the excess entropy of activation to the frequency factor we see that the smaller the absolute value of $\Delta S^* < 0$, the smaller A.

"These results suggest that the probability of adiabatic reaction may be of the order of unity, at least for these systems. The frequency factors of Class I reactants provide the most direct measure of this probability factor.

Many of the hydrated metal cations, which are mainly of Class II, tend to hydrolyze and form other complexes easily. In a careful investigation Silverman and Dodson [59] unraveled the rate constant of the $Fe^{+3,+2}$ system... The frequency factor was very small."

NOTE that the term $-\boldsymbol{\partial w^r/\partial T}$ of $\boldsymbol{\Delta S^*}$ in Eq. (3.21) can be influenced by the ionic strength, for instance by the high acidity needed to minimize hydrolysis, because w^r is influenced by the ionic strength (see earlier).

"Baker, Basolo, and Neumann [56] have reported a most interesting result of a very high frequency factor, 5×10^{16} cc mole^{-1} sec^{-1} for a Class III system,**(2)** $Co(phen)_3^{+3,+2}$.

"A similarly high frequency factor has been reported for another Class III system [63], $Cr(H_2O)_6^{+3,+2}$."

"The frequency factors of the electrochemical systems... Randles and Somerton [61] found them to be typically in the range 3×10^2 to 3×10^4 cm sec^{-1}. The reactants were primarily hydrated metal cations and so were mostly of Class II. In one system, $Cr(CN)_6^{-3,-4}$, a rather low frequency factor (10 cm sec^{-1}) was found. The electrostatic repulsion between ion and electrode was estimated to be large for this system, and the low A-value was attributed to a small *transition probability factor* because of the increased R." [4, p. 160]

Marcus indicates now two other parameters which are influenced by R:

"However, we see from Eq. (3.35):

$$\lambda = m^2 e^2[(1/a) - (1/R)][(1/D_{op}) - (1/D_s)]$$

that there is also an *image term* present. This term would tend to favor small R's. An alternative explanation for the A-value could be given on the basis of the usual sign of $-\partial w^r/\partial T$. When two reactants of like sign approach each other, the enhanced electrostatic field polarizes the solvent more strongly and causes a negative ΔS^*.".

See above.

"Activation energies for several electrochemical electron transfers have been measured [61]...

The results tentatively indicate the activation energies of the electrode reactions to be *more than one-half* those of the corresponding exchange processes. Should this result prove to be generally true (at least for Class I and Class II reactants) one possible interpretation would be that the distance between the ion in the activated complex and its electrical image *exceeds twice* the ionic radius."

We have seen that $\lambda_{ex} \cong 2\lambda_{el}$. This is valid in the hypothesis $R = 2a$. If $R > 2a$ because of the presence of a layer of molecules adjacent to the electrode which the ion cannot penetrate, then λ_{el} is greater than for $R = 2a$ and recalling the formula for λ_{el} we can explain the earlier experimental result [4, pp. 159–161].

3.21.4. *Salt Effects*

"No detailed study of salt effects for simple electron transfers at electrodes appears to have been reported" until 1958. "Therefore it is not yet possible to adequately test the usual assumption (one not made here) that the work required to transport an ion from the body of the solution to the electrode, w^r or w^p, equals the ionic charge multiplied by the difference of potential at the initial and final positions of the ion. This assumption is clearly valid when the ionic charge is so small that it does not perturb the configuration of the remaining ions. It is also clearly invalid at the *point of zero charge* [17, 20]. At this point, the work term estimated on the basis of the above assumption is zero, whereas it actually equals the free energy of interaction of the ion with the image. For infinite dilution, the latter term is $-q^2/2D_sR$, q being the ionic charge, while for dilute salt solutions, an approximate estimate of it is $(-q^2/2D_sR)\exp(-\kappa R)$, κ being the usual Debye κ." [4, p. 161]

NOTE that the interaction energy is half of what it would be between two real charges [27]. [4, p. 161]

3.21.5. *Solvent Effects*

"Solvent effects for simple electron transfers will occur, according to the theory, whenever there is a change in dielectric constant, refractive index, ionic radius, or standard free energy of reaction. Changes in composition of the coordination shell, which may include the solvent, naturally alter a, ΔF^0, and the ease of intramolecular compression or stretching. In addition there are effects which were not incorporated in the theory, effects such as changes in ion pairing, selective solvation of solvent mixtures, and a change of the mechanism itself."

"Heavy water has been used as a solvent in several studies [64, 65]. Hudis and Dodson [64] have demonstrated the importance, when hydrolyzed species and other complexes are present, of determining the various equilibrium constants so that meaningful electron transfer rate constants for the different species can be obtained and solvent effects evaluated. The rate constant for $Fe^{+3,+2}$ was smaller by a factor of two in D_2O [64], while that of $NpO_2^{+2,+1}$ was smaller by a somewhat smaller factor [65]. A qualitative interpretation of solvent effects can be given in terms of atom transfer [64, 65] or electron transfer [1] theories. Thus far, [1958], however, comparative studies in the two solvents do not permit either mechanism to be distinguished from the other. This situation results from the fact that ions have different solvation energies in the two media [66], so that water does not behave as an inert solvent.

The rate of the $Co(NH_3)_6^{+3,+2}$ reaction appears to be faster in liquid ammonia [67] than in water, where it was immeasurably slow [57]. The activation energy was high, being in the range found for two other Class III compounds [56, 63]. Possible explanations for the solvent effect include a change of mechanism, such as a dissociation [67] or possibly a hydrolyzed intermediate, $Co(NH_3)_5NH_2^{+2}$." [4, pp. 161–162]

3.21.6. *Effects of Electrode Material*

"A dependence of rate constant on electrode material will occur if there is any change in surface contamination [61] and possible metal–solvent binding, or if electrode charge density at a given η_a changes sufficiently to alter w^r or w^p. The rate constants measured for one system with several solid electrodes underwent no great variation (within a factor of 10) but were much smaller than that found for a mercury electrode [61]." [4, p. 162]

3.21.7. *The Image Force Law and Its Implications*

"The role of the electrostatic image has been generally ignored in electrochemical theories. A recent investigation [68] of the quantum limitations of the image force law for a vacuum is reassuring. It has been applied by the writer (M.) to dielectric media and to electrochemical theory.

The effect of this image is to partially neutralize the field of the ion and to reduce, thereby, the *configurational rearrangement free energy* needed to satisfy the energy restriction. It is noteworthy that its calculated effect remains, even at salt concentrations sufficiently large as to make w^r and w^p negligible. This is because it is impossible for the ionic atmosphere in the activated complex to neutralize the *ion–image interactions* of the two different hypothetical charge distributions at the same time. A *compromise configuration* results. A similar behavior exists in the homogeneous case, where the $1/r$ term remains even if w^r and w^p are zero.

At the point of zero electric charge [17, 21], there is a net, shielded Coulombic attraction between the ion and its image, which is quite large for very dilute solutions. According to the theory a *positive* ΔS^* would result, since the attraction lowers the entropy of solvation."

See above for the increase of entropy when two ions of opposite charge approach each other.

"A measurement of the frequency factor in this region would permit a determination of this ΔS^*"

"If this prediction is verified for Class I reactants, it will be interesting to compare ΔS^* with the theoretical estimates" [4, p. 162]

3.22. Comparison of Isotopic Exchange Rate and Corresponding Electrochemical Exchange Current

"A comparison of $\sqrt{k_{ex}/10^{11}}$ and $k_{el}/10^4$ on the basis of the existing experimental data is given in Table I." of Ref. [6], "All rate constants are pseudo-rate constants, their use being justified under the conditions cited [6]. The qualitative trend in both k_{el} and k_{ex} is seen to be the same, and the values... are relatively close to each other, considering the fact that approximations in the theory enter exponentially (a fairer comparison would be of $\Delta F^*_{ex}/2$ and ΔF^*_{el}), that stationary electrodes (with their absorption problems) were usually necessary, and that the work terms may not have been negligible." [6, p. 854]

3.23. Comparison of Chemical and Electrochemical Oxidation–Reduction Rates of a Series of Related Reactants

"In this comparison we shall consider systems in which *a constant reagent* is used in the chemical system, and *a constant electrode potential* in the electrochemical one, to oxidize or reduce a series of related compounds. In a series of *a given charge type*, the work terms are either exactly or roughly constant in each of these two systems. Furthermore, if the ΔF^*'s are in the region where they would depend linearly on ΔF^0, then according to Eqs. (3.1, 3.78, 3.83, 3.25), the ratio:

$$\frac{k_{\text{soln}}}{k_{el}} \tag{3.95}$$

should be the same for each member of the series: in both cases, the terms λ_1, ΔF^0, and, at a constant E, $\eta(= E - E^0)$, will normally vary from member to member. λ_2 refers to the constant reagent. However, since $\Delta F^0 = -nFE^0 + \text{const}$ in the series, one sees from Eqs. (3.78, 3.83, 3.25), that these variations in λ_1, ΔF^0, and E^0 cancel when one compares values of $(\Delta F^*_{\text{soln}} - \Delta F^*_{el})$, that is of k_{soln}/k_{el}. Vlcek [69] has recently observed (1961) that the electroreduction and the $Co(dipy)_3^{+3}$ reduction [69] of $Co(NH_3)_6^{+3}$ and $Co(NH_3)_5(H_2O)^{+3}$ has essentially the same k_{soln}/k_{el} for both compounds [6]. This experimental result is in agreement with the earlier theoretical deductions. Presumably both E^0 and λ_1 differed in the two compounds."

"Similarly the ratio:

$$\frac{k^{\text{a}}_{\text{soln}}}{k^{\text{b}}_{\text{soln}}} \tag{3.96}$$

for each member of the series oxidized or reduced by two reagents, a and b, should be constant. This result was found experimentally for the $Co(NH_3)X$ compounds reduced by $V(H_2O)_6{}^{+2}$ and $Cr(dipy)_3{}^{+3}$, respectively, with X being NH_3, H_2O, and Cl^- [70] (Table II of Ref. [6]).

The restriction to a given *charge type* will not be important if the work terms are relatively minor.

The comparison involving V^{+2} should be accepted with some reserve since the V(II) reaction is not necessarily an 'outer sphere' one, as Taube has pointed out." [6, p. 855]

3.24. Early Comparisons of Theoretical and Experimental Results

(a) Excess Free Energy of Activation

The very first application of the theory consisted in comparisons of experimental and theoretical ΔF^*'s [1, p. 870].

The experimental ΔF^*_{expt} are known very accurately because they depend only on the logarithm of k_{bi} and so a factor of 2 in k_{bi} introduces an error of only 0.4 kcal mole^{-1} in ΔF^*_{expt}.

ΔF^*_{expt} is found measuring k_{bi} while Eq. (3.17); (or, more simply, Eq. 3.4) is used to compute ΔF^*_{calc}.

In most instances, M. found an encouraging agreement between experimental and calculated ΔF^*'s, considering that his theory is free from adjustable parameters. The agreement was in particular excellent for compounds of class I ("tightly knit covalently bound ions"), but was not as good, even if reasonable, for small hydrated cations [3, p. 426].

The reasons given to explain differences between experimental and calculated ΔF^*'s are the following:

(i) It is possible that the *innermost solvation layer* of the ions is *not completely dielectrically saturated* as assumed. In this case, there would be an additional contribution to ΔF^* arising from any changes which may have to occur in interatomic distances in the innermost solvation layer, that is, within the sphere of radius a, prior to the electronic jump, in order to form the activated state.

(ii) The *electron tunneling factor* may be somewhat less than unity. Note that this factor is not temperature dependent and therefore does not enter into any comparison between experimental and calculated values of the activation energy.

(iii) For reactions in which there is uncertainty of mechanism, say between two successive one-electron transfer and one two-electron transfer it is necessary to know the mechanism before calculating ΔF^*_{calc}.

From calculations applied to concrete ET cases one finds that the relative magnitude of the two contributions to ΔF^* in Eq. (3.4), the Coulombic repulsion and the solvent reorganization free energy are generally of the same order of magnitude. From this, it may be inferred that no simple correlation between the reaction rate and the size of the Coulombic term would be expected.

(b) Excess Entropy of Activation

From the carefully studied temperature dependence of the ferrous–ferric reaction, the experimental value of the excess entropy of activation was computed with the aid of Eq. (3.7). It was $\Delta S^*_{\text{expt}} = -23\,\text{cal mole}^{-1}\,\text{deg}^{-1}$ at 0°C. "The theoretical value ΔS^*_{calc} is found from Eq. (3.15) to be $-14\,\text{cal mole}^{-1}\,\text{deg}^{-1}$. Considering the experimental errors that always accompany a measurement of ΔS^* and considering the assumptions of the theory, the experimental and calculated values agree reasonably well." [1, p. 871]

3.25. Reaction Rates in Heavy Water—A Possible Criterion of Mechanism

The way in which the reaction rate changes when D_2O is used as solvent instead of H_2O is rather subtle. Any changes of the interatomic O–H distances in the innermost hydration layer needed to form the activated state are easier for the O–H bonds than for the O–D bonds since the former have a higher zero-point energy and so they span a larger range of interatomic distances. Note that the influence of D_2O on ET can only happen through the differences in atomic polarization in the dielectrically saturated innermost solvation layer since the two solvents do not differ appreciably in their D_s or D_{op} and so the difference of rates in the two solvents cannot arise from the dielectrically unsaturated part of the medium.

M. proposed [1, p. 871] an experimental method to help distinguishing between a simple outer-sphere ET mechanism and an atom transfer mechanism in a series of reactions. If one measures the reaction rates for the reactions Fe^{+2}–Fe^{+3}, Fe^{+2}–$FeOH^{+2}$ and Fe^{+2}–$FeCl^{+2}$ in H_2O and in D_2O, one should find a comparable D_2O–H_2O isotope effect on the reaction rates if these reactions have a small-overlap electron transfer mechanism, that is, if there is but a small overlap of the electronic orbitals of the two reactants in the activated complex, a basic assumption of the theory. Since the Fe^{+2}–Fe^{+3} and Fe^{+2}–$FeOH^{+2}$ rate constants were twice as great in H_2O as they were in D_2O, this effect would predict an effect of similar magnitude for the Fe^{+2}–$FeCl^{+2}$ reaction.

If a reaction has, on the other hand, an atom transfer mechanism, then a D_2O–H_2O effect would probably be expected only if the atom were hydrogen. Thus an isotope effect could occur for the Fe^{+2}–Fe^{+3} and Fe^{+2}–$FeOH^{+2}$ reactions, but not for the Fe^{+2}–$FeCl^{+2}$ reaction if it involves a chlorine atom transfer. This expected absence of isotope effect in a chlorine atom transfer process can, of course, be tested by measuring the rate constant of the atom transfer Cr^{+2}–$CrCl^{+2}$ reaction in the two solvents (vide supra).

3.26. Two Case Studies in Detail (Verbatim with Glosses) [2]

(a) OXIDATION OF HYDROQUINONES BY FERRIC IONS

Mechanism

"The overall reaction for the oxidation of a hydroquinone by ferric ions is given by Eq. (3.97)

$$2\mathrm{Fe}^{+3} + Q\mathrm{H}_2 \xrightarrow{k_1} 2\mathrm{Fe}^{+2} + Q + 2\mathrm{H}^+ \quad K \tag{3.97}$$

where QH_2 and Q denote the hydroquinone and quinone, respectively. The rate of this reaction has been measured as a function of the concentrations of the various reactants and products. One mechanism consistent with the data involves the ionization of QH_2 to QH^-, followed by an electron transfer between QH^- and Fe^{3+}:

$$\mathrm{Fe}^{+3} + Q\mathrm{H}^- \rightleftarrows \mathrm{Fe}^{+2} + Q\mathrm{H}$$

Where QH denotes the semiquinone. This step was in turn followed by the ionization of QH to Q^- and then by the latter's oxidation to

Q [1]. The four elementary steps of the mechanism are then:

$$QH_2 \rightarrow H^+ + QH^- \quad K_1 \tag{3.98}$$

$$Fe^{+3} + QH^- \xrightarrow{k} Fe^{+2} + QH \tag{3.99}$$

$$QH \rightarrow H^+ + Q^-$$

$$Fe^{+3} + Q^- \rightarrow Fe^{+2} + Q$$

On the right side of the reactions, I have written the symbol of the corresponding equilibrium constants. k_1 and k are symbols of the rate constants.

"We shall be concerned here with the experimental and theoretical value of k, the rate constant for the forward reaction in Eq. (3.99)."

As we know Marcus' theory applies to the elementary redox step of an overall redox reaction.

"The experimental k can readily be computed from the known overall bimolecular rate constant k_1 and the known first ionization constant of QH_2, K_1."

For convenience of the reader, I explicitly show how this is done following a note on p. 874 of Ref. [2].

The rate constant of the elementary redox step (3.82) can be written as:

$$v = k[Fe^{+3}][QH^-]$$

It was also written [51] in terms of a pseudo-rate constant k_1:

$$v = k_1[Fe^{+3}][QH_2]$$

Consider reaction (3.98) and the associated chemical equilibrium:

$$\frac{[H^+][QH^-]}{[QH_2]} = K_1$$

whence:

$$QH^- = K_1\frac{[QH_2]}{[H^+]}$$

and

$$v = k[\mathrm{Fe}^{+3}]K_1\frac{[Q\mathrm{H}_2]}{[\mathrm{H}^+]} = k_1[\mathrm{Fe}^{+3}][Q\mathrm{H}_2]$$

For the pseudo-constant k_1, we have:

$$k_1 = \frac{kK_1}{[\mathrm{H}^+]}$$

and we see that it is inversely proportional to $[\mathrm{H}^+]$.

The various quantities needed for the theoretical calculation of k are computed in the following sections.

3.27. Standard Free Energy Change of the Redox Step, ΔF^0

The standard free energy ΔF^0 of reaction (3.99) has not been measured directly but can be estimated from K_1, and from K and K_S, the equilibrium constants of reactions (3.97) and (3.100), respectively,

$$Q\mathrm{H}_2 + Q \rightarrow 2Q\mathrm{H} \quad Ks \tag{3.100}$$

It is readily verified that ΔF^0 is given by

$$\Delta F^0 = -RT\ln(KK_S/K_1^2)^{\frac{1}{2}} \tag{3.101}$$

The values of K and K_1 have been determined experimentally [51, 70]. Numerous data on the formation constants of semiquinones K_S have also been obtained. When, as in reaction (3.100), all three compounds in this equilibrium are uncharged, or have the same charge, the free energy of this reaction is found to be practically independent of the chemical structure of these compounds (this can be inferred from a detailed analysis of data on the on the pH dependence of the first and second oxidation potential of many hydroquinone-like compounds). The standard free energy change of reaction (3.100), $-RT\ln K_S$, is found to be about 2.3 kcal mole^{-1}.

With these values of K, K_1, and K_S, the values of ΔF^0 given in Table I were calculated from Eq. (3.101).

Table I. Kinetic and thermodynamic data for $Fe^{+3} + QH^-$ reaction at 25°C[a]

Hydroquinone	$k \times 10^{-9}$ $\left(\frac{\text{liter}}{\text{mole} \times \text{sec}}\right)$	ΔF^0	ΔF^*_{expt}	ΔF^*_{calc}	$(\Delta F^*_x - \Delta F^*_y)^b$ Expt	Calc
2,6-dichloro-	0.023	−9.3	7.7	7.5	0	0
benzo-	3.3	−13.2	4.8	5.9	2.9	1.6
tolu-	24	−14.9	3.6	5.2	4.1	2.3
duro-	270	−20.5	2.1	3.2	5.6	4.3

[a] All free energy units are in kcal mole^{-1}.
[b] x denotes the 2,6-dichloro compounds and y denotes any other compound.

3.28. Effective Radii

"In the derivation of Eq. (3.4) for ΔF^*, each ion was treated as being surrounded by a sphere of radius a inside of which the dielectric medium, that is, the solvent, is saturated and outside of which it is unsaturated. Similar models have been used extensively in calculating the free energy of solvation of the ions," see Refs. in [2].

As discussed in Chapter 2 "if, as it is usually assumed, the innermost hydration layer around monoatomic ions is largely saturated, it will not contribute to ΔF^*, and a then equals the sum of the crystallographic radius of the ion and of the diameter of a water molecule. Polyatomic ions such as $Fe(CN)_6^{-3}$ and MnO_4^-, being rather large, would not be expected to cause dielectric saturation of the solvent as readily as the smaller monoatomic ions, since the electric field of the ion, which is responsible for the saturation, falls off roughly as the square of the distance from the center of the ion. Thus one would expect that a for the cited polyatomic ions would be approximately the actual crystallographic radii. This is also consistent with the assumption that only the first layer of the water molecules about a monoatomic ion is saturated, since a hydrated monoatomic ion has about the same radius as these polyatomic ions.

With a slightly different for an ion when it is a reactant and when it is a product it was suggested that a mean value for a be adopted." "The crystallographic radii of Fe^{+2}, Fe^{+3}, and H_2O are [2] 0.75, 0.60, and 0.72. The mean of the first two is 0.68 Å."

The refinements of the theory will take into account the effect of the radii variations on the reaction rate.

When organic ions are considered one has to consider two new features. "First, the ion is far from being spherical and, second, it is possible that that the *effective radius* a could be quite different when this particle is a charged reactant and when it is an uncharged product. This is discussed later, where it is inferred from entropy data that the *effective polarizing radius* of a hydroquinone-like ion such as $HOC_6H_4O^-$ is about the same as that of the hydroxyl ion. It is further suggested that the dielectric saturation around this group when it is a charged particle be neglected as a first approximation, and that a equals the crystallographic radius of the oxygen group 1.4 Å. This approximation can be removed by a refinement of the theory."

3.29. Excess Free Energy of Activation

"Experimental values of ΔF^* were calculated from the rate constant k given in Table 1, using Eq. (3.1) and setting Z equal to 10^{13} liter mole^{-1} sec^{-1} in that equation. These values are reported in Table I.

Using the effective radii and the ΔF^0's deduced in the preceding section and setting $D_{op} = 1.8$ and $D_s = 78.5$ at 25°C, values of ΔF^*_{calc} were obtained with the aid of Eqs. (3.2)–(3.4), and are given in Table I. These results will be discussed in detail later."

An important note: M. later [71] used for Z the value 10^{11} liter mole^{-1} sec^{-1} instead of 10^{13} liter mole^{-1} sec^{-1} of Ref. [72].

3.30. Excess Entropy of Activation

"Experimental values for the excess entropy of activation of reaction (3.99), ΔS^*_{expt}, were calculated from the data."

It may be instructive to report, from note 26 in Ref. [2], how this calculation was done.

The pseudo-rate constant k_1 of Eq. (3.97) has an activation energy E_1, say, and a frequency factor A_1 so that $k_1 = A_1 \exp(-E_1/RT)$. Values of A_1 were determined experimentally. The rate constant k, Eq. (3.99), has a frequency factor A, say, which can be calculated from the known A_1 [51] and the known [70] entropy of ionization of QH_2, ΔS_1 say. It is found that $A = A_1[H^+]\exp(-\Delta S_1/R)$. According to Eq. (3.7), ΔS^*_{expt} is then found by setting $A = Z\exp(\Delta S^*/R)$, Z being in Ref. [2] equal to 10^{13} liter mole^{-1} sec^{-1} [72]. But see the previous note on the correct value of Z.

"The values of ΔS^*_{calc} were computed from Eq. (3.9) using the previously determined radii and ΔF^0 and using the calculated ΔS^0." From note 27 in Ref. [2]: "$\Delta S^0 = -\partial \Delta F^0/\partial T$ and therefore according to Eq. (3.101), $\Delta S^0 = \frac{1}{2}\Delta S + \frac{1}{2}\Delta S_S - \Delta S_1$ where ΔS and ΔS_S are the standard entropy changes in reactions (3.97) and (3.100), respectively [51, 70]. Since the sum of the translational, rotational, and vibrational entropies of the reactants and of the products of reaction should be about the same, it may be assumed that ΔS_S is essentially zero."

"In this way ΔS^*_{expt} was found to be 46, 57, 47, and 44 cal mole^{-1} deg^{-1}, and ΔS^*_{calc} to be 36, 38, 36, and 31 entropy units, for the 2,6-dichloro, benzo, tolu, and duro hydroquinones, respectively. The average of the former group of values is 49 and that of the latter is 35. Values of ΔS^*_{calc} were also computed from the approximate equation, Eq. (3.10). They agreed well with the exact calculated values, within about one and a half entropy units."

(b) AEROBIC OXIDATION OF THE LEUCOINDOPHENOLS

3.31. Excess Free Energy of Activation

"The overall reaction of the leucoindophenols with dissolved oxygen is represented by Eq. (3.102), where QH_2 denotes a leucoindophenol such as $HOC_6H_6NHC_6H_6OH$, and Q denotes the corresponding indophenol $HOC_6H_4NC_6H_4O$.

$$QH_2 + O_2 = Q + H_2O_2 \tag{3.102}$$

A mechanism consistent with the data [51] involved the ionization of QH_2 to QH^-, which then transferred an electron to O_2:

$$QH^- + O_2 \xrightarrow{k} QH + O_2^- \tag{3.103}$$

This step was followed by reactions of QH and O_2^-. As in the preceding reaction series discussed in Ref. [2], the rate constant k for the forward reaction in Eq. (3.103) can be inferred from the overall bimolecular rate constant k_1 and the first ionization constant K_1 of QH_2 ($k = k_1[H^+]/K_1$). The k's and the values of ΔF^*_{expt} calculated from them using Eq. (3.1) are given in Table II.

It can be shown that the ΔF^0 of reaction (3.103), needed for the estimation of ΔF^*_{calc}, is given by Eq. (3.104):

$$\Delta F^0 = -F(E^0_{QH_2} - E^0_{O_2^-}) + RT \ln K_1/K_S^{\frac{1}{2}} \tag{3.104}$$

Table II. Kinetic and thermodynamic data for $O_2 + QH^-$ reaction at 30°C[a]

Indophenol	$K\left(\dfrac{\text{liter}}{\text{mole} \times \text{sec}}\right)$	ΔF^0	ΔF^*_{expt}	ΔF^*_{calc}	$(\Delta F^*_y - \Delta F^*_x)$[c] Expt	Calc
Unsubstituted	47	$15.9 - \alpha$	15.7	23.7[b]	0	0
o-chloro	5.1	$17.7 - \alpha$	17.0	24.9[b]	1.3	1.2
o-bromo	6.1	$17.4 - \alpha$	16.9	24.7[b]	1.2	1.0
2,6-dibromo	0.23	$19.7 - \alpha$	18.9	26.7[b]	3.2	2.5

[a] All free energy units are in kcal mole^{-1}.
[b] There is an appreciable uncertainty in $-E^0_{O_2^-}$ and therefore in ΔF^0, of an amount α, where α may be about 0 to +5 kcal mole^{-1}, according to Latimer [73]. This uncertainty introduces a corresponding uncertainty in the ΔF^*_{calc}, which may be about 3 kcal mole^{-1} less than those reported in this table.
[c] x denotes the unsubstituted compound and y denotes the substituted compound.

where $E^0_{QH_2}$ is the standard oxidation potential of QH_2 ($QH_2 = Q + 2H^+ + 2e$), $E^0_{O_2^-}$ is that of O_2^- ($O_2^- = O_2 + e$).".

Note that M. is using here Latimer's standard oxidation potentials whose values are opposite of IUPAC's standard reduction potentials [73].

"F is the Faraday, and K_S is the equilibrium constant of reaction (3.100) for the formation of semiquinones or indophenols. The ΔF^0's of Table II were estimated using the known [74] $E^0_{QH_2}$, the estimated [73] $E^0_{O_2^-}$ the value of $-RT \ln K_S$ discussed in the previous section (2.3 kcal mole^{-1}) and the known [47] K_1."

Introducing into Eq. (3.4), these ΔF^0's and the effective radii later deduced, the ΔF^*_{calc} values of Table II were computed.

3.32. Excess Entropy of Activation

In this section, M. shows how it is possible to estimate unknown entropy values.

"All the entropy data needed for the estimation of ΔS^*_{expt} and ΔS^*_{calc}, the experimental and the calculated excess entropy of activation of the electron transfer step, Eq. (3.103), are not available. However, some estimate of the undetermined entropy values may be made.

For example, just as in the case of the ferric-hydroquinone reaction, ΔS^*_{expt} can be calculated from the frequency factor of the pseudo-rate constant and the entropy of ionization of QH_2, ΔS_1.

Assuming that ΔS_1 equals the value for the hydroquinones and water (which are later shown to be equal) one finds in this way that $\Delta S^*_{\text{expt}} \cong 7$ cal mole^{-1} deg^{-1}.

The term ΔS^*_{calc} is seen from the approximate equation, Eq. (3.10), to be about $\Delta S^0(1 + \Delta F^0/\lambda)/2$ since $e_1 e_2$ and $e_1^* e_2^*$ are each equal to zero in reaction (3.103). In reaction (3.103), it is also expected that the translational, vibrational, and rotational entropies of the reactants are each about equal to those of the products."

Because of the very similar chemical structure of reagents and products.

"Accordingly, if the entropy change ΔS^0 of reaction (3.103) were appreciable, the main contributions to it would arise from possible differences in the entropy of solvation of QH^- and O_2^-. This would not be expected to be large so that $\Delta S^0/2$ should be relatively small. Further, $\Delta F^0/\lambda$ is calculated to be about 0.3, so we conclude that ΔS^*_{calc} is small, in agreement with the estimated small value of ΔS^0_{expt}."

3.33. Discussion

"A comparison of Tables I and II shows that the rate constants of the redox step in the two series of reactions considered here differ from each other by a factor 10^9 on the average. This difference stems from the considerable difference in the standard free energy change ΔF^0 of the redox step in the two cases, one being about—13 kcal mole^{-1}, the other perhaps lying between +17 and + 12 kcal mole^{-1}, depending on the correct value of $E^0_{O_2^-}$. This difference in turn is related principally to the differences between standard oxidation potentials of the ferrous ion and of the oxygen molecule ion.

According to the theory developed in Part I, the standard free energy change affects the reaction rate in the following way. During an ET step, there is *first a reorganization* of the solvent molecules about the reacting ions *prior to the jump* of an electron from one reactant to the other. Now, for a given ΔS^0, the more positive ΔF^0 is, the greater would be the energy of the state of the system *just after the jump*. Therefore, the energy of the system *just before the jump*, which is *equal* to this, would also have to be greater. This simply means that there is a greater energy barrier to forming from the isolated reactants a *suitable collision complex in which* the electron can jump. Conversely, the more negative ΔF^0, the less the energy barrier."

This is pictorially described in Fig. 5* (c) and (d) of Chapter 1 in terms of PESs.

"In spite of the extremely favorable value of ΔF^0, the redox step of the ferric ion–hydroquinone ion reaction is seen from Table I not to proceed at every collision."

As a matter of fact the most rapid reactions in solution are those *encounter controlled* or *microscopic diffusion controlled* such as the recombination reaction of H^+ and OH^- in water with a rate constant of 1.4×10^{11} liter mole^{-1} sec^{-1} at 25°C [75] while the k for the Fe^{+3} — benzo-hydroquinone anion is only 3.3×10^9 liter mole^{-1} sec^{-1}.

"According to the theory, this is because of the preliminary solvent reorganization prior to the electron jump. However, it is of interest that the rate constant of the redox step in the ferric ion–hydroquinone reaction is much larger, on the average, than those of the isotopic exchange electron transfer reactions having zero standard free energy change considered in the preceding paper [1]. The major reasons for this difference lie in:

(i) The large negative value of ΔF^0 in the former reaction as compared with the zero value of the latter.
(ii) The Coulombic *attraction* of the Fe^{+3}–QH^- reactants, as compared with the Coulombic *repulsion* of the reactants in those isotopic exchange reactions, and moreover:

"While the general agreement between the calculated and experimental results is satisfactory, the type of agreement obtained in Table II for the *absolute* value of ΔF^* in the reaction of ferric and hydroquinone ions is partly fortuitous. Two *compensating approximations* were employed: the a value chosen for the iron ion assumed complete dielectric saturation in the innermost hydration layer of this ion and as in [1] tends to make ΔF^*_{calc} too small."

The reason for this statement is that if the dielectric saturation in the innermost hydration layer is not complete, a reorganization energy contribution of this layer to ΔF^*_{calc} should also be considered.

"The a value for the oxygen group correctly assumed no dielectric saturation around the uncharged reactant but made the same assumption when it was charged. This tends to make ΔF^*_{calc} too large."

Because the innermost hydration layer of the ion is supposed completely unsaturated and so it contributes to ΔF^*_{calc}.

"In the oxygen-leucoindophenol reaction, only the second of these approximations was involved and therefore there is no compensation. This may be the reason why the absolute value of ΔF^* is somewhat larger than ΔF^*_{expt} in Table II.

In Table I, it is observed that the rate constant for the electron transfer step of the durohydroquinone is very high, namely 2.7×10^{11} liter mole^{-1} sec^{-1}. This value is about the maximum value that a rate constant can have in solution. The maximum corresponds to the situation in which the probability of reaction per collision is so high that the slow process in the reaction becomes the diffusion of the reactants toward each other" (vide supra). "Using a formula of Debye [76], we estimate that the rate constant for a diffusion-controlled reaction between two ions of charges +3 and −1 is about 5×10^{10} liter mole^{-1} sec^{-1}. Within the error of the various determinations, this is about equal to the rate constant k of the electron transfer step for the durohydroquinone reaction."

3.34. Ionic Radii

"The negative charge on a hydroquinone ion such as $HOC_6H_4O^-$ or $HOC_6H_4NHC_6H_4O^-$ is largely on the oxygen. Thus it is this atom which polarizes the dielectric. Accordingly, the *appropriate polarizing radius a* to be used for this *charged center* may be the same as that for another negatively charged oxygen, such as the OH^- ion. It is true that the organic residue will prevent the close approach of some of the solvent molecules and hence reduce their polarization. On the other hand, *this residue is itself polarized by the charged oxygen*, atomic polarization being induced, although it is less strongly polarized than the solvent. In this way, *the organic residue and the solvent play analogous roles.*

A similarity between the hydroquinone ion and the hydroxyl ion in their *extent of solvation* and therefore in their *effective polarizing radius a*, can be inferred from the standard entropy change of reaction:

$$\mathrm{H}Q\mathrm{H} + \mathrm{OH}^- = \mathrm{H}Q^- + \mathrm{HOH} \tag{3.105}$$

In such a reaction, the translational and rotational entropies of each of the two products would be expected to be about the same as those of the corresponding two reactants. Moreover, the sum of the vibrational entropies of the products should be about equal to the sum of those of the reactants.

If there is an appreciable entropy change in the reaction, it would be expected to arise from differences in the ability of the OH^- and the $\mathrm{H}Q^-$ ions to polarize the solvent molecules and therefore to vary in their entropy of solvation"

There is of course a close relation between the ability of solvent polarization by an ion and its entropy of solvation.

"Now the standard entropy change of reaction (3.105) is readily shown to equal the difference in entropy of ionization of water and of $Q\mathrm{H}_2$. The entropy of ionization of $Q\mathrm{H}_2$ for the various hydroquinones in Table I is [70], in the respective order in which they occur in the table, −26, −32, −29, and −25 entropy units, with an average value of −28. The entropy of ionization of water is [77] −26.7 eu. Accordingly, the entropy change of reaction (3.105) is seen to be zero within the experimental error. This therefore provides some basis for assuming that $Q\mathrm{H}^-$ and OH^- have about the same polarizing radius a.

Another question discussed in the text concerned the relative abilities of $Q\mathrm{H}^-$ and OH^- to dielectrically saturate the neighboring water molecules. The $Q\mathrm{H}$ molecule, being uncharged, cannot saturate the dielectric. The former, being charged, could. Thus a would apparently be different for $Q\mathrm{H}^-$ and $Q\mathrm{H}$, contrary to an assumption made in the derivation of Eq. (3.4) for ΔF^*. However, it is possible that the dielectric saturation by the anion is relatively small. For example, studies [78, 79] of the dielectric constant of aqueous salt solutions indicate *less dielectric saturation in the vicinity of anions, as compared with cations.* It is of interest, too, though not necessarily significant, that the 'experimental' free energy of solvation of anions agree reasonably well [80, 81] with those calculated from the Born formula [82]. The Born formula assumes that the dielectric is unsaturated.

In the interest of simplicity, we shall assume that as a first approximation there is little dielectric saturation in the vicinity of the $Q\text{H}^-$ ion. Accordingly it follows that a will be the crystallographic radius of the oxygen group, which is about 1.4 Å."

M. will later remove the assumption of absence of dielectric saturation.

"The O_2^- may be regarded as an ellipsoid of revolution whose [83] semimajor axis was found crystallographically to be 2.02 and whose semiminor axis is 1.51 Å. As a first approximation, this molecule will be treated as a sphere of radius 1.7 Å which has the same volume as the ellipsoid. This 'crystallographic' radius will be taken as the value of a for O_2^-, for the same reason given previously for the choice of the crystallographic radius as the a value of the oxygen atom in the hydroquinone ion" [2, p. 877].

NOTES

1. **M**: "κ is the probability of electron tunneling from one electronic configuration to the other, the rate constant is then proportional to κ. If one wants to phrase that in terms of the Eyring theory, since κ is not an energy dependent factor, that would contribute to the energy independent part of the free energy of activation. The energy independent part is the entropy of activation."
2. **M**: "For a typical bimolecular frequency, where we don't have to worry about charges, the number would be of the order of 10^{14},

so this is higher than that by a factor of 500, it is a strange because the charges would orient more, would reduce below the entropy, making the number below 10^{14}. Now, of course, you have all sorts of ion pairing effects. . ."

References

1. R. A. Marcus, *J. Chem. Phys.* **26**, 867, (1957).
2. R. A. Marcus, *J. Chem. Phys.* **26**, 872, (1957).
3. R. A. Marcus, *Trans. N.Y. Acad. Sci.* **19**, 423, (1957).
4. R. A. Marcus, *Can. J. Chem.* **37**, 155, (1959).
5. R. A. Marcus, "A Theory of Electron Transfer Processes at Electrodes," in *Transactions of the Symposium on Electrode Processes*, p. 239, E. Yeager, Editor, Wiley, New York (1961).
6. R. A. Marcus, *J. Phys. Chem.* **67**, 853, 2889, (1963).
7. R. A. Marcus, "Electron Transfer at Electrodes," in *Encyclopaedia of Electrochemistry*, p. 529, C. A. Hampel, Editor, Reinhold, New York (1964).
8. R. A. Marcus, P. Siddarth, in *Photoprocesses in Transition Metal Complexes, Biosystems and Other Molecules: Experiment and Theory*, Vol. 376, pp. 49–88, Kluwer Academic Publishers (1992).
9. R. A. Marcus, *Angew. Chem. Int. Ed. Engl* **32**, 1111–1121, (1993).
10. R. A. Marcus, in *Lecture Notes, International Summer School on the Quantum Mechanical Aspects of Electrochemistry*, pp. 1–75, P. Kirkov, Editor, Ohrid, Yugoslavia (1971).
11. R. A. Marcus, *Ann. Rev. Phys. Chem.* 15, 155–196, (1964).
12. R. A. Marcus, *Phys. Chem. Sci. Res. Rep.* **1**, 477, (1975).
13. R. A. Marcus, "Theory and Applications of Electron Transfers at Electrodes and in Solution," in *Special Topics in Electrochemistry*, p. 161, P. A. Rock, Editor, Elsevier Scientific Publishing Company, Amsterdam, Oxford, New York (1977).
14. M. Maimonides, *The Guide of the Perplexed*, Translated and with an Introduction and Notes by Shlomo Pines, Introductory Essay by Leo Strauss, 2 volumes, The University of Chicago Press, Chicago and London (1963).
15. S. Glasstone, K. J. Laidler, H. Eyring, *The Theory of Rate Processes*, McGraw-Hill, New York and London (1941).
16. R. A. Marcus, *Discuss. Faraday Soc.* **29**, 21, (1960); (a) R. A. Marcus, *Int. J. Chem. Kinet.* **13**, 865, (1981).
17. A. J. Bard, L. R. Faulkner, *Electrochemical Methods, Fundamentals and Applications*, John Wiley and Sons, Inc., New York, Chichester, Brisbane (1980).
18. R. A. Marcus, *J. Chem. Phys.* **43**, 679–701, (1965).

19. R. A. Marcus, "Electrostatic Free Energy and Other Properties of States Having Non-Equilibrium Polarization. II. Electrode Systems", Office of Naval Research, Technical Report No. 11 Contract Nonr 839(09), Task No. 051-339, in *Special Topics In Electrochemistry*, p. 210, P. A. Rock, Editor, Elsevier Scientific Publishing, Amsterdam, Oxford, New York (1977).
20. R. A. Marcus, "On the Theory of Overvoltage for Electrode Processes Possessing Electron Transfer Mechanisms. I", Office of Naval Research, Technical Report No. 12 Contract Nonr 839, Task No. 051-339, in*Special Topics in Electrochemistry*, p. 180, P. A. Rock, Editor, Elsevier Scientific Publishing Company, Amsterdam, Oxford, New York (1977).
21. J. O'M. Bockris, A. K. N. Reddy, A. M. Gamboa-Aldeco, *Modern Electrochemistry 2A, Fundamentals of Electrodics*, 2nd Edition, Kluwer Academic/ Plenum Publishers, New York, Boston, Dordrecht (2000).
22. K. J. Vetter, *Electrochemical Kinetics, Theoretical Aspects*, Academic Press, New York, London (1967).
23. P. Delahay, *Double Layer and Electrode Kinetics*, Interscience Publishers, New York, London, Sidney (1965).
24. L. Eberson, *Electron Transfer Reactions in Organic Chemistry*, Springer-Verlag, Berlin, Heidelberg, New York (1987).
25. H. C. Brown, *J. Phys. Chem.* **56**, 868, (1952).
26. H. Taube, *Can. J. Chem.* **37**, 129, (1959).
27. D. J. Griffiths, *Introduction to Electrodynamics*, 3rd Edition, Prentice Hall International Inc., (1999).
28. W. K. H. Panofsky, M. Phillips, *Classical Electricity and Magnetism*, 2nd Edition, Addison-Wesley, Reading, Massachusetts (1962).
29. M. Mason, W. Weaver, *The Electromagnetic Field*, Dover Publications, Inc., New York (1929).
30. R. P. Feynman, R. B. Leighton, M. Sands, *The Feynman Lectures on Physics*, Vol. II, Addison-Wesley Publishing Company, Reading, Massachusetts, Menlo Park, California, London (1964).
31. J. J. Lingane, *Electroanalytical Chemistry*, 2nd Edition, Interscience Publishers, Inc., New York (1958).
32. L. Meites, *Polarographic Techniques*, 2nd Edition, Interscience Publishers, New York, London, Sidney (1965).
33. J. Heyrovský, J. Kůta, *Principles of Polarography*, Academic Press, New York and London (1966).
34. R. McWeeny, *Coulson's Valence*, 3rd Edition, Oxford University Press (1979).
35. S. S. Shaik, H. B. Schlegel, S. Wolfe, *Theoretical Aspects of Physical Organic Chemistry. The S_N2 Mechanism*, John Wiley and Sons, New York (1992).
36. V. G. Levich, in *Physical Chemistry, An Advanced Treatise*, Vol. 98, H. Eyring, D. Henderson, W. Jost, Editors, Academic Press, New York (1970).

37. R. R. Dogonadze, in *Reactions of Molecules at Electrodes*, p. 135, N. S. Hush, Editor, Wiley, New York (1971).
38. R. A. Marcus, *J. Chem. Phys.* **24**, 979–989, (1956).
39. P. Lorrain, D. R. Corson, *Electromagnetic Fields and Waves*, 2nd Edition, p. 68, Freeman and Co., San Francisco (1970), Eq. (2.92).
40. J. A. Stratton, *Electromagnetic Theory*, p. 203, McGraw-Hill Book Company, New York and London (1941), Eq. 12.
41. P. P. Schmidt, *Electrochemistry Spec. Period. Rep. Electrochem.* **5**, 21, (1975).
42. N. Sutin, *Ann. Rev. Phys. Chem.* **17**, 119–142, (1966).
43. N. Sutin, *Acc. Chem. Res.* **1**, 225–231, (1968).
44. H. Taube, *J. Chem. Ed.* **45**, 452, (1968).
45. H. Taube, *Electron Transfer Reactions of Complex Ions in Solution*, p. 30, Academic Press, New York and London (1970).
46. H. Taube, E. L. King, *J. Am. Chem. Soc.* **76**, 4053, (1954).
47. H. Taube, H. Meyers, *J. Am. Chem. Soc.* **76**, 2013, (1954).
48. H. Gerischer, *Z. Elektrochem.* **54**, 366, (1950).
49. N. S. Hush, *J. Chem. Phys.* **28**, 962, (1958).
50. H. Rubin, F. C. Collins, *J. Phys. Chem.* **58**, 958, (1954).
51. J. R. Baxendale, H. R. Hardy, *Trans. Faraday Soc.* **50**, 808, (1954).
52. J. H. Baxendale, S. Lewin, *Trans. Faraday Soc.* **42**, 126, (1946).
53. J. C. Sheppard, A. C. Wahl, *J. Am. Chem. Soc.* **79**, 1020, (1957).
54. A. C. Wahl, C. P. Deck, *J. Am. Chem. Soc.* **76**, 4054, (1954).
55. H. A. Laitinen, P. Kivalo, *J. Am. Chem. Soc.* **75**, 2198, (1953).
56. B. R. Baker, F. Basolo, H. M. Neumann, Paper presented at the Symposium on Mechanisms of Inorganic Reactions in Solution. Northwestern University, Chicago, Ill. 1958.
57. W. B. Lewis, C. D. Coryeli, J. W. Irvine, *J. Chem. Soc.* **2**, S386, (1949).
58. L. E. Orgel, *Report to Xe Inst. Intern. Chim. Solvay, Conseil chim.* 289, (1956).
59. J. Silverman, R. W. Dodson, *J. Phys. Chem.* **56**, 846, (1952).
60. A. W. Adamson, K. S. Vorres, *J. Inorg. Nucl. Chem.* **3**, 206, (1956).
61. J. E. B. Randles, K. W. Somerton, *Trans. Faraday Soc.* **48**, 937, (1952).
62. H. A. Laitinen, M. W. Grieb, *J. Am. Chem. Soc.* **77**, 5201, (1954).
63. A. Anderson, N. A. Bonner, *J. Am. Chem. Soc.* **78**, 3826, (1954).
64. J. Hudis, R. W. Dodson, *J. Am. Chem. Soc.* **78**, 1911, (1956).
65. J. C. Sullivan, D. Cohen, J. C. Hindman, *J. Am. Chem. Soc.* **79**, 3672, (1957).
66. E. Lange, J. Martin, *Z. Elektrochem.* **42**, 662, (1936).
67. J. J. Grossman, C. S. Garner, *J. Chem. Phys.* **28**, 268, (1958).
68. R. G. Sachs, D. L. Dexter, *J. Appl. Phys.* **21**, 1304, (1950).
69. A. A. Vlcek, in *Sixth International Conference on Coordination Chemistry*, p. 590, S. Kirschner, Editor, The Macmillan Co., New York, NY, 1961.
70. J. H. Baxendale, H. R. Hardy, *Trans. Faraday Soc.* **49**, 1140, (1953).
71. R. A. Marcus, *Int. J. Chem. Kinet.* **13**, 865, (1981).

72. A. A. Frost, R. G. Pearson, *Kinetics and Mechanism*, 2nd Edition, John Wiley and Sons, Inc., New York, London, Sydney (1961).
73. W. M. Latimer, *Oxidation Potentials*, Prentice Hall, Inc., New York (1952).
74. *International Critical Tables*, Vol. 6, p. 333, McGraw-Hill Book Company, Inc., New York (1929).
75. R. Parsons, *Modern Aspects of Electrochemistry*, Chap. 3, J.O'M. Bockris, Editor, Academic Press, Inc., New York, 1954.
76. P. Debye, *Trans. Electrochem. Soc.* **82**, 265, (1942).
77. H. S. Harned, B. B. Owen, *The Physical Chemistry of Electrolytic Solutions*, Chap. 15, Reinhold Publishing Corporation, New York (1950).
78. J. B. Hasted, D. M. Ritson, C. H. Collie, *J. Chem. Phys.* **16**, 1, (1948).
79. D. H. Everett, C. A. Coulson, *Trans. Faraday Soc.* **36**, 633, (1940).
80. W. M. Latimer, K. S. Pitzer, C. M. Slansky, *J. Chem. Phys.* **7**, 108, (1939).
81. W. M. Latimer, *J. Chem. Phys.* **25**, 90, (1955).
82. M. Born, *Z. Phys.* **1**, 45, (1920).
83. A. Helms, W. Klemm, *Z. anorg. u. allgem. Chem.* **241**, 97, (1939).

CHAPTER 4

Theory of Electrochemical Electron Transfer

The first and fundamental formulation of Marcus theory for homogeneous ET reactions (redox reactions) was summarized in Chapter 2. In this chapter, an abridged version is presented of its extension to electrode systems from Refs. [1, 2, and 2a]. The two Office of Naval Research reports were written by Marcus in 1957 but were published in 1977 in a book edited by Peter A. Rock. In the following, the symbol M. is for Marcus.

4.1. Electrostatic Free Energy and Nonequilibrium Dielectric Polarization for Electrode Systems

In ET at electrodes, the total dielectric polarization $\mathbf{P}(\mathbf{r})$ is, like in the homogeneous case, the sum of two terms:

$$\mathbf{P}(\mathbf{r}) = \mathbf{P}_e(\mathbf{r}) + \mathbf{P}_u(\mathbf{r})$$

We remind the reader that the electronic polarization $\mathbf{P}_e(\mathbf{r})$ is the portion of the electric polarization which is in electrostatic equilibrium with the electric field strength $\mathbf{E}(\mathbf{r})$, that is, the polarization is determined by the field through the electronic polarizability α_e:

$$\mathbf{P}_e(\mathbf{r}) = \alpha_e \mathbf{E}(\mathbf{r})$$

The E-type polarization is related to D_{op}, the square of the refractive index:

$$4\pi\alpha_e = D_{op} - 1$$

An important point is that $\mathbf{E}(\mathbf{r})$ depends on the charge distribution *and on* $\mathbf{P}_u(\mathbf{r})$. On the other hand, $\mathbf{P}_u$ is the portion of dielectric polarization which is in general not in equilibrium with $\mathbf{E}$. When it is in equilibrium, it can be expressed in terms of $\mathbf{E}$ and of the atomic–orientational polarizability α_u:

$$\mathbf{P}_u(\mathbf{r}) = \alpha_u \mathbf{E}(\mathbf{r})$$

When it is not, it appears as a solution of a variational equation.

The U-type polarization depends on both D_{op} and D_s:

$$4\pi\alpha_u = D_s - D_{op}$$

4.2. The Electrochemical System

After having briefly reminded the reader of the properties of the polarization vector function, I describe now the simple model of electrode system considered by Marcus.

An oriented, thin, and nonpolarizable solvent layer is supposed to exist at the electrode–solution interface ("inner layer" of the electrical double layer, see, e.g., Ref. [3]). The adsorbed layer produces a fixed potential difference χ across the interface. The magnitude of χ depends on the solvent, on the nature of the electrode and on the temperature. χ is first supposed to be independent of the average electrode charge density. The ions taking part in ET are designated as "central ions." When the solution is not infinitely diluted, the central ions will be surrounded by an ionic atmosphere of other ions. If a salt is present in solution whose ions do not exchange electrons with the electrode (inert electrolyte), the ionic atmosphere will be made up also from ions of the inert electrolyte.

The picture of the system is now complete: metal electrode, solvent layer, central ions taking part in the ET process, ionic atmospheres surrounding them [1, p. 211].

4.3. Electrostatic Potential

As in the case of homogeneous ET, "each central ion is treated as a sphere having a surface charge density $\sigma(\mathbf{r})$ equal to the ionic charge divided by its area. The remaining ions are described in terms of a volume charge density $\rho(\mathbf{r})$. There will also be a surface charge distribution on the electrode surface." As in Chapter 2 (see Ref. [4]), "the potential can be expressed in terms of contributions from the charges *and from the polarized volume elements.*" The potential arises then from the following contributions:

(i) Volume charges of the ions in ionic atmospheres
(ii) Surface charges on the central ions
(iii) Surface charges on the surface of the M electrode
(iv) Polarized volume elements
(v) Potential difference at the interface

This potential "is the same as the 'inner potential' $\varphi(\mathbf{r})$ described by Parsons [5] and Lange [6]."

"We have:

$$\phi(\mathbf{r}') = \int \frac{\rho(\mathbf{r})}{|\mathbf{r}-\mathbf{r}'|}dV + \int_{\text{ions}} \frac{\sigma(\mathbf{r})}{|\mathbf{r}-\mathbf{r}'|}dS + \int_{M} \frac{\sigma(\mathbf{r})}{|\mathbf{r}-\mathbf{r}'|}dS + \mathbf{P}(\mathbf{r}) \cdot \nabla_r \frac{1}{|\mathbf{r}-\mathbf{r}'|}dV + \beta \quad (4.1)$$

where ∇_r is the gradient operator and where

$$\begin{aligned} &\beta = \chi \text{ (on the electrode)} \\ &\text{and} \\ &\beta = 0 \text{ (in solution)} \end{aligned} \quad (4.2)$$

The volume integrals are over the entire volume of the dielectric (the solution). The surface integral is over the surface of each central ion and over the surface of the electrode M:

$$\int \frac{\sigma(\mathbf{r})}{|\mathbf{r}-\mathbf{r}'|}dS = \int_{\text{ions}} \frac{\sigma(\mathbf{r})}{|\mathbf{r}-\mathbf{r}'|}dS + \int_{M} \frac{\sigma(\mathbf{r})}{|\mathbf{r}-\mathbf{r}'|}dS$$

It is convenient to define a function $\psi(\mathbf{r})$ which is identical with the potential employed in Marcus' Part I (Ref. [4]):

$$\psi(\mathbf{r}) = \varphi(\mathbf{r}) - \beta \tag{4.3}$$

For the electric field strength $E(\mathbf{r})$, we have $E(\mathbf{r}) = -\nabla\varphi = \nabla\psi$" [1, p. 212].

Equation (4.1) is like Eq. (2.11a) in Chapter 2 with the addition of the $\int_M \frac{\sigma(\mathbf{r})}{|\mathbf{r}-\mathbf{r}'|}dS$ and β terms, characteristic of the electrode surface and of the electrode–solution interface.

4.4. Electrode Charge Distribution and the Method of Electrical Images

The term $\int_M \frac{\sigma(\mathbf{r})}{|\mathbf{r}-\mathbf{r}'|}dS$ is expressed by M. using the method of electrical images.

"In systems having equilibrium polarization the method of images [7] can be used to determine the charge distribution" [1, p. 217].

M. briefly discusses the quantum limitations to the image force theory in reference 9 of Ref. [1] deducing from Ref. [8] that the percent error in the energy change computed from the image force theory is small.

"The same method is also suitable for systems having nonequilibrium polarization. For simplicity, consider a planar electrode, that is, any electrode whose radius of curvature is appreciably greater than the thickness of the double layer. Let the electrode–dielectric interface be situated in the plane $x = 0$ and let the dielectric occupy the semi-infinite region $x > 0$. As before, the coordinates of any point in the dielectric will be specified by the vector r drawn from any arbitrary origin to the point. The *coordinates of the 'image point,'* having the same y and z and differing only in the sign of x, will be specified by a vector $\mathrm{r_{im}}$. Mirror image functions $\bar{\rho}(\mathbf{r}_{im})$, $\bar{\sigma}(\mathbf{r}_{im})$, and $\bar{\mathbf{P}}(\mathbf{r}_{im})$ will then be defined in *the region of negative* x, according to Eq. (4.4). They are undefined in the region of positive x, i.e., in the dielectric.

$$\bar{\rho}(\mathbf{r}_{im}) = \rho(\mathbf{r}), \quad \bar{\sigma}(\mathbf{r}_{im}) = \sigma(\mathbf{r}) \tag{4.4}$$
$$\bar{\mathbf{P}}_x(\mathbf{r}_{im}) = -\mathbf{P}_x(\mathbf{r}), \quad \bar{\mathbf{P}}_y(\mathbf{r}_{im}) = \mathbf{P}_y(\mathbf{r}), \quad \bar{\mathbf{P}}_z(\mathbf{r}_{im}) = \mathbf{P}_z(\mathbf{r})$$

The last three expressions relate the components of the vector $\bar{\mathbf{P}}$ to those of $\mathbf{P}$. In Eq. (4.4), σ refers only to the surface charge density on the central ions and not to that on the electrode."

Figure 1 shows schematically the electrical image of a central ion, represented by a small dashed circle. The dashed shell around it represents the image of its ionic cloud. The region between the solid line representing the electrode–solution interface and the dashed line parallel to it is the region of the electrode where the images of most of the ions of the electrical double layer are to be found.

Note that $\bar{\rho}(\mathbf{r}_{im}) = \rho(\mathbf{r})$ because *both charge and volume* in the *image* charge density are opposite to those in the charge density,

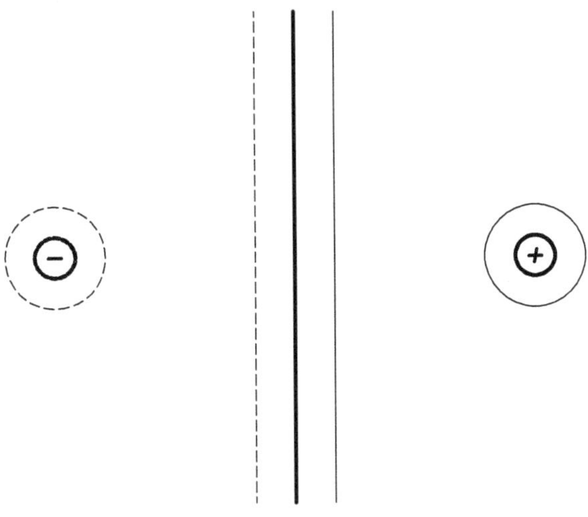

Fig. 1*.

that is, $dV = dxdydz$ in the dielectric but $dV = (-dx)dydz$ in the metal. Notice that not only the charge but also the polarization vector function has an image.

"Applying the method of images, the electrode charge distribution which satisfies the condition of constant potential on the electrode M and in the body of the solution obeys the following equation:

$$\int_M \frac{\sigma(\mathbf{r})}{|\mathbf{r}-\mathbf{r}'|}dS = -\int_{\text{ions}} \frac{\bar{\sigma}(\mathbf{r}_{im})}{|\mathbf{r}_{im}-\mathbf{r}'|}dS_{im} - \int \frac{\bar{\varrho}(\mathbf{r}_{im})}{|\mathbf{r}_{im}-\mathbf{r}'|}dV_{im} - \int \bar{\mathbf{P}}(\mathbf{r}_{im})\cdot\nabla\frac{\mathbf{1}}{|\mathbf{r}_{im}-\mathbf{r}'|}dV_{im} \tag{4.5}$$

In Eq. (4.5), $\mathbf{r}$ and $\mathbf{r}'$ denote any points on the electrode's Msurface and in the dielectric medium, respectively. The first surface integral is over M and the second over the mirror images of each central ion. The volume integrals are over the entire mirror image of the volume of the dielectric."

It appears in this way very clearly that the charge induced on the surface of the electrode can be expressed in terms of the dielectric images: "The total charge in the dielectric is equal and opposite to the total image charge, which in turns equals the total electrode charge."[1]

The integrals extended to the ions can be evaluated treating the ions as spheres with surfaces of uniform charge densities. "It can be readily shown (see Chapter 2) that:

$$\int_{\text{ion}} \frac{\sigma dS}{|\mathbf{r}-\mathbf{r}'|} = \frac{\int_{\text{ion}} \sigma dS}{r} = \frac{q}{r}$$

Where r is the distance from the ion to the field point, see Fig. 2, and q is the ionic charge.

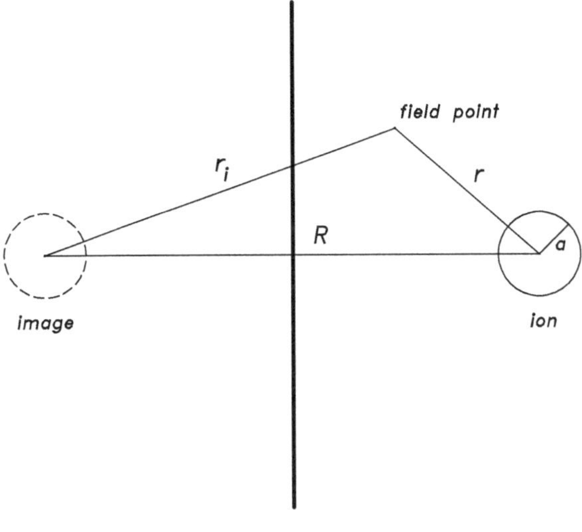

Fig. 2*.

Similarly, it can be shown that:

$$\int_{\text{ion}} \frac{\bar{\sigma}(\mathbf{r}_{im})dS_{im}}{|\mathbf{r}_{im} - \mathbf{r}'|} = \frac{q}{r_{im}}$$

where r_{im} is the distance from the field point to the center of the electrical image of this ion" [1, p. 218].

4.5. Mixing and Electrostatic Entropy and Free Energy

I report here the Note added in proof in p. 983 of Ref. [10]: "The charge distribution at the end of Stage I of the charging process, denoted by ρ^0 and σ^0, is a fictitious distribution used to produce the specified U-type polarization $\mathbf{P}_u(\mathbf{r})$. By contrast, the charge distribution at the end of Stage II is the actual distribution of charges on the ions of the system. The complete charging process is performed at fixed configurations of these ions, though they may be uncharged in Stage I and charged up in Stage II. Thus the reversible work given by Eq. (4.6)

$$F = \frac{1}{2}\int \left\{ \frac{\mathbf{E}_c \cdot \mathbf{E}_c}{4\pi} - \mathbf{P}\cdot\mathbf{E}_c + \mathbf{P}_u \cdot \left(\frac{\mathbf{P}_u}{\alpha_u} - \mathbf{E} \right) \right\} dV \tag{4.6}$$

was also performed at fixed configuration of these ions. There is therefore an additional free energy term which should be considered, namely the entropy term associated with the preliminary formation of this given uncharged configuration of the ions from a random configuration. Let e_i, $c_i(\mathrm{r})$, and c_i^0 denote the charge, the concentration, and the average concentration of ions of the ith species which make up the continuous volume charge distribution $\rho(\mathrm{r})$; $\rho(\mathrm{r}) = \sum_i e_i c_i(\mathrm{r})$. Then, this additional entropy term associated with the formation of the given configuration from a random one is the well-known excess entropy of mixing and is given by Eq. (4.6a):

$$-\boldsymbol{k}\sum_i \int \boldsymbol{c}_i \cdot (\ln \boldsymbol{c}_i - \ln \boldsymbol{c}_i^{\mathbf{0}}) \boldsymbol{d}\boldsymbol{V} \tag{4.6a}$$

(typo in the original). The total electrostatic contribution to the free energy includes this term *and* F given by Eq. (4.6). It is the sum of these two terms, rather than F alone, which is the electrostatic free energy, F_e, say

$$F_e = F + kT\sum_i \int c_i(\ln c_i - \ln c_i^0) dV \tag{4.6b}$$

where F is given by Eq. (4.6). In a very dilute solution c_i equals c_i^0 and F_e equals F.

Similarly, the total electrostatic contribution to the entropy is the sum of the term in Eq. (4.6a) and of S given by Eq. (4.7) in the following section

$$S = -\left(\frac{\partial W_I}{\partial T} \right)_{E_c^0}. \qquad (4.7)\text{”}$$

Altogether the total electrostatic entropy will be:

$$S_e = S - k\sum_i \int c_i(\ln c_i - \ln c_i^0) dV \tag{4.8}$$

F is computed by the two stages charging process. The work done during each stage can be calculated from the equation:

$$W = \iint \varphi^{\lambda} \frac{d\rho}{d\lambda} d\lambda dV + \iint \varphi^{\lambda} \frac{d\sigma}{d\lambda} d\lambda dS \tag{4.9}$$

where λ is a *charging parameter* which is increased from 0 to 1 during each charging stage and where φ^{λ} denotes the value of φ at any λ (Notice that $d\rho dV$ is a charge and that $\varphi d\rho dV$ is then a product potential $\times$ charge, that is, electrical work. The same for $\varphi d\sigma dS$, compare Eq. (4.9) in Ref. [10]). Introducing Eqs. (4.2) and (4.3) in this expression for W, that is, going from treating ET in solution to the parallel treatment of ET at the electrode, Eq. (4.9) becomes:

$$W = \iint \psi^{\lambda} \frac{d\rho}{d\lambda} d\lambda dV + \iint \psi^{\lambda} \frac{d\sigma}{d\lambda} d\lambda dS + \chi \iint_{M} \frac{d\sigma}{d\lambda} d\lambda dS \tag{4.10}$$

Eq. (4.10) differs from the corresponding expression used for the redox case [10] only in the last term. For this reason, the formula deduced for F by the two stages charging process in Ref. [10], see Chapter 2, will differ from the one for the electrode system only in last term:

$$F = \frac{1}{2} \int [-\mathbf{P} \cdot \mathbf{E}_c - \mathbf{P}_u \cdot (\mathbf{E} \cdot \mathbf{P}_u/\alpha_u) + \mathbf{E}_c \cdot \mathbf{E}_c/4\pi] dV$$
$$+\chi \int_{M} \sigma dS \tag{4.11}$$

where $\mathbf{E}_c(\mathbf{r})$ is the electric field which the given *ionic and electrode* charge distribution would exert in a vacuum:

$$\mathbf{E}_c(\mathbf{r}') = -\nabla_{r'} \left[\int \frac{\rho(\mathbf{r})}{|\mathbf{r} - \mathbf{r}'|} dV + \int \frac{\sigma(\mathbf{r})}{|\mathbf{r} - \mathbf{r}'|} dS \right]$$

On the electrode, this $\sigma(\mathbf{r})$ arises from all the induced charges, including those induced by the *polarized dipoles*. M. defines then a function $\mathbf{E}_v(\mathbf{r})$ which "depends only on the ionic charges in the solution and on the surface charge density $\sigma_v(\mathbf{r})$ which they would induce in a vacuum." The potential $\psi_v(\mathbf{r}')$ in that system must be a constant on the electrode.

$\mathbf{E}_v(\mathbf{r}')$ and $\psi_v(\mathbf{r}')$ are given by:

$$\mathbf{E}_v(\mathbf{r}') = -\nabla_{r'}\psi_v(\mathbf{r}')$$

$$\psi_v(\mathbf{r}') = \int \frac{\rho_v(\mathbf{r})}{|\mathbf{r} - \mathbf{r}'|} dV + \int \frac{\sigma_v(\mathbf{r})}{|\mathbf{r} - \mathbf{r}'|} dS$$

where $\rho_v(\mathbf{r}) = \rho(\mathbf{r})$ and where $\sigma_v(\mathbf{r})$ and $\sigma(\mathbf{r})$ are equal on the surface of each central ion but differ on the electrode. They differ there by an amount equal to the surface charge density induced in the electrode by the polarized dielectric, that is, by $\mathbf{P}(\mathbf{r})$.

M. shows that Eq. (4.12) can be rewritten in terms of $\mathbf{E}_v$:

$$F = \frac{1}{2}\int [-\mathbf{P}\cdot\mathbf{E}_v - \mathbf{P}_u\cdot(\mathbf{E} - \mathbf{P}_u/\alpha_u) + \mathbf{E}_v\cdot\mathbf{E}_v/4\pi]dV + \chi\int \sigma dS$$

For electrode systems, Eq. (4.8) will be:

$$S_e = -\left(\frac{\partial W_I}{\partial T}\right)_{\mathbf{E}^0_{v,\chi}} - k\sum_i \int c_i \ln(c_i/c_i^0)dV$$

where $\mathbf{E}^0_{v,\chi}$ stands for $\mathbf{E}^0_c$ and the internal energy will be $U = F_e + TS_e$.

4.6. Evaluation of the F_e of an Equilibrium System

Equations (4.11) and (4.12) give the *general expressions* for F_e *valid for equilibrium and nonequilibrium* dielectric polarization systems. F_e will have its minimum value F_e^{eq} at equilibrium [1, p. 215]. Fluctuations $\delta\mathbf{P}_u$ and δc_i carry the system away from equilibrium and F_e^{eq} moves to corresponding nonequilibrium values $F_e^{\text{eq}} + \delta F_e$. In order to get the particular expression for F_e at equilibrium, M. minimizes F_e setting $\delta F_e = 0$ and considering two equations of restraint. For the variation δF_e, we have from Eq. (4.6b):

$$\delta F_e = \delta F + kT \sum_i \int \ln(c_i/c_i^0)\delta c_i \delta V$$

In δF the term $\chi \int_M d\sigma dS$ will appear.

In the minimization process, one should consider two equations of constraint of immediate physical interpretation:

(i) $\int \delta c_i(\mathbf{r})dV = 0$ for each i

[2.215] because the total number n_i of ions of species i is fixed in solution [(2)] and:

(ii) $\int_M d\sigma dS = -\sum_i \int e_i \delta c_i dV$

[2.220] which means that if the charge density on the electrode varies by σ, the charge in the solution must vary by an equal and opposite amount because of the electroneutrality of the whole electrode and solution system.

Solving the variational problem for δF_e subject to the earlier constraints, M. shows that at equilibrium $\mathbf{P}_u(\mathbf{r})$ is determined by $\mathbf{E}(\mathbf{r})$(as it was intuitively to be expected):

$$\mathbf{P}_u(\mathbf{r}) = \alpha_u \mathbf{E}(\mathbf{r}) \tag{4.14}$$

and that $c_i(\mathbf{r})$ depends on $\phi^{\mathrm{eq}}(\mathbf{r})$ around a central ion:

$$c_i = n_i \exp(-e_i \phi^{\mathrm{eq}}/kT)/V_i \tag{4.15}$$

$$V_i = \int \exp(-e_i \phi^{\mathrm{eq}}/kT)dV$$

[1, p. 215], that is, the concentration $c_i(\mathbf{r})$ of volume charges of ions of species i around a central ion depends on the potential $\phi^{\mathrm{eq}}(\mathbf{r})$ and is Boltzmann weighted by the ratio:

$$\frac{\exp(-e_i\phi^{\text{eq}}/kT)}{\int \exp(-e_i\phi^{\text{eq}}/kT)dV}$$

Notice, moreover, that the integral in the denominator has the dimension of a volume and so there is a concentration in both members of Eq. (4.15).

The equilibrium electrostatic free energy (4.16a) can be deduced directly from Eqs. (4.6b) and (4.10) [2, p. 215]

$$F_e^{\text{eq}} = -\frac{1}{2}\int \rho\varphi dV + \frac{1}{2}\int \sigma\varphi dS + \chi\int_M \sigma dS + kT\sum_i n_i \ln V/V_i \tag{4.16a}$$

An equivalent expression is:

$$F_e^{\text{eq}} = \int_{\sigma=0}^{\sigma=\sigma}\int_S \phi^{\text{eq}} d\sigma dS$$

[2.222] where $d\sigma = \frac{dq}{dS}$. The formula is of transparent physical meaning: the equilibrium free energy of the electrodic system is the energy necessary to charge up all of the surfaces present in it. If the system is supposed to be made up of

(i) A central ion
(ii) Mobile ions
(iii) An electrode M
(iv) A medium of dielectric constant D_s the potential ϕ^{eq} is built up from the contributions of (i), (ii), and (iii) [2, pp. 196, 197]:

$$\phi^{\text{eq}} = \frac{q}{D_s}\left(\frac{1}{r} - \frac{1}{r_i}\right) + \phi_\rho^q + \beta$$

"where r and r_i denote the distances from the field point to the center of the ion and to the center of its electrical image, respectively. ϕ_ρ^{eq} is the contribution from the mobile ions and from the electrode charges they induce. On the electrode surface, ϕ^{eq} is a constant, $\beta = \chi$. On the surface of the central ion of radius a, see Fig. 2, and surface $4\pi a^2$, the surface density $d\sigma$ equals $dq/4\pi a^2$." It can be shown that (see Ref. [4, p. 974] with note on the Born charging formula):

$$\int_{\text{ion}} \frac{q}{D_s}\left(\frac{1}{r} - \frac{1}{r_i}\right) d\sigma dS = \frac{q^2}{2D_s}\left(\frac{1}{a} - \frac{1}{R}\right)$$

where R is the distance from the center of the ion to its electrical image as in Fig. 2.

We have then:

$$F_e^{\text{eq}} = \frac{1}{4\pi a^2}\int_{\text{ion}}\int_{q=0}^{q=q} \phi_\rho^q dq dS + \frac{q^2}{2D_s}\left(\frac{1}{a} - \frac{1}{R}\right) + \chi \int_M \sigma dS \tag{4.16b}$$

φ_ρ^q is, on its turn, the sum of three independent contributions, when the ion is outside the double layer region:

(1) "ϕ_{atm}^q, arising from a spherically symmetric ionic atmosphere about the central ion," see Fig. 1.
(2) "A contribution due to the electrode charge density induced by this atmosphere. It is symbolized by a dashed spherical shell in Fig. 1. Since the atmosphere is concentric with the central ion, spherically symmetric, and has a total charge of $-q$, it can be shown that the same electrode charge density would be introduced by a point charge $-q$ situated at the center of the central ion. The image of this charge is q and its contribution to the potential is therefore $q/D_s R$." This contribution to the potential will give a contribution $q^2/2D_s R$ to F_e^{eq} that will cancel the contribution $-q^2/2D_s R$ in Eq. (4.16b). This means that the contributions of the image of the central ion and of the image of its ionic atmosphere cancel each other.
(3) "φ_S arising from the ions of the electrical double layer together with the electrode charges they induce," see Fig. 1.

"Remembering that both $\phi_{\rm atm}^{q}$ and φ_S are constant over the surface of this ion, we obtain:

$$F_e^{\rm eq} = \int_{q=0}^{q=q} \phi_{\rm atm}^{q} dq + q\phi_S + \chi \int_M \sigma dS + q^2/2D_s a \qquad (4.17)$$

Since $\int_{q=0}^{q=q} \phi_{\rm atm}^{q} dq = kT \ln \gamma_q$, where γ_q is the electrostatic contribution to the activity coefficient of the ion of charge q in the body of the solution, Eq. (4.17) becomes:

$$F_e^{\rm eq} = kT \ln \gamma_q + q\phi_S + \chi \int_M \sigma dS + q^2/2D_S a. \qquad (4.18)\text{"}$$

4.7. Theory of Overvoltage for Electrode Processes Possessing ET Mechanism

4.7.1. *Introduction*

Among the mechanisms for electrode processes, there may be simple ET or more complicated ones in which the ET process "is preceded or followed by chemical equilibrium or by other chemical reactions. From such studies similarities have been noted between the rate with which certain species are oxidized or reduced by electrochemical and by chemical means." Randles [11] and Vlcek [12] "interpreted the rates of *electrochemical* electron transfers in terms of concepts related to those used for electron transfer in *chemical* reactions. A brief qualitative comparison of the *relative rates* of several isotopic exchange reactions and of the corresponding electrochemical systems" was given by M. in Ref. [13] and is here reported in Chapter 3 [2, pp. 181, 182].

The theory presented in this chapter for ET in electrode processes is a continuation of that for homogeneous ET reactions. Like the theory for homogeneous (redox) ET reactions, even this theory "proceeds from first principles plus assumptions that appear reasonable on *a priori* grounds and is free from arbitrary assumptions and adjustable parameters. As before, the theory is not applicable to atom transfer mechanisms (hydrogen overvoltage, for example)" [2, p. 182].

4.8. The ET Rate Constants

"Equations describing any *overall* electrochemical process may include terms for

(i) The transport of ions to the electrochemical double layer region at the electrode for any chemical reaction of the electrochemically active species
(ii) The work required to penetrate the double layer (if necessary)
(iii) The actual ET

Only under certain conditions can these facts be disentangled in a relatively simple way and simple ET rates defined: The double layer region should be sufficiently thin that

(1) No chemical reaction occurs in it
(2) Diffusion across it is sufficiently rapid so that the concentration of an ion at any point in the double layer is related to its concentration just outside it by the work required to transport it to that point.

At appreciable salt concentration, the double layer thickness appears to be only of the order of several Ångstroms [14]. Under the earlier conditions, one can describe the entire electrochemical process by force-free differential equations [14] (containing any chemical reaction terms [16] if necessary) outside of the double layer region, while *the boundary condition*, at a boundary surface S drawn *just outside the double layer*, *contains rate constants* which depend only on the ET process itself. This boundary condition is that the flux of any ion through this surface S equals its net rate of disappearance by ET. If A and B represent the electrochemically active ion before and after its electron transfer with the electrode, the ET step may be written as:

$$A + ne \underset{k_b}{\overset{k_f}{\rightleftarrows}} B \tag{4.19}$$

where n is the number of electrons lost by the electrode.

If C_A^S and C_B^S denote the concentrations of A and B just outside the double layer, then the ET constants, k_f and k_b, will be defined by the relation:

Net ET per unit area = $k_f C_A^S - k_b C_B^S$" [2, pp. 182–183].

4.9. Basic Assumptions

The basic assumptions used in this extension of the Marcus theory to the electrode systems are analogous to those made in the oxidation–reduction theory for homogeneous systems [4].

The assumptions are

(a) "In the activated state of reaction (4.19), the spatial overlap between the electronic orbitals of the two 'reactants' is assumed to be small. The 'reactants' are now the electrode and the discharging ion or molecule (to be referred as 'central ion')." This assumption was previously discussed in Chapter 1.
(b) The central ion in this 1957 early form of the theory is treated as "a sphere within which no changes in interatomic distances occur during reaction (4.19) and outside of which the solvent is treated as a dielectrically unsaturated continuum. For complex ions or hydrated monoatomic cations, the sphere includes the first coordination shell of the central atom [4, 17]." [2, p. 183] The concept of "effective size" for unsymmetric organic molecules has been described in Chapter 3. In a refinement of the theory, changes, if any, in interatomic distances during ET were considered by Marcus, and were briefly mentioned in Chapter 3 and will be further dealt with later in Chapter 6.

4.10. Nature of the Transition State

As in the case of homogeneous ET, a successful ET between the central ion and the electrode proceeds via *two successive intermediate states*, X^* and X, both participating in the ET process. X^* represents the state of the reactants *just before* ET, X that of the products *just after* ET while *a linear combination of the two wave function of the states X^* and X represents the TS*. "These states have the same atomic configuration and the same total energy but in X^* the electronic configuration in the central ion is that of A and in X that of B. The electronic configuration of the electrode undergoes a corresponding change" because of the Franck–Condon principle as was already observed in Chapter 1. "By arguments similar to those employed in the redox theory" dealt with in the preceding Chapters, "it then follows that the actual ET, that is, the formation of X from X^*, must be *preceded* by a reorientation of the solvent molecules *in the vicinity of the discharging ion and nearby area of the electrode*. For similar reasons, a change in ionic atmosphere in this region *also precedes* ET. The new configuration of the solvent *and* ionic atmosphere, which is the same in X^* and X, will prove to be *intermediate* between that of the initial state and that in the final state. As such, it is not that which is *predictable by the charge distribution* in X^* or by that in X. That is, it is not in electrostatic equilibrium with either charge distribution, and its properties cannot be described by the usual electrostatic expressions. Instead, expressions which take this nonequilibrium behavior into account must be used. Once again [4] there are an infinite number of pairs of thermodynamic states, X^* and X, satisfying the constant energy-atomic configuration restriction and we are interested in determining the properties of *the most probable pair*, that is, the one with the *minimum free energy of formation from the initial state*. This is done by *minimizing* the expression for the free energy of formation subject to the restriction that X^* and X have the same atomic configuration and the same energy" [2, pp. 183–184].

4.11. Reaction Scheme

"From the preceding discussion, we may write for the mechanism of reaction (4.19) the following scheme, analogous to that in Ref. [4]" and here reported in Chapter 2:

$$A \underset{k_{-1}}{\overset{k_1}{\rightleftarrows}} X^* \tag{4.20}$$

$$X^* \underset{k_{-2}}{\overset{k_2}{\rightleftarrows}} X \tag{4.21}$$

$$X \xrightarrow{k_3} \mathrm{B} \tag{4.22}$$

"Here, a central ion just outside the double layer is denoted, in its initial electronic state, by A and in its final state by B. As noted earlier, step (4.20) involves a *suitable reorganization* of the solvent and ionic atmosphere and (if necessary) a *suitable penetration of the electrical double layer.* Step (4.21) is the actual ET itself, and step (4.22) involves *a reversion of configuration* of solvent and atmosphere to one in equilibrium with the new charge on the central ion. Step (4.22) also involves a motion away from the electrode. As in the redox theory, the reverse of Eq. (4.22) occurs but it needs not be considered in the computation of k_f. Steady-state considerations for c_{X^*} and c_X" (where the c's are concentrations) "lead to the relation [4, p. 969]:

$$k_f = k_1/[1 + (1 + k_{-2}/k_3)k_{-1}/k_2] \tag{4.23}$$

As discussed later, when the probability of ET in the lifetime of the intermediate state X^* (10^{-13} sec) is large, k_f is about half of k_1. k_1 depends on the free energy of formation of X^* from the initial state in reaction (4.20). We proceed first to the calculation of this free energy change and later to a discussion of the evaluation of the rate constants."

The treatment given here by Marcus in 1957 is today considered by him only as a formalism that helped him to get started, just a step in an evolution, "a thing that may be interesting from the point of view of how thoughts developed but not for presentation of a way of thinking. That was history, I'm not thinking anymore in those terms." (M). I shall then not discuss the rate constants of the elementary steps and I refer the reader to Chapter 6 for the modern treatment.

4.12. Free Energy of Formation of State X^*

"The electrostatic contribution, ΔF^*, to the free energy of formation of state X^* from state A will be different from zero because of

(1) The work which may be required *to transport* the central ion from a point just outside the double layer to some particular point in it (if necessary)
(2) The work required *to reorient* the solvent molecules *and* the ionic atmosphere to a nonequilibrium configuration, for this position of the central ion

Let F_e^A, F_e^*, F_e, and F_e^B denote the electrostatic free energy of the electrode system when the system is in the states A, X^*, X, and B, respectively (throughout, asterisks will be used to designate the properties of X^* and will be omitted in designating the properties of X). We have then for ΔF^*:

$$\Delta F^* = F_e^* - F_e^A$$

F_e^* is given by the general expression (4.24) and F_e is given by a similar equation, minus the asterisks:

$$F_e^* = \frac{1}{2}\int [\mathbf{E}_v^* \cdot \mathbf{E}_v^* - \alpha_e \mathbf{E}^* \cdot \mathbf{E}_v^* - \mathbf{P}_u \cdot (\mathbf{E}^* + \mathbf{E}_v^* - \mathbf{P}_u/\alpha_u)]$$

$$dV + kT \sum_i \int c_i \ln c_i/c_i^0 dV + \chi \int_M \sigma^* dS \qquad (4.24)$$

where $\mathbf{E}_v^*(\mathbf{r})$ = electric field at point $\mathbf{r}$ exerted in a vacuum by all the ionic charges in the state X^* and by those electrode charges which they would induce in a vacuum.

$\mathbf{E}^*(\mathbf{r})$ = electric field in state X^*. It equals $-\nabla\varphi^*$.
$\varphi^*(\mathbf{r})$ = inner potential in state X^* (cf. Eq. (4.1))
α_e = E-type polarizability = $(D_{op} - 1)/4\pi$
α_u = U-type polarizability = $(D_s - D_{op})/4\pi$
$c_i(\mathbf{r})$ = concentration of ions of type i in states X^* and X
c_i^0 = average concentration of ions of type i. It equals their number n_i in the solution divided by the latter's volume V
$\mathbf{P}_u(\mathbf{r})$ = U-type polarization in states X^* and X
χ = potential drop at electrode–solution interface due to an oriented solvent dipolar layer"

Note that $c_i(\mathbf{r})$ and $\mathbf{P}_u(\mathbf{r})$ in state X^* are the same as those in state X because of the constant atomic configuration restriction.

F_e^A is obtained from the general Eq. (4.24) introducing in it the equilibrium expressions (4.14) and (4.15) for $\mathbf{P}_u$ and c_i.

"F_e^B is given by a similar equation, but the functions in the integrand now refer to state B.

Equation (4.24) and subsequent equations treat χ as being independent of the average electrode charge density." However, M. shows that "the final equations" (4.31)–(4.33) in the following "are unchanged even if χ were a function of this quantity" [2, pp. 186–187].

4.13. Constraint Imposed by the Constant Energy — Constant Atomic Configuration Restriction

"The free energy of formation of X^* from A consists of the electrostatic contribution ΔF^* and of a term, described later, associated with the localization of the center of gravity of the central ion in a narrow region near the electrode.[(3)] The free energy of formation of X from B contains an electrostatic term $F_e - F_e^B$, ΔF say, and a center of gravity term equal to that noted previously. As in the redox theory, it follows from assumption (a) given earlier that no energy change and no configurational entropy change accompany the formation of state X from state X^*. The electronic entropy change arising from any possible change in the electronic degeneracy of the central ion and of the electrode is zero or negligible. Accordingly, X^* and X have the same free energy."

Note that the equality of the free energies of X^* and X have been deduced from the equality of their energies *and of their entropies.*

"The net free energy change in forming B from A is therefore

$$F^* - F_e^A + (F_e^B - F_e) = \Delta F^* - \Delta F \tag{4.25}$$

Independently, this free energy change of reaction (4.19) can be written as the sum of the following terms:

(a) The change in chemical potential of ions A and B and of the electrons in electrode M. This is $\mu_B - \mu_A - n\mu_e^M$ where n is the number of electrons transferred in reaction (4.21).
(b) The change in free energy due to the transfer of charge $e - e^*$ from a metal of inner potential φ_M to a solution of inner potential φ_S, e^* and e denoting the charges of ions A and B, respectively. This term is $(e^* - e)(\varphi_M - \varphi_S)$ which we shall denote by $(e^* - e)\Delta\varphi$".

"If electrode M were *in electrochemical equilibrium* with the *actual* concentrations of A and B just outside the double layer, then the sum of these two terms would be zero."

Denoting the corresponding value of $\Delta\varphi$ as $\Delta\varphi'$ we would then have:

$$\mu_B - \mu_A - n\mu_e^M + (e^* - e)\Delta\varphi' = 0 \tag{4.26}$$

For $\Delta\phi \neq \Delta\phi'$, there wouldn't be such an equilibrium and the difference:

$$\Delta\varphi - \Delta\varphi' = \eta_a$$

defines the *activation overvoltage.*

For the net free energy change $\Delta F^* - \Delta F$ written as sum (a) + (b) we have:

$$\Delta F^* - \Delta F = \mu_B - \mu_A - n\mu_e^M + (e^* - e)\Delta\varphi$$

Introducing $\Delta\varphi'$ in the above equation using Eq. (4.26) we have:

$$\Delta F^* - \Delta F = (e^* - e)(\Delta\varphi - \Delta\varphi') = (e^* - e)\eta_a \tag{4.27}$$

"It may be remarked that $\mu_B - \mu_A$ is related to its standard value $\mu_B^0 - \mu_A^0$ and $\Delta\varphi$ to its standard value $\Delta\varphi^0$, by Eqs. (4.28) and (4.29), where the f's denote activity coefficients and the c^S's denote concentrations just outside the double layer:

$$\mu_B - \mu_A = \mu_B^0 - \mu_A^0 + kT\ln(f_B/f_A)(c_B^S/c_A^S) \tag{4.28}$$

$$(e - e^*)\Delta\varphi = (e - e^*)\Delta\varphi^0 + kT\ln(f_B/f_A)(c_B^S/c_A^S)\text{''} \tag{4.29}$$

[2, pp. 187–188].

4.14. Minimization of ΔF^* Subject to the Constraint Imposed by Eq. (4.27)

The Final Theoretical Equations for ΔF^*

We shall not delve in the details of the minimization process of ΔF^* which is analogous to the one described in Chapters 2 and 3 for the redox theory. I only remark that Marcus considers now not only the configuration of the solvent molecules in the activated complex, but also the configuration of the ionic atmosphere so that ΔF^* is minimized now not only with respect to arbitrary variations $\delta\mathbf{P}_u(\mathbf{r})$ of the orientation–atomic polarization, but also with respect to arbitrary variations $\delta c_i(\mathbf{r})$ of ionic concentrations in the ionic atmosphere. There is then now a new condition of constraint, that is, besides the condition:

$$\delta\Delta F^* - \delta\Delta F = 0 \tag{4.30}$$

there is the new condition of fixed number of ions n_i in the solution, expressed as:

$$\delta n_i = \int \delta c_i dV = 0$$

There are correspondingly two Lagrange multipliers: the m, already met in Chapter 2, and a new one, $-\ln l_i$. In terms of the multipliers, the *solvent* and *ionic* configuration of the pair of intermediate states X^* and X is obtained so that:

$$\mathbf{P}_u = \alpha_u \left[\mathbf{E}^* + m(\mathbf{E}^* - \mathbf{E})\right]$$

and $c_i = c_i^0 l_i \exp\left\{-e_i\left[\varphi^* + m(\varphi^* - \varphi)\right]/kT\right\}$

where $l_i = V / \int \exp\left\{-e_i\left[\varphi^* + m(\varphi^* - \varphi)\right]/kT\right\}$

so that $c_i = n_i \dfrac{\exp\left\{-e_i\left[\varphi^* + m(\varphi^* - \varphi)\right]/kT\right\}}{\int \exp\{-e_i\left[\varphi^* + m(\varphi^* - \varphi)\right]/kT\} dV}$

We see that the concentration of ions at distance r from the central ion is obtained by Boltzmann weighting the total number of ions n_i with the expression in the fraction which depends on the inner potential $\varphi^* + m(\varphi^* - \varphi)$.

Notice that Marcus uses here the condition of constraint (4.30) instead of the condition $\delta F^* - \delta F = 0$ which was used in the redox case but the two conditions are really the same because $\Delta F^* = F_e^* - F_e^A$, $\Delta F = F_e - F_e^B$, and $\delta F_e^A = \delta F_e^B = 0$.

Marcus evaluates then ΔF^* introducing what he will later name in Ref. [20] the *equivalent equilibrium system* (e.e.s). Such a system is introduced here as a hypothetical system having *fictitious* charges on the central ion and on the electrode which are *in equilibrium* with the configuration of solvent and ionic atmosphere characterized by the same $\mathbf{P}_u(\mathbf{r})$ and $c_i(\mathbf{r})$ as those of states X^* and X and which were not in equilibrium with the *real* charges of X^* and X. The electrostatic free energy $F_e^\dagger$ of this hypothetical system is determined and used to calculate the final theoretical equations for ΔF^*. The hypothetical equilibrium system is characterized by having an inner potential $\varphi^\dagger = \varphi^* + m(\varphi^* - \varphi)$, ionic surface charge $\sigma^\dagger = \sigma^* + m(\sigma^* - \sigma)$ and, likewise, $E^\dagger = \mathbf{E}^* + m(\mathbf{E}^* - \mathbf{E})$, $\mathbf{P}_u^\dagger = \alpha_u \mathrm{E}^\dagger = \alpha_u(\mathbf{E}^* + m(\mathbf{E}^* - \mathbf{E})$, and $e^\dagger = e^* + m(e^* - e)$, where $e^\dagger$ is the charge on the central ion in the hypothetical system.

Note that it is not $F_e^\dagger = F_e^* + m(F_e^* - F_e)$.

The final theoretical equations for ΔF^* are:

$$\Delta F^* = w^* + \frac{m^2}{2}\lambda \tag{4.31}$$

"where m satisfies the equation

$$-\left(m + \frac{1}{2}\right)\lambda = (e^* - e)\eta_a + w - w^* \tag{4.32}$$

and where

$$\lambda = (\Delta e)^2 \left(\frac{1}{a} - \frac{1}{R}\right)\left(\frac{1}{D_{op}} - \frac{1}{D_S}\right) \tag{4.33}$$

In these equations, $\Delta e = e^* - e$, the charge transferred to the electrode in reaction (4.19); a denotes the effective radius of the central ion, discussed earlier; $R/2$ is the distance of the ion to the electrode surface in the state X^*; w^* and w denote the work required to transport the central ion from the body of the solution to a distance $R/2$ from the electrode surface when the ion is its initial and final state, respectively; the remaining quantities have been defined previously" [2, pp. 190–191].

I want to stress that $\Delta F^* = \Delta F^*(R)$, i.e., ΔF^* depends, through λ, on the distance of the central ion from the electrode.

"The term w^* can be evaluated by solving the usual Poisson–Boltzmann equation [3] when the central ion is in the body of the solution, and then introducing these solutions for the electrostatic potential into the appropriate equations for the electrostatic free energy of the system (Eqs. (4.16) and (4.18)). The difference in electrostatic free energy is w^*. Similarly, w can be computed from analogous equations for the central ion B.

However, a considerably simpler though less rigorous procedure has generally been assumed for calculating the work required to transport an ion from the body of the solution to some distance $R/2$ from the electrode. The assumption is generally made that this work equals the charge of the central ion multiplied by the difference in the value of the electrostatic potential at $R/2$ and in the body of the solution (i.e., outside the double layer), *this potential being computed in the absence of the central ion.* This particular potential can be inferred from measurements on the electrical double layer by various methods [14, 21]" [2, p. 192].

4.15. Dependence of χ on Mean Electrode Charge Density

In the earlier theory, "it has been assumed for simplicity that χ, the potential drop due to electrode–solvent and solvent–solvent interactions at the interface, was independent of the mean electrode charge density, $\overline{\overline{\sigma}}$ say. Recent work suggests, however, that the degree of orientation of the solvent molecules next to the electrode (and hence χ) may vary with this quantity [22, 23]."

Marcus considers then the possible modification of the theory for the case in which χ is not just a constant, but a function $\chi(\overline{\overline{\sigma}})$ and concludes that "when a few electrons are transferred in some process" as in our case "the average electrode charge density is affected only to a negligible extent. In such cases, χ is a constant for all states in the process." M. then shows that χ considered either constant or depending on $\overline{\overline{\sigma}}$ does not appear in the final Eqs. (4.31)–(4.33).

"Accordingly, we infer that any effect of a change in degree of orientation in the solvent layer next to the electrode is a more indirect one. It might affect the 'dielectric constant' in the vicinity of the electrode, particularly the contribution from orientation polarization. But, we observe from Eq. (4.33), unless D_S is close to D_{op}, changes in the former have very little effect on ΔF^*.

Again, it might affect to some extent the distance of closest approach of the central ion. In some studies of the equilibrium properties of the electrode double layer, Grahame [22, 23] has shown that a self-consistent interpretation of the data can be obtained assuming such an effect. Extremely interesting inferences were drawn about the behavior of the solvent in this region toward ions of different size. It is clear that an analogous study of the electrode kinetics of simple electron transfers at various electrode charge densities should be very interesting. At certain charge densities, it appeared from the equilibrium studies, there is no oriented solvent layer. The interpretation of kinetic data obtained under such conditions would be correspondingly simplified" [2, pp. 204–205].

4.16. Presence of Fixed, Adsorbed Ions in the Electrical Double Layer

If fixed adsorbed ions are present in the double layer [3], one should consider their contribution to the electrostatic free energy F_e^{eq}.

"If q_k denotes the charge of the kth fixed ion and if r_k and r_k^i denote the distance from the kth ion and from its image to the field point, then the contribution to the potential arising from all fixed ions is, in an equilibrium polarization system,

$$\sum_k \frac{q_k}{D_S}\left(\frac{1}{r_k} - \frac{1}{r_k^i}\right), \tag{4.34}$$

the summation being over all fixed ions.

This can be shown to add the following terms to F_e^{eq} in Eq. (4.16):

$$\sum_k \int_{\text{ion}} \int_{q_k=0}^{q_k=q_k} \frac{\phi_\rho^{q_k} dq_k}{4\pi a_k^2} dS + \sum_k \frac{q_k^2}{2D_S}\left(\frac{1}{a_k} - \frac{1}{R_k}\right)$$
$$+ \sum_j \sum_{k \neq j} \frac{q_k q_j}{2D_S}\left(\frac{1}{r_{jk}} - \frac{1}{R_{jk}}\right) + \sum_k \frac{q q_k}{D_S}\left(\frac{1}{\rho_k} - \frac{1}{\rho_k^i}\right) \tag{4.35}$$

where a_k is the radius of the kth fixed ion; R_k, r_{jk}, R_{jk}, ρ_k, and ρ_k^i denote the distance from this ion to its electrical image, to the jth fixed ion, to the latter's electrical image, to the central ion and to the latter's electrical image, respectively. The first term in Eq. (4.35) describes the interaction of the mobile ions with the fixed ions.

In Eqs. (4.17) and (4.18), φ_S now represents the potential in the body of the solution due to all ions, fixed and mobile, of the double layer and so includes Eq. (4.34). Accordingly, $q\varphi_S$ in Eqs. (4.17) and (4.18) now includes the last term of (4.35). The remaining three terms of Eq. (4.35) should be added to Eqs. (4.17) and (4.18)" [2, pp. 203–204].

M. demonstrates that "because of a cancellation of the newly added terms" the equation for ΔF^* remains unchanged "except for the new significance of $\boldsymbol{\varphi}_{\mathrm{s}}$" [2, p. 203].

4.17. Summary and Final Remarks

At the end of this chapter, it may be useful to highlight some pivotal points of Marcus' treatment.

M. begins with an extension to the electrode system of his general equation for the electrostatic free energy F_e of systems with nonequilibrium dielectric polarization. In the electrode system, he explicitly considers that the central ions have ionic atmospheres. Moreover, a potential drop χ is present at the electrode–solution interface. The expression for F_e is obtained by the two stages charging process that has been described in detail, for a simplified model, in Chapter 3. The expression for F_e in a nonequilibrium polarization system is then obtained from F_e^{eq} by a variational calculation considering that F_e obtains from F_e^{eq} when, because of fluctuations (variations) $\delta\mathbf{P}_u$ and δc_i, F_e departs from its equilibrium value. M. found an expression for F_e^{eq} in which $\mathbf{P}_u = \alpha_u\mathbf{E}$—as expected for a system in electrostatic equilibrium—and an expression for $c_i(\mathbf{r})$ is given as function of $\varphi(\mathbf{r})$. M. considers here also explicitly for the first time the hypothetical equilibrium system, called equivalent equilibrium system, whose

$\mathbf{P}_u$ is the same as that in X^* and X but in electrostatic equilibrium with fictitious charges on the ions. The treatment of the equivalent equilibrium distribution in Ref. [20] will later extend this concept.

The ionic charge on the electrode is described in terms of the electrical images of the ions in the body of the solution and at the electrical double layer. With a second variational calculation, Marcus finds the expression for the minimum value of ΔF^* corresponding to the activation free energy for the ET process.

Formulas (4.31)–(4.33) for ΔF^*, λ, and m are very similar to those for the redox systems and were discussed in detail in Chapter 3. A unifying point of view of the formulas for redox and electrode systems consists in reducing the ET in the electrode case to that between an ion and its electrical image in the electrode.

It is finally demonstrated that the presence of χ can be neglected in deriving the final equations because of cancellations. For the same reason, the presence of adsorbed ions in the double layer can also be overlooked.

The contribution of the ionic atmosphere to the free energy of activation has also been demonstrated to be small. In Chapter 6, this last point will be taken up again and an expression for the contribution of ΔF^*_{atm} to ΔF^* will be given.

NOTES

1. **M**: "The charge on the electrode is an image charge, it is not uniformly distributed over the electrode. It is on the surface of the electrode but it behaves as though is situated a certain distance into the electrode, that's the image."
2. **M**: "If you have a certain species of ions you may change a local concentration keeping the total number of ions of that species constant."
3. **M**: "We have, in the case of a reaction at an electrode, a localization analogous to that considered by Sutin when he considers the reactants to come together and to form a complex."

References

1. R. A. Marcus, "Theory and Applications of Electron Transfers at Electrodes and in Solution," in *Special Topics in Electrochemistry*, p. 161, P. A. Rock, Editor, Elsevier, New York (1977).
2. R. A. Marcus, "Electrostatic Free Energy and Other Properties of States Having Nonequilibrium Polarization. II. Electrode Systems. (ONR Report No. 11, dated 1957)," in *Special Topics In Electrochemistry*, p. 210, P. A. Rock, Editor, Elsevier, New York (1977); (a) R. A. Marcus, "On the Theory of Overvoltage for Electrode Processes Possessing Electron Transfer Mechanisms. I. (ONR Report No. 12 dated 1957)," in *Special Topics in Electrochemistry*, p. 180, P. A. Rock, Editor, Elsevier, New York (1977).
3. A. J. Bard, L. R. Faulkner, *Electrochemical Methods, Fundamentals and Applications*, John Wiley & Sons, Inc., New York, Chichester, Brisbane (1980).
4. R. A. Marcus, *J. Chem. Phys.* **24**, 966–978, (1956).
5. R. Parsons, in *Modern Aspects of Electrochemistry*, Chap. 3, J. O'M. Bockris, Editor, Academic Press, Inc., New York (1954).
6. Compare Reference 5 for bibliography.
7. M. Mason, W. Weaver, *The Electromagnetic Field*, Dover Publications, Inc., New York (1929).
8. R. G. Sachs, D. L. Dexter, *J. Appl. Phys.* **21**, 1304, (1950).
9. J. O'M. Bockris, A. K. N. Reddy, *Modern Electrochemistry 1*, 2nd Edition, Plenum Press, New York and London (1998).
10. R. A. Marcus, *J. Chem. Phys.* **24**, 979–989, (1956).
11. J. E. B. Randles, *Trans. Faraday Soc.* **48**, 828, (1952).
12. A. A. Vlcek, *Coll. Czech. Comm.* **20**, 894 (1955).
13. R. A. Marcus, *Trans. N.Y. Acad. Sci.* **19**, 423, (1957).
14. D. C. Grahame, *Chem. Rev.* **41**, 441, (1947).
15. Compare reviews by (a) J. O'M. Bockris, *Modern Aspects of Electrochemistry*, Chap. 4, Academic Press, Inc., New York, 1954; (b) P. Delahay, *New Instrumental Methods in Electrochemistry*, Interscience Publishers, Inc., New York, 1954; (c) A. E. Remick, *J. Chem. Ed.* **33**, 564, (1956).
16. Compare J. Koutecky, *Collect. Czech. Chem. Commun.* **18**, 183, (1953).
17. R. A. Marcus, *J. Chem. Phys.* **26**, 867, (1957).
18. J. J. Lingane, *Electroanalytical Chemistry*, 2nd Edition, Interscience Publishers, Inc., New York (1958).
19. Compare J. O'M. Bockris, *Modern Aspects of Electrochemistry*, Chap. 3, p. 167, Academic Press Inc., New York (1954).
20. R. A. Marcus, *Discuss. Faraday Soc.* **29**, 21, (1960).
21. A. Frumkin, *Trans. Faraday Soc.* **36**, 117, (1940).
22. D. C. Grahame, *J. Chem. Phys.* **25**, 364, (1956).
23. D. C. Grahame, *J. Am. Chem. Soc.* **79**, 2093, (1957).

CHAPTER 5

Statistical Mechanical Theory of Electron Transfer, Equivalent Equilibrium Distribution, Reaction Coordinate for ET, Inverted Free Energy Effect, Interactions in Polar Media

In this chapter, I have summarized the theory on potential energy expressions, on interparticle interaction energy, and on interactions in polar media which can be found in Refs. [1–5].

In the pivotal paper written for the *Discussions of the Faraday Society* of 1960 M. introduced for the first time a treatment of his theory in terms of potential energy surfaces and of statistical mechanics. This generalization of the earlier macroscopic treatment in terms of nonequilibrium solvent polarization was needed in order to treat in a consistent manner the contributions of both the inner coordinates of the molecules (in particular of the coordination shells, when present) and the outer solvent coordinates in the electron transfer process. The new concept of equivalent equilibrium distribution (e.e.d.), already briefly mentioned and used in Ref. [5a] (see Chapter 4), was formulated, "inverted free energy effect" was here predicted for the first time and, finally, the contribution of the internal molecular coordinates of the central ions or molecules, in particular that of the coordination shells, to the reaction rates was considered in some detail. Such topics as well as a more rigorous and extensive formulation of the theory were also the object of Marcus' great paper of 1965, Ref. [6], and they will be dealt with in Chapter 6.

5.1. Many-Dimensional Potential Energy Surfaces

"(i) No electronic interactions

In discussions of ET, problems which have frequently arisen and have occasioned some uncertainty and confusion concern the charge distribution in the transition state, the mode of calculating its interaction with surrounding molecules, and the mechanism of ET itself. To treat these problems, we first consider a hypothetical case where *no electronic coupling* between the redox orbitals of the reactants occurs, so that no ET is possible ....

In this case, we have two distinctly different electronic states — one having the electronic structure of the reactants, the other having that of the products. The lowest electronic state of each chemical pair has its own potential energy surface in a many-dimensional atomic configuration space, whose coordinates are those of all the atoms of the two reactants, of the solvent, and of any electrolyte.

The two surfaces each have their own valleys but the two sets of valleys occur in quite different regions of the space, reflecting differences in stable bond lengths, solvent orientations, and so on. The surfaces intersect, usually along some upper reaches of each, and form thereby a surface of one less degree of freedom" [1].

In Fig. 1 (from Ref. [1]), the N-dimensional potential energy surfaces are schematically represented.(1)

"The intersection surface can be reached by any suitable fluctuation of atomic coordinates to produce some atomic configuration which is usually a compromise between the stabler ones of the two electronic states. Because of the absence of electronic interaction of the redox orbitals, such a fluctuation does not cause any ET. The system merely stays on the surface corresponding to the original electronic configuration on passing through the intersection. Fluctuations of this nature(2) involve *simultaneous* changes in orientation,

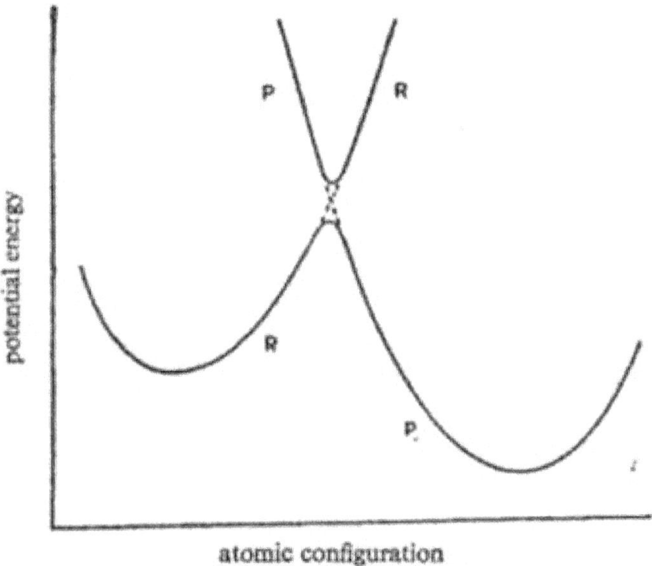

Fig. 1. Profile of N-dimensional Potential energy surface plotted against an Atomic configurational coordinate of the entire system. Curve R denotes reactants ($ox_1 + red_2$); curve P, products ($red_1 + ox_2$). Dotted lines show intersection of surface (zero electronic interaction case) and solid lines indicate the splitting for the case of weak interaction. From Ref. [1].

position, and atomic polarization of the solvent molecules, in internuclear distances in the coordination shell, in relative motion of the reactants, and in configuration of the ionic atmosphere.

(ii) Adiabatic and nonadiabatic mechanisms for ET

Consider next the weak electronic interaction between the redox orbitals which occurs, for example, when the reactants are not too far apart. Their interaction leads to the usual splitting of the surfaces, as indicated in Fig. 1. For sufficient electronic interaction, a system passing across the intersection during a fluctuation will always stay on the lowest surface. We see from Fig. 1 therefore, that the products have been formed from the reactants *adiabatically* (in the quantum mechanical sense) as a result of this atomic motion. This motion, then, is one which produces an atomic configuration of the system more favorable to the electronic charge distribution of the products.

When the electronic interaction is *extremely weak*, on the other hand, for example, *when the reactants are far apart*, the system tends to retain its original electronic configuration on passing across the intersection, that is, the system "jumps" to the upper surface at such times and jumps back on its return. Each time no ET tends to occur. There is, nevertheless, in such cases, a small probability of "transition." For this system, we have thereby a "nonadiabatic" mechanism for ET. As long as the interaction is not too strong, the splitting is relatively small, and little error is made in regarding the correct potential energy at the "intersection" surface as being essentially equal to that for the zero-interaction system. *Thus, both the potential energy and the probability distribution on the intersection surface can be computed for the weak interaction system by the simple expedient of regarding the system as being the conceptually simpler zero-interaction*[(3)] *one*. Moreover, it may be emphasized here that in the computation, the charge distribution for the *zero-interaction* case should be used. It is the one for the reactants (or products) and not some compromise.

In both cases, adiabatic and nonadiabatic, it is necessary for the system to pass through the intersection surface. In the first approximation, the theoretical rate expression deduced below for the adiabatic mechanism will apply to a nonadiabatic one if, in the latter case, it is multiplied by some factor denoting an average probability per passage through the intersection region. (Nuclear tunneling through the barrier in Fig. 1 is neglected here in both cases.)

(iii) Equations for rate and for intersection surface

We shall use an equation for rate of passage through a surface in many-dimensional space, in our case the intersection surface. It is similar to the usual transition state theory .... If $\Delta F^{\ddagger}$ denotes the difference in the free energy of the reactants when they are considered to exist on this $(N-1)$ dimensional surface compared with their existing in all atomic configurations, the rate constant, k_r is

$$k_r = (kT/h)\exp(-\Delta F^{\ddagger}/kT). \tag{5.1}$$

For defining the intersection surface in terms of molecular properties, we introduce the following notation:

k = Any atomic configuration of the entire system in N-dimensional space
p = Superscript to designate throughout a property of the products (a change of notation from Part 1)
$\mathcal{E}_k$ = Potential energy of reactants in configuration k
$\Delta\mathcal{E}$ = Difference between electronic energy of the lowest electronic state of the products and that of the reactants when each is at its own zero of potential energy

NOTE: M. uses in this paper the same symbol for the atomic configuration and the Boltzmann constant, which might be confusing.

Since the electronic energies of the reactants and products are equal along the intersection surface, the latter obeys the relation

$$\mathcal{E}_k = \mathcal{E}_k^p + \Delta\mathcal{E} \tag{5.2}$$

at its intersection.

The potential energy of the reactants on the intersection surface, $\mathcal{E}_k^{\ddagger}$, say, equals $\mathcal{E}_k$ and because of Eq. (5.2) could also be written as

$$\mathcal{E}_k^{\ddagger} = \mathcal{E}_k + m(\mathcal{E}_k - \mathcal{E}_k^p) - m\Delta\mathcal{E}, \tag{5.3}$$

where m is any constant.(4) " [1]

NOTE: An all-important point is here that at the intersection, where Eq. (5.1) is valid, that is, where $\mathcal{E}_k - \mathcal{E}_k^p = \Delta\mathcal{E}$, $\mathcal{E}_k^{\ddagger}$ is the energy at the transition state TS, and there $\mathcal{E}_k^{\ddagger}$ is equal to the value of $\mathcal{E}_k$ at the intersection when the energy difference between the minima is a certain *fixed* $\Delta\mathcal{E}$. Marcus defines then a *shift coordinate* Γ for a *variable* $\Delta\mathcal{E}$, designated as

$$\Gamma \equiv \mathcal{E}_k - \mathcal{E}_k^p - \Delta\mathcal{E} \tag{5.4}$$

So that

$$\mathcal{E}_k^{\ddagger} \equiv \mathcal{E}_k + m\Gamma, \tag{5.5}$$

$\mathcal{E}_k^{\ddagger}$ defines *energies of intersections* obtained by lowering or raising the P surface with respect to the R surface by an amount Γ, so that $\mathcal{E}_k^{\ddagger}$ defines a potential energy surface which will be described in the following.

5.2. Potential Energy in Microscopic Terms

In his first treatment of ET reactions, M. discussed the reorganization of configurations of solvent molecules before and after the ET act. The contribution of the solvent reorganization to the free energy of activation for the ET process was calculated using a *macroscopic* treatment for the medium having nonequilibrium dielectric polarization.

In some ET processes, there are also changes in the coordination shell as well, as we have already seen in Chapter 3. Their contribution to the free energy of the activated complex was estimated in microscopic terms. In order to include both contributions in a consistent manner, there is the need of treating also the solvent in microscopic terms, that is, as if made up of molecules and not of polarized volume elements.

In his first and most simple treatment of a system in which central ions and solvent molecules are treated on the same footing as "particles" [1], the jth particle is considered as possessing only

(i) A permanent dipole moment μ_j
(ii) An isotropic polarizability α_j
(iii) A charge e_j

some of which may be zero.

In the "all particles" treatment "we assume that the potential energy of the reactants is essentially the sum of two contributions:

$$\mathcal{E}_k = \mathcal{E}_{k^i} + \mathcal{E}_{k^o} \tag{5.6}$$

where $\mathcal{E}_{k^i}$ depends on the internal coordinates, k^i, of the coordination shells alone ($\mathcal{E}_{k^i}$ being defined as zero at the equilibrium values of these coordinates), and $\mathcal{E}_{k^o}$ depends on all other coordinates, k^o, of the entire system. Thus, k, the totality of all coordinates, is an abbreviation for $k^i + k^o$ ('inner' and 'outer') [1].

M. introduced the following additional notations:

$\mathbf{E}$ = Electric field strength at any point, arising from all the ionic charges and from the permanent and induced dipoles.
ϕ = Potential arising directly from all ionic charges $= \sum_j e_j/r_j$.
$\mathbf{D}$ = Contribution to $\mathbf{E}$ arising directly from all ionic charges. $\mathbf{D} = -\nabla\varphi$.
$\mathbf{D}_\mu$ = Contribution to $\mathbf{E}$ arising solely from the permanent dipoles.
Ω_{k^o} = Van der Waals' potential energy of interaction of all the particles (repulsive, dispersive, permanent dipole–dipole). Ω_{k^o} is taken to depend only on k^o.
j = Subscript to also denote fields at particle j minus the latter's contribution."

M. shows that E_{k^o} is given by:

$$\mathcal{E}_{k^o} = \Omega_{k^o} + \Sigma_j(e_j\phi_j/2 - \mu_j \cdot \mathbf{D}_j - \alpha_\mathbf{j}\mathbf{E}_j \cdot (\mathbf{D}_j + \mathbf{D}_{\mu j})/2) \tag{5.7}$$

To establish Eq. (5.7), arguments related to those in Section 2.4 of Chapter 2 (from Appendix IV of Ref. [7]) may be used (cf. also Ref. [8]). Equation (5.1) is the sum of the following interactions (the 1/2 in some of the expressions in the following avoids counting the interactions twice):

$\sum_j e_j\varphi_j/2$: between charges
$-\sum_j \mu_j \cdot \mathbf{D}_j$: between charges and permanent dipoles (cf. Ref. [9])

$-\sum_j \alpha_j \mathbf{E}_j \cdot \mathbf{D}_{\mu j}$: induced dipole–permanent dipole ($\alpha_j \mathbf{E}_j$ is an induced dipole)
$-\frac{1}{2}\sum_j \alpha_j \mathbf{E}_j \cdot \left(\mathbf{E}_j - \mathbf{D}_j - \mathbf{D}_{\mu_j}\right)$: induced dipole–induced dipole.

In the above expression, the contribution to the field given by the *induced* dipoles is obtained by subtracting the field due to the charges and that given by the *permanent* dipoles from the total field $\mathbf{E}_j$.

$\sum_j \alpha_j \mathbf{E}_j^2/2$: energy stored up in the induced dipoles
Ω_{k^o} : the term includes the permanent dipole–permanent dipole term

$$-\sum_j \mu_j \cdot \mathbf{D}_{\mu j}/2$$

"$\mathbf{E}_j$ obeys the relation:

$$\mathbf{E}_j = \mathbf{D}_j + \mathbf{D}_{\mu j} - \nabla_j \sum_{m \neq j} \alpha_m \mathbf{E}_m \cdot \nabla_m \frac{1}{r_{jm}} \tag{5.8}$$

(cf. Ref. [10] for nonpolar and Ref. [8] for polar molecules between parallel electrodes in a nonelectrolyte system). Unlike the $\mathbf{D}$ in Ref. [10], say, ours is not the dielectric displacement, a quantity with little molecular significance here, but the microscopic equivalent of $\mathbf{E}_c$ [7]." [1] Compare Chapter 2.

5.3. Potential Energy for Transition State

In Ref. [7], his first paper on ET theory, M. observed that the U-type polarization appropriate to the TS was given by $\mathbf{P}_u = \alpha_u\{\mathbf{E}^* + (\mathbf{E}^* - \mathbf{E})m\}$, that is, by a polarization determined by a field strength intermediate, in the sense expressed by the formula, between that characteristic of reagents and products. If it could then be possible to produce a field strength given by $\{\mathbf{E}^* + (\mathbf{E}^* - \mathbf{E})m\}$, this field strength would on its turn produce a $\mathbf{P}_u$ polarization in equilibrium with the field strength and the configurational distribution of the TS would be the same as that of this equilibrium state. In this section, it is shown how such a field could be produced. As a matter of fact, introducing the earlier quantities (5.6) and (5.7) into Eq. (5.2) and using Eq. (5.8), it is possible to express the potential energy of the TS, that is, at the crossing of the R and P surfaces, in terms of electrostatic properties and of outer and inner coordinates which are in between those of reactants and products. We obtain:

$$\mathcal{E}_k^{\ddagger} = \mathcal{E}_{k^i}^{+} + \mathcal{E}_{k^o}^{+} + C, \tag{5.9}$$

where

$$\mathcal{E}_{k^o}^{+} = \Omega_{k^o} + \Sigma_j [e_j^{+}\phi_j^{+}/2 - \mu_j \mathbf{D}_j^{+} - \alpha_j \mathbf{E}_j^{+} \cdot (\mathbf{D}_j^{+} + \mathbf{D}_{\mu j})/2], \tag{5.10}$$

ϕ_j^{+}, $\mathbf{D}_j^{+}$, $\mathbf{E}_j^{+}$ and $\mathcal{E}_{k^i}^{+}$ are abbreviations for functions of the type

$$X^{+} = X + m(X - X^{p})$$

and

$$C = m(m+1)\Sigma_j[\alpha_j(\mathbf{E}_j - \mathbf{E}_j^{p}) \cdot (\mathbf{D}_j - \mathbf{D}_j^{p})/2 - (e_j - e_j^{p})(\phi_j - \phi_j^{p})/2] - m\Delta\mathcal{E} \ldots.$$

Erratum: In the last equation, two typos present in the equation in Ref. [1] have here been corrected.

5.4. The Equivalent Equilibrium Distribution (e.e.d.) and the Reaction Coordinate

"It is instructive, for evaluating $\Delta F^{\ddagger}$, to first compare the transition state, in which $\exp(-\mathcal{E}_k^{\ddagger}/kT)$ is integrated *over the intersection surface*, with a state in which this factor is integrated *over all space*. We shall term the configurational distribution of the latter state the "equivalent equilibrum distribution" (e.e.d.).[(5)]

Comparison of Eq. (5.9) with Eq. (5.6) and of Eq. (5.10) with Eq. (5.7) reveals that the e.e.d. is one which would be obtained in a corresponding *equilibrium* system in which the charges of the two central ions were e_n^{+}, i.e., $e_n + m(e_n - e_n^{p})$, $(n = 1, 2)$,[(6)] and which had $\mathcal{E}_{k^i}^{+}$ as a potential function for the coordination shell.

The transition state of the zero-interaction system differs from this system in only two respects: the charges of the two central ions are those of the reactants, and it has one less dimension of freedom than an N-dimensional system.

For exact evaluation of the surface integral, we should examine in detail the motion along the surface, for example, by examining the atomic motion *normal* to it, i.e., the *reaction coordinate.*[(7)] We hope to analyze this dynamical problem at a later date. For the present, we use instead the following procedure.

By a suitable choice of m, the e.e.d. is made to center on the intersection surface, and thereby to die away fairly rapidly along the normal. Since Eq. (5.9) applies both to e.e.d. and to the transition state, we may then set the surface integral over $\exp(-\mathcal{E}_k^{\ddagger}/kT)$ equal to the volume integral for the e.e.d. divided by a partition function along the *normal*... Most of the likely motions along the *normal* to the transition state surface... have a "frequency" of motion of about 10^{13} sec^{-1}. We anticipate, therefore, that the partition function just noted, which may be written as $kT/h\nu$, will be of the order of unity..." [1].

5.5. Calculation of the Free Energy of Activation Using the e.e.d.

The inner coordinates k^i appear in the potential energy (5.9) of the e.e.d. The vibrational potential energy in the TS which is there represented by $\mathcal{E}_{k^i}^{+}$ is here designated as $\mathcal{E}_{k^i}^{\ddagger}$. This vibrational energy in the TS is made up of the difference $\Delta\mathcal{E}_i^{\ddagger}$ between the minima of this energy in the TS with respect to that in the ground state, plus the vibrational excitation $E_v^{\ddagger}$ with respect to the minimal vibrational energy in the TS state, where v stands for the totality of vibrational quantum numbers. The vibrational energy in the TS is then $\mathcal{E}_{k^i}^{\ddagger} = \Delta\mathcal{E}_i^{\ddagger} + \mathcal{E}_v^{\ddagger}$. For the difference between the free energies of the TS and the ground states, we then have the following formula [7a, pp. 45, 116] in which the free energy of the transition state is expressed in terms of its potential energy:

$$\exp(-\Delta F^{\ddagger}/kT) = \frac{\exp(F/kT)}{h^{N_0}} \int .. \int \sum_{v=0}^{\infty} \exp[-(\mathcal{E}_v^{\ddagger} + \Delta\mathcal{E}_i^{\ddagger} + \mathcal{E}_{k^o}^{\ddagger} + C + K)/kT]\, d\tau_o \tag{5.11}$$

where F denotes the free energy of the reactants and "where K is the kinetic energy of the N_o outer coordinates and $d\tau_o$ is their volume element in phase space. Summation over all v immediately yields the vibrational partition function, $Q^{\ddagger}_{vib}$, for the inner coordinates, and integration over the N_o moments cancels a corresponding momentum integral in an expression for F (as does the h^{N_o} factor). The residual integrand depends only on the relative 'outer coordinates' (positional and orientational) of all the particles. We next hold all of these relative coordinates fixed within two fairly large spheres,[8] one about each central ion (large enough so that the long range ion–ion–solvent interactions are negligible on their surfaces).

We then integrate over the coordinates of the center of gravity of these two ions and over the orientations of their line of centers, translating and rotating, respectively, the entire system within the large spheres to ensure constancy of the important relative coordinates during integration. Holding the distance r between the ions fixed, we next integrate over all other outer coordinates, the integral being denoted later by $\exp(-F_o^{\ddagger}(r)/kT)$. In integrating finally over r, we first note two factors which favor small rs in spite of any Coulombic repulsion: the solvent reorganization barrier is smaller there (cf. below) and, at the larger rs, the electronic interaction becomes so weak that out there the integral should be multiplied by some small nonadiabatic transition probability. We presumably err relatively little if we simply take r as the distance of closest approach and set the corresponding r-partition function equal to unity, that is, $kT/h\nu_r \sim 1$. We obtain after some cancellation[9]:

$$k_r = Z \exp[-(\Delta F_i^{\ddagger} + \Delta F_o^{\ddagger}(r))/kT], \tag{5.12}$$

where

$$\Delta F_i^{\ddagger} = \Delta \mathcal{E}_i^{\ddagger} - kT \ln Q_{vib}^{\ddagger}/Q_{vib} \tag{5.13}$$

$$\Delta F_o^{\ddagger} + F_o = F_o^{\ddagger}(r) = -kT \ln \int .. \int \exp(-\mathcal{E}_{k^o}^{\ddagger}/kT) dk_o'. \tag{5.14}$$

In these equations, Q_{vib}, *the vibrational partition function of the inner coordinates of reactants, was extracted from F*: $F_o^{\ddagger}(r)$ is the configurational free energy of the reactants due to all ion–solvent–ion interactions in Eq. (5.14), at fixed positions of the central ions; dk_o' is the configurational volume element of the remaining $(N - 6)$ outer coordinates. Z is the same as the usual collision frequency of two nonpolar molecules in solution (probably about 10^{11} liter mole^{-1} sec^{-1} rather than the value suggested in Ref. [6a]. . . .

5.6. Evaluation of $\Delta F_0^{\ddagger}$ and $\Delta F_i^{\ddagger}$

It follows from Section 5.5 that Eq. (5.14) for $F_0^{\ddagger}(r)$

$$\Delta F_o^{\ddagger} + F_o = F_o^{\ddagger}(r) = -kT \ln \int .. \int \exp\left(-\mathcal{E}_{k^o}^{\ddagger}/kT\right) dk_o' \tag{5.15}$$

is simply the free energy of a system having the charges of the reactants a distance r apart but a distribution of orientations of solvent molecules and of positions of ions in the ionic atmosphere which would be in equilibrium with the hypothetical charges

$$e_n + m(e_n - e_n^p), \quad (n = 1, 2),$$

on the two central ions. It is at this point that we introduce the macroscopic expression for the free energy of this type of nonequilibrium system. We obtain

$$\Delta F_0^{\ddagger} = w + m^2 \lambda \ldots$$

5.7. Equation for *m*

The equation for m is obtained by equating the difference between free energies of activation for the forward and reverse reaction to the standard free energy of reaction at the prevailing electrolyte concentration, $\Delta F^{0\prime}$. In the process, we tacitly set the free energy of the reactants on the intersection surface equal to that of the products there... and so satisfy the energy condition (A.1) in Appendix 1...

$$\langle \mathcal{E}_k \rangle_{av} = \langle \mathcal{E}_k^p + \Delta\mathcal{E} \rangle_{av} \quad \text{(A.1)}$$

Thereby the e.e.d. is made to center on the intersection surface....[10]

APPENDIX 1 Properties of the e.e.d.[11]

If the e.e.d. is indeed 'centered' on the intersection surface, a system having the e.e.d. and the electronic configuration of the reactants would have the same average energy as a system having the same e.e.d. and the electronic configuration of the products. That is, it would satisfy

$$\langle \mathcal{E}_k \rangle_{av} = \langle \mathcal{E}_k^p + \Delta\mathcal{E} \rangle_{av}, \tag{A.1}$$

each averaged over the e.e.d. (They would have the same free energy too.)

It is easy to show that such an e.e.d. exists.[12] Consider the expression $-kT \ln \int .. \int \exp(-\mathcal{E}_{k^o}^{\ddagger}/kT)dk_o'$ as a function of m and minimize it with respect to m. One obtains immediately, using Eq. (3.1.3):

$$\frac{\int .. \int \exp(-\mathcal{E}_k^{\ddagger}/kT)\mathcal{E}_k d\tau_k}{\int .. \int \exp(-\mathcal{E}_k^{\ddagger}/kT)d\tau_k} = \frac{\int .. \int \exp(-\mathcal{E}_k^{\ddagger}/kT)(\mathcal{E}_k^p + \Delta\mathcal{E})d\tau_k}{\int .. \int \exp(-\mathcal{E}_k^{\ddagger}/kT)d\tau_k}. \tag{A.2}$$

This is the desired property. Thus, there is an e.e.d. centered on the intersection surface. It has the property that the ln term, i.e. the free energy, is a minimum with respect to m ....

We now examine in more detail the approximation of replacing the $(N-1)$-dimensional surface integral by an N-dimensional volume integral over the e.e.d., N being large. A *coordinate system* is introduced as follows[(13)]:

We note first that the intersection of the two potential surfaces in Fig. 1 defines a surface of $(N-1)$ degrees of freedom and that shifting the potential energy surface of the products vertically by an amount Γ without change in shape, produces a different intersection which defines a new $(N-1)$-dimensional surface parallel to the first one. In this way, a family of parallel surfaces can be generated, each member associated with a particular value of Γ and obeying (A.3)...

$$\mathcal{E}_k = \mathcal{E}_k^p + \Delta\mathcal{E} + \Gamma. \tag{A.3}$$

Let σ denote the totality of $(N-1)$ orthogonal curvilinear coordinates defining position on any given surface and let γ be the coordinate normal to the family of surfaces, so that k in E_k consists of σ and γ. Let the origin of γ be at the actual intersection surface, for which $\Gamma = 0$. $\Gamma(\gamma)$ is a strictly monotonic function of γ and $\Gamma(0) = 0$.

Consider now the volume integral over the e.e.d.,

$$\int \cdots \int \exp(-\mathcal{E}_{\sigma\gamma}^{\ddagger}/kT)d\sigma d\gamma,$$

where $\mathcal{E}_{\sigma\gamma}^{\ddagger}$ satisfies Eq. (A.4) and m was selected so that Eq. (A.5) is satisfied (cf. (A.1) using (A.3)).

$$\mathcal{E}_{\sigma\gamma}^{\ddagger} = \mathcal{E}_{\sigma\gamma} + m\Gamma(\gamma), \tag{A.4}$$

$$\int \cdots \int \Gamma\exp(-\mathcal{E}_{\sigma\gamma}^{\ddagger}/kT)d\sigma d\gamma = 0.\text{"} \tag{A.5}$$

The equation is valid because Γ is an odd function while the e.e.d., being a Gaussian, is an even function. Note that $\exp(-\mathcal{E}_{\sigma\gamma}^{\ddagger}(\gamma)/kT)$ defines the *e.e.d.* of atomic configurations corresponding to the potential energy surface $E^{\ddagger}_{\sigma\gamma}(\gamma)$. $\mathcal{E}^{\ddagger}_{\sigma\gamma}(\gamma)$ is supposed to be of a parabolic shape — an approximation — and so $\exp(-\mathcal{E}^{\ddagger}_{\sigma\gamma}(\gamma)/kT)$ is a Gaussian, with its maximum at $\mathcal{E}^{\ddagger}_{\sigma\gamma}(0)$. It is schematically shown in Fig. 1* later discussed with Marcus in the **NOTES** section. The e.e.d. dies away fairly rapidly along the normal γ.

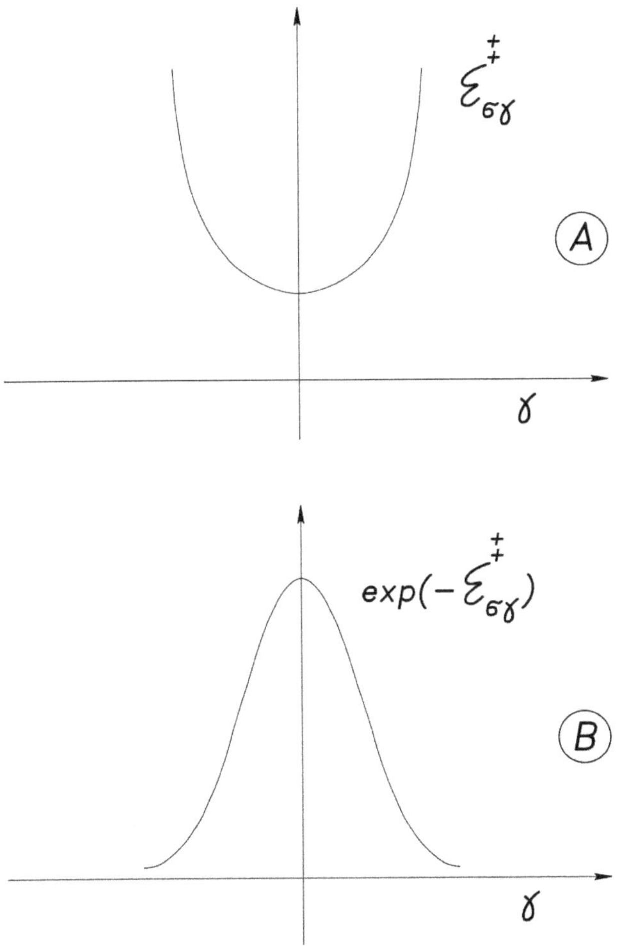

Fig. 1*.

"In the vicinity of $\gamma = 0$, Γ equals $\gamma(d\Gamma/d\gamma)_0$, the derivative being then nonzero. Equation (A.5) becomes

$$\int_{-\infty}^{+\infty} \gamma \exp(-F^{\ddagger}_{(\gamma)}/kT)d\gamma = 0 \tag{A.6}$$

where $F^{\ddagger}_{\gamma}$, the free energy of a system constrained to exist on the surface γ, is given by

$$\exp(-F^{\ddagger}_{(\gamma)}/kT) = \int \ldots \int \exp(-\mathcal{E}^{\ddagger}_{\sigma\gamma}/kT)d\sigma\text{"} \tag{A.7}$$

Note: consider that Γ can be written as $\Gamma(\gamma) = \Gamma(0) + \gamma(d\Gamma/d\gamma)_0$, with $\Gamma(0) = 0$.

"Since Eq. (A.6) is applicable to all T,[(14)] we infer that $F^{\ddagger}_{\gamma}$ is an even function of γ and write therefore the Taylor series (A.8), retaining only terms up to γ^2 for physical reasons based on Fig. 1:

$$F^{\ddagger}_{(\gamma)} = F^{\ddagger}_{(0)} + (\gamma^2/2!)F''_{(0)} + \cdots. \tag{A.8}$$

Using Eq. (A.8), the N-dimensional volume integral becomes

$$\int_{-\infty}^{+\infty} \exp(-F^{\ddagger}_{(\gamma)}/kT)d\gamma = \exp(-F^{\ddagger}_{(0)}/kT)\sqrt{2\pi kT/F''_{(0)}} \tag{A.9}$$

But

$$\exp(-F^{\ddagger}_{(0)}/kT) = \int \ldots \int \exp(-\mathcal{E}_{\sigma 0}/kT)d\sigma,$$

using Eqs. (A.7) and (A.4) at $\gamma = 0$, and so equals the desired ($N - 1$)-dimensional surface integral. Thus, if the 'vibrational-like partition function'[(15)] $\sqrt{2\pi kT/F''_{(0)}}$ is of the order of unity, the basic approximation enunciated in Section 5.4 is seen from Eq. (A.9) to be justified."

5.8. The Meaning of m, the "Inverted" Chemical Behavior

The meaning of the Lagrangian multiplier m in the dielectric continuum treatment of the theory was discussed in Chapter 2. It will be discussed in Chapter 6 in terms of potential energy surfaces and statistical mechanics. We anticipate here that $m = 0$ if the crossing of the R and P surfaces is at the bottom of the R surface, that is, if the TS resembles the reactants. The TS would resemble the products if it were $m = -1$ and the crossing would happen at the bottom of the P curve. $m = -1/2$ if the TS would resemble the reactants as much as the products being intermediate between reactants and products.

"If ΔF^0 becomes too negative, intersection of the two surfaces becomes possible only at high potential energies, unless in such cases a more favorable reaction mechanism is found." [1, p. 28]. The crossing of the R and P parabolas happen to the left of the bottom of the P parabola and the m's become positive.

5.9. Interactions in Polar Media

5.9.1. *Introduction*

In the previous treatment of the dielectric properties of condensed phases, M. considered each molecule as having a permanent dipole plus an induced dipole proportional to the reacting field acting on it. This was in accord with theoretical analyses reviewed in Refs. [8, 10–15]. In an effort toward generalization, M. subsequently devised a more general model of polar interactions in condensed media developing a Hartree method to compute *the polar contribution to the interparticles interaction energy*, using permanent plus induced charge distributions on particles. The *total* interparticle interaction energy is taken to be the sum of this *polar* term and of an interparticle *correlation* term including effects due to interparticle London dispersion and to exchange repulsion forces. It can be shown that a number of idealized models of molecules, or an entire medium or an electrode, satisfy the same formal equations. For this reason, use of this model permitted the development of a *unified* theory for electron transfer rates in solution and at electrodes.

5.10. Interparticle Interaction Energy

"We may consider the macrosystem to be composed of 'particles,' each particle having a specified set of nuclei in a fixed configuration, which is later permitted to vary. Each *particle* may, for example, be a molecule or any collection of molecules (a whole electrode for example)" [2].

The *total electronic energy* of the system at any specified nuclear configuration is the sum of the electronic energies of the particles considered as isolated, at infinite distance from each other, each having a specified nuclear configuration, and the interparticle interaction energy.

"It is desired to calculate this *interparticle term*, that is, to calculate *the work required to bring the particles together from infinite...* we assume that the interparticle interaction energy is the sum of a polar term and an interparticle electron correlation term. (Among the contributions to the second would be those arising from interparticle London dispersion and exchange repulsion interactions). We *define* this *polar term* to be the interaction energy calculated when the electronic wave function of the macrosystem is assumed to be *the best product* of antisymmetrized many-electron wave functions, one for each particle" [2].

M. shows that a polar term so defined is equal to the *classical electrostatic* interaction energy of charge densities on the particles. A Hartree-like treatment is employed in which the component wave functions are many-electron ones rather than the usual one-electron functions of the atomic Hartree method. The best wave function is the one which minimizes energy.

"Let X be the electronic wave function of the system and X_i the antisymmetrized electronic wave function of particle i. Let each X_i be normalized to unity.

$$X = \prod_i X_i \tag{5.16}$$

When the interparticle interaction is removed, each X_i reduces to the exact wave function X_i^0 for the isolated i particle....

Let H_i^0 be the Hamiltonian for an *isolated* i and let E_i^0 be the corresponding electronic energy:

$$H_i^0 X_i^0 = E_i^0 X_i^0 \tag{5.17}$$

Let H_{ij} be the Coulombic interaction of the electrons and nuclei of i with those of j, (H_{ij} is the sum of the pertinent $1/r$ terms). The Hamiltonian H of the entire system is therefore

$$H = \sum_i H_i^0 + \frac{1}{2}\sum_{i \neq j}\sum_j H_{ij}$$

We shall use the symbol (,) to denote the pertinent inner product, for example, (X, HX) is $\int X^* HX d\mathbf{r}$, where one integrates over all electron coordinates $\mathbf{r}$ of the entire system....

Minimizing (X, HX) with respect to variations in each X_i, subject to constancy of (X, X), X_i is found in a standard way to be the solution of Eq. (5.18):

$$(H_i^0 + V_i - E_i)X_i = 0 \tag{5.18}$$

where E_i is a Lagrangian multiplier, and V_i is the sum of Coulombic attraction terms of the electrons *of particle* i with the nuclei of all j, $\sum_i v_j^i$, and of v_{ij}, the internuclear and interelectron repulsion of i and j, averaged over the electronic distribution of all $j \neq i$, and of

$\sum_j v_i^j$, the Coulombic attraction of the nuclei of i with electrons of all $j \neq i$, averaged over the electronic distribution of the j's

$$V_i = \sum_j v_j^i + \sum_j (X_j, [v_{ij} + v_i^j] X_j), \tag{5.19}$$

$$E = \sum_i E_i - \frac{1}{2} \sum_{i \neq j} \sum_j (X_i X_j, v_{ij} X_i X_j) - \sum_{i \neq j} \sum_j (X_j, v_i^j X_j) \tag{5.20}$$

(**typo** corrected from the original) where we have normalized X, that is, have set $(X, X) = 1$. E^{pol}, the polar contribution to the interparticle interaction energy is given by Eq. (5.21):

$$E^{pol} = E - \Sigma_i E_i^0 \tag{5.21}$$

that is, the *polarization energy* is the difference between the energy of the whole system when the particles interact with each other and the sum of the energies stored in them when they do not interact.

We wish to obtain an expression for E^{pol} in terms of charge densities. If particle i has N electrons (coordinates $\mathbf{r}_1, \ldots, \mathbf{r}_N$) and nuclei $a, b, \ldots$ (coordinates $\mathbf{r}_a, \mathbf{r}_b, \ldots$ and charges $Z_a e, Z_b e, \ldots$), then the total charge density of particle i, $\rho_i(\mathbf{r})$, is the sum of electronic and nuclear terms $\rho_i^e(\mathbf{r})$ and $\rho_i^n(\mathbf{r})$:

$$\rho_i(\mathbf{r}) = Ne \int \ldots \int X_i^*(\mathbf{r}, \mathbf{r}_2, \ldots, \mathbf{r}_N) \times X_i(\mathbf{r}, \mathbf{r}_2, \ldots, \mathbf{r}_N) d\mathbf{r}_2 \ldots d\mathbf{r}_N + \sum_a Z_a e \delta(\mathbf{r} - \mathbf{r}_a) = \rho_i^e(\mathbf{r}) + \rho_i^n(\mathbf{r}) \tag{5.22}$$

(**typo** corrected from the original), where δ is the Dirac δ function. N and a should bear a subscript but we omit it for brevity. The charge density of the isolated particle i is

$$\rho_i^0(\mathbf{r}) = Ne \int \ldots \int \frac{X_i^{0*}(\mathbf{r}, \mathbf{r}_2, \ldots \mathbf{r}_N) X_i^0(\mathbf{r}, \mathbf{r}_2, \ldots \mathbf{r}_N) d\mathbf{r}_2 \ldots d\mathbf{r}_N}{(X_i^0, X_i^0)} + \sum_a Z_a e \delta(\mathbf{r} - \mathbf{r}_a). \tag{5.23}$$

The *induced charge density* ρ_i^{in} is the difference:

$$\rho_i = \rho_i^0 + \rho_i^{in}. \tag{5.24}$$

Equation (5.19) for V_i can be written in terms of these ρ_i's:

$$V_i(\mathbf{r}_1 \ldots \mathbf{r}_N) = e \sum_{m=1}^{N} \Phi_i(\mathbf{r}_m) + \sum_{\mathbf{j}} [\boldsymbol{v}_{\mathbf{ij}} + (X_j, v_i^j X_j)] \tag{5.25}$$

(**typo** corrected from the original) where [2]

$$\Phi_i(\mathbf{r}) = \sum_{j \neq i} \int \frac{\rho_j(\mathbf{r}')}{|\mathbf{r} - \mathbf{r}'|} d\mathbf{r}' \text{''} \tag{5.26}$$

"When ρ_i^{in} is linear in Φ_i, that is, when the electronic response of i is linearly dependent on the perturbing potential field" [2], it can be demonstrated that:

$$E^{pol} = \frac{1}{2} \sum_i \rho_i^0 \Phi_i(\mathbf{r})$$

which is the E^{pol} in terms of charge densities.

5.11. Comparison with the Electrostatic Calculations

"The values of both ρ_i^0 and ρ_i^{in} themselves can only be determined quantum mechanically. . . .

The electrostatic energy of interaction between the particles can be regarded as the sum of

(i) Interaction of the permanent charge distributions,

$$\frac{1}{2}\sum_{i\neq j}\sum_{j}\iint \frac{\rho_i^0(\mathbf{r})\rho_j^0(\mathbf{r}')d\mathbf{r}d\mathbf{r}'}{|\mathbf{r}-\mathbf{r}'|}$$

the ½ avoids double counting.

(ii) Interaction of the permanent and induced charge distributions,

$$\sum_{i\neq j}\sum_{j}\iint \frac{\rho_i^0(\mathbf{r})\rho_j^{in}(\mathbf{r}')d\mathbf{r}d\mathbf{r}'}{|\mathbf{r}-\mathbf{r}'|}$$

(iii) Interaction of the induced charge distributions,

$$\frac{1}{2}\sum_{i\neq j}\sum_{j}\iint \frac{\rho_i^{in}(\mathbf{r})\rho_j^{in}(\mathbf{r}')d\mathbf{r}d\mathbf{r}'}{|\mathbf{r}-\mathbf{r}'|}$$

(iv) Energy stored in the induced charge distribution [2],

$$\sum_{i}\int_{\lambda=0}^{1}\int \left(\rho_i^{in\lambda}-\rho_i^{in}\right)\Phi_i d\mathbf{r}d\lambda."$$

The meaning of $\rho_i^{in\lambda}$ and of ρ_i^{λ} is explained in the following.

The first three components parallel analogous interactions "on a macroscale" in Chapter 2, see Ref. [7]]. Contribution (iv) needs an extra explanation.

Imagine of building up the ρ_i^{in} starting from ρ_i^0 and adding to it $\rho_i^{in\lambda} = \rho_i^{in}\lambda$ with λ a charging parameter varying from 0 to 1. $\rho_i^{in\lambda}$ equals zero for $\lambda = 0$ and ρ_i^{in} for $\lambda = 1$. Term (iv) is then the work for the charging process in which the charge density on i varies from ρ_i^0 to $\rho_i^0 + \rho_i^{in}$.

Quantum-mechanically we can write E_i in Eq. (5.18) as:

$$E_i = (X_i, (H_i^0 + V_i)X_i)$$
$$\text{or} \quad E_i = (X_i, H_i^0 X_i) + (X_i, V_i X_i)$$

and term (iv) is:

$$(\text{iv}) = \sum_i [(X_i H_i^0 X_i) - E_i^0]$$
$$\text{or} \quad (\text{iv}) = \sum_i [(X_i, H_i^0 X_i) - (X_i^0, H_i^0 X_i^0)]$$

that is, it is the difference between the expectation values of H_i^0 when the wave function is the X_i of the perturbed system and that, X_i^0, of the unperturbed system.

5.12. Interactions in Polar Media

In describing the polar contribution to the potential energy of interaction between particles in a polar medium — in particular in an electrolytic solution — M. has used the following models:

(1) "In Part I of this series, [Ref. [7]: 'Electrostatic Free Energy and Other Properties of States Having Nonequilibrium Polarization. I.'].... Certain molecules under special consideration, which will be termed central species, were regarded as being *electronically* polarizable regions (continua) imbedded in the remainder of the system treated as an *electronically* and *orientationally* polarizable continuum. Examples of the central species are two reacting ions, a reacting ion and an electrode, and a fluorescing solute molecule....

(2) The treatment in Part I can be extended considerably by using a recent and more general description of the central species [2]: One may regard them as being particles, each of which represents a single molecule or any collection of molecules at any specified configuration of the particles' nuclei. These particles are assigned general, permanent, and induced charge distributions and can be

embedded, as before in the remainder of the system and treated as a dielectric continuum. (A statistical mechanical description of nonequilibrium polarization systems, based on this model of a particle, is given in Part III of this series [5].)....

In the present paper [3], we describe this extension of Part I. Because of the general nature of these particles, electrode systems are automatically included, and thereby Part I is further generalized since it was concerned with homogeneous systems only. Electrode systems had to be described separately.

An expression is obtained for the polar contribution to the free energy of systems in which the medium surrounding the particles is treated as an orientationally and electronically polarizable dielectric continuum having nonequilibrium dielectric polarization. It will then be shown that this free energy can be expressed as a sum of free energies of hypothetical systems having equilibrium dielectric polarization. An advantage of this relation is that literature expressions for equilibrium polarization systems can be immediately used to compute the free energy of nonequilibrium ones....

It will be useful to regard each ion and its inner coordination shell as forming a single particle. In electrochemical systems, it will also be useful to treat the electrode and any strongly bound adjacent layer of molecules or ions as another particle... dielectric continuum theory is used in this paper [4] to calculate the polar contribution to the free energy arising from the mutual interaction of the particles and from their interactions with the medium. The calculation is made at any specified nuclear configuration of the particles. One may then integrate over these nuclear configurations, suitably weighted as in statistical mechanics. We consider first dielectrically unsaturated systems and then deduce Eq. (5.47) for partially saturated systems as well" [3].

(3) "In Part I [Ref. [2]: 'Interactions in Polar Media. I. Interparticle Interaction Energy'] equations were derived for the polar contribution to the potential energy of interaction between

particles. Each particle represented a single molecule or any collection of molecules (a whole electrode, for example) at any specified nuclear configuration. The equations involved multipolar permanent and induced charge distributions of the particles and constituted a generalization of earlier potential energy expressions in which the particles were restricted to being single molecules interacting via permanent and induced dipoles only."[(16)] [8].... In the present paper [4], it is shown that the equations of Part I are also formally satisfied by a *common model of a particle*, wherein it is treated as an electronically polarizable dielectric continuum containing some or no fixed charge distribution", see Ref. [12] for example.... (An important example of such a model occurs in the standard electrostatic treatment of polar interactions between electrodes and particles in solution: The electrode is regarded as being a continuum of infinite dielectric constant.).... Because of computational difficulties in the statistical mechanical treatment of positions of the nuclei, many literature calculations are based on a continuum treatment of part of the orientation and electronic polarization in the system.... Each of the two specialized models has its merits, the induced dipole approximation being more appropriate for *small* molecules and the electronically polarizable continuum model being suited to *very large* ones, a crystal for example. In particular, the method of approximating an induced charge distribution by its leading term, the induced dipole, does not explain the dissymmetry of light scattering by large particles [16]. However, the dissymmetry is explained by the continuum model because of the appearance of terms higher than the dipolar one. Moreover, even a uniform static field sometimes produces induced moments higher than the dipole one" [4].

The dielectric continuum approximation "can be regarded as a limiting case of the statistical mechanical one in which the ratio of size of the central species to that of each molecule composing the medium becomes extremely large" [5].

5.13. Molecular Treatment of Particles

"Let $\phi_i(\mathbf{r})$ be the electrostatic potential arising from all parts of the system other than from the charge density $\rho_i(\mathbf{r})$ on particle i. Equation (5.26) for ϕ_i is easily extended to the case where, besides the particles, a medium is present, having a polarization $\mathbf{P}(\mathbf{r})$:

$$\phi_i(\mathbf{r}') = \sum_{j \neq i} \int \frac{\rho_j(\mathbf{r})d\mathbf{r}}{r} + \int \mathbf{P}(\mathbf{r}) \cdot \nabla r^{-1} d\mathbf{r} \tag{5.27}$$

where integration is over the entire volume of the system, $1/r$ denotes $1/|\mathbf{r} - \mathbf{r}'|$, $\mathbf{P}(\mathbf{r})$ vanishes inside the region V_i occupied by any particle i and $\rho_i(\mathbf{r})$ is given by:

$$\rho_i(\mathbf{r}) = \rho_i^0(\mathbf{r}) + \rho_i^{in}(\mathbf{r}) \tag{5.28}$$

where ρ_i^0 is the charge distribution on i when it is isolated" [4]. It may be demonstrated that "the induced charge distribution ρ_i^{in} is related to ϕ_i via a linear Hermitian operator A_i:

$$\rho_i^{in}(\mathbf{r}) = A_i(\mathbf{r}, \mathbf{r}')\phi_i(\mathbf{r}') \tag{5.29}$$

When only particles are present,[(17)] each having a specified nuclear configuration, the polar contribution to the interparticle potential energy U^{pol} is given by:

$$U^{pol} = \frac{1}{2} \sum_i \int \phi_i \rho_i^0 d\mathbf{r} \tag{5.30}$$

where ϕ_i is given by Eq. (5.27) with $\mathbf{P} = 0$.

When *both* particles *and* the polarized medium are present, it is the polar contribution to the free energy F^{pol} which is of interest. When $\mathbf{P}(\mathbf{r})$ is in electrostatic equilibrium with the charge equilibrium, for example, when it is given by

$$\mathbf{P}(\mathbf{r}) = -\chi(\mathbf{r})\nabla\psi(\mathbf{r}) \tag{5.31}$$

where χ is the dielectric susceptibility of the medium and ψ is the electrostatic potential arising from all parts of the system, then F^{pol}

is given by the RHS of Eq. (5.30) with ϕ_i given by Eq. (5.27)" [4]. A straightforward proof of Eq. (5.30) for U^{pol} in the particle-only case, and for F^{pol} one, can be given simultaneously. In both cases it equals the work needed to change every ρ_i^0 from 0 to its final value ρ_i ... The value of F^{pol} for a particle-medium system when $\mathbf{P}(\mathbf{r})$ is not in electrostatic equilibrium with the ρ_i is given later.

Equations (5.27) to (5.30) and an equation for $\mathbf{P}(\mathbf{r})$ are the basic equations. We also make use of Eqs. (5.32) and (5.33) for the electrostatic potential $\psi(\mathbf{r})$ due to *all* sources and for the potential $\psi_{is}(\mathbf{r})$ in a system containing an *isolated* particle i (This isolated system is denoted by "is"):

$$\psi(\mathbf{r}') = \sum_{\mathrm{i}} \int \frac{\rho_{\mathrm{i}}(\mathbf{r})}{r} d\mathbf{r} + \int \mathbf{P}(\mathbf{r}) \cdot \nabla r^{-1} d\mathbf{r} = \phi_i(\mathbf{r}') + \int \frac{\rho_i(\mathbf{r})}{r} d\mathbf{r} \tag{5.32}$$

$$\psi_{is}\left(\mathbf{r}'\right) = \int \frac{\rho_i^0(\mathbf{r})}{r} d\mathbf{r} \tag{5.33}$$

5.14. Electronically Polarizable Continuum Model of a Particle

Each particle i is treated as a *region* having electronic dielectric susceptibility $\chi_i(\mathbf{r})$ and having a fixed charge density $\rho_i^v(\mathbf{r})$. Both $\chi_i(\mathbf{r})$ and $\rho_i^v(\mathbf{r})$ can vary with position r inside the molecule i, and both vanish outside V_i, the volume occupied by i. The potential $\psi(\mathbf{r}')$ is the sum of contributions of the charge distributions ρ_i^v and from all polarized volume elements, those inside each V_i, $(-\chi_i \nabla \psi)$, and those outside:

$$\psi(\mathbf{r}') = \sum_i \int \left\{ \frac{\rho_i^v}{r} - \chi_i \nabla \psi \cdot \nabla r^{-1} \right\} d\mathbf{r} + \int \mathbf{P}(\mathbf{r}) \cdot \nabla r^{-1} d\mathbf{r} \tag{5.34}$$

where $\mathbf{P}(\mathbf{r})$ vanishes inside each V_i"....

For an isolated i we have:

$$\psi_{is}(\mathbf{r}') = \int \frac{\rho_i^v}{r} d\mathbf{r} - \int \chi_i \nabla \psi_{is} \cdot \nabla r^{-1} d\mathbf{r} \ldots . \quad (5.35)$$

U^{pol} or F^{pol} is obtained by subtracting the reversible work to charge the *isolated i*' s from that to charge them in the actual system. U^{pol} applies to the particles-only system and F^{pol} to the particles-medium system:

$$U^{pol}(\text{or } F^{pol}) = \sum_i \int \int_{\rho_i^v=0}^{\rho_i^v=\rho_i^v} \psi d\rho_i^v(\mathbf{r}) d\mathbf{r} - \sum_i \int \int_{\rho_i^v=0}^{\rho_i^v=\rho_i^v} \psi_{is} d\rho_i^v(\mathbf{r}) d\mathbf{r}. \quad (5.36)$$

Equation (5.41) may be integrated. . . . We find

$$U^{pol}(\text{or } F^{pol}) = \frac{1}{2} \sum_i \int (\psi - \psi_{is}) \, \rho_i^v d\mathbf{r}. \quad (5.37)$$

Electronically Polarizable Continuum Model Satisfies Eqs. (5.27)–(5.30).

It is shown now that this model satisfies Eqs. (5.27)–(5.30) with special choice of A_i" [4]. Among the necessary conditions "for it to satisfy these equations, are

(i) ψ_{is} and ψ given by Eqs. (5.35) and (5.34) equal those given by Eqs. (5.32) and (5.30), respectively
(ii) Identical values for F^{pol} or U^{pol}" [4]

M. shows that "one can find a molecular description with a $\rho_i^0 \ldots$ expressed entirely in terms of continuum properties of i, that is, in terms of $\chi_i(\mathbf{r})$, ρ_i^v, and geometry of V_i and (a crucial point in the argument) independent of the presence of any other $j\,(\neq i)$, of any external field and of the value of χ. . . . One may then conclude that a particular molecular description has been found which is mathematically equivalent to the electronically polarizable continuum model of this molecule *insofar as polar interactions are concerned*, that is, that Eqs. (5.27–5.29) are satisfied by this continuum model" [4].

5.15. Systems with Nonequilibrium Polarization

"We consider now the case where the polarization of the medium $\mathbf{P}(\mathbf{r})$ in a particle-medium system is not in electrostatic equilibrium with the ρ_i^0. $\mathbf{P}(\mathbf{r})$ no longer equals $-\chi\nabla\psi$. Instead it can be written as

$$\mathbf{P}(\mathbf{r}) = \mathbf{P}_0(\mathbf{r}) - \chi_e(\mathbf{r})\nabla(\psi - \psi_0) \tag{5.38}$$

where $\mathbf{P}_0(\mathbf{r})$ and ψ_0 are quantities characterizing the nonequilibrium state of the medium surrounding the particles i, and where χ_e is the dielectric *electronic* susceptibility of the medium.

The terms $\mathbf{P}_0$ and ψ_0 are independent of all ρ_i^0 and ρ_i^{in}." "χ_e is related to D_{op}, the square of the refractive index of the medium, $4\pi\chi_e = D_{op} - 1$. In an equilibrium polarization system $\psi = \psi_0$ and $\mathbf{P}_0 = -\chi\nabla\psi_0$, so that Eq. (5.38) reduces to the value given earlier, Eq. (5.31), for equilibrium polarization systems" [4].

"Equations 5.30 and 5.42 for F^{pol} no longer apply" for the nonequilibrium case. "The correct expressions are obtained as follows: let the earlier $\mathbf{P}_0(\mathbf{r})$ be the polarization that the medium would have if, for its *given* orientation polarization, it *were* in electrostatic equilibrium with *a* charge distribution. The properties of this *hypothetical state and charge distribution* are designated by a subscript 0" [4]. M. then shows that "The value of F^{pol} for the *molecular-type* and the *continuum-type* models of the particles is given by Eqs. (5.39) and (5.40), respectively:

$$F^{pol} = F_0^{pol} + \sum_i \int (\phi_i + \phi_{i0})\left(\rho_i^0 + \rho_{i0}^0\right) d\mathbf{r} \tag{5.39}$$

$$F^{pol} = F_0^{pol} + \sum_i \int (\psi + \psi_0)\left(\rho_i^v + \rho_{i0}^v\right) d\mathbf{r} \tag{5.40}$$

The equality of these expressions is also shown. . . " [4].

5.16. Error of Induced Dipole Approximation

"The approximation of regarding the *induced charge distribution* as being an *induced dipole* at a point in a molecule will break down when the molecule becomes sufficiently large or the field sufficiently nonuniform. In fact, this breakdown has permitted the estimate of molecular size through light-scattering experiments [16]" [4]. "When the ratio of particle size to wavelength is of the order of 0.1, a dissymmetry occurs in the scattered light. Estimates of the molecular size have been made thereby, particularly in the Rayleigh–Gans region [16], using in effect the continuum model" [4, note 24].

5.17. The Total Interaction Potential

"The total interaction potential is usually taken to be the sum of the *polar term*, a *dispersion term*, and for shorter distances, a *repulsive term*. For large r only the first two need to be considered. Certain possible contributions of an inelastic nature,[(18)] such as *charge transfer* or *electronic excitation* can be minimized by use of ions of low electron affinity (e.g., monovalent alkali metal cations) and of low energy" [4, p. 465]. A list of intermolecular forces can be found on p. 5 of the recent Ref. [17] where a brief discussion of intermolecular forces in a medium is also given on p. 147.

5.18. Free Energy of Nonequilibrium Polarization Systems

5.18.1. *Homogeneous and Electrode Systems*

In the two stages charging process of Ref. [7], the free energy of a nonequilibrium polarization system was calculated using an intermediate fictitious equilibrium system at the end of charging stage I. In the more refined method of Ref. [3], which is also extended to "a variety of systems," for instance to electrode systems, *four equilibrium polarization states* are considered:

"If $(e_a, e_b, \ldots, e_j, \ldots)$ denotes the charges the charges of the reactants A and B and of any other ions j present before electron transfer, and if $(e_a^p, e_b^p, \ldots, e_j^p, \ldots)$ denotes the charges after, then $e_j^p = e_j$."

Only the charges of the reacting ions change after the ET reaction.

State [1] is the initial state, the equilibrium polarization system having the e_i's existing before ET, and the permanent charge densities ρ_i^0 of the ions before the ET process happens. Such charge densities are symbolized as $\rho_{i_1}^0$ and $\mathbf{P}(\mathbf{r})$ is in equilibrium with them. Note that the state [1] is referred to in the sub-subscript 1 of $\rho_{i_1}^0$. The notation will also be used in relation to the following states.

State [0] has charge densities $\rho_i^0 = \rho_{i_0}^0$, different from those in the activated complexes X and X^* *and in between.* Such charge densities are the ones which would be in equilibrium with the orientational contribution of $\mathbf{P}(\mathbf{r})$ at the transition states $X$and X^*. They are the charge densities at the end of stage I of the charging process in Ref. [7]. This system has been denoted by M. as the "equivalent equilibrium system" (e.e.s.) [1].

State [1–0] has $\rho_i^0 = \rho_{i_{1-0}}^0$ where $\rho_{i_{1-0}}^0 = \rho_{i_1}^0 - \rho_{i_0}^0$ and having $\mathbf{P}(\mathbf{r})$ in equilibrium with $\rho_{i_{1-0}}^0$'s.

State [1–0, op] has $\rho_i^0 = \rho_{i_{1-0}}^0$ and is *responding to those ρ_i^0 only via the electronic polarization.*

State [1–0, op] is the equilibrium state which exists only for the very short time during which the electronic polarization is already in equilibrium with the charges $(-m\Delta e, m\Delta e, \ldots, 0, \ldots)$ but the atomic-orientational polarization has still not responded to the charges and is then equal to zero.

In order to have a less abstract, more visible picture of the earlier states, let us describe them not in terms of charge densities but, rather, in terms of charges on the ions.

"If $(e_a, e_b, \ldots e_j, \ldots)$ denotes the charges of the reactants A and B and of any ions j present before ET, and if $(e_a^p, e_b^p, \ldots, e_j^p, \ldots)$ denotes the charges after, then $e_j^p = e_j$," that is, the *noncentral* ions do not take part in ET. State [1] has the e_i's existing before ET and can be represented as:

$$\text{State [1]} \equiv (e_a, e_b, \ldots, e_j, \ldots).$$

It has been shown [5b, 1] that in the activated complex, the medium has an orientation polarization which would be in dielectric equilibrium only with the hypothetical charges $e_i + m(e_i - e_i^p)$ $(i = a, b,$ or $j)$ where m satisfies a given equation. State [0] has therefore the charge distribution described by:

$$\text{State [0]} \equiv (e_a + m\Delta e, e_b - m\Delta e, ..., e_j, \ldots).$$

"Hence the state [1–0] is given by:

$$\text{State [1–0]} \equiv (-m\Delta e, m\Delta e, \ldots, 0, \ldots)."$$

In the Notes a synopsis of the five states is given[(19)] with their polarizations and charge densities.

I shall now briefly show how the free energies of the earlier systems are calculated.

"We let each particle i have a *total* charge density $\rho_i(\mathbf{r})$, the sum of a *permanent* charge density $\rho_i^0(\mathbf{r})$ (the charge density when i is isolated) and an *induced* charge density $\rho_i^{in}(\mathbf{r})$. The latter is related to $\phi_i(\mathbf{r}')$, the potential due to all sources but i, via a linear Hermitian operator A_i.

$$\rho_i(\mathbf{r}) = \rho_i^0(\mathbf{r}) + \rho_i^{in}(\mathbf{r}) \tag{5.41}$$

$$\phi_i(\mathbf{r}') = \sum_{j \neq i} \int \frac{\rho_i(\mathbf{r})}{r} dr + \int \mathbf{P}(\mathbf{r}) \cdot \nabla \frac{1}{r} dr \tag{5.42}$$

where $\mathbf{P}(\mathbf{r})$ is the polarization of the medium (it vanishes inside the volume V_i occupied by any particle i) and $1/r$ is $1/|\mathbf{r} - \mathbf{r}'|$. Integration in (2) is over the entire volume of this system and ρ_i vanishes outside V_i" [3].

Notice that the $\rho_i^{in}(\mathbf{r})$ is inside the formula for $\phi_i(\mathbf{r}')$. Moreover, M. demonstrates the validity of the relation:

$$\rho_i^{in}(\mathbf{r}) = A_i(\mathbf{r}, \mathbf{r})\phi_i(\mathbf{r}) \tag{5.43}$$

where A_i is a linear operator. For the sake of simplicity, I shall not make use of the earlier formula and of others of the same kind showing the *linear dependence* of charge densities on potentials and vice versa.

"Let $\chi(\mathbf{r})$ and $\chi_e(\mathbf{r})$ denote the *static* and *electronic* dielectric susceptibilities of the medium, respectively, at any point r in the medium.

$$\chi(\mathbf{r}) = (D_s - 1)/4\pi, \quad \chi_e(\mathbf{r}) = (D_{op} - 1)/4\pi,$$

where $D_s(\mathbf{r})$ and $D_{op}(\mathbf{r})$ denote the static dielectric constant and the square of the refractive index of the medium, respectively. (Usually the χ's and the D's are constant scalars, but we include the possibility that they are variable tensors.) Let $\psi(\mathbf{r})$ denote the potential at any point r due to all parts of the system:

$$\psi(\mathbf{r}) = \sum_i \int \frac{\rho_i}{r} d\mathbf{r} + \int \mathbf{P} \cdot \nabla \frac{1}{r} d\mathbf{r},$$

with ρ_i given by Eq. (5.41). When appropriate subscripts are added, these equations refer to any of the five systems described earlier. $\mathbf{P}(\mathbf{r})$ for the nonequilibrium polarization system differs from $\mathbf{P}$ of the [0] system, $\mathbf{P}_0(\mathbf{r})$, *only in electronic polarization.* Assuming the electronic polarization to be dielectrically unsaturated,[(11)] we have [2, 3 note 12]:

$$\mathbf{P} - \mathbf{P}_0 = -\chi\nabla(\psi - \psi_0).\text{"} \tag{5.44}$$

After these premises on potentials and charge densities, M. goes on calculating the free energies for the earlier systems considering that "The *free energy change* ΔF, when the system is charged from one set of charges on the particles to another, equals the *reversible work* done to form one system from the other[15]:

$$\Delta F = \int \int_{\lambda=0}^{1} \phi_i^{\lambda} d\rho_i^{0^{\lambda}} d\mathbf{r} \qquad (5.45)$$

where λ is a charging parameter.

Note 15: 'Initially, in using Eq. (5.45) to derive Eq. (5.46),

$$\Delta F = \int \int_{\lambda=0}^{1} \phi_i^{\lambda} d\rho_i^{0\lambda} d\mathbf{r} \tag{5.45}$$

$$F = F_1 + F_{1-0}^{op} - F_{1-0} \tag{5.46}$$

the electrostriction which may result when this work is performed at constant pressure is ignored. It can be concluded by taking into account any dependence of $\chi(r)$ and $\chi_e(r)$ on λ. We note that this electrostriction is the one which occurs outside the inner coordination shell of each ion. In the derivation of Eq. (5.47),

$$F = F_1 + F_{1-0}^{op\prime} - F_{1-0}^{\prime} \tag{5.47}$$

the electrostriction on forming system [0] is not neglected. Rather, one makes the normally milder assumption of neglecting the additional electrostriction on forming system [1] from [0]. Systems [1–0]' and [1–0, op]' have the $\chi(r)$ and $\chi_e(r)$ of system [0].'[20]

Starting with system [0] and holding the orientation polarization of the medium fixed, we can form the nonequilibrium polarization system by changing the $\rho_i^{0^{\lambda}}$ according to Eq. (5.48)

$$\rho_i^{0^{\lambda}} = \rho_{i_0}^{0} + \lambda(\rho_{i_1}^{0} - \rho_{i_0}^{0}) \tag{5.48}$$

Since $\mathbf{P}_0$ is being held fixed" [3] it is possible to apply a linear relation between charge densities and potentials and conclude that

$$\phi_i^{\lambda} = \phi_{i0} + \lambda(\phi_{i1} - \phi_{i0}) \ldots$$

(typo in the original)

For equilibrium polarization

$$F = \frac{1}{2}\sum_i \int \phi_i \rho_i^0 d\mathbf{r}. \tag{5.49}$$

Equations for F for the four equilibrium polarization systems described earlier are obtained from Eq. (5.49) by adding appropriate subscripts and superscripts. For the [1–0] and [1–0, op], we thus have

$$F_{1-0} = \frac{1}{2}\sum_i \int \phi_{i_{1-0}}(\rho_{i_1}^0 - \rho_{i_0}^0) d\mathbf{r}, \tag{5.50a}$$

$$F_{1-0}^{op} = \frac{1}{2}\sum_i \int \phi_{i_{1-0^{op}}}(\rho_{i_1}^0 - \rho_{i_0}^o) d\mathbf{r} \ldots. \tag{5.50b}$$

As ... $\rho_i^{0\lambda}$ is varied, ... $\mathbf{P}^\lambda$ changes from $\mathbf{P}_0$ according to Eq. (5.48), $\mathbf{P}^\lambda$ being linearly dependent on ρ^0:

$$\mathbf{P}^\lambda - \mathbf{P}_0 = \chi\nabla(\psi^\lambda - \psi_0)[3] \tag{5.51}$$

It is finally demonstrated the formula giving the free energy of the *nonequilibrium system* in terms of the free energies of the *systems in equilibrium*:

$$F = F_1 + F_{1-0}^{op} - F_{1-0} \tag{5.52}$$

5.19. Some Applications of Eq. (5.51)

I. Electron Transfers in Solution

"If A and B are treated as spherical, and if electrostatic image effects are ignored (they contribute about 10% to the free energy change [1, note p. 27]), then the *polar contribution* to the free energy of the activated complex minus that of an equilibrium polarization system with charges e_a, e_b, and... e_j ... *in the same specified positions* is given by $F - F_1$, i.e., by $F_{1-0}^{op} - F_{1-0}$. However, using a well-known formula, F_{1-0} equals the sum of the *Born charging terms* at infinity plus the *work* required to bring the ions from ∞ to some mean distance r apart. (It will be recalled that the F's do not include interactions within the inner coordination shells.) Since the 1–0 system has no ions other than A and B, we then find:

$$F_{1-0} = -\frac{1}{2a_1}\left(1 - \frac{1}{D_s}\right)(m\Delta e)^2 - \frac{1}{2a_2}\left(1 - \frac{1}{D_s}\right)(m\Delta e)^2 + \frac{(m\Delta e)(-m\Delta e)}{D_s r}$$

and

$$F_{1-0}^{op} = -\frac{1}{2a_1}\left(1 - \frac{1}{D_{op}}\right)(m\Delta e)^2 - \frac{1}{2a_2}\left(1 - \frac{1}{D_{op}}\right)(m\Delta e)^2 + \frac{(m\Delta e)(-m\Delta e)}{D_{op} r}$$

where a_1 and a_2 denote the radii of the reactants, each including any inner coordination shell. Accordingly, this contribution to the free energy change is [3]:

$$F - F_1 = (m\Delta e)^2\left(\frac{1}{D_{op}} - \frac{1}{D_s}\right)\left(\frac{1}{2a_1} + \frac{1}{2a_2} - \frac{1}{r}\right). \quad \textbf{(21)}\text{"}$$

II. Electron Transfer at Electrodes

"Remarks similar to the earlier apply, but using the appropriate equations for F_{1-0} and F_{1-0}^{op} for electrode systems. If a reacting ion A is about to undergo ET with the electrode, one finds:

State $[1] \equiv (e_a, \ldots, e_j, \ldots)$" [3]

Notice that e_b is here missing and so will it be also in the notation for the following states:

"State $[0] \equiv (e_a + m(e_a - e_a^p), \ldots, e_j, \ldots)$

Hence,

State $[1\text{–}0](-m(e_a - e_a^p), \ldots, 0, \ldots)$.

In state [1–0] any change on the electrode, other than the electrostatic image charge, $m(e_a - e_a^p)$, vanishes. For such a system [1–0] having all $e_j = 0$, F_{1-0} is the sum of the Born charging term when the hypothetical ion and electrode are far apart" [3] plus the work required to bring the center of the ion from ∞ distance from the electrode to some mean distance r from the electrode surface and $2r$ from its electrical image:

$$F_{1-0} = \frac{1}{2a}\left(1 - \frac{1}{D_s}\right)(m\Delta e)^2 - \frac{(m\Delta e)^2}{2rD_s},$$

"Accordingly we obtain [3]

$$F - F_1 = F_{1-0}^{op} - F_{1-0} = \left(\frac{1}{2a} - \frac{1}{2r}\right)\left(\frac{1}{D_{op}} - \frac{1}{D_s}\right)(m\Delta e)^2\text{"}$$

We see that on applying the general compact formula $F - F_1 = F_{1-0}^{op} - F_{1-0}$ (5.51) M. gets immediately his two formulas for the homogeneous and electrode cases.

5.20. Partially Dielectrically Saturated Nonequilibrium Polarization Systems

"Any solvent present in the inner coordination shell of an ion will be at least *partially* dielectrically saturated. Partly for this reason the inner coordination shells are treated separately and by a noncontinuum method [1, 6]. However, partial saturation can also occur in the solvent *just outside* this shell if the ion is *highly charged*, and in this section this latter saturation will be considered.

Equation (5.48) can be derived for systems having this partial dielectric saturation, provided the change in *permanent* charges of the particles on going from the e.e.s. [0] to the nonequilibrium one is not too large. These changes are, in fact, normally small. In ET reactions, for example, they usually involve a change of $\frac{1}{2}$ a charge unit on each reacting particle" [3].

Such a change is given by $m\Delta e$ and m is usually about — 0.5 (cf. Refs. [1, 6]).

"There are two types of dielectric saturation to be included, electronic and orientational. We can consider both at once. It will be assumed that the change in charge of any species on going from the e.e.s. [0] to the nonequilibrium one and, therefore, to the equilibrium system [1] is *sufficiently small* that although the linear Eqs. (5.41), (5.43), (5.49) may not be applicable, each may be replaced by a linear relation between increments:

$$\delta\rho_{\mathrm{i}}^{\mathrm{in}} = \mathrm{A}_{\mathrm{i}}'\delta\phi_{\mathrm{i}} \ldots$$

For example, it is assumed that when the plot of **P** versus $\nabla\psi$ becomes *curved* due to *dielectric saturation,* we may compute the change $\delta\mathbf{P}$ from the tangent, $-\chi'$, and from $\delta\nabla\psi$" [3].

That is, one supposes that

$$\delta\mathbf{P} = -\chi'\delta\nabla\psi.$$

M. demonstrates the validity of the equation (5.47)

$$F = F_1 + F_{1-0}^{op\prime} - F_{1-0}'$$

in the case of partially dielectric saturation systems where in the primed F's χ is substituted by χ', χ_e by χ_e' and A_i by A_i'.

5.21. Free Energy of Nonequilibrium Polarization Systems

Statistical Mechanics of Homogeneous and Electrode Systems

"*Polar molecular interactions* play a role not only in the usual dielectric properties [18], but also in a variety of other phenomena, such as solvent effects on the spectra of polar solutes [19], homogeneous and electrochemical electron transfers [20], intramolecular charge transfers [21], and properties of polarons in semiconductors and other materials [22]. These phenomena usually involve systems with a nonequilibrium dielectric polarization [23].

In theories of these processes, the calculation of the free energy of the system often plays a central role" [5].

In the preceding sections, in which I summarized the theory in Ref. [3], it was shown "by dielectric continuum theory that the polar contribution to the free energy of a nonequilibrium dielectric polarization system equals the sum of free energies of related equilibrium ones.... However, for *noncontinuum* discussions, a statistical mechanical derivation of the free energy relation would be desirable. This derivation is given in the present paper [5] and is summarized in the following. 'The analysis also provides some further insight into the continuum description' [5].

5.22. Potential Energy Function

"The *macrosystem* may be considered to be composed of *particles*, each of which represents a *single molecule* or a *collection of molecules*, an *entire electrode* for example. For some purposes, it suffices to compute functions of the *interparticle* potential energy. Consequently, detailed assumptions about *intraparticle* behavior are then unnecessary and can be avoided.

We consider the behavior of the macrosystem at a *specified configuration* of the nuclei of the '*central species*' the remainder of the system being called the '*medium*.' Each of these central species s is treated as a separate particle, and the medium is treated as one giant particle M. Each central species consists of any *molecule* or *electrode* undergoing a transformation in the phenomenon under investigation. If the macrosystem is analyzed *at specified positions of the other ions*, the central species will be defined *to include* these ions as well. The subscript i will be used to denote both s and M.

Examples of central species are a pair of reacting molecules, an electrochemically active ion, an electrode, a fluorescing molecule, and so on. When one of the central species is an ion, this ion plus its inner coordination shell will be regarded as a single particle. When one of the central species is an electrode, the electrode plus any strongly bound adsorbed ions will be also treated as one particle.

In this paper [5], we are primarily concerned with the *nonequilibrium statistically distributed configurations* which can arise in a system. We explore in some detail the behavior of the medium *outside* of these inner coordination shells of the central species. The vibrational motion inside the latter shells can be handled by more standard methods. For this reason, we consider the behavior of the system for any *given* value of the coordinates *in these coordination shells* as well as for any *given positional* and *orientational* coordinates *of the central species*. ... Let $\boldsymbol{\tau}^f$ denote the coordinates held *fixed*[(22)] and $\boldsymbol{\tau}$ denote the other coordinates. We treat the $\boldsymbol{\tau}$ coordinates classically" [5].

Notice that the medium is considered as a single giant molecule which can have intraparticle coordinates and the $\boldsymbol{\tau}$ coordinates comprise all of the coordinates of the medium and the interparticle coordinates between the central particles and the medium particles.

"For any given $\boldsymbol{\tau}^f$ and $\boldsymbol{\tau}$, the electronic energy U_{tot} is the sum of an *intraparticle* and of an *interparticle* one. The former is the total electronic energy [24, p. 3] when the particles are isolated from each other and have the intraparticle configuration contained in this value of $(\boldsymbol{\tau}^f, \boldsymbol{\tau})$.

The second term is the change on bringing the particles together, the interparticle configuration specified by this $(\boldsymbol{\tau}^f, \boldsymbol{\tau})$. This electronic energy serves, of course, as a potential energy function for nuclear motion, in the adiabatic approximation, in the standard way [5, Note 9].

We may therefore write U_{tot} as

$$U_{tot} = U + U^f$$

where U is the sum of the*intraparticle* term *for the medium* and of all *interparticle* terms. U^f is the intraparticle term *for the central species* and is a function of $\boldsymbol{\tau}^f$ alone.

The interparticle electronic energy itself will be taken to be *the sum* of *polar* and '*nonpolar*'[(23)] or, more precisely, '*electron correlation*' terms, the latter including repulsion and London dispersion energies of interaction *between* the particles [2]. We then expand the *interparticle* energy in terms of the ρ_i^0's, the charge densities on the isolated particles at the intraparticle configuration given by $(\boldsymbol{\tau}^f, \boldsymbol{\tau})$, retaining terms up to *second powers* in the ρ_i^0's [5, Note 9a]. These terms contain only zero and second powers of the ρ_i^0's [5, Notes 7(a) and 7(b)]. They may be conveniently classified further accordingly as they contain zero, first, or second powers *of* the central species, ρ_s^0, namely as $U(0)$, $U(1)$, and $U(2)$, respectively [5]:

$$U = U(0) + U(1) + U(2).\text{"} \tag{5.53}$$

Note that M. expands U in terms of ρ_i^0's which are ρ_s^0 for the central species and ρ_M^0 for the medium. The ρ_i^0's are supposed to appear in U as products, the numbers 0, 1, and 2 referring to the exponents of the ρ_s^0's and the sum of the exponents with which ρ_s^0 and ρ_M^0 appear in the three terms in U is 2. This means that in $U(2)$ there is the factor $(\rho_s^0)^2$, the factor $\rho_s^0 \rho_M^0$ is in $U(1)$, and the factor $(\rho_M^0)^2$ in $U(0)$. So that:

"$U(2)$ must be independent of ρ_M^0, $U(1)$ must contain first powers of ρ_M^0 and hence vanish when either ρ_M^0 or ρ_s^0 vanishes, and $U(0)$ contains both *zero* and second powers of ρ_M^0. The part of $U(0)$ which has zero powers of ρ_M^0 consists of the intraparticle term for M and of the *interparticle nonpolar* term. Each term in Eq. (5.53) depends not only on $\boldsymbol{\tau}$ but also on $\boldsymbol{\tau}^f$."

The earlier partition of U separates the polar interactions between central species and between central and noncentral species — which is $U(2)$ and in $U(1)$ respectively — from the nonpolar interactions among the same species and the polar interactions between noncentral species, which is left in $U(0)$.

5.23. Statistical Mechanics of Equilibrium and Nonequilibrium Polarization Systems

"In the systems considered here [5], the *configurational distribution of the 'medium'* may or may not be the equilibrium one for the *specified electronic state of the central species*." As in Refs. [7, 3], "these systems are called equilibrium and nonequilibrium polarization systems, respectively.

For a medium having an equilibrium configurational distribution, the *configurational contribution of the $\boldsymbol{\tau}$ coordinates* to the free energy of the system at the *specified* $\boldsymbol{\tau}^f$ is:

$$F_c = -kT \ln \int \exp\left(\frac{-U}{kT}\right) d\boldsymbol{\tau}, \tag{5.54}$$

where, for brevity, we have omitted the usual product of factorials that takes care of indistinguishability of like molecules. (This product cancels in all of the free-energy differences in this paper [5].)"

Note that the *total* configurational contribution to the free energy would be given by Eq. (5.53) considering in it not only U but $U + U^f$. This point will be taken up shortly at the end of this chapter.

"The *polar contribution* to F_c is defined as F_c *minus* its value $F_c(0)$ when all ρ_s^0 vanish. Denoting this polar term by F we thus obtain from Eqs. (5.53) and (5.54):

$$F = F_c - F_c^0 = -kT \ln\left(\int \exp\left(\frac{-U}{kT}\right) d\boldsymbol{\tau} \Big/ \int \exp\left[\frac{-U(0)}{kT}\right] d\boldsymbol{\tau}\right) \tag{5.55}$$

We also require the *configurational free energy* of a system in which *the medium* responds to the ρ_s^0 of the central species via an *electronic polarizability* rather than via an adjustment of its nuclear configurational distribution. The *distribution function for the $\boldsymbol{\tau}$ coordinates* of this 'equilibrium *optical* polarization system' is the same as the one in which the ρ_s^0 vanish, and so is proportional to $\exp[-U(0)/kT]$. The *polar contribution F^{op} to the free energy* of this system is the *polar energy of interaction* of the polar species *with each other and with the medium*, $U(1) + U(2)$, averaged with respect to this distribution:

$$F^{op} = \int [U(1) + U(2)] \times \exp\left[\frac{-U(0)}{kT}\right] d\boldsymbol{\tau} \Big/\!\!\int \exp\left[\frac{-U(0)}{kT}\right] d\boldsymbol{\tau}$$

Considering *nonequilibrium polarization systems* next, those that we have investigated thus far have a *configurational distribution functions* appropriate to *some* charge distribution *on the central species*, but not to the existing one. However, the *electronic* polarization of the system, which has a very short relaxation time, is that which is appropriate *to the existing* ρ_s^0*'s and to the existing orientation–atomic polarization of the medium.* We use a subscript 0 to denote the properties of the state for which the configurational distribution *of the noncentral species would be the equilibrium one* at the *specified* configuration of the *central species.* This state was called the 'e.e.s.' [1].

An example of a nonequilibrium polarization system is one in which a central species has just absorbed or emitted light: the configurational distribution of the *surrounding* molecules is appropriate to the molecular state *just before but not after* the transition. Again, *in the transition state* of ET processes, the configurational distribution of molecules *near the reacting pair* (or near a reacting ion and electrode) is appropriate to some *hypothetical* charge distribution, *a compromise between that of the reactants and that of the products*, but not to the existing one [1].

The $\boldsymbol{\tau}$ contribution to the free-energy difference $F_c^{non} - F_{c0}$ between the nonequilibrium state and *its e.e.s.* equals the energy difference, since *both have the same configurational distribution* and, thereby, the same entropy:

$$F_c^{non} - F_{c0} = \int (U_1 - U_0) \times \exp\left(\frac{-U_0}{kT}\right) d\boldsymbol{\tau} \Big/ \int \exp\left(\frac{-U_0}{kT}\right) d\boldsymbol{\tau}, \tag{5.56}$$

where U_1 and U_0 denote the potential energy U when the ρ_i^0's are those in the nonequilibrium polarization system, $\rho_{i_1}^0$, say, and those in the e.e.s., $\rho_{i_0}^0$, respectively."

Note that the above "energy difference" is the *internal energy* difference of the systems. Moreover, the nonequilibrium system and *its* e.e.s. differ by theirs ρ_s^0's but the charge densities of the noncentral particles are the same, that is, $\rho_{M_1}^0 = \rho_{M_0}^0$.

"Because of the assumed additivity of the interparticle *polar* and *nonpolar* terms, $U_1 - U_0$ in Eq. (5.56) can be written as the sum of two differences, one due to *the difference in polar properties* of the central species, $U_1(1) + U_1(2) - U_0(1) - U_0(2)$, the other due to *the difference in nonpolar properties,* $U_1(0) - U_0(0)$" [5]. That is:

$$U_1 - U_0 = (U_1(1) + U_1(2) - U_0(1) - U_0(2)) + U_1(0) - U_0(0)$$

"We may define the *polar contribution* to Eq. (5.56) to be given by Eq. (5.57), write it as $F^{non} - F_0$, and thereby define F^{non}.

$$F^{non} - F_0 = \langle U_1(1) + U_1(2) - U_0(1) - U_0(2)\rangle_0 \tag{5.57}$$

where [5]

$$\langle f\rangle_0 = \int f \exp\left(\frac{-U_0}{kT}\right) d\tau \bigg/ \int \exp\left(\frac{-U_0}{kT}\right) d\tau."$$

5.24. Dielectrically Unsaturated Systems

Partial Dielectric Saturation

"Dielectric unsaturation(24) is characterized by a *linear response of the medium* to the ρ_i^0 *of the central species*. If the coordinates of any *ions in the medium* are included in $\boldsymbol{\tau}$, then the introduction of a *linear response approximation* will give rise not only to dielectric unsaturation but also to a Debye–Hückel approximation for salt effects as well. To avoid the second of these, we hold the ionic coordinates fixed by including them among the $\boldsymbol{\tau}^f$ (i.e., every ion is regarded as a central species). This constraint need not be imposed in the milder approximation of partial dielectric unsaturation.

Because of the linear response(25) of the medium to the ρ_s^0 's, the polar contribution to the free energy at given $\boldsymbol{\tau}^f$ is quadratic in the ρ_s^0 's" [5].

Just consider the $(\rho_s^0)^2$ in $U(2)$.

"Accordingly, this approximation may be introduced by multiplying each ρ_s^0 by a parameter λ, then expanding the free energy in a Taylor series in λ about $\lambda = 0$ retaining terms up to λ^2, and finally setting $\lambda = 1$. [In Eq. (5.53), the $U(1)$ and $U(2)$ are therefore multiplied by λ and λ^2, respectively.] In this way, Eqs. (5.54) and (5.55) yield [5]

$$F = \langle U(1) + U(2)\rangle - (1/2kT)[\langle U(1)^2\rangle - \langle U(1)\rangle^2] \qquad (5.58)$$

$$F^{op} = \langle U(1) + U(2)\rangle \qquad (5.59)$$

where

$$\langle f\rangle = \int f \exp\left[\frac{-U(0)}{kT}\right] d\boldsymbol{\tau} \Big/ \int \exp\left[\frac{-U(0)}{kT}\right] d\boldsymbol{\tau}.\text{''} \qquad (5.60)$$

Using the linear response approximation, M. demonstrates in Ref. [5] that the equation

$$F^{non} - F_1 = F^{op}_{1-0} - F_{1-0}$$

which was derived on the basis of the dielectric continuum approximation, can also be derived in the framework of a noncontinuum, "particles" description of the entities involved in the theory of ET. In the dielectric continuum approximation, the F's were considered as free energies of charging written in terms of electric potentials and charge densities. In the statistical mechanical treatment of Ref. [5], they are written in terms of potential energy functions and distribution functions.

"*Partial dielectric saturation* associated with the coordinates $\boldsymbol{\tau}$ can occur outside the inner coordination shell of any *highly charged ion*. The inner coordination itself is largely 'saturated,' but it depends on $\boldsymbol{\tau}^f$ and so is unaffected by any unsaturation approximation" [5].

Even for the case of partial dielectric saturation M. derives in Ref. [5], on the basis of statistical mechanics, Eq. (47) that was previously derived on the basis of continuum methods.

5.25. Relation of the τ—to Total Configurational Contribution to Free Energy

"The total configurational contribution to the free energy of equilibrium systems is:

$$-kT \ln \int \exp\left[\frac{-(U^f + U)}{kT}\right] d\tau d\tau^f$$

$$= -kT \ln \int \exp\left[\frac{-\left(U^f + F_c\right)}{kT}\right] d\tau^f$$

where F_c is given by Eq. (5.54) [5].

NOTES

1. **M**: "The abscissas of the profiles refer to instantaneous values of the coordinates."
2. Ref. 1 p. 22 bottom: "Fluctuations of this nature involve simultaneous changes in orientation, position, and atomic polarizations of the solvent molecules, in internuclear distances in the coordination shell, in relative motion of the reactants and in configuration of the ionic atmosphere."

   Q: Doesn't it sound almost impossible that all these coordinates move simultaneously in the right way to the configurations of the TS?

   M: Oh, no, they don't have to move in the right direction, they're fluctuating all over, they're fluctuating like mad over all sorts of wrong directions, but then the idea is that in the fluctuation you want to cross an $N - 1$ dimensional surface, that still gives you a wide latitude of fluctuations possible, it's just one dimension less than the regular thing. You're trying to reach a point on that $N - 1$ dimensional surface. As long as you reach anyplace on that surface it's OK, so there's a huge number of configurations that're OK. For a single point it would be impossible. It would be a miracle and miracles are better left out of this paper.

3. **M**: In the "zero-interaction system," the surfaces are not distorted by the electronic coupling, those are the surfaces we're using in the calculation of the free energy. You know, we don't make any correction, although one could, for a small resonance energy. The zero-interaction system is compared with the actual system. The main thing is that we want to use the diabatic free energies, *diabatic free energies are the free energies of the zero-interaction system.*
4. In **M30**, Part IV (1960), you introduce an m, in the identity

$$\mathcal{E}_k^{\ddagger} = \mathcal{E}_k + m(\mathcal{E}_k - \mathcal{E}_k^p) - m\Delta\mathcal{E} \tag{5.3}$$

and then, surprisingly, this m, "any constant" in the identity above, happens now to be nothing but our acquaintance of old, the m in $\Delta F_0^{\ddagger} = w + m^2\lambda$. Are the two m's the same? From what you write here apparently the old one is obtained out of the condition that the difference of the activation free energies for reactants and product is equal to ΔG^0, that is, from a purely thermodynamic equation.

M: They are not, if you are not at the intersection surface then m will not be that which appears in Eq. $\Delta F_0^{\ddagger} = w + m^2\lambda$. You get the usual m from $(2m + 1)\lambda =$ etc. If you're not satisfying that, then m can be regarded as *a flexible parameter.* But remember that this paper was really preparation for the 1965 paper, for it I had much more time, then entered more detail, and so on.

M: "Apparently the old one is obtained out of the condition that the difference of the activation free energies for reactants and product is equal to ΔG^0."
That's when you are on the intersection surface, yes.

5. **M**: I introduced those charges because that is the way of generating a nonequilibrium polarization. In other words, they describe the polarization but those charges are *fictitious charges.*

6. Ref. 1 p. 25: "(iv) EQUIVALENT EQUILIBRIUM DISTRIBUTION

It is instructive, for evaluating $\Delta F^{\ddagger}$, to first compare the transition state, in which $(-\mathcal{E}_k^{\neq}/kT)$ is *integrated over the intersection surface*, with a state in which *this factor is integrated over all space.* We shall term the configurational distribution of the latter state the 'e.e.d.' Comparison of Eq. (5.9) with (5.6) and of Eq. (5.10) with (5.7)

$$\mathcal{E}_k^{\neq} = \mathcal{E}_{k^i}^{+} + \mathcal{E}_{k^o}^{+} + C \tag{5.9}$$

$$\mathcal{E}_k = \mathcal{E}_{k^i} + \mathcal{E}_{k^o} \tag{5.6}$$

$$\mathcal{E}_{k^o}^{+} = \Omega_{k^o} + \sum_j [e_j^{+}\phi_j^{+}/2 - \mu_j \cdot \mathbf{D}_j^{+} - \alpha_j \mathbf{E}_j^{+} \cdot (\mathbf{D}_j^{+} + \mathbf{D}_{\mu j})/2] \tag{5.10}$$

$$\mathcal{E}_{k^o} = \Omega_{k^o} + \sum_j [e_j\phi_j/2 - \mu_j \cdot \mathbf{D}_j - \alpha_j \mathbf{E}_j \cdot (\mathbf{D}_j + \mathbf{D}_{\mu j})/2] \tag{5.7}$$

reveals that the e.e.d. is one which would be obtained in a corresponding *equilibrium* system in which the charges on the two central ions were e_n^{+}, that is, $e_n + m(e_n - e_n^{p})$, $(n = 1, 2)$, and which had $\mathcal{E}_{k^i}^{+}$ as a potential function for the coordination shell.

The transition state of the zero-interaction system differs from this system in only two respects: the charges of the two central ions are those of the reactants, and it has one less dimension of freedom than an N-dimensional system."

Q: (1) How can you integrate over all space the factor $(-\mathcal{E}_k^{\neq}/kT)$ where the atomic configuration k is belonging to the intersection surface? From what you write in Appendix 1 you vary the k's shifting the P surface with respect to the R surface and so you change the crossing points. Is it so?

(2) Please check my understanding: if the charges on the ions are changed from those of the reactants to those of the e.e.d., the potential

energy curves will in general differ. To a certain atomic configuration, k different energies will correspond when the charges on the ions are different. But near the TS $\mathbf{P}_u$ will by definition be the same for systems with the charge distribution of the reactants and systems with the charge distribution of the e.e.d. This means that the atomic configuration near the TS will be similar for systems with the reactants and the e.e.d. charge distributions. Is it so? Apparently not, why so? because in Appendix 1 you only shift the P surface with respect to the R one with their shapes remaining unchanged.

(3) In [1] the starting point of your theory is the equation:

$$\mathcal{E}_k^{\neq} = \mathcal{E}_k + m(\mathcal{E}_k - \mathcal{E}_k^p) - m\Delta\mathcal{E} \qquad (5.3)$$

The equation was derived for the case in which "The potential energy of the reactants on the intersection surface, $\mathcal{E}_k^{\neq}$ say, equals $\mathcal{E}_k$" and "where m is any constant." I have two sub-subquestions:

(i) Is m any constant even in the e.e.d. if we don't require the e.e.d. to be centered on the intersection surface?
(ii) How do we know that the e.e.d. is really an equilibrium one?

M: We shall term the configurational distribution of the latter state the "e.e.d.,"... in other words the e.e.d. is centered on that intersection surface but in a sort of like a stable thing, you know, on the intersection surface in, say, *at its lowest point, its lowest part* and is just fluctuating hypothetically *perpendicular to that* $(N-1)$*-dimensional surface.* The actual system of course is unstable, the reactants go one way and the products go the other way, if it existed on that surface. So this is a *hypothetical system* which has *a distribution identical with the TS distribution* but is different in that we got *at a coordinate perpendicular to that which is stable,* it's a fictitious thing.

Sort of like a *harmonic potential, that is existing perpendicular to that* $(N-1)$*-dimensional surface.*

(1) Well, in your integration, let's say, you have say 10^{23} variables? All right, you have $10^{23}-1$ in the integration if you're integrating over

the transition state. If you're integrating over the e.e.d., you put in a fictitious harmonic oscillator, let's say, situated on the intersection surface and then you can integrate over it easily.

"From what you write in Appendix 1 you vary the k's shifting the P surface with respect to the R surface and so you change the crossing points. Is it so?"

Yes, that's right, and *I use that vertical shift as a kind of reaction coordinate, so that, for example, if you start off at the reactants then shift a products curve . . . consider a value of the shift coordinate* Γ *at which the product curve intersects the reactants curve, at that crossing point you have the reactants, now slowly raise the products curve and eventually raise at a point where exists the transition state, the reaction system, that would give you the value of the coordinate* Γ *at the transitino state.* The idea was evolving there, I was doing in a way much less clear, than I did in later papers, on choosing it as a reaction coordinate. . . there was an evolution in my choice of reaction coordinate so that's really a beginning of it. . . but I state the theory more carefully in later papers.

The reaction coordinate is a vertical difference between the two potential energy surfaces,all points that have the same vertical difference have the same value of the reaction coordinate.

The Γ *is a vertical shift and is a coordinate* for the $10^{23} - 1$ dimensional space. It's a global coordinate. That really for me was sort of a breakthrough. I learned much later that somebody by the name of Lax, Norbert Lax, in 1952 had used something like that for the treatment not of reactions but of some spectral properties, of harmonic oscillators. I learned it much later but it was restricted, I mean he considered harmonic oscillators. This is really general, doesn't matter if it is oscillators, rotators, anything."

(2) "Near the TS $\mathbf{P}_u$ will by definition be the same for systems with the charge distribution of the reactants and systems with the charge distribution of the e.e.d."

Yes, because they're exactly the same system except for 1 dimension.

"This means that the atomic configuration near the TS will be similar for systems with the reactants and the e.e.d. charge distributions. Is it so?"

Yes, as long as you are on the intersection surface they will be the same. When you're out the intersection surface then the TS doesn't exist there.

Q: Yes, but what troubles me is that in Appendix 1 you shift the P surface with respect to the R and the shapes of the R and P curves remain the same.

Yes, that's right... *all I'm doing is I'm trying to get intersection at different points, as a reaction coordinate, so the initial intersection, although I may not say this, would be intersecting the bottom of the reactants potential energy surface, then I would gradually raise it to where it really is in the transition state, and the shift coordinate is the reaction coordinate.*

Q: But if you use different charges on the ions, don't you have different shapes of the curves?

M: Oh, I see what you're saying, *the energy would be different, polarization would be different. Near the intersection I realize that they are the same, by definition, but away from the intersection?* Well, it may be that if you look at the intersection, as you move along the reaction coordinate, in other words with the intersections occurring at different points, that the distribution of configurations on those intersections has very different arrangements as you're going along there, they don't have the same solvent arrangement, all of them are changing, as you go from one crossing to another crossing raising the curve, in that new crossing all of the orientations can be very different from how they were in the old.

Q: Yes, OK, but would the two parabolas remain the same?

M: They are not parabolas, they're surfaces in many dimensions, and you're occupying different regions and different parts, so all is changing.

Q: OK, because you just shift P with respect to R.

M: Think of to the other $10^{23} - 1$ dimensions, that cross section there could look totally different from the starting cross section. In other words, the shape of the potential energy curve in one part of the many dimensional surface can be very different from the shape in another part.

Q: So this is just a profile that has not to be taken too seriously...

M: Yes, well... that's right in the sense that it's a cross section, and so it's a cross section though of something that has 10^{23} dimensions, and *all they have in common is that they have the same vertical distance, the same value of the reaction coordinate*, but if you look at different parts of that profile, of that cross section, properties would look very different, just the energy difference is the same, but the orientations of the solvent molecules be very different.

(3) (i) "m determines just where the e.e.d. exists, just where that intersection occurs."

For different values of m, you get different values of the intersection. In other words, think of raising the products curve vertically, you get a set of different ... points, each point would correspond to a different value of m.

(3) (ii) *By definition. You're assuming that in transition state theory with respect to every coordinate but for reaction coordinate the distribution in the transition state is an equilibrium distribution.*

Q: The book of Eyring has been your Bible.

M: Oh, yes, absolutely, in fact he used the term activated complex and for the longest time I used activated complex until I finally went over to the other Polanyi's term, TS, which is what everybody uses nowadays.

Q: But aren't they really two slightly different concepts in the sense that one concept refers to a state and one refers to an objective molecule at the TS?

M: Yes, probably neither terminology is great because *the state doesn't exist, is a fictitious state and a fictitious arrangement somehow.* Yes, . . . *it's an arrangement, is a particular arrangement in the* 10^{23} *dimensional space, it's a hypersurface.* So, that's not fictitious but *typically you don't observe it.*

Q: But aren't there modern techniques with lasers such that it is possible to observe the TS?

M: Neumark, at Berkeley took a system and photoexcited the electron from it and the actual state that he formed then was a neutral system which actually was at the TS for a different reaction, so in a sense he observed it, but. . . I think there're few examples of it, those are really the only examples. . . one observed that but I think that was really very indirect. By the way, Polanyi was another one who spoke about observing transition states, but I think that too was a stretch of the imagination.

Q: But nonetheless those atoms must go through some kind of intermediate. . .

M: Oh, yes, they do, the only question is what's your definition of observing it. And the one who came closer to it I think it was Neumark. Because by photoexciting an electron he left himself on top of an unstable situation, which is a transition state.

7. Ref. 1 p. 25 bottom: "For exact evaluation of the surface integral, we should examine in detail the *motion along the surface*, for example, by examining the atomic motion *normal* to it, that is, the *reaction*

coordinate... By a suitable choice of m ... the e.e.d. is made to center on the intersection surface, *and thereby to die away fairly rapidly along the normal.*"

Q: Please explain this last point: how do we recognize from your formalism this dying away fairly rapidly? Please look at my Fig. 1* where on panel A I have shown how qualitatively the energies of systems with the atomic configurations k differ by ΔE when the charges are those of the reactants and when they are those of the e.e.d. depending on the distance between the configuration k and the one on the TS along the coordinate γ perpendicular to the TS intersection surface. In panel B, I show the probability of the configurations of the systems with the charge of the reactants close to the configuration of the system with the e.e.d. charge as a function of the distance of the configurations along the coordinate γ perpendicular to the TS intersection surface. Is there some truth in this intuitive figure?

M: What dies away is the population in the distribution, in the e.e.d., it dies when you get away from the surface because the fictitious harmonic potential is very high. Yes, that's right, see, on the upper figure you have the fictitious potential, perpendicular to the intersection surface, you have it harmonic, which is localized in the system, the population dies away from there, yes, you were fine, you fine, very good, congratulations, that's a good figure to put in the book.

Q: In panel B, I showed the probability of the configuration of the system with the charge of the reactants being similar with the configuration of the system with the e.e.d. charge as a function of the difference Δk of the configurations.

M: Yes, in the e.e.d., if you look at the direction perpendicular to the TS, I imagine that except in one dimension the shape of the potential energy curve is identical as you move along that, you know, the e.e.d. as you move along in a one-dimensional surface, and *the only difference is in the coordinate perpendicular to the TS, to the* $(N-1)$-*dimensional surface*, then otherwise the configurations are identical, and so *the fictitious system differs from the TS only in introducing this coordinate that's perpendicular and making the system stable with respect to that coordinate. This is a very important point* yes, *you made it stable by using that harmonic potential centered in the upper panel, yes, that's right.*

Q: And also panel B is OK?

M: Yes, panel B is all right because if you have that, thermally you'll have a Gaussian distribution.

8. Ref. 1, p. 26: "We next hold all of these relative coordinates fixed within two fairly large spheres, one about each central ion (large enough so that the long range ion–ion–solvent interactions are negligible on their surfaces). We then integrate over the coordinates of the center of gravity of these two ions and overall orientations of their lines of centers, translating and rotating, respectively, the entire system within the large spheres to ensure constancy of the important relative coordinates during integration.

$$k_r = Z \exp[-(\Delta F_i^{\neq} + \Delta F_o^{\neq})/kT] \tag{5.12}$$

Q: (1) Are the spheres containing most of the solvent molecules and of ionic atmosphere defining the reacting system? Do you know the order of magnitude of the molecules contained in such spheres or their radius?

M: "Certainly there's a correlation function which sort of reaches unity, you know, if one plots a density correlation functions, in statistical mechanics, then you see a set of wiggles, and sort of oscillating about the value at large distances, and I suppose that the correlation function dies off after four or five shells, . . . layers of solvation, but with the ions of course the effects are longer range, and there of course the ions feel each other, at finite concentrations, and there I got to see, in terms of correlations, at what point the ionic effect sort of dies off, and I don't have to worry about it, but in computer simulations, they were worried about pretty long range interactions.

The theory extends well beyond those spheres because you want any Coulombic effects to be included and they die off very slowly, so you should really consider a pair of reactants, an encounter complex, in, say, a mole of water."

(2) Do you get the collision frequency Z by the above integrations? I remind you that in your first paper on electron transfer you obtained Z using partition functions:

$$\frac{k_1}{k_{-1}} \cong \frac{(2\pi(m_1+m_2)kT/h^2)^{\frac{3}{2}}(8\pi^2\mu R^2 kT/h^2)\exp(-\Delta F^*/kT)}{(2\pi m_1 kT/h^2)^{\frac{3}{2}}(2\pi m_2 kT/h^2)^{\frac{3}{2}}}$$

$$= \frac{h}{kT}(8\pi kT/\mu)^{\frac{3}{2}}$$

$$k_1 = (8\pi kT/\mu)^{\frac{1}{2}} R^2 \exp(-\Delta F^*/kT)$$

Are the two methods equivalent? If so, please explain.

M: Yes, that's right, in other words if you take the line of centers of the two reactants, and now imagine that line of centers can have any orientation, ... translate all and do the integration over all the other coordinates afterward, then you'll get a $4\pi R^2$, where R is sort of the collision parameter, collision distance... and so that will give you the geometric factor that comes into the collision frequency.

Q: In your first paper on ET, you obtained Z using partition functions.

M: Yes, and in a sense this is using partition functions, it's just that now the partition function is a 10^{23}-dimensional one and in the first part of the integration I'm holding all relative coordinates constant and I'm just rotating the axes joining the line of centers, and I'm integrating over each first, and then integrate over the other coordinates then. Since we are using dielectric continuum theory before, we didn't have to say about the 10^{23} coordinates....

Q: Are the two methods equivalent?

M: One is dielectric continuum theory and the other one is statistical mechanics.

Q: But they give the same Z.

M: Well, one gives an expression in terms of dielectric properties and the other gives an expression in terms of statistical mechanical properties. In other words, when you do the statistical mechanics only later do I then try to identify some of the results that we obtained using dielectric properties. Here I use the statistical mechanical method.

Q: But we are talking about the Z.

M: Well, the Z part, there in the dielectric continuum theory I treated the few coordinates involving Z entirely independently of the other coordinates, essentially.

Q: Like in the gas phase.

M: Like in the gas phase except that at some point later on I used a pair distribution function to correct the gas phase value, whereas here I'm operating in two stages, I'm holding all the other coordinates fixed before integrating over them and I'm just rotating the whole system because that gives me the line of centers orientations in the TS.

9. Following: "We obtain after some cancellation:

$$k_r = Z\exp[-(\Delta F_i^{\ddagger} + \Delta F_o^{\ddagger}(r))/kT], \tag{5.12}$$

where

$$\Delta F_i^{\ddagger} = \Delta \mathcal{E}_i^{\ddagger} - kT\ln Q_{vib}^{\ddagger}/Q_{vib} \tag{5.13}$$

$$\Delta F_o^{\ddagger}(r) + F_o = F_o^{\ddagger} = -kT\ln\int..\int\exp(-\mathcal{E}_{ko}^{\ddagger}/kT)dk_o'." \tag{5.14}$$

In Ref. 1 p. 23 we find $k_r = (kT/h)\exp(-\Delta F^{\ddagger}/kT)$ (5.1)

M: "F_0 is the starting free energy, $Q_{vib}^{\ddagger}$ is the vibrational partition function for the inner coordinates in the TS obtained by summing over the all v's, Q_{vib} is the analogous partition function for the reactants and dk_o' is the configurational volume element of the $N_o - 6$ outer coordinates obtained from the total N_o subtracting the coordinates of the center of gravity of the couple of central ions and their rotational coordinates, and r is the distance of closest approach."

Q: How do you go from Eq. (5.1) with the preexponential factor (kT/h) to Eq. (5.12) with the preexponential factor Z?

M: (1) Yes, introducing into the ΔF^* a partition function related to the rotation of the activated complex and the translation of the reactants, that converts from one to the other. From Eqs. (5.12) to (5.1).

Q: The partition function converts the kT/h to a Z.

M: Yes. . . . terms in the free energy.

Q: In the 1975 paper **M118**, ("Electron Transfer in Homogeneous and Heterogeneous Systems," in *Physical and Chemical Sciences Report*) you write on p. 480: "The net result for an activated complex theory expression for the rate constant k_{rate} is:

$$k_{rate} = \kappa(kT/h)\exp(-\Delta G^{\ddagger}/kT)\text{"} \qquad (3.1)$$

on p. 486: "The contribution of the vibrations of the reactants to $\Delta G^{\ddagger}$ was included in a straightforward way. . . as was the free energy change arising from the formation of a collision pair. (The latter proves to be $-kT \ln Zh/kT\ldots$) . . . One obtains Eq. (4.1) for bimolecular reactions in solution. . .

$$k_{rate} = \kappa Z\exp(-\Delta G^{\ddagger}/kT) \qquad (4.1)$$

ΔG^* is the contribution to $\Delta G^{\ddagger}$ arising from the solvent polarization. . . . "

Q: I believe in Eq. (4.1) we should substitute a * for a ‡.

M: One of them should be starred. *The one in Eq.* (4.1) *should be starred.* Yes, that's right, that should be starred. Absolutely. In other words, $\Delta G^{\ddagger}$ *equals a number of partition functions plus* ΔG^*. Yes. You found a mistake in notation and in Eq. (4.1) there should be ΔG^*. In Eq. (4.1), there is a misprint, it should be $k_{rate} = \kappa Z\exp(-\Delta G^*/kT)$.

Q: I'm reading your papers rather carefully....

M: And I'm glad you do. Maybe when I'll get back to them I'll read them carefully too.

Q: Which one is the difference between $\Delta G^{\ddagger}$ and ΔG^{*}?

M: ΔG^{*} *has a different meaning, some of the properties in* $\Delta G^{\ddagger}$ *had been in the partition function in there, certain ones had been extracted, and one is left with the* ΔG^{*}. Equations (5.70) and (4.1) are correct, no mistakes.

Q: In your **M53** 1965 paper, you write on p. 685, 2nd column bottom:

$$k_{\text{bi}} = \kappa \rho Z_{\text{bi}} \exp(-w^{r}/kT) \exp[-\Delta F^{*}(R)/kT] \qquad (31)$$

In this case, $\exp(-w^{r}/kT)$ is separated from $\exp[-\Delta F^{*}(R)/kT]$.

Q: w^{r} is apparently contained in ΔG^{*}.

M: Yes, that's right, contained in the ΔG^{*}in the Eq. (4.1), that's right.

Q: But if you look at your Eq. (31) which is taken from your paper, your great paper of 1965...

M: There I used ÄF^{*}, so ÄF^{*} isn't the same as the ΔG^{*}.

Q: Isn't the same?

M: No, no. ΔG^{*} would really have to include a w^{r}, for example.

Q: So, no contradiction.

M: No, a change of notation.

10. From **M30**, Part IV, p. 27 top and earlier reported in p. 12 in this Chapter: "It follows from §3.4 that (4.2.4) for $F_{0}^{\ddagger}(r)$

$$\Delta F_{o}^{\ddagger} + F_{o} = F_{o}^{\ddagger}(r) = -kT \ln \int .. \int \exp(-\mathcal{E}_{k^{o}}^{\ddagger}/kT) dk'_{o} \quad (4.2.4)$$

Is simply the free energy of a system having the charges of the reactants a distance r apart but a distribution of orientations of solvent molecules and of positions of ions in the ionic atmosphere which would be in equilibrium with the hypothetical charges

$$e_{n} + m(e_{n} - e_{n}^{p}), \quad (n = 1, 2),$$

on the two central ions. It is at this point that we introduce the macroscopic expression for the free energy of this type of nonequilibrium system. We obtain

$$\Delta F_0^{\ddagger} = w + m^2\lambda\text{"} \tag{4.3.3}$$

Following p. 27 bottom: "(iv) Equation for m. The equation for m is obtained by equating the difference between free energies of activation for the forward and reverse reaction to the standard free energy of reaction at the prevailing electrolyte concentration, $\Delta F^{0\prime}$. In the process, we tacitly set the free energy of the reactants on the intersection surface equal to that of the products there... and so satisfy the energy condition (A.1) in Appendix 1...

$$\langle \mathcal{E}_k \rangle_{av} = \langle \mathcal{E}_k^p + \Delta\mathcal{E} \rangle_{av} \tag{A.1}$$

Thereby the e.e.d. is made to center on the intersection surface."

Q: In **M30**, Part IV (1960), you introduce an m, in the identity

$$\mathcal{E}_k^{\ddagger} = \mathcal{E}_k + m(\mathcal{E}_k - \mathcal{E}_k^p) - m\Delta\mathcal{E} \tag{5.3}$$

and then, surprisingly, this m, "any constant" in the identity earlier, happens now to be nothing but our acquaintance of old, the m in Eq. (4.3.3) above. Are the two m's the same? From what you write here apparently the old one is obtained out of the condition that the difference of the activation free energies for reactants and product is equal to ΔG^0, that is, from a purely thermodynamic equation. In the present case, you consider instead something new, that is, the potential energy hypersurface, and you impose the condition of equal average energy on the hypersurface with the energies there averaged following the distribution function obtained from the e.e.d. of potential energies. Is it so?

M: That *av* is average on the intersection surface. That you have at the TS. You can always define a magnitude like in Eq. (3.1.3) anyplace, but if you want it defined at the TS then you have to be set to satisfy Eq. A.1.

11. Following: "PROPERTY OF THE e.e.d.

If the e.e.d. is indeed 'centered' on the intersection surface, a system having the e.e.d. and *the electronic configuration of the reactants* would have the same average energy as a system having the same e.e.d. and *the electronic configuration of the products.* That is, it would satisfy

$$\langle \mathcal{E}_k \rangle_{av} = \langle \mathcal{E}_k^p + \Delta\mathcal{E} \rangle_{av} \tag{A.1}$$

each averaged over the e.e.d. (They would have the same free energy too.)

It is easy to show that such an e.e.d. *exists.* Consider the expression

$$\frac{\int .. \int \exp(-\mathcal{E}_k^{\neq}/kT)\mathcal{E}_k d\tau_k}{\int .. \int \exp(-\mathcal{E}_k^{\neq}/kT)d\tau_k} = \frac{\int .. \int \exp(-\mathcal{E}_k^{\neq}/kT)(\mathcal{E}_k^p + \Delta\mathcal{E})d\tau_k}{\int .. \int \exp(-\mathcal{E}_k^{\neq}/kT)d\tau_k} \tag{A.2}$$

This is the desired property. Thus there is an e.e.d. centered on the intersection surface. *It has the property that the ln term, i.e., the free energy, is a minimum with respect to m.* (4.4.1), the equation in the text for m,

$$-(2m+1)\lambda + \Delta\mathcal{E}_i^{\neq} - \Delta\mathcal{E}_i^{p\neq} = \Delta F^{0\prime} - \Delta F_{vib}^{0} + w^p - w \tag{4.4.1}$$

satisfies the above equation."

Q: (1) (i) For every atomic configuration *k on the intersection surface* we have that

$$\mathcal{E}_k = \mathcal{E}_k^p + \Delta\mathcal{E} \tag{5.2}$$

From that you deduce Eq. (A.1). Do the averaged energies represent the energy of the activated complex X^* of your first paper?

(ii) Whereas you have considered earlier an e.e.d. in equilibrium with the *fictitious* charges $e_n + m(e_n - e_n^p)$, you consider now two e.e.d.'s, with real charges, in one in which the charges are those of the reactants and in the other the charges are those of the products. Is it so? We have now then three e.e.d.'s, one with the electronic configuration of the reactants, one with that of the products and one with the fictitious intermediate electronic configuration.

(2) Please explain why the free energy is a minimum with respect to m, which defines the e.e.d. with the fictitious charges.

M: (1) (i) Ehm... well, remember this is a statistical mechanical distribution, while the other is dielectric continuum theory. But yes, the average energy of the reactants on the TS or in the e.e.d., the average energy is that of the products, yes.

(ii) No, it will be the same e.e.d., it will be the same e.e.d., ... in every respect except the charges of the reactants and the charges of the products. But the actual distribution of coordinates will be identical for the two, so the e.e.d. is identical, yes, the distribution of distances, the distribution of angles, all, just the only difference is the charges on the reactants versus the charges on the products, but independently of that every coordinate is identical in the two e.e.d.'s, it's one e.e.d.

(2) Oh... different values of m will correspond to different intersections, but there's only... I have to think about that. It's a good question, let me just think for a moment... you see, you change m by shifting one surface with respect to the other *but only one of those shifts corresponds to the exact intersection*. And of course it is when you shift it above, when you shift it below in one way it is differing from the exact intersection, because it's no longer the exact intersection, and somehow there's a minimization, that I'd have to think that through, I don't know the answer right now, a good question.

12. **M30,** p. 28 bottom: "If the e.e.d. is indeed 'centered' on the intersection surface, a system having the e.e.d. and the electronic configuration of the reactants would have the same average energy as a system having the same e.e.d. and the electronic configuration

of the products. That is, it would satisfy

$$\langle \mathcal{E}_k \rangle_{av} = \langle \mathcal{E}_k^p + \Delta\mathcal{E} \rangle_{av} \tag{A.1}$$

each averaged over the e.e.d. . . .

It is easy to show that such e.e.d. exists. Consider the expression $-kT \ln \int .. \int \exp(-\mathcal{E}_k^{\ddagger}/kT) d\tau_k$ as a function of m and minimize it with respect to m. One obtains immediately, using Eq. (3.1.3):

$$\mathcal{E}_k^{\ddagger} = \mathcal{E}_k + m(\mathcal{E}_k - \mathcal{E}_k^p) - m\Delta\mathcal{E} \tag{3.1.3}$$

$$\frac{\int .. \int \exp(-\mathcal{E}_k^{\neq}/kT)\mathcal{E}_k d\tau_k}{\int .. \int \exp(-\mathcal{E}_k^{\neq}/kT) d\tau_k} = \frac{\int .. \int \exp(-\mathcal{E}_k^{\neq}/kT)(\mathcal{E}_k^p + \Delta\mathcal{E}) d\tau_k}{\int .. \int \exp(-\mathcal{E}_k^{\neq}/kT) d\tau_k} \tag{A.1}$$

Q: Please check my derivation of Eq. (A.1):

I first rewrite Eq. (3.1.3) in the form:

$$\begin{aligned} \mathcal{E}_k^{\ddagger} &= \mathcal{E}_k + m(\mathcal{E}_k - \mathcal{E}_k^p) - m\Delta\mathcal{E} = m\mathcal{E}_k + \mathcal{E}_k - m(\mathcal{E}_k^p + \Delta\mathcal{E}) \\ &\Rightarrow d\mathcal{E}_k^{\ddagger}/dm = \mathcal{E}_k - (\mathcal{E}_k^p + \Delta\mathcal{E}) \Rightarrow \text{ if } d\mathcal{E}_k^{\ddagger}/dm = 0 \\ &\Rightarrow \mathcal{E}_k = (\mathcal{E}_k^p + \Delta\mathcal{E}). \end{aligned}$$

Now for the derivative of $-kT \ln \int .. \int \exp(-\mathcal{E}_k^{\ddagger}/kT) d\tau_k$ with respect to m:

$$\begin{aligned} \frac{d[-kT \ln \int .. \int \exp(-\mathcal{E}_k^{\ddagger}/kT) d\tau_k]}{dm} &= -kT \frac{\int .. \int \frac{d[\exp(-\mathcal{E}_k^{\ddagger}/kT)]}{dm} d\tau_k}{\int .. \int \exp(-\mathcal{E}_k^{\ddagger}/kT) d\tau_k} \\ &= 0 \Longrightarrow \end{aligned} \tag{A.2}$$

M: The e.e.d. exists as a hypothetical quantity. How you get 0? Can you refresh my memory?

Q: I'm deriving Eq. (A.1).

M: OK, all right, let me see. Why do you minimize with respect to m? Why do I write that? OK, fine, I'm delighted, I just assume it's correct. You derived, that's great...

13. Following: "We now examine in more detail the approximation of replacing the $(N-1)$-dimensional surface integral by an N-dimensional volume integral over the e.e.d., N being large. A coordinate system is introduced as follows.

We note first that the intersection of the two potential surfaces in Fig. 1 defines a surface of $(N-1)$ degrees of freedom and thus shifting the potential energy surface of the products vertically by an amount Γ without change of shape, produces a different intersection which defines a new $(N-1)$-dimensional surface parallel to the first one. In this way, a family of parallel surfaces can be generated, each member associated with a particular value of Γ and obeying Eq. (A.2)

$$\mathcal{E}_k = \mathcal{E}_k^p + \Delta\mathcal{E} + \Gamma \tag{A.2}$$

Let σ denote the totality of $(N-1)$ orthogonal curvilinear coordinates defining positions on any given surface and let γ be the coordinate normal to the family of surfaces, so that k in E_k consists of σ and γ. $\Gamma(\gamma)$ is a strictly monotonic function of γ and $\Gamma(0) = 0$.

Consider now the volume integral over the e.e.d.

$$\int \cdots \int \exp(-\mathcal{E}_{\sigma\gamma}^{\neq}/kT)d\sigma d\gamma,$$

where $\mathcal{E}_{\sigma\gamma}^{\neq}$ satisfies Eq. (A.3) (cf. Eq. (3.1.3))

$$\mathcal{E}_k^{\neq} = \mathcal{E}_k + m(\mathcal{E}_k - \mathcal{E}_k^p) - m\Delta\mathcal{E} \tag{3.1.3}$$

and m was selected so that Eq. (A.5) *is satisfied (cf. Eq.* (A.2), *using* (A.3)).

$$\mathcal{E}_{\sigma\gamma}^{\neq} = \mathcal{E}_{\sigma\gamma} + m\Gamma \tag{A.3}$$

$$\int \cdots \int \Gamma \exp(\mathcal{E}_{\sigma\gamma}^{\neq}/kT)d\sigma d\gamma = 0\text{"} \tag{A.4}$$

Q: Please check my derivation of Eqs. (A.3) and (A.4), considering that $k = \sigma\gamma$:

(3.1.3) $\overset{\text{yields}}{\longrightarrow} \mathcal{E}_k^{\ddagger} = \mathcal{E}_k + m\left(\mathcal{E}_k - \mathcal{E}_k^{p} - \Delta\mathcal{E}\right) \Rightarrow$ if $\mathcal{E}_k - \mathcal{E}_k^{p} - \Delta\mathcal{E} \equiv \Gamma$ $\Rightarrow$ (A.4) and $\mathcal{E}_k^{p} + \Delta\mathcal{E} = \mathcal{E}_k - \Gamma$ $\Rightarrow$ substituting $\mathcal{E}_k - \Gamma$ for $\mathcal{E}_k^{p} + \Delta\mathcal{E}$ in the numerator on the RHS of Eq. (A.1)

$$\frac{\int .. \int \exp(-\mathcal{E}_k^{\neq}/kT)\mathcal{E}_k d\tau_k}{\int .. \int \exp(-\mathcal{E}_k^{\neq}/kT) d\tau_k} = \frac{\int .. \int \exp(-\mathcal{E}_k^{\neq}/kT)(\mathcal{E}_k^{p} + \Delta\mathcal{E}) d\tau_k}{\int .. \int \exp(-\mathcal{E}_k^{\neq}/kT) d\tau_k} \tag{A.1}$$

one gets Eq. (A.4).

M: k is $k = \sigma\gamma$. Now, what is σ?

Q: It's the set of coordinates over the surface.

M: Oh yes, OK, all right.

Q: And γ is the perpendicular coordinate

M: Yes, all right, OK... It's best that one derives, by all means.

Q: Why is the integral in Eq. (A.5) equal to zero?

M: I really have to think about that... the most probable value of a Gaussian, the average value is zero. It's that, if Γ is the shift from the true TS potential energy surface, in other words the one that really does intersect it, and Γ is the shift up and down from there, then this is related to there being a Gaussian situated there, so I think it's related to that and so is your previous question. That Γ is a function of what? Is a function of σ?

Q: *It's a function of* γ

M: It's a function of γ, right, oh, I see.... It may be that if you look at the vertical difference of the surfaces, as you move along the $10^{23} - 1$ dimensions, sometimes one's higher than the other and sometimes one is lower than the other, you know, these are many-dimensional surfaces, so just goes higher at one point of the $10^{23} - 1$ and doesn't mean that the surface is always higher than the other ones, it can be lower than the other ones, and the main point is that on the average it's as much higher as it is low, compared with the other, I think that's associated with that minimization business. That otherwise it would tend to be much higher around the average and much lower on the average, the one on the average is sort of as much as high as low then you get zero. You asked good questions, your figures were completely right.

14. Following: "In the vicinity of $\gamma = 0$, Γ equals $\gamma(d\Gamma/d\gamma)_0$, the derivative being nonzero. Equation (A.4) thus becomes

$$\int_{-\infty}^{+\infty} \gamma \exp\left(-F^{\neq}_{(\gamma)}/kT\right) d\gamma = 0 \tag{A.5}$$

where $F^{\neq}_{\gamma}$, the free energy of a system constrained to exist on the surface γ, is given by Eq. (A.6)

$$\exp(-F^{\neq}_{(\gamma)}/kT) = \int \dots \int \exp(-\mathcal{E}^{\neq}_{\sigma\gamma}/kT) d\sigma \tag{A.6}$$

Since Eq. (A.7) is applicable to all T, we infer that $F^{\neq}_{\gamma}$ is an even function of γ and write therefore the Taylor series (A.7), retaining only terms up to γ^2 for physical reasons based on Fig. 1:

$$F^{\neq}_{(\gamma)} = F^{\neq}_{(0)} + (\gamma^2/2!) F''_{(0)} + \cdots . \tag{A.7}$$

M: Well... that statement that I made then there, it is not correct, it is equal to zero only when $\gamma = 0$, not in the vicinity of $\gamma = 0$. It's only when $\gamma = 0$, when $\gamma \neq 0$ of course Γ it doesn't equal zero. So, it's only when $\gamma = 0$ then that is equal zero. So, in other words then I shouldn't have said "in the vicinity," I should have said: when $\gamma = 0$, $\Gamma(0) = 0$, but clearly when $\gamma \neq 0$, then that quantity is not zero.

Q: (1) Why "Since Eq. (A.5) is applicable to all T, we infer that $F^{\neq}_{\gamma}$ is an even function of γ"? Which is the influence of temperature?

(2) Which ones are the physical reasons you refer to in the last sentence?

M: (1) (2) Yes, so, what you have in Eq. (A.5) is γ times some function of temperature, and, regardless of the temperature, the integral still zero, and the temperature sort of means that it can weight one type of condition, it can weight another type of condition, you know, it's all possible conditions, so that means you're covering probably quite a range of some function space, because you have always different functions of temperature, you know, there is one value one temperature, another value of temperature, yet the integrals are always zero, and so when I saw something like that, I figured that, well, even though it may have in some region of temperature some cubic terms, but to cover such a wide range of that would have to be really a quadratic function of temperature. It's not a rigorous argument but... anyway it was because of the fact that that property is so universal that way, that there is such a wide range of the values of the function, because you can vary temperature wildly, that I felt that I didn't see how there could be any cubic terms that would dominate some temperature region or another, otherwise that integral wouldn't be zero for all temperatures. So it is an intuitive argument.

Q: The mathematical reason why $F^{\ddagger}_{(\gamma)}$ is to be an even function is that the integral in Eq. (A.5) can only be equal to zero if $\exp(-F^{\ddagger}_{(\gamma)}/kT)$ is an even function of the variable and the simplest way to have that is if we write $F^{\ddagger}_{(\gamma)}$ as in Eq. (A.7). But which is the physical reason for that? You refer for that to Fig. 1.

M: It is something like... if you regard the $1/kT$ as a Laplace transform parameter, then basically when you equate two integrals you're equating their Laplace transforms, then you take the inverse, to get what's inside, and under certain conditions when two Laplace transforms are equal, then the objects inside are also equal. Yes, it's sort of based upon thinking like: OK, I have a Laplace transform, take the inversion, so the two inversions are equal. So, I think it's related to that. Probably that's what I had in mind, but I didn't say it. In fact, as you can regard $1/kT$ as a Laplace transform parameter, one might be able to phrase it in those terms and in that case it would be more illuminating to the audience. In other words, when two Laplace transforms are equal for all values of the parameter, if the inverse of the Laplace transform is unique, then it means that the objects inside have to be equal. So I think there is really some sort of a theorem there. I write that sort of intuitively, but, you know, I forgot... I did sit for two years of mathematics courses, maybe I remembered more than then I do now, I think. The physical reason in my mind is this: *think of the two parabolas, and now at the point of intersection draw another parabola, I think of that parabola as being the potential energy for the e.e.d.* That's the physical reason. It's really mathematical, that was on my mind.

15. Section 4.1, p. 25 bottom: "...we may then set the surface integral over $\exp(-\mathcal{E}_k^\ddagger/kT)$ equal to the volume integral for the e.e.d., divided by a partition function along the normal, as found in Appendix 1 ... Most of the likely motions along the normal to the transition state surface, ... have a 'frequency' of motion of about 10^{13} sec^{-1}. We anticipate, therefore, that the partition function just noted, which may be written as $kT/h\nu$, will be of the order of unity... "

p. 26 bottom: "In integrating finally over r, we first note two factors which favor small rs in spite of any Coulombic repulsion: the solvent reorganization barrier is smaller there ... and, at large rs, the electronic interaction becomes so weak that out there the integral should be multiplied by some small nonadiabatic transition probability. We presumably err relatively little if we simply take r as the distance of closest approach and *set the corresponding r-partition function equal to unity, that is,* $kT/h\nu_r \sim 1$ (*cf. also Section* 4.1).

p. 30 top: "Using Eq. (A.7)

$$F^\ddagger_{(\gamma)} = F^\ddagger_{(0)} + (\gamma^2/2!)F''_{(0)} + \cdots . \tag{A.7}$$

the N-dimensional volume integral becomes

$$\int_{-\infty}^{\infty} \exp(-F^{\ddagger}_{(\gamma)}/kT)d\gamma = \exp(-F^{\ddagger}_{(0)}/kT)\sqrt{2\pi kT/F''_{(0)}}. \quad \text{(A.8)}$$

But $\exp(-F^{\ddagger}_{(0)}/kT)$ equals $\int \cdots \int \exp(-\mathcal{E}_{\sigma 0}/kT)d\sigma$, using Eqs. (A.6) and (A.3) at $\gamma = 0$,

$$\exp(-F^{\ddagger}_{(\gamma)}/kT) = \int \ldots \int \exp(-\mathcal{E}^{\ddagger}_{\sigma\gamma}/kT)d\sigma \quad \text{(A.7)}$$

$$\mathcal{E}^{\ddagger}_{\sigma\gamma} = \mathcal{E}_{\sigma\gamma} + m\Gamma \quad \text{(A.3)}$$

and so equals the desired $(N-1)$-dimensional surface integral. Thus, if the 'vibrational-like partition function' $\sqrt{2\pi kT/F''_{(0)}}$ is of the order of unity, the basic approximation enunciated in Section 4.1 is seen from Eq. (A.9) to be justified."

Q: The basic approximation was: "By a suitable choice of m, the e.e.d. is made to center on the intersection, and thereby to die away fairly rapidly along the normal. Since Eq. (3.3.1)

$$\mathcal{E}^{\neq}_{k} = \mathcal{E}^{+}_{k^i} + \mathcal{E}^{+}_{k^o} + C \quad \text{(3.3.1)}$$

applies both to e.e.d. and to the transition state, we may then set the surface integral over $\exp(-E^{\neq}_{k}/kT)$ equal to the volume integral for the e.e.d., divided by a partition function along the normal, as found in Appendix 1."

Please explain: (1) why the suitable choice of m, (2) The centering point, (3) How you see the dying away fairly rapidly along the normal, (4) Why one gets the "vibrational-like partition function."

M: (1) Well, you know... given by the intersection, ... that's the suitable choice of m.

(2) For a certain value of m, the free energy of the e.e.d. is, apart from vibrational partition function, identical with the free energies of the reactants and the products, and only for that value of m.

(3) Well, I think that for the e.e.d. the distribution about that intersection is Gaussian, the distribution is dying away rapidly as you get away from the intersection, a Gaussian function.

(4) Because I wanted to be centered there, and if it is centered in some region and occupies a region near, that's vibrational-like, it's not rotation or anything, it's not translation.

Q: Can you explain more in detail?

M: Yes, that's right, yes, because is localized, you see, I mean, that's one way of localizing a system, you give a distribution function that has an effective parabolic-type potential along the direction normal to the intersection, it has to be vibrational-like.

Q: (1) From statistical mechanics, we know that for high temperatures $z_{vib} \sim \frac{kT}{h\nu}$ and that $z_{\text{transl}} = \sqrt{2\pi mkT L_1}/h$. Isn't then the above "vibrational-like partition function" $\sqrt{2\pi kT/F''_{(0)}}$ more similar to a translational partition function than to a vibrational partition function? The $\sqrt{2\pi kT}$ expression is present in both.

(2) Which is the physical meaning of the "vibrational like partition function"? Why, in other words, we have that: volume integral = surface integral × partition function? What does it mean?

(1) "From statistical mechanics we know that for high temperatures $z_{vib} \sim \frac{kT}{h\nu}$."

M: No, z is $\sqrt{2\pi kT/F''_{(0)}}$ and that's no relation to $z_{vib} \sim \frac{kT}{h\nu}$.

Q: So $\sqrt{2\pi kT/F''_{(0)}}$ is a translational.

M: Yes, one is translational, the other is vibrational.

Q: Why do you call it vibrational-like when it is translational?

M: Ehm, if it is $\frac{kT}{h\nu}$, that implies a frequency, that's a vibration. So, vibrational-like.

Q: But your formula is very similar to z translational.

M: Yes, that's right, and the z there refers to the translational. In other words, if you take the $\frac{kT}{h\nu}$ and you combine with a certain partition function you can get the z out. So, it's a result I derived, from the $\frac{kT}{h}$ expression. It's actually not dated, just derived.

Q: And that's why you call it vibrational-like, OK?

M: It's $\frac{kT}{h\nu}$, that's vibrational-like.

Q: Yes but the square root...

M: Yes, that's translational-like, yes.

Q: It looks like a mix-up of the two.

M: Oh... well, let me see... that $\frac{kT}{h\nu}$ it's like if you treat a collision in liquids *as a half collision*, if you treat it as a vibration then one would have a partition function of $\frac{kT}{h\nu}$, but if you treat it as a half vibration, you still have that but strictly you should divide by 2 then. In other words, *a collision in liquids is like a half vibration,* let these molecules in solution as vibrating with respect to each other.

Q: But the mathematical form is $\sqrt{2\pi kT/F''_{(0)}}$ is similar to a translational.

M: Yes, that's right, but has no connection with that $\frac{kT}{h\nu}$, that's a derivation where you take the free energy, and you put in the appropriate partition function, and you come out with a Z, starting from the $\frac{kT}{h}$, I actually derived it, it has not to do directly with the $\frac{kT}{h\nu}$. That "vibrational-like" is a kind of hand-waving statement, the other you may call a rigorous derivation, given certain assumptions. In my kinetics courses, I always give it.

M: "Why, in other words, we have that: volume integral = surface integral × partition function?" In other words, a volume is equal to a surface times a distance.

Q: And the partition function takes care of the distance.

M: Yes, that's right, that's right, in other words, you have a volume integral as a surface integral times a $d\gamma$, effectively a distance, then you integrate over γ.

Q: And you have a partition function.

M: Yes. I just mentioned how I did that and I'll try to dig it out here. What I did really, you know, there is bimolecular collision theory and there is an expression for it, all I did was put in the partition function, that appears in the numerator and denominator, and, lo and behold, out comes the collision frequency expression.

16. From **M44**: p. 460, 1st column top: "In Part I... The equations involved multipolar permanent and induced charge distributions on the particles and constituted a generalization of earlier potential energy expressions in which the particles were restricted to being single molecules interacting via permanent and induced dipoles only."

Q: (1) In your previous paper, there was no multipolar expansion. Was it somehow implicit in the equations? If so, how was it?

(2) Is an induced charge distribution automatically comprehensive also of induced multipoles?

M: (1) No, I just had it as a functional relationship.

Q: Is it implicit?

M: Yes, it's implicit, I just didn't make the explicit expansion. In other words, with the multipolar terms there I still have a linear relation between some charges distribution and a potential, so I just wrote it as a linear, a formal linear relationship.

Q: So if you have an induced charge distribution, it is automatically comprehensive of induced multipoles?

M: Well, if it's linear, yes. See, the key question is whether the relationship between the output and the input is linear or not. If it is linear, the relation between the output and the input, then you can substitute an operator without putting in all the explicit expressions.

I mean, it's just a way of avoiding doing a lot of writing which may not be necessary at that point, unless you want to do some calculation explicitly, but you can still derive relationships.

Q: One of the marks of your being a very clever theoretician is that of avoiding of getting involved in extremely difficult computations

M: Yes, yes, that's right, it's easier to do it that way. *But smarter too* I don't know about that, it's easier. You know, I think maybe that's one characteristic that came out of the being at the Courant Institute, sure I was there for two years, I did a modest teaching at Brooklyn Poli during the second year, so there I stayed for two years, and you know, this idea of the kind of minimization of equations, you know, I think that it's possible I got that, extracted that from the experience at the Courant Institute, I'm not sure. Nirenberg would come in and have his own proofs, discuss them with the professors.

17. From **M44**: p. 461, 2nd column middle: (1) "When only particles are present, each having a specified nuclear configuration, the polar contribution to the interparticle potential energy U^{pol} is given by Eq. (4)

$$U^{pol} = \frac{1}{2}\sum_i \int \phi_i \rho_i^0 d\mathbf{r} \tag{4}$$

where ϕ_i is given by Eq. (1)

$$\phi_i(\mathbf{r}') = \sum_{j \neq i} \frac{\rho_j(\mathbf{r})d\mathbf{r}}{r} + \int \mathbf{P}(\mathbf{r}) \cdot \nabla \mathbf{r}^{-1} d\mathbf{r} \tag{1}$$

with $\mathbf{P} = 0$.

When both particles and polarized medium are present, it is the polar contribution to the free energy F^{pol} which is of interest. When $\mathbf{P}(\mathbf{r})$ is in electrostatic equilibrium with the charge equilibrium, e.g., when it is given by $-\chi(\mathbf{r})\nabla\psi(\mathbf{r})$, where χ is the dielectric susceptibility of the medium and ψ is the electrostatic potential arising from all parts of the system, then F^{pol} is given by the RHS of Eq. (4), with ϕ_i given by Eq. (1)."

(2) p. 463, 1st column top: "U^{pol} applies to the particles-only system and F^{pol} to the particle-medium system"

Q: I was thinking of the symbol m... you do not always give it the same meaning.

M: All right, yes, I mean that that is a Lagrangian multiplier and that there is a way of regarding it instead as a reaction coordinate and it takes some value that was obtained from Lagrangian multiplier when you want to be at the TS. And you could use in the sense that $\Delta F^* = m^2\lambda$, use m as a reaction coordinate that starts at $m = 0$ and goes to the $m = -1$ for the products...

(1) By the way, I used that form, now I'm trying to remember about 60 years back, but it was that form that I found in Mason and Weaver, that I found so appealing because it separated the charge distribution and the polarization distribution, what I was looking for, you know, I looked as I mentioned probably in some place, maybe the Nobel talk, I looked through all the books on electrostatics and magnetism that I could find in the Polytechnic library, there were 11 of them, and the one I liked the best and I made more use of, was the one by Mason and Weaver, and they happened to be, or certainly Mason, and I think Weaver too, happened to be at Caltech at the time he wrote the book, yes.

Q: Or at the University of Chicago, are you sure?

M: Well, Mason definitely was at Caltech, and Weaver I think that became famous and he may have been at Caltech too but Mason definitely was, as faculty member. But I didn't, you know, I didn't make any connection with that... but that book turned out to be for me the best of the 11 books and in particular I remember drawing an equation like Eq. (4) or Eq. (1) from the book because it had polarization explicitly in a such a simple way, that gave a sort of a real physical feeling to doing what I was trying to do.

(2) The free energy is the work required to produce, for example, a dipole. A dipole interacting with a field has an energy $-\mu \cdot \mathbf{E}$, where

E is the field, but the free energy associated with the whole thing if the dipole is built up from nothing, is $-\frac{\mu\cdot\mathbf{E}}{2}$, because $-\mu\cdot\mathbf{E}$ is the energy and the interaction with the dipole, but it took work to produce the dipole and that was $\frac{\mu\cdot\mathbf{E}}{2}$, so there's some partial cancellation, you get for the free energy $-\frac{\mu\cdot\mathbf{E}}{2}$. So, there can be situations like that in here, I can't tell just from what was given. There is usually a difference from what we call the energy of an *existing* system which has a certain dipole, and the free energy, the work required to *produce* it and if that dipole isn't a permanent dipole, if it had to be produced, then that free energy causes $-\mu\cdot\mathbf{E}$ to be divided by two.

Q: But I realized that you often use the energy for particles systems and free energy for the medium.

M: Well if you only had the particles alone you just really have the energy, you are not normally producing it from some process, so you don't have a process, but when you have the particles and the solvent around them, then normally you charge up the particles and the solvent and you have a process.

Q: You have a process but with a process goes an entropy somehow?

M: Yes, that's right, there is an entropy, yes.

Q: Let me ask you a very simple question, imagine you have a capacitor, and you charge up the capacitor...

M: Now, does it have a dielectric in there? Because the situation is different if you don't have the dielectric, then there is a distinction between energy and free energy.

Q: In vacuum anyway the energy is a free energy.

M: Yes, when you have a dielectric and you charge it up, then there is an entropy change in the dielectric.

Q: If there was no dielectric....

M: Then the energy and the free energy would be the same.

18. From **M44**, p. 465, 2nd column bottom: "Certain possible contribution of an inelastic nature, such as charge transfer or electronic excitation..."

Q: (1) If charge transfer inside a molecule is like an electronic excitation and is considered the result of an inelastic collision, is the electron transfer process, with an electron jumping from one molecule to another, a reactive collision? Or is it an inelastic collision for the self-exchange reactions (same reactants and products) and a reactive collision for the others?

(2) Could one speak of electron absorption or electron emission instead of speaking of electron transfer?

M: (1) Well... somebody thinks that an inelastic collision is a process that changes the internal degrees of freedom, ... all sorts of vibrational changes and rotational changes in solution, so, you know, there's so much going on that one no longer really uses the word inelastic for them. Inelastic is really a term more used for some collisions in the gas phase, that are not perfectly elastic collisions.

Q: So you consider them reactive collision.

M: This is the term for that, because all of chemical reaction would be inelastic, so why use it?

(2) No, because electron emission refers to emitting into a vacuum or into some dielectric continuum and not into a specific molecule, ET is more... I mean, you can have an electron... you can have a molecule that emits an electron, ... but if it is going directly to another reactor without existing in the vacuum, then you don't call that electron emission.

Q: So the transfer implies something taking it up.

M: Yes.

19. From **M43**, p. 1861, 2nd column top: "If $(e_a, e_b, \ldots, e_j, \ldots)$ denotes the charges of the reactants A and B and of any other ions j

present before electron transfer, and if $(e_a^p, e_b^p, \ldots, e_j^p, \ldots)$ denotes the charges after, then $e_j^p = e_j$. It has been shown that in the activated complex, the medium has an orientation–atomic polarization which would be in dielectric equilibrium only with the hypothetical charges $e_i + m(e_i - e_i^p)$ $(i = a, b, \ or \ j)$ where m satisfies a given equation. State [0] has, therefore, the charge distribution described by:

$$\text{State [0]} \equiv (e_a + m\Delta e, e_b - m\Delta e, \ldots, e_j \ldots), \tag{27}$$

where $\Delta e = e_a - e_a^p = e_b^p - e_b$.

State [1], which consists of the equilibrium polarization system having the same nuclear configuration of i's and having the e_j's existing before electron transfer, is represented by Eq. (28). Hence the state [1–0] is given by Eq. (29).

$$\text{State [1]} \equiv (e_a, e_b, \ldots, e_j, \ldots), \tag{28}$$

$$\text{State [1 - 0]} \equiv (-m\Delta e, m\Delta e, \ldots, 0, \ldots), \tag{29}$$

i.e., F_{1-0} is the free energy of an equilibrium polarization having *no charged particles* other than those undergoing electron transfer."

Q (1): The system [0] is the e.e.s. Giving $m$successively the values 0, $-1/2$ and -1 one has three [0] systems, one of which is the initial one, one is the final and one is the activated complex. Imagine now of having the reaction

$$Fe^{3+} + Fe^{*2+} = Fe^{2+} + Fe^{*3+}.$$

Can we say that the e.e.s. corresponding to the activated complex is the fictitious state in which the charges on the ions are given by:

$$Fe^{2,5+} + Fe^{*2,5+}?$$

(2): I had previously symbolized your hypothetical systems the following way, indicating for each of them charge densities, $\mathbf{P}_u$ and $\mathbf{P}_e$

polarizations

$$[\text{non eq.}]: \mathbf{P}_{neq}, \rho_{i1}^{0} \quad \mathbf{P}_{neq} = \mathbf{P}_{u,neq} + \mathbf{P}_{e,eq1}$$

$$[0]: \mathbf{P}_{eq0} = \mathbf{P}_{neq}, \rho_{i0}^{0} \qquad \mathbf{P}_{eq0} = \mathbf{P}_{u,neq} + \mathbf{P}_{e,eq0}$$

$$[1]: \mathbf{P}_{eq1}, \rho_{i1}^{0} \qquad \mathbf{P}_{eq1} = \mathbf{P}_{u,eq1} + \mathbf{P}_{e,eq1}$$

$$[1-0]: \mathbf{P}_{eq1-0}, \rho_{i1-0}^{0} = \rho_{i1}^{0} - \rho_{i0}^{0} \quad \mathbf{P}_{eq1-0} = \mathbf{P}_{u,eq1-0} + \mathbf{P}_{e,eq1-0}$$

$$[1-0, \text{op}]: \mathbf{P}_{e}, \rho_{i1-0}^{0} = \rho_{i1}^{0} - \rho_{i0}^{0}$$

From above, we see that [1 – 0] = [1] – [0] as far as charge densities and polarizations are concerned. But if one considers the free energies, you distinguish between $F_1 - F_0$ and F_{1-0} because from equations in **M43**:

$$F_{1-0} = \frac{1}{2}\sum_i \int \phi_{i1-0}(\rho_{i1}^{0} - \rho_{i0}^{0})d\mathbf{r} \tag{17}$$

and

$$F_1 - F_0 = \frac{1}{2}\sum_i \int (\phi_{i1} + \phi_{i0})(\rho_{i1}^{0} - \rho_{i0}^{0})d\mathbf{r} \tag{22}$$

where

$$\phi_{i1-0} = \phi_{i1} - \phi_{i0} \tag{19}$$

we see that

$$F_{1-0} = \frac{1}{2}\sum_i \int (\phi_{i1} - \phi_{i0})(\rho_{i1}^{0} - \rho_{i0}^{0})d\mathbf{r}$$

so that the reason to consider the [1–0] system independent from the [1] and the [0] is the charging energy. Is it so?

M: (1) I've already looked at page 1, and the answer to your question for Part 1 is yes, but it looks as though you would have repulsions between those fictitious charges, which you don't have. But you noticed that in the theoretical expression is $F - F_1 = F_{1-0}^{op} - F_{1-0}$ so that horrible repulsion between fictitious charges cancels in that difference. So that horrible repulsion actually never appears, it appears twice really, but cancels. Yes, because remember F_{1-0} doesn't appear alone, only appears in connection with F^{op}_{1-0}.

(2) **M**: Yes, as far as the charges yes, but... Yes, the charges that'd be the same, but not the result... yes, there is a cross term actually that appears when you have [1–0].

20. From **M43**, p. 1860, 1st column "The *free energy change* ΔF, when the system is charged from one set of charges on the particles to another, equals the *reversible work* done to form one system from the other[15]:

$$\Delta F = \int \int_{\lambda=0}^{1} \phi_i^{\lambda} d\rho_i^{0\lambda} d\mathbf{r} \tag{7}$$

where λ is a charging parameter."

Note 15: "Initially, in using Eq. (7) to derive Eq. (5.46),

$$\Delta F = \int \int_{\lambda=0}^{1} \phi_i^{\lambda} d\rho_i^{0\lambda} d\mathbf{r} \tag{7}$$

$$F = F_1 + F^{op}_{1-0} - F_{1-0} \tag{23}$$

the electrostriction which may result when this work is performed at constant pressure is ignored. It can be included by taking into account any dependence of $\chi(r)$ on $\chi_e(r)$ on λ. We note that this electrostriction is the one which occurs outside the inner coordination shell of each ion. In the derivation of Eq. (23′),

$$F = F_1 + F^{op\prime}_{1-0} - F'_{1-0} \tag{23′}$$

the electrostriction on forming system [0] is not neglected. Rather, one makes the normally milder assumption of neglecting the additional electrostriction on forming system [1] from [0]. Systems [1–0]' and [1–0, op]' have the $\chi(r)$ and $\chi_e(r)$ of system [0]."

Q5: Please explain the whole story... in particular why doesn't one neglect the electrostriction for system [0] but one neglects it for system [1].

M: I really did enjoy trying expressing in terms of equilibrium free energies, yes.

Oh, well, the idea is that supposing in charging up to [0] there is some electrostriction, and supposing that when one charges going from [0] to [1] there is further electrostriction, but if one neglects that further electrostriction, then between [0] and [1] you have a kind of linearity.

Q: And why are you allowed to do this?

M: Well, it is not that you are allowed to, it's just that it's a less restricted assumption to assume a linearity over that region.

21. "If A and B are treated as spherical, and if electrostatic image effects are ignored (they contribute about 10% to the free energy change [1, note p. 27]), then the *polar contribution* to the free energy of the activated complex minus that of an equilibrium polarization system with charges e_a, e_b, and... e_j ...*in the same specified positions* is given by $F - F_1$, i.e., by $F^{op}_{1-0} - F_{1-0}$.

However, using a well-known formula, F_{1-0} equals the sum of the *Born charging terms* at infinity plus the *work* required to bring the ions from ∞ to some mean distance r apart. (It will be recalled that the F's do not include interactions within the inner coordination shells.) Since the 1–0 system has no ions other than A and B, we then find:

$$F_{1-0} = -\frac{1}{2a_1}\left(1 - \frac{1}{D_s}\right)(m\Delta e)^2 - \frac{1}{2a_2}\left(1 - \frac{1}{D_s}\right)(m\Delta e)^2 + \frac{(m\Delta e)(-m\Delta e)}{D_s r}$$

and

$$F_{1-0}^{op} = -\frac{1}{2a_1}\left(1-\frac{1}{D_{op}}\right)(m\Delta e)^2 - \frac{1}{2a_2}\left(1-\frac{1}{D_{op}}\right)(m\Delta e)^2 + \frac{(m\Delta e)(-m\Delta e)}{D_{op}r}$$

where a_1 and a_2 denote the radii of the reactants, each including any inner coordination shell. Accordingly, this contribution to the free energy change is [3]:

$$F - F_1 = (m\Delta e)^2\left(\frac{1}{D_{op}} - \frac{1}{D_s}\right)\left(\frac{1}{2a_1} + \frac{1}{2a_2} - \frac{1}{r}\right)."$$

M: Notice that the expression for $F - F_1$ is identical with the 1956 expression, so all those other charges that appear in F_{1-0} and in F_{1-0}^{op} cancel to give the 1956 expression. Remember, F_{1-0} doesn't appear alone, only appears in connection with F_{1-0}^{op}. The observable is the free energy of activation and that's the difference of F_{1-0}^{op} with F_{1-0}.

Q: Are you sure that it never comes alone?

M: No, I'm positive, not in the final expression, not in the observable. In the observable is the free energy of activation, and that's with a difference of the F_{1-0}^{op} and the F_{1-0} terms.

22. From **M45**, p. 1735, 1st column middle: "Let $\boldsymbol{\tau}^f$ the coordinates held fixed and $\boldsymbol{\tau}$ denote the other coordinates. We treat the $\boldsymbol{\tau}$ coordinates classically. For any given $\boldsymbol{\tau}^f$ and $\boldsymbol{\tau}$, the electronic energy U^{tot} is the sum of an intraparticle term and of an interparticle one. The former is the total electronic energy when the particles are isolated from each other and have the *intraparticle* contained in this value of $(\boldsymbol{\tau}^f, \boldsymbol{\tau})$."

Q1: I believe the $\boldsymbol{\tau}^f$ coordinates are just the internal ones. Is it so?

M: It may be the position of, say, an ion, that's not an internal coordinate, and may also denote the position of the ion and some charges inside the ion, yes. It all depends on how I use it, one can tell from how it's used.

23. Following: "The interparticle electronic energy itself will be taken to be the sum of polar and 'nonpolar' or, more precisely, 'electron correlation' terms, the latter including repulsion and London dispersion energies of interaction *between* the particles."

Q: Why are the correlation terms nonpolar?

M: Well. . . in the case of the London dispersion interaction, you have an instantaneous electron distribution, you know, think of it sort of classically for the moment, on each particle, and there is a feeble interaction between those distributions, . . . really what amounts to if you take the instantaneous interactions between the two ions, say, or two molecules, and you use second order perturbation theory, which means you're taking into account the effect of one on the wave function of the other, and calculate the result for that effect, then that's an example of a dispersion interaction and, you know, you may call that nonpolar because you reserve the term polar for the interaction of the fixed charge distribution. I think I'm using polar for the fixed charge distributions. In connection with the Pauli repulsion, you know, that's a strange interaction, and that's not an interaction of the charge distribution versus charge distribution it's an interaction of the distributions due to Pauli exclusion, the result of Pauli principle that gives rise to an exponential, so again it's not the result of just two fixed charge distributions interacting with each other. I meant to separate those two kinds of effects, some effect will be treated by simple electrostatics and all the rest. . . .

24. From **M45**, p. 1737, 1st column top: "Dielectric unsaturation is characterized by a linear response of the medium to the ρ_i^0 of the central species. If the coordinates of any ions in the medium are included in $\boldsymbol{\tau}$, then the introduction of a linear response approximation will

give rise not only to dielectric unsaturation but also to a Debye–Hückel approximation for salt effects as well. To avoid the second of these, we hold the ionic coordinates fixed by including them among the $\boldsymbol{\tau}^f$."

M: When the dielectric polarization is proportional to the field, that's unsaturation, when it becomes independent of the field, that's saturation. Partial saturation is in between.

Q1: When and why do you want the ions of the ionic atmosphere to belong to $\boldsymbol{\tau}$ or rather to the $\boldsymbol{\tau}^f$ coordinates?

M: Yes... well... we knew that there're limitations to the Debye–Hückel, don't we? Nevertheless, Debye–Hückel is a linear approximation, and if you want to include... the response to the atmosphere, that the atmosphere is responding in a nonequilibrium way, it tends satisfy the charges, the initial charges and the final charges so it's adapted to, and if you wanted to include that, then you treated it the same way that you treating the polarization. Make it part of that response term, then hold the fixed charges.

25. Following: "Because of the linear response of the medium to the ρ_s^0's the polar contribution to the free energy at a given $\boldsymbol{\tau}^f$ is quadratic in the ρ_s^0's."

Q2: May we make the following analogy? Consider Hooke's law: $F = -k(x - x_0) \Rightarrow V = \frac{1}{2}kx^2$. If instead of a mechanical force, a linear displacement and a potential energy I consider an electric force, a charge density, and a free energy may I write $E = k(\rho - \rho_0)$ and $F \propto k\rho^2$?

M: Yes, I think... I've looked at that question too, I think if you wrote $F = k(\rho - \rho^0)^2$, the difference squared, see, the one is a difference, you're starting at ρ^0, then the free energies have to start at ρ^0 also. The analogy is OK, yes.

Q: These analogies are important to understand things.

M: You're right.

Q: Well Dr. Marcus, thanks a lot. The right title for the book should be: Marcus comments on his own theory. When did your ancestors come to America?

M: Let's see, my father was born here in 1895, he was the oldest, so my guess is they came around 1882–1883. My mother's side they went instead to England and she didn't come over until 1921 or so, surely after the First World War. She came with her sister and brother-in-law and niece.

Q: Was she born in England?

M: She was born in England, yes, her parents were born in, I assume, Russia, the grandfather changed his name, so we never knew where he came from. I mean, dad and children knew where he came from but I never did. But he maybe had a long name and he simplified it to what to him was a well-known, namely Cohen, and that's how my mother became a Cohen.

Q: But Cohen is a name used very much also by Jews here in Italy.

M: Yes, Cohen is sort of an honor name and I knew as a child that somehow we weren't real Cohen's, but I never knew why. Much later I learned that when my grandfather came to England they changed his name to Cohen, but I didn't know that at the time, I just knew we weren't real Cohens.

Q: If you were a real Cohen, you would have some special responsibility.

M: Yes, that's right, that's right.

Q: They are priests.

M: Yes, that's right, that's right, maybe my grandfather wanted to be one. Anyway it was known in the family that we weren't real.

Q: I personally don't see in you a very great vocation for priesthood.

M: No, no, although, you know, being a teacher even of Chemistry, you are a little bit of a priest.

Q: Oh, yes, yes, that's for sure.

M: Yes, and sometimes with all that... that some priests have.

Q: Anyway you are the Pope of the electron transfer.

M: That's right, and I'm looking around for a few Cardinals, I think if there is anybody I can raise to sainthood that I'd be glad to, but I really need first a batch of Cardinals first.

References

1. R. A. Marcus, *Discuss. Faraday Soc.* **29**, 21–31, (1960).
2. R. A. Marcus, *J. Chem. Phys.* **38**, 1335, (1963).
3. R. A. Marcus, *J. Chem. Phys.* **38**, 1858, (1963).
4. R. A. Marcus, *J. Chem. Phys.* **39**, 460, (1963).
5. R. A. Marcus, *J. Chem. Phys.* **39**, 1734, (1963); (a) R. A. Marcus, "On the Theory of Overvoltage for Electrode Processes Possessing Electron Transfer Mechanisms." I. (ONR Report No. 12, dated 1957). In *Special Topics in Electrochemistry*, p. 181, P. A. Rock, Editor, Elsevier, New York (1977); (b) R. A. Marcus, *J. Chem. Phys.* **24**, 966, (1956).
6. R. A. Marcus, *J. Chem. Phys.* **43**, 679–701, (1965); (a) R. A. Marcus, *J. Chem. Phys.* **26**, 867, (1957); (b) R. A. Marcus, *J. Chem. Phys.* **26**, 872, (1957).
7. R. A. Marcus, *J. Chem. Phys.* **24**, 979–989, (1956); (a) D. A. McQuarrie, *Statistical Mechanics*, Harper & Row (1973).
8. M. Mandel, P. Mazur, *Physica.* **24**, 116, (1958).
9. J. O' M. Bockris, A. K. N. Reddy, *Modern Electrochemistry 1*, 2nd Edition, Plenum Press, New York and London (1998).
10. J. Kirkwood, *J. Chem. Phys.* **4**, 592, (1936).
11. C. J. F. Böttcher, *Theory of Electric Polarization*, Elsevier Scientific Publishing Company, Amsterdam, London, New York (1973).
12. H. Frohlich, *Theory of Dielectrics*, Oxford University Press, London (1950).
13. W. F. Brown, Jr., *Encyclopedia of Physics*, Vol. 17, S. Flügge, Editor, Springer Verlag, Berlin (1956).
14. W. F. Brown, Jr., *Physica.* **24**, 695, (1959).
15. Permanent multipoles have been included by L. Jansen, *Phys. Rev.* **110**, 661, (1958).
16. See, for example, H. C. van de Hulst, *Light Scattering by Small Particles*, John Wiley & Sons, Inc., New York (1957).
17. A. J. Stone, *The Theory of Intermolecular Forces*, Clarendon Press, Oxford (2002).

18. See reviews: W. F. Brown, *Handbuch der Physik*, Vol. 17, p. 1, S. Flügge, Editor, Springer-Verlag, Berlin (1956); (a) R. H. Cole, *Ann. Rev. Phys. Chem.* **11**, 149, (1960).
19. Y. Ooshika, *J. Phys. Soc. Japan.* **9**, 594, (1954); E. Lippert, *Z. Naturforsch.* **10a**, 541, (1955); E. G. Mc Rae, *J. Phys. Chem.* **61**, 562, (1957); R. Platzmann, J. Franck, *Z. Phys.* **138**, 411, (1954); R. A. Marcus, paper presented at 133rd National American Chemical Society Meeting, San Francisco (1958); Tech. Rept. No. 4, NSF Grant G-3690, (1958).
20. R. A. Marcus, *J. Chem. Phys.* **24**, 966, (1956); *Can. J. Chem.* **37**, 155, (1959); *Trans. Symp. Electrode Processes Philadelphia, PA.* 1959, 239, (1961); *Discuss. Faraday Soc.* **29**, 21, (1960); V. C. Levich, R. R. Dogonadze, *Collection Czech. Chem. Commun.* **26**, 193, (1961) (O. Boshko translator, University of Ottawa, Ontario), compare *Dokl. Akad. Nauk SSSR* **133**, 1368, (1960); N. S. Hush, *Trans. Faraday Soc.* **57**, 557, (1961).
21. See H. M. McConnell, *J. Chem. Phys.* **35**, 508, (1961).
22. For example, S. I. Pekar, *Untersuchungen über die Elektronentheorie der Kristalle*, Akademie Verlag, Berlin (1954); (a) T. Holstein, *Ann. Phys.* **8**, 325, 343, (1959).
23. See, R. A. Marcus, (a) *J. Chem. Phys.* **24**, 979, (1956) (Part I); (b) *J. Chem. Phys.* **38**, 1858 (Part II).
24. R. McWeeny, *Coulson's Valence*, 3rd Edition, Oxford University Press (1979).

CHAPTER 6

Unified General Theory of Electron Transfer Reactions

In this chapter, Ref. [1] is summarized in which M. presented a unified treatment for homogeneous and electrode reactions. This Reference is # VI in the series of papers "On the Theory of Electron Transfer Reactions" and is the "summa" of M. theory up to 1965. Many topics dealt with in the preceding chapters are taken up again and more extensively dealt with in the framework of Marcus' generalization of the activated complex theory of reaction rates. In this theory, generalized coordinates are used in both ordinary and mass-weighted configuration space. The elements of the generalized activated complex theory necessary to understand the application to ET theory are briefly described in Sections 6.4–6.7.

6.1. General Characters of ET Reactions

Range of their Rate Constants

A series of characteristic properties and unusual features make the ET reactions "quite different from other types of reactions in the literature":

(a) "No rupture or formation of chemical bonds" are involved "in the elementary electron transfer step"
(b) "Weak electronic interactions of the reactants" and consequently possible nonadiabaticity

(c) "Nonequilibrium solvent polarization of the solvent medium"[2]
(d) "Unusual reaction coordinate"
(e) They allow "an approximate calculation of the reaction rate without use of arbitrary adjustable parameters"

"Applications of the theoretical equations were made in several subsequent papers (**M 129, 28, 33**). The mechanism of electron transfer was later examined in more detail in Part IV using potential energy surfaces (PESs) and statistical mechanics (**M30**). (In Part I, the solvent medium outside the inner coordination shell of each reactant has been treated as a dielectric continuum. The free energy of reorganization of the medium, accompanying the formation of the activated complex having nonequilibrium dielectric polarization, was computed by a continuum method.) In Part IV, changes in bond lengths in the inner coordination shell of each reactant were also included, and the statistical mechanical term for the free energy change in the medium outside was replaced only in the final step by its dielectric continuum equivalent.

A number of predicted quantitative correlations among the data were made on the basis of Part IV. They have received some measure of experimental support, described in Part V and in recent review articles (**M41, 48**). A more general basis for these correlations is described in the present paper, which also presents a unified treatment of chemical and electrochemical transfers.

The form of the final equations for the rate constants is comparatively simple, a circumstance which leads almost at once to the earlier correlations. (It permits extensive cancellations in computed ratios of rate constants.) This simplicity has resulted from several factors: (1) Some of the more complex aspects of the rate problem[(1)] are rephrased so that they affect only a preexponential factor (ρ) appearing in the rate constant, a factor that appears to be close to unity. (2) Little error is found to be introduced when the force constants of reactants and products are replaced by symmetrical reduced force constants. (3) An important term (λ) in the free energy of activation is essentially an additive function of the properties of the two redox systems in the reaction.

The electron transfer rate constants can vary by many orders of magnitude: for example, known homogeneous electron exchange rate constants vary by factors of more than 10^{15} from system to system, and electrochemical rate constants derived from electrochemical exchange currents vary by about 10^8 at any given temperature[3]. (An electron exchange reaction is one between ions differing in their valence state but otherwise similar). Thus, small factors of 2 or 3 are of relatively minor importance in any *theory which is intended to cover this wide range of values*. Some approximations in this paper are made with this viewpoint in mind.

In the present paper, classical statistical mechanics is employed for those coordinates which vary appreciably during the course of the reaction. This classical approximation is a reasonable one for orientational and translational coordinates at the usual reaction temperatures and, in virtue of the above remark, for the usual low-frequency vibrations in inner coordination shells. Because of *cancellations which occur in computations of ratios of rate constants, this approximation could be weakened* for deriving the predicted correlations, even when the quantum corrections would not be small.

In calculations of absolute values of the electron transfer rate constants, a classical approximation will introduce some error when *the necessary changes in bond lengths to affect ET* are so small as to be comparable with *zero-point fluctuations*. However, in the latter case, the vibrational contribution to the free energy of activation is itself small and does not account for any large differences in reaction rates in redox reactions which have been investigated experimentally. Hence, for our present purpose and, in the interest of simplicity, this particular possible quantum effect may be ignored"[1].

Author's Comment: The theory of Marcus is intended to cover a wide range of values of ET rates. M. very cleverly avoids points of the theory that would be at the same time difficult and of minor importance for his stated purpose. He follows in the theory the example set in experimental physics by Galileo when, in order to avoid difficult and at the same time unessential effects in his study of the motion of heavy bodies, he used the inclined plane.

6.2. Individual and Overall Rate Constants

"Many chemical and electrochemical redox reagents are ions which possess inner coordination shells and which may form complexes with ions of opposite sign. Any such complex is 'inner' or 'outer' according as the latter ions do or do not enter the inner coordination or shell of the reactant.(2) To a greater or lesser extent, all such complexes normally contribute to the measured rate of the redox process. For this reason both a rate constant for the overall reaction, involving all complexes, and a rate constant for each individual step, involving a specific pair of complexes in a bimolecular step, have been defined in the literature. They equal the overall rate divided by the stoichiometric concentration (or product of such concentrations in the bimolecular case), in the case of an overall rate constant, and the reaction rate divided by the concentration of the particular complex (or product of such concentrations in the bimolecular case), in the case of an individual rate constant. Often the individual rate constants are measured experimentally. Frequently, however, only the overall rate constant is determined in the experiment.

The derivation up to and including Section 6.6 applies to overall rate constants as well as to individual rate constants. The Sections 6.7–6.17 apply only to individual rate constants. To calculate the overall rate constant from the expression derived for the individual ones in these latter sections, one must take cognizance of any reactions leading to the formation and destruction of the complexes and must average over the behavior of all complexes, as in Appendix VIII" [1].

6.3. On Potential Energy Surfaces

6.3.1. *A More Extensive Treatment*

"The potential energy of the system is a function of the translational, rotational, and vibrational coordinates of the reacting species and

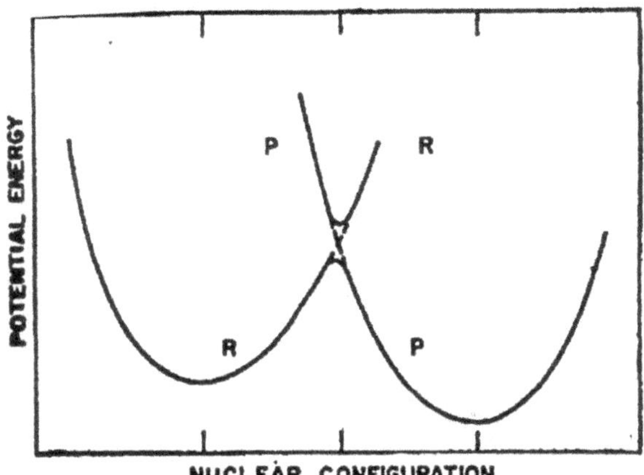

Fig. 1. Profile of potential energy surface of reactants *R* and products *P* plotted versus configuration of all the atoms in the system. The dotted lines refer to a system having no electronic interaction of the reacting species. The adiabatic surface is indicated by a solid line (From Ref. [1]).

of the molecules in the surrounding medium" [1]. Profiles of PESs are given in Figs. 1 and 2 for homogeneous and electrochemical ET reactions.

"The abscissa, a line drawn in the above many-dimensional coordinate space, represents any *concerted* motion of the above types leading from one spatial configuration (of all atoms) that is *suited* to the electronic structure *of the reactants* to that of the *products*.[(3,4)] Surface R denotes the potential energy profile when the reacting species have *the electronic structure of the reactants*, and surface *P* corresponds to their having *the electronic structure of the products*. If the *distance* between the reacting species is sufficiently small, there is the usual *splitting* of the two surfaces in the vicinity of this intersection of *R* and *P*. If the *electronic interaction causing the splitting* is sufficient, the system will always remain on the lowest surface as it moves from left to right in Fig. 1. Thus, the system has moved from surface *R* to surface *P adiabatically*, in the usual sense that the corresponding motion of the atoms in the system is treated by a quantum-mechanical adiabatic method. On the other hand, if the

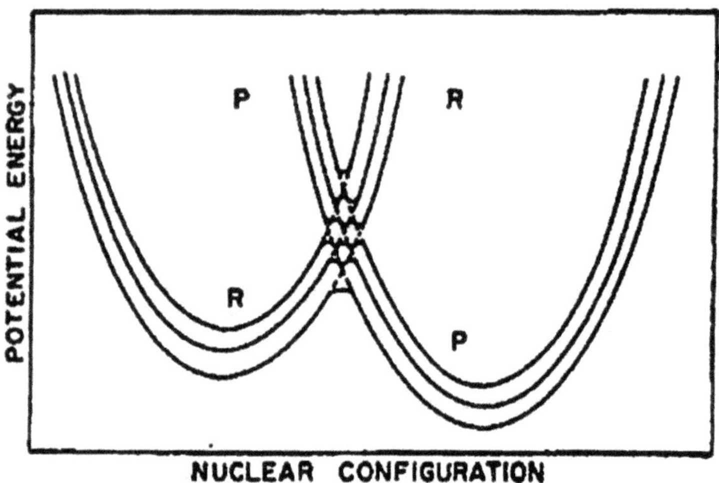

Fig. 2. Same plot as Fig. 1 but for an electrode reaction. The finite spacing between the many-electron levels of a finite electrode is enormously magnified, and only three of them are indicated. The splitting differs from level to level (From Ref. [1]).

electronic interaction causing the splitting is *very* weak, a system initially on curve *R will tend to stay on R* as it passes to the right across the intersection. The probability that as a result of this nuclear motion the system ends up on curve *P* is then calculated treating this motion nonadiabatically[(5)][4].

It should be noted that the system can undergo this ET either by surmounting the barrier if it has enough energy or by tunneling of the atoms of the system through it if it has not. We confine our attention to the case where the systems surmount this barrier. Some atom tunneling calculations have been made, however [5].

Since the abscissa in Fig. 1 is *some combination* of translational, rotational, and vibrational coordinates, this 'reaction coordinate' is rather complex: the surfaces *R* and *P* intersect, and the set of configurations describing this intersection form a hypersurface in configuration space. The exact motion *normal* to this hypersurface depends *on the part being crossed*. In some parts, it involves *changes in bond distances* in the inner coordination shells of the reactants, in other parts, it involves a *change of separation distance* of the reactants, and in still others, it involves *reorientation of polar molecules in the medium*.

Analogous remarks apply to electrode reactions except that the intersection region is more complex because of the presence of many electronic energy levels in the metal. A blown-up portion of this region is indicated in Fig. 2. The diagram consists of many PESs, each for a many-electron state of the entire macrosystem. All the surfaces are parallel since they differ only in the distribution of electrons among 'single-electron quantum states' in the metal. (Only one distribution of the electrons among these single-electron quantum states correspond to each surface in Fig. 2 if the energy level of the entire macrosystem is nondegenerate. It corresponds to several distributions in the case of degeneracy.) There is a probability distribution of finding the macrosystem in any many-electron energy level indicated in Fig. 2.[(6)] As a consequence of a Fermi–Dirac distribution of the electrons in the metal, most electrons which are transferred to or from the many-electron energy levels in the metal will behave as though they go into or from a level which is *within kT* of some *mean energy level*, and hence practically equal to it. Thus, except for the calculation of the transition probability associated with the transition from surface R to surface P in the intersection region, the situation is in effect very similar to that in Fig. 1" [1].

This point will be taken up again and discussed in Section 6.8.

6.4. *Some Concepts from Marcus's Generalization of Activated Complex Theory of Reaction Rates

In order to better understand the following sections, it may be useful to excerpt some concepts and notation from Refs. [6–8] whose results were used in Ref. [1] to obtain the ET reaction rates.

"For the reaction to occur, some $n-1$ dimensional hypersurface in the n-dimensional configuration space must be crossed. The *hypothetical system* constrained to exist on this surface is the 'activated complex' [7]. Or else '*The set of points in configuration space defining this hypersurface is the activated complex*' [8]." The surface is called the "reaction hypersurface... The equation of this hypersurface S may be written as $f(q^1, \ldots, q^N) = 0$" [7] where the $q^{i's}$ are generalized coordinates $q^i = q^i(x^1, \ldots, x^N)$.

"A choice of coordinates may be made so that... one of them, q^N ... is *constant* on S, and can be taken as zero on it. This surface will be a q^N coordinate hypersurface both in mass-weighted and in ordinary configuration space" [7]. This coordinate is the "reaction coordinate" [8]. "On the one side of S (the reactants') q^N is negative and on the other is positive." "The reaction rate is the net rate at which systems cross S" [7].

"It will be supposed that some distance from this hypersurface, closer to the 'reactants' region' of configuration space, in a small interval the motion along the reaction coordinate can be described classically and the probability of finding a system moving in that interval in the forward direction (i.e. toward the products) has approximately the equilibrium value." In "*the immediate TS neighborhood*... tunneling may occur... The reaction coordinate q^N leading from the cited interval, through TS, to products, is to be chosen in a way which facilitates solution of the dynamical equations approximately or exactly *in the neighborhood of the TS*" [8].

"In a dilute gas, an activated complex may be regarded as an isolated particle. In a liquid or dense gas, its motion may be strongly coupled to those of the surrounding molecules. In the latter case, it will be useful to *consider as the activated complex a macroscopic subsystem* near the center of which is the actual reactant or pair of reactants and on the boundary of which the correlation of the motions of the solvent molecules with those of the reactants is negligible. This *subsystem* is regarded as *imbedded* in the *remainder* of the infinite (or practically infinite) system. For homogeneous reactions, rigid translations or rotations will later be performed on the subsystem, and the solvent molecules of the remaining part will be permitted to continuously adjust themselves. For heterogeneous systems, rigid translations of the *macroscopic activated complex* parallel to the interface will be performed with a similar adaptation of the remaining molecules occurring.

The activated complex of a homogeneous reaction in a gas or liquid, defined above, has as coordinates three translations, (x, y, z), two rotations (ϑ, ϕ) of an axis fixed in the complex, and $N-5$ other coordinates, which will be called the *internal coordinates* of the complex, though one of them, rotation about the body-fixed axis) has a property analogous to the five 'external' ones: The potential energy of the entire system is invariant to changes in the five external coordinates.

In the case of a heterogeneous reaction on a *uniform* interface, the potential energy function for the activated complex is invariant to the two Cartesian coordinates x and y parallel to the interface of the two phases. Presumably, such a case occurs in electrochemical electron transfers to a good approximation when the reactant is not adsorbed. In reactions involving *localized adsorption* on *perfect crystals*, the potential energy is a periodic function of x and y. For any heterogeneous reaction, the remaining $n-2$ are the internal ones of the activated complex, though in the particular case of a nonuniform surface, the potential energy $U^{\ddagger}$ of the activated complex and $q^r \ldots$ depend on *all* the n coordinates" [7].

6.5. *Use of Generalized Curvilinear Coordinates

Marcus refers, in his generalized theory, to the n-dimensional Schrödinger equation in the form given by Pauli [8a]

$$\sum_{r,s=1}^{n} (g)^{-\frac{1}{2}} \frac{\partial}{\partial q_r}\left(g^{\frac{1}{2}} g^{rs} \frac{\partial \psi}{\partial q_s}\right) + k^2(\alpha_1 - V)\psi = 0$$

"where k^2 equals $2/\hbar^2$, α_1 is the energy E, V is the potential energy, and ψ the wave function for the entire system. g^{rs} is reciprocal to g_{rs}. The latter appear in the fundamental line element ds in a mass-weighted space having the q^r as generalized coordinates and are defined in Eq. (6.2)

$$ds^2 = \sum_{r,s=1}^{n} g_{rs} dq^r dq^s, \tag{6.1}$$

If the coordinates q^r are the *ordinary Cartesian coordinates* x^r, then g_{rs} equals $m_r\delta_{rs}$, where m_r is the mass of the atom whose coordinates include this x^r and where δ_{rs} is the Kronecker delta function. If the coordinates q^r are any other ones, the corresponding g_{rs} and g^{rs} are then computed from the expressions

$$g_{rs} = \sum_{i=1}^{n} m_i \frac{\partial x^i}{\partial q^r}\frac{\partial x^i}{\partial q^s}, \quad g^{rs} = \sum_{i=1}^{n} \frac{1}{m_i}\frac{\partial q^r}{\partial x^i}\frac{\partial q^s}{\partial x^i} \tag{6.2}$$

For example, if the q^r are *mass-weighted Cartesian coordinates*, that is, if they equal $(m_r)^{\frac{1}{2}}x^r$, then g_{rs} equals δ_{rs}.

Robertson considered systems having orthogonal coordinates ($g^{rs} = 0$ if $r \neq s$). More precisely, when the coordinates are orthogonal the quantities reciprocal to g^{rs}, g_{rs}, are diagonal. However, since gg^{rs} is the cofactor of g_{rs}, where g *is the* $n \times n$ *determinant of the* g_{rs}, and since g_{rs} is diagonal, so is g^{rs}. For example, see Ref. [8c], Eq. (6.14), p. 44" [8b].

6.6. *Hamiltonian and Other Properties

"The line element in *mass-weighted configuration space ds* is given by

$$ds^2 = \sum_{k=1}^{n} m^k (dx^k)^2 = \sum_{i,j=1}^{n} g_{ij} dq^i dq^j \tag{6.3}$$

where the x^k are space-fixed Cartesian coordinates of the atoms ($m^{(3r-3)+1} = m^{(3r-3)+2} = m^{(3r-3)+3}$ is the mass of the rth atom. (NOTE: The original formula in the paper has been corrected). The

q^i are generalized coordinates, and g_{ij} is a symmetric, covariant second-order tensor [8c, p. 24], given by

$$g_{ij} = \sum_{i=1}^{n} m^k \frac{\partial x^k}{\partial q^i} \frac{\partial x^k}{\partial q^j}. \tag{6.4}$$

(a typo present in the original paper has been corrected)

The contravariant tensor conjugate to g_{ij} is g^{ij}:

$$g^{ij} = \sum_{k=1}^{n} \frac{1}{m^k} \frac{\partial q^i}{\partial x^k} \frac{\partial q^j}{\partial x^k} \tag{6.5}$$

$$\sum_{j=1}^{n} g^{ij} g_{jk} = \sum_{j=1}^{n} g_{kj} g^{ji} = \delta_k^i, \tag{6.6}$$

where δ_k^i is 0 or 1 according as $i \neq k$ or $i = k$.

The kinetic energy Tequals $\frac{1}{2}(ds/dt)^2$ and so is given by Eq. (6.7) in terms of the generalized velocities $\dot{q}^i$

$$T = \frac{1}{2} \sum_{i,j=1}^{n} g_{ij} \dot{q}^i \dot{q}^j. \tag{6.7}$$

Some of the $\dot{q}^i$'s are usually rotations, with the result that many of the g_{ij} are then neither diagonal nor constant. Since the generalized momentum p_i equals $\partial(T - U)/\partial \dot{q}^i$, where $U(q^1, \ldots, q^n)$ is the potential energy, p_i is given by Eq. (6.8). From Eqs. (6.6) to (6.8), Eq. (6.9) is obtained for H, the Hamiltonian of the system:

$$p_i = \sum_{i,j=1}^{n} g_{ij} \dot{q}^j \tag{6.8}$$

$$H = \frac{1}{2} \sum_{i,j=1}^{n} g^{ij} p_i p_j + U(q^1, \ldots, q^n). \tag{6.9}$$

We also need the line element ds in *ordinary configuration space*:

$$ds^2 = \sum_{k=1}^{n} (dx^k)^2 = \sum_{i,j=1}^{n} a_{ij} dq^i dq^j, \tag{6.10}$$

where a_{ij} is a covariant tensor. The contravariant tensor a^{ij} is conjugate to it. Both are defined in Eq. (6.11):

$$a_{ij} = \sum_{k=1}^{n} \frac{\partial x^k}{\partial q^i} \frac{\partial x^k}{\partial q^j}; \quad a^{ij} = \sum_{k=1}^{n} \frac{\partial q^i}{\partial x^k} \frac{\partial q^j}{\partial x^k}, \tag{6.11}$$

$$\sum_{j=1}^{n} a^{ij} a_{jk} = \sum_{j=1}^{n} a_{kj} a^{ji} = \delta_k^i. \tag{6.12}$$

We now make use of some results on determinants. Because of the product rule, Eq. (6.13) follows from Eq. (6.4), and Eq. (6.14) from Eq. (6.11), see Note 11 in Ref. [7]:

$$\underset{i,\, j\, =\, 1}{\overset{n}{\det}} g_{ij} = \prod_{k=1}^{n} m^k \left(\underset{i,\, j\, =\, 1}{\overset{n}{\det}} \frac{\partial x^i}{\partial q^j} \right)^2, \tag{6.13}$$

where the indexes i and j in the determinant go from 1 to n

$$a = \underset{i,\, j\, =\, 1}{\overset{n}{\det}} a_{ij} = \left(\underset{i,\, j\, =\, 1}{\overset{n}{\det}} \frac{\partial x^i}{\partial q^j} \right)^2. \tag{6.14}$$

The *volume element* in mass-weighted configuration space and that in ordinary configuration space are denoted by $d\tau$ and dV, respectively [8c]:

$$d\tau = (\det g_{ij})^{\frac{1}{2}} \prod_{i=1}^{n} dq^i \tag{6.15}$$

$$dV = (\det a_{ij})^{\frac{1}{2}} \prod_{i=1}^{n} dq^i = \left(\det_{i,j=1}^{n} \frac{\partial x^i}{\partial q^j} \right) \prod_{i=1}^{n} dq^i, \qquad (6.16)$$

where $\det \partial x^i/\partial q^j$ is understood to be the positive square root of $(\det \partial x^i/\partial q^j)^2$.

Because of Eq. (6.13) one obtains Eq. (6.17) from Eq. (6.15):

$$d\tau = \left[\prod_{k=1}^{n} (m^k)^{\frac{1}{2}} \right] \det \frac{\partial x^i}{\partial q^j} \prod_{i=1}^{n} dq^i = \left[\prod_{k=1}^{n} (m^k)^{\frac{1}{2}} \right] dV. \qquad (6.17)$$

The *area element* of a coordinate hypersurface on which q^N is a constant will be denoted by $d\sigma$ and by dS for mass-weighted and ordinary configuration space, respectively. These area elements are the volume elements in an $n-1$ dimensional space in which dq^N is zero, see **M47**. Hence

$$d\sigma = (\det_{i,j \neq N} g_{ij})^{\frac{1}{2}} \prod_{i \neq N} dq^i \qquad (6.18)$$

$$dS = (\det_{i,j \neq N} a_{ij})^{\frac{1}{2}} \prod_{i \neq N} dq^i \qquad (6.19)$$

Since g^{NN} and $\det_{i,j\neq N} g_{ij}$ are each the cofactor of g_{NN} in g, they are equal. From Eq. (6.18) one then obtains Eq. (6.20). Equation (6.21) follows similarly from Eq. (6.19), since both a^{NN} and $\det_{i,j\neq N} a_{ij}$ are the cofactor of a_{NN} in a:

$$d\sigma = (g g^{NN})^{\frac{1}{2}} \prod_{i \neq N} dq^i \qquad (6.20)$$

$$dS = (a a^{NN})^{\frac{1}{2}} \prod_{i \neq N} dq^i. \quad [7]\text{''} \qquad (6.21)$$

6.7. *Effective Mass

Area Element versus Volume Element

"An effective mass $m^{\ddagger}$ *for motion normal to* S in ordinary n-dimensional configuration space may be defined in several ways. A definition suited to our purpose is the following: When the momentum $\mathbf{p}$ is normal to S in this configuration space the proportionality factor of $p^2/2$ in the kinetic energy is designated by $1/m^{\ddagger}$." Marcus demonstrates then [7, p. 2627] that

$$m^{\ddagger} = a^{NN}/g^{NN}$$

where the indexes *NN* have been used instead of *rr*.

"The area element $d\sigma$ of S_N ... equals the n-dimensional volume element $d\tau$ divided by δ, the perpendicular distance between S_N and another q^N-coordinate hypersurface for which q^N differs by dq^N. If $\boldsymbol{\delta}$ is a vector normal to S_N and having a length δ its contravariant component along q^N must be equal to dq^N by definition. Some manipulation then shows that $\delta = dq^N/(g^{NN})^{\frac{1}{2}}$. The value for $d\sigma$ then follows" [8b, p. 609].

We then have $d\sigma = d\tau/\delta$, that is, $d\tau = d\sigma\delta \Longrightarrow d\tau = (g^{NN})^{-\frac{1}{2}} d\sigma dq^N$.

6.8. Expression for the Rate Constant

"We consider any particular pair of reactants (or *a* reactant, in the case of *intramolecular* ET). These 'labeled' reactants may be any two given molecules in solution or one molecule and the electrode, and each may form complexes to various extents with other ions and molecules. In effect, we need to calculate the probability that the vibrational–rotational–translational coordinates of the entire system are such that the system is *in the vicinity* of the many-dimensional *intersection hypersurface* in configuration space.

It is assumed in the following that the distribution of systems in the vicinity of the intersection region of Figs. 1 or 2 is an equilibrium one.(7) The usual equilibrium-type derivation of the rate of a homogeneous or heterogeneous reaction in the literature employs a special form for the kinetic energy, a form consistent with the set of configurations of the activated complex being describable by a hyperplane in configuration space.(8) A more general curvilinear formulation has been given recently [7]. Upon integrating over a number of coordinates which leave the potential energy invariant one obtains Eqs. (6.22), (6.23), and (6.24) for homogeneous bimolecular reactions, homogeneous unimolecular reactions, and heterogeneous reactions, respectively [7]." The U in the Eqs. (6.22)–(6.24) is denoted there by $U^{\ddagger}$.

$$k_{\mathrm{bi}} = (8\pi kT)^{\frac{1}{2}} \int_S \frac{\exp(-U/kT)R^2(m^{\neq})^{-\frac{1}{2}}dS}{Q} \tag{6.22}$$

$$k_{\mathrm{uni}} = \left(\frac{kT}{2\pi}\right)^{\frac{1}{2}} \int_S \frac{\exp(-U/kT)(m^{\neq})^{-\frac{1}{2}}dS}{Q} \tag{6.23}$$

$$k_{\mathrm{het}} = \left(\frac{kT}{2\pi}\right)^{\frac{1}{2}} \int_s \frac{\exp\left(-U/kT\right)\left(m^{\neq}\right)^{-\frac{1}{2}}dS}{Q} \tag{6.24}$$

Note 13 in Ref. [1]: "In these equations, S is an abbreviation, made for brevity of notation, for S_{int} where the subscript means *internal*, since several integrations over 'external coordinates' have been performed and there remains only the integration over a hypersurface in internal coordinate space [7]. Similarly, the symbols S', V, and V' discussed later should bear a subscript int, which is omitted here for brevity."

"In these equations, $m^{\ddagger}$ is the effective mass for motion normal to the hypersurface S, R is the distance between the two reactants (normally between their centers of mass)."

Q is the configuration integral [9, p. 116] for the reactants when they are far apart [7]

$$Q = \int e^{-U/kT} dV_{\mathrm{int}}$$

dS_{int} and dV_{int} are the area and volume elements in a many-dimensional internal coordinate space. "Both $m^{\ddagger}$ and R may vary over S_{int}.[(9)]" U is the potential energy function.[(10)] dS_{int} will be abbreviated in the following with dS.

"In adapting these equations to ET reactions, one should consider the possibility of the reaction occurring nonadiabatically[(11)] and, in the case of electrodes, should consider the existence of many levels which may accept or donate an electron to a reactant in solution. In the framework of a *classical*[(12)] treatment of the motion of the nuclei in (1) to (3), a factor κ can be shown to appear in the integrand. . . . κ is a *momentum-weighted average*[(13)] of the *transition probability* from the R to P surface *per passage* through the intersection region. (It is momentum weighted since the transition probability depends on the momentum.) κ can vary over S. Normally, we take κ as approximately equal to unity *when the reactants are near each other*, introducing thereby the assumption that the reaction is adiabatic.

In the case of Eq. (6.24), the situation is somewhat more complex because of the presence of the many electrode levels. At present (1965) there is, in the literature, no theoretical calculation of the transfer probability from a level R to a continuum (essentially) of levels P, per passage through the intersection range, for the entire range of transfer probabilities from 0 to 1. Such a calculation would take into account the fact that in an unsuccessful passage through the intersection region the system can also revert to other R levels different from the original one. At present (1965) only the limiting case of very small transfer probability has been considered in the literature [10]. In this case, transfers to and from each of the levels have been treated independently using perturbation theory; they do not interfere at this limit.

When the transfer probability in electrode reactions is fairly large *when ion and electrode are close*, a different approach must be employed [1, note 15]. Here, we take advantage of the fact that for a metal electrode most of the electron transfers occur to and from levels near the Fermi level [1, note 15]: In the terminology of a one-electron model, most of the levels several *kT* below the Fermi level are fully occupied and cannot accept more electrons. The Boltzmann factor discourages transfer to the rather unoccupied levels several *kT* above the Fermi level. Conversely, transfers from the occupied levels below the Fermi level are discouraged by a higher overall energy barrier to reaction while transfer from a higher level is discouraged by the fact that most of the higher levels are unoccupied. To illustrate this point more precisely, let $n(\varepsilon)$ be the density of the "one-electron model levels" for the electrode and $f(\varepsilon)$ the Fermi–Dirac distribution,

$$f(\varepsilon) = \{\exp[(\varepsilon - \bar{\mu}_e)/kT] + 1\}^{-1}$$

where ε is the energy of one of these levels and where $\bar{\mu}_e$ is the electrochemical potential of electrons in the metal. Both ε and $\bar{\mu}_e$ depend on the electrostatic potential of the metal ϕ:

$$\varepsilon = \varepsilon(0) - e\phi \quad \bar{\mu} = \mu - e\phi$$

where $\varepsilon(0)$ is the value of ε at $\phi = 0$ and μ is the chemical potential [11, p. 299].

The probability that electron transfer from the electrode to the ion or molecule in solution will occur from a "one-electron model" level of energy ε would be expected to depend on ε by a factor roughly proportional to

$$n(\varepsilon) f(\varepsilon) \exp(\varepsilon/2kT) \tag{6.26}$$

the third factor arising in the region where the "electrochemical transfer coefficient" is 0.5, a common value [3]. Since $n(\varepsilon)$ is a weak function of ε, the last two factors in Eq. (6.26) largely determine *the most probable value of* ε. The maximum of Eq. (6.26) is then easily shown to occur at $\varepsilon = \bar{\mu}_e$. Similarly, contribution to ET from an ion in solution to a particular level ε would be expected to vary with ε as in

$$n(\varepsilon)[1 - f(\varepsilon)] \exp(-\varepsilon/2kT),$$

which also has a maximum at $\varepsilon = \bar{\mu}_e$, of course.

Because of this circumstance (that most contributions arise from levels ε near $\bar{\mu}_e$), we approximate the situation in Fig. 2 by replacing the set of surfaces R by one surface and P by another surface, corresponding to an electronic energy in the electrode given by $\bar{\mu}_e$ as above [1, note 15]. If ET accompanies each passage through the intersection region in Fig. 1, the reaction is referred to as "adiabatic," purely by analogy with the term in the homogeneous reaction. The reaction rate is given by Eq. (6.24), where the equation of S depends on the electrostatic potential. On the other hand, when the transfer probability per passage is very weak, a term κ should be introduced in the integral, κ being a velocity-weighted transition probability appropriately summed over all energy levels in the electrode. A value for κ in this weak interaction limit has been discussed elsewhere [12]. When a complete calculation for the transfer probability from and to a continuum of electrode levels becomes available it can be used to estimate κ. Normally, however, we assume the electrode reaction to be "adiabatic" and so take $\kappa \cong 1$ on the average.

6.9. *Relation of the Surface Integrals (6.22) to (6.24) to Volume Integrals

Although some deductions can be made from the surface integrals in Eqs. (6.22) to (6.24) when the equation of the intersection surface S is simple, we find it convenient to express the surface integral in terms of a volume one. The same aim was followed in Part IV but in a less precise way. The principal equation derived in this section is Eq. (6.27), which is later used in conjunction with Eqs. (6.22) to (6.24) to obtain an expression for $\boldsymbol{k}_{\mathbf{rate}}$.

Let U^r be the potential energy function for the reactants and U^p be that for the products. As mentioned earlier, the intersection of R and P surfaces forms a hypersurface in configuration space.

This hypersurface is called the "*reaction hypersurface.*" Its equation is given by Eq. (6.28):

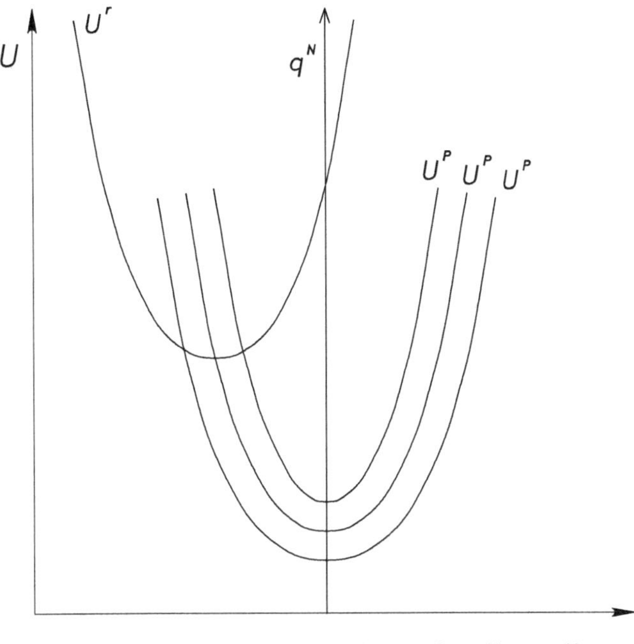

Fig. 1.*

$U^r - U^p = 0$ (for points on reaction hypersurface) (6.28)

This surface is a member of a family of hypersurfaces in configuration space represented by Eq. (6.29) where c is a constant:

$$U^r - U^p = c \tag{6.29}$$

The surface (6.29) can be obtained from surface (6.28) by lowering the P surface in Fig. 1 by an amount c.

We employ a coordinate system q^1 to q^n used in the derivation of Eqs. (6.22) to (6.24) and recall that one coordinate, q^N, in the internal coordinate space was chosen to be *a coordinate constant on the hypersurface* (6.28). Let q^N be zero there. In fact, each member of the family of hypersurfaces (6.29) is made a coordinate hypersurface for q^N" [1]. See Fig. 1*.

"We consider any of the integrals in Eqs. (6.22) to (6.24), include the factor κ in the integrand, and write dS as $dS' dR$. The factor κ

depends primarily on R. In the following expression, the same symbol κ is used to denote this κ, averaged over S'.

Note 18: The κ appearing in Eq. (6.30) is now a symbol representing

$$\int \kappa (m^{\ddagger})^{-\frac{1}{2}} \exp\left(\frac{-U}{kT}\right) dS' \Big/ \int (m^{\ddagger})^{-\frac{1}{2}} \exp\left(\frac{-U}{kT}\right) dS'$$

where κ is the original kappa.

Each of the integrals in (1) to (3) can be rewritten as

$$\int_R \kappa R^{\alpha} \left[\int_{S'} \exp\left(-\frac{U}{kT}\right) (m^{\neq})^{-\frac{1}{2}} dS' \right] dR,^{(14)} \tag{6.30}$$

where α is 2, 0, or 0, according to whether Eqs. (6.22), (6.23), or (6.24) is the equation involved.

We wish to relate *the earlier integral over* S' to a volume integral (6.31) over the internal coordinate space *at fixed* R as in Eq. (6.38) and finally as in Eq. (6.27):

$$\int_{V'} \exp\left(-\frac{U*}{kT}\right) dV',^{(15)} \tag{6.31}$$

where U^* is a function to be determined; $R^{\alpha} dV'$ is an element of volume of this internal coordinate space at fixed R.

Note 19: These 'internal coordinates' were defined as those coordinates for which integration was not performed in obtaining Eqs. (6.22) to (6.24).

To establish Eq. (6.38), we first note from Appendix II that the distribution in volume which is centered on S' (but not confined to S', of course), is f^* given by Eq. (6.32)

$$f^* = \exp\left(-\frac{U^*}{kT}\right) \Big/ \int \exp\left(-\frac{U^*}{kT}\right) dV' \tag{6.32}$$

where

$$U^* = U^r + m(U^r - U^p)^{(16,17)} \tag{6.33}$$

and m is a parameter which varies with the coordinate R and which is determined in Section 6.16.[16-18] On S', one sees from Eq. (6.3), U^* equals U^r for any given R.

We then recall from Ref. [7] that dV' and dS' are related by Eq. (6.34), and we introduce a quantity $I(q^N, R)$ defined by Eq. (6.35)

$$dV' = (a^{NN})^{-1} dS' dq^N \quad [19,20] \tag{6.34}$$

$$\exp\left[-I\left(q^N, R\right)\right] = \int \exp\left(-\frac{U^*}{kT}\right)\left(a^{NN}\right)^{-\frac{1}{2}} dS' \quad [21] \tag{6.35}$$

where a^{NN} is conjugate to an element a_{NN} in the line element of the many-dimensional configurational space. On recalling from Ref. [7] that $m^\ddagger$ equals a^{NN}/g^{NN}, where g^{NN} is conjugate to an element g^{NN} in the line element of the corresponding mass-weighted configuration space, the S' integral in Eq. (6.30) can then be rewritten as in Eq. (6.36), where $\langle (g^{NN})^{\frac{1}{2}} \rangle$ is a suitable average over S'.

$$\int \exp\left(-\frac{U}{kT}\right)(m^\ddagger)^{-\frac{1}{2}} dS' = \langle (g^{NN})^{\frac{1}{2}} \rangle \exp\left[-I(0, R)\right]. \quad [21] \tag{6.36}$$

In deriving Eq. (6.36), we have also used the fact that U^* equals U^r on S'.

Finally, the integral in Eq. (6.31)

$$\int_{V'} \exp\left(-\frac{U^*}{kT}\right) dV'$$

can be rewritten as $\int \exp\left[-I\left(q^N, R\right)\right] dq^N$ in virtue of Eqs. (6.34) and (6.35). On the basis of a *Gaussian expansion*, Eq. (6.37) can be derived

$$\int \exp\left[-I\left(q^N, R\right)\right] dq^N = \left[2\pi / I''(0, R)\right]^{\frac{1}{2}} \exp[-I(0, R)]. \quad [22] \tag{6.37}$$

(**Typo** corrected in the original equation)

where $I''(0, R)$ is $d^2I\left(q^N, R\right)/dq^{N2}$, evaluated on S' (and hence at $q^N = 0$). One then obtains from Eq. (6.30),

$$\int_R \kappa R^\alpha \left[\int_{S'} \exp\left(-\frac{U}{kT}\right)(m^\ddagger)^{-\frac{1}{2}} dS'\right] dR$$
$$= \int_R \frac{\kappa R^\alpha \langle (g^{NN})^{\frac{1}{2}}\rangle \exp[-F^*(R)/kT]}{[2\pi/I''(0, R]^{\frac{1}{2}}} dR \qquad (6.38)$$

where $F^*(R)$ is the configurational free energy of a system having the potential energy function U^* for this separation distance R

$$\exp\left(-\frac{F^*(R)}{kT}\right) = \int \exp\left(-\frac{U^*}{kT}\right) dV'^{(23)} \qquad (6.39)$$

To complete the proof of Eq. (6.38), we must verify Eq. (6.37).

We recall from the definition of q^N that $U^r - U^p$ depends only on q^N on any hypersurface (6.29). To ensure centering of the system on S', that is, at $q^N = 0$, $m(R)$ is to be selected so that Eq. (6.40) is satisfied:

$$\langle U^r \rangle = \langle U^p \rangle \qquad (6.40)$$

where $\langle\ \rangle$ denotes average with respect to the distribution function f^*. On using Eqs. (6.32), (6.34), and (6.35), Eq. (6.40) becomes

$$\int \exp\left[-I\left(q^N, R\right)\right]\left(U^r - U^p\right) dq^N = 0 \qquad (6.41)$$

Because of the centering of f^*, expansion of $I\left(q^N, R\right)$ about $q^N = 0$ is permissible, as is one of $U^r - U^p$

$$I\left(q^N, R\right) = I(0, R) + q^N I'(0, R) + \left[\left(q^N\right)^2 \Big/ 2!\right] I''(0, R) + \ldots \qquad (6.42)$$

$$U^r - U^p = 0 + q^N (U^r - U^p)' + \ldots, \qquad (6.43)$$

where $'$ indicates a derivative with respect to q^N, evaluated at $q^N = 0$. We retain only leading terms in each case. Insertion of Eqs. (6.42) and (6.43) into Eq. (6.41) followed by integration reveals that $I'(0, R)$ vanishes. Introduction of Eq. (6.42) into the left-hand side of Eq. (6.37) then establishes Eq. (6.37).

Some of the terms in Eq. (6.38) can be expressed in terms of quantities of more immediate physical significance. It may be shown from Eqs. (6.32), (6.34), (6.35), (6.42), and the vanishing of $I'(0, R)$ that

$$I''(0, R) = \langle (\delta q^N)^2 \rangle^{-1}, \tag{6.44}$$

where $\langle (\delta q^N)^2 \rangle$ is the mean square deviation of q^N.[22] The mean square deviation of the *perpendicular distance s* from the reaction hypersurface is given by Eq. (6.45) for small s's:

$$\left\langle (\delta s)^2 \right\rangle = \left\langle \left(a^{NN}\right)^{-1} \left(\delta q^N\right)^2 \right\rangle = \left\langle \left(a^{NN}\right)^{-1} \right\rangle \left\langle \left(\delta q^N\right)^2 \right\rangle, \tag{6.45}$$

where $\langle (a^{NN})^{-1} \rangle$ is a suitable average of $(a^{NN})^{-1}$.[23]

We make use of the fact that $\langle (g^{NN})^{\frac{1}{2}} \rangle \langle (a^{NN})^{-1} \rangle^{\frac{1}{2}}$ has units of $(\text{mass})^{-\frac{1}{2}}$, and denote it by $(m^*)^{-\frac{1}{2}}$, and that the integrand in Eq. (6.38) has a maximum at some value of R, denoted in Eq. (6.45) by R. [When R becomes large κ tends to zero and when R is small the Van der Waals' repulsion makes $F^*(R)$ large.] On treating the integrand as a Gaussian function of $R^{(24)}$, Eq. (6.38) becomes

$$\int_R \kappa R^\alpha \left[\int_{S'} \exp\left(-\frac{U}{kT}\right) (m^\ddagger)^{-\frac{1}{2}} dS' \right] dR$$
$$= \kappa \rho R^\alpha (m^*)^{-\frac{1}{2}} \exp\left[-F^*(R)/kT\right] \tag{6.27}$$

where κ is evaluated at this value of R and where ρ is a ratio (6.46) whose value should be of the order of magnitude of unity:

$$\rho = \left[\left\langle (\delta R)^2 \right\rangle / \left\langle (\delta s)^2 \right\rangle\right]^{\frac{1}{2}} \tag{6.46}$$

where $\langle(\delta R)^2\rangle$ is the mean square deviation in the value of R; ρ[25] and κ can be calculated from more specific models when the various integrals defining them can be evaluated.

6.10. Rate Constants in Terms of ΔF^*

Let F^r be the *configurational free energy* associated with the Q of Eqs. (6.22) to (6.24). Thereby, it is the free energy contribution for an equilibrium contribution of 'V' coordinates' when the reactants are *very far apart* but *fixed in position*,

$$F^r = -kT \ln Q. \tag{6.47}$$

Let $F^r(R)$ be the corresponding quantity when the reactants are a distance R apart. We then have

$$w^r = F^r(R) - F^r, \tag{6.48}$$

where w^r can be called the reversible work to bring the reactants from *fixed* positions *infinitely* far apart to the cited *separation distance*. We also introduce $\Delta F^*(R)$:

$$\Delta F^*(R) = F^*(R) - F^r(R) \tag{6.49}$$

Equations (6.22) to (6.24) for k_{rate} now yield Eqs. (6.50) to (6.52), when Eqs. (6.27) and (6.47) to (6.49) are used,

$$k_{\text{bi}} = \kappa\rho Z_{\text{bi}} \exp(-w^r/kT) \exp[-\Delta F^*(R)/kT] \tag{6.50}$$

$$k_{\text{uni}} = \kappa\rho (kT/2\pi m^*)^{\frac{1}{2}} \exp(-\Delta F^*/kT) \tag{6.51}$$

$$k_{\text{het}} = \kappa\rho Z_{\text{het}} \exp(-w^r/kT) \exp[-\Delta F^*(R)/kT] \tag{6.52}$$

where Z_{bi} and Z_{het} are given by

$$Z_{\text{bi}} = (8\pi kT/m^*)^{\frac{1}{2}} R^2 \text{[26]} \quad Z_{\text{het}} = (kT/2\pi m^*)^{\frac{1}{2}} \tag{6.53}$$

[In Eq. (6.51) ΔF^* is simply $F^* - F^r$, there being only one reactant.] Z_{bi} is in fact the collision number of two uncharged species in solution when they have unit concentration, when their reduced mass is m^*, and when their collision diameter is R. Z_{het} is the collision number of an uncharged species with unit area of an interface (here, the electrode), when it has unit concentration and when its mass is m^*.

F^* and F^r in Eqs. (6.50) to (6.52) involve an integration over the orientation of each reactant.[(27)] The integrand in F^r is independent of these coordinates and, in the case of the 'outer sphere electron transfer mechanism' the integrand in F^* is assumed to be independent of them also.[(28)] (For purposes of deriving many of the correlations in Section 6.19, this assumption could be weakened because of cancellations.)[(29)] Integration over these coordinates is regarded as having been performed in Eqs. (6.50) to (6.53), since the orientational factors now cancel in $\Delta F^*(R)$.[(30)] Thus, in the subsequent calculation of F^* and F^r each reactant may be regarded as fixed not only in position, as before, but in orientation also[(31)]" [1].

6.11. Distribution Function and the Free Energy

"Eq. (6.39) for $F^*(R)$ can be rewritten as in Eq. (6.54), with the aid of Eqs. (6.32), (6.33), and (6.40):

$$F^*(R) = \langle U^r \rangle + kT \langle \ln f^* \rangle \tag{6.54}$$

where

$$\langle U^r \rangle = \int U^r f^* dV'^{(32)}, \quad \langle \ln f^* \rangle = \int (\ln f^*) f^* dV' \tag{6.55}$$

Since $-k \langle \ln f^* \rangle$ is the *configurational entropy* of the system having the distribution function f^* and since $\langle U^r \rangle$ is the mean potential energy of a nonequilibrium system having a potential energy function U^r but a distribution function of f^* *inappropriate* to this U^r, we see that $F^*(R)$ is also equal to the *configurational free energy* of this *nonequilibrium system.*

In obtaining an expression for $F^*(R)$, it is convenient to divide, as one usually does in related problems, the internal coordinates *at the given R* into two groups: V_i' coordinates describing the positions of the atoms *in the inner coordination shells of the reactants*, and V_o' coordinates describing the positions of the atoms *of the medium* relative to each other *and to those in the inner coordination shells.* It is also convenient to write U as the sum of two terms, U_i and U_o, one describing the *intramolecular interactions* of the atoms in each coordination shell, the other describing the interactions of the atoms of the medium with each other *and with those of the inner coordination shells*. Thus, U_i depends only on the V_i' coordinates; U_o depends primarily on the V_o' coordinates, *but also depends on the* V_i' *ones*,

$$U = U_i + U_o \tag{6.56}$$

The quantities U_i^* and U_o^* are defined in terms of U_i^r, and so on, to be given by Eq. (6.33), with i and o subscripts, respectively. Then, U^* is the sum of U_i^* and U_o^*.

The volume element dV' is written as

$$dV' = dV_i' dV_o' \tag{6.57}$$

where dV_i' is defined as the product of the differentials $\left(\prod_i dq^i\right)$ of the V_i' coordinates. Thereby, dV_o' contains the Jacobian appearing in dV'. It may vary, therefore, with the V_i' coordinates.

In calculating F^* and F^r, we may evaluate the integrals appearing in them by first integrating over the V_o' coordinates and then over the V_i' ones. This procedure is convenient since *the* V_i' *ones perform small oscillations* while *the others can undergo considerable fluctuations.* With this procedure in mind, we define new quantities f_i^* and f_o^*, the former depending only on the V_i' coordinates, the latter depending on the V_o' coordinates, and *parametrically* on the V_i' ones.

$$f_o^* = \exp\left[\left(\chi_o^* - U_o^*\right)/kT\right] \tag{6.58}$$

$$f_i^* = \exp\left(-\frac{\bar{U}_i^*}{kT}\right) \bigg/ \int \exp\left(-\frac{\bar{U}_i^*}{kT}\right) dV_i' \tag{6.59}$$

Where

$$\bar{U}_i^* = U_i^* + \chi_o^{*\,(\mathbf{33})} \tag{6.60}$$

$$\exp\left(-\frac{\chi_o^*}{kT}\right) = \int \exp\left(-\frac{U_o^*}{kT}\right) dV_o'. \tag{6.61}$$

One then obtains

$$f^* = f_o^* f_i^* \tag{6.62}$$

Quantities f_o^r, f_i^r ... can be defined, by replacing the * by an r superscript in Eqs. (6.58) to (6.61). However, χ_0^r is simply F_o^r, the V_o' contribution to the configurational free energy of the reactants *for the given value of the* V_i' *coordinates*

$$\exp\left(-\frac{F_o^r}{kT}\right) = \int \exp\left(-\frac{U_o^r}{kT}\right) dV_o'.$$

We also introduce F_o^*, defined by

$$F_o^* = \left\langle U_o^r \right\rangle_{*o} + kT\left\langle \ln f_o^* \right\rangle_{*o} \tag{6.63}$$

where the average $\langle\ \rangle_{*o}$ is computed with respect to f_o^*. F_o^* is the V_o' contribution to the free energy of the nonequilibrium system having the potential energy function U^r and the distribution function f_o^*. The first and second term in Eq. (6.63) are the energy and entropy contributions, respectively.

To obtain an expression for $\bar{U}_i^*$, the function largely controlling the V_i' coordinates distribution, we first obtain Eq. (6.64) by introduction of Eq. (6.58) into Eq. (6.63) and by use of Eq. (6.33) with subscripts o's added. Equations (6.60), (6.56), (6.56) and, with subscripts i's added, (6.33), then yield Eq. (6.64), since U_i^r and U_i^p are independent of the V_o' coordinates:

$$\chi_0^* = F_o^* + m\langle U_o^r - U_o^p\rangle_{*o} \tag{6.64}$$

$$\bar{U}_i^* = U_i^r + F_o^* + m\langle U^r - U^p\rangle_{*o} \tag{6.65}$$

Equations (6.54), (6.56), (6.62), and (6.63) yield Eq. (6.66a), when one notes that U_i^r and f_i^* are independent of the value of the V_o' coordinates. Equation (6.66b) then follows from Eqs. (6.40), (6.65), and (6.66a):

$$F^*(R) = \langle U_i^r + F_o^* \rangle_{*i} + kT \langle \ln f_i^* \rangle_{*i} \tag{6.66a}$$

$$= \langle \bar{U}_i^* + kT \ln f_i^* \rangle_{*i} \tag{6.66b}$$

where the average $\langle\ \rangle_{*i}$ is computed with respect to the distribution function f_i^*.

The free energy $F^r(R)$, given by

$$-kT \ln \int \exp\left(-\frac{U^r}{kT}\right) dV'$$

evaluated at R, can also be shown to be given by expressions similar to Eq. (6.65) but with the asterisks replaced by r's

$$F^r(R) = \langle U_i^r + F_o^r \rangle_{ri} + kT \langle \ln f_i^r \rangle_{ri}. \tag{6.67}$$

To evaluate k_{rate}, we compute ΔF^* from Eqs. (6.49), (6.66), and (6.67) and use Eq. (6.65)

The similarity of Eqs. (6.66) and (6.67) and later of Eqs. (6.69) and (6.70) is an example of the fact that the properties of the [$\boldsymbol{r}$] system can be obtained from those of the [$*$] system by setting $m = 0$. The origin of this behavior is seen in the original Eq. (6.33) defining the [$*$] system.

6.12. Vibrational Contribution to $\Delta F^*(R)$

While it is not necessary to introduce the harmonic approximation, the expressions are appreciably simplified by it. There is evidence that the approximation is adequate for many reactions of interest.

It is recalled that the generalized coordinates were denoted by q^j. Let the first n_i of these be *vibrational coordinates* of the reacting species, that is, the V_i' coordinates, and let q_*^j denote the value of the jth vibrational q^j occurring *at the minimum* of $\bar{U}_{\mathbf{i}}^*$. We have

$$\bar{U}_i^* = \bar{U}_i^*(\mathbf{Q}_*) + \frac{1}{2}\sum_{j,k=1}^{n_i} f_{jk}^*(q^j - q_*^j)(q^k - q_*^k)$$

$$= \bar{U}_i^*(\mathbf{Q}_*) + \frac{1}{2}(\mathbf{Q}^T - \mathbf{Q}_*^T)\cdot\mathbf{F}^*(\mathbf{Q} - \mathbf{Q}_*) \qquad (6.71)$$

where $\mathbf{Q} - \mathbf{Q}_*$ denotes a column vector whose elements are $q^j - q_*^j$. $\mathbf{F}^*$ denotes a square matrix whose elements are f_{ij}^*. "The f_{ij}^* are force constants." The superscript T denotes a transpose (a row vector here), and the dot indicates the scalar product of this row vector with the column vector $\mathbf{F}^*(\mathbf{Q} - \mathbf{Q}_*)$.

On multiplying numerator and denominator of Eq. (6.59) by $\exp[\bar{U}_i^*(\mathbf{Q}_*)/kT]$ (**Typo** in the original equation), introducing this expression for f_i^* into Eq. (6.66b), then using Eq. (6.71), integrating, and finally introducing an expression for $\bar{U}_i(\mathbf{Q}_*)$ [Eq. (6.65) evaluated at $\mathbf{Q} = \mathbf{Q}_*$], Eq. (6.72) follows:

$$F^*(R) = U_i^r(\mathbf{Q}_*) + F_o^*(\mathbf{Q}_*) + m\langle U^r(\mathbf{Q}_*) - U^p(\mathbf{Q}_*)\rangle_{*_o}$$
$$-\frac{1}{2}kT\ln\left[(2\pi kT)^{n_i}\Big/\left|f_{jk}^*\right|\right]^{(34)} \qquad (6.72)$$

where $\left|f_{ik}^*\right|$ is the determinant of the f_{ik}^*'s.

If q_r^j denotes the value of a vibrational q^j occurring at the minimum of U_i^r, it can be shown that F^r is given by Eq. (6.73) after a quadratic expansion of $\bar{U}_i^r(\mathbf{Q})$ about $\mathbf{Q}_r$,

$$F^r = U_i^r(\mathbf{Q}_r) + F_o^r(\mathbf{Q}_r) - \frac{1}{2}kT\ln\left[(2\pi kT)^{n_i}\Big/\left|f_{jk}^r\right|\right]^{(35)} \qquad (6.73)$$

where

$$f_{ik}^r = (\partial^2\bar{U}_i^r/\partial q^i\partial q^k) \quad \text{at} \quad \mathbf{Q} = \mathbf{Q}_r. \qquad (6.74)$$

Equation (6.32) is then obtained from Eqs. (6.72) and (6.73) by noting that $\langle U^r(\mathbf{Q}_*) - U^p(\mathbf{Q}_*)\rangle_{*_o}$ vanishes (Appendix V)," (recall that $\langle U^r \rangle = \langle U^p \rangle$) "and that $\bar{U}_i^r(\mathbf{Q}_*)$ can be expanded about the value of $\bar{U}_i^r$ at $\mathbf{Q}_r$:

$$F^*(R) - F^r(R) = \frac{1}{2}\left(\mathbf{Q}_*^{\mathbf{T}} - \mathbf{Q}_r^{\mathbf{T}}\right) \cdot \mathbf{F}^r\left(\mathbf{Q}_* - \mathbf{Q}_r\right) + \Delta F_o^*\left(\mathbf{Q}_*\right)$$
$$+\frac{1}{2}kT \ln\left(\left|f_{jk}^*\right| \Big/ \left|f_{jk}^r\right|\right) \tag{6.75}$$

Where $\Delta F_o^*(\mathbf{Q}) = F_o^* - F_o^r(\mathbf{Q})$ (at any given R)" [1].

In Eq. (6.71), the force constants f_{jk}^* appear, defined as $f_{jk}^* = \partial^2\widetilde{U}_i^*/\partial q^i \partial q^k$. Marcus shows that the force constants for the transition state can be written in terms of the force constants for reactants and products, that is,

$$f_{jk}^* = \partial^2\widetilde{U}_i^*/\partial q^i \partial q^k = (m+1)f_{jk}^r - mf_{jk}^p$$

where $f_{jk}^r = (\partial^2\widehat{U}_i^r/\partial q^j \partial q^k)$ at $\mathbf{Q} = \mathbf{Q}_r$. The symbol q_r^j denotes the value of a vibrational q^j occurring at the minimum of U_i^r. The f_{jk}^p's are analogously defined. It is then convenient to introduce symmetrical and antisymmetrical functions of the force constants:

$$k_{jk} = 2f_{jk}^r f_{jk}^p \Big/ \left(f_{jk}^r + f_{jk}^p\right) \tag{6.76}$$

$$l_{jk} = \left(f_{jk}^r - f_{jk}^p\right) \Big/ \left(f_{jk}^r + f_{jk}^p\right) \tag{6.77}$$

"The first of these quantities was chosen so as to have dimensions of a force constant and the second of these so as to be dimensionless" [1]. Use of the reduced force constants k_{jk} and of the antisymmetrical functions l_{jk} will be made in the following sections.

6.13. Orientation and Other Contributions to ΔF^*

"For purposes of generality we employ the particle description of the potential energy in a macrosystem [13, 14]" [1]. See the preceding chapter. Notice that the notation in Ref. [14] and in the preceding chapter differs somewhat from that in Ref. [1] and in this chapter. τ, U, U^f, and ρ_s^0 there become V_o', U_o, U_i, and ρ_a^0 here. The particle description "introduces fewer assumptions than those normally used in condensed polar media. Because of its comparative generality, it also permits *a simultaneous formulation* of the theory of *homogeneous intermolecular* electron transfer, electron transfers *at electrodes*, and *intramolecular* electron transfers. In this description, the system consists of particles each of which is a reacting molecule or any electrode present, the latter including as part of it any strongly bound layer of ions or solvent. The remainder of the system, the medium, can be regarded as one giant particle.

The potential energy is the sum of an intraparticle term (the energy when the particles are isolated, each having the given intraparticle coordinates) and an interparticle term (the energy change when the particles are brought together *for the given values of the intraparticle coordinates.* The solvent particle possesses a 'cavity' for each reactant particle, which the latter fills when they are brought together.

The intraparticle terms below contain the electronic and potential energy of the reactants and of the medium. The interparticle term is, in the first approximation, the sum of interparticle polar terms and of interparticle electron correlation (i.e., exchange repulsion and London dispersion) energies [14]" [1], see Chapter 5. "It can then be expanded in powers of the permanent charge density ρ_a^0 of the reactants. The usual approximations in the literature correspond to neglect of powers higher than the second, together with the assumption of specific forms for these terms [13].

In terms of the symbols U_i and U_o introduced earlier, we have

$$U = U_i + U_o \tag{6.78}$$

where

$$U_o = U(0) + U(1) + U(2). \tag{6.78a}$$

In Eq (6.78), U_i is the intraparticle term for the reactants and U_o is the sum of the intraparticle term for the medium and of the interparticle term. $U(0)$, $U(1)$, and $U(2)$ depend functionally on zeroth, first and second powers of ρ_a^0 and, respectively, on second, first, and zeroth powers of ρ_M^0, the permanent charge density of the medium [13]. $U(0)$ also contains the intraparticle term for the medium and the electron correlation interparticle term. U_i and ρ_a^0 depend only on the intraparticle coordinates, V_i', of the reactants, and ρ_M^0 depends only on those of the medium, V_o'.

The distribution function f_o^* defined in Eq. (6.58) can be shown to be similar to that which occurs when the permanent charge distribution on a reactant A is $\rho_a^{0\,*}$, given by Eq. (6.79) for all A:

$$\rho_a^{0\,*} = \rho_{ar}^0 + m(\rho_{ar}^0 - \rho_{ap}^0) \tag{6.79}$$

where ρ_{ar}^0 is the permanent charge distribution of molecule A when it is actually a reactant and ρ_{ap}^0 is that when it is a product. . . . Normally, as will be seen later, m will be close to $-\frac{1}{2}$.

The V_o' contribution to the free energy of formation of a system with a nonequilibrium V_o' distribution, ΔF_o^*, *at any given R and at any given* **Q**, has been evaluated elsewhere on the basis of the particle description described above and of an assumption of (at most) partial dielectric saturation [14]:

$$\Delta F_o^* = F_{m(r-p)}^{op} - F_{m(r-p)}\ ^{\mathbf{(36)}} \tag{6.80}$$

In Eq. (6.80), F^{op} and F denote the *polar contributions* to the free energies of two *hypothetical equilibrium and dielectrically unsaturated* systems, each having a ρ_a^0 equal to $m(\rho_{ar}^0 - \rho_{ap}^0)$ on each reactant. The first system is an '*optical polarization*' system [13]," [1] (see Chapter 5) "i.e., a system whose medium responds to these ρ_a^0's only via an electronic polarization. The second system responds via all polarization terms. Both F^{op} and F are quadratic functions of the $m(\rho_{ar}^0 - \rho_{ap}^0)$'s.

It can be shown[26] that $F^{op} - F$ depends on the square of the permanent charge distribution *on the reactants*, in this case $m(\rho_{ar}^0 - \rho_{ap}^0)$.

Note 26: According to Eqs. (10) and (11) of Ref. [15] $F^{op} - F = [\langle U(1)^2\rangle] - \langle U(1)\rangle^2]/2kT$. The latter depends only on the second power of the charge distribution, since $U(1)$ is a linear functional of the first power [14, note 27]" [1].

Note that from a statistical mechanical point of view

$$\Delta F_o^* = F^{op} - F = [\langle U(1)^2\rangle - \langle U(1)\rangle^2]/2kT.$$

has the nature of a *fluctuation* and ΔF_0^* depends on $m^2(\rho_{ar}^0 - \rho_{ap}^0)^2$. Note that as in Chapter 5, the free energy is expressed both in terms of potential energy functions and of charge distributions.

"We may then describe the dependence of ΔF_0^* by

$$\Delta F_o^* = m^2\lambda_o, \tag{6.81}$$

where

$$\lambda_o = \left\langle \left[U_o^r(1) - U_o^p(1) - \left\langle U_o^r(1) - U_o^p(1)\right\rangle\right]^2\right\rangle \Big/ kT^{(37,38)} \tag{6.82}$$

and the averaging function is:

$$\exp\left(-\frac{(U_o^r + U_o^p)}{2kT}\right) dV_o' \Big/ \int \exp\left(-\frac{(U_o^r + U_o^p)}{2kT}\right) dV_o'' [1].$$

For a discussion of this function and of other possible choices see note 27 in Ref. [1].

"To use Eq. (6.81) and those derived earlier, an expression is needed for m. It is derived below after some preliminary analysis involving the standard free energy of reaction, the electrochemical cell potential, and the activation overpotential.

6.14. Standard Free Energy of Reaction

The configurational free energy of the system when the *reacting* species are *labeled* reactant molecules, fixed in position but far apart, was denoted by F^r. The corresponding quantity when the *pair* refers to labeled product molecules was denoted by F^p. The *momentum and translational contributions* of each member of the reacting pair *to the free energy* of the initial state *cancels* that in the final state in these reactions involving no change in the total number of moles of redox species. Thus the difference $F^p - F^r$ is equal to the free energy of reaction when a *pair* of labeled reactants form a pair of labeled products in the prevailing medium.[39]

This free energy of reaction in the prevailing medium can be expressed in terms of "standard" chemical potentials. The chemical potential μ_i can be written as $\mu_i^{0\prime} + kT\ln c_i$, where $\mu_i^{0\prime}$ is the "standard" chemical potential, defined here as the value of μ_i at $c_i = 1$. Because of the labeling, $F^p - F^r$ does not contain a contribution from entropy of mixing of the reactants. Since it is these mixing terms which contribute the $kT\ln c_i$ to μ_i, we therefore have

$$F^p - F^r = \sum_p \mu_p^{0\prime} - \sum_r \mu_r^{0\prime}, \tag{6.83}$$

where $\sum_p$ and $\sum_r$ denote summation over products and reactants. There are *one* or *two* terms in each sum, according as the reaction is *unimolecular* or *bimolecular*.

The right-hand side of Eq. (6.83) is $\Delta F^{0\prime}$, the "standard" free energy of reaction for the prevailing medium, temperature and pressure. Hence,

$$F^p - F^r = \Delta F^{0\prime}. \tag{6.84}$$

It equals $-kT \ln K$, where K is the equilibrium "constant" measured under these conditions. Both $\Delta F^{0\prime}$ and K can vary with *electrolyte concentration*, with temperature, and with pressure.

6.15. Activation Overpotential and Electrode–Solution Potential Difference

For electrode systems, the counterpart of Eq. (6.84) is obtained by considering the free energy of reaction (6.85) for a labeled molecule at any *fixed* position in the body of a solution, but *far from the electrode*, M

$$\text{Red} + M = \text{Ox} + M(ne) \tag{6.85}$$

where Ox and Red denote the oxidized and reduced forms of the labeled molecule in the body of the solution. This free energy change, which accompanies the transfer of n electrons from the ion or molecule to the electrode at a *mean energy level* discussed in Section 6.3, *has a number of contributions* "such as

(i) One from the change in electronic energy
(ii) One from the change in ion–solvent interactions in the vicinity of the ion
(iii) One from the change in vibrational energy

Let F^r now denote the configurational free energy of the system containing the electrode and a labeled reactant, the latter fixed in a position far outside the electrode double layer.

Let F^p denote the corresponding quantity when the labeled molecule is a product, the electrode having gained n electrons as in Eq. (6.85).

The term $F^p - F^r$ is linear in the metal–solution potential difference, as may be seen from the discussion in Section 6.8, and thereby in the half-cell potential E. (E is defined to be the half-cell potential corrected for any ohmic drop and concentration polarization.) We have then

$$F^p - F^r = \Delta + neE \tag{6.86}$$

where Δ is independent of E, and where we have used a standard convention regarding the sign of E. [This convention is one which makes reaction (6.85) increasingly spontaneous with increasing positive $E^{0\prime}$, a quantity defined later.]

Because of the labeling the entropy-of-mixing term of the oxidized molecules and that of the reduced molecules are again absent in F^r and F^p. When the system is at electrochemical equilibrium and when the probability of finding the labeled species as a reactant is the same as that for finding as a product, $F^p - F^r$ must vanish. Also, E then has its equilibrium value, which is E_0' for the case of equal concentrations of the labeled species. [E_0' is the 'standard' oxidation potential or, as it is sometimes called, the formal oxidation potential of the half-cell; E_0' is defined in terms of the equilibrium half-cell potential E_e by Eq. (6.87) for any ratio of concentrations (Red)/(Ox)]

$$E_e = E_0' + (kT/ne)\ln[(\mathrm{Red})/(\mathrm{Ox})]. \tag{6.87}$$

One then obtains, from Eq. (6.86),

$$0 = \Delta + neE_0'. \tag{6.88}$$

Hence

$$F^p - F^r = ne\left(E - E_0'\right) \tag{6.89}$$

We observe from Eq. (6.89) that $E - E_0'$, rather than the activation overpotential $E - E_e$, plays the role of the 'driving force' in these reactions. The same role is played by $\Delta F^{0\prime}$ in the homogeneous reactions.

In terms of formal electrochemical potentials of the product and reactant ions and in terms of the *electrochemical potential of the electrons in the electrode* we have, incidentally, for reaction (6.85),

$$F^p - F^r = \bar{\mu}_p^{0\prime} - \bar{\mu}_r^{0\prime} + n\bar{\mu}_e \tag{6.89}$$

6.16. Equation for *m*

We first note that $\Delta F^{0\prime}$ can be written as the algebraic sum of the following terms:

(i) The free energy change when the reactants are brought together to the separation distance R, w^r;
(ii) The free energy of reorganization of the reacting system at this R, ΔF^*;
(iii) The free energy difference of reactants and products *in this reorganized state*,[40] which equals

$$\left(\left\langle U^p + kT\ln f^*\right\rangle - \left\langle U^r + kT\ln f^*\right\rangle\right)$$

because of *cancellation of momentum and of translational contributions*;

(iv) Minus the free energy of reorganization of the product system at this R to the above state, $-\Delta F^{*p}$;
(v) And minus the free energy change when the products are brought together to the separation distance R, $-w^p$.

Thus Eq. (6.91) is obtained when Eq. (6.40) is used,
(Homogeneous)

$$\Delta F^{0\prime} = w^r + \Delta F^*(R) - \Delta F^{*p}(R) - w^p \qquad (6.91)$$

The electrochemical equation corresponding to Eq. (6.91) is (6.92), as one may show from Eq. (6.89),
(Electrochemical)

$$ne\left(E - E_0^{\prime}\right) = w^r + \Delta F^*(R) - \Delta F^{*p}(R) - w^p \qquad (6.92)$$

Here, ΔF^{*p} is obtained from ΔF^{*r}, and w^p from w^r by interchanging r and p superscripts and, at the same time, interchanging $-m$ and $m+1$. To establish this result, it suffices to note from Eq. (6.33) that U^* and all its associated properties are unaffected by such a transformation, but the properties of the reactants become those of the products" [1].

Summary of Final Equations

"On using the results of Appendix IV and referring to Eqs. (6.50) to (6.52), it is found that the rate constant for a bimolecular homogeneous reaction or a unimolecular electrochemical reaction is given by

$$k_{\text{rate}} = \kappa\rho Z \exp\left(-\Delta F^*/kT\right) \quad (6.50), (6.52)$$

where Z is given by Eq. (6.53), ΔF^* by Eqs. (6.93) and (6.94)" [1]. "The rate constant of an *intramolecular* electron transfer reaction, on the other hand, is given by Eq. (6.51), with ΔF^* given by Eq. (6.93) but with the work terms w^r and w^p omitted:

Homogeneous:

$$\Delta F^* = \frac{w^r + w^p}{2} + \frac{\lambda}{4} + \frac{\Delta F^{0\prime}}{2} + \frac{\left(\Delta F^{0\prime} + w^p - w^r\right)^2}{4\lambda} \tag{6.93}$$

Electrochemical

$$\Delta F^* = \frac{w^r + w^p}{2} + \frac{\lambda}{4} + \frac{ne\left(E - E_0'\right)}{2} + \frac{\left(neE - neE_0' + w^p - w^r\right)^2}{4\lambda} \tag{6.94}$$

In Eqs. (6.93) and (6.94) λ is given by

$$\lambda = \lambda_i + \lambda_o \tag{6.95}$$

where λ_i is given by Eq. (6.96) at $\mathbf{Q} = \mathbf{Q}_*$. On introducing the symmetrical force constants, one finds $\mathbf{Q}* = \mathbf{Q}_r + m(\mathbf{Q}_r - \mathbf{Q}_p)$. Since λ_o depends but weakly on $\mathbf{Q}_*$ and since m is usually close to $-\frac{1}{2}$, it suffices to evaluate λ_o at $\mathbf{Q}* = \frac{1}{2}(\mathbf{Q}_r + \mathbf{Q}_p)$ in the typical case. This result is used in deriving Eq. (6.96), (**typo** in the original)

$$\lambda_i = \frac{1}{2} \sum_{j,k} k_{j,k} \Delta q_j \Delta q_k. \tag{6.96}$$

The reduced force constants k_{jk} are defined in Eq. (6.76) and the Δq_j's are *differences in equilibrium values of bond coordinates* (e.g., independent bond lengths and angles), $q_j^r - q_j^p$. It is expected that typically ρ should be about unity. As noted earlier, Z is essentially[41] the collision number, being about 10^{11} liter mole$^{-1}\cdot$ sec^{-1} for homogeneous and electrochemical reactions, respectively.

In Ref. [3], the earlier equations were written in an equivalent form

$$\Delta F^* = w^r + m^2 \lambda \tag{6.98}$$

$$\begin{aligned} &\text{Homogeneous:} && -(2m+1)\lambda = \Delta F^{0\prime} + w^p - w^r \\ &\text{Electrochemical:} && -(2m+1)\lambda = ne\left(E - E_0'\right) + w^p - w^{r\prime\prime}[1] \end{aligned} \tag{6.99}$$

"... the final equations obtained when Eq. (6.99) is introduced into Eq. (6.98) are identical with Eqs. (6.93) and (6.94).

According to Eqs. (6.93) and (6.94), ΔF^* depends on $\Delta F^{0\prime}$ or on neE according to Eqs. (6.100a) and (6.100b) *when w^r and w^p are held constant.*

$$\left(\partial \Delta F^*/\partial \Delta F^{0\prime}\right)_w = \frac{1}{2} + (1/2\lambda)\left(\Delta F^{0\prime} + w^p - w^r\right) \quad (6.100a)$$

(**Typo** in the original equation)

$$(\partial \Delta F^*/\partial neE)_w = \frac{1}{2} + (1/2\lambda)\left(neE - neE_0' + w^p - w^r\right) \quad (6.100b)$$

We refer to these slopes as *'transfer coefficients at constant work terms.'* The second term in Eqs. (6.100a) and (6.100b) can be calculated when λ is known, and *this in turn can be estimated from the experimental value of ΔF^* at $\Delta F^{0\prime} = 0$, or at $E = E_0'$ using Eq. (6.93) or (6.94), when the work terms can be estimated or are negligible.* Typically this second term is found to be small, so that these 'transfer coefficients' are then 0.5.

Equations (6.100a) and (6.100b) are based on the neglect of the antisymmetrical functions l_{jk} defined in Eq. (6.77). When these functions are not neglected, the transfer coefficient is not exactly 0.5 for zero $\left(\Delta F^{0\prime} + w^p - w^r\right)/\lambda$ or zero $\left(neE - neE^{0\prime} + w^p - w^r\right)\lambda$ but is given instead by Eq. (6.101i) in Appendix IV. When these two sources of deviation from a 0.5 value are small, we may add them and so obtain Eqs. (6.100c) and (6.100d) instead of Eqs. (6.100a) and (6.100b):

$$\left(\partial \Delta F^*/\partial \Delta F^{0\prime}\right) w = \frac{1}{2} + (1/2\lambda)\left(\Delta F^{0\prime} + w^p - w^r + \frac{1}{2}\lambda_i \langle l_s \rangle\right) \quad (6.100c)$$

$$(\partial \Delta F^*/\partial neE)_w = \frac{1}{2} + (1/2\lambda)\left(neE - neE_0' + w^p - w^r + \frac{1}{2}\lambda_i \langle l_s \rangle\right). \quad (6.100d)$$

As noted in Appendix IV, the $\langle l_s \rangle$ term could cause a deviation from the 0.5 value by 0.04 when the force constants in the products are all *twice as large* (or all *twice as small*) as the corresponding ones in the reactants and when λ_i/λ is about $\frac{1}{2}$. Smaller differences in force constants would lead to even smaller deviations than 0.04. This source of deviations would be difficult to detect experimentally, since there are other sources of deviation as well. In the case of homogeneous reactions, force constants *on one reactant* may stiffen and those *in the other* weaken, so that the *average* value of $\langle l_s \rangle$ may be even less than that for the above case, and the deviation from the 0.5 value arising from this source correspondingly smaller.

In summary, the transfer coefficient at constant w's is expected to be close to $\frac{1}{2}$, reflecting *a type of symmetry* of the *R* and *P* surfaces in the vicinity of the reaction hypersurface' see Section 6.20. A source of deviation from this symmetry arises from a difference in corresponding force constants in reactants and products. It appears as an $\langle l_s \rangle$ term in Eq. (6.100) and has been shown to be small. A second source of deviation arises when the *R* or *P* surface is appreciably lower than the other, and is reflected in the presence of the $\Delta F^{0\prime}$ and $ne(E - E_0')$ terms in Eq. (6.100). This source of deviation, too, is normally small. The leading term in Eq. (6.100), $\frac{1}{2}$, arises from *the quadratic nature* of both V_0' and V_i' contributions to ΔF^*.[(42)]"

6.17. APPENDIX IV. Simplification of Eq. (6.75) and the Equation for k_{rate}

"We introduced the quantities k_{jk} and l_{jk} defined in Eqs. (6.76) and (6.77). The first was chosen so as to have dimensions of a force constant, and the second so as to be dimensionless.

Principally it is the *diagonal stretching contributions* which are usually important. Purely for simplicity of argument we confine our attention to the diagonal terms. We denote the new force constants by f_s^r, f_s^p, and their symmetric and antisymmetric combinations cited above by k_s and l_s. In terms of k_s and l_s, f_s^r equals $k_s/(1 - l_s)$ and

f_s^p equals $k_s/(1+l_s)$. To make use of the symmetry of the resulting equations, we use the parameter ε, equal to $(m+\frac{1}{2})$.

We obtain Eq. (6.101a) from (6.75) and $\Delta \mathrm{Q} = \mathrm{Q}_p - \mathrm{Q}_r$:

$$\Delta F_i^* = \frac{1}{2}\left(\varepsilon - \frac{1}{2}\right)^2 \sum_s k_s \left(\Delta q_s^0\right)^2 (1-l_s)(1+2\varepsilon l_s)^{-2} + \frac{1}{2}kT \sum_s \ln\left[(1+2\varepsilon l_s)\,/\,(1+l_s)\right] \tag{6.101a}$$

where ΔF_i^* is defined as $\Delta F^*(R) - \Delta F_o^*(R)$. Similarly, we find Eq. (6.101b) by noting that it is obtained from Eq. (6.101a) by replacing $-m$ by $m+1$ and interchanging r and p subscripts (see Section 6.16)

$$\Delta F_i^{*p} = \frac{1}{2}\left(\varepsilon + \frac{1}{2}\right)^2 \sum_s k_s \left(\Delta q_s^0\right)^2 (1+l_s)(1+2\varepsilon l_s)^{-2} + \frac{1}{2}kT \sum_s \ln\left[(1+2\varepsilon l_s)\,/\,(1-l_s)\right], \tag{6.101b}$$

where ΔF_i^{*p} is $\Delta F^{*p}(R) - \Delta F_0^{*p}(R)$.

In terms of ε, Eq. (6.91) can be written as Eq. (6.101c), upon introducing Eq. (6.81) for ΔF_0^* and its counterpart for $\Delta F_o^{*p}[= (m+1)^2\lambda_o]$

$$-2\lambda_o + \Delta F_i^* - \Delta F_i^{*p} = \Delta F_R^{0\prime} \tag{6.101c}$$

where

$$\Delta F_R^{0\prime} = \Delta F^{0\prime} + w^p - w^r \tag{6.101d}$$

Most of the data are obtained in the vicinity of $\Delta F_R^{0\prime} = 0$ [3, 12]. We consider this region first. Near the point ($l_s = 0$, $\Delta F_R^{0\prime} = 0$) one readily verifies from the equations below that ε is close to zero and that it vanishes at that point." Where $m = -\frac{1}{2}$. "We let η denote ε or

l_s, and O denote the 'order of.' (η is a small quantity in the vicinity of this point.) One then finds from Eqs. (6.101a) to (6.101c)

$$-2\varepsilon\lambda - \frac{1}{2}\lambda_i \langle l_s \rangle + O\left(\eta^3\right) = \Delta F_R^{0\prime} + kT \sum_s l_s \qquad (6.101e)$$

where

$$\begin{aligned} \lambda_i &= \frac{1}{2}\sum_s k_s \left(\Delta q_s^0\right)^2, \quad \lambda = \lambda_0 + \lambda_i \\ \langle l_s \rangle &= \sum_s k_s \left(\Delta q_s^0\right)^2 l_s \Big/ \sum_s k_s \left(\Delta q_s^0\right)^2 \end{aligned} \qquad (6.101f)$$

Furthermore, according to Eq. (6.91) ΔF^* equals $\Delta F^{*p} + \Delta F_R^{0\prime}$. Hence,

$$\Delta F^* = \frac{1}{2}(\Delta F^* + \Delta F^*) = \frac{1}{2}(\Delta F^* + \Delta F^{*p}) + \frac{1}{2}\Delta F_R^0.$$

On introducing Eqs. (6.101a) and (6.101b) one finds

$$\Delta F^* = \frac{1}{2}\Delta F_R^{0\prime} + \lambda\left(\varepsilon^2 + \frac{1}{4}\right) + \lambda_i O\left(\eta^4\right) + \frac{1}{4}kT \sum_s \left(4\varepsilon l_s + l_s^2\right) \qquad (6.101g)$$

On introducing Eq. (6.101e) for ε one finds that Eq. (6.101g) becomes

$$\begin{aligned} \Delta F^* = {} & \frac{1}{2}\Delta F_R^{0\prime} + \frac{1}{4}\lambda + (1/4\lambda)\left(\Delta F_R^{0\prime} + \frac{1}{2}\lambda_i \langle l_s \rangle\right) \\ & + \frac{1}{4}kT\left[\sum_s l_s^2 - (kT/\lambda)\left(\sum_s l_s\right)^2\right] + O\left(\eta^4\right) \end{aligned} \qquad (6.101h)$$

The same expression obtains for electrode processes, with the $\Delta F^{0\prime}$ in $\Delta F_R^{0\prime}$ replaced by $ne(E - E^{0\prime})$.

In an isotopic exchange reaction which involves mere *interchange of charges* in the electron transfer step, the term $\langle l_s \rangle$ vanishes by symmetry. In other reactions, there will be some tendency for it to vanish, for while l_s increases on one reactant on going from state R to state P (due to an increase of charge), it will tend to decrease on the other. As a somewhat extreme case involving no compensation, consider two reactants one of which has vanishing l_s and also vanishing contribution to λ_i. (Hence we include the possibility that this 'reactant' is an electrode.) For the other molecule, let the force constants k_s^r and k_s^p differ by as much as a factor of 2. Then one finds $\langle l_s \rangle \approx \frac{1}{3}$.

If $\lambda_i/\lambda \approx \frac{1}{2}$ then $\lambda_i^2 \langle l_s \rangle^2/16\lambda$ is only about 1% of $\lambda/4$, the main term at $\Delta F_R^{0\prime} = 0$. When $\lambda/4$ has *its usual value of 10–20* kcal mole^{-1}, say 10, and when the reactant has a coordination number of six, then the kT term in Eq. (6.101h), is estimated to be about 4% of the $\lambda/4$ term at room temperature.

We consider next the effect of nonvanishing $\langle l_s \rangle$ on the derivative $(\partial \Delta F^r/\partial \Delta F^{0\prime})_{\lambda_0 \lambda_i}$ at $\Delta F_R^{0\prime} = 0$, the region of greatest interest. This derivative equals

$$\frac{1}{2} + (\lambda_i/4\lambda)\,\langle l_s \rangle + O\left(\eta^2\right) \tag{6.101i}$$

In the case cited earlier, the $\lambda_i \langle l_s \rangle/4\lambda$ term is about + 0.04. Thus, the derivative differs by only 8% for this case. Hence, the $\langle l_s \rangle$ term may be neglected when ε (and hence $|\Delta F_R^{0\prime}/\lambda|$) is small. When $|\Delta F_R^{0\prime}/\lambda|$ is not small, one finds that Eq. (6.101h) should be replaced by Eq. (6.101j), to terms correct to first order in the l_s

$$\Delta F^* = \frac{1}{2}\Delta F_R^{o\prime} + \frac{1}{4}\lambda + \frac{1}{4}(\Delta F_R^{o\prime})^2 + (\Delta F_R^{o\prime}/4\lambda)\lambda_i \langle l_s \rangle [1 - (\Delta F_R^{o\prime}/\lambda)^2] \tag{6.101j}$$

The term containing $\langle l_s \rangle$ is still small: A fairly extreme case is one where the activated complex resembles the reactants ($\boldsymbol{m} = 0$) or the products ($\boldsymbol{m} = 1$). *At each extreme* $|\Delta F_R^{0\prime}/\lambda|$ *is about unity*, since the expression for $\varepsilon(= -\Delta F_R^{0'}/2\lambda)$ is but slightly affected and since $|\varepsilon|$ equals $\frac{1}{2}$ when m is 0 or 1. In the interval

$$0 \leq \left|\Delta F_R^{0'}/\lambda\right| \leq 1$$

the last term in Eq. (6.101j) has a maximum at

$$\left|\Delta F_R^{0'}/\lambda\right| = \left(\frac{1}{3}\right)^{\frac{1}{2}}.$$

At this point, it equals about 1 kcal mole^{-1} for the values of $\langle l_s \rangle$, λ_i/λ, and $\lambda/4$ cited earlier. When one does not neglect second and higher order terms in l_s and solves Eqs. (6.101a) to (6.101c) numerically in this region one obtains the same result: The l_s terms may be neglected" [1].

6.18. Properties of the Reorganization Term λ

"For use in subsequent correlations, we examine an additivity property of λ and the relation between the values of λ in related homogeneous and electrochemical systems. We consider first the (hypothetical) situation when R is *very large*, so large that *the force field from one reactant* does not influence the other. On noting that λ_0 is given by Eq. (6.82) and that the *fluctuations* around each reactant are now *independent* (large R), λ_0 can be written as the sum of two independent terms, one per reactant. It then follows that when R is large, the value of λ_0 for a reaction between reactants from two different redox systems A and B, λ_0^{ab}, is the arithmetic mean of the values λ_0^{aa} and λ_0^{bb} of the respective systems:

$$\lambda_0^{ab} = \frac{1}{2}(\lambda_0^{aa} + \lambda_0^{bb}) \quad (R \text{ large}).^{(43)} \qquad (6.102a)$$

Furthermore, *in the electrochemical case, there is only a contribution from one ion* (assuming that any distortion of atomic structure of the electrode yields only a relatively minor contribution to ΔF_0^*). Denoting the values of λ_0 for *the electrochemical redox system A* and *the homogeneous redox system A* by λ_0^{el} and λ_0^{ex}, respectively, we have

$$\lambda_o^{ex} = 2\lambda_o^{el} \quad (R \text{ large}) \tag{6.102b}$$

Relations similar to Eqs. (6.102a) and (6.102b) also hold for λ_i, *independent of R*, as may be seen from Eq. (6.96): Part of the sum for λ_i is over the bonds of the first reactant and the remainder is over those of the second one. While the k_{jk}'s of one reactant in the activated complex depends *slightly* on the fact that there is a neighboring reactant, this influence is taken to be weak.

In the absence of *specific interactions*, Eqs. (6.102a) and (6.102b) would also hold for smaller R since in the equation for λ_0 *each ion would merely see another charge*, $-m\Delta e$, and the surrounding medium, in both the homogeneous and electrode cases" [1].

Note that *the last sentence describes a nonspecific interaction.*

"In the homogeneous case, the $-m\Delta e$ is centered on the other ion. In the electrode case, it is an image charge on the electrode [1, note 28]. To obtain some estimate of deviations from Eq. (6.71a) due to differences in *ion size* (one type of 'specific effects'), we examine in the next section the evaluation of λ_0 in the dielectric continuum approximation.

6.19. Dielectric Continuum Estimate at ΔF_0^*

The present section on *a continuum estimate* of ΔF_0^* is included partly for what it can reveal *approximately* about certain aspects of the *statistical mechanical value* for ΔF_0^* and partly for making some *approximate numerical calculations*. It does not form a necessary part of the *present* electron transfer theory itself, of course, for the latter rests on statistical mechanics.

We note that ΔF_0^* can be regarded as *the sum of two contributions*, ΔF_{sol}^* and ΔF_{atm}^*. ΔF_{sol}^* is defined as the contribution if the atmospheric ions have *not adapted* themselves to the change $m(\rho_{ar}^0 - \rho_{ap}^0)$," [1], i.e. to that charge density on the reactants, "and

ΔF^*_{atm} is defined as the contribution due to their *adaptation* (*'reorganization'*)" [1].

NOTE: Here M. defines the term "reorganization."

"ΔF^*_{sol} in an *electrolyte medium* will not have exactly the same value it has at infinite dilution, since *the local dielectric properties near the reactants* will be altered somewhat by the presence of salt" [1].

These two contributions are estimated in Appendix VII by treating

(i) The *medium* as a dielectric continuum
(ii) The *ion atmosphere* as a continuum
(iii) The reactants as spheres
(iv) And by neglecting dielectric image effects [1, note 29]

We obtain Eqs. (6.103) and (6.104) for the value of ΔF^*_{sol} for a medium treated as *dielectrically unsaturated continuum* outside the inner coordination shell of each reactant. If *partial saturation* occurs, Eq. (6.80) still applies [14]. If one then introduces "differential" rather than "integral" dielectric constants, as later defined, and treats them approximately as constants, Eqs. (6.103) and (6.104) again apply but now D_{op} and D_s *are mean values* of these *differential constants.*

NOTE: The differential and the integral dielectric constants were not defined in the Appendix. Marcus defines them here in the following **NOTE**: "The dielectric constant normally comes in $\mathbf{D} = \varepsilon\mathbf{E}$. That's the usual thing. And now let us suppose that the dielectric constant is a function of the electric field, in other words it is not a constant, so you can regard that ε being a function of $\mathbf{E}$, what I mean for differential is that if for some $\mathbf{E}$. I change it to $\delta\mathbf{E}$, change it by a little amount, and the dielectric displacement changes to $\delta\mathbf{D}$, then the proportionality between those two I would call the differential dielectric constant. And I would call integral the usual one, I don't think 'integral' is a good term for it. The usual dielectric constant is a constant but the actual response, the detailed response, of $\mathbf{D}$ to $\mathbf{E}$

or of **E** to **D** may not be constant, especially when one gets strong fields and so you have a variable dielectric constant."

Homogeneous:

$$\Delta F^*_{sol} = m^2(ne)^2\left(\frac{1}{2a_1} + \frac{1}{2a_2} - \frac{1}{R}\right)\left(\frac{1}{D_{op}} - \frac{1}{D_s}\right) \tag{6.103}$$

where ne is Δe, the charge transferred from one reactant to the other; a_1 and a_2 are the radii of the two reactants computed at intramolecular coordinates $q^i = q^i_*$ (the radii are of spheres each of which includes any inner coordination shell); R is taken to be the sum of $a_1 + a_2$; D_{op} is the square of the refractive index of the medium; and D_s is the static dielectric constant of the medium

Electrochemical:

$$\Delta F^*_{el} = \frac{m^2(ne)^2}{2}\left(\frac{1}{a_1} - \frac{1}{R}\right)\left(\frac{1}{D_{op}} - \frac{1}{D_s}\right) \tag{6.104}$$

(**Error** corrected in the subscript of the original)

where $2R$ is twice the distance from the center of the ion to the electrode surface and a_1 is the radius of the reactant (*and hence of the product*) computed at $q^i = q^i_*$.

The value obtained in Section 6.24 for ΔF^*_{atm} in the *electrically unsaturated region* (i.e., in the Debye–Hückel region for the atmosphere around the ion and, in the electrode system, for the diffuse part of the double layer) is given by Eq. (6.105) for homogeneous systems for the case of $a_1 = a_2 (= a)$, and by Eq. (6.106) for electrode systems. *The value of ΔF^*_{atm} for partially electrically saturated systems can also be obtained from Eq. (6.80).* Once again, one introduces *"differential" quantities.* If the latter are replaced by *"mean" values near the central species,* Eqs. (6.105) and (6.106) are again obtained, but *with D_s and κ reinterpreted*; κ is given by Eq. (6.106) in APPENDIX VII, Section 6.24. A more reliable procedure, however, would be to use the position-dependent value of κ in solving this particular linearized Poisson–Boltzmann equation, since the electric fields in electrolyte media die out fairly rapidly, namely as $r^{-1}\exp(-kr)$. Equations (6.105) and (6.106) are based on the solution of a linearized Poisson–Boltzmann equation with a *local mean* κ

Homogeneous:

$$\Delta F^{*}_{am} = \frac{m^2(ne)^2}{D_s R} \times \left[\frac{\kappa R + \exp[-\kappa(R-a)]\left(1+\kappa^2 a^2/2\right)}{1+\kappa a + \exp[-\kappa(R-a)]\kappa^2 a^3/3R} - 1\right] \tag{6.105}$$

Electrode:

$$\Delta F^{*}_{\text{atm}} = \frac{1}{2}[\text{RHS of (6.105)}] \tag{6.106}$$

Calculated as above, ΔF^{*}_{atm} is *much smaller* than ΔF^{*}_{sol} and is also expected to be less than the salt effects on w^r and w^p. Even at high κ it is only $m^2(ne)^2 D_s R$, which is only about 2% of ΔF^{*}_{sol}. Parenthetically, we note that this term arising from Eqs. (6.105) and (6.106) just cancels the D_s term in Eqs. (6.103) and (6.104), respectively.

Added electrolyte can influence the rate constant, we conclude, *principally* by affecting w^r, w^p, and (by affecting dielectric properties) ΔF^{*}_{sol}.[(44)] Comparison of Eq. (6.103) with Eq. (6.104) reveals that λ_0 for an isotopic exchange reaction has twice the value of λ_0 for an electrode reaction involving this redox couple when the value of R is the same in each case. It is recalled that R *is the value for which* $\kappa\rho\exp(-\Delta F^*/kT)$ *had a maximum.* If one presumes this R to be the distance of closest approach of the "hard spheres" and assumes the reactant to just *touch* the electrode, then R is the same in each case. In Eq. (6.103) $a_1 = a_2$ for an isotopic exchange reaction since *these are the radii evaluated for* $q = q_*$, it is recalled, and since typically the transition state should be symmetrical in this respect. (From the equation cited, the actual q^j_*'s can be computed and the presumed symmetry verified for typical conditions.)

It may be seen from Eq. (6.103) that λ_0 is *essentially equal* to the sum of two terms, one per reactant, and that for the same R the value of λ_0 for homogeneous reaction in any *redox system a* (**typo** in the original) equals twice its value for the electrochemical case. While *the presence of the R term* makes λ_0 not quite additive, the deviation from additivity can be shown to be small: On denoting the radii for ions of the two systems by a and b we obtain Eq. (6.107).

$$\lambda_o^{ab} - \frac{1}{2}\left(\lambda_o^{aa} + \lambda_o^{bb}\right) = \frac{[1-(b/a)]^2}{4b[1+(b/a)]}(ne)^2\left(\frac{1}{D_{op}} - \frac{1}{D_s}\right) \tag{6.107}$$

Even if b/a is $\frac{1}{2}$, a fairly extreme case, the ratio of the earlier difference to λ_0^{bb} is

$$\frac{(1-b/a)^2}{2(1+b/a)}$$

that is, $\frac{1}{12}$. In virtue of this result, λ_0 has been treated as an additive function in applications [3, 12] of the equations of this paper.

6.20. Significance of *m*

The parameter m was chosen… so as to satisfy the *centering condition* (6.40), a condition which led to the vanishing of $I'(0)$." "On differentiating $I(q^N R)$ given in Eq. (6.35) and setting $q^N = 0$ one finds:

$$m = -\left\langle \frac{\partial U^r}{\partial q^N} \right\rangle \Big/ \left\langle \frac{\partial}{\partial q^N}\left(U^r - U^p\right)\right\rangle^{(45)} \tag{6.108}$$

where the average $\langle\ \rangle$ is over the distribution function *on the reaction hypersurface* S' at the given R,

$$\exp\left(\frac{-U^r}{kT}\right) \Big/ \int \exp\left(\frac{-U^r}{kT}\right)\left(a^{NN}\right)^{\frac{1}{2}} dS'$$

From Eq. (6.108), $-m$ is seen to be the mean slope at the reaction hypersurface S', $\langle \partial U^r/\partial q^{\mathrm{N}}\rangle$, of the Rsurface, for the given separation distance R, divided by the sum of the mean slopes $\langle \partial U^r/\partial q^{\mathrm{N}}\rangle$ and $\langle -\partial U^p/\partial q^{\mathrm{N}}\rangle$ of the R and P surfaces at S'. If the intersection surface S' at this R passed through the stable configurations of the reactants, on the average, then $\langle \partial U^r/\partial q^N \rangle$ would be zero. If it passed through those of the products instead, $\langle \partial U^p/\partial q^N \rangle$ would be zero. In these two cases, one sees from Eq. (6.108) that m would be 0 and -1, respectively. When in the vicinity of S' the R and Psurfaces are, on the average, mirror images of each other about S', $\langle \partial U^r/\partial q^N \rangle$, equals $\langle -\partial U^p/\partial q^N \rangle$ and one sees that $m = -\frac{1}{2}$.**(46)** Values of m close to $-\frac{1}{2}$ are typical [3, 12] and are to be expected, one sees from Eq. (6.99), when $\Delta F^{0\prime}$ is near zero or when E is close to E_0' (typically of the order of or less than 10 kcal mole^{-1} or 0.25 V, respectively).

6.21. Deductions from the Final Equations

Equations (6.50) to (6.52), together with the additivity property of λ (Section 6.18), and the relation between the electrochemical and chemical λ's described earlier lead to the following deductions, if κ and ρ are about unity, or if they satisfy milder conditions in some cases.[30]

Note 30: For example, it suffices for some of the deductions that $\kappa\rho$ be constant in a given *series of reactions* or that it have a geometric mean property.

(a) The rate constant of a homogeneous 'cross-reaction,' k_{12} is related to that of the two electron exchange reactions, k_{11} and k_{22}, and to the equilibrium constant k_{12}, in the prevailing medium by Eq. (6.109), *when the work terms are small or cancel*,

$$\mathrm{Ox}_1 + \mathrm{Red}_2 \overset{k_{12}}{=} \mathrm{Red}_1 + \mathrm{Ox}_2$$

$$k_{12} = (k_{11}k_{22}K_{12}f)^{\frac{1}{2}}, \tag{6.109}$$

where

$$\ln f = \frac{(\ln K_{12})^2}{4\ln\left(k_1 k_{22}/Z^2\right)}. \tag{6.110}$$

Frequently, f is close to unity.**(47)**

(b) The electrochemical transfer coefficient at metal electrodes is 0.5 for small activation overpotentials [1 note 31a] (i.e., if $|nF\eta| < |\Delta F_0^*|$, where ΔF_0^* is the free energy of activation for the exchange current[48]), [1, note 31b], when the work terms are negligible.[49]

(c) When a substituent in the coordination shell of a reactant is *remote* from the central metal atom and is varied *in a series*, a plot of the free energy of activation ΔF^* versus the 'standard' free energy of reaction in the prevailing medium $\Delta F^{0\prime}$ will have a slope of 0.5, if $\Delta F^{0\prime}$ is not too large (i.e., if $|\Delta F^{0\prime}|$ is less than the intercept in this plot at $\Delta F^{0\prime} = 0$).
In this series, for a sufficiently remote substituent, λ and the work terms are constant[50] but $\Delta F^{0\prime}$ varies, as in Eq. (6.100a). The slope of the ΔF^* -versus $-\Delta F^{0\prime}$ plot has been termed the *chemical transfer coefficient* [3], by analogy with the electrochemical terminology.

(d) When a *series of reactants* is oxidized (reduced) by *two* different reagents, the *ratio* of the two rate constants is the same for all members of the series in the region of the region of chemical transfer coefficient equal to 0.5[51] [i.e., in the region where $|\Delta F^{0,}| < (\Delta F^*)_{\Delta F^{0\prime}=0}$ in each case].

(e) When the series of reactants in (d) is oxidized (reduced) *electrochemically* at a given metal–solution potential difference, the *ratio* of the *electrochemical* rate constant to either of the *chemical* rate constants in (d) *is the same for all members of the series*, in the region where the chemical and (work corrected) electrochemical transfer coefficient is 0.5.[52]

(f) The rate constant of a (chemical) electron exchange reaction, k_{ex}, is related to the electrochemical rate constant at zero activation overpotential, [1, nota 31a] k_{el}, for this redox system, according to Eq. (6.111) when the work terms are negligible[53]:

$$(k_{\text{ex}}/Z_{\text{soln}})^{\frac{1}{2}} \cong k_{\text{e}}/Z_{\text{el}} \quad (6.111)$$

[54]

where Z_{soln} and Z_{el} are collision frequencies, namely about 10^{11}mole^{-1}. sec^{-1} and 10^4cm sec^{-1}. When the work terms are not negligible, or do not cancel in the comparison, the deductions which depend on this condition refer to rate constants, to K_{12} and to an electrochemical transfer coefficient corrected for these terms. Again, a minor modification of the transfer coefficients from the value of $\frac{1}{2}$ in (b) or (c) can also arise from the antisymmetrical force constant term $\langle l_s \rangle$ in Eq. (6.101i).

It is shown in Appendix VII that under certain conditions these expected correlations apply to *overall* rate constants as well as to those involving only one pair of reactants" [1].

6.22. Generalization and Other Improvements

In Section 19 of Ref. [1] (Part VI of the series of papers I to VI "On the Theory of Electron Transfer Reactions"), M. lists the "extensions and improvements... compared with the earlier ones " to be found in Part VI with respect to the earlier Parts I–V.

(1) "Use is made of a more general expression for the reaction rate as the starting point.
(2) A more detailed picture of the mechanism of electron transfer is given for the electrochemical case.
(3) The derivation is now given for both electrode and homogeneous reactions, and in a single treatment.
(4) The statistical-mechanical treatment of polar interactions, based in Part IV on the interactions of permanent and induced dipolar molecules in the medium, was replaced by a more general particle description of polar interactions, through the use of the potential energy function (6.78) and (6.78a).
(5) The equivalent equilibrium distribution (e.e.d.) made plausible in Part IV was proved more rigorously here.
(6) The functional form Eq. (6.81) for ΔF_0^*, obtained in Part IV only by subsequently treating the medium as a dielectric

continuum, was derived here using a statistical-mechanical treatment of nonequilibrium polarization systems.

(7) The basic equation for k_{rate} has been converted to a simple form [e.g., Eqs. (6.50) and (6.93)], a form used in Part V, by neglecting the antisymmetrical function of the force constants, a neglect which has only a minor effect numerically.

(8) The symmetry arguments used in Part IV to convert the kT/h and a portion of a ΔF^* term to the Z factor in Eq. (6.50) have now been given more rigorously.[55]

(9) The ion atmospheric reorganization term was but mentioned in Part IV. It is now incorporated into ΔF_0^*. The *nonpolar contribution* to w^r and w^p is also formally included.

(10) The contribution of *a range of separation distances* to the rate constant is included.

The results in the present paper may be compared with earlier papers in this series. In Part I, ΔF_0^* was computed for homogeneous reactions at zero ionic strength, and dielectric continuum theory was used. Equation (6.103) was obtained. The actual mechanism of electron transfer was discussed there, but without the *detailed description* which the use of many-dimensional PESs provides. The latter was used in later papers of this series, a use which added to the *physical picture*. The counterpart of Part I for electrode systems was also derived and applied to the data in a subsequent paper [15].

In the earliest papers, *the dielectric continuum equivalent of the e.e.d.* was derived by a method apparently different from that used in the present paper.

The distribution selected was the one that minimized the free energy subject to the constraint embodied in Eq. (6.40), or really embodied in the dielectric continuum counterpart of Eq. (6.40)" [1].

In Ref. [1], Appendix II and Appendix IX, M. shows the equivalence of the two methods.

6.23. Nonadiabatic Electron Transfers

"Several estimates are available for the probability of nonadiabatic reaction, per passage through the intersection region of two PESs, and have been referred to and discussed in Ref. [12]. In each case, *the motion along the reaction coordinate* was assumed to be dynamically *separable* from the remaining motions. (For conditions on separability see, for example, Ref. [16] and references cited therein.) The probability of electron transfer per passage through the intersection region in Fig. 1 will depend in the first approximation on the momentum p_N conjugate to the reaction coordinate q^N, as, for example, in the Landau–Zener [4] equation. While the value of κ is not so simply represented in more rigorous treatments, we simply write it as $\kappa(p_N)$. In the above treatments, the reaction coordinate was assumed to be orthogonal to the others in *mass-weighted* configuration space, so that g_{Ni} vanishes for $i \neq N$ (and so, therefore, does g^{Ni}) in the kinetic energy. On recalling the derivation of Eqs. (6.22) and (6.23) in Ref. [7] and on introducing the above assumptions, the rate constant is given, one can show, by Eqs. (6.22) or (6.23), but with the integrand multiplied by κ:

$$\kappa = \frac{\int_{p_N=0}^{\infty} \kappa\,(p_N)\, g^{NN}\, p_N \exp\left(-\frac{g^{NN} p_N^2}{2kT}\right) dp_N}{\int_0^{\infty} g^{NN}\, p_N \exp\left(-\frac{g^{NN} p_N^2}{2kT}\right) dp_N}$$

This κ can depend on all the other coordinates, $q^i (i \neq N)$ at the given value of q^N characterizing the intersection surface S. The denominator in the above equation is easily shown to equal kT, and so to be independent of the q^i. In the discussions of $\kappa(p_N)$ in the literature, the derivation of the Landau–Zener equation, for example, the reaction coordinate has been assumed to be rectilinear; g^{NN} is then a constant and the integral in the numerator then becomes independent of the q^i and may be removed from the integral in Eqs. (6.22) and (6.23).

There appears to be no treatment in the literature for nonadiabatic reactions involving many closely spaced energy surfaces as in Fig. 2 covering the range of $\kappa(p_N)$ from 0 to 1. If $\kappa(p_N)$ is sufficiently small, the transition to each P surface from the initial R surface may be assumed to be independent, as mentioned earlier, and the reverse transition to the initial R surface during this passage may also be neglected. In this case only does the method of Levich and coworkers in this connection become appropriate. (For references, see Ref. [12].) In this case, the above κ appears in the integrand of Eq. (6.23) and care is taken to sum over all levels in an appropriate fashion, as done by Levich *et al.* (see Ref. [12] for bibliography). One can then evaluate the κ appearing in Eq. (6.52) and defined earlier. Usually, however, we assume that $\kappa(p_N)$ is close to unity (within some small numerical factor, say) for the p_N's of interest" [1].

6.24. APPENDIX VII. Calculation of ΔF_0^* in Continuum Approximation

"When *dielectric* unsaturation and *electric* unsaturation[(56)] prevail, there is, respectively, a *linear response of the solvent polarization* and *of the charge density of ions* in an ion atmosphere to *the charging* of the central ion (or ions), and not merely to *a small change* in its charge. In real systems, some partial dielectric saturation outside of the coordination shells may occur and, at appreciable concentration of added electrolyte, the response of the atmospheric ions is certainly nonlinear. (The region of linear response of an ion atmosphere to a charging of the 'central ion' is confined to the Debye–Hückel region.)

We introduce the *partial saturation approximation*, wherein only a linear response to *a small change* in charge of the central ion (ions) is assumed. The special case of unsaturation is automatically included, therefore. We are interested, typically, in changes of magnitude $m\Delta e$, that is, about $\frac{1}{2}$ *an electronic charge unit*. Equation (6.80)

$$\Delta F_o^{**} = F_{m(r-p)}^{op} - F_{m(r-p)} \qquad (6.80)$$

was derived for both the *partially saturated* and for *unsaturated* systems, but in the former case the definition of $F^{op}_{m(r-p}$ and $F_{m(r-p)}$ has to be interpreted carefully.

To calculate $F_{m(r-p)}$ appearing in Eq. (6.80) and to take partial saturation into account, one considers *two* charge distributions: (i) *The original charge distribution* of the reactants and the medium for the cited R. (ii) *A hypothetical charge distribution* in which the reactants' charge distribution is altered from (i) by an amount $m(\rho^0_{ar} - \rho^0_{ap})$, in a hypothetical system which has responded linearly to this change. To obtain the properties of the hypothetical system in $F_{m(r-p)}$, one subtracts the above two charge distributions on the reactants and also subtracts the portions of the remaining charge distributions, induced or otherwise, which did not respond. *One now has in this hypothetical* $[m(r-p)]$ *system reactants which have permanent charges given by the distribution* $m\left(\rho^0_{ar} - \rho^0_{ap}\right)$ and are imbedded in a medium of solvent and atmospheric ions which has linear 'response function' describing the above response. For example, if we use a continuum model, then the *effective* dielectric susceptibility of the solvent is the proportionality constant $\chi(\mathbf{r})$ in Ref. [17]

$$\delta\mathbf{P}(\mathbf{r}) = -\chi(\mathbf{r})\delta\mathbf{E}(\mathbf{v}), \tag{6.105}$$

where $\delta\mathbf{P}$ and $\delta\mathbf{E}$ are the change in polarization and in electric field at $\mathbf{r}$. The *effective* dielectric constant describing the response to this $\delta\mathbf{E}$ is $D_s(\mathbf{r})$ equal to $1 + 4\pi\chi(\mathbf{r})$. The quantities $\chi(\mathbf{r})$ and $D_s(\mathbf{r})$ can be tensors.

Then again, if $\rho(\mathbf{r})$ is *the charge distribution in the ion atmosphere* and, if one wishes, *in the electrical double layer* at the electrode–solution interface, and if $\rho(\mathbf{r})$ *is approximated by a continuum expression*

$$\rho(\mathbf{r}) = \sum_{\mathbf{i}} c_i^{\infty} e_i \exp(-e_i\psi/kT)$$

where e_i is the charge of species i in this atmosphere, C_i^{∞} is its concentration at infinity, and ψ is the potential at $\mathbf{r}$ relative to the value at infinity, then

$$\delta\varrho(\mathbf{r}) = -\left(\sum_i \frac{c_i^\infty e_i^2 e^{-e_i\psi/kT}}{kT}\right)\delta\psi(\mathbf{r}).$$

On recalling that *the Debye kappa* is defined as *the square root of the proportionality constant of* $\rho(\mathbf{r})$ *and*$-\psi(\mathbf{r})$ *in linear systems*, the quantity which plays the same role in this hypothetical system is $\kappa'(\mathbf{r})$.

$$\kappa'(\mathbf{r}) = \left[\sum_i c_i^\infty e_i^2 \exp(-e_i\psi/kT)\right]^{\frac{1}{2}} \tag{6.106}$$

To calculate $F^{op}_{m(r-p)}$, we recall that this system *responds only via the electronic polarization of the medium,*[(57)] and so κ' vanishes for this system and $\chi'(\mathbf{r})$ becomes $\chi_e(\mathbf{r})$, the proportionality constant replacing $\chi(\mathbf{r})$ in Eq. (6.105). The medium in this hypothetical system behaves as though it had a dielectric constant $D'_{\text{op}}(\mathbf{r})$ equal to $1 + 4\pi\chi_e$.

If we take D'_{op} to be *approximately a constant*, for simplicity, then $F^{\text{op}}_{m(r-p)}$ is easily calculated. We neglect dielectric image effects. $F^{\text{op}}_{m(r-p)}$ is the sum of the free energy of solvation of the central species when they are far apart, plus the free energy change when they are brought together in this 'op' medium. The former is given by the Born formula (it is not the free energy of solvation of the bare ion, but of the coordinated ion) and the latter by the Coulombic term. Hence,

$$F^{\text{op}}_{m(r-p)} = -\left[\frac{(m\Delta e)^2}{2a_1}\left(1 - \frac{1}{D'_{\text{op}}}\right) + \frac{(m\Delta e)^2}{2a_2}\left(1 - \frac{1}{D'_{\text{op}}}\right)\right] - \frac{(m\Delta e)^2}{D'_{\text{op}}R} \tag{6.107}$$

The $F_{m(r-p)}$ term is the sum of its value when the ion atmosphere does not respond

$$F_{m(r-p)} = -\left[\frac{(m\Delta e)^2}{2a_1}\left(1-\frac{1}{D'_s}\right)+\frac{(m\Delta e)^2}{2a_2}\left(1-\frac{1}{D'_s}\right)\right]-\frac{(m\Delta e)^2}{D'_s R} \tag{6.108}$$

and the contribution due to their response via $\boldsymbol{\kappa}'(\mathbf{r})$, that is, ΔF^*_{atm}. On taking $\boldsymbol{\kappa}'$ to be approximately a constant near the central series the leading terms of the second contribution are

$$\Delta F^*_{\text{atm}} = -\frac{(m\Delta e)^2}{D'_s R} \times \left[\frac{\kappa' R + \exp\left[-\kappa'(R-a)\right]\left(1+\kappa'^2\right)a^2/2)}{1+\kappa' a + \exp\left[-\kappa'(R-a)\right]\kappa'^2 a^3/3R} - 1\right] \tag{6.109}$$

when $a_1 = a_2$.

The difference of Eqs. (6.107) and (6.108) is the value of $F^{\text{op}} - F$ when the atmosphere does not respond, and was called ΔF^*_{sol}, that is, $\Delta F^*_{\text{sol}} = F^{\text{op}} - F$. In Eqs. (6.103) to (6.106), we have omitted the prime superscripts for brevity.

In the case of *electrode systems*, there is only one ion, but there is also the image charge of opposite sign*in* the electrode.

Instead of Eqs. (6.107) to (6.109) one finds,

$$(electrode) F^{\text{op}}_{m(r-p)} = -\frac{(m\Delta e)^2}{2a}\left(1-\frac{1}{D'_{\text{op}}}\right)-\frac{(m\Delta e)^2}{2D'_{\text{op}}R} \tag{6.110}$$

and that $F_{m(r-p)}$ is the sum of Eq. (6.111) and one-half (6.109),

$$-\frac{(m\Delta e)^2}{2a}\left(1-\frac{1}{D'_s}\right)-\frac{(m\Delta e)^2}{2D'_s R}. \tag{6.111}$$

In this way Eqs. (6.105) and (6.106) of the text were obtained" [1].

NOTES

1. **M**: "By more complex aspects I mean *the details* of what's going on, how much of the reaction coordinate is a vibration, how much is orientation, how much is the ions sort of coming together, all of those properties involve big or small portion of a reaction coordinate. But... to do proper justice to everything, you have to take into account, in a kind of rigorous way and...that's almost an impossible task, to do it that way. *An approximate theory is necessary* and so I lumped everything together, in a sense."
2. **M**: "For example in the ferrous–ferric isotopic exchange reactions... Cl^- and OH^- ions may well go into the coordination shell of the ferric ion. Dick Dodson studied the effects of those ions on the rate. I was very conscious at the time with that sort of results."
3. **M**: "Imagine I pick a particular profile. If I take this 10^{23}-dimensional space and draw now a different vertical plane through it, then I'd have a different combination of coordinates. *The reaction coordinate of ET reactions is really the difference of the energies of the two PESs. This reaction coordinate is the abscissa of the diagram in which you have the free energy of the system as ordinate.* And the free energy curves, two parabolas, one for the reactants and one for the products show a splitting at the crossing which is some average splitting of the potential energy curves."
4. Q: One often considers the lowest energy path leading from reactants to products.

   M: "The lowest energy path doesn't describe the whole reaction coordinate, you need a global coordinate."
5. Q: There is an interval of possible values of the splitting, from zero to a maximum... To which range of this interval does your theory applies? From zero interaction to how much?

   M: "If the interaction were fantastic, tremendous, then the splitting would be large and calculating just the free energy of activation at the crossing of diabatic states would give you a big error, because the actual TS is lowered quite a bit by the splitting, so that's a factor, that depends on what error you're willing to tolerate.... For the simple application of the theory you want the splitting to be enough, for example, let's say kT, so that your error is maybe of the order of kT which is small, you know, compared with the usual values. On the other hand, you don't want it to be too small, too small a splitting, because then you got to bring in a nonadiabatic probability for transfer, so you want it some place in between. I just say kT because that's a typical value, I mean if you make an error of a kT in your activation energy, by picking a point lower than the intersection, because of the splitting, then you made an error of $e^{kT/kT}$, you made an error of a factor of 2.7, that's not a big deal. If you modify the theory by putting in a preexponential factor that allows a nonadiabatic process, then you can treat all sorts of situations, of high and low nonadiabaticity, but if you go to the other limit where the interaction is so strong that the splitting is great, then it depends on what a bound of error you want to tolerate. If the splitting is a kT of one, one has a factor of 3, say, 2.7, if the splitting is $2kT$ one gets a factor of 10, you see."

6. **M**: "If you take one of those curves of the reactants and you add or subtract an electron, you get one of those curves of the products. Supposing the electron can go into several different possible states, then from anyone of those curves on the left it can go into a number of states on the right, as long as there is an unoccupied level it can go into, so it can go into an infinite number of them, just go to the right level, the right intersection."
7. **M**: "That would be the e.e.d., that's a fictitious concept to discuss the centering concept... well, it's as theoretical as TS theory."
8. **M**: "If S is your TS, different parts of the TS correspond to one separation distance, other parts of the TS hypersurface may

correspond to another. Electron transfer could occur at a range of distances."

9. **M**: "You can regard R as one of the coordinates involved in the space S. As long as you have independent coordinates, even if you start with Cartesian coordinates you can still change coordinates, so one of them becomes R."
10. **M**: "U is dependent on all the coordinates that are included in S."
11. **M**: "$\kappa = \kappa(R)$ and when $1/R$ is small there might be some reaction where there is very little electronic coupling. If that's the case, then κ would be small."
12. **M**: "The quantum treatment of course would be different, you wouldn't use both coordinates and momentum variables, it's only classically that you can specify coordinates and momenta as independent variables."
13. **M**: "Momentum-weighted because we have integrated over momenta and lumped everything into the κ."
14. Writing the coordinates more explicitly:

$$\int_R \kappa R^\alpha \left[\int_{S'} \exp\left(-\frac{U(S', R, q^N = 0)}{kT}\right)(m^\ddagger)^{-\frac{1}{2}}\, dS'\right] dR$$

15. Writing the coordinates more explicitly:

$$dV' = (a^{NN})^{-\frac{1}{2}} dS' dq^N \Rightarrow \int_V (a^{NN})^{-\frac{1}{2}} \exp\left(-\frac{U^*(S', R, q^N)}{kT}\right) dS' dq^N$$

16. **M**: "Think of U^r and U^p fixed, and determined by ΔG^0. m is a fixed quantity which is not variable but is determined by the

bottoms of the two PESs, it's determined by ΔG^0. For each value of m you get a different e.e.d."

17. For $m = -1/2$, U^* represents the potential surface

$$U^* = \frac{U^r + U^p}{2}$$

It is interesting to see the form of this curve in the simplest case. Imagine that U^r and U^p are two parabolas of the same form, U^r with the vertex at $x = x_1$, U^p with the vertex at $x = x_2$ and with $x_1 = -x_2$ and with $\Delta G^0 = 0$. We then have $U^r = k(x - x_1)^2 = k(x + x_2)^2$ and $U^p = k(x - x_2)^2$. We then have: $U^* = \frac{U^r+U^p}{2} = \frac{k(x+x_2)^2+k(x-x_2)^2}{2} = \frac{k(2x^2+2x_2^2)}{2} = k(x^2 + x_2^2)$. Such parabola has its minimum value at $x = 0$, where it has the common value of the parabolas U^r and U^p at the crossing, that is, $U^* = kx_2^2$ and has the value $U^*(x_1) = U^*(-x_2) = 2kx_2^2 = U^*(x_2)$. We then have the interesting result that curve U^* for $m = -1/2$ represents a parabolic PES with its minimum value at the crossing point of the of the parabolas U^r and U^p and this minimum value is exactly the energy at the TS. The potential energy curve $U^* = \frac{U^r+U^p}{2}$ is the one for the e.e.d. centered on the intersection surface.

M: "It gives an extra representation, a concrete representation to U^* and you see that U^* is like an effective potential with a minimum. This is what my intention was actually, but this makes it very concrete."

18. **M**: "In the 1956 paper, you have a variational parameter m, and then only in the last step you find its value,by a free energy balance equation, and that value of m turns out to be the one you get in the transition state."
19. **M**: "S' and R together form S, together they are 10^{23}–1 coordinates, and q^N is a third coordinate, the remaining coordinate of the system of 10^{23} coordinates. Then you integrate over q^N in the fictitious system, I mean the e.e.d., q^N goes from $-\infty$ to

∞ but the potential energy in the fictitious system goes down rapidly about $q^N = 0$."

20. **M**: "q^N is related to γ of Part IV but doesn't have to be γ, it's just an Nth coordinate."
21. For the sake of clarity writing explicitly the range of integrations one can rewrite the integrals (6.35) and (6.36) as:

$$\exp\left[-I\left(q^N, R\right)\right] = \int_{S(q^N)} \left(-\frac{U^*}{kT}\right)\left(a^{NN}\right)^{-\frac{1}{2}} dS' \qquad (6.10)$$

$$\int_{S(q^N=0)} \exp\left(-\frac{U}{kT}\right)\left(m^{\ddagger}\right)^{-\frac{1}{2}} dS' = \left\langle \left(g^{NN}\right)^{\frac{1}{2}} \right\rangle \exp\left[-I(0, R)\right] \qquad (6.11)$$

22. **M**: "$I(q^N, R)$ divided by kT is an effective surface free energy."
23. **M**: "That F^* is a free energy of an N-dimensional system which has been extended from the $N-1$ dimensional system, considering sort of a hypothetical U on either side of the actual hypersurface. So, we convert it in that way to a full N-dimensional conventional free energy rather than a free energy with the restriction of being on a hypersurface. The system is spread because in one case you have the free energy of a restricted system, existing on a single q^N and in the other case you have the free energies of an extended system, it means that you're not confined to q^N, but you have a hypothetical distribution."
24. **M**: "The idea is this: that for the integral as a whole there's some advantage in the reactants being close to each other because the reorganization energy is less and the tunneling probability is greater: you reduce R, there is some advantage being close. On the other hand, if you get too close you run into big repulsion, you know, pushing the reactants against each other. So that means there's some nearly optimal R when you take all of those factors into account. In one respect, there is an advantage in making R as small as possible, but you can't squash the reactants too close

together, there's a huge potential energy so there's an optimum R, and I treated it as a Gaussian about that optimum R,that's an approximation. I mean, the main thing is that it's falling off rapidly in one direction, falling off rapidly in another, it's probably not exactly a Gaussian, but that's an approximation. The main property is that it is localized."

25. **M**: "That's a magnitude I didn't want to try to evaluate. The main problem is that I had some terms that were of the size of maybe a few tenths of an Ångstrom divided by another term a few tenths of an Angstrom squared, so I assumed: OK, ballpark unity, that's all. In other words, a tremendous amount of details . . . only with computer simulations could you be able to get some result, and then you treat the whole situation in a bit different way, and the details would be tremendous. . . it wasn't worth it, now I wouldn't get much out of it anyway. Well, now I was getting closer to experiment in the sense that I was getting more experimental detail, and so I thought it was important to bring out R as a variable some way or another. The perpendicular distance now was just a matter of convenience, I think I got a little bit tired of all of those Riemannian geometries, and so now I was prepared to be a little careless."
26. **M**: "m^* is an effective reduced mass for that type of problem. If you are crossing a hypersurface, then in terms of kinetic energy there is some effective mass associated with it."
27. **M**: "The integration over all orientations in each reactant is implicit in the integrations in Eq. (6.27) and in the calculation of Q. As long as the integration variables include all of the variables then that's automatically the case, using the full dimensionality of everything."
28. **M**: "If you think of F^r, the free energy of the reactants, and imagine rotating the reactants separately. . . then your free energy isn't changing, if the free energy of one orientation is going to be the same for the other orientations, in fact normally you

integrate over all orientations. But if you hold the orientation fixed, then it wouldn't matter what the orientation of the reactants is when it's out at infinity, it'd have the same free energy. And I'd have the same free energy also when you bring them together if they're sufficiently far apart so that the interactions don't depend on the mutual orientations. *This is unlike most reactions, which involve really pressing one thing against another, and they are very much dependent on orientations.* But if ET is occurring at a distance such that it is not depending on orientations, then you can integrate over them. But there are some ET reactions that depend on chirality and orientations."

Q: But you wrote a paper later on about orientations.

M: Yes, that's right, we did with Bob Cave.

29. **M**: "Suppose we included the effective orientation. . . for a whole series of reactants and suppose that the orientation affected the reactions in about the same way, the orientation then wouldn't affect, say, the effective standard free energy of reaction. If the orientation occurred, for whatever reason, both for the cross-reaction and for the self-exchange reactions, in other words if there was a certain orientation of the reactants that would be favorable not only for the cross-reaction but also for the exchange reactions, and maybe because of the electron centers being concentrated near one part of the molecule, one would get a faster rate no matter what the self-exchange or the cross-reaction, in that case there could be some cancellation."
30. **M**: "Cancelled by the previous assumption, cancelled exactly because there's no dependence on orientation, because at infinity the results are independent of orientation, so you can integrate over the orientations there. The different orientations have the same free energy, so you can integrate over orientations to get the full free energy, which includes the orientational degrees of

freedom. If an integral is independent of R, then you can integrate over that and you'll get the collision frequency out of that. In other words, imagine you save for the last step any integration over capital R. If you integrate over the angles, that is what you do for hard spheres collisions. If, in the TS, R is localized around some region and if you have some little distribution perpendicular to that, you focus on that region, then what comes out of the contact distance, for getting all the way to R, appears as a collision frequency as in hard spheres collisions, and you now multiply by sort of like a partition function about that R and that's one of the magnitudes that cancels in the ρ business, you know..."

31. **M**: "You want to have as few variables as you can, you want to reduce the number of variables as much as you can and if the results are independent of orientation, then you fix the orientation. You want to integrate over the irrelevant variables, so in this case in which the orientation is irrelevant, in other words everything is the same for all of the orientations, pick anyone orientation, calculate what you want and do the integration over orientations analytically."
32. **M**: "f^* extends the F which is constrained to the hypersurface where $U^r = U^p$ to what happens when you go off that hypersurface in this e.e.d. On the hypersurface, you start at U^r and what you start at is extended to an effective PES when you go off the hypersurface, so that the system you get is a stable distribution that I called in 1960 the e.e.d. The region around the point at the bottom contributes most, and regions away from that contribute less, in a Boltzmann way."
33. Q: $\bar{U}_i^* = U_i^* + \chi_o^*$: is a potential expressed as a mix of electronic energy and free energy: the first term is an electronic energy of the internal coordinates and the second one is a free energy which

is a mean over the solvent coordinates. Are you allowed to mix up electronic energy and free energy?

M: "If a particular free energy depends parametrically on some coordinates, then you're allowed to mix them up."

34. *Demonstration of Eq. (6.72): The steps described in words are here made explicit one after the other:

(i) I multiply numerator and denominator of f_i^*, Eq. (6.59), by $\exp[-\bar{U}_i(\mathbf{Q}_*)/kT]$:

$$f_i^* = \exp\left(-\frac{U_i^*}{kT}\right)\exp\left[\bar{U}_i^*(\mathbf{Q}_*)/kT\right] \Big/ \exp\left[\bar{U}_i^*(\mathbf{Q}_*)/kT\right]$$
$$\times \int \exp\left(-\frac{U_i^*}{kT}\right) dV_i' \Rightarrow f_i^* = \exp\left(-\frac{(\bar{U}_i^* - \bar{U}_i^*(\mathbf{Q}_*))}{kT}\right)$$
$$\Big/ \int \exp\left(-\frac{(\bar{U}_i^* - \bar{U}_i^*(\mathbf{Q}_*))}{kT}\right) dV_i'$$

(ii) I introduce the above result first into Eq. (6.66a) and then in Eq. (6.66b):

$$F^*(R) = \left\langle U_i^r + F_o^*\right\rangle_{*i} + kT\left\langle \ln\left[\exp\left(-\left(\frac{(\bar{U}_i^* - \bar{U}_i^*(\mathbf{Q}_*))}{kT}\right)\Big/\right.\right.\right.$$
$$\left.\left.\left.\times \int \exp\left(-\frac{(\bar{U}_i^* - \bar{U}_i^*(\mathbf{Q}_*))}{kT}\right) dV_i'\right)\right]\right\rangle_{*i}$$
$$= \left\langle \bar{U}_i^* + kT \ln\left[\exp\left(-\frac{(\bar{U}_i^* - \bar{U}_i^*(\mathbf{Q}_*))}{kT}\right. \Big/\right.\right.$$
$$\left.\left.\times \int \exp\left(-\frac{(\bar{U}_i^* - \bar{U}_i^*(\mathbf{Q}_*))}{kT}\right) dV_i'\right)\right]\right\rangle_{i_i}$$

(iii) I introduce Eq. (6.71) in the above equation:

$$\left\langle \bar{U}_i^* + kT \ln\left[\exp\left(-\frac{\frac{1}{2}\sum_{j,k=1}^{n_i} f_{jk}*\left(q^j - q^{*j}\right)\left(q^k - q^{*k}\right)}{kT}\right.\right.\right.$$

$$\left.\left.\left./\int \exp\left(\frac{-\frac{1}{2}\sum_{j,k}^{n_j}\left(q^j - q^{*j}\right)\left(q^k - q^{*k}\right)}{kT}\right) dV_i'\right)\right]\right\rangle_{*i}$$

$$= \left\langle \bar{U}_i^* + kT \ln\left[\exp\left(-\frac{\frac{1}{2}(\mathbf{Q}^{\mathrm{T}} - \mathbf{Q}_*^{\mathrm{T}}) \cdot \mathbf{F}^*(\mathbf{Q} - \mathbf{Q}_*)}{kT}\right.\right.\right. /$$

$$\left.\left.\left.\times \int \exp\left(-\frac{\frac{1}{2}(\mathbf{Q}^{\mathrm{T}} - \mathbf{Q}_*^{\mathrm{T}}) \cdot \mathbf{F}^*(\mathbf{Q} - \mathbf{Q}_*)}{kT}\right) dV_i'\right)\right]\right\rangle_{*i} \qquad (*)$$

(iv) I begin integrating the above equation evaluating separately two integrals:
$\left\langle \bar{U}_i^* \right\rangle_{*i} = \left\langle U_i^r \right\rangle_{*_i} + \left\langle F_0^* \right\rangle_i^* + m\left\langle \langle U^r - U^p \rangle_{*0} \right\rangle_{*_i}$ from Eqs. (6.65) and Eqs. (6.66a) and (6.66b):

$$\bar{U}_i^* = U_i^r + F_o^* + m\left\langle U^r - U^p \right\rangle_{*o} \qquad (6.112)$$

$$F^*(R) = \left\langle U_i^r + F_o^* \right\rangle^{*i} + kT\left\langle \ln f_i^* \right\rangle_{*i} = \left\langle \bar{U}_i^* + kT \ln f_i^* \right\rangle_{*i} \qquad (6.66\text{a,b})$$

The result is the first part (first three terms) of Eq. (6.72)

$$F^*(R) = U_i^r(\mathbf{Q}_*) + F_o^*(\mathbf{Q}_*) + m\left\langle U^r(\mathbf{Q}_*) - U^p(\mathbf{Q}_*) \right\rangle_{*_o} - \frac{1}{2}kT \ln\left[(2\pi kT)^{n_i}/|f_{jk}^*|\right] \qquad (6.72)$$

Q: Why $\langle U_i^r \rangle_{*i} = U_i^r(\mathbf{Q}_*)$, $\langle F_o^* \rangle_{*i} = F_o^*(\mathbf{Q}_*)$ and so why $\langle \bar{U}_i * \rangle_{*i} = \langle \bar{U}_i^*(\mathbf{Q}_*) + \frac{1}{2} \sum_{j,k=1}^{n_i} f_{jk}^*(q^j - q^{*j})(q^k - q^{*k}) \rangle_{*i} = \bar{U}_i^*(\mathbf{Q}_*)$? Why does the average value of the distribution with respect to f_i^* equal the value of the potential corresponding to the equilibrium values of the vibrational coordinates? This means that $\left\langle \frac{1}{2} \sum_{j,k=1}^{n_i} f_{jk}^* \left(q^j - q^{*j}\right)\left(q^k - q^{*k}\right) \right\rangle_{*i} = 0.$

M: To demonstrate that

$$\left\langle \bar{U}_i^* \right\rangle_{*i} = \int \left[\bar{U}_i^* \exp\left(-\frac{\bar{U}_i^*}{kT}\right) \Big/ \int \exp\left(-\frac{\bar{U}_i^*}{kT}\right) dV_i' \right] \times dV_i' = \bar{U}_i^*\left(\mathbf{Q}_*\right)$$

One uses the method of the steepest descent: Supposing you have the integral $\int f(x) e^{-g(x)} dx$ and $e^{-g(x)}$ is some Gaussian. Then to evaluate the integral you consider that $e^{-g(x)}$ has its maximum value at the minimum of $g(x)$, you evaluate then the preexponential function at that x at which the exponential has its maximum value and that minimum value of the preexponential is here exactly $\bar{U}_i^*(\mathbf{Q}_*)$.

(v) I just show at this point the integral in the second part of Eq. (*):

$$\left\langle kT \ln \left[\exp\left(-\frac{\frac{1}{2}\left(\mathbf{Q}^{\mathrm{T}} - \mathbf{Q}_*^{\mathrm{T}}\right) \cdot \mathbf{F}^*(\mathbf{Q} - \mathrm{Q}_*)}{kT} \right. \Big/ \int \exp\left(-\frac{\frac{1}{2}\left(\mathbf{Q}^{\mathrm{T}} - \mathbf{Q}_*^{\mathrm{T}}\right) \cdot \mathbf{F}^*(\mathbf{Q} - \mathbf{Q}_*)}{kT} \right) dV_i' \right) \right] \right\rangle_{*i}$$

In Note 24 at the bottom of p. 687, you cite Bellman's *Introduction to Matrix Analysis*. I have the Second Edition, so the pages numbers don't agree, but I believe you were referring to Eq. (6.1) p. 97 in my book:

$$I_N = \int_{-\infty}^{\infty} \cdots \int_{-\infty}^{\infty} e^{-(x,Ax)} dx$$

where $dx = dx_1 dx_2 \cdots dx_N$. The result is:

$$I_N = \frac{\pi^{N/2}}{|A|^{1/2}}.$$

In our case we would have:

$$\int \exp\left(-\frac{\frac{1}{2}\left(\mathbf{Q}^{\mathrm{T}} - \mathbf{Q}_*^{\mathrm{T}}\right) \cdot \mathbf{F}^*(\mathbf{Q} - \mathbf{Q}_*)}{kT} \right) dV_i'$$

$$= \frac{(2\pi kT)^{n_i/2}}{|\mathbf{F}^*|^{1/2}} = \frac{(2\pi kT)^{n_i/2}}{\left|f_{jk}^*\right|^{1/2}}. \qquad (*)$$

Substituting inside the above equation:

$$\left\langle kT \ln \left[\exp\left(-\frac{\frac{1}{2}\left(\mathbf{Q}^{\mathrm{T}} - \mathbf{Q}_*^{\mathrm{T}}\right) \cdot \mathbf{F}^*(\mathbf{Q} - \mathbf{Q}_*)}{kT} \right) \Big/ \frac{(2\pi kT)^{n_i/2}}{\left|f_{jk}^*\right|^{1/2}} \right] \right\rangle_{*i}$$

$$\Rightarrow kT \int \left\{ \begin{array}{l} \ln \left[\exp\left(-\frac{\frac{1}{2}\left(\mathbf{Q}^{\mathrm{T}} - \mathbf{Q}_*^{\mathrm{T}}\right) \cdot \mathbf{F}^*(\mathbf{Q} - \mathbf{Q}_*)}{kT} \right) \Big/ \right. \\ \left. \frac{(2\pi kT)^{n_i/2}}{\left|f_{jk}^*\right|^{1/2}} \right] \left[\exp\left(-\frac{\frac{1}{2}\left(\mathbf{Q}^{\mathrm{T}} - \mathbf{Q}_*^{\mathrm{T}}\right) \cdot \mathbf{F}^*(\mathbf{Q} - \mathbf{Q}_*)}{kT} \right) \Big/ \right. \\ \left. \int \exp\left(-\frac{\frac{1}{2}\left(\mathbf{Q}^{\mathrm{T}} - \mathbf{Q}_*^{\mathrm{T}}\right) \cdot \mathbf{F}^*(\mathbf{Q} - \mathbf{Q}_*)}{kT} \right) \right] dV_i' \end{array} \right\} dV_i'$$

$$kT \int \left\{ \frac{\ln\left[\exp\left(-\dfrac{\frac{1}{2}\left(\mathbf{Q}^{\mathrm{T}} - \mathbf{Q}_*^{\mathrm{T}}\right)\cdot \mathbf{F}^*\left(\mathbf{Q} - \mathbf{Q}_*\right)}{kT}\right) \Big/ \dfrac{(2\pi kT)^{n_i/2}}{\left|f_{jk}^*\right|^{1/2}}\right]}{\left[\exp\left(-\dfrac{\frac{1}{2}\left(\mathbf{Q}^{\mathrm{T}} - \mathbf{Q}_*^{\mathrm{T}}\right)\cdot \mathbf{F}^*\left(\mathbf{Q} - \mathbf{Q}_*\right)}{kT}\right) \Big/ \dfrac{(2\pi kT)^{n_i/2}}{\left|f_{jk}^*\right|^{1/2}}\right]} \right\} dV_i'$$

$$kT \int \left\{ \frac{\left[-\dfrac{\frac{1}{2}\left(\mathbf{Q}^{\mathrm{T}} - \mathbf{Q}_*^{\mathrm{T}}\right)\cdot \mathbf{F}^*\left(\mathbf{Q} - \mathbf{Q}_*\right)}{kT} - \ln \dfrac{(2\pi kT)^{n_i/2}}{\left|f_{jk}^*\right|^{1/2}}\right]}{\left[\exp\left(-\dfrac{\frac{1}{2}\left(\mathbf{Q}^{\mathrm{T}} - \mathbf{Q}_*^{\mathrm{T}}\right)\cdot \mathbf{F}^*\left(\mathbf{Q} - \mathbf{Q}_*\right)}{kT}\right) \Big/ \left(\dfrac{(2\pi kT)^{n_i/2}}{\left|f_{jk}^*\right|^{1/2}}\right)\right]} \right\} dV_i'$$

$$kT \int \left\{ \left[\begin{array}{l} -\dfrac{\frac{1}{2}\left(\mathbf{Q}^{\mathrm{T}} - \mathbf{Q}_*^{\mathrm{T}}\right)\cdot \mathbf{F}^*\left(\mathbf{Q} - \mathbf{Q}_*\right)}{kT} \\ \exp\left(-\dfrac{\frac{1}{2}\left(\mathbf{Q}^{\mathrm{T}} - \mathbf{Q}_*^{\mathrm{T}}\right)\cdot \mathbf{F}^*\left(\mathbf{Q} - \mathbf{Q}_*\right)}{kT}\right) \Big/ \\ \dfrac{(2\pi kT)^{n_i/2}}{\left|f_{jk}^*\right|^{1/2}} - \ln \dfrac{(2\pi kT)^{n_i/2}}{\left|f_{jk}^*\right|^{1/2}} \\ \exp\left(-\dfrac{\frac{1}{2}\left(\mathbf{Q}^{\mathrm{T}} - \mathbf{Q}_*^{\mathrm{T}}\right)\cdot \mathbf{F}^*\left(\mathbf{Q} - \mathbf{Q}_*\right)}{kT}\right) \Big/ \dfrac{(2\pi kT)^{n_i/2}}{\left|f_{jk}^*\right|^{1/2}} \end{array} \right] \right\} dV_i' \quad (**)$$

(vi) I evaluate now the above integral beginning with the one in the last line:

$$-kT \ln\left[(2\pi kT)^{n_i/2} / \left|f_{jk}*\right|^{1/2}\right]$$
$$\times \int \left[\exp\left(-\frac{\frac{1}{2}\left(\mathbf{Q}^{\mathrm{T}} - \mathbf{Q}_*^{\mathrm{T}}\right)\cdot \mathbf{F}^*\left(\mathbf{Q} - \mathbf{Q}_*\right)}{kT}\right) \Big/ \right.$$
$$\left. \times \frac{(2\pi kT)^{n_i/2}}{\left|f_{jk}^*\right|^{1/2}} \right] dV_i'$$

$$= -kT \ln\left[(2\pi kT)^{n_i/2} / \left|f_{jk*}\right|^{1/2}\right]$$

$$= -\frac{1}{2}kT \ln\left[(2\pi kT)^{n_i} / \left|f_{jk*}\right|\right]$$

which is the last addend of Eq. (6.72).

I write now the integral of the first line in Eq. (**):

$$\int \left[-\frac{1}{2}\left(\mathbf{Q}^{\mathrm{T}} - \mathbf{Q}_*^{\mathrm{T}}\right) \cdot \mathbf{F}^* \left(\mathbf{Q} - \mathbf{Q}_*\right)\right.$$
$$\left. \times \exp\left(-\frac{\frac{1}{2}\left(\mathbf{Q}^{\mathrm{T}} - \mathbf{Q}_*^{\mathrm{T}}\right) \cdot \mathbf{F}^* \left(\mathbf{Q} - \mathbf{Q}_*\right)}{kT}\right)\right] dV_i'.$$

This integral approximately vanishes for the same reason previously seen. Even here we have the evaluation of the integral by the method of steepest descent.

35. In $F^r = U_i^r(\mathbf{Q}_r) + F_o^r(\mathbf{Q}_r) - \frac{1}{2}kT \ln\left[(2\pi kT)^{n_i} \Big/ \left|f_{jk}^r\right|\right]$; the dependence on the force constants has the form $-\frac{1}{2}kT \ln\left[(2\pi kT)^{n_i} / \left|f_{jk}^r\right|\right]$. This form reminds one of a configurational entropy, that is, it says that the configurational entropy depends on the force constants.

    M: "They're related to the entropy. Imagine... having the statistical mechanical expression for the free energy of a harmonic oscillator and of taking minus the derivative of the free energy with respect to temperature to get the entropy..."
36. **M**: "A little bit earlier than the 1965 paper we obtained a very simple way of expressing the nonequilibrium free energy of the system in terms of the equilibrium free energies, and we had $F_{non} - F_1 = F_{1-0}^{op} - F_{1-0}$. Now, that equation is fine, and that expression applies even if you put into both systems all the atmospheric charges, as long as you don't start rearranging the

charges... Even though you put into the system all of the atmospheric charges that were absent in the 1956 paper, there we just focused on the two ions, and not the atmosphere. So, even in this expression if you put in all those ions, but don't allow them to rearrange, if you keep them fixed, so that the only rearrangement be in the reorganization of the system, the orientation of the solvent molecules, and so on, then you get exactly the same expression, but in a much simpler way than I did in the 1956 paper, in other words all those extra charges cancel out on the right-hand side, as indeed they should if the expression is correct.... In the actual systems or the hypothetical systems I leave all the charges in there, the electrolyte, and so on, charges that I never had in the 1956 paper. All cancels, so that one achieved the same expression as I did in 1956, when I derived it in a very different way. And what I wanted to do in this particular case, is that I wanted to express the result in terms of certain hypothetical equilibrium systems without having to go through the terrible gyrations of going through the nonequilibrium, and it was important that what didn't appear in the 1956 paper should not appear in this, and indeed they cancel out exactly in the new formulation. Now, in an Appendix VII we also talk about what happens when you now allow some of those ions, the ions in the atmosphere, to adjust themselves, that I didn't have in the 1956 paper. What I have in the Appendix is in the linear approximation, when you are treating ions interactions and solvent effects under typical conditions you have to go beyond linear. Linear is Debye–Hückel."

37. λ_o is a *fluctuation term*. As a matter of fact, expanding the square and then taking the average one has:

$$\lambda_o = \Big\langle [U_o^r(1) - U_o^p(1)]^2 - [U_o^r(1) - U_o^p(1)]\langle U_o^r(1) - U_o^p(1)\rangle$$
$$+ \langle U_o^r(1) - U_o^p(1)\rangle^2 \Big\rangle / kT \Rightarrow$$

$$\lambda_o = \left(\left\langle\left[U_o^r(1) - U_o^p(1)\right]^2\right\rangle - 2\left\langle U_o^r(1) - U_o^p(1)\right\rangle \right.$$

$$\left. \times \left\langle U_o^r(1) - U_o^p(1)\right\rangle + \left\langle U_o^r(1) - U_o^p\right\rangle^2 \right)/kT$$

$$\Rightarrow \lambda_o = \left(\left\langle\left[U_o^r(1) - U_o^p(1)\right]^2\right\rangle - \left\langle U_o^r(1) - U_o^p\right\rangle^2\right)/kT.$$

where the statistical fluctuation nature of λ_o is now apparent.

38. **M**: "Eq. $\lambda_o = (\langle[U_o^r(1) - U_o^p(1)]^2\rangle - \langle U_o^r(1) - U_o^p(1)\rangle^2)/kT$ is the statistical mechanical counterpart of Eq. $\lambda_i = \frac{1}{2}\sum_j k_j (\Delta q_j)^2$ which is very straightforward or Eq. $\lambda_o = (ne)^2 \left(\frac{1}{2a_1} + \frac{1}{2a_2} - \frac{1}{R}\right)\left(\frac{1}{D_{\text{op}}} - \frac{1}{D_{\text{s}}}\right)$, much less straightforward because you have to go to continuum theory."
39. The free energy is a macroscopic thermodynamic quantity. How is it related to the couple of central ions, or more in general central particles exchanging electrons in solution?

    M: "The microscopic part comes with all the solvent around and imagine that you have many replicas of the system, here is one system, there is another, there is another, there is another... the point is that for dilute solutions you can always take that pair of particles and look at their interaction with the surroundings and if you take the whole of such couples of interacting particles then you have a molar dilute solution."
40. **M**: "The activation free energy includes a work term w in it, in fact a couple of w's, and the reorganization energy doesn't. *That's the big difference between the activation free energy and the reorganization energy.* In other words, the activation free energy is w^r plus the reorganizational term and the reorganizational term has a w in it anyway. $\frac{\ddot{e}}{4}$ is a reorganizational term ... the reorganization free energy ΔG^* has a work term plus a reorganization term, that reorganization term has a $\Delta G^0(R)$ which in turn has a w in it.

    Any transition state is reorganized from the reactants. But one should relate the TS to the whole reorganization and remember that $\Delta G^0(R)$ has a ΔG^0 plus a difference of work terms."

41. **M**: "The actual collision number consists of repeated collisions separated by getting into a sort of a collision shell, or something, so the whole terminology would be loose then. In solution you don't have collisions as you have in the gas phase, instead there is a sort of wandering of molecules toward each other, and then when they are in contact they are in contact for an all set of rapid fired collisions and then they separate. When you average over the rapid fire collisions and the small diffusion, then together you would get the Z."
42. Q: The leading term $\frac{1}{2}$ in Eq. (6.100a) arises really from the $\frac{\Delta F^{0\prime}}{2}$ in the RHS of Eq. (6.93):

$$\Delta F^{*}=\frac{w^{r}+w^{p}}{2}+\frac{\lambda}{4}+\frac{\Delta F^{o\prime}}{2}+\frac{\left(\Delta F^{o^{\prime}}+w^{p}-w^{r}\right)^{2}}{4\lambda} \tag{6.93}$$

On the other hand, we have the Equations:

$$\Delta F^{*}(R)=F^{*}(R)-F^{r}(R) \tag{6.49}$$

$$F^{*}(R)-F^{r}(R)=\frac{1}{2}\left(\mathbf{Q}_{*}^{T}-\mathbf{Q}_{r}^{T}\right)\cdot\mathbf{F}^{r}(\mathbf{Q}_{*}-\mathbf{Q}_{r}) +\Delta F_{o}^{*}(\mathbf{Q}_{*})+\frac{1}{2}kT\ln\left(\left|f_{jk}^{*}\right|\Big/f_{jk}^{r}\right) \tag{6.75}$$

$$F^{*}(R)-F^{r}(R)=\frac{1}{2}\sum_{j,k=1}^{n_i}f_{jk}^{*}(q^{j}-q_{*}^{j})(q^{k}-q_{*}^{k}) +\Delta F_{o}^{*}(\mathbf{Q}_{*})+\frac{1}{2}kT\ln\left(\left|f_{jk}^{*}\right|/f_{jk}^{r}\right)$$

$$\lambda_{i}=\frac{1}{2}\sum_{j,k}k_{j,k}\Delta q_{j}\Delta q_{k} \tag{6.76}$$

$$\Delta F_{o}^{*}=m^{2}\lambda_{o} \tag{6.81}$$

$$\Delta F_o^* = F^{\text{non}} - F_1 = F_{1-0}^{\text{op}} - F_{1-0}\text{``}$$

$$F_{1-0}^{\text{op}} = \frac{1}{2}\sum_i \int \phi_{i1-0}^{\text{op}}(\rho_{i1}^0 - \rho_{i0}^0)d\mathbf{r} \quad (5.52\text{b}),$$

$$F_{1-0} = \frac{1}{2}\sum_i \int \phi_{i1-0}(\rho_{i1}^0 - \rho_{i0}^0)d\mathbf{r} \quad (5.52\text{a})$$

and so we see that the activation energy depends on three terms, $\lambda_i = \frac{1}{2}\sum_{j,k} k_{j,k}\Delta q_j \Delta q_k$, $F_{1-0}^{\text{op}} = \frac{1}{2}\sum_i \int \phi_{i1-0}^{\text{op}}(\rho_{i1}^0 - \rho_{i0}^0)d\mathbf{r}$, and $F_{1-0} = \frac{1}{2}\sum_i \int \phi_{i1-0}(\rho_{i1}^0 - \rho_{i0}^0)d\mathbf{r}$ of a quadratic nature and each one with a $\frac{1}{2}$ in front of each. But which one is the connection with the $\frac{1}{2}$ in Eq. (6.100a) which depends on Eq. (6.93)?

M: "I think the $\frac{1}{2}$ is more connected with the assumed symmetry. In other words, for the leading term to be $\frac{1}{2}$ in that derivative there is an assumption that I made about symmetrization of the PESs, but if I hadn't done that, then that leading term wouldn't be $\frac{1}{2}$. You get $\frac{1}{2}$ if the corresponding force constants are equal, but that's too stringent a condition, and it's in fact not true, so that's why I introduced, in the 1965 paper, symmetrization, and expanded in terms of symmetric and antisymmetric combinations of the force constants."

43. **M**: "The equation is true when R is large, it's also approximately true when R is not large.
    I wrote it just by thinking about the physical situation, only thinking about the physical situation. When reactants are far apart, you still have the reorganizations but clearly they're independent of each other."
44. **M**: "The dielectric constant of a salt solution is not exactly as that of water."
45. **M**: "It's best to think of U^r and U^p fixed, and determined by ΔG^0. The e.e.d. is $U^* = U^r + m(U^r - U^p)$. Suppose we introduce m that way, m is determined by ΔG^0."

Q: Which is basically the difference between the bottoms of the PESs.

M: Roughly, that's right. For each value of m, you get a different e.e.d. If one thinks of a different value of q^N, then to describe that q^N you need a different value of m, so m does depend on q^N. q^N *labels the hypersurfaces when you move the* U^p *up and down.* If the barrier is equal to $m^2\lambda$, then when $m = 0$ the parabola of the products on the average crosses the bottom of the other, in other words, the TS is at the bottom of the first parabola, that of the reactants. So $m = 0$. If the activation free energy is $m^2\lambda$, then one of the parentheses crosses on the average the bottom of the other, then there is no activation energy, so $m = 0$.So, there $\partial U^r / \partial q^N = 0$. When $m = 0$, $U^r$is on the average at its minimum, so that derivative is zero. m *becomes positive in the inverted region.*"

46. **M**: "If you have two parabolas and if they look identical, one for the reactants and one for the products, then clearly there's a kind of symmetry in the way the surroundings see those reactants and products. If you think of the one-dimensional profiles, and if you have that symmetry, then you can see physically that the slopes are equal and of opposite sign at the transition state. The $\frac{1}{2}$ is when at the transition state the slopes are equal enough and opposite. That's only when you have the two parabolas the same with$\Delta G^0 = 0$. Now you have the same parabolas, the one displaced from the other, and if you have $\Delta G^0 = 0$, you see, if those two slopes are equal to each other, if you look at Eq. (6.108) then you have $m = -\frac{1}{2}$, by making the two slopes equal one to each other, then the denominator becomes twice the numerator.
47. **M**: "When $\Delta G^0/\lambda$ is small, $f$is close to unity and K_{12} is close to 1, but you don't have to go to the extreme limit, that's when $\Delta G^0/\lambda = 0$. But when $\Delta G^0/\lambda$ is small, then f is close to unity as long as you're in the linear regime. In other words, you have a

quadratic expansion in $\Delta G^0/\lambda$, you expand it, and as long as you can neglect the quadratic term you are in the linear regime. That corresponds, in this other language, to f being close to unity, that is, f measures the deviation from the quadratic term."

48. **M**: "That corresponds to the $\lambda/4$."
49. **M**: "When an ion approaches an electrode, there may be some work term to get there, and the definition of the work term in a sense is that it can't be changed by the standard free energy of reaction nor can it be changed by the overpotential. That's Debye's assumption."
50. **M**: "For example, in experiments with phenanthrolines... if you have substituents on some edge of one of those rings and if it's not affecting the electronic structure of the ion–ligand interaction too much, then it's just acting as something which changes the ΔG^0 and doesn't change the λ, maybe because the λ that's involved is largely dependent on the metal-immediate ligand atom change of those distances and some remote substituent may not have a large effect on changes of those distances between oxidized and reduced form. That's the assumption."
51. **M**: "When a series of reactants is oxidized (reduced) by two different reagents, the ratio of the two rate constants is the same for all members of the series in the region of chemical transfer coefficient equal to 0.5. In other words, that's the linear regime, due to neglect of the quadratic. If you get a linear slope in the free energy plot, with a slope of 0.5, you are in that region. And you may remember that it was difficult to get into the nonlinear region, one really had to change by a large amount the standard free energy of reaction. When you remember in the electrochemical case, Campus tried to go where the overpotential was enough that he could get into the nonlinear regime, but the trouble was that then the reaction became diffusion controlled, then finally Chidsey in 1991 did that famous Science experiment (Christopher E. D. Chidsey, *Science* **251**, (4996), 919–922) where he

attached the -SH to a gold electrode, then he had a large fixed distance and he was able to vary the standard potential a lot and see the curvature. A famous paper.

52. Q: Which ones are the distinguishing features of an electrode with respect to a molecule with the same oxidizing or reducing power?

    M: "The electrode has a surface, so some of the details are different, there are double layer questions that come in and how you handle that. One should also consider that with a metal you don't have an inverted effect, for reasons that are clear. And if the reaction is nonadiabatic, the specific coupling term depends upon what level of the metal you're going into."
53. **M**: "The work terms refer to what happens when you bring the reactant up to a distance at which it undergoes electron transfer."
54. **M**: "The reason for the $\frac{1}{2}$ is that in one case one has the reorganization of two species, in the other case there is just the reorganization of one, so the free energy barrier is roughly half, and that gives rise to that $\frac{1}{2}$ in that relationship.
55. Q: In Part IV, you wrote two different expressions for k_r:

$$k_r = (kT/h)\exp\left(-\Delta F^{\ddagger}/kT\right) \tag{5.1}$$

and

$$k_r = Z\exp\left[-\left(\Delta F_i^{\ddagger} + \Delta F_o^{\ddagger}\right)/kT\right] \tag{5.12}$$

Which are the symmetry arguments by which you go from one to the other of the earlier two expressions and which are the more rigorous arguments?

M: "That's about integration over the external coordinates, the 'outer' coordinates (position and orientational) of all the particles. The $\Delta F^{\ddagger}$ in Eq. (5.1) is different from the terms in Eq. (5.12), the $\Delta F^{\ddagger}$ in Eq. (5.1) also includes some of those external coordinates that one integrates over, and integrating over that you get Z. The symmetry arguments.... You write that $\Delta F^{\ddagger}$ in terms of external coordinates and internal coordinates, and then try to do some integration and you take a theoretical expression for $\Delta F^{\ddagger}$. We started from the symmetry with respect to rotating the line of centers and the symmetry with rotating the system within two large spheres, one about each central ion."

56. **M**: "Electric unsaturation means that the response of the ions is linear and because then the response of the ions is linear, you know, the Poisson–Boltzmann equation simplifies, so you have a linear differential equation."
57. Q: Do you mean that the "op" medium is a system without ionic atmosphere?

    M: "No, it may have the ionic atmosphere but the ionic atmosphere isn't responding. Just like it has solvent molecules but other than electronically the solvent molecules aren't responding. And what I don't know, I would have to think about it, is: in what configuration are the ions nevertheless, ... in the TS configuration or are they in the reactants configuration? And it may well be that since there is an m there, they're probably in the TS configuration but in terms of the interaction they are not responding to the charges per se but only by their polarizability, their *electronic* polarizability. So, in other words, *when you're looking at the optical response, it's as though you do have the nuclei there and there may be some dipole arrangement, but those nuclei aren't responding, they stay fixed, whatever the distribution they're in.*

References

1. R. A. Marcus, *J. Chem. Phys.* **43**, 679–701, (1965).
2. R. A. Marcus, *J. Chem. Phys.* **94**, 979–989, (1956).
3. R. A. Marcus, *J. Phys. Chem.* **67**, 853, 2889, (1963).
4. See, for example, L. Landau, *Phys. Z. Sowjetunion.* **1**, 88, (1932); **2**, 46, (1932); C. Zener, *Proc. R. Soc. (London)* **A137**, 696, (1932); **A140**, 660, (1933); C. A. Coulson, K. Zalewski, *Proc. R. Soc. (London)* **A268**, 437, (1962). The present

(1965 *e.n.*) situation has been summarized in Ref. [12], where the definition of nonadiabaticity was also discussed. References should also have been made there to the work of E. C. G. Stueckelberg, *Helv. Phys. Acta.* **5**, 369, (1932); Compare H. S. W. Massey, in *Encyclopedia of Physics*, Vol. 36, p. 297, S. Flügge, Editor, Springer-Verlag, Berlin (1956).

5. N. Sutin and M. Wolfsberg, quoted by N. Sutin, *Ann. Rev. Nucl. Sci.* **12**, 285, (1962). These authors discussed the possibility of tunneling of the atoms in the inner coordination shell.

   Possible quantum effects which include atom tunneling in the medium outside this shell have been treated by V. G. Levich, R. R. Dogonadze, *Proc. Acad. Sci. USSR, Phys. Chem. Sec.* [English transl. **133**, 591, (1960)]; *Collect. Czechoslov. Chem. Comm.* **26**, 193, (1961) [transl. O. Boshko, University of Ottawa, Ontario.] Any conclusions concerning the contribution of atom tunneling depend in a sensitive way on the assumed values for the bond force constants and lengths in the inner coordination shell, properties on which data are now becoming available, and on the assumed value for a mean polarization frequency for the medium. [Atom tunneling is different from electron tunneling, the latter being a measure of the splitting in Fig. 1 (Ref. [12]).]
6. R. A. Marcus, *J. Chem. Phys.* **41**, 2614, (1964).
7. R. A. Marcus, *J. Chem. Phys.* **41**, 2624, (1964).
8. R. A. Marcus, *J. Chem. Phys.* **43**, 1598, (1965); (a) W. Pauli, Jr., in *Handbuch der Physik*, Vol. 5, p. 39, S. Flügge, Editor, Springer-Verlag, Berlin (1958); (b) R. A. Marcus, *J. Chem. Phys.* **41**, 603–609, (1964); (c) C. E. Weatherburn, *Riemannian Geometry*, Cambridge University Press, New York (1957).
9. D. A. McQuarrie, *Statistical Mechanics*, Harper & Row (1973).
10. R. R. Dogonadze, Y. A. Chizmadzhev, *Proc. Acad. Sci. USSR, Phys. Chem. Sec.English Transl.* **144**, 463, (1962), **145**, 563, (1962); V. G. Levich, R. R. Dogonadze, *Intern. Comm. Electrochem. Thermodyn. Kinet.*, 14th Meeting, Moscow (1963), preprints. This work is reviewed in Ref. [12].
11. E. A. Guggenheim, *Thermodynamics*, North-Holland (1977).
12. R. A. Marcus, *Ann. Rev. Phys. Chem.* **15**, 155, (1964).
13. R. A. Marcus, *J. Chem. Phys.* **38**, 1335, (1963).
14. R. A. Marcus, *J. Chem. Phys.* **39**, 1734, (1963).
15. R. A. Marcus, ONR Tech. Rept. No. 12, Project NR 051-331, (1957); Compare *Can. J. Chem.* **37**, 155, (1959) and *Trans. Symp. Electrode Processes, Phila, PA.* 1959, 239–245, (1961).
16. R. A. Marcus, *J. Chem. Phys.* **41**, 603, (1964).
17. R. A. Marcus, *J. Chem. Phys.* **38**, 1858, (1963).

CHAPTER 7

Theory of Electron Transfer Reactions of Solvated Electrons and of Chemiluminescent Electron Transfer Reactions

In this chapter is summarized the work of Marcus on the theory of ET reactions of solvated electrons and on the chemiluminescent ET reactions, **M61** [1], **55** [2], **57** [3], **57 Erratum** [3], **60** [4]. The text, very close to Marcus' papers, is interspersed with notes and remarks. Marcus' Notes are collected at the end of the chapter.

From **M61**: R. A. Marcus, "**Theory of Electron Transfer Reactions and Related Phenomena.**" *Proceedings of Symposium on Exchange Reactions*, p. 1, Atomic Energy Agency (1965).

7.1. Summary of the Assumptions, Approximations, and Consequences of the M. Theory of ET Reactions

In **M61**, M. gives a very terse summary of the elements of his theory. I report it almost verbatim often listing the items for didactical convenience.

"Among the *principal assumptions* are the following [5, 6]:

(1) The usual assumption of using *equilibrium statistical mechanics* to calculate the *chance* that the system will form an 'activated complex'(1) and of assuming that the rate is the rate of '*first passages*' through the activated complex configurations.

(2) *Classical statistical mechanics* for calculating the principal contributions to the free energy of activation.
(3) *Weak overlap* of the redox orbitals.
(4) Reaction is usually quantum-mechanically *adiabatic*.
(5) Partial dielectric unsaturation outside coordination shells.
In addition, several *approximations* were introduced to simplify the equations considerably for easier comparison with the experimental data:
(6) *Harmonic* vibrations inside coordination shells.
(7) '*Symmetrized*' potential energy surfaces (PESs) of reactants and products.
(8) In some applications, it is assumed that effects due to *specific interactions cancel* in certain comparisons and *can be ignored* in others.

Assumption (7) and, except when large changes in bond length occur, assumption (6), can be shown to introduce little error.

Assumptions (1) to (7) lead to the following simple expression for the rate constant k of chemical or electrochemical ET reactions [5–9]:

$$k = \kappa\rho Z e^{-\Delta F^*/kT} \tag{7.1}$$

$$\Delta F^* = w^r + \frac{\lambda}{4}(1 + \Delta/\lambda)^2 \tag{7.2}$$

where $\kappa \cong 1$ (assumption (4)), $\rho \cong 1$, $Z \approx 10^{11}\,\mathrm{M^{-1}\,sec^{-1}}$ (homogeneous) or $10^4\,\mathrm{cm\,sec^{-1}}$ (heterogeneous)."

The value of Z was later adjusted by **M** to $Z = 10^{12}\,\mathrm{M^{-1}\,sec^{-1}}$ in **M170**.

"Δ equals $\Delta F^{0\prime} - w^r + w^p$ for homogeneous reactions and equals $ne(E - E_0') - w^r + w^p$ for electrode reactions. The remaining symbols are defined as follows: w^r, w^p = work to bring reactants together or products together, respectively, to *mean* separation distance R in the activated complex[2];

Note 2: In an electrode reaction, the electrode is one of the 'reacting species.' The R now refers to the ion–electrostatic image distance, that is, to twice the ion–electrode distance [5–7, 9].

$\Delta F^{0\prime}$ = 'standard' free energy of the homogeneous reaction for the *prevailing*[(2)] *temperature and medium*"; n = number of electrons transferred in electrode reaction; $E - E_0'$ = cell potential minus "standard" cell potential (corrected for iR drop, etc.); λ = *intrinsic reorganization factor*, which depends on differences of orientation polarization and of bond lengths in the two valence states of each reacting species. λ is essentially an additive function $\left[= \frac{1}{2}(\lambda_a + \lambda_b)\right]$ of homogeneous reactants $A$and B, λ for the electrode reaction of species $A(= \lambda_a^{el})$ equals $\frac{1}{2}\lambda_a$ when the distance R in the homogeneous reaction is twice the ion–electrode distance in the electrode reaction [5–7, 9].

From these equations, a number of *predictions* follow involving *relations between the rate constants* of cross-reactions, homogeneous exchanges, and electrode reactions [6, 9]. In turn, these tests provide some insight into the applicability of various assumptions as well as into the mechanism of the reactions. The predictions are summarized elsewhere and compared there with the data [8, 9].

A direct *consequence* of assumptions (5), (6), and (7) is

(i) The *simplicity* of the final results (e.g., Eq. (2) and
(ii) The *additivity* of λ as well as
(iii) The fact that for *small* Δ's ($\Delta \ll \lambda$): $\partial \Delta F^* / \partial \Delta = 0.5$"
*Demonstration:
Expanding the square in parenthesis in Eq. (7.2):

$$\left(1 + \frac{\Delta}{\lambda}\right)^2 = 1 + \frac{2\Delta}{\lambda} + \frac{\Delta^2}{\lambda^2}.$$

$$\text{If } \Delta \ll \lambda \Rightarrow \left(1 + \frac{\Delta}{\lambda}\right)^2 \cong 1 + \frac{2\Delta}{\lambda} \Rightarrow \Delta F^*$$

$$= w^r + \frac{\lambda}{4} + \frac{1}{2}\Delta \Rightarrow \frac{\partial \Delta F^*}{\partial \Delta} = 0.5.$$

"Evidence for this numerical value of the derivative for *small* Δ's exists for both homogeneous and electrochemical reactions as is discussed elsewhere [8, 9].

(iv) Another consequence of assumptions (5) to (7) is the existence of *a relation between the rate constants* k_{ab} for $A_{\text{ox}} + B_{\text{red}} \to A_{\text{red}} + B_{\text{ox}}$ and those for the electron exchanges (k_{aa} and k_{bb}) and the equilibrium constant K_{ab} [8]:

$$k_{ab} \cong \sqrt{k_{aa}k_{bb}K_{ab}f} \tag{7.3}$$

where f is a certain function of k_{aa}, k_{bb}, and K_{ab}, and is usually close to unity.

(v) A reflection of the λ additivity *and* of the 0.5 value for the $\partial\Delta F^*/\partial\Delta$ is the *constancy of the ratios* k_a^{el}/k_{ab} and k_{ab}/k_{ac} when A is varied in a series of reactions with an electrode at a *given* cell potential or with a reagent B and with another reagent C (cf. reductions of $Co(NH_3)_5X^{\text{III}}$ for different X's). Evidence for such constancy is discussed elsewhere [8, 9].

Some tests of Eq. (7.4) are available. If k is written as an Arrhenius expression:

$$k = A\exp(-E_a/kT)$$

then

$$A \approx \kappa\rho Z\exp(\Delta S^*/k) \tag{7.4}$$

where

$$\Delta S^* = -\partial\Delta F^*/\partial T$$

and where the temperature dependence of κ on T, if any, has been neglected. A principal contribution to ΔS^* is the entropy change *associated with w^r and w^p* in the case of ionic reactions. The increased or decreased *local* electric fields in the vicinity of reactants, caused by bringing together ions of like or unlike signs, respectively, causes increased or decreased order in the solvent molecules. In *electrode* *reactions* such an effect occurs as well (ion–electrode repulsion or attraction) but is usually assumed to be swamped by high salt concentrations.[3] In that case,

$$A/Z \cong \kappa \quad \text{since} \quad \rho \cong 1.$$

Evidence in support of $A/Z \cong 1$ (and so $\kappa \approx 1$) is available for electrode reactions. For homogeneous reactions, however, it may be difficult to completely swamp these electrostatic effects because of the high charge of *both* reactants with most reactions. *Bridging effects* occur and further complicate the value of ΔS^*. However, $A/Z \approx 1$ for $MnO_4^- - MnO_4^=$ under certain conditions."

Author's Note: M. follows very closely with his theory the most recent experimental work. He also suggests new experiments. Here is the suggestion following the earlier theoretical considerations:

"Studies on electron exchange reactions of uncharged and slightly charged species would be highly desirable."

NOTE: For example, reactions such as:

$$A + A^* \rightarrow A^- + A^+ \quad \text{or} \quad A_1^- + A_2 \rightarrow A_1 + A_2^-$$

7.2. Caveats About the Applicability of the Theory

"There are several occasions where some breakdown of Eq. (7.2) would be expected:

(a) Changes of bond length are so large that approximation (6) should be omitted.
(b) Approximation (8) breaks down.
There are several cases where relations deduced from Eq. (7.2) are false unless carefully calculated:
(c) Equation (7.3) would be false if one of these reactions involved an *excited electronic state* and the others did not, unless all constants referred to the same excited state.

(d) Separation distance in electrode reaction may be larger than that used to deduce

$$\lambda_a^{el} \cong \frac{1}{2}\lambda_a$$

Condition (c) might lead to enormous deviations in expected ratios of rate constants, while (a) and (b) should lead to smaller errors. (An error of only 3 kcal mole^{-1} in ΔF^*_{calc} at 25°C due to (a) or (b) causes an error of 100 in k.)"

3 kcal mole^{-1} $= 4.3 \times 10^{-2} \times 3 = 12.9 \times 10^{-2} = 0.13$ eV.

"In three reactions, all involving cobalt ions and summarized in the cited review [9], there were considerable deviations in the theoretical and experimental value of

$$k_{ab}/\sqrt{[k_{aa}k_{bb}K_{ab}f]}$$

suggesting that excited states may be involved in some of these cobalt reactions." Author's Note: Here M. suggests some experiments:

"A study of reactions of the $Co(NH_3)_6^{2+,3+}$, $Co(phen)_3^{2+,3+}$, and $Co^{2+,3+}$ systems with a wide variety of reactants would be desirable for localizing the source of these deviations.

The λ's obtained from chemical and electrochemical rate constants have an *intrinsic* interest. For example, *differences of equilibrium bond lengths in the two valence states* of a reactant A influence the value of λ_a and reflect ligand field and other effects.(4) The wide variance in λ_a's from system to system is expected and is presumably *the principal source* of the *large range* of k's for electron exchanges (15 orders of magnitude) and electron exchange currents (8 orders of magnitude) [8]. The λ's also have other applications, for example, to calculations for *reactions of solvated electrons* [2, 4] and for *chemiluminescent reactions* arising from ET [3].

A theoretical expression for the rate constant of electron transfer reactions of solvated electrons has been derived elsewhere [2, 4]. It takes cognizance of the *sensitivity of the wave function* of the solvated electron to fluctuations in oriented polarization of the medium, fluctuations which are *required* for formation of the activated complex. For very fast reactions with a substance A_{ox}, the theoretical expression for the *activation-controlled* rate constant k_{act} reduces to Eqs. (7.1) and (7.2), with $\lambda = \lambda_a + \lambda_e$, where λ_e depends on the properties of the solvated electron, and with $\Delta F^{0\prime}$ replaced by a $\Delta F^{0\prime}_{int}$, from which it differs by an amount equal to the standard state translational free energy of the solvated electron.[(5)] The 'effective' standard reduction potential, used to calculate $\Delta F^{0\prime}_{int}$ from E^0's of the two reactants ($\Delta F^{0\prime}_{int} = -ne(E^0_e - E^0_a)$, was estimated to be about 2.9 volts and the value of λ_e was estimated to be ca. 15 kcal mole^{-1}" (=0.64 eV).

"The possibility of *diffusion control* must be considered in these fast reactions of solvated electrons; the rate constant is given by:

$$\frac{1}{k_{obs}} = \frac{1}{k_{act}} + \frac{1}{k_{diff}}$$

where k_{diff} is the theoretical diffusion-controlled rate constant. When k_{diff} is much greater than k_{act}, k_{obs} simply equals k_{act}, and conversely. Several calculations of the rate constants of solvated electron reactions are described elsewhere [2, 4]. They are based on λ's determined from chemical or electrochemical exchange data.

Chemiluminescent ET reactions have been the subject of several recent experimental investigations cited in Ref. [4]. They would be *very exothermic* if the product had been formed in its ground state instead of in an excited one, a circumstance which has some immediate consequences; some insight into the relative rates of formation of product in the two electronic states can be obtained from Eq. (7.2). *One finds that when a* reaction is so exothermic that $-\Delta F^{0\prime}/\lambda > 1$ for the ground state, the value of ΔF^* for formation of the ground state can become relatively high, while that for the formation of an excited state (calculated by use of the appropriate $\Delta F^{o\prime}$ and λ for formation of that state) can be relatively small. Such conditions are favored by low λ's. (Compare the 'inverse' $\Delta F^{o\prime}$ effect in Ref. [6].)"

Notice that $\lambda > 0$; if $\Delta F^{o\prime}/\lambda = -1$, that is, if $\Delta F^{o\prime} - w^r + w^p = -\lambda$ in Eq. (7.2) $\Delta F^* = w^r$. If w^r and w^p are negligible, then

$\Delta F^* \cong \frac{\lambda}{4}(1 + \Delta F^{o\prime}/\lambda)^2$. Wherefrom if $\Delta F^{o\prime} = 0 \Rightarrow \Delta F^* = \frac{\lambda}{4}$; if $\Delta F^{o\prime} = -\lambda \Rightarrow \Delta F^* = 0$; if $\Delta F^{o\prime} > -\lambda \Rightarrow \Delta F^* > 0$ *again* and the P profile crosses the R profile *to the left* of the minimum of the R profile.

"Chemical exchange data show that the λ is low for aromatic molecules. Calculations for these reactions and for possible chemiluminescence of solvated electrons reactions have been given elsewhere [2, 4]. For this purpose, chemical or electrochemical λ's were employed. Arguments are also given there to show that *chemiluminescent reactions would not be expected to occur on metal electrode surfaces.* (Formation of an excited state of the electrode is relatively easy, and it would *internally convert* to its ground state.) Chemiluminescence can occur homogeneously from electrochemically generated species, however."

As, for example, in the case of anthracene ions electrochemically generated and then reacting according to the reactions [10]:

$$A^+ + A^- \rightarrow A^* + A$$

$$A^* \rightarrow A + h\nu$$

"The dependence of orientation polarization on the charge of a species is one source of the λ term. This same dependence is also a contributing factor to the marked *spectral shifts* observed for certain molecules when the polarity of the solvent is changed. A detailed discussion of this phenomenon has been given elsewhere, where reference to earlier work are also cited [11]."

From **M55**: R. A. Marcus, "**Theoretical Study of Electron Transfer Reactions of Solvated Electrons.**" *Adv. Chem. Ser.* **50**, 138, (1965).

In **M55** [2] M. briefly describes once more the electron transfer mechanism with some novel considerations about what he means for *profiles* of a PES and for *activated complex* of a simple ET reaction. He also distinguishes between *intersection surface* and *intersection region* and for the first time introduces the general concept of *bond* *distances in a reactant versus* the until now used more restrictive *bond distances in an inner coordination shell.*

7.3. Electron Transfer Mechanism

The theory of ET reactions of solvated electrons "stems from the writer's work on simple ET reactions of *conventional* reactants [7, 9]. A simple ET reaction is *defined* as one in which *no bonds are broken or formed* during the *redox step*; such a reaction might be preceded or followed by *bond-breaking* or *bond-forming steps* in a several-steps reaction mechanism. Other chemical reactions involve *rupture* or *formation* of one or several chemical bonds, and only a *few coordinates* suffice to establish their essential features. In simple ET in solution, on the other hand, *numerous coordinates* play a role. One cannot use the usual two-coordinate potential energy *contour diagram* [12] to visualize the course of the reaction, but must resort to some other pictorial method such as the use of *profiles* of a PES."

NOTE the concept of *contours* versus *profiles*. See in the following M.'s definition of profiles.

"In the postulated mechanism of simple ET reactions, a *weak coupling* of the '*redox orbitals*' of the two reactants is assumed [7, 9], and *fluctuations* of *translational*, *rotational*, and *vibrational* coordinates leading from those characterizing *stable configurations of reactants* to *stable configurations of products* is described in terms of PESs. A surface for the reactants, plotted *versus* the *many relevant coordinates* of the system, intersects one for the products, and a *profile* of these surfaces is given in Fig. 1. The intersection is split in the usual quantum-mechanical manner by a coupling of the redox orbitals [7, 9].

Accordingly, reaction can occur if

(1) The system reaches the *intersection region* during a fluctuation.
(2) If the *electronic coupling* is large enough to cause the system to remain on the lowest surface during the crossing of the intersection.

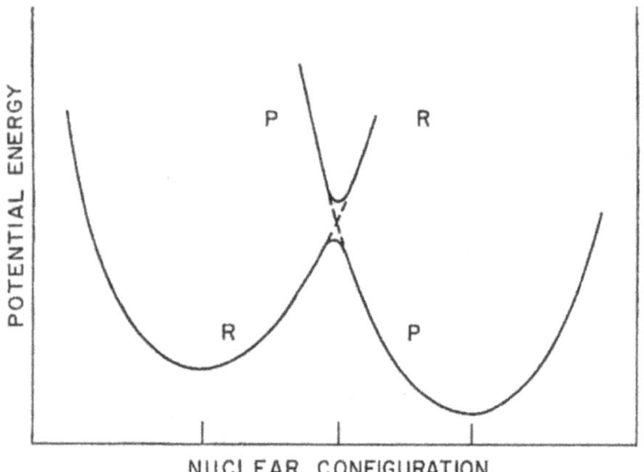

Fig. 1. Profile of potential energy surface of reactants (R) and that of products (P) plotted versus configuration of all the atoms in the system. The dotted line refer to a system having zero electronic interaction of the reacting species. Each adiabatic surface is indicated by a solid line.

The *configurations* occurring *at the intersection* constitute the '*activated complex*' for the reaction. They define a *hypersurface* in this many-dimensional configuration space. The theoretical problem of *calculating* the *reaction rate* is largely that of *calculating* the *chance of crossing* this hypersurface *in unit time.* The*coordinates* undergoing some or appreciable change during the simple ET include

(i) The *separation distance* of the reactants.
(ii) *Bond distances in a reactant* (those in an inner coordination shell for example).
(iii) *Orientation* of solvent molecules.

All of these coordinates contribute to the '*reaction coordinate*' leading from reactants through activated complex to products.

7.4. Special Features for Solvated Electron Reactions

In addition to the earlier features which ET reactions of *solvated electrons* and *ordinary ETs* have in common, those of the solvated electron possess several novel aspects:

(1) The electronic wave function of a solvated electron, spread over several solvent molecules, should be *very sensitive* to *orientation fluctuations* of these molecules, unlike that of an ordinary reactant.
(2) The solvated electron 'disappears' into the other reactant so that there is a change in number of reacting particles.
(3) Unlike many of the conventional ETs, many reactions of the solvated electron are *diffusion controlled.*"

The solvated electron disappears in a way similar to that in which a photon disappears in, say, the reaction $A + h\nu \rightarrow A^*$.

"Because of (1) the *mean kinetic energy* of the solvated electron changes appreciably during the orientation fluctuations required for the system to reach the intersection hypersurface. Such changes are included in Fig. 1, which represents a plot of the *total electronic energy* of the system as a function of the atomic coordinates.

Item (2) contributes to the calculated free energy of formation of the product from a system coming *from the intersection.*

Because of (3), the observed rate constant k_{obs} is given by

$$\frac{1}{k_{\text{obs}}} = \frac{1}{k_{\text{diff}}} + \frac{1}{k_{\text{act}}} \tag{7.5}$$

where k_{diff} and k_{act} are diffusion-controlled and activation-controlled rate constants. According as $k_{\text{diff}} \gg k_{\text{act}}$ or $k_{\text{act}} \gg k_{\text{diff}}$, k_{obs} equals k_{act} or k_{diff}, respectively [6].

7.5. Assumptions

The mathematical derivation of the theoretical expression for k_{act} for solvated electron transfers has been given elsewhere [4]. The following assumptions were the principal ones made, of which (a) to (c) are standard in activated complex theory:

(a) The adiabatic (Born–Oppenheimer) approximation is used to treat the electron–nuclear motion. (In the vicinity of the intersection hypersurface, nonadiabatic effects arise only if the splitting is *very small*. The factor κ in Equation (7.6) is then less than unity and is calculated by various nonadiabatic methods. Normally, the reaction is assumed to be adiabatic [$\kappa \approx 1$].)"

If there is no splitting, the system remains on the single PES of the reactants and there is no ET reaction. If the splitting is large, the system remains on a single PES which has initially the character of the PES of the reactants and finally the character of the PES of the products. The reaction, happening on a single PES, is an adiabatic reaction, see Chapter 1.

(b) "Classical equilibrium statistical mechanics is used to calculate the probability of reaching the intersection hypersurface. Any vibrational quantum effects which occur are treated approximately in the usual way for activated complex theory.
(c) The *rate* is given by the number of '*first passages*' of a system across the intersection surface per unit time.
(d) Highly *specific interactions* of the two reactants (e.g., *steric effects*) are assumed *to contribute primarily to the work* w required to bring the reactants together.
(e) The solvent polarization (outside any inner coordination shell) caused by the two reactants is treated by dielectric continuum theory. (Statistical mechanics was used for conventional ETs and can be used to replace the continuum theory when this refinement becomes appropriate.[(6)])
(f) In *hydrogen-bonded solvents* of present interest, the electron is assumed to be in a *cavity-free* medium. Arguments were given to suggest that when a solvent cavity for the electron occurs, it will primarily affect the *numerical value* of λ_e^∞ in Eq. (7.7) rather than the *functional form* of those final equations [4].
(g) The vibrations of the second reactant are treated as harmonic oscillators.
(h) A *symmetrized vibrational* potential energy function is used for reactants and products. The minor error introduced thereby has been estimated elsewhere [5, 6]."

NOTE that M. collects the specific effects in the work term.

"A solvated electron in a polar solvent has a high classical frequency of motion in its 'orbit.'

Accordingly, the rotating or librating solvent molecules of the system see it primarily as some diffuse charge distribution. The molecules orient themselves toward this diffuse charge in a way *consistent* with their *thermal motion* and with their *bonding* to other molecules. The valence and inner electrons of these molecules are polarized by the instantaneous field of the solvated electron *when they are some distance* from the electron. If they are too close to it, they cannot follow the motion of the instantaneous field, as uncertainty principle arguments show [4]. A quantum-mechanical continuum estimate of the radius of this *dynamical sphere of exclusion* of electron polarization and of the contribution to the interaction energy was made [4]. When the radius is small relative to the *circumference of the 'orbit'* of the electron, the radius does not influence the numerical value of λ_e^∞ in Eq. (7.7). More refined models for the solvated electron would also recognize any *vibronic effect*, since the estimated frequency of electronic motion [4] is not far from that of polar OH vibrations of the medium."

7.6. Calculation of k_{act}

"The rate constant k_{act} is given by Eq. (7.6), on the basis of assumptions (a) to (c) and the introduction of certain comparatively minor approximations [7, 9, *A*, 13]. (In the present paper, a reference labelled by a letter refers to a comment in the Appendix.)

$$k_{\mathrm{act}} = Z\kappa\rho e^{-\Delta F^*/kT} \tag{7.6}$$

where ΔF^* is the free energy of formation from reactants of a system *centered* on the intersection hypersurface and Z is a collision frequency between uncharged species. (Z is about 10^{11} liters mole^{-1} sec^{-1}. The charge effects on collision frequency[(7)] are included in ΔF^*.)

NOTE: The value of Z was changed to 10^{12} in **M170.**

ΔF^* is computed for the most probable separation distance contributing to reaction. ρ is a ratio of certain mean square displacements and is taken to be about unity [7, 9, *A*, 13]."

"Statistical mechanics was used to calculate the *probability of occurrence* of the necessary fluctuations in vibrational coordinates. Continuum theory was used to calculate the free energy of formation of any *nonequilibrium state of polarization* of the medium in the presence of the electron, and solvation free energy of the electron owing to changes in solvent polarization [4]. In this way, the free energy was calculated for a system having the electronic charge distribution of the reactant and having *Boltzmann-type polarization and vibrational distribution functions.* Similarly, for *the same distribution of coordinates*, the free energy was calculated for a system having the electronic configuration of the product. However, as Fig. 1 illustrates, these two free energies are equal when each system is *constrained* to be *centered* on the intersection hypersurface:

(1) The potential energy, *averaged over the given distribution*, is the same because of the intersection.
(2) The entropy of a system is determined only by the configurational distribution and so must also be the same for two systems having the same such distribution.
(3) The kinetic energy of each nucleus is also the same in the two systems."

Such are the three components of free energy.

"Minimization of the free energy of this arbitrary state of a system containing the reactants, subject to the condition of equality of the two free energies, yields an expression for the free energy of the reactants in this *centered distribution* and, thereby, for ΔF^*. The functional form of the equation for ΔF^* is given by Eq. (7.7), and that for

ΔF^{*p}, the free energy of formation of the *centered distribution* from the product, is given by Eq. (7.8) [4]:

$$\Delta F^* = w + m^2 \left[\left(1 - \frac{m^2}{2} \right) \lambda_e^\infty + \lambda_2^\infty + \Delta\lambda_R \right] \quad (7.7)$$

$$\Delta F^{*p} = (m+1)^2 [(1 - m^2)\lambda_e^\infty + \lambda_2^\infty + \Delta\lambda_R] \quad (7.8)$$

where w is the work required to bring the reactants together to a *mean separation distance R in the activated complex.* (Both Coulombic and non-Coulombic terms can contribute to w.) λ_e^∞ is an '*intrinsic reorganization factor*,' which can be expressed in terms of the properties of the solvated electron [4]. λ_2^∞ is an 'intrinsic reorganization factor' of the second reactant and depends only on the properties of that reactant, such as differences in equilibrium bond lengths and orientation polarization in the oxidized and reduced forms [4]."

Author's Note: The λ^∞'s of Marcus's theory correspond, because of their additivity property, to the additivity of single electrode standard potentials in the theory of Nernst. But Nernst's additivity of electrode potentials is a rigorous thermodynamic property whereas Marcus' kinetic theory is approximate at finite distances between the reactants. Moreover, in the case of Marcus' λ^∞'s, we have free energies instead of Nernst's electrostatic potentials.

"The ∞ superscript indicates that the quantities are evaluated for reactants far apart, and $\Delta\lambda_R$ is the change in the sum of intrinsic reorganization factors from $(\lambda_e^\infty + \lambda_2^\infty)$ to $(\lambda_e^R + \lambda_2^R)$." That is,

$$\Delta\lambda_R = (\lambda_e^R + \lambda_2^R) - (\lambda_e^\infty + \lambda_2^\infty).$$

"In terms of continuum theory $\Delta\lambda_R$ for the one-electron transfer equals $-e^2(D_{\text{op}}^{-1} - D_s^{-1})/R$."

In the earlier formula, the term $e^2(D_s^{-1})/R$ is the normal electrostatic energy at R.

"where D_{op} and D_s denote the optical and static dielectric constants, respectively, and where e is the electronic charge [4]. The quantity m is the solution of

$$\Delta F^* - \Delta F^{*p} = \Delta F^{o\prime}_{int} - w."$$

$\Delta F^{o\prime}$ is here divided in an intrinsic part and in the work term.

"In this last equation, $\Delta F^{o\prime}_{int}$ is the 'standard' free energy of reaction $\Delta F^{o\prime}$ in the prevailing medium, corrected for the *translational free energy loss* when the '*oriented center*,' in which the electron formerly resided, disappears during the formation of product from the centered distribution on the hypersurface. This corrected $\Delta F^{o\prime}$ constitutes the 'driving force' for reaction at the mean separation distance R:

$$\begin{aligned}\Delta F^{o\prime}_{int} &= \Delta F^{o\prime} - \Delta F^{o}_{trans}\\ &= \Delta F^{o\prime} - RT\ \ln[(2\pi m_p kT)^{3/2}1000/h^3 N_a]\end{aligned}$$

for a standard state of 1 M, where m_p is the *effective mass* for translation of the *solvated electron* (the '*polaron*') and N_a is the Avogadro's number. The value of ΔF^{o}_{trans} is 5.3 kcal mole^{-1} if m_p is 3/N_a grams/molecule [*B*].

The functional form of Eqs. (7.7) and (7.8) differs somewhat from that found earlier [7, 9] for conventional ETs, the difference arising from the sensitivity of the wave function to changes in solvent polarization. Equations (7.7) and (7.8) simplify on close examination: These reactions are extremely rapid because the solvated electron is a *very strong reducing agent*, so that $\Delta F^{o\prime}$ is very negative. In this case, ΔF^* is very small and, therefore, m^2 is seen from Eq. (7.7) to be small. In fact, for the usual reactions of the solvated electron, *a posteriori* numerical calculations from the observed rates show that $m^2 \ll 1$, and so m^2 can be neglected in the coefficients of λ_e^∞ in Eqs. (7.7) and (7.8)."

This means that $(1 - m^2/2) \cong 1$ in Eq. (7.7) and consequently the $\lambda_e^\infty + \lambda_2^\infty$ there cancels with the $-(\lambda_e^\infty + \lambda_2^\infty)$ in $\Delta\lambda_R$.

"The equations then become

$$\Delta F^* = w + m^2\lambda \tag{7.9}$$

$$m = -\frac{1}{2}\left(1 + \frac{\Delta F^{\circ\prime}_{\text{int}} - w}{\lambda}\right) \tag{7.10}$$

where

$$\lambda = \lambda_e^R + \lambda_2^R$$

These equations are now similar to those derived earlier for conventional ET reactions [7, 9].

The value for λ_2^R is the same as that for this same reactant in an ordinary homogeneous or electrochemical ET occurring at the same R and can be estimated from them, as described later [6].

$\Delta F^{\circ\prime}_{\text{int}}$ is known for many reactions of the solvated electron, and w can be estimated approximately. Accordingly, a theoretical value of ΔF^* can be calculated from Eq. (7.9) once λ_e^R is known. Either λ_e^R can be calculated from other sources (it depends on *the model* of the solvated electron) or a value can be used which best fits data on k_{act} for several reactions, or both. In making such calculations, it should be noted that ΔF^* is not very accurately given by Eq. (7.9), because of the various approximations. It is more realistic to compare theoretical and experimental values for ΔF^*, therefore, rather than those for k_{act}.[(8)]

The work term w makes a relatively minor contribution to ΔF^*. In numerical calculations, it is usually assumed to be electrostatic in origin and to be given roughly by the shielded Coulombic formula,

$$w^r = (e_1 e_2/DR)\exp(-\kappa R)$$

where e_1 and e_2 are the charges of the reactants, D is the dielectric constant and κ is the Debye kappa. (In very dilute solutions $\exp(-\kappa R) \cong 1$.)

7.7. Dependence of k_{act} on $\Delta F^{\circ\prime}$

Possible Chemiluminescence

To explore *the behavior of* ΔF^* *with* $\Delta F^{\circ\prime}$, it is convenient to rewrite Eqs. (7.9) and (7.10) as

$$\Delta F^* = w + \frac{\lambda}{4}\left[1 + \frac{\Delta F^{0\prime}_{\text{int}} - w}{\lambda}\right]^2 .\text{''} \tag{7.11}$$

Rewriting Eq. (7.11) as

$$\Delta F^* - w = \frac{\lambda}{4}\left[1 + \frac{\Delta F^{0\prime}_{\text{int}} - w}{\lambda}\right]^2$$

we see that "The value of $\Delta F^* - w$ is seen to decrease at first as $\Delta F^{\circ\prime}$ becomes increasingly negative, to pass through a minimum at

$$(\Delta F^{\circ\prime}_{\text{int}} - w) = -\lambda,$$

and then to increase as $\Delta F^{\circ\prime}$ becomes still more negative. The *physical origin* of the behavior is seen from Fig. 1: As $\Delta F^{\circ\prime}$ is made more negative, the *product surface* in Fig. 1 is *lowered* relative to the R surface, and ΔF^* becomes smaller at first, because the intersection occurs at lower energies on the R surface. This effect of $\Delta F^{\circ\prime}$ on ΔF^* is the *normal* one, and the configurations at the intersection are seen to be a *compromise* between the stable ones of the initial state and the stable ones of the final state. When $\Delta F^{\circ\prime}$ becomes *still more negative*, the intersection is seen to occur *to the left of the minima* of the R and Psurfaces in Fig. 1, at higher and higher parts of the initial R surface as $\Delta F^{\circ\prime}$ becomes increasingly negative. The configurations at the intersection are *no longer compromise ones.* This latter $\Delta F^{\circ\prime}$ region might be called the *abnormal* $\Delta F^{\circ\prime}$ region.

When $\Delta F^{\circ\prime}$ is sufficiently negative ΔF^* becomes large, and

(i) Either the reaction with the solvated electron should become very slow or

(ii) A reaction should occur by other paths, two of which are the following:

(a) *Formation of electronically excited states* and possible chemiluminescence [3]:
Although the intersection of the R surface with the surface for the electronic ground state of the product occurs at high ΔF^*'s when $\Delta F^{\circ\prime}/\lambda$ becomes very negative (slightly more negative than -1), the calculations given in the following indicate that an intersection with a surface in which the product is excited may then occur at *low* ΔF^*; chemiluminescence may therefore result:

For formation of the *excited state of the product* ΔF^* is again given by Eq. (7.11), *with* λ_2^R *and* $\Delta F_{\text{int}}^{\circ\prime}$ *now referring to formation of this excited state.* For example, in a reaction for which $\Delta F_{\text{int}}^{\circ\prime}$ is as negative as -4 eV, and in which the fluorescence occurs at, say, 6000 A. (2 eV), the $\Delta F_{\text{int}}^{\circ\prime}$ for the formation of the excited state is then only -2 eV or even less. If $\lambda/4$ for *both* reactions is about 0.4 eV and if, for present purposes, we neglect w, then ΔF^* for a reaction leading to the ground and excited state of the product is 0.9 and 0.025 eV, respectively," the ΔF^* having been calculated from Eq. (7.11). "In the first case, the reaction is much too slow for measurement in these systems and in the second case, it would occur with ease, the barrier ΔF^* being about 0.6 kcal mole^{-1}.

(b) *Atom transfer (AT) reaction*: In principle, it is possible that an AT reaction such an Eq. (7.12) can occur when the second reactant is an AT acceptor or is otherwise reactive to the atom. To be sure, there is no evidence as yet for such reactions of the solvated electron. By suitable choice of reactant they could be avoided.

$$e(H_2O)^- + Fe(CN)_6^{-3} \rightarrow OH^- + Fe(CN)_5CNH^{-3} \tag{7.12a}$$

$$\text{Fe(CN)}_5\text{CNH}^{-3} \rightarrow \text{Fe(CN)}_6^{-4} + \text{H}^{+}\text{''} \quad (7.12b)$$

NOTE that in reaction (7.12a) there is a H transfer. The successive reaction is a normal acid dissociation.

7.8. Estimating $\Delta F^{0\prime}_{\text{int}}$ and λ

From measured forward and reverse rate constants of the reaction of a solvated electron with water Baxendale [14] estimated the standard potential for the solvated electron."

From $k_{\rightarrow}$ and $k_{\leftarrow} \Rightarrow K_{eq} \Rightarrow E^{\circ}$.

"Use of a more recent rate constant [15] and correction [4] for a certain omitted entropy change yields a value, $E_e^{\circ\prime} = +2.7$ volts, for the *standard oxidation potential* of the *solvated electron.* To calculate $\Delta F^{\circ\prime}_{\text{int}}$ for a reaction from a difference of the standard oxidation potentials of the two reactants, $E_e^{\circ\prime} - E_2^{\circ\prime}$, the $\Delta F^{\circ\prime}$ must be corrected for the $\Delta F^{\circ}_{\text{trans}}$ of about 5 kcal mole^{-1}. This correction can be made [4] by taking the effective $E_e^{\circ\prime}$, $E^{\circ\prime}_{\text{eff}}$ for a solvated electron to be 2.9 volts.

$$\Delta F^{\circ\prime}_{\text{int}} = -eF(E^{\circ\prime}_{\text{eff}} - E_2^{\circ\prime})\text{''}$$

NOTE that 5 kcal mole^{-1} $\times\, 4.33 \cdot 10^{-2} = 0.2$ eV, this is how to get 2.9 from 2.7.

"In homogeneous (electron exchange) reactions between two species differing only in their valence states ΔF^{*} is given by Eq. (7.11) with λ equal to $2\lambda_2^R$ and $(\Delta F^{\circ\prime}_{\text{int}} - w)$ replaced by $\Delta F^{\circ\prime} - w^r + w^p$ [7, 9]. (w^r and w^p denote the work required to bring the reactants together to the mean separation distance R, and the products to this R, respectively.) $\Delta F^{\circ\prime}$ is zero for a simple electron exchange reaction *and w^r equals w^p for it,* since the products are chemically indistinguishable from the reactants.

$$\Delta F^{*}_{\text{ex}} = w^r + \frac{\lambda_2^R}{2} \quad (7.13)$$

The separation distance R should affect *primarily* the *orientation polarization contribution* to λ rather than the *vibrational contribution* from the inner coordination shell [7, 9]. If R is about the same for this reaction as it is for reaction with the solvated electron, then λ_2^R is the same. The difference in λ_2^R would probably be relatively minor in any case for typical R's.

Since $\Delta F_{ex}^* = -RT \ln(k_{ex}/10^{11}\,\text{M}^{-1}\,\text{sec}^{-1})$, λ_2^R can be estimated from the electron exchange rate constant k_{ex} when correction of ΔF_{ex}^* is made for w^r or when w^r is small enough to be neglected. Values of ΔF_{ex}^* have also been obtained indirectly from measurements of rate constants of other redox reactions involving the species, particularly in Ref. [16], where tests of this evaluation are described.

In *electrochemical* ET reactions, the value of the rate constant k_{el} at *zero activation overpotential* yields a value of $\Delta F_{el}^* = -RT \ln(k_{el}/10^4\text{cm} \cdot \text{sec}^{-1})$, and the theoretical expression for ΔF_{el}^* is [7, 9]

$$\Delta F_{el}^* = \frac{w^r + w^p}{2} + \frac{\lambda_2^R}{4} \tag{7.14}$$

Correction of this ΔF_{el}^* for the work terms then yields a value of λ_2^R. The consistency of Eqs (7.13) and (7.14) has received some experimental support [17], though further work is desirable. Examples of some approximate values of λ_2^R computed in these ways are [C]: $Co(NH_3)_6^{+2,3}$ (~60 kcal mole^{-1}), $Fe(phen)_3^{+2,3}$ (~15 kcal mole^{-1}), $Eu^{+2,+3}$ (~40 kcal mole^{-1}).

The value of λ_e^R for the solvated electron has been estimated in several ways [4, 17]. By assuming that diffusion of the solvated electron in water occurs as a site-to-site ET and using an expression for ΔF^* for a unimolecular ET reaction [7, 9], Sutin estimated [17] a *lower bound* of 5 kcal mole^{-1} for λ_e^R from the known diffusion constant. The writer has estimated [4] a value of roughly 15 kcal mole^{-1} to fit the rate constant for reaction of the solvated electron with Sm^{+3}, assuming that λ_2^R was about the same as that for another rare earth, Eu^{+3}. An estimate of λ_e^R can also be made from *spectral* and *solvation* data, but depends on the *detailed model* used for the solvated electron [4] and neglects the '*electron affinity*' *of the first excited state* of the solvated electron. (This electron affinity describes *local interactions*; these are not covered by simple polaron theory.)

The latter estimated value of λ_e^R is rough but is consistent with the value just cited and with a value estimated *a priori* [4]. At the same time, this second estimate from the data yields a rough and perhaps not reliable value of the *'electron affinity' of the ground state* of the solvated electron [4].

From Eq. (7.11), an estimate can be made of an *error* in calculated ΔF^* owing to an error in λ. If the errors are denoted by δ's we have

$$\delta \Delta F^* = -\left(1 + \frac{\Delta F_{\text{int}}^{o\prime} - w}{\lambda}\right) \frac{\Delta F_{\text{int}}^{o\prime}}{2\lambda} \delta\lambda \qquad (7.15)$$

For example, an error of 5 kcal mole^{-1} in λ introduces an error of 0.6 kcal mole^{-1} in ΔF^*, when $\Delta F_{\text{int}}^{o\prime}/\lambda \cong -0.6$. Similarly, the error in ΔF^* owing to an error in $\Delta F_{\text{int}}^{o\prime}$ is

$$\delta \Delta F^* = \left(1 + \frac{\Delta F_{\text{int}}^{o\prime} - w}{\lambda}\right) \frac{\delta \Delta F_{\text{int}}^{o\prime}}{2} \qquad (7.16)$$

An error of 2 kcal mole^{-1} in $\Delta F_{\text{int}}^{o\prime}$ introduces an error of 0.9 kcal mole^{-1} in ΔF^* when $\Delta F_{\text{int}}^{o\prime}/\lambda \cong -0.6$. However, one sees from Eqs. (7.15) and (7.16) that the sensitivity of ΔF^* to a change in λ or $\Delta F_{\text{int}}^{o\prime}$ depends on the value of $\Delta F_{\text{int}}^{o\prime}/\lambda$. We have selected a typical value.

7.9. Further Applications

The prediction [3] of possible chemiluminescence at *suitable* $\Delta F_{\text{int}}^{o\prime}/\lambda$ has already been discussed. Applications of the equations have also been made to calculations of rate constants for reactions of solvated electrons, using λ_2^R's estimated as earlier [4, 17]. If, when $\Delta F_{\text{int}}^{o\prime}$ is *very negative*, the calculated rates are *too low*, the explanation may lie in the *formation of excited states* or, in some cases, in *ATs*. A search for the *predicted chemiluminescence* is under way [18]. It would be favored by a reaction for which λ_2^R is small enough that $\Delta F_{\text{int}}^{\circ\prime}/\lambda$ is quite negative for formation of the ground state of the product."

NOTE: In M.'s writing, "quite negative" means a number negative and of rather great absolute value.

"Interestingly enough, it is possible to vary ΔF° systematically *without varying* λ, simply by *varying a substituent* in a large organic ligand[(9)] [16]. Thus *a control of the relative values* of ΔF^*'s for reactions leading to ground and excited states becomes possible and so, thereby, does the *yield* of any chemiluminescence.

To investigate reactions of solvated electrons in the *borderline region* of *diffusion* and *activation control*, there are *two regions* of $E^{\circ\prime}$ for the *second reactant* which are of interest, for typical λ's. Estimated from Eq. (7.11), neglecting w, these are $E_2^{\circ\prime} > ca.$ 1.5 to 2 volts [4] and $E_2^{\circ\prime} < ca.$ -0.5 to -1.0 for a typical λ of about 55 kcal mole^{-1}."

NOTE that 55 kcal mole^{-1} = 2.28 eV. Moreover, in $\Delta F_{\text{int}}^{\circ\prime} = -eF(E_{\text{eff}}^{\circ\prime} - E_2^{\circ\prime})$ the $E_2^{\circ\prime} < 0$ is subtracted from $E_{\text{eff}}^{\circ\prime}$ and $\Delta F_{\text{int}}^{\circ\prime}$ is negative and small in absolute value.

"The former region, particularly, would be expected to yield the most reliable empirical values of λ_e^R, since they involve *compromise configurations* in the activated complex, just as k_{ex} and k_{el} do. The resulting λ_e^R's *may* be compared with those obtained from the second $E_2^{\circ\prime}$ region. In cases which involve *appreciable changes* in bond lengths of a reactant, it may be necessary to replace the harmonic oscillator approximation (g) by use of anharmonic potential energy functions. (Approximation (h) is automatically replaced at the same time.) For this purpose, one may replace Eq. (7.11) by an equation derived earlier for ordinary ET reactions [6] as the derivation of Eq. (7.11) from Eq. (7.7) and as the similarity of Eq. (7.11) to the equation for conventional ET both indicate.

7.10. Appendix

(A) In activated complex theory, the rate constant can be expressed in terms of the free energy $F^{\neq}$ of a system *hypothetically* constrained to exist on a certain hypersurface, the 'activated complex,' cf. Marcus, R. A., *J. Chem. Phys.* **41**, 2624, (1964).**(10)** $F^{\neq}$ can be expressed in terms of the free energy F^* of a system *centered* on that hypersurface [5, 6].**(11)** The difference between $F^{\neq}$ and F^* contribute a factor to ρ. A *second factor* in ρ arises from the *fluctuations* in the separation distance of the reactants in the activated complex [7, 9].

(B) Theoretical values of m_p and λ_e^{∞} would depend on the model of a solvated electron. The value of m_p used in the text is at best rough [4], but $\Delta F^{\circ}_{\text{trans}}$ is relatively insensitive to it in the region of interest. A theoretical value of $\Delta F^{\circ\prime}$ would also depend on the model, but fortunately its experimental value is known instead.

(C) Values of λ_2^R for $Co(NH_3)_6^{+2,+3}$ and for $Eu^{+2,+3}$ were estimated from k_{el} in Ref. [8], neglecting w because of the high salt concentrations. The value of $Fe(phen)_3^{+2,+3}$ was estimated from the value assigned to k_{ex} in Ref. [16], neglecting w. Only a lower limit for k_{ex}, quoted in Ref. [16], is known. A possible non-Coulombic source of w for certain $Fe(phen)_3^{+3}$ reactions is noted in Refs. [16] and [8] and, if correct, may apply to reactions between e(aq) and $Fe(phen)_3^{+3}$. Data on other k_{ex}'s and k_{el}'s are given in Refs. [16] and [8] and in various other articles."

From **M57**: R. A. Marcus, "**On the Theory of Chemiluminescent Electron-Transfer Reactions.**" *J. Chem. Phys.* **43**, 2654, (1965) a. **Erratum**: **52**, 2803, (1970).

7.11. On the Theory of Chemiluminescent Electron Transfer Reactions

"A variety of AT chemiluminescent reactions are known, but more recently a number of chemiluminescent ET reactions have been reported in the literature [19]. We apply a theory of electron transfers [6–9] to these reactions and examine some consequences for experiment.

We consider a PES for the reactants and one for the products, each plotted as a function of all the translational, rotational, and vibrational coordinates in the system [6–9]. In the zeroth approximation of no electronic coupling of the redox orbitals of the ion and electrode, these two PESs intersect. In the next approximation of *some* electronic coupling, the intersection is removed by the usual quantum-mechanical splitting, as in Fig. 1(a). A *suitable fluctuation* of coordinates involving approach of the reactants, reorientation of solvent molecules, and change in bond lengths of reactants, permits the system to cross the original intersection region in a coordinate region (small separation distance) where the electronic coupling is appreciable. In this way electron transfer has occurred, adiabatically if the splitting is sufficient and nonadiabatically otherwise. The system has moved from the *R* to the *P* surface [Fig. 1(a)].

In the usual thermal ET reactions, the *products* are formed in their *ground electronic states*, for only these are usually conveniently *accessible energetically*. Nevertheless, the PES of the reactants will 'cross' a surface of the products in which one or more products is *electronically excited*, in some other region of configuration space. If this latter intersection region is easily accessible (*energetically* and *entropically*), a reaction to form an excited product can occur. The excited product may either emit light or react. An example of such a reaction is a possible triplet–triplet annihilation to form an excited singlet state which later fluoresces [19(a)].

In a *sufficiently exothermic* [20] ET reaction, the region where the surface of *unexcited products* 'intersects'(12) that of unexcited reactants is not readily accessible and becomes less accessible with increasing exothermicity (see Ref. [6] where the postulated effect was called an inverted free energy effect). Under such conditions, the region where the PES involving an *excited product* intersects that for the unexcited reactants *may* be readily accessible. The formation of an excited product can then occur easily.

A *numerical estimate* of the rate to form *an unexcited* or *an excited product* can be made as follows: The ET rate constant to form *a particular electronic state* of the products is given by Eq. (7.6) according to an ET theory [7, 9]:

$$k_{\text{act}} = Z\kappa\rho e^{-\Delta F^*/kT} \tag{7.6}$$

where Z is about 10^{11} liter mole$^{-1}\cdot$sec^{-1}, κ is a factor close to unity unless the splitting is extremely small, $\rho \approx 1$, and ΔF^* *reflects the accessibility* of the intersection region:

$$\Delta F^* = w^r + (\lambda/4)[1 + (\Delta F_R^{0\prime}/\lambda)]^2 \tag{7.17}$$

$$\Delta F_R^{\circ\prime} = \Delta F^{\circ\prime} + w^p - w^r.\text{”}$$

Note that: (1) if $\Delta F^{\circ\prime} = -\lambda \Rightarrow \Delta F^* = 0$ (neglecting w); (2) if $\Delta F^{\circ\prime} = \lambda \Rightarrow \Delta F^* = \lambda$ (always neglecting the work term); (3) if $\Delta F^{\circ\prime} = 0 \Rightarrow \Delta F^* = \lambda/4$.

“In these equations, w^r is the work required to bring the reactants together *to the most probable separation distance R* in the *intersection region*, that is, in the ‘*activated complex*’; w^p is the corresponding work to bring the products together, each w referring to the given *state of excitation*. $\Delta F^{\circ\prime}$ is the ‘standard’ free energy of reaction in the prevailing medium. λ is a *reorganization term* expressible in terms of

(i) *Differences* in *equilibrium bond lengths* of each reacting species in its *initial* and *final* electronic states, that is the electronic states *shortly prior to and after* ET, and in terms of
(ii) Dielectric properties related to *differences* in *equilibrium orientation polarization* in these electronic states

Equation (7.17) displays an interesting behavior as $\Delta F^{\circ\prime}$ becomes increasingly negative for a reaction leading to the given electronic state[5] of the products. At small $\Delta F^{\circ\prime}$’s the ‘barrier’ ΔF^* decreases with increasingly negative $\Delta F^{\circ\prime}$, the expected slope of 0.5 being

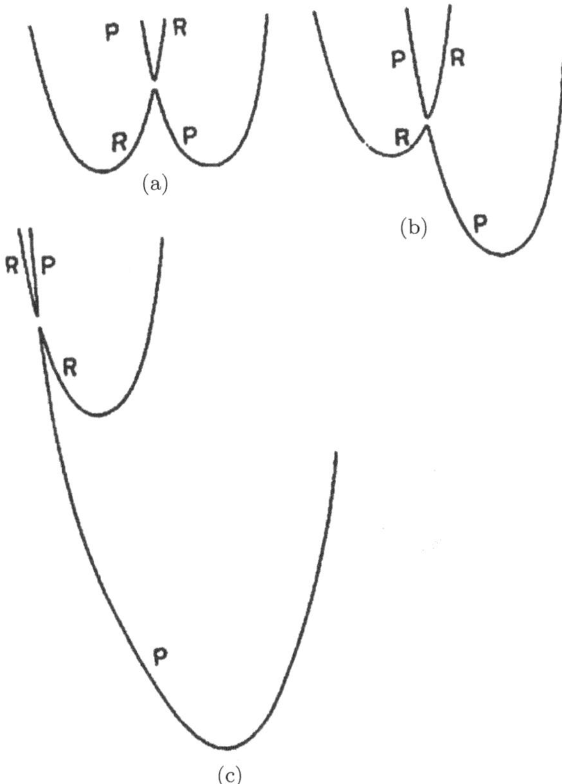

Fig. 1. Profile of potential energy surface of reactants (R) and that of products (P) plotted versus configuration of all the atoms in the system. The adiabatic surfaces are indicated by solid lines for the case of (a) $\Delta F^{\circ\prime} \cong 0$, (b) $\Delta F^{\circ\prime} < 0$, and (c) $\Delta F^{\circ\prime} \ll 0$.

supported by recent experimental studies [21]. In this region of $\Delta F^{\circ\prime}$, a typical region in fact for chemical reactions, the likely configurations of coordinates at the intersection region represent some *compromise* between *stable* coordinate configurations of the reactants and those of the products [Figs. 2(a) and (b)] [6–9].

When $\Delta F^{\circ\prime}$ becomes *so negative that it equals* λ (neglecting the work terms w^r and w^p for the moment) ΔF^* vanishes. For *still more negative* $\Delta F^{\circ\prime}$'s, ΔF^* *begins to increase with increasingly negative* $\Delta F^{\circ\prime}$, as one sees from Eq. (7.17). In this *new* $\Delta F^{\circ\prime}$ *region*, the intersection region does not occur at compromise configurations [Fig. 2(c)]."

NOTE: Fig. 2(c) is wrong and will be corrected in the following Erratum by Marcus where the correct Fig. 1 there shows how Fig. 2(c) should have been.

"With increasingly negative $\Delta F^{\circ\prime}$, the intersection's configurations become increasingly different from both the stable configurations of the reactants and those of the products for the given electronic states [Fig. 2(c), corrected in the following erratum]. The above $\Delta F^{\circ\prime}$ *regions* can be called the '*normal*' and '*abnormal*' $\Delta F^{\circ\prime}$ *regions.*

When the value of λ for reaction leading to a ground state and that leading to an excited state is about the same, the difference of ΔF^*'s for the two reactions is determined by the respective values of $\Delta F^{\circ\prime}/\lambda$ [Eq. (7.17)]. The magnitude $\Delta F^{\circ\prime}/\lambda$ for a typical reaction of interest can be determined as follows: We consider a reaction leading to *an excited state* for which the excitation energy is about 3 eV. (The calculations are easily amended for a different excitation energy.) As a first approximation (neglecting $T\Delta S^{\circ\prime}$), the $-\Delta F^{\circ\prime}$ for the ground-state reaction is then about -70 kcal mole^{-1}."

1 eV corresponds to 23.0609 kcal mole^{-1} $\Longrightarrow$ the above 70 kcal mole^{-1}.

"The value of λ can be estimated from appropriate homogeneous or electrochemical ET rates of related reactions when they are available [6–9]. For example, the λ appearing in Eq. (7.17) is essentially the mean of the λ's of the two-electron exchange reactions (7.18) and (7.19), [6–9]

$$A_{\text{ox}} + A_{\text{red}} \rightarrow A_{\text{red}} + A_{\text{ox}} \tag{7.18}$$

$$B_{\text{ox}} + B_{\text{red}} \rightarrow B_{\text{red}} + B_{\text{ox}} \tag{7.19}$$

for the cited electronic states. That is,

$$\lambda_{ab} = \frac{1}{2}(\lambda_{aa} + \lambda_{bb}). \tag{7.20}$$

Alternatively, if λ_{aa}^{el} denotes the λ for the electrochemical exchange rate, the theoretical λ_{aa}^{el} equals

$$\lambda_{aa}^{\mathrm{el}} = \frac{\lambda_{aa}}{2}$$

essentially and

$$\lambda_{ab} = \lambda_{aa}^{\mathrm{el}} + \lambda_{bb}^{\mathrm{el}}$$

Note 7: When λ is very small, λ may arise largely from change in orientation polarization and so be more sensitive to specific influences. $\lambda_{\mathrm{el}} = \frac{1}{2}\lambda_{\mathrm{ex}}$ if the mean separation distance of the reactants in the activated complex of the exchange reaction equals the distance between the ion and its electrostatic image in the electrode reaction, and if specific effects affect the λ's but slightly.

(We restrict our considerations to systems in which adsorption at an electrode is absent.) Evidence supporting these ideas has been given recently [8, 9].

When A_{red} is an *aromatic anion* and A_{ox} an *aromatic molecule*, an estimate of λ can be made from available measurements of the reaction rate constant for Eq. (7.17). The value of k is about 10^7–10^9 liter mole^{-1} sec^{-1} [22]. (However, the results are complicated by at least partial alkali cation bridging."

The cation bridging is another specific effect besides adsorption.

"Certain peculiar factors suggest that further study of the mechanism is desirable [23].) Analogous data without *cation bridging* are available for another aromatic compound, Wurster's blue and its positive ion, for which k is about 10^8 liter mole^{-1} sec^{-1} [24]. If work terms are neglected for simplicity and if these data provide a rough measure of k for an unbridged reaction, then

$$\frac{\lambda_{aa}}{4}$$

becomes about 4 kcal mole^{-1} for an aromatic system, a very small value."

4 kcal mole^{-1} = 0.17 eV.

Note that in the earlier electron exchange reaction $\Delta F_R^{\circ\prime} = 0$ and so we see from Eq. (7.6) that from k one gets ΔF^* and from Eq. (7.17) that neglecting work terms one obtains directly $\lambda/4$.

"The value of λ_{bb} depends on the nature of B. When λ_{bb} is also small, roughly equal to λ_{aa}, then $\lambda/4$ becomes about 4 kcal mole^{-1} and $-\Delta F^{\circ\prime}/\lambda$" becomes about 4.4."

NOTE: $\lambda_{aa}/4$ was supposed to be 4 kcal mole^{-1} and so $\lambda_{aa} \approx$ 16 kcal mole^{-1}. From $\lambda_{aa} \approx \lambda_{bb}$ and from Eq. (7.20), one sees that $\lambda_{ab} = \lambda \approx 16$ and so $\lambda/4 \approx 4$ kcal mole^{-1}. M. now tacitly supposes that $\Delta F^{\circ\prime} = -70$ kcal mole^{-1}. In this way dividing by $\lambda \approx 16$ one gets $-\Delta F^{\circ\prime}/\lambda \approx 4.4$.

"The barrier ΔF^* leading to an unexcited product then equals $3.4\lambda/4$ kcal mole^{-1} according to Eq. (7.17)."

Typo: From $-\Delta F^{\circ\prime}/\lambda \approx 4.4$ one sees that $1 + (\Delta F_R^{\circ\prime}/\lambda) = -3.4$ but in Eq. (7.17) there is a square, so it should be $(3.4)^2\lambda/4$.[(13)]

"Such a reaction has a rate constant which is extremely small, the reaction proceeding only at one every 10^{10} collisions."

Note that from $k = \kappa\rho Z e^{-\Delta F^*/kT}$ one sees that if $\kappa \approx 1$ and $\rho \approx 1 \Rightarrow k \approx Z e^{-\Delta F^*/kT}$ and if $\Delta F^* = 0 \Rightarrow k \approx Z$.

"On the other hand, for this small λ, the reaction leading to *formation of an excited state* occurs with a *high rate* if that state of the products is accessible. For example, according to whether $\Delta F^{\circ\prime}$ for formation of an excited state is 0 or -10 kcal mole^{-1}, assumption of *the same* λ yields a ΔF^* of $\frac{1}{4}\lambda$ or $\frac{9}{94}\left(\frac{1}{4}\lambda\right)$, respectively. The reaction then proceeds at one in every 10^3 or one in every three collisions respectively.

As calculations based on Eq. (7.17) readily verify, a large difference in rates leading to formation of an excited state versus ground state of a product, and favoring the former, exists for this very exothermic system even for larger λ_{bb}'s, though not for very large λ_{bb}'s. It should be noted, however, that λ for an elementary step depends on the *change of equilibrium bond lengths of each reacting species in that step*, and so may differ in two competitive reactions. Knowledge of the electronic states often permits an at least qualitative insight into these differences in the absence of exact quantitative calculations.

By introduction of *suitable substituents* in the aromatic ring, some *manipulation* of the $\Delta F^{o\prime}$'s becomes possible, making formation of an excited state difficult or easy and suggesting thereby further experiments. Extensive variations of $\Delta F^{o\prime}$ of inorganic complexes containing heterocyclic aromatic ligands has been achieved in this way for ordinary ET reactions [21]. Under suitable conditions, the $\Delta F^{o\prime}$'s for ground-electronic-state reactions are available from polarographic half-wave potentials, even for some electrochemically irreversible systems.

The above remarks apply to homogeneous ET reactions. The mechanism of *electrochemical* ETs can also be analyzed in terms of motion on PESs [6–9]. A major difference now arises, however: For any electronic state of the reactant there are many PESs, each shifted vertically from one another and each involving a different *many-electron quantum state* of the electrode. In terms of *one-electron states*, each many-electron state corresponds to a different *distribution of electrons* of the metal among *one-electron states*. Thus, the surface R in Fig. 2(a) now denotes *one* of these many surfaces for the *given* electronic state of the reactant and for a *given* many-electron state of the electrode. Similarly, the surface P in Fig. 2(a) now denotes a surface for the given electronic state of the product *and* that of the electrode.(14)

The splitting at the intersection of an R and P surface depends on the particular *surfaces*, varying from zero to an appreciable amount: In terms of one-electron quantum states of the electrode, there is an appreciable splitting *at an 'intersection'* of an R and P surface for a one-electron transfer if the distribution of electrons among the one-electron quantum states is *the same* for the R and P states *except* for*the electron* undergoing transfer. We term such a pair of surfaces as *suitable* for occurrence of ET.

Although there is now almost a *continuum* of such R and P surfaces, most ETs under the usual electrochemical conditions occur to and from one-electron electrode quantum states which are within kT of the Fermi level [9]: in the case of ET from ion to electrode, transfer to a suitable but high P surface is unlikely because of the greater activation energy needed to reach the intersection region. Transfer to a low P surface is difficult because the probability of finding an *otherwise suitable* surface having an unoccupied one-electron orbital is small. Most of the one-electron states that are more than kT below the Fermi level are already occupied. A quantitative description has been given elsewhere [9].

An equation derived for the electrochemical rate constant is given by Eq. (7.6), [6–9] where Z is now about $10^4\,\text{cm}\,\text{sec}^{-1}$, λ is the *electrochemical* λ, which has a value of about one-half the value of λ for a homogeneous electron exchange reaction such as Eq. (7.18) [6–9], and $\Delta F^{\circ\prime}$ is replaced by $ne(E - E_o')$, n is the number of electrons transferred, usually equal to one in an elementary step. E and E_o' are the *electrode–solution* potential and the 'standard' *electrode–solution* potential drop for the given electronic states of reactant and product.

We consider now a very exothermic [20] electrode process involving transfer of an electron from the ion to the electrode, that is, a process that *would be* very exothermic *if* transfer occurred to the Fermi level of the electrode. There are now many suitable potential energy P surfaces that cross the original R surface at an accessible region, and each corresponds to the electron going into some high unoccupied orbital of the metal, near the top of the *unfilled half of the conduction band* if need be. Because of the large width of this band in metals [25], the system can easily *reduce the exothermicity*, by ET in such an unfilled level.

Similar remarks apply to a very exothermic ET from an electrode to a molecule or ion in solution. In this case, the electron can *relieve* the exothermicity by coming from one of the lower levels in the *filled half* of the conduction band.

Even when the exothermicity for transfer to or from the Fermi level is of the order of 3 eV, it can be alleviated [26] because the widths of the unfilled and filled halves of the conduction band are each at least this amount [25]. Transfer then proceeds *to or from* one-electron states in the electrode for which the ΔF^* in Eq. (7.17) is small, in the case of adiabatic transfers. [The $e(E - E_o')$ which replaces ΔF^o there for a one-electron transfer electrode reaction is in turn now replaced by $e(E - E_o') \pm (\varepsilon - \varepsilon_F)$ where $\varepsilon - \varepsilon_F$ is the energy of this one-electron quantum state in the metal relative to that of the Fermi level. The sign depends on the convention used for E_o'.]

Thus, unless the formation of the *excited state of an ion* by ET *to or from* the electrode has a very small ΔF^*, it cannot compete with the above process which involves formation of the ground state of an ion and an *excited state of the electrode.*[(15)]

In agreement with these observations, no chemiluminescence has been observed when an aromatic anion comes in contact with a positively charged metal electrode [19d]."

Which means that no reaction like $A^- + M(+) \rightarrow A + M + h\nu$ is observed.

"Again, when an aromatic anion is formed at an electrode and the electrode is suddenly made very positive [19a, c] no chemiluminescence occurs until the supporting electrolyte is oxidized by the electrode."

Same case as before but now the A^- is formed at the electrode following the reaction $A + M(e) \rightarrow A^-$.

"A homogeneous chemiluminescent reaction between the aromatic anion and the resulting radicals was then presumed to occur.

It would be interesting to investigate theoretically and experimentally the possibility of chemiluminescence with electrodes for which the conduction band width is small. Semiconductor electrodes offer such a possibility. ET theory has been applied to these electrodes [27]. The fact that transfer can occur with *both the valence and the conduction bands* would, of course, be taken into consideration."

From "**Erratum**" **M57**: R. A. Marcus "**Erratum: On the Theory of Chemiluminescent Electron Transfer Reactions**

7.11a

"The curve drawn in Fig. 1(c) of the earlier article indicates the 'reaction path' leading to formation of *ground-state* products in a highly exothermic ET. The adiabatic curves are given instead by Fig. 1 in the Erratum. Thereby, the transition from the lower segment of the *R* (reactants') curve is nonadiabatic. Thus, the argument that the rate constant for the case of Fig. 1(c) for formation of the electronic ground states of products is small, made there on general grounds, is now *further strengthened*: the rate constant is decreased by a transition probability factor for nonadiabatic reactions."

NOTE: While until now the nonadiabatic transition would *reduce* the adiabatic ET because some systems would leak passing through the PESs splitting from *R* to *R*, here we have a *nonadiabatic ET* in which the ET happens *because* of a transition *R* to *P* through the splitting.

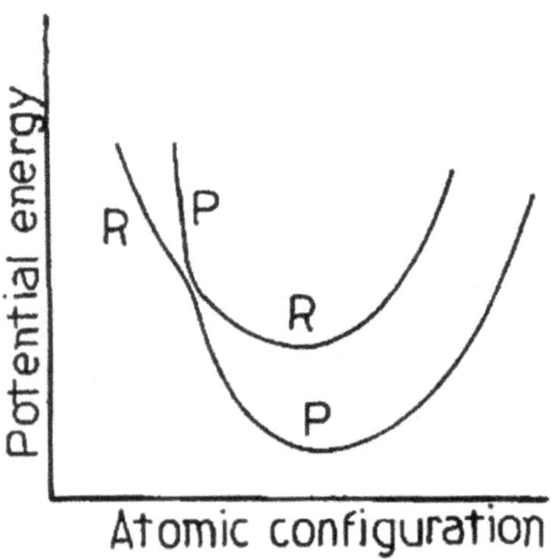

Fig. 1. Profile of potential energy surface of reactants (R) and that of products (P) plotted versus configuration of all the atoms in the system, for the case of very negative ΔF°.

From **M60**: R. A. Marcus, "**Theory of Electron-Transfer Reaction Rates of Solvated Electrons.**" *J. Chem. Phys.* **43**, 3477, (1965).

7.12. Introduction

"In the present paper, a reaction rate theory is formulated for electron transfer reactions of solvated electrons. Consideration is restricted to 'pure' ETs, i.e., to reactions for which chemical bond *rupture* does not occur or occurs *subsequently* to ET as in Eqs. (7.21) and (7.22).

$$e(\text{aq}) + \text{Ox} \rightarrow \text{Red} \tag{7.21}$$

$$\text{Red} \xrightarrow{\text{fasr}} \text{Products.} \tag{7.22}$$

An expression for the reaction rate is obtained in terms of

(i) Standard reduction potentials of the two reactants
(ii) Spectral data for the solvated electron
(iii) Reorganization parameters found from rate data on chemical or electrochemical ET when available.

When the calculated rate constant is very high, it is not used directly but appears, instead, in a boundary condition in a differential equation for mutual diffusion of the two reactants. For very high rate constants, the reaction becomes diffusion controlled.

A major purpose of the present paper is to relate rate constants of solvated electron reactions to other rate constants and to quite different properties. Many rate constants for reactions of solvated electrons have now been measured in water and in alcoholic solvents [29, 30]. More precisely, rate constants have been measured for reaction with *some species* which, measurements of salt effects in the diffusion-controlled region suggest, has a unit negative charge [31], and which has an *absorption spectrum* in water and alcoholic solvents shifted to the blue from that in liquid ammonia [29, 30].

At present, tests of the theoretical rate expression derived in the present paper are somewhat limited by the fact that many of the common redox reactants, for which conventional ET rates are known, are *reduced very quickly* by the solvated electron because of the latter's *extremely high reduction potential.* Thus, many of the reactions of the solvated electron are *diffusion controlled,* so that only lower bounds for the corresponding activation-controlled rate constants are available from the data. For many reactants whose rates of reaction with the solvated electron are not diffusion controlled, the corresponding redox reactions, chemical or electrochemical, are slow."

That is, species that are not easily reduced by the solvated electron are also not easily reduced by conventional reducing agents.

"Nevertheless, by suitable choice of solvents, reactants, and techniques, one may anticipate, this sparsity of comparable data will be reduced.

PESs and Reaction Mechanism

The concepts of ET theory presented elsewhere [5–9, 32] remain applicable here when the overlap of the *electronic orbital of the solvated electron* and that of the second reactant is weak: One may draw a PES as before, the electronic structure of the *reacting species* being that of the reactants and the surface being plotted as a function of the position of all the atoms in the system. Similarly, a second PES in this many-dimensional configuration space may be drawn, the electronic structure of the reacting species being that of the products."

Note that both reactants and products are "reacting species."

"When there is zero *interaction of the redox orbitals* of the two reacting species, these two potential energies intersect.

In the absence of an interaction of the redox orbitals, a fluctuation of coordinates from ones describing *stable spatial configurations of the atoms in the initial system* to those characterizing the *intersection region* and finally to those describing *stable spatial configurations of the products* cannot lead to reaction. It merely represents a fluctuation of coordinates, *fluctuations which occur continually.* The presence of some electronic *interaction between* the *redox orbitals splits the surfaces at each point of intersection* in the usual quantum-mechanical manner. If the '*redox interaction*' is appreciable so is the splitting. A fluctuation of the earlier type then causes ET to occur as the system passes through the *region of configuration space* characterizing the intersection region. The process occurs adiabatically if the splitting is sufficiently large, and nonadiabatically if it is not [5–9, 32].

The coordinates of this many-dimensional configuration space *include* ones describing

(a) The *vibrations* of *each* molecule present
(b) Its *translations*
(c) Its *orientations*

Fluctuations leading to reaction *include* those of the

(d) *Separation distance* of the reactants
(e) *Reorientation* of the dielectrically polarized solvent molecules
(f) *Vibrations* in the second reactant"

In the present theory, the *first reactant* is the solvated electron.

"In fact, any favorable changes in a coordinate which has a somewhat different *equilibrium distribution* in the initial and final states, or which permits the reactants to come close together, helps facilitate the reaction, i.e., in the present case permits the system to *reach and pass through* the intersection region in a place *where the redox interaction is appreciable.* To be sure, related remarks apply to all chemical reactions, but in the usual chemical reactions one concentrates mainly on the chemical bonds that are broken and those that are formed. The present step (7.21) involves no such bonds for the reaction being considered.

The *electronic wave function of the solvated electron* is *very sensitive to fluctuations* of the solvent molecules, unlike the electronic wave functions of a conventional reactant. This circumstance leads to one principal quantitative difference between the present treatment of its reactions and that given earlier [5–9, 32] for more conventional ET reactions. A second difference lies in the fact that the *number of particles* changes in the simple ET step (7.21), while in those considered previously there was no such a change. Normally in reactions, the number of particles changes only because of formation or rupture of new chemical bonds. In the present situation, however, it occurs by *absorption* of the electron into the second reactant and *deorientation* of the solvent molecules formerly oriented about it.

For purposes of simplicity in the present paper, we formulate the theory using *dielectric-continuum* theory for the solvent polarization. Any vibrational changes in the other reactant are treated in *molecular terms* however. The *functional form* of the resulting rate equation suggests a functional form for a statistical mechanically derived expression, much as it did in the formulation of an ET theory: a *continuum treatment* was given first [5] and was followed by a *statistical mechanical* treatment [6, 7]. Both had a rate expression of the *same functional form.*

A qualitative summary of the assumptions and of the principal results of the present paper have been given in a recent monograph [2]. A glossary of the principal symbols employed is given in Appendix III.

7.13. Theory

Since the orientation polarization has a much lower *characteristic frequency* than the frequency of motion [33] of the solvated electron, the orientation polarization 'sees' a *smeared-out* charge distribution of the electron. When the environment is in thermal equilibrium with the solvated electrons, the *polarization function* and *wave function* are found by *minimizing a certain functional*, the sum of the kinetic energy of the electron and the free energy of the polarized system [34]. No constraint is imposed in the minimization (other than that of normalization of the wave function).

The activated complex for a weak-overlap ET reaction is *a species* having *the set* of spatial configurations that occur at the intersection of the two PESs described earlier. It has the potential energy of the intersection for each point of the set and has an *equilibrium distribution of coordinates* within this set [9]."

NOTE: The possible configurations of the activated complex are those corresponding to the intersection surface. Their distribution is a Boltzmann distribution among the energies corresponding to the different molecular configurations of the activated complex.

"The properties of the activated complex have been related elsewhere to those of *two other constrained systems* [5–9, 32]: One of these two systems has the electronic structure of the reactants and the other has that of the products.**(16)** Each system is *simpler* than the activated complex in that each is *constrained to be centered* on the intersection region rather than *confined* to it. Each has the same thermodynamic energy (the same as that of the activated complex) and the same Boltzmann distribution of configurations. Since the two systems have the same distribution of coordinates in configuration space they also have the same entropy and, thereby, the same free energy. They have different electronic structures, and the condition that they have the same free energy at all temperatures constitutes an *equation of constraint*."

Here M. refers to equivalent equilibrium distributions.

"Let ΔF^* be the free energy of formation, from reactants fixed in position far apart, of a system in which the system is centered in the above intersection region, as described earlier, and in which the reactants are fixed in position a distance R apart. Then, it has been shown, the bimolecular rate constant is given by Eq. (7.6)

$$k_{\text{act}} = Z\kappa\rho e^{-\Delta F^*/kT} \tag{7.6}$$

[5, 9], where Z is a collision frequency $(8\pi\kappa T/m^*)^{\frac{1}{2}}R^2$ calculated for the most probable separation distance R between the centers of the reactants in the activated complex. Typically, R is the sum of the radii

of the reactants, any uncertainty in R leading to a minor uncertainty in k; m^* is approximately the reduced mass for the reaction coordinate; κ is unity for an adiabatic reaction (as we usually assume it to be) and less than unity for a nonadiabatic one; ρ is a factor usually close to unity: it is the ratio of the root-mean-square fluctuation of R in the activated complex to that of s, the displacement normal to the 'intersection hypersurface' in the centered distribution [7].

We let $F^{*r}(R)$ denote the free energy of a system of reactants, fixed in position, a distance R apart and having the constrained equilibrium distribution of configurations described earlier,**(17)** let $F^{*p}(R)$ denote that of the product in a system with the same constrained distribution, and let $F^r(\infty)$ denote the free energy of an unconstrained system of the reactants when they are fixed in position but far apart. We have

$$\Delta F^* = F^{*r}(R) - F^r(\infty) \tag{7.23}$$

The value of ΔF^* is determined as the solution of the *variational problem* (7.24) and (7.25)

$$\delta F^{*r}(R) = 0, \quad \text{subject to (7.25)} \tag{7.24}$$

$$F^{*r}(R) = F^{*p}(R). \tag{7.25}$$

The variation in Eq. (7.24) is to be performed with respect to *parameters describing the configurational distribution* of the system: The *orientation polarization* is characterized by a continuous parameter, a *polarization function*, and the *distribution of vibrational coordinates* of the *second reactant* is characterized by several parameters, q_*^i the *most probable value* of each vibrational coordinate *in the centered distribution*."

Three contributions to ΔF^* are considered.

(1) "The vibrational contribution to ΔF^* is then given by [35]

$$\Delta F^*_{\text{vib}} = \sum_{i,j} k_{i,j}(q_*^i - q_r^i)(q_*^j - q_r^j), \tag{7.26}$$

where q_r^i and q_p^i denote the most probable (unconstrained) values of q^i for the reactant and product, respectively; $k_{i,j}$ is a symmetrical function,

$$k_{i,j} = 2k_{ij}^r k_{ij}^p/(k_{ij}^r + k_{ij}^p),$$

of the corresponding force constants, k_{ij}^r and k_{ij}^p, of the reactant and product. The vibrational contribution to the free energy of formation of the activated complex *from the separated products* is the same as Eq. (7.26) but with q_r^i replaced by q_p^i.

(2) Another term in $F^{*r}(R)$ is the kinetic energy of the excess electron,

$$\left(-\frac{\hbar^2}{2\mu}\right)\int \psi * \nabla^2 \psi d\mathbf{r}.$$

On integration by parts it becomes

$$\hbar^2 \int |\nabla\psi|^2 d\mathbf{r_1}/2\mu,$$

where $\mathbf{r_1}$ and μ denote the position and *effective mass* of the electron [34]. Hence another contribution to ΔF^* is the change in this kinetic energy arising from a change in ψ."

*Demonstration: Recall $\int_a^b u'vdx = uv|_a^b - \int_a^b uv'dx$; here $\psi* = u$ and $\nabla\psi = v$. The $\psi * \nabla\psi$ term vanishes because the function ψ is supposed to be an even function (an s-like orbital, say) and so its derivative is an odd function. Then $\psi * (\mathbf{r})\nabla\psi(\mathbf{r}) = -\psi * (-\mathbf{r})\nabla\psi(-\mathbf{r})$.

(3) "A final contribution to $F^{*r}(R)$ is the free energy of polarization of the system. It has been suggested [36] that cavity formation does not occur in strongly hydrogen-bonding solvents such as water and alcohols, unlike liquid ammonia. In the present paper, we assume cavities to be absent—and so perhaps restrict attention to the strongly hydrogen-bonding solvents. Use of a more elaborate expression for the free energy of polarization would permit the inclusion of cavities."

In the liquids with strong hydrogen bonds, the molecules are strongly bound to the neighbors so that no cavities are left among them.

"The usual polaron-type theory [34, 37, 38] is employed to describe the polarization free energy of the system. However, we do not wish to tie the present reaction rate mechanism too closely to any particular model for the solvated electron. A simple approximate analysis of the effect of using alternative models is described in a later section, therefore.

For each value of $\mathbf{r}_1$, the free energy of polarization of the system is shown in Appendix I to be Eq. (7.27), where $\mathbf{P}(\mathbf{r})$ is a function of the orientation polarization at $\mathbf{r}$:

$$F_{\text{pol}}(\mathbf{r}_1) = -\int \mathbf{P}\cdot\mathbf{D}d\mathbf{r} + \frac{2\pi}{c}\int \mathbf{P}^2 d\mathbf{r} - \frac{e_2^2}{2a_2}\left(1-\frac{1}{D_{\text{op}}}\right) + \frac{e_1 e_2}{D_{\text{op}}|\mathbf{r}_1-\mathbf{r}_2|} \tag{7.27}$$

where c is given by Eq. (7.28), $\mathbf{D}$ is the field due to the permanent charges and is given by Eq. (7.29), and the radius a_2 for the second reactant is that of a sphere which includes any inner coordination shell. (One can weaken the spherical assumption, however; then, the final rate expression contains an orientational factor.) Changes in values of the coordinates in that shell contribute to ΔF^*_{vib}.

$$c = D_{\text{op}}^{-1} - D_s^{-1} \tag{7.28}$$

$$\mathbf{D} = \mathbf{D}_1 + \mathbf{D}_2 \tag{7.29}$$

$$\mathbf{D}_1 = -\nabla(e_1/|\mathbf{r}-\mathbf{r}_1|), \quad \mathbf{D}_2 = -\nabla(e_2/|\mathbf{r}-\mathbf{r}_2|) \tag{7.30}$$

In Eq. (7.30), e_2 is the charge of the second reactant, which is centered at $\mathbf{r}_2$. D_{op} and D_{s} denote the optical and static dielectric constant of the medium.

The total free energy of polarization is obtained by multiplying Eq. (7.27) by $|\psi|^2 d\mathbf{r}_1$ and integrating over $\mathbf{r}_1$.

If we denote by $\mathscr{F}_{\mathbf{o}}^{\mathbf{r}}$, the sum of the electron kinetic energy and the free energy of polarization of the *unconstrained* system of reactants when they are far apart then ΔF^* becomes

$$\Delta F^* = \Delta F_{\text{vib}}^* + \frac{\hbar^2}{2\mu}\int |\nabla\psi|^2 d\mathbf{r_1} + \int F_{\text{pol}}^r(\mathbf{r_1})|\psi|^2 d\mathbf{r_1} - \mathscr{F}_{\mathbf{o}}^{\mathbf{r}} \tag{7.31}$$

with ΔF_{vib}^* and F_{pol}^r given by Eqs. (7.26) and (7.27) with $\mathbf{D}$, e_1 and e_2 all bearing r superscripts to denote properties of reactants.[18]

We consider next the corresponding free energy of formation of the *constrained* state with the electronic structure of the product, from an *unconstrained* state, ΔF^{*p}. Because of the manner in which $F_{\text{pol}}^p(R)$ and $F_{\text{pol}}^r(R)$ were defined ('reacting species fixed in position'), there is no translational contribution to this

$$\Delta F^{*p} = F^{*p}(R) - F^p(\infty).$$

By arguments analogous to those used to derive Eq. (7.31) we obtain

$$\Delta F^{*p} = \frac{1}{2}\sum_{i,j} k_{ij}(q_*^i - q_p^i)(q_*^j - q_p^j) + \frac{2\pi}{c}\int \mathbf{P}^2 d\mathbf{r} - \int \mathbf{P}\cdot\mathbf{D}^p d\mathbf{r} - \frac{e_2^2}{2a_2}\left(1 - \frac{1}{D_{\text{op}}}\right). \tag{7.32}$$

The expression is somewhat simpler than Eq. (7.31): The charge e_1^p vanishes. The electron kinetic energy term is missing, since the electron now resides in the second reactant" (in its reduced form) "where its kinetic energy is insensitive to solvent fluctuations and does not contribute to ΔF^{*p}, therefore.

For any given R, we then minimize ΔF^* with respect to the quantities $\mathbf{P}$ and q_*^i, subject to the constraint imposed by Eq. (7.25). When $\mathbf{P}$ is varied, ψ in Eq. (7.31) also varies. However, since ψ is determined as that quantity which minimizes the sum of the kinetic energy and $\int F_{\text{pol}}|\psi|^2 d\mathbf{r_1}$ (and hence, which minimizes ΔF^*) for any given $\mathbf{P}$, the variation of ΔF^* with ψ at fixed $\mathbf{P}$ is zero. In minimizing

ΔF^* with respect to **P**, therefore, ψ can be regarded as constant and, in the final step, is set equal to the ψ appropriate to the given **P**."

At the end of a minimization process that follows in the steps of the process seen in Chapter 2, after the introduction of the Lagrangian multiplier m , M. obtains "the properties of the *constrained state*, the '*centered distribution*':

$$\mathbf{P} = [(m+1)\int |\psi|^2 \mathbf{D}^r d\mathbf{r}_1 - m\mathbf{D}^p]c/4\pi$$

$$q_*^i = (m+1)q_r^i - mq_p^i. \tag{7.33a}$$

Note that Eq. (7.33a) may be written in a more transparent form as $q_*^i = q_r^i + m(q_r^i - q_p^i)$.

The expression for the ΔF^* and ΔF^{*p} of the reaction obtained by the earlier minimization process are

$$\begin{aligned}\Delta F^* = {} & \frac{\hbar^2}{2\mu}\int |\nabla\psi|^2 d\mathbf{r}_1 + \frac{m^2 c}{8\pi}\int (\mathcal{D}^r - \mathbf{D}^p)^2 d\mathbf{r} \\ & - \frac{c}{8\pi}\int \mathcal{D}^{\mathbf{r}2} - \frac{e_2^{r2}}{2a_2}\left(1 - \frac{1}{D_{\text{op}}}\right) \\ & - \frac{ee_2^r}{D_{\text{op}}}\int \frac{|\psi|^2 d\mathbf{r}_1}{|\mathbf{r}_1 - \mathbf{r}_2|} + m^2\lambda_i - \mathcal{F}_{\mathbf{o}}^{\mathbf{r}},\end{aligned} \tag{7.33b}$$

where $-e$ is the charge of the electron and

$$\lambda_i = \frac{1}{2}\sum_{i,j} k_{i,j}(q_p^i - q_r^i)(q_p^j - q_r^j)$$

$$\mathcal{D}^r = \int \mathbf{D}^r |\psi|^2 d\mathbf{r}_1.$$

'Analogously, ΔF^{*p} is found to be given by (7.33c):

$$\begin{aligned}\Delta F^{*p} = {} & \frac{(m+1)^2 c}{8\pi}\int (\mathcal{D}^r - \mathbf{D}^p)^2 d\mathbf{r} - \frac{c}{8\pi}\int \mathbf{D}^{p2} d\mathbf{r} \\ & - \frac{e_2^{p2}}{2a_2}\left(1 - \frac{1}{D_{\text{op}}}\right) + (m+1)^2\lambda_i - \mathcal{F}_{\mathbf{o}}^{\mathbf{p}}.\end{aligned} \tag{7.33c}$$

To determine the parameter m in terms of the known quantities, we next express $F^{*r} - F^{*p}$ in terms of the 'standard' free energy of reaction.

The 'standard' free energy of reaction for the elementary ET step (7.21) at the prevailing pressure, temperature, and reaction medium is denoted by $\Delta F^{\circ\prime}$, to distinguish it from the standard one ΔF° at 25°C, 1 atm, and unit activities. Let the standard-state translational free energy of the solvated electron in step (7.21) be $F^{\circ}_{\text{transl}}$. If we subtract from $\Delta F^{\circ\prime}$ this translational contribution, F^{0}_{transl}, (**Typo** in the original symbol) and note that the rotational contribution of Ox is the same as that of Red, we obtain the internal contribution, $\Delta F^{\circ\prime}_{\text{int}}$. The latter is the free energy of formation of products from reactants when all reacting particles are fixed in position:

$$\Delta F^{0\prime}_{\text{int}} = \Delta F^{0\prime} - F^{0}_{\text{transl}}{}^{(\mathbf{19})}, \tag{7.34}$$

(**Typo** in the original equation). (We note parenthetically that, in contrast to $\Delta F^{0\prime}$, $\Delta F^{0\prime}_{\text{int}}$ is independent of the choice of concentration units in the standard state).

With $\Delta F^{\circ\prime}_{\text{int}}$ one may now rewrite Eq. (7.25). The free energy change $\Delta F^{\circ\prime}_{\text{int}}$ on going from reactants *fixed in position* to a product *also fixed in position* can be regarded as the sum of three terms:

(1) Formation of constrained state of the reactants in activated complex, ΔF^{*};
(2) The change due to a change in electronic structure in the constrained state, $F^{*p}(R) - F^{*}(R)$, i.e., zero because of Eq. (7.25);
(3) Formation of product from the constrained state of the product, $-\Delta F^{*p}$. Thus, we obtain

$$\Delta F^{0\prime}_{\text{int}} = \Delta F^{*} - \Delta F^{*p}, \tag{7.34a}$$

where $\Delta F^{0\prime}_{\text{int}}$ is given by Eq. (7.34).

For any given m, the function ψ is determined as the ψ which minimizes ΔF^*. To determine the parameter m, Eqs. (7.33b) and (7.33c) are then introduced into Eq. (7.34a).

For completeness, we note that k_{-1}, the unimolecular rate constant for the reverse step of reaction (7.21), can be obtained from ΔF^{*p} (7.33c) and from the equilibrium constant for reaction (7.21). One finds:

$$k_{-1} = Z\kappa\rho \exp(-F^{\circ}_{\text{transl}}/kT) \exp(-\Delta F^{*p}/kT) \tag{7.35}$$

where Z is about $10^{14}\,\text{cc} \cdot \text{mole}\,\text{sec}^{-1}$." See **M170** for a value of Z later corrected by M.

7.14. Calculation for Large *R*

M. considers first the form ΔF^* and ΔF^{*p} would take "if the separation distance R were large in the activated complex." "In this situation, the 'polaron' is isolated and one may make use of known solutions of the polaron problem to evaluate the various integrals."

M. denotes with $\mathscr{F}^{\mathbf{r}}_{\mathbf{o}}$ "the sum of the electron kinetic energy and the free energy of polarization of the unconstrained system of reactants when they are far apart" (see earlier) and with $\mathscr{F}^{\mathbf{r}}_{\mathbf{o1}}$ and $\mathscr{F}^{\mathbf{r}}_{\mathbf{o2}}$ the contributions of the isolated polaron and that of the isolated second reactant to $\mathscr{F}^{\mathbf{r}}_{\mathbf{o}}$.

They happen to be:

$$\mathscr{F}^{\mathbf{r}}_{\mathbf{o2}} = e_2^{r^2}/2a_2[1 - 1/D_s]$$

for the second reactant and

$$\mathscr{F}^{\mathbf{r}}_{\mathbf{o1}} \equiv H_o = \frac{\hbar^2}{2\mu}\int |\nabla\psi_o|^2 d\mathbf{r_1} - \frac{c}{8\pi}\int \mathscr{D}^{r^2}_{o1} d\mathbf{r}. \tag{7.36}$$

for the isolated unconstrained polaron where ψ_o is its wave function and

$$\mathscr{D}^r = \int \mathbf{D}^r |\psi|^2 d\mathbf{r_1}.$$

"The wave function ψ_0 is determined by minimizing H_0 [39]."

NOTE: $-\frac{c}{8\pi}\int \mathscr{D}_{01}^{r^2} d\mathbf{r}_1$ is an electrostatic energy density, see [39a, p. 288].

"Since ψ is determined by minimizing ΔF^* at a given m, it is also obtained by minimizing that part of ΔF^* which depends on ψ at this m, namely H:

$$H = \frac{\hbar^2}{2\mu}\int |\nabla\psi|^2 d\mathbf{r_1} - \frac{c(1-m^2)}{8\pi}\int \mathscr{D}_1^{r^2} d\mathbf{r}.\text{"} \qquad (7.37\text{a})$$

Note that the interaction term is given here by $\frac{cm^2}{8\pi}\int \mathscr{D}_1^{r^2} d\mathbf{r}$.

Once ψ has been determined by minimizing ΔF^* *at a given m* "H is obtained from the well-known values of H_o in the literature merely by replacing c in the formula for H_o by $c(1-m^2)$."

It is then shown that

$$\Delta F^* = H - H_o + m\lambda_2^\infty$$

where

$$\lambda_2^\infty = \lambda_i + \lambda_o^\infty$$

and

$$\lambda_o^\infty = \frac{c}{8\pi}\int (\mathbf{D}_2^r - \mathbf{D}_2^p)^2 d\mathbf{r} = -(e_2^r - e_2^p)^2 c/2a_2$$

Note that the contribution to ΔF^* by the solvated electron is $H - H_o$. Moreover, the λ_0^∞ term is the energy density of the electric field of reactant 2 due to the charge which is transferred.

"A useful virial theorem can be derived for the solution of the nonlinear wave equation for ψ_0, $\delta H_0 = 0$, using a scalar factor technique [40] similar to the one [41] used to establish the virial theorem for the usual (linear) Schrödinger equation." M. shows that [40]:

$$\frac{\hbar^2}{2\mu}\int |\nabla\psi_o|^2 d\mathbf{r_1} = -H_o$$

and that

$$-\frac{c}{8\pi}\int D_{o1}^{r^2} d\mathbf{r} = 2H_o\text{“} \tag{7.37b}$$

Here $\frac{\hbar^2}{2\mu}\int |\nabla\psi_o|^2 d\mathbf{r_1} = -H_o$ is the kinetic energy and $-\frac{c}{8\pi}\int D_{o1}^{r^2} d\mathbf{r} = 2H_o$ is the potential energy.

"Similarly, the first term of Eq. (7.37a) is $-H$ and the second is $2H$. One can always assure satisfaction of this virial theorem by introduction of a scale factor as one of the variational parameters, as one may see from the above proofs [39, 40] of the theorem."

M. shows then that H_o "can be written as Eq. (7.37b), where A is a dimensionless constant.

$$H_o = -A\mu e^4 c^2/\hbar^2 \tag{7.38}$$

It is then shown that "H is given by

$$H = (1 - m^2)^2 H_o. \tag{7.39}$$

It is finally shown that "Under typical conditions where these reactions are very fast... m^2 is small... One the finds that $m^2 \ll 1$, so that":

$$\Delta F^* = m^2(-2H_o + \lambda_2^{\infty}) \tag{7.40}$$

$$\Delta F^{*p} = (m+1)^2(-2H_0 + \ddot{e}_2^{\infty}) \tag{7.41}$$

"and m is the solution of Eq. (7.42):

$$-(2m+1)(-2H_o + \lambda_2^{\infty}) = \Delta F_{\text{int}}^{o\prime}. \tag{7.42}$$

These equations now have the same functional form as those derived for more conventional ET reactions. (In those cases, however, $\Delta F^{o\prime}_{\text{int}}$ equaled $\Delta F^{o\prime}$.)

7.15. Evaluation of H_o

Variational functions of the type (7.43) have been used by Pekar to evaluate H_0

$$\psi_o = B(1 + \alpha r + \beta r^2)e^{-\alpha r}, \tag{7.43}$$

the best result for A in Eq. (7.38) being -0.0544 [34]. Jortner has used instead a $1s$ wave function, $B\exp(-\gamma r)$, which yields a slightly worse result: $A = -0.0488$, [36, 42]. Quantitative calculations show that the βr^2 is hardly necessary in Eq. (7.38) but that the αr is useful, since the *effective potential* for the polaron is found to be of a *harmonic oscillator* nature near the origin rather than Coulombic [34]. Results of other variational calculations have been summarized by Allcock [43]. They include the use of a harmonic oscillator wave function by Pekar *et al.* (1948, 1954) and by Feynman (1955), which yields $A = -0.053$, and the use of a $1s$ wave function by Frohlich (1954), which yields $A = -0.0488$. On using a $2p$ wave function for the excited state Pekar [34] found the transition energy to be... $-1.28H_o$.

Since this transition energy is about 1.7 eV when the solvent is water, H_o is about -1.3 eV in that solvent"

NOTE: Because $1.7\,\text{eV} = -1.28H_o \Rightarrow H_o \approx -1.3\,\text{eV}$.

"This particular result is independent of the value used for c, as long as the same c is used in the ground and $2p$ state. Nevertheless, such an evaluation of H_o is open to some question: one should subtract from 1.3 eV an amount $\Delta\mathscr{A}/1.28$ where $\Delta\mathscr{A}$ is some difference in 'electron affinity' (i.e., in short-range interactions) of the solvent molecules for the solvated electron in its ground and excited states. Perhaps, then, this magnitude 1.3 eV is too high.

A comparison in Appendix II" (Section 7.25) "of spectral and solvation data permits one to eliminate the unknown 'electron affinity' of the ground state, $\mathscr{A}_g$. When that of the excited state can be neglected, the values of $-H_o$ and of $\mathscr{A}_g$ are estimated to be roughly about 0.7 and 0.8 eV, respectively, but they are quite uncertain since they depend on a small difference (~5) between fairly large quantities (~35 kcal mole^{-1}). The actual value of $-H_o$ is larger, and that of $\mathscr{A}_g$ smaller, if the neglected electron affinity of the excited state is positive.

A direct calculation of H_o made by assuming μ = electron mass and $D_{\mathrm{op}} = 1.33$ [30], yields 0.45 for $-H_o$. To explain a particular rate constant, we later take $-H_o$ to be some compromise—about 0.65 eV—though the value must be regarded as highly tentative pending accumulation of more data.

The above results suggest that part of the large spectral difference of the solvated electron in ammonia and in water may be due to a difference in $\mathscr{A}_g$, if the affinity is greater for water.

7.16. Case of Finite *R*

On introducing Eq. (7.90) into Eq. (7.33b), we find

$$\begin{aligned}\Delta F^* &= \frac{\hbar^2}{2\mu}\int |\nabla\psi|^2 d\mathbf{r}_1 + \frac{(m^2-1)c}{8\pi}\int \mathscr{D}_1^{r^2}\, d\mathbf{r}\\ &\quad - \frac{ee_2^r}{D_s}\int \frac{|\psi|^2 d\mathbf{r}_1}{|\mathbf{r}_1-\mathbf{r}_2|} + \frac{m^2ce^2}{2a_2}\\ &\quad - m^2ce^2\int \frac{|\psi|^2 d\mathbf{r}_1}{|\mathbf{r}_1-\mathbf{r}_2|} + m^2\lambda_i - H_o \qquad (7.44)\end{aligned}$$

where H_o is given by Eq. (7.38).

The first term involving $|\mathbf{r}_1 - \mathbf{r}_2|^{-1}$ and, when m^2 is small, the second $|\mathbf{r}_1 - \mathbf{r}_2|^{-1}$ term also, are small in comparison with the electronic energy of the polaron [44]. The *distortion* of ψ from the form it has when these terms are absent is assumed small, therefore. In this case, ψ is seen from Eq. (7.44) to satisfy the same variational equation, for any given m, as the ψ in the previous section, i.e., as the ψ for the *isolated but constrained* polaron, for only the first two terms in Eq. (7.44) now depend on ψ."

NOTE: For $|\mathbf{r}_1 - \mathbf{r}_2| \to \infty$ the 3rd and 4th terms in Eq. (7.44), go to zero, compare Eq. (7.37a).

"Thus, the first two terms on the RHS of Eq. (7.44) equal $-H$ and $2H$ as before, H being given by Eq. (7.39).

The overlap of the polaron wave function and the charge of the second reactant was assumed earlier to be small. In this case, the term $\int |\psi|^2 d\mathbf{r}_1/|\mathbf{r}_1 - \mathbf{r}_2|$ becomes R^{-1} because of the spherical nature of $|\psi|^2$ in this nondistorted ground state.

We let w^r denote the reversible work required to bring the reactants together. It is the free energy of formation of an *unconstrained state* at the above R from an *unconstrained state* at $R = \infty$. This term is obtained from Eq. (7.44) by setting $m = 0$, as one may see by recalling the origin of m as a Lagrangian multiplier. (When m vanishes, so does the equation of constraint.)

$$\Delta F^* = w^r - m^2(2 - m^2)H_o + m^2\{\lambda_i + e^2c[(1/2a_2) - (1/R)]\},$$

where H_o is given by Eq. (7.38) and

$$w^r = -ee_2^r/D_sR.$$

Similarly one finds

$$\begin{aligned}\Delta F^{*p} = &-2(m+1)^2(1 - m^2)H_0 \\ &+ (m+1)^2\{\lambda_i + e^2c[(1/2a_2) - (1/R)]\}\text{''}\end{aligned}$$

Where obviously the work term w^p doesn't appear.

"And m is then the solution of Eq. (7.34a).

If m is small we again may set $m^2 \ll 1$ and write:

$$\Delta F^* = w^r + m^2\lambda, \tag{7.45}$$

$$\Delta F^{*p} = (m+1)^2\,\lambda, \tag{7.46}$$

where m is given by Eq. (7.47) and λ by Eq. (7.48).

$$-(2m+1)\lambda = \Delta F^{o\prime}_{\text{int}} - w^r \tag{7.47}$$

$$\lambda = -2H_o + \lambda_i + e^2[(2a_2)^{-1} - R^{-1}](D_{\text{op}}^{-1} - D_s^{-1}). \tag{7.48}$$

If we define λ_e by Eq. (7.49) and call it the λ *for the polaron* and write λ_2 as in Eq. (7.50), Eq. (7.51) follows:

$$\lambda_e = -2H_o - (e^2/2R)(D_{\text{op}}^{-1} - D_s^{-1}), \tag{7.49}$$

$$\lambda_2 = \lambda_i + \tfrac{1}{2}e^2(D_{\text{op}}^{-1} - D_s^{-1})(a_2^{-1} - R^{-1}) \tag{7.50}$$

$$\lambda = \lambda_e + \lambda_2 \tag{7.51}$$

This λ_2 appears as a characteristic parameter for the second reactant in ordinary chemical and electrochemical transfers. The R may vary somewhat from reaction to reaction but correction can be made for this variation.

A value for $\Delta F^{o\prime}_{\text{int}}$ is given in the next section."

NOTE: λ_2 may be better compared to λ_e if it is written as:

$$\lambda_2 = \lambda_i + \frac{e^2}{2a_2}(D_{\text{op}}^{-1} - D_{\text{s}}^{-1}) - (e^2/2R)(D_{\text{op}}^{-1} - D_{\text{s}}^{-1})$$

7.17. Standard Potential and $\Delta F^{o\prime}_{\text{int}}$

"From an estimate of the forward and reverse rate constants of reaction (7.52), Baxendale estimated the latter's equilibrium constant [45a].

$$e(\text{aq}) + \text{H}_2\text{O} = \text{H} + \text{OH}^-. \tag{7.52}$$

This constant was then combined with the autoprotolysis constant of water and with the equilibrium constant for $2H = H_2$ to yield a ΔF° of $-61.5\,\text{kcal mole}^{-1}$ for reaction (7.48) and, hence, an E° of $+2.7\,\text{V}$

$$e(\text{aq}) + \text{H}^+(\text{aq}) = \frac{1}{2}\text{H}_2. \tag{7.53}$$

A review of the argument employed reveals two errors [45b]. On making corrections for them and on using very recent data for the equilibrium constant of Eq. (7.52) then ΔF° for Eq. (7.53) is found to be $-62.7\,\text{kcal mole}^{-1}$ and the new E° to be $+2.7\,\text{eV}$.

To calculate F°_{transl} for the electron in reaction (7.52):

$$e(\text{aq}) + \text{Ox} \rightarrow \text{Red} \tag{7.21}$$

we may proceed as follows:

On the right side of the equilibrium, the excess electron *resides* on the reduced reactant and its motion is correspondingly *restricted*, while on the left-hand side this electron has three translational degrees of freedom. On recalling the definition of $\Delta F^{\circ\prime}_{\text{int}}$ and denoting the *translational partition function* of the solvated electron by $(pf)_{\text{transl}}$, the equilibrium constant of reaction (7.21), K, is given by Eq. (7.54) (we have canceled the translational partition functions of Red and Ox).

$$K = \exp(-\Delta F^{\circ\prime}_{\text{int}}/kT)(\text{pf})^{-1}_{\text{transl}}. \tag{7.54}$$

Any volume change is supposed to be included in $\Delta F^{\circ\prime}_{\text{int}}$ through electrostriction. That is, $\Delta F^{\circ\prime}_{\text{int}}$ is a 'Gibbs free energy' change.

The energy of the polaron, $H(\upsilon)$, calculated by solving the Schrödinger equation for a dynamic polaron moving with velocity $\tilde{o}$, is assumed to have the form [34]

$$H(\upsilon) = H_0 + \frac{1}{2}m_p\upsilon^2, \tag{7.55}$$

where m_p is the reduced mass for the polaron and H_o is the energy (or really, in part, free energy) of the *stationary* polaron.

We now obtain

$$(\mathrm{pf})_{\mathrm{transl}} = (2\pi m_p kT)^{\frac{3}{2}} V/\hbar^3, \tag{7.56}$$

where V is 1 cc when the units of K are molecules per cubic centimeter. On recalling the definition of $F^{\circ}_{\mathrm{transl}}$ by Eq. (7.34):

$$\Delta F^{0\prime}_{\mathrm{int}} = \Delta F^{0\prime} - F^{0}_{\mathrm{trasl}} \tag{7.34}$$

we obtain

$$\exp(-F^{\circ}_{\mathrm{transl}}/kT) = (\mathrm{pf})_{\mathrm{transl}}. \tag{7.57}$$

At present, the value for m_p is somewhat uncertain. It does not seem quite appropriate to use a value computed for a solid-state system [46]. On the other hand, the estimated *diffusion constant* [47] of $\sim 10 \times 10^4\,\mathrm{cm}^2\,\mathrm{sec}^{-1}$ for the polaron is substantially higher than that of a water molecule ($2.45 \times 10^4\,\mathrm{cm}^2\,\mathrm{sec}^{-1}$), indicating a considerable *freedom of motion*. Perhaps a value for m_p of the order of $\frac{1}{4}$ to 10 molecular weight units would be appropriate.

On using Eqs. (7.58) and (7.59), we obtain $-5.3\,\mathrm{kcal\,mole}^{-1}$ for $F^{\circ}_{\mathrm{transl}}$ when $m_p \sim 3$ and the standard state, in Eq. (7.59) and for K, is 1 mole liter^{-1}.

It is convenient to define an effective $E^{\circ\prime}_1$, $E^{\circ\prime}_{\mathrm{eff}}$, for the solvated electron by means of Eq. (7.60), since $\Delta F^{\circ\prime}_{\mathrm{int}}$ (and, hence, $E^{\circ\prime}_{\mathrm{eff}} - E^{\circ\prime}_2$) describes the effective driving force of the reaction, according to Eqs. (7.45) to (7.47).

$$\Delta F^{\circ\prime}_{\mathrm{int}} = -eF(E^{\circ\prime}_{\mathrm{eff}} - E^{\circ\prime}_2), \tag{7.60}$$

where $E^{\circ\prime}_2$ is the 'standard' reduction potential of the second reactant. Using the above value for $F^{\circ\prime}_{\mathrm{transl}}$ and Eq. (7.34), we find

$$E^{\circ\prime}_{\mathrm{eff}} = +2.9\,\mathrm{V}.$$

7.18. Activation and Diffusion Control

The observed rate constant of a reaction can be expressed in terms of D, the *sum of the diffusion coefficients of the reactants*, and in terms of the activation-controlled rate constant k_{act} as in Eq. (7.61) [6].

$$k_{obs}^{-1} = k_{act}^{-1} + k_{diff}^{-1} \tag{7.61}$$

where

$$k_{diff} = 4\pi D / \int_R^\infty e^{w/kT} \frac{dr}{r^2}$$

The activation-controlled constant k_{act} is the one calculated by Eq. (7.6). When $k_{act} \ll k_{diff}$, k_{obs} equals k_{act}, and when $k_{diff} \ll k_{act}$, k_{obs} equals k_{diff}."

In chemical kinetics, the important step is the slowest one, that is, the one with a lower rate constant...

"The diffusion-controlled constants for the solvated electron reactions are quite high, a result which is attributed to a high diffusion coefficient ($\sim 10^{-4}\,cm^2\,sec^{-1}$) for the solvated electron [47], *comparable to that of H_3O^+*. These rate constants ranged from $13 \times 10^{10} M^{-1} \cdot sec^{-1}$for reactions with a large, positively charged cation [$Cr(en)_3^{3+}$], where en is ethylene diamine, through 2.0×10^{10} for a neutral species (O_2) to 0.3×10^{10} for a reaction with a negative species [$Fe(CN)_6^{3-}$] [30]."

The reaction is faster with a positive ion, less so with a neutral reagent and less still with a negative ion.

7.19. Calculations of k_{act}

"For water as solvent at room temperature $ce^2/2R$ is $90/R$ kcal mole^{-1}, with R in angstroms."

Recall that $c = D_{op}^{-1} - D_s^{-1}$ (7.28) and that

$$\lambda_e = -2H_o - (e^2/2R)(D_{op}^{-1} - D_s^{-1}) \tag{7.49}$$

"Thus, if R is about 6 Å, this *contribution to* λ_e in Eq. (7.49) is about 15 kcal mole^{-1}. [Such a choice of R is a possible one since the radius for e(aq), it has been suggested [48a], is about 2.5–3.0 Å, while the radius of a small hydrated cation is typically about 3.5 Å.] However, the value for H_o itself is presently uncertain, as we saw earlier, and the value for λ_e is tentatively taken as 15 kcal mole^{-1} to explain the Sm^{3+} data described later. [This λ_e corresponds to an H_o of -15 kcal mole^{-1}, according to Eq. (7.49).]"

NOTE: $H_0 = -15 \Rightarrow -2H_0 = 30 \Rightarrow 30 - 15 = 15$ for λ_e.

"A lower bound to λ_e appears to be about 6 kcal mole^{-1} [48b]. Calculations based on a λ_e of 45 kcal mole^{-1} are also given for comparison.

The values of λ_2 may be estimated from electrochemical or chemical ET rates, though in some cases the magnitude of the correction of the rate for a work term is somewhat uncertain. According to a recent tabulation [49], we find λ_2 to be about 35, 40, and 60 kcal mole^{-1} for a single ion in the $Fe^{2+,3+}$, $Tl^{3+,2+}$, and $Co(NH_3)_6^{2+,3+}$ systems, for example. (λ_2 values for many other ions are available.) On recalling that the E_2°'s for these reactants are -0.77, $\sim +0.34$, and $\sim -0.1\,V$ one then finds from Eq. (7.47)

$$-(2m+1)\lambda = \Delta F_{\text{int}}^{o\prime} - w^r \tag{7.47}$$

that the calculated ΔF^*'s are small, regardless of whether λ_e is taken to be 15–20 or 45 kcal mole^{-1}: For a λ_e of 15 kcal mole^{-1}, the calculated ΔF^*'s are about 1, -2, and -1.5 kcal mole^{-1}, respectively, and for a λ_e of 45 kcal mole^{-1} they are about -2, -1, and $+2.5$ kcal mole^{-1}. The $Co(NH_3)_6^{3+}$ reaction has been investigated and is diffusion controlled.

We consider next a hydrated cation with a λ_2 of about 40 kcal mole^{-1} and inquire as to how positive its standard reduction E_2° potential should be in order that the reaction no longer be diffusion controlled. When its E_2° is about 2.0 V, $m^2\lambda$ is estimated to be about 6 or 12 kcal mole^{-1} according as λ_e is 15–20 or 45 kcal mole^{-1}, neglecting w^r."

NOTE: Recall $\Delta F^* = w^r + m^2\lambda$. (7.45)

"Even with a favorable w^r of several kilocalories per mole, this reaction should be activation controlled. Similar remarks apply to an E_2^0 of 1.5, except that the reaction should be partially in the diffusion-controlled region if the lower λ_e is used. Interesting from this point of view is a recent study of reactions with rare earths [50]. E°'s ($M^{3+,2+}$) for Eu^{3+}, Yb^{3+}, and Sm^{3+} are known. They are about +0.43, +1.15, and +1.55 V, respectively [51]. The corresponding E°'s for the others are not known but are probably higher since these other rare earths do not form stable divalent states."

For modern values of standard potentials, see the book of Bard, Parsons, and Jordan on standard potentials. It is, in particular, −0.35 for $Eu^{3+} + e^- = Eu^{2+}$. In the European convention, the numbers would be negative.

"Reactions for the first two and perhaps for the third appear to be diffusion controlled ($k = 6.1, 4.3$, and 2.5×10^{10} M^{-1} sec^{-1}, respectively). The rate constants of the other rare earths are definitely activation controlled, the rate constant for Tm^{3+} being 3×10^9 M^{-1} sec^{-1}, for example.

Judging from an electrochemical rate constant, the value of λ_2 for $Eu^{2+,3+}$ is about 40 kcal mole^{-1} [49]. In the cases of Eu^{3+} and Yb^{3+}, the theoretical rate is so high that the reaction is predicted to be diffusion controlled for a λ_e of 15–20 kcal mole^{-1}. (For Yb^{+3} ΔF^* is 4 kcal mole^{-1} if λ_e is 45 kcal mole^{-1}.) In the case of Sm^{3+}, ΔF^* is estimated to be about 1, 1.5, or 7 kcal mole^{-1} according as λ_e is taken to be 15, 20, or 45 kcal mole^{-1}. (For w^r a Coulombic term, $-3e^2/D_s R$, is assumed.) If k_{diff} in Eq. (7.61) is taken to be

$6.1 \times 10^{10}\,M^{-1}\,sec^{-1}$ (the value for Eu^{3+}) then the experimental k_{act} is estimated to be $4 \times 10^{10}\,M^{-1} \cdot sec^{-1}$ for Sm^{3+}, which corresponds to a ΔF^* of about 0.5 kcal mole^{-1}."

NOTE: From the knowledge of k_{diff} and of k_{rate}, it is possible to know $k_{act} \Rightarrow \Delta F^*$.

"Thus, a value for λ_e of around 15 kcal mole^{-1} seems slightly favored at this time, but the various uncertainties in the λ's and in the E°_{eff} for the solvated electron must be borne in mind in such comparisons with the data. In turn, further kinetic and equilibrium data involving the solvated electron and further kinetic data on ordinary chemical and electrochemical exchanges should provide more precise E°_{eff}, λ_e, and λ_2's. An important region for such investigations would be one with E°_2's in the vicinity of 1.5–2.0 V."

This range is justified because for $E^0 > 2$ the reactions are diffusion controlled.

NOTE how M. suggests new experiments after a careful theoretical analysis.

"These calculations are further discussed in Ref. [32].

7.20. Possibility of Chemiluminescence

With a strong reducing agent such as the solvated electron, $\Delta F^{\circ\prime}_{int}$ can be extremely negative.

For example, when $E^{\circ\prime}_2$ is as negative as -1.9 V, $\Delta F^{\circ\prime}_{int}$ is -110 kcal mole^{-1}. In such cases, there is a possibility of formation of the product in an excited state, and then chemiluminescence becomes possible. The possibility of forming an excited state is enhanced when reaction leading to the ground state of the product is made less favorable. In principle such a circumstance can arise when the reorganization parameter λ is made sufficiently small by suitable choice of reactant: When $|\Delta F^{\circ\prime}_{int}/\lambda|$ is large (somewhat greater than unity) the *conventional type of crossing* of the two potential surfaces becomes difficult. The surfaces no longer cross at configurations which are *compromises* between reactants and products, and then the value of the *reorganizational barrier* $m^2\lambda$ actually increases as $\Delta F^{\circ\prime}_{\text{int}}$ becomes more negative [6].

For example, when λ_2 is 20 or 40 kcal mole^{-1} and $\Delta F^{\circ\prime}_{\text{int}}$ is -110 kcal mole^{-1}, $m^2\lambda$ is about 35 or 15 kcal mole^{-1}, respectively, if λ_e is 15 kcal mole^{-1}. It is still large even if λ_e is 45 kcal mole^{-1}, if λ_2 is 20. Then, $m^2\lambda$ equals 8 kcal mole^{-1}. If there is a *readily accessible* excited state of the product *available*, the product might then be formed in the excited state and, under favorable conditions, fluoresce. The rate of reaction (7.21) to form the product in an excited state, Red*, (**Erratum** the original symbol Ox*) is again given by Eq. (7.6), but $\Delta F^{\circ\prime}_{\text{int}}$ is now the associated free energy change for the reaction leading to Red* (**Erratum** corrected). Other things being equal, low λ_2's are favored by ligands such as $o-$phenantroline and bipyridyl or by having an aromatic molecule as one of the reactants. For example, λ_2 for an iron o-phenantroline ion appears to be about 15–20 kcal mole^{-1}, or less [52]."

From $\Delta F^* = w^r + m^2\lambda$ neglecting w^r one has $\Delta F^* = m^2\lambda$ and from $m = -\frac{1}{2}\left(1 + \frac{\Delta F^{\circ\prime}_{\text{int}} - w}{\lambda}\right)$ knowing $\Delta F^{\circ\prime}_{\text{int}}$ and $\lambda \Rightarrow \Delta F^*$.

7.21. Use of Alternative Models for the Polaron

"Comparison of Eqs.

$$\Delta F^* = m^2(-2H_o + \lambda_2^{\infty}) \quad (7.4) \quad \text{and} \quad \Delta F^* = w^r + m^2\lambda, \quad (7.45)$$

for the values of ΔF^* at $R = \infty$ and at finite R reveals one feature identical with that in the equations derived earlier [5–8] for conventional ET reactions: The expression for ΔF^* has the form of m^2 times an *intrinsic term* (the value of λ at $R = \infty$) plus m^2 times a term e^2c/R."

NOTE: Consider the Eqs. (7.45) and (7.48). One sees that λ is made up from the intrinsic term $-2H_0 + \lambda_i + \frac{e^2c}{2a_2}$ plus the term $-\frac{e^2c}{R}$.

"As the derivation for the conventional transfers reveals, this particular c is a *property of the medium outside* of the region occupied by the two reactants, regardless of what value one may eventually use for c appearing in the *expressions for* H_o and for λ at $R = \infty$.

Accordingly, use of some other model for the polaron is expected to lead to Eqs. (7.45) and

$$\lambda = -2H_o + \lambda_i + e^2\left[(2a_2)^{-1} - R^{-1}\right]\left(D_{\text{op}}^{-1} - D_s^{-1}\right) \quad (7.48)$$

for ΔF^*, but with a somewhat different expression for part of λ, namely the $-2H_o$ term. That is, we have

$$\lambda_e = -2H_o - (e^2/2R)(D_{\text{op}}^{-1} - D_s^{-1}) \quad (7.49)$$

$$\text{and} \quad \lambda_2 = \lambda_i + \frac{1}{2}e^2(D_{\text{op}}^{-1} - D_s^{-1})(a_2^{-1} - R^{-1}) \quad (7.50)$$

as before but with a value of H_o which may differ somewhat from

$$H_o = -A\mu e^4 c^2/\hbar^2 \quad (7.38)$$

Further analysis of the *dynamical polaron* may well again lead to an expression of the type

$$H(\upsilon) = H_0 + \frac{1}{2}m_p\upsilon^2, \quad (7.55)$$

in the first approximation but with a value for the effective mass of the polaron derived for the behavior of polarons in polar liquids. The exact value for m_p can also affect somewhat the preexponential factor ρ in Eq. (7.6). This factor is a ratio of the root-mean-square fluctuation of R to that of the *perpendicular displacements* from the intersection region in the centered distribution, it is recalled. A small value of m_p may lead to a somewhat larger $[(\delta R)^2]^{\frac{1}{2}}$ although this could be partially offset by a larger $[(\delta s)^2]^{\frac{1}{2}}$. The exact value of ρ depends on the nature of the typical motion along the reaction coordinate [7].

7.22. Further Discussion

In the derivation of expression (7.26) for ΔF^*_{vib}:

$$\Delta F^*_{\text{vib}} = \sum_{i,j} k_{i,j}(q^i_* - q^i_r)(q^j_* - q^j_r)$$

the vibrations of the reactant were treated as harmonic oscillators. In these strongly negative $\Delta F^{\circ\prime}_{\text{int}}$ systems the activated complex resembles the reactants much more than the products, so that the second reactant may have a *vibrational configuration* considerably strained from that which it will later have as a product. In such cases, the harmonic oscillator approximation may be less accurate than it normally is. One can readily avoid the approximation, though the resulting expressions become more complex, as in Ref. [6]. Alternatively, some estimate of the error can be made *a posteriori* by comparing a harmonic term,

$$\frac{1}{2}\sum k^p_{ij}(q^i_* - q^i_p)(q^j_* - q^j_p),\text{''}$$

NOTE: For the calculation of $\Delta F^{*p}_{\text{vib}}$.

"With the corresponding change of vibrational potential energy for the actual product molecule when the q^i are stretched from q^i_p to q^i_*.

The factor

$$Z\exp(-F^{\circ}_{\text{transl}}/kT)$$

appearing in the *unimolecular* rate constant

$$k_{-1} = Z\kappa\rho\exp(-F^{\circ}_{\text{transl}}/kT)\exp(-\Delta F^{*p}/kT) \tag{7.35}$$

is of some interest. If, for the sake of simplicity, the *reaction coordinate* is supposed to be one purely of *relative motion* of the polaron and the second reactant, then because m_p is small Z becomes effectively

$$Z = (8\pi kT/m_p)^{\frac{1}{2}}R^2$$

and

$$Z\rho\exp(-F^{\circ}_{\text{transl}}/kT) \Rightarrow (kT/h)(8\pi IkT/h^2)\rho$$

where

$$I = m_p R^2$$

and $\rho = 1$. This factor in the unimolecular rate constant becomes the usual $10^{13}\,\text{sec}^{-1}$ factor multiplied by a rotational partition function. To be sure, if $m_p R^2$ were small this rotational partition function would be replaced by its quantum value and a corresponding correction would be made in Eq. (7.35): the latter would be multiplied by be replaced by its quantum value and a corresponding correction would be made in Eq. (7.35): the latter would be multiplied by

$$\sum_{j=0}^{\infty}(2j+1)\exp(-J^2\beta)\beta$$

where $\beta = \hbar^2/2IkT$.

However, R^2 is very large.

Finally, it may be noted, the value of R in Eq. (7.48) for λ:

$$\lambda = -2H_o + \lambda_i + e^2[(2a_2)^{-1} - R^{-1}](D_{\text{op}}^{-1} - D_s^{-1})\text{''}$$

Note that λ depends on R.

"Finally, it may be noted, the value of R in Eq. (7.48) is by no means established and, in the case of solvated electron reactions, might depend on the reaction. When

$$\Delta F_{\text{int}}^{\circ\prime}/\lambda$$

or, more precisely, when

$$-(\Delta F_{\text{int}}^{\circ\prime} - w^r)/\lambda$$

becomes more positive than 1, ΔF^* actually increases as $-\Delta F_{\text{int}}^{\circ\prime}$ increases [Eqs. (7.45) and (7.47)]. This situation occurs only at very negative $\Delta F_{\text{int}}^{\circ\prime}$'s. In such a case, the most favorable R might be that which makes λ a little larger, namely by being larger itself (Eq. [7.48]) and hence makes $|\Delta F_{\text{int}}^{\circ\prime}/\lambda|$ a little smaller. However, too large an R would make κ too small in Eq. (7.6). Again, it should be emphasized, we have considered only weak overlap ETs. We have not permitted the solvated electron wave function to overlap the orbital of the second reactant appreciably, partly because the resulting *desolvation* might not be economical and partly because the theoretical calculations become more involved. The present work is intended to be a first approximation for comparing with and interpreting the experimental data.

An alternative AT mechanism for some reactions is discussed in Ref. [32]."

Author's Note: M.'s theory is a first approximation in the same sense in which $PV = nRT$ is an approximation.

7.23. Electron Polarization of the Solvent

"An aspect of the theory of the *unconstrained* polaron of particular interest involves the extent of *electron polarization of the solvent*. Purely from the viewpoint of *the frequency of motion of the solvated electron alone* (about $4 \times 10^{14}\ \text{sec}^{-1}$) [33] there would appear to be on the surface no difficulty in the solvent electrons following the motion of the solvated electron: When the frequency of a light wave is 1.5, 3, 4, 5, and $13 \times 10^{14}\,\text{sec}^{-1}$, the refractive index of water at 25°C is quite high [53]: $n = 1.30, 1.32, 1.33, 1.33$, and 1.38 respectively. ($D_{\text{op}} = n^2$). In fact, refractive index dispersion data for typical solvents can be interpreted [41] by regarding the *valence and inner electrons of the solvent as having a mean frequency of about* $3 \times 10^{15}\,\text{sec}^{-1}$, which is much larger than that of the solvated electron. Their corresponding angular frequency ω is 2π times this value.

However, an *exclusion* of solvent electron polarization nevertheless must occur in the *immediate vicinity* of the solvated electron. This *particular exclusion* has not explicitly been discussed in the literature from the viewpoint of continuum theory, but can be treated by applying to the electron polarization analogous arguments [55] made in the literature for high-frequency lattice polarization. We consider first a system free of solvent orientation and vibrational polarization and consider the qualitative behavior and then in Eq. (7.62)

$$\mathscr{E} = -e^2/2d[1 - (1/D_{\text{op}})] \tag{7.62}$$

the quantitative result.

Electrons of the solvent which are too close to the solvated electron cannot respond instantaneously to its motion no matter how high their natural frequency ω, namely electrons within a distance [56] $d \cong \upsilon/\omega$, where υ is the velocity of the solvated electron."

This d is then the distance covered by the solvated electron while the electron of the electronically polarized solvent molecule performs an entire oscillation.

"This d is small when υ is small. Yet, υ cannot be very small and at the same time the electron be localized with a distance d less than $\sim \hbar/\mu\upsilon$ according to the uncertainty principle. Thus, we can at best have $\upsilon/\omega \cong \hbar/\mu\upsilon$ as a condition defining a distance outside of which the solvent electrons are polarized by the solvated electron. That is, $\upsilon \cong (\hbar\omega/\mu)^{\frac{1}{2}}$ and, hence $d \cong (\hbar/\omega\mu)^{\frac{1}{2}}$."

*Demonstration.

The last Equation is from $\Delta x \Delta p \approx \hbar \Rightarrow d \cdot \mu\upsilon \cong \hbar \Longrightarrow \frac{\upsilon}{\omega} \cdot \mu\upsilon \cong \hbar \Longrightarrow \frac{\upsilon^2}{\omega} \cong \frac{\hbar}{\mu} \Longrightarrow \frac{\upsilon^2}{\omega^2} \cong \frac{\hbar}{\omega\mu} \Longrightarrow d \cong \left(\frac{\hbar}{\omega\mu}\right)^{\frac{1}{2}}$

"The *energy of polarization* is then of the order of $-(e^2/2d)(1 - 1/D_{\text{op}})$ (Born formula).

A quantitative solution of the Schrödinger equation for the motion of the solvated electron and of a polarizable continuum of angular frequency ω leads in fact to Eq. (7.62) for the *energy* in this *high-frequency limit*:

$$\mathscr{E} = -e^2/2d[1 - (1/D_{\text{op}})] \tag{7.62}$$

where

$$d \cong (\hbar/\omega\mu)^{\frac{1}{2}}. \tag{7.63}$$

This result can be derived from that given by Allcock [43, 57] by noting that in the present case of zero orientation polarization and a frequency ω for motion of the solvent electrons, the formalism for the problem is identical with the one given there in the high-frequency approximation but that the D_s and D_{op} appearing there should now be replaced by D_{op} and unity, respectively."

Compare the preceding formula (7.62) with

$$\lambda = -2H_o + \lambda_i + e^2[(2a_2)^{-1} - R^{-1}](D_{\text{op}}^{-1} - D_s^{-1}). \tag{7.48}$$

"When $\omega/2\pi$ has the value cited earlier of 3×10^{15} sec^{-1}, d equals 0.55 Å. Related remarks concerning an exclusion sphere of solvent electron polarization should apply in systems containing orientation and vibrational polarization. The size of this *exclusion sphere* in either case is relatively small: The radius of the $1s$ polaron orbit [43] is about 1.80 Å $(16\hbar^2/5ce^2\mu)$ so that the circumference of the orbit is about 11 Å or about 10 times the diameter of the exclusion sphere. Thus, one is inclined to suspect that this exclusion sphere, which occurs both in orientation free and in orientation polarization systems, has relatively little effect on H_o, which is the *free energy of formation* of the *orientation polarization system* from the *orientation-free one.*

With the framework of this continuum approximation this problem could of course be investigated precisely: A Schrödinger equation could be set up for this dynamical motion of the solvent electrons, the solvated electron, and the lattice polarization. The lattice polarization would be treated in the low-frequency approximation and the electron polarization in the high-frequency approximation. The foregoing arguments suggest, however, that the net change from the simple polaron theory calculation of H_o is small, although a molecular treatment of the electron polarization might yield a somewhat different conclusion. Since d is smaller than the 'lattice' distance, the earlier continuum estimate cannot be an accurate one.

7.24. *Appendix I. Derivation of Eq. (7.27) for $F_{\text{pol}}(r_1)$

It is convenient to decompose the total polarization at $\mathbf{r}$, $\mathbf{P}_{\text{tot}}(\mathbf{r})$, into the sum of two parts [37]:

$$\mathbf{P}_{\text{tot}}(\mathbf{r}) = \mathbf{P}_e + \mathbf{P}_u$$

where

$$\mathbf{P}_e(\mathbf{r}) = (D_{\text{op}} - 1)\mathbf{E}(\mathbf{r})/4\pi\,,\text{''}$$

Compare [39a] p. 452: $k = n^2$ and p. 454: $\mathbf{P} = \frac{k-1}{4\pi}\mathbf{E}$.

"And $\mathbf{E}(\mathbf{r})$ is the electric field at $\mathbf{r}$. If the orientation polarization is held fixed and a charge is changed, the change in polarization is in fact $(D_{\text{op}} - 1)/4\pi$ times the change in $\mathbf{E}(\mathbf{r})$. Thus, the change in $\mathbf{P}_e(\mathbf{r})$ describes the *response* at *fixed orientation polarization.* Therefore, $\mathbf{P}_u(\mathbf{r})$ is some function of the orientation polarization:

It is automatically held fixed when the latter is held fixed, since the change in **P** is fully accounted for by that in $\mathbf{P}_e$.

If one neglects dielectric image effects it can be shown that $\mathbf{E}(\mathbf{r})$ depends on the field directly due to the charges, $\mathbf{D}(\mathbf{r})$, *and* on $\mathbf{P}_u(\mathbf{r})$ according to Eq. (7.64):

$$\mathbf{E} = (\mathbf{D} - 4\pi\mathbf{P}_u)/D_{\text{op}} \tag{7.64}$$

In this case, the reversible work required *to charge* the system to any given nonequilibrium state described by functions $\mathbf{D}(\mathbf{r})$ and $\mathbf{P}_u(\mathbf{r})$ is [37]

$$W_{rev} = -\frac{1}{8\pi}\left(1 - \frac{1}{D_{\text{op}}}\right)\int \mathbf{D}^2 d\mathbf{r} - \int \mathbf{P}\cdot\mathbf{D}d\mathbf{r} + 2\pi c\int \mathbf{P}^2 d\mathbf{r}, \tag{7.65}$$

where **P** denotes $\mathbf{P}_u(\mathbf{r})/D_{\text{op}}$ and is again fixed once the orientation polarization is specified. The term $F_{\text{pol}}(\mathbf{r}_1)$ is obtained by subtracting from Eq. (7.65) its value, $W_{rev}(\infty, eq)$, when the reactants are far apart and when the orientation polarization vanishes. Thus

$$W_{\text{rev}}(\infty, \text{eq}) = -\frac{1}{8\pi}\left(1 - \frac{1}{D_{\text{op}}}\right)\int \mathbf{D}^2 d\mathbf{r} \tag{7.66}$$

calculated at $R = \infty$. In this manner one obtains Eq. (7.27)

$$F_{\text{pol}}(\mathbf{r}_1) = -\int \mathbf{P}\cdot\mathbf{D}d\mathbf{r} + \frac{2\pi}{c}\int \mathbf{P}^2 d\mathbf{r} - \frac{e_2^2}{2a_2}\left(1 - \frac{1}{D_{\text{op}}}\right) + \frac{e_1 e_2}{D_{\text{op}}|\mathbf{r}_1 - \mathbf{r}_2|} \tag{7.27}$$

for F_{pol}. We again neglect dielectric image effects in evaluating integrals. They are small [5].

7.25. *Appendix II. Estimate of H_o and Electron Affinity from Solvation and Spectral Data

A rough estimate of these quantities can be attempted from spectral and solvation data as follows. Noyes [58] has suggested a value of 108.3 kcal mole^{-1} for the ΔF° of the following hypothetical process:

$$\frac{1}{2}H_2(g) \rightarrow H^+(aq) + [e^-](aq), \tag{7.67}$$

where $[e^-]$(aq) denotes a *hypothetical electron* at the bulk electrostatic potential of the solvent, and having zero entropy and zero energy. For a standard state of 1 M for H_2 this ΔF° becomes 102.9 kcal mole^{-1}. Subtraction of this value from that for the actual process (7.53)

$$e(aq) + H^+(aq) = \frac{1}{2}H_2 \tag{7.53}$$

yields a value of 40.2 kcal mole^{-1} for the free energy of a solvated electron."

Summing Eqs. (7.67) and (7.53) one obtains:

$$e(aq) \rightarrow [e^-](aq) \Rightarrow [e^-](aq) \rightarrow e(aq).$$

"If the translational free energy of this solvated electron is about −5.3 kcal mole^{-1}, as discussed earlier, the solvation portion is −34.9 kcal mole^{-1}. The difference of this value and the spectral transition energy (1.72 eV) is about 4.7 kcal mole^{-1}. It is independent of the electron affinity of the electron in the ground state and equals 0.28 H_o when the electron affinity of the excited state can be neglected. In that case, H_o would be about 0.7 eV, which is close to the value estimated from rate data in the text, and the electron affinity of the ground state would be about 0.8 eV. However, the considerable uncertainties in these values should be emphasized."

A Glossary of Principal Symbols follows.

where $[e^-](aq)$ denotes a hypothetical electron at the bulk electrostatic potential of the solvent, and having zero entropy and zero energy. For a standard state of $1M$ for H_2, this ΔF° becomes 102.9 kcal mole^{-1}. Subtraction of this value from that for the actual process (54) yields a value of 40.2 kcal mole^{-1} for the free energy of a solvated electron. If the translational free energy of this solvated electron is about -5.3 kcal mole^{-1}, as discussed earlier, the solvation portion is -34.9 kcal mole^{-1}. The difference of this value and the spectral transition energy (1.72 eV) is about 4.7 kcal mole^{-1}. It is independent of the electron affinity of the electron in the ground state and equals 0.28 H_o when the electron affinity of the excited state can be neglected. In that case, H_o would be about 0.7 eV, which is close to the value estimated from rate data in the text, and the electron affinity of the ground state would be about 0.8 eV. However, the considerable uncertainties in these values should be emphasized.

APPENDIX III. GLOSSARY OF PRINCIPAL SYMBOLS

$F^{*r}(R)$	Free energy of system when reactants are fixed in position a distance R apart and are in the distribution centered on the intersection region
$F^{*p}(R)$	Corresponding term for product fixed in position and in the above distribution centered on intersection region
$F^r(\infty)$	Free energy of system when reactants are fixed in position far apart and are otherwise unconstrained
$F^p(\infty)$	Free energy of system when product is fixed in position and is otherwise unconstrained
ΔF^*	$F^{*r}(R)-F^r(\infty)$
ΔF^{*p}	$F^{*p}(R)-F_p(\infty)$
w^r	Work required to bring reactants together, a contribution to ΔF^*
Z	Collision frequency calculated for the above R and for uncharged and otherwise noninteracting species (charge effects, etc., are included in w^r)
q^i	ith vibrational coordinate of reactants (and products), whose equilibrium value in reactant is q_r^i, in product is q_p^i, and whose most probable value in the centered distribution is q_*^i
ΔF^*_{vib}	Vibrational contribution to ΔF^*
k_{ij}^r, k_{ij}^p	Force constant of ij cross term in vibrational potential energy of reactant and of product, respectively [reduced force constant k_{ij} is $2k_{ij}^r k_{ij}^p/(k_{ij}^r+k_{ij}^p)$]
ψ, ψ_o	Wavefunction of electron when the reactants are in the centered distribution and when they are unconstrained plus far apart, respectively
$F_{\text{pol}}(\mathbf{r}_1)$	Polarization free energy as a function of the electron position $\mathbf{r}_1$
$\mathbf{P}(\mathbf{r})$	A particular function of the orientation polarization of the medium
a_2	Radius of second reactant, including any inner coordination shell
e_1, e_2	Charge of electron and of second reacting species, respectively (superscript r indicates reactants; p, products; hence, $e_1^p=0$, $e_1^r=-e$)
D_{op}, D_s	Optical and static dielectric constants of the medium, respectively
c	$(D_s-D_{\text{op}})/D_sD_{\text{op}}$
$\mathbf{D}_1$, $\mathbf{D}_2$	Field due directly to the electron at $\mathbf{r}_1$ and to the second reactant at $\mathbf{r}_2$, respectively [Eq. (11)] $\mathbf{D}=\mathbf{D}_1+\mathbf{D}_2$
$\mathfrak{D}_i^r$	Value of $\mathbf{D}_i$, averaged over $\lvert\psi\rvert^2$ [Eq. (20)] $\mathfrak{D}^r=\mathfrak{D}_1^r+\mathfrak{D}_2^r$
$\mathfrak{F}_o^r$	Sum of electron kinetic energy and of polarization free energy of system of separated reactants
$H_o\equiv\mathfrak{F}_{o_1}^r$	$\mathfrak{F}_o^r$ minus contribution of second reactant
$\mathfrak{F}_o^p$	Polarization free energy of system containing the product
μ, m_p	Effective mass of electron and of polaron, respectively
m	A Lagrangian multiplier, determined by Eq. (23)
$\Delta F^{o\prime}$, K	"Standard" free energy of Reaction (1) at prevailing temperature and for prevailing medium, K being the corresponding equilibrium constant
F°_{transl}, $(\text{pf})_{\text{transl}}$	Translational free energy and translational partition function of the solvated electron, calculated for a state of unit concentration
$\Delta F^{o\prime}_{\text{int}}$	The internal contribution to $\Delta F^{o\prime}$ [Eq. (22)]

Symbol	Meaning
λ_i	A factor in ΔF^*_{vib}, $\Delta F^*_{vib} = m^2\lambda_i$
λ_o^{m}	An intrinsic orientation polarization reorganization term for second reactant when the reactants are far apart [Eq. (28) ff.]
λ_t^{m}	$\lambda_i + \lambda_o^{m}$
H	The kinetic energy of the electron plus the contribution of the electron and its environment to the polarization free energy in the centered distribution when the reactants are far apart
λ	An intrinsic reorganization term [Eq. (46)]
λ_e, λ_2	"Contribution" of the electron and of the second reactant to λ [Eq. (49) ff.]
$E_2^{0\prime}$	"Standard" reduction potential of second reactant
$E^{0\prime}_{eff}$	Effective standard reduction potential of solvated electron (~2.9 V)
k_{diff}	Diffusion controlled rate constant, Eq. (61)
k_{act}	Activated controlled rate constant, Eq. (3)
k_{obs}	Observed rate constant, Eq. (60)
ω	Classical angular frequency of solvated electron in its orbit.

NOTES

M61, Theory of ET Reactions and of Related Phenomena. *Proceedings of Symposium on Exchange Reactions*, p.1, Atomic Energy Agency (1965).

1. Marcus uses initially the term "activated complex," later he uses "transition state."
2. Q: What does "prevailing" mean?

   M: It just does mean the conditions considered in the experiment.
3. Q: Does the above mean that in electrode reactions $\Delta S^* = 0$? Or is it very small?

   M: Only as far as the work terms are concerned.
4. Q: Which ones?

   M: Other effects related for instance to bond lengths. The ligand field theory is limited to considering t and e orbitals, and so on.
5. Q: Is it somehow like the difference between $\Delta G^{\neq}$ and ΔG^*?

   M: Yes, the same with the translational partition function...

M55, Theoretical Study of ET Reactions of Solvated Electrons. *Adv. Chem. Ser.* **50**, 138, (1965).

6. **M**: I was induced to turn to statistical mechanics in order to make the theory more general. So, this is the only reason for refinement.
7. Q: Where are these charge effects on collision frequencies dealt with?

   M: In the work terms.
8. **M**: There is amplification because of the exponential.
9. Q: The part of a molecule which determines λ will be the part more directly involved in ET, I believe. But is there a well-defined boundary between that part and the part which determines $\Delta F°$?

   M: No sharp boundary. As a matter of fact the constant is to be considered as a first approximation in the absence of other information. See papers of Ed. S. Lewis of Rice University on transfer of methyl between organic molecules.
10. **M**: A transition state is always constrained, by definition.... In ρ there is also a correction for going from the N dimensions to $N - 1$ dimensions. There is not a single separation distance for ET to happen, that would be too much to expect, so there is a range of separation distances. ρ contains that range of distances plus the correction from N dimensions to $N - 1$ dimensions.
11. Q: Please explain the difference between the system *on* the hypersurface and the one *centered* on the hypersurface. The system centered on the hypersurface must extend beyond it, toward the R and P parts of the PES. There must be some kind of distribution, a Gaussian distribution centered on the hypersurface, say?

    M: OK for the system extending toward R and toward P. OK also for the Gaussian distribution. If you have transition state theory, you don't have just a distribution *centered* on it but a distribution that *is* on it (the hypersurface).

M57, On the Theory of Chemiluminescent ET Reactions. *J. Chem. Phys.* **43**, 1598, (1970).

12. Q: Why "intersects" in quotes?

    M: It means readily accessible energetically. If there is no intersection, like in "touching," if the distance between R and P is considerable, because of strong coupling, one should consider the tunneling.
13. Q: Equation (2) is $\Delta F^* = w^r + (\lambda/4)[1 + (\Delta F_R^{\circ\prime}/\lambda)]^2$. From $\Delta F^{\circ\prime}/\lambda = -4.4$, we see that $1 + \Delta F^{\circ\prime}/\lambda = -3.4$. The effect of squaring this number is then only that of making it positive. There must be an error.

    M: There is. It should be $(3.4)^2$.
14. Q: Did anybody ever compute such PESs for electrodes and solutions?

    M: Levich and Dogonadze were the first to introduce the Fermi–Dirac distribution within this topic. See the 1965 paper. They integrate over the many levels. Because of the many excited states available there is no inverted effect with the metal electrodes. The story is different with semiconductor electrodes.
15. **M**: The concept of the excited state of an electrode is simple. When an electron occupies (goes into), a higher metal orbital one has an excited state of an electrode. There you have a continuum of levels, and so no inverted effect. It may be different with semiconductors because there, there may be narrow bands.

M60, Theory of ET Reaction Rates of Solvated Electrons. *J. Chem. Phys.* **43**, 3477, (1965).

16. **M**: They are, I believe, the precursor and successor complexes.
17. **M**: In the case of activated complex theory, there is the hypersurface which *is* the activated complex. Here we have polarization and all N degrees of freedom are involved in it instead of only $N-1$. We have then a distribution of which we can geometrically consider a center. But this is a fictitious set of configurations whereas in the case of the activated complex the configurations on the hypersurface are *real even if fleeting*. In a

sense it is fictitious because the system passes instantaneously through it. Note that using macroscopic variables does not tell anything about dimensionality. If there is this centered distribution dividing it by some width... we have a density of states along the reaction coordinate... it is like a concentration...

18. Q: Can we say that the second term is the kinetic energy of the solvated electron and the third is its potential energy in the polarization field?

    M: Yes. Here we have quantum mechanics for the kinetic energy and quantum mechanics + statistical mechanics for the potential energy, which is really a polarization free energy taking in consideration also same entropy.

19. Q: It looks like there are two sign errors in the last lines. They should probably read: "If we subtract from $\Delta F^{\circ\prime}$ this translational contribution, $F^{\circ}_{\mathrm{transl}}$, and note that the rotational contribution of Ox is the same as that of Red, we obtain the internal contribution, $\Delta F^{\circ\prime}_{\mathrm{int}}$. The latter is the free energy of formation of products from reactants when all reacting particles are fixed in position:

$$\Delta F^{\circ\prime}_{\mathrm{int}} = \Delta F^{\circ\prime} - F^{\circ}_{\mathrm{transl}}\text{"} \tag{7.68}$$

    Moreover, can we put it this way: If we consider the $\Delta F^{\circ\prime}$ of the reaction $e(\mathrm{aq}) + \mathrm{Ox} \rightarrow \mathrm{Red}$ we can write it as $\Delta F^{\circ\prime} = F^{\circ\prime}(\mathrm{Red}) - F^{\circ\prime}(\mathrm{Ox}) - F^{\circ\prime}(e(\mathrm{aq}))$. The initial potential energy of the solvated electron transforms itself in kinetic energy, the solvated electron reacts, disappears, and if we put $F^{\circ\prime}(e(\mathrm{aq})) = F^{\circ}_{\mathrm{transl}}$ and $\Delta F^{\circ\prime}_{\mathrm{int}} = F^{\circ\prime}(\mathrm{Red}) - F^{\circ\prime}(\mathrm{Ox})$ we recover Eq. (7.68). Is it OK?

    M: OK and we have two sign errors above.

References

1. R. A. Marcus, *Proceedings of Symposium on Exchange Reactions*, p. 1, Atomic Energy Agency (1965).

2. R. A. Marcus, *Adv. Chem. Ser.* **50**, 138, (1965).
3. R. A. Marcus, *J. Chem. Phys.* **43**, 2654, (1965), **52**, 2803, (1970).
4. R. A. Marcus, *J. Chem. Phys.* **43**, 3477, (1965).
5. R. A. Marcus, *J. Chem. Phys.***24**, 966–978, (1956). (**typo** in the original reference)
6. R. A. Marcus, *Discuss. Faraday Soc.* **29**, 21–31, (1960). (The term involving $(\Delta e)^2(D_{\text{op}}^{-1} - D_{\text{s}}^{-1})/2a_1$ there (**typo** in the original formula) is to be replaced by $\lambda_{\text{e}}^{\infty}$ and the Δq_i's now refer only to the second reactant.)
7. R. A. Marcus, *J. Chem. Phys.* **43**, 679–701, (1965).
8. R. A. Marcus, *J. Phys. Chem.* **67**, 853, 2889, (1963).
9. R. A. Marcus, *Annu. Rev. Phys. Chem.* **15,** 155–196, (1964).
10. F. Di Giacomo, *Z. Phys. Chem. Neue Folge.* **122** (1), 1–13, (1980).
11. R. A. Marcus, *J. Chem. Phys.* **43**, 1261, (1965).
12. S. Glasstone, K. J. Laidler, H. Eyring, *The Theory of Rate Processes*, McGraw-Hill, New York and London (1941).
13. R. A. Marcus, *J. Chem. Phys.* **41**, 2624, (1964).
14. J. H. Baxendale, *Radiat. Res. Suppl.* **4**, 139, (1964).
15. L. M. Dorfman, M. S. Matheson, *Prog. React. Kinet.* **3**, 237, (1965).
16. R. J. Campion, N. Purdie, N. Sutin, *Inorg. Chem.* **3**, 1091, (1964).
17. Sutin, N. in "*Symposium on Exchange Reactions*," International Atomic Energy Agency, Brookhaven (June 1965).
18. N. Sutin, H. A. Schwarz, private communication.
19. (a) D. M. Hercules, *Science.* **145**, 808, (1964); E. A. Chandross, F. I. Sonntag, *J. Am. Chem. Soc.* **86**, 3179, (1964), and references contained therein; (c) K. S. V. Santhanan, A. J. Bard, *J. Am. Chem. Soc.* **87**, 139, (1965); (d) G. J. Hoitjink (private communication); (e) This possibility was suggested to the writer by G. J. Hoitjink: In some cases, polarographic data on aromatic hydrocarbons compounds suggest that a reaction may be sufficiently exothermic to form a triplet state of an aromatic molecule but not quite enough to form an excited singlet. The triplets may then phosphoresce in rigid media, or in solution they may annihilate each other or react with other solutes. Polarographic data are summarized by G. J. Hoitjink, *Ind. Chim. Belge* **12**, 1371, (1963).
20. In the present paper, we use the word "exothermic" for convenience to designate a reaction with a large negative standard free energy of reaction rather than specifically one with a large negative heat of reaction. However, for these reactions $|T\Delta S^{\circ}| \ll |\Delta H^{\circ}|$ normally.
21. For example, Ref. [16] and references cited therein.
22. (a) R. L. Ward, S. I. Weissman, *J. Am. Chem. Soc.* **79**, 2086, (1957); (b) P. J. Zandstra, S. I. Weissman,*J. Am. Chem. Soc.* **84**, 4408, (1962); (c) T. Layoff, T. Miller, R. N. Adams, H. Fah, A. Horsfield, W. Proctor, *Nature.* **205**, 382, (1965).

23. For example, Ref. [6], the activation energies are anomalously high, considering the high rate constants. As a result, the frequency factor A for the apparent cation free path are too high by many orders of magnitude: $A = 10^{17}$, 10^{17}, and 10^{23} liter mole^{-1} · sec^{-1} for naphtalene–naphtalene anion in three solvents, rather than the expected 10^{11} liter mole^{-1} sec^{-1} or somewhat less.
24. A. D. Britt, *J. Chem. Phys.* **41**, 3069, (1964).
25. This width of the filled half of the conducting band is about 3.7, 5.8, 7.1, 4.2, and 3.0 eV for Fe, Ni, Cu, Ni, and Na, respectively. See soft X-ray emission studies of E. M. Gyorgy, G. G. Harvey, *Phys. Rev.* **93**, 365, (1954); Compare D. Pines, *Solid State Phys.* **1**, 368, (1955); C. Kittel, *Introduction to Solid State Physics*, 2nd Edition, p. 310, John Wiley & Sons, Inc., New York (1961).
26. An example where the exothermicity is not alleviated occurs in the neutralization of gaseous He^+ by metal electrodes. Here, the exothermicity is the difference between the ionization potential (24.47 eV) and the work function of, say a molybdenum electrode (4.3 eV). It is extremely large. On the other hand, the difference between the "ionization potential" of the 2 ^{3}S excited state of helium (4.5 eV) and this work function is very small. Thereby, helium atoms are formed in metastable states. See H. S. W. Massey, E. H. S. Burhop, *Electronic and Ionic Impact Phenomena*, p. 570 ff, Oxford University Press, London (1952).
27. Compare the theoretic studies of electrode reactions by Levich, Dogonadze, and Chimadzev and by Gerischer, described or noted in Ref. [27]. ETs at semiconductor electrodes have also been discussed by these authors, and by Dewald [6].
28. For example, see Ref. 12, p. 149; E. E. Nikitin, in *Chemische Elementarprozesse*, pp. 43–77, H. Hartmann Editor, Springer-Verlag, New York (1968).
29. For review see E. H. Hart, J. K. Thomas, S. Gordon, *Radiat Res. Suppl.* **4**, 24, (1964).
30. For review see L. M. Dorfman, M. S. Matheson, *Prog. React. Kinet.* (1965).
31. G. Czapski, H. A. Schwarz, *J. Phys. Chem.* **66**, 471, (1962); E. Collinson, F. S. Dainton, D. R. Smith, S. Tazuke, *Proc. Chem. Soc.* **1962**, 140 and more recent references.
32. The assumptions of this ET theory and of the additional ones used in the present investigation are summarized by the author in Ref. [2].
33. The frequency of motion of the electron ν_e is approximately $\Delta E/h$ where ΔE, the transition energy in water, is about 1.7 eV. This is about 4×10^{14} sec^{-1}.
34. S. I. Pekar, *Untersuchungen über die Elektronentheorie der Kristalle*, Akademie Verlag, Berlin (1954).

35. This result utilizes the experience of the derivation in Ref. [7], where it was found that the vibrational coordinates in the centered distribution have a Boltzmann distribution characterized by parameters q_*^i and that the expression can be considerably simplified with little error by introduction of the k_{ij}'s.

Instead, one can include in $F^{*r}(R)$ the vibrational free energy $\int \rho_i(h_i^r + kT \ln \rho_i) d\tau_i$, where ρ_i, h_i^r, and τ_i are the vibrational phase space density, sum of vibrational kinetic and potential energy, and vibrational phase space volume element. The corresponding vibrational term with h_i^r replaced by h_i^p can be included in $F^{*p}(R)$. One then solves the constrained variational problem (7.24) and (7.25), the variations being $\delta\rho$ and $\delta\mathbf{P}$, and eventually introduces methods such as those employed in Ref. [7] to obtain Eq. (7.26), with q_*^i given by Eq. (7.33a), thus achieving the same result as before.

36. J. Jortner, *Radiat. Res. Suppl.* **4**, 24, (1964); *Mol. Phys.* **5**, 257, (1962).
37. R. A. Marcus Refs. [5], [37a], [37b]. See the appendix of Ref. 4 for a notational change; (a) R. A. Marcus, *J. Chem. Phys.* **38**, 1858, (1963); (b) R. A. Marcus, *J. Chem. Phys.* **39**, 1734, (1963).
38. A recent survey of polaron theory is given in *Polarons and Excitons*, C. G. Kuper, G. D. Whitfield, Editors, Oliver and Boyd, Ltd., Edinburgh (1963).
39. We note, incidentally, that this H_0 is also the same as that used for the isolated, unconstrained polaron by Pekar in Ref. [34] and by Jortner in Ref. [36]. (However, Jortner assumed $D_{op} = 1$ for the ground-state calculation.); (a) G. Joos, *Theoretical Physics*, Dover Publications, Inc., New York (1986).
40. Ref. [34, pp. 38–41] and Eqs. (8.10) and (8.11). Our LHS of Eq. (7.37b') is his $\overline{V}/2$.
41. For example, W. Kauzmann, *Quantum Chemistry*, pp. 229–232, Academic Press Inc., New York (1957).
42. This particular results is unaffected by the fact that Jortner assumes $c = 1 - (1/D_s)$ for the ground state.
43. G. R. Allcock, *Adv. Phys.* **5**, 412, 450, (1956).
44. For example, when m^2 is small Eqs. (7.37a), (7.37b), and (7.39) show that the second term in Eq. (7.44) is approximately $2H_0$, that is, about 2.6 eV according to the value derived earlier for H_o (1.3 eV...). The third term in Eq. (7.44) is about $-ee_2^r/D_s R$, that is, about -0.03 eV in water when $R = 6$ Å and $e_2^r = e$. The fifth term in Eq. (7.36) is about $-m^2ce^2/R$, that is, about -0.15 eV when m^2 has a typical value of about 0.1 (or less) in these very fast reactions.
45. (a) J. H. Baxendale, *Radiat. Res. Suppl.* **4**, 139, (1964); (b) The value used for the equilibrium constant of water should be $(10^{-14}/55)M$ instead of $10^{-14}M^2$, since 55 M is the molarity of water. [Alternatively, in a water medium a pseudo first order rate constant for the forward step in Eq. (7.52) can be used and the ordinary autoprotolysis constant employed.] Secondly, to obtain proper cancellation of units the ΔF° used for $H \rightleftarrows \frac{1}{2}H_2$ should be that for a standard state of H atoms of 1 M rather than of 1 atm. (c) M. S. Matheson,

Advan. Chem. Ser. **50**, 45 (1965): These latest values of the rate constants of forward and reverse reactions in Eq. (7.52) are $16 \pm 1\,\text{M}^{-1} \cdot \text{sec}^{-1}$ and $1.8 \pm 0.6 \times 10^7\,\text{M}^{-1} \cdot \text{sec}^{-1}$, respectively.

46. The mass m_p for the solid state system equals $0.020\ \alpha^4$, where α equals $e^2c(\mu/2\omega\hbar^3)^{\frac{1}{2}}$. (Cf. Ref. [34] or survey by G. R. Allcock, Ref. [43]). The value of m_p depends, therefore, on the magnitude of the angular frequency of the "lattice polarization," ω. If $\omega/2\pi$ were $10^{13}\,\text{sec}^{-1}$ and γ equaled the ratio of μ to the electron mass, α for such a model would equal $18c\gamma^{\frac{1}{2}}$. If one chose a value of $c\gamma^{\frac{1}{2}}$ of 0.65 (to fit the H_o of Appendix II (Section 7.25) then m_p would be about one quarter of a molecular weight unit, and would be larger if the appropriate frequency for the liquid motion were less.
47. H. A. Schwarz, *Radiat. Res. Suppl.* **4**, 89, (1964).
48. (a) Ref. [36]. In Ref. [34, pp. 35–38], Pekar estimates a different radius on the basis of a different criterion, which leads to a radius of 8 Å when $c = 0.65$ [46] and $\mu =$ electron mass; (b) N. Sutin, in *Symposium on Exchange Reactions* (International Atomic Energy Agency, Brookhaven, New York, June 1965) calculated this value of λ_e by assuming that the diffusion of the solvated electron in water occurs as a site-to-site ET (jump distance of 2 Å) and by using an expression for ΔF^* for a unimolecular ET reaction. The value is an upper bound, since the mean jump distance l may be shorter than 2 Å for then the *solvent reorganization barrier* would be smaller. Indeed, if the diffusion constant depends on l roughly as $l^2 \exp[-\Delta F^*(l)/RT]$, the most probable value of l is that which maximizes this expression.
49. R. A. Marcus, *J. Phys. Chem.* **67**, 853, (1963) in Ref. [8]. The value of λ_2 equals $\lambda/2$ in Eq. (2) there, when Eq. (2) refers to a chemical electron exchange reaction, and to λ when it refers to electrochemical exchange. (λ_2 is the λ per ion and there are two reactants in the chemical and one in the electrochemical exchange.) Values of λ are estimated from the data in Table I there, with the aid of this Eq. (2). (Both the λ_2 and the E_2^o for the $Tl^{3+,2+}$ are somewhat uncertain since there are known to be two consecutive one-electron transfers in the electrochemical reaction $Tl^{3+,1+}$: The value given is obtained from a geometric mean of the rate constants. The latter k's are close together, and therefore we have assumed for simplicity that the E_2^o's for the reaction probably are also.)
50. J. K. Thomas, S. Gordon, E. J. Hart, *J. Phys. Chem.* **68**, 1524, (1964); Compare J. H. Baxendale *et al.* *Nature.* **201**, 468, (1964).
51. These values are those listed in Ref. [50] and may be compared with polarographic half-wave potentials (0.43, 0.93, and 1.56, respectively) given by I. M. Kolthoff, J. J. Lingane, *Polarography*, Vol. 2, p. 440, Interscience Publishers, Inc., New York (1952).

52. There is some uncertainty as the exact value of w^r when two organic-like ions react or when an organic-like ion reacts with an inorganic one. Compare data of G. Dulz, N. Sutin, *Inorg. Chem.* **2**, 917, (1963).
53. H. H. Landolt, R. Börnstein, in *Zahlenwerte und Funktionen*, 6th Edition, Vol. 2, Part 8, pp. 5-562–5-566, K. H. Hellwege, A. M. Hellwege, Editors, Springer-Verlag, Berlin (1962).
54. Ref. [43], p. 692.
55. H. Frohlich, in Ref. [38, pp. 6–7].
56. The solvent electrons outside of a sphere of radius d, centered at the solvated electron, will see the solvated electron as a static charge if the fractional change of field D is small in the time required ω^{-1} for appreciable change of electron polarization. This change in D at a point on this sphere is small if in time ω^{-1} the path length of the moving solvated electron subtends only a small angle at that point, i.e., if $(\upsilon/d)\omega^{-1} \leq 1$.
57. This result, together with higher order terms which are small in our case (α there is about 0.5 for electron polarization) has been obtained for lattice polarization by S. W. Tiablikov (1952, 1954) T. D. Lee, F. E. Low, and D. Pines (1953), M. Guari (1953), G. Höhler (1955), and R. P. Feynman (1955). The results are summarized largely in Ref. [43] and, in part, in Ref. [38].
58. R. M. Noyes, *J. Am. Chem. Soc.* **86**, 971, (1964).

CHAPTER 8

Rate Constants, Barriers, and Brønsted Slopes, Electron Transfer at Electrodes and in Solution: Theory versus Experiment—Electrode Reactions of Organic Compounds

In this chapter, I have followed very closely the content of **M69**, **M71**, and **M77**, Refs. [1, 2, and 3]. Notes and remarks are interspersed throughout the text, Marcus' notes are collected at the end.

M69. Theoretical Relations among Rate Constants, Barriers, and Brønsted Slopes of Chemical Reactions, *J. Phys. Chem.* **72**, 891, (1968).

8.1. Theoretical Relations among Rate Constants, Barriers, and Brønsted Slopes of Chemical Reactions

In this paper, M. considers the possible extension of his theory for *weak-overlap* electron transfer (ET) reactions, that is, for ET reactions in which the *resonance splitting* between potential energy surfaces (PESs) is small, to reactions "with considerable resonance splitting, such as atom transfers (ATs), proton transfers (PTs), and *strong-overlap* ETs."

Introduction

"A simple relation has been derived for the free energy barrier and rate constant of weak-overlap ET reactions [4, 5]

$$k = Z\exp(-\Delta F^*/RT) \tag{8.1}$$

$$\Delta F^* = w^r + \lambda(1 + \Delta F_R^{o\prime}/\lambda)^2/4 \tag{8.2}$$

where Z is a *bimolecular collision frequency in solution* ($\cong 10^{11}$ liter mole^{-1} sec^{-1}),"

NOTE: the value was later corrected to 10^{12}, see **M170**.

"$\Delta F_R^{o\prime} \equiv \Delta F^{o\prime} + w^p - w^r$, $\Delta F^{o\prime}$ is the 'standard' free energy of the reaction for the prevailing medium and temperature, w^r (or w^p) is the work required to bring the reactants (or products) together to the *mean separation distance in the activated complex*, and λ for a cross-reaction is the mean of that for two electron exchange (EE) reactions [6].

$$\lambda_{12} = (\lambda_{11} + \lambda_{22})/2 \tag{8.3}$$

As a consequence of Eqs. (8.1–8.3) one finds [7, 8]

$$k_{12} \cong (k_{11}k_{22}K_{12}f_{12})^{1/2} \tag{8.4}$$

where k_{12} and K_{12} are the rate constant and equilibrium constant of the *cross-reaction*, k_{11} and k_{22} are the rate constants of the EE reaction of the two different redox systems, and

$$\ln f_{12} = -\Delta F^{o\prime 2}/2\lambda RT,$$

is given by Eq. (8.5).

$$\ln f_{12} \cong (\ln K_{12})^2/4\ln(k_{11}k_{22}/Z^2) \tag{8.5}$$

Equation (8.4) has been applied in the literature to weak-overlap ETs [9]. Recently, as a conjecture, it was applied to a few examples of AT reactions [10]. Equations (8.1 and 8.2) have been similarly used to calculate Brønsted slopes in atom- and proton-transfer reactions [11]. In each case, the results were encouraging, but more extensive application is needed.

A semiempirical, bond energy–bond order (BEBO) method has been used to calculate activation energies of gas-phase ATs [12]. The *potential energy form of* Eq. (8.2), slightly modified in a way expected for ATs, permitted [13] the calculation of potential energy barriers for some 45 cross-reactions (CRs) from those of 10 exchange reactions (EEs), with a reasonable agreement of about 2 kcal mole^{-1}. (There were 90 CRs, but only 45 were independent.) These results are given in Appendix I."

Two remarks:

(i) In this paper, M. discusses the relation between equations representing PESs and equations of related free energies. An equation in terms of free energies is here expressed in terms of potential energy.

(ii) Two other well-known cross relations are: (1) The one represented by the electrode potentials which allow one to know the potential difference of all possible galvanic cells built by combining the electrode potentials; (2) The Pauling postulated relation between the bond energy of the molecule A–B from the bond energies of A–A and B–B.

Note, moreover, that $\Delta F^{\circ\prime 2}/2\lambda = -RT \ln f_{12}$, $f_{12} = \exp(-\Delta F^{\circ\prime 2}/2\lambda RT) = K_{12}^{\Delta F^{\circ\prime}/2\lambda}$.

If $\Delta F^{\circ\prime} = 0 \Rightarrow f_{12} = 1$ and if $\Delta F^{\circ\prime} = -2\lambda \Rightarrow f_{12} = K_{12}^{-1} \Rightarrow k_{12} \cong (k_{11}k_{22})^{\frac{1}{2}}$. In this case, the rate constant for the cross-reaction is the geometric mean of those for the two exchange reactions and we have a relation similar to that of Pauling's *postulate of the geometric mean* (The Nature of the Chemical Bond, p. 83).[(1)]

"In the present paper, these equations are discussed for '*strong-overlap' reactions*, such as ATs, PTs, and strong-overlap ETs, and various consequences are noted. In some respects, the present discussion is a quantitative treatment of the common notion in the literature [14] that the Brønsted slope reflects the extent to which the activated complex resembles the reaction products (e.g., our Eq. (8.23))."

8.2. A Modification of Equation 8.2

"Because of an expected difference in PESs, discussed in Appendix II, any applicability of Eq. (8.2) to gas-phase AT reactions is expected to be limited to $|\Delta F_R^{\circ\prime}| \leq \lambda$. Outside that interval, Eq. (8.6) is to be used. The same remarks apply to reactions in solution (Appendix II) if *most* of the reorganization comes from the bonds being broken and formed, rather than from all the other coordinates [15].

$$\begin{aligned} \Delta F^* &\cong w^r(-\Delta F_R^{\circ\prime} \geq \lambda) \\ \Delta F^* &\cong \Delta F^{\circ\prime} + w^p(\Delta F_R^{\circ\prime} \geq \lambda)\text{''} \end{aligned} \tag{8.6}$$

NOTE: Recall that $\Delta F_R^{\circ\prime} = \Delta F^{\circ\prime} + w^p - w^r$. If $-\Delta F_R^{\circ\prime} = \lambda$ in Eq. (8.2) $\Rightarrow \Delta F^* \cong w^r$.

"The λ in Eqs. (8.2) and (8.6) is seen from Eq. (8.2) to equal approximately $4\Delta F_0^*$":

$$\lambda \cong 4\Delta F_0^* \tag{8.6$'$}$$

$$\text{or} \quad \frac{\lambda}{4} \cong \Delta F_0^*$$

"where ΔF_0^* *is the value of* ΔF^* *at* $\Delta F^{\circ\prime} = 0$."

The following potential energy *counterparts* (for gas-phase reactions) of Eqs. (8.2) and (8.6) were used to calculate the energy barriers for gas-phase ATs mentioned earlier. (The work terms, w^r and w^p, *usually coulombic*, are absent now.) Let E_{ij} be the *potential energy barrier* of the AT Eq. (8.7).

$$A_iB + A_j \rightarrow A_i \cdots B \cdots A_j \rightarrow A_i + BA_j \tag{8.7}$$

and let ΔE° be the net potential energy change when $i = 1$ and $j = 2$. Then

$$E_{12} = E(1 + \Delta E^\circ/4E)^2 \quad (|\Delta E^\circ| \leq 4E) \tag{8.8a}$$

$$
\begin{aligned}
E_{12} &= 0 \qquad (-\Delta E^0 = 4E) \\
E_{12} &= \Delta E^0 \quad (\Delta E^0 = 4E)\text{''}
\end{aligned}
\tag{8.8b}
$$

Two **Errata** in last two conditions in parenthesis in the original paper have been corrected.

NOTE: To better understand the last equations compare them with Eq. (8.6) and remember that M. sets here the work terms equal to zero. Notice also that E_{12} plays the role of ΔF^* and E that of $\lambda/4$. As a matter of fact, if $\Delta E^\circ = 0 \Rightarrow E_{12} = E$ which parallels $\Delta F^* = \frac{\lambda}{4}$ if $\Delta F_R^{\circ\prime} = 0$.

"In Eqs. (8.8a) and (8.8b)

$$E = (E_{11} + E_{22})/2\text{''} \tag{8.8c}$$

that E_{11} and E_{22} correspond to the reactions (8.7) with $i = j = 1$ or 2, respectively.

8.3. Outline of Treatment

Before proceeding with detailed derivations, aspects of the paper are first reviewed.

In *weak-overlap ET reactions*, it may be recalled, reactants experience work terms w^r, which are of coulombic and, in some cases [7], of 'hydrophobic–hydrophilic' origin (*solvent structural effects*). These reactants also experience a readjustment of bond lengths, of bond angles when appropriate, and of orientations of solvent molecules outside the reactants' coordination shells. Typically, the need for making these readjustments, more than w^r, constitutes the *principal barrier* to reaction. They occur because the system not only has to undergo ET, but also has to eventually adopt values of these coordinates which are appropriate to the reaction products.

In the present paper, the arguments are *extended* to *reactions in which bonds are broken and formed.* Initially, a very simple BEBO model for gas-phase reactions (8.9) is considered and Eq. (8.10) is derived. The main purpose of using Eq. (8.10) is to provide a simple, plausible vehicle for considering ATs and for comparing with Eq. (8.8), not for making a detailed calculation for these reactions. More elaborate quantum mechanical calculations of E_{12}, E_{11}, and E_{22} would be useful for testing Eq. (8.8) or for testing Eq. (8.15), a result derived from Eq. (8.10).

For some purposes, it is not necessary to employ an equation which contains the *specific assumptions* present in Eq. (8.10) or (8.8). A more general, functional Eq. (8.11) is therefore introduced to generalize a portion of the subsequent treatment. Equation (8.11) includes Eqs. (8.8) and (8.10) as special cases and, like them

(i) Relates barriers of CRs to exchange reactions and
(ii) Serves also as a basis for a discussion of Brønsted slopes.

A comparison of these equations is then given. *At low* $\Delta E°/4E$, they are found to give *exactly the same first-order term for the barrier*, and the latter term is found to contain no *intrinsic asymmetry*. When Eq. (8.8) and a symmetrized Eq. (8.10) are compared *at arbitrary* $\Delta F_R^*/\lambda$, they are found to give fairly similar results."

"Reactions in solution are considered next. The coordinates of the reactants and solvent molecules in such systems can be roughly grouped as follows:

(i) Several *bond distances*, for bonds undergoing *rupture or formation*, and
(ii) Coordinates describing *more minor adjustments* in *bond lengths* and *angles* in reactants, *of which the solvent may be one*, and *orientations* of solvent molecules.

A simple treatment of group 1 could parallel the one used to obtain Eq. (8.8) and, simultaneously, group 2 could be treated by the method used to obtain Eq. (8.2). The details of the latter would differ somewhat from that for weak overlap ET [16]. A discussion of steric and statistical factors is also given. (They were absent in weak-overlap ETs).

Equation (8.4) relating rate constants of CRs to those of ERs is next examined, and the implications of the preceding results are given. The discussion suggests that Eq. (8.4) comes through fairly intact, particularly *at low* $\Delta F^{\circ\prime}/4\Delta F_0^*$."

Notice the critical role of *the same condition* expressed in terms of PESs, $\Delta E^\circ/4E$, and in terms of free energies, $\Delta F^{\circ\prime}/4\Delta F_0^*$.

"In several subsequent sections, the meaning and magnitude of the Brønsted slope are considered, as is its relation to a kinetic isotope effect. The various findings are then summarized and, in a concluding section, the classification of activation free energies into *intrinsic* and *extrinsic* contributions is noted.

8.4. One Model for Atom Transfers

In the BEBO method [17], the *energy* of the A_iB *bond* in a gas-phase reaction (8.7) is written empirically as $-V_i n_i^{p_i}$, where n_i is the *instantaneous* bond order of A_iB, V_i is the A_iB bond energy when $n_i = 1$, and p_i is a quantity determined from BEBO relations in the literature; p_i is quite close to unity. Thus, E_f, the *potential energy of formation* of the system *from the initial configuration* is Eq. (8.9) when $n_1 = 1$ and $n_2 = 0$ initially.

$$E_f = (-V_1 n_1^{p_1} - V_2 n_2^{p_2}) + V_1 \tag{8.9}$$

In the cited model of the gas-phase reaction, it is then assumed that the total bond order $n_1 + n_2$ is constant along the *reaction path*; it is unity in the present case. By setting

$$\frac{dE_f}{dn_2} = 0$$

at the energy maximum along the reaction path, the activation energy is calculated, from empirically known V_i's and p_i's."

NOTE: It may be useful, to get an intuitive grasp of the meaning of Eq. (8.9), to compute the initial and final energy of the system in the reaction (8.7). Initially we have $n_2 = 0$ and $n_1 = 1$, so that $E_\mathrm{f} = 0$. At the end of the reaction, $n_1 = 0$, $n_2 = 1$, and $E_\mathrm{f} = V_1 - V_2$.

Notice that if $p_1 = 1$, one has $n_1 = 2$ for a double bond and $n_1 = 3$ for a triple bond. Because a double bond has an energy which is not exactly that of two single bonds, the exponent p_1 takes care of this fact.

"To later compare Eqs. (8.8) with (8.9), we first note that the former depends on only three quantities, E_{11}, E_{22}, and ΔE°, while the latter depends on four, V_1, V_2, p_1, and p_2. We may remove this difference as follows. Since $p_i \cong 1$, a Taylor expansion of Eq. (8.9) can be made and powers of $(p_i - 1)$ higher than the first neglected. The instantaneous bond energy of A_iB is then

$$-V_i n_i^{p_i} \cong -V_i[n_i + (p_i - 1)n_i \ln n_i]\text{''} \tag{8.9a}$$

*Demonstration: write the earlier as $-V_i n_i n_i^{p_i - 1} \cong -V_i n_i[1 + (p_i - 1)\ln n_i]$. Now remember that $a^x = 1 + x \log_e a + \cdots$ and we see that the term in parenthesis is made up by the first two terms of the expansion of $n_i^{p_i - 1}$.

"Maximization of Eq. (8.9) using Eq. (8.9a) yields $n_i = \frac{1}{2}$ for an exchange reaction [18]. One finds

$$E_{ii} = V_i(p_i - 1)\ln 2 \tag{8.9b}$$

$$E_\mathrm{f} = n_2 \Delta E^\circ - E_{11}\frac{n_1 \ln n_1}{\ln 2} - E_{22}\frac{n_2 \ln n_2}{\ln 2}\text{''} \tag{8.10}$$

NOTE: E_f now depends only on three properties ΔE°, E_{11}, and E_{22} like Eq. (8.8a). If we use Eq. (8.10) to determine the potential energy barrier of a gas-phase CR, we see that Eq. (8.10) doesn't depend on n_1 and n_2 because $n_1 + n_2 = 1$ and n_2 is determined by $\mathrm{d}E_{\mathrm{f}/}/\mathrm{d}n_2 = 0$.

"It is useful to introduce E, the symmetric combination of E_{11} and E_{22} given by Eq. (8.8c), and an antisymmetric combination ε defined by

$$\varepsilon = \frac{E_{11} - E_{22}}{E_{11} + E_{22}} = \frac{E_{11} - E_{22}}{2E}$$

The terms *intrinsic asymmetry*, measured by ε, and *extrinsic asymmetry*, measured by

$$\frac{\Delta E^\circ}{4E}$$

will be employed throughout this paper."

NOTE: The intrinsic asymmetry is then the ratio of the difference between the two potential energy barriers of the gaseous exchange reactions to twice the potential energy barrier of the cross-reaction, while the extrinsic asymmetry is the ratio of the potential energy change in the cross-reaction to four times the potential energy barrier.

8.5. A More General Equation

"Equations (8.8) and (8.10) are special cases of a more general one

$$E_{\mathrm{f}} = n\Delta E^\circ + \frac{1}{2}E_{11}g_1(n) + \frac{1}{2}E_{22}g_2(1-n) \qquad (8.11)$$

$$\mathrm{d}E_{\mathrm{f}}/\mathrm{d}n = 0 \quad (n = n^{\neq})$$

where n is some *degree of reaction parameter*, being zero initially, unity finally, and $n^{\neq}$ for the activated complex. . . . The term $g_i(n)$ is any function [19, 20] of n, normalized so that $g\left(\frac{1}{2}\right) = 1$.[(2)]

In the region of *small intrinsic and extrinsic asymmetry*, Eq. (8.11) may be expanded in powers of $n - \frac{1}{2}$ (**typo** corrected)[(3)] and terms beyond the second power neglected. One obtains

Eq. (8.12).

$$E_f = E + \frac{1}{2}\Delta E^0 - \frac{x^{\neq 2}}{4}[E_{11}g_1'' + E_{22}g_2''] \qquad (8.12)$$

where $x^{\neq}\left(= n^{\neq} - \frac{1}{2}\right)$ is

$$x^{\neq} = 2\left[-\Delta E^{\circ} - \frac{1}{2}(E_{11}g_1' - E_{22}g_2')\right] \Big/ (E_{11}g_1'' + E_{22}g_2'') \qquad (8.13)$$

The primes denote derivatives evaluated at $n = \frac{1}{2}$."

NOTE: Recall that $E = (E_{11} + E_{22})/2$.

8.6. Relation among Equations (8.8), (8.10), and (8.11)

"It has already been noted that Eqs. (8.8) and (8.10) are special cases of Eq. (8.11). Unlike Eqs. (8.8), (8.10), and (8.11) do have an intrinsic asymmetry term. However, a rather striking result can be proven when the intrinsic $\left(\varepsilon = \frac{E_{11}-E_{22}}{E_{11}+E_{22}} = \frac{E_{11}-E_{22}}{2E}\right)$ and extrinsic asymmetry ($\Delta E^{\circ}/4E$) are both small: the intrinsic asymmetry makes no contribution to the first-order term in those equations for E_f .... Thus, all three equations agree at low $\Delta E^{\circ}/4E$ ...

We next consider the relation between Eqs. (8.8) and (8.10) at *any value* of $\Delta E^{\circ}/4E$. In both equations, E_f becomes 0 or ΔE°, accordingly, as $\Delta E^{\circ}/4E$ becomes very negative or very positive. For other $\Delta E^{\circ}/4E$, Eq. (8.8) is most easily compared with Eq. (8.14), the symmetrized form of Eq. (8.10), since Eq. (8.8) contains no intrinsic asymmetry.

$$E_f = n\Delta E^{\circ} - E[n \ln n + (1 - n)\ln(1 - n)]/\ln 2 \qquad (8.14)$$

The value of n which solves $\mathrm{d}E_f/\mathrm{d}n = 0$ is found and inserted into Eq. (8.14). Manipulation yields [21]

$$E_f = E + \frac{1}{2}\Delta E^{\circ} + \left(\frac{1}{2}\Delta E^{\circ}\Big/ y\right)\ln\cosh y \qquad (8.15)$$

where $y = (\Delta E^\circ/2E)\ln 2$. The $\Delta E^{\circ 2}/16E$ term in Eq. (8.8a) can be written, for comparison, as $\frac{1}{2}\Delta E^\circ y/4\ln 2$.

When y tends to $\pm\infty$, $\ln\cosh y$ tends to $\pm y - \ln 2$. Thereby, E_f in Eq. (8.15) tends to 0 or ΔE°, according as ΔE° tends to $-\infty$ or to $+\infty$, respectively, in agreement with Eq. (8.8b). When $\Delta E^\circ = 2E$, which is midway between the extremes of small $\Delta E^\circ/4E$ and of $\Delta E^\circ/4E \sim 1$, $(1/y)\ln\cosh y$ is about $\frac{1}{3}$, while $y/4\ln 2$ is $\frac{1}{4}$. The difference of Eqs. (8.8a) and (8.15) is, therefore, $\frac{1}{24}\Delta E^\circ$, which is small.

E, defined by Eq. (8.8c), is sometimes of the order of 10 kcal mole^{-1}, so that a ΔE° of about $2E$ is then about 20 kcal mole^{-1}.

8.7. Reactions in Solution

Two quite different gas-phase models of a reaction obeyed Eq. (8.11), one being BEBO and the other (see Appendix II) having *a pair of intersecting potential energy parabolas*. In *the free energy analog* of Eq. (8.11), *to be used for reactions in solution*, E_f, ΔE° and E_{ii} are replaced by their analogs *at a mean separation distance*, R, in the activated complex, $\Delta F^* - w^r$, $\Delta F_R^{\circ\prime}$, and $\Delta F_{ii}^* - w_{ii}$

$$\Delta F^* - w^r = n\Delta F_R^{\circ\prime} + \frac{1}{2}(\Delta F_{11}^* - w_{11})g_1(n) + \frac{1}{2}(\Delta F_{22}^* - w_{22})g_2(1-n) \tag{8.16}$$

$$\partial\Delta F^*/\partial n = 0 \quad (n = n^{\neq})$$

where $w_{ii} = w_{ii}^r = w_{ii}^p$ and $\Delta F_R^{\circ\prime} = \Delta F^{\circ\prime} + w^p - w^r$.

The quadratic expression (8.2) can in fact be written as in Eq. (8.16), with $g_1(n) = g_2(n) = 4n(1-n)$ and $\Delta F_{ii}^* - w_{ii} = \lambda_{ii}/4$. In Appendix III, the results of Ref. [22] (**typo** in the original reference) are used to show that weak-overlap ETs in solution obey Eq. (8.16), even before some approximations present in Eq. (8.2) are introduced.

As noted earlier, the nuclear coordinates in ET reactions are of *two types*

(i) *Vibrational coordinates* (bond lengths and angles) in reactants, including those of any solvent molecules in the coordination shell, and
(ii) *Orientational coordinates* of solvent molecules outside the coordination shell.

For the former, a quadratic potential energy function is appropriate. For the latter, it is not. Instead, the statistical mechanical equivalent of dielectric unsaturation for partial saturation was introduced into the free energy expression for the solvent system [22–24]. The *free energy* of the solvent then became *a quadratic function of fluctuations in solvent polarization*, just as the harmonic potential energy for vibrational coordinates is a quadratic function of fluctuations in those coordinates. The total of the *two contributions* to ΔF^* leads, as noted in Appendix III, to Eq. (8.16).

In the case of an atom, proton, or *strong-overlap* ET reactions in solution, it was noted that there are

(i) Bonds being broken or formed, including any involving addition or removal of a solvent molecule to or from a reactant;
(ii) Vibrational coordinates undergoing small changes; and
(iii) Orientations of solvent molecules changing their distribution because of a change in charge distribution in the reactants.

The first group might be treated as in Eq. (8.10) or (8.11), the second as in the preceding paragraph, and the third by the statistical mechanical dielectric unsaturation method noted there. The final result for the latter two contributions would differ somewhat from that found for weak-overlap ETs, because the *change in charge distribution* as the system moves along the reaction coordinate is now *less abrupt*. However, since *quite different models* lead to Eq. (8.11), there is little doubt that a reasonable theory consistent with Eq. (8.16) can be formulated."

Note that one is accustomed to consider ΔF^* at $n^{\neq}$ but here M. considers how the activation barrier varies with n. Note moreover that $\Delta F^* - w^r$ is the *configurational free energy barrier*, that is, the barrier before the fluctuation. Here, M. makes explicit the concept that in the gas-phase one uses E_f while in liquid phase one uses ΔF^*.

8.8. Steric and Statistical Factors

"In atom or PTs, steric and statistical factors may contribute to the rate constant in a manner, which depends on *details* of the PES. Several models can be considered in such a way as to permit the preceding formalism to be utilized intact, as for example the following.

(i) The reactants come together, requiring the coulombic or other work term, w^r. They reorient with a steric and statistical factor of S^r and s^r, respectively [25].
(ii) The system undergoes the pertinent changes of bond lengths and solvation and so reaction occurs.
(iii) The products separate, the relevant terms for the *reverse process* being w^p, S^p, and s^p.

The free energy change in steps (i) and (iii) is $w^r - RT \ln S^r s^r$ and $w^p - RT \ln S^p s^p$, respectively, and we write

$$W^r = w^r - RT \ln S^r s^r\text{''}$$

Note that in M. theory the steric effects appear in these "generalized" work terms.

Compare for instance in K. J. Laidler's "Theory of Chemical Reaction Rates," McGraw-Hill Book Company, (1969) the concepts of steric and statistical factors.

"The overall 'standard' free energy of reaction in steps (i)–(iii) is $\Delta F^{\circ\prime}$, so that in step (ii) is

$$\Delta F_R^{\circ\prime} = \Delta F^{\circ\prime} + W^p - W^r.$$

The *configurational* free energy barrier to form the activated complex is

$$\frac{1}{4}\lambda(1 + \Delta F_R^{0\prime}/\lambda)^2$$

when $|\Delta F_R^{\circ\prime}/\lambda| \leq 1$, according to the arguments which led to Eq. (8.2); λ has the additivity in Eq. (8.3).

The '*translational*' contribution [26] to the free energy of activation $\Delta F^{\neq}$ is

$$-kT \ln(hZ/kT),$$

where Z is the collision number of uncharged species in solution ($\sim 10^{11}$ liter mole^{-1} sec^{-1})."

NOTE: The value 10^{11} was later corrected to 10^{12}, see **M170**.

"Thus from

$$k_{\text{rate}} = (kT/h)\exp(-\Delta F^{\neq}/RT),$$

one again obtains Eqs. (8.1) to (8.3), where now the w^r and w^p are replaced by W^r and W^p. For example

$$\Delta F^* = W^r + \frac{\lambda}{4}(1 + \Delta F_R^{\circ\prime}/\lambda)^2$$

$|\Delta F_R^{\circ\prime}/\lambda| \leq 1$, where $\Delta F_R^{\circ\prime} = \Delta F^{\circ\prime} + W^p + W^r$ now."

NOTE: The *free energy barrier* is made up by the *work term* + the *configurational* free energy barrier. Notice the difference between ΔF^* and $\Delta F^{\neq}$.

8.9. Remarks on Equation (8.4)

"In this section, a modification of Eq. (8.4) based on the *free energy analog* of Eq. (8.15) is first given. We also consider a case where Eq. (8.4) could break down.

The free energy analog of Eq. (8.15) yields Eq. (8.4), but with

$$f_{12} = K_{12}^{(1/y \ln \cosh y)} \tag{8.17a}$$

where

$$y = (\ln K_{12})(\ln 2)/\ln(k_{11}k_{22}/Z^2) \tag{8.17b}$$

For comparison, Eq. (8.5) can be rewritten as

$$f_{12} = K_{12}^{y/4 \ln 2} \tag{8.18}$$

When, as in a previous section, $\Delta E°/2E \sim 1$ and so $y = \ln 2$, and when $K_{12} = 10^{-12}$, the $f_{12}^{\frac{1}{2}}$'s in Eqs. (8.17a) and (8.18) differ only by a factor of 3.

The breakdown of Eq. (8.4) can be investigated by examining the breakdown of the first-order term in Eq. (8.8a), $E + \frac{1}{2}\Delta E°$, because of the related theoretical origin of both equations.[4] Equation (8.4) rests on the dependence of all terms in the energy change on any degree of reaction variable n.[5], [6] Even a natural asymmetry in the PES did not affect this equation in the region of small $\Delta E°/4E$, because of a compensation.[7], [8]

Correspondingly, some breakdown will occur in these equations when an important term in the free energy barrier does not vary with n.[9] For example, in the case of Eq. (8.11), let a fraction c of the $\Delta E°$occur before the principal reorganization.[10] Then E_f is zero initially, and is given by Eq. (8.11), subsequently, with $n\Delta E^\circ$ replaced by

$$c\Delta E^\circ + (1-c)n\Delta E^\circ.$$

Manipulation as before leads to

$$E_\mathrm{f} = E + \frac{1}{2}(1+c)\Delta E^0 - \frac{x^{\ddagger 2}}{2}Eg''\left(\frac{1}{2}\right) + \cdots\text{''} \quad (8.19)$$

where $x^{\neq} = n^{\neq} - \frac{1}{2}$ and M. shows that

$$x^{\neq} = 2\left[-(1-c\Delta E^\circ) - \frac{1}{2}(E_{11}g_1' - E_{22}g_1')\right] \Big/ (E_{11}g_1'' + E_{22}g_1'')$$

where the primes denote derivatives evaluated at $n = \frac{1}{2}$ and "$g_1(n) = g_2(n)$ for simplicity of illustration.

From Eq. (8.19), or really from its free energy counterpart, one finds

$$k_{12} \cong (k_{11}k_{22}K_{12}^{1+c})^{\frac{1}{2}} \quad (8.20)$$

When c is small, $K_{12}^{\frac{c}{2}}$ becomes a second-order term. E.g., if $c = 0.1$ and $K_{12} = 10^{10}$, $K_{12}^{\frac{c}{2}} \cong 3$."

NOTE: Because $(10^{10})^{0,1} = 10, \Rightarrow 10^{\frac{1}{2}} \cong 3\ldots$

"When steric and statistical effects are included, Eq. (8.4) becomes

$$k_{12} \cong (k_{11}k_{22}K_{12}f_{12})^{\frac{1}{2}}(\zeta_{12}^2\,\zeta_{11}\zeta_{22})^{\frac{1}{2}} \quad (8.21)$$

where ζ for a reaction is

$$\zeta = (S^r s^r S^p s^p)^{\frac{1}{2}}$$

and f_{12} is

$$\ln f_{12} = (\ln K_{12})^2/4\ln(k_{11}k_{22}/\zeta_{11}\zeta_{22}Z^2)$$

(Similarly, Eq. (8.17b) can be corrected for the ζ's by dividing $k_{11}k_{22}$ by $\zeta_{11}\zeta_{22}$).

8.10. Meaning of the Brønsted Slope

The Brønsted slope, α, is

$$\alpha = \frac{\partial \Delta F^*}{\partial \Delta F^{\circ\prime}} \tag{8.22}$$

and, according to Eq. (8.16), equals

$$n + \left(\frac{\partial \Delta F^*}{\partial n}\right)_{\Delta F^{\circ\prime}} \left(\frac{\partial n}{\partial \Delta F^{\circ\prime}}\right),$$

evaluated at $n = n^{\neq}$. Since $(\partial \Delta F^*/\partial n)_{\Delta F^{\circ\prime}}$ vanishes at $n = n^{\neq}$, we obtain

$$\alpha = n^{\neq} \tag{8.23}$$

(i) For the BEBO model described earlier for a gas-phase reaction $n^{\neq}$, and hence α, is the *bond order of the bond being formed.*

(ii) For weak ETs in solution, discussed in Appendix III and involving *no bond ruptures*, the *distribution of coordinates in the activated complex* is determined by a potential energy function

$$(1 - n^{\neq})U^r + n^{\neq} U^p; \tag{8.23$'$}$$

That is, $n^{\neq}$ and thereby α represents the *products' contribution* to this function.

(iii) For an atom or PT or strong-overlap ET in solution, if some method, such as that briefly touched on in a preceding section, were employed to calculate the terms in Eq. (8.16), $n^{\neq}$ and hence α would characterize the *product-like character* of *both* types of coordinates.

As exemplified by the equations of the following section, and in accordance with the usual notion, the Brønsted plot should be curved when $\Delta F^{\circ\prime}$ is varied over a *sufficiently wide range*.

For any compound the α and, hence, the $n^{\neq}$ equals the *instantaneous slope* of this plot at the given $\Delta F^{\circ\prime}$ for this compound."

Note the triple meaning of α: (1) $\alpha = \frac{\partial \Delta F^*}{\partial \Delta F^{o\prime}}$; (2) $\alpha = n^{\neq}$; (3) $\alpha =$ bond order of the molecule in the activated complex.

8.11. Magnitude of the Brønsted Slope

Since α is $\partial \Delta F^*/\partial \Delta F^{o\prime}$, Eq. (8.2) yields

$$\alpha = \frac{1}{2}(1 + \Delta F_R^{o\prime}/\lambda)\text{''} \tag{8.24}$$

Note that $\alpha = 0.5$ when $\Delta F_R^{o\prime} = 0$.

"when $|\Delta F_R^{o\prime}| \leq \lambda$. If ΔF_0^* denotes the *intercept at* $\Delta F^{o\prime} = 0$ of a plot of ΔF^* versus $\Delta F^{o\prime}$, then

$$\Delta F_0^* = w^r + \frac{\lambda}{4}[1 + (w^p - w^r)/\lambda]^2 \cong \frac{w^r + w^p}{2} + \frac{\lambda}{4}\text{''} \tag{8.25}$$

Compare Eq. (8.2) with $\Delta F^{o\prime} = 0$.

"since $\lambda \gg (w^p - w^r)$. Thus [27]

$$\alpha \cong (1 + \Delta F_R^{o\prime}/4\Delta F_0^*)/2 \tag{8.26}$$

When statistical and steric factors are included, Eq. (8.11.3) again follows, but with w^r and w^p in $\Delta F_R^{o\prime}$ and ΔF_0^* replaced by W^r and W^p.

For a reaction whose *functional dependence* of ΔF^* on $\Delta F^{o\prime}$ is given instead by Eq. (8.15) with E_f and ΔE° replaced by ΔF^* and $\Delta F_R^{o\prime}$, etc.) α would be

$$\alpha = (1 + \tanh y)/2 \tag{8.27}$$

where $y = (\Delta F_R^{o\prime}/2\Delta F_0^*)\ln 2$. When $\Delta F_R^{o\prime}/\Delta F_0^*$ is 2, this α is 0.8, while that obtained from Eq. (8.26) is 0.75.

Comparisons of experimental plots of ΔF^* versus $\Delta F^{\circ\prime}$ with these equations are appropriate when λ is *constant for all points in the plot.*

8.12. Kinetic Isotopic Effect k_H/k_D and Brønsted Slope

A maximum in plots of k_H/k_D versus pK's has sometimes been reported. A useful simple explanation has been given [28], based on *little isotopic effect* on ΔpK and on the reaction's forward or reverse step becoming fast (diffusion controlled) at either extremity of the plot [28]. With the aid of Eq. (8.2), we can formulate the suggestion in quantitative terms. We consider, then, reactions for which there is *little isotopic effect on* $\Delta F^{\circ\prime}$ "(and W^r or W^p). The barrier difference for hydrogen and deuterium isotopes is then found from Eq. (8.2) to be [29]

$$\Delta F_H^* - \Delta F_D^* \cong \frac{1}{4}(\lambda_H - \lambda_D)\left[1 - \left(\frac{\Delta F_R^{\circ\prime}}{\lambda_H}\right)^2\right] \tag{8.28}$$

when $|\Delta F_R^{\circ\prime}| \leq \lambda_H$. We have used the fact that $\lambda_H \lambda_D \cong \lambda_H^2$. The barrier difference is seen to pass through a maximum. Since

$$\alpha - \frac{1}{2} = \Delta F_R^{\circ\prime}/2\lambda_H,\text{''}$$

From Eq. (8.24) "the last factor in Eq. (8.28) is $\left[1 - 4\left(\alpha - \frac{1}{2}\right)^2\right]$.

8.13. Summary of Findings for Cross-Relation and for Brønsted Slope

We summarize our findings regarding Eq. (8.4).

(i) Except as noted in Appendix II, the application of Eq. (8.4) to *atom* or *PTs* is probably limited to $|\Delta F^{\circ\prime}/4\Delta F_0^*| < 1$, a normally minor limitation.

(ii) Several models including the BEBO one and a generalization thereof, lead to Eq. (8.4) in the vicinity of small $\Delta F^{\circ\prime}/4\Delta F_0^*$. Intrinsic asymmetry does not alter this equation when $\Delta F^{\circ\prime}/4\Delta F_0^*$ is small.

(iii) The difference of the BEBO and quadratic approximations, Eqs. (8.15) and (8.8), is rather small where we have tested it.

(iv) Since Eq. (8.4) rests ultimately upon a continuous dependence of the main configurational free energy change on some reaction parameter, the equation will break down when some appreciable fraction of the total free energy change becomes independent of n, to the extent given by Eq. (8.20).

(v) When steric and statistical effects occur, several models can be considered, one of which replaces Eqs. (8.4) by (8.21).

(vi) Any effects which are specific only for the cross-reaction or only for one of the exchange reactions are excluded in Eq. (8.11) and so tend to cause Eq. (8.4) to break down.[11]

The Brønsted slope, α, is expected to be 0.5, when $\Delta F^{\circ\prime} = 0$. However, deviations would occur

(i) If $c \neq 0$ in Eq. (8.20), that is, if an appreciable fraction of the free energy change were independent of the reaction parameter n,[12] or

(ii) if $g_i' \neq 0$ in Eq. (8.13) and, at the same time, the asymmetry ε is large. Equation (8.11.3) is a simple approximate expression for α for any $\Delta F^{\circ\prime}$ for which $|\Delta F^{\circ\prime}/4\Delta F_0^*| < 1$. It yielded a value (0.75) close to that obtained (0.8) from *a quite different model* (Eq. 8.27) even when the $\Delta F^{\circ\prime}/\Delta F_0^*$ had the fairly large value of 2. An *interpretation of* α for a certain *class of models*, those for which Eq. (8.16) is applicable, is given by Eq. (8.23) and the associated discussion.

8.14. Remark on Intrinsic Barriers

Reaction barriers have been considered here in terms of

(i) Intrinsic (λ_{ii}) and
(ii) Thermodynamic ($\Delta F^{\circ\prime}$) *contributions*, as well as of
(iii) Steric (ζ) and
(iv) Statistical ones.

Some test of the *intrinsic–extrinsic separation* can be made with data on Brønsted slopes and with the cross relation, e.g., with tests of Eqs. (8.4) and (8.26).

In interpretations of Brønsted slopes, λ has been assumed *constant* as a conjecture for a *reaction series* when the substituent is not part of the *reaction site* [9, 11]. When a similar assumption is valid for other reactions, such as those involved in the Hammett $\sigma\rho$ relation, there is an interesting consequence. In the region of $\alpha = \frac{1}{2}$, σ and ρ then depend only on variations in $\Delta F^{\circ\prime}$. For other α's they depend on variations in $\Delta F^{\circ\prime}(1 + \Delta F^{\circ\prime}/2\lambda)$ at *constant* λ. Then, with λ estimated from the data, discussions of substituent effects reduce purely to a discussion of effects on $\Delta F^{\circ\prime}$ and so fall within a broader class of problems concerning the effect of substituents on thermodynamic properties.

8.15. Appendix I. Comparison of BEBO Calculations and Equation (8.8)

Equation (8.8) relates the barriers of CRs ($i \neq j$ in Eq. 8.7) to ERs ($i = j$ in Eq. 8.7).

Table 1 gives a comparison with calculations made for a BEBO *model* more complicated than Eq. (8.9). The reactions in this table are hydrogen ATs, that is, B in Eq. (8.7) is H now. The diagonal elements in Table 1 provide the values of the *exchange barriers* E_{ii}. Values of ΔE° are obtained by subtracting the dissociation energies of the

A_iH used in the BEBO calculations, D_{A_iH} given in the last column. Use of Eq. (8.8) then permits the nondiagonal elements of the table to be computed. They are given in parentheses. For brevity, only values below the diagonal are given. Those above the diagonal refer to the *reverse* reaction and are not independent of the former.

The values based on Eq. (8.8) are seen to be *close to the BEBO ones*. The two sets of results show about the same agreement with the experimental data.

8.16. Appendix II. Potential Energy Surfaces in Reactions

If the potential energy of a reaction *along the reaction coordinate* involves *a pair of intersecting parabolas*, Eq. (8.8a) is obtained.(13) In this case, the *predominant* motion along the reaction coordinate is a vibrational or a *pseudo-vibrational* one.(14) In the case of a weak-overlap ET in solution, numerous coordinates are involved and this plot is a profile of the potential energy along a reaction coordinate in many dimensional configuration space (e.g., Figure 1 of Ref. [30]). The reaction coordinate is expected to be associated in part with vibrational motion of the ligands and dielectric relaxation of the solvent polarization [22].

In a gas-phase AT, the reaction coordinate involves a *concerted motion* of a compressing of one bond and a stretching of the other. Here, using the usual *potential contour diagrams*, one can plot the potential energy along the reactants' valley, up to the saddle point, and down to the products' valley. Here, the curve, potential energy versus reaction coordinate, is initially constant, then rises to a maximum, like an Eckart barrier, and then falls to another constant value [31]. One can no longer, therefore, obtain the "inverted chemical effect" [30] possible with Eq. (8.2) at *large* $|\Delta F^{\circ\prime}/\lambda|$, and so the added Eq. (8.6) is imposed.(15)

When the *bond rupture-bond formation* in atom or PTs in solution is the *principal contributor* to the reaction coordinate, as one would expect, the remarks of the last paragraph apply here as well. However, when *most* of the reorganization is associated with coordinates not involved in bond rupture or formation, Eq. (8.6) would no longer be applicable. The "inverted chemical effect" could again occur, and the unrestricted Eq. (8.2) would again be relevant.

8.17. Appendix III. Weak Overlap Electron Transfers and the Free Energy Analog of Equation (8.11)

We denote by ΔF_R^* the *configurational* contribution to ΔF^* at any separation distance, that is

$$\Delta F_R^* = \Delta F^* - w^r \tag{8.29}$$

Similarly, for formation of activated complex from products at a separation distance R, we write

$$\Delta F_R^{*p} = \Delta F^{*p} - w^p \tag{8.30}$$

In Refs. [30] and [22], it was noted

(i) That the *distribution* of *activated complex configurations* for weak—overlap reactions was *centered* at the intersection of two potential energy functions U^r and U^p in many dimensional configuration space.
(ii) That the distribution can be expressed in terms of an *equivalent equilibrium distribution* for which the configurations are distributed in accordance [32] with the function

$$f^* = A\exp\{-[(1-n)U^r + nU^p]/kT\} \tag{8.30a}$$

where A is a normalizing constant, and
(iii) That f^* is unchanged when the symbols (r, p, and n) are changed to (p, r, and $1-n$), respectively."

NOTE: For $n = 0$, f^* is the same as that for the reagents, for $n = 1$ is the same as that for the products, for $n = \frac{1}{2}$ U^r and U^p contribute by the same amount.

"ΔF_R^* can be written as

$$\Delta F_R^* = (1-n)\Delta F_R^* + n\Delta F_R^{*p} + n\Delta F_R^{\circ\prime}\,^{(\mathbf{16})} \tag{8.30b}$$

since

$$\Delta F_R^{\circ\prime} = \Delta F_R^* - \Delta F_R^{*p}.$$

There is now a useful symmetry property. Because of property iii, examination of an expression for ΔF_R^* shows at once that when (r, p, and n) is replaced by (p, r, and $1-n$), respectively, ΔF_R^* goes over to ΔF_R^{*p}, ΔF_R^{*p} goes over to ΔF_R^*, and, thus, $(1-n)\Delta F_R^* + n\Delta F_R^{*p}$ remains unchanged. This result is used later to obtain Eq. (8.32).

We shall first show that

$$\frac{\partial \Delta F_R^*}{\partial n} = 0.^{(\mathbf{17})}$$

From the definition of F_R^*, i.e.

$$F_R^* = \langle U^r \rangle + kT\langle \ln f^* \rangle, \tag{8.30c}$$

(Eq. 35, Ref. [22]), where $\langle\ \rangle$ denotes average with respect to f^*, and from the equation reflecting the central distribution

$$\langle U^r \rangle = \langle U^p \rangle$$

(Eq. 20, Ref. [22], **erratum** in the original citation), one finds $\partial F_R^*/\partial n = 0$."

NOTE: From Eqs. (8.30a) and (8.30c), we see how the free energy depends through statistical mechanical averages, on the PESs for reagents and products.

"We next derive Eq. (8.32).

If for the moment, we regard the ions as far apart, the distribution function f^* can be *factored* into *two parts, one for each ion and its environment.*(**18**) So can that of the initial reactants. Thereby, the expression

$$(1-n)\Delta F_R^* + n\Delta F_R^{*p}$$

can also be written as the *sum* of two functions $f_{11}(r, p, n)$ and $f_{22}(r, p, n)$, one for each reactant."

NOTE: It is so because f^* appears in the expression of F^* under the ln sign. Notice moreover that the f_{11} and f_{22} are the *free energies* obtained from the factoring.

"Thus, we may write

$$\Delta F_R^* = n\Delta F_R^{o\prime} + f_{11}(A_1^{ox}, A_1^{red}, n) + f_{22}(A_2^{red}, A_2^{ox}, n) \quad (8.31)$$

since the reactants are A_1^{ox} and A_2^{red}. In virtue of the symmetry property proven earlier, the last term also equals

$$f_{22}(A_2^{ox}, A_2^{red}, 1-n).$$

For brevity, we denote this term by $f_{22}(1-n)$ and the term before it in Eq. (8.31) by $f_{11}(n)$

$$\Delta F_R^* = n\Delta F^{o\prime} + f_{11}(n) + f_{22}(1-n) \quad (8.32)$$

We consider now the case where the ions are close together in the activated complex rather than separated. Their vibrational contribution to U^r and U^p are largely additive, so this contribution to the distribution function f^* factors as before. The solvent polarization contributions from the two ions *are not independent*. However, as noted on p. 693 of Ref. [22], deviation from an *apparent* additivity can be estimated to be small. Thus, Eq. (8.31) applies also for interacting cases.

For an exchange reaction (ER)

$$f_{22}(n) = f_{11}(n).$$

For these systems, according to the analysis

$$n = \frac{1}{2}.$$

(There is no *double maximum*[19] for barriers versus reaction coordinate in this treatment.) Thus, for an exchange we have

$$\Delta F_{ii}^* = 2f_{ii}\left(\frac{1}{2}\right). \quad (8.33)$$

If we denote $f_{ii}(n)/f_{ii}\left(\frac{1}{2}\right)$ by $g_i(n)$, the free energy analog of Eq. (8.11) follows.

In summary, there are two basic ingredients to this free energy analog of Eq. (8.11)

(i) A dependence of ΔF^* on a parameter n, and
(ii) An additivity, which is partly apparent and partly actual,[20] of intrinsic barrier for the changes of configuration around each center, A_1 and A_2, i.e., for the changes that would occur if $\Delta F_R^{o\prime}$ were zero."

NOTE: That is, if $\Delta F_R^* = \Delta F_R^{*p}$.

A Glossary of Principal Symbols Follows.

M71: Electron Transfer at Electrodes and in Solution: Comparison of Theory and Experiment, *Electrochim. Acta* **13**, 995, (1968).

8.18.

From the Abstract: "Recent theoretical work has predicted certain *quantitative correlations* between rates of

(i) Crossed-redox reactions and rates of isotopic exchange, and between
(ii) Homogeneous and electrochemical rates.

Experimental tests of these predictions yield insight into 'intrinsic' and 'driving force' factors.

The *intrinsic factor* is related to *differences in properties* of oxidized and reduced species (e.g., differences in corresponding bond lengths and differences in solvent orientation polarization).

The *driving force term* is related to the standard free energy of reaction in the homogeneous reaction and to the activation overpotential in the electrode reaction.

Measurements of *temperature coefficients* of rates in dilute solution provide some information about *adiabatic* and *dielectric-saturation* effects. Absolute rates, in conjunction with knowledge of bond-length differences and bond-force constants, provide some insight into the overall picture, (intrinsic, adiabatic, unsaturation factors, etc.). Study of the cited quantitative correlations permits the cancellations of many effects, and so can reveal others.[21]

The present state of experimental information on these topics is described."

8.19. Introduction, Assumptions, and Theory

"We shall review some of our results on the theory of ET reactions in solution and at electrodes [22, 33], which we have compared [33] with other treatments [34, 35], and then consider ways of testing experimentally various aspects of the theory.

The ET process in solution or at electrodes is considered in terms of a PES for the *entire* system. That surface is plotted as a function of all coordinates of the system.

These coordinates include bond lengths of the reactants, orientational coordinates of the solvent molecules, bond lengths and intermolecular distances of the latter, distance between the two reacting species and between each of them and the other molecules, and so on. The two 'reactants' can either be two species in solution *or* one species and an electrode. Similar remarks apply to the 'products.'

A PES is *first* drawn for a system containing the two reactants and the rest of the system, *without including the electronic coupling* of the reactants (Fig. 1). A surface is also drawn for the two products, again *without including coupling* (Fig. 1). The two surfaces intersect at certain values of the coordinates. If there are N coordinates initially, this intersection set forms an $N - 1$ dimensional subspace, which must be crossed for reactions to occur."

Notice that each point of the $N - 1$ dimensional subspace has N coordinates when immersed in the N—dimensional space and that there is there an Nth coordinate common to all the points of the $N - 1$ dimensional subspace.

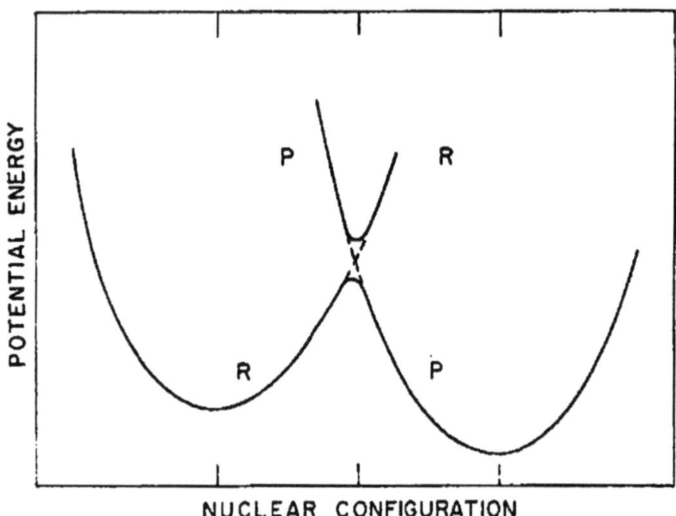

Fig. 1. Profile of potential energy surface of reactants (R) and that of products (P), plotted versus nuclear configuration of all the atoms in the system. - - -, surface for zero electronic interaction of the reacting species. —, adiabatic surface.

"When the electronic coupling is present the above surfaces are split at their intersection in a well-known quantum mechanical manner (Fig. 1, solid curves), yielding thereby a surface for a quantum mechanical adiabatic reaction.

When the system undergoes a suitable fluctuation of coordinates from values appropriate to reactants to values appropriate to the intersection region, it reaches the latter region and, one sees from Fig. 1, the ET occurs if the coupling is strong enough. With strong enough coupling, the system continues to reside on the lower surface, which is R initially and P finally. If the coupling is not strong enough, the system jumps from the lower R surface to the upper R surface, simply *by retaining its original electronic configuration.* The chance that the system remains on the lower adiabatic surface on passing through the intersection region, and so yield a successful ET is then small. The reaction in this case can be said to have nonadiabatic aspects.

The coupling is enhanced by decreasing separation distance between the two reactants. Thus, for ET one needs

(i) A *suitable fluctuation* of coordinates and
(ii) An *appropriately small separation distance* of reactants.

When an electrode reaction is involved, there are many *R* and *P* PESs to be considered, *each* corresponding to *different distribution* of the electrons among the quantum states of the electrode.

However, because of the Fermi distribution, most of the ETs occur to and from levels within *kT* of the Fermi level of the metal. For *diagrammatic purposes* in Fig. 1, therefore, the transfer can be *visualized* in terms of a simple *average* level [22].

The details of the *calculation of the ET rate* have been given both for solution reactions and electrode reactions [22].

(i) Part of the calculation involves obtaining an expression for the *probability density* of finding the system *in the intersection region* (per unit length along the abscissa of Fig. 1 in many-dimensional configuration space).
(ii) Part involves use of a suitable expression for calculating the tendency of the system to remain on the *lowest* surface, and
(iii) Part involves the introduction of suitable approximations that simplify the theoretical expressions and permit their comparison with the experimental data.

The motion along the abscissa of Fig. 1 (in many-dimensional configuration space) has been treated classically [34]. (Thus, any *nuclear tunneling* through this barrier is ignored. It is normally regarded as minor except at low enough temperatures.) The reactions treated were those which did not involve rupture of a bond in the elementary step. For simplicity, the potential energy for stretching of bonds in the coordination shell of each reactant was treated as a *quadratic function* of the coordinates. In the statistical mechanical calculation of the free energy of the ion–solvent and solvent–solvent interactions throughout the reaction, a dielectric unsaturation (or at most partial saturation) treatment was used. Each reacting species was taken to have its coordination ligands *intact*, so that *bridged activated complexes* were not considered. A quasi-equilibrium distribution was used for computing the probability of finding the system in the intersection region.

8.20. Theoretical Equations

For an *electrode* reaction (8.34) or a *homogeneous* reaction (8.35) the expression given by (8.36) was obtained for the rate constant,

$$\mathrm{Ox} + ne \rightarrow \mathrm{Red} \tag{8.34}$$

$$\mathrm{Ox}_1 + \mathrm{Red}_2 \rightarrow \mathrm{Red}_1 + \mathrm{Ox}_2 \tag{8.35}$$

$$k = Z\kappa\rho \exp(-\Delta F^*/RT), \tag{8.36}$$

where ΔF^* is given by Eq. (8.37) for an electrode reaction and by Eq. (8.38) for a solution one

$$\Delta F^* = \frac{w^r + w^p}{2} + \frac{\lambda_{\mathrm{el}}}{4} + \frac{nF(E - E_0')}{2} + \frac{[nF(E - E_0') + w^p - w^r]^2}{4\lambda_{\mathrm{el}}} \tag{8.37}$$

$$\Delta F^* = \frac{w^r + w^p}{2} + \frac{\lambda}{4} + \frac{\Delta F^{\circ\prime}}{2} + \frac{(\Delta F^{\circ\prime} + w^p - w^r)^2}{4\lambda}\text{,''} \tag{8.38}$$

NOTE: Some observations about the equations.

The same equations describe the reactions in solutions and those at the electrode.

In all of them we have

$$\Delta F^* = \Delta F^*(w^r, w^p, \Delta F^{\circ\prime}, \lambda).$$

In Eq. (8.37) appears the *activation overpotential* $E - E^{\circ\prime}$ which takes the role of $\Delta F^{\circ\prime}$ in the electrochemical case. Notice moreover that the work terms appear in the formulas first as a sum and then as a difference.

"Here, w^r is the work required to bring the reactants together until their separation distance R is R_0, the *average R for those systems which react* (i.e., R_0 is the separation distance *in the activated complex*); w^p is the corresponding work term for the products for the same separation distance R_0 (R_0 is usually taken to be the distance of closest approach, but could be greater than that when the repulsions of reactants or products are large; λ and λ_{el} are *intrinsic 'reorganization factors,' describing in energy terms the reorganization of coordinates needed when* $E = E_0'$ *or when* $\Delta F^{o\prime} = 0$, expressions for them have been given [22, 33]; n is the number of electrons transferred *in the elementary step*; E_0' is the 'standard' potential of the half-cell for the prevailing temperature, medium, electrolyte, etc.; E is the potential of the half-cell; Z is the collision number, and equals *ca* 10^{11} liter mole^{-1} sec^{-1} for a solution reaction and *ca* 10^4 cm sec^{-1} for an electrode reaction"

NOTE: The value 10^{11} was later better estimated as 10^{12} in **M170**.

"κ is an *averaged probability of remaining on the lowest surface* on passage through the intersection region ($\kappa = 1$ for an adiabatic reaction); ρ is a *ratio of mean displacements*:

ρ = mean fluctuation in R_0 to mean fluctuation in 'activated complex region' along the abscissa in Fig. 1, being usually taken to be about unity. We have given [22] a more precise definition of ρ.

Sometimes reactions (8.34) and (8.35) are preceded or followed by other elementary steps, but all properties in Eqs. (8.36) to (8.38) refer explicitly to step (8.34) or step (8.35).

8.21. Comparisons of Rate Constants

We have summarized the deductions arising from Eqs. (8.36) to (8.38) [22, 33].

(i) The rate constant of a homogeneous 'cross-reaction,' k_{12}, is related to those of the two exchange-reactions, k_{11} and k_{22}, and to the equilibrium constant K_{12}, in the prevailing medium by

Eq. (8.40), *when the work terms are small or cancel*

$$Ox_1 + Red_2 \underset{\longleftarrow}{\overset{k_{12}}{\longrightarrow}} Red_1 + Ox_2 \tag{8.39}$$

$$k_{12} = (k_{11}k_{22}K_{12}f)^{\frac{1}{2}} \tag{8.40}$$

where

$$f = \frac{(\ln K_{12})^2}{4\ln(k_{11}k_{22}/Z^2)}. \tag{8.41}$$

Frequently, f is within an order of magnitude of unity.

(ii) The electrochemical transfer coefficient at metal electrodes is 0.5 for small activation overpotentials (i.e., if

$$|nF(E - E_0')| < |\Delta F_0^*|,$$

where ΔF_0^* is the value of ΔF^* for the exchange current, *when the work terms are negligible*.

See Eq. (87) of Ref. [22] for a more general expression for this transfer coefficient.

(iii) When a *substituent* in the coordination shell of a reactant is *remote* from the *central metal* atom and is varied in a *series*, a plot of the free energy of activation ΔF^* versus the 'standard' free energy of reaction in the prevailing medium $\Delta F^{\circ\prime}$ will have a slope of 0.5, if $\Delta F^{\circ\prime}$ is not too large (i.e., if $|\Delta F^{\circ\prime}|$ is less than the intercept in this plot at $\Delta F^{\circ\prime} = 0$). In this series, for a sufficiently remote substituent, λ *and the work terms are constant* but $\Delta F^{\circ\prime}$ varies. The slope of the ΔF^* versus $\Delta F^{\circ\prime}$ plot has been termed the *chemical transfer coefficient* [22], by analogy with the electrochemical terminology. See Eq. (87) of Ref. [22] for a more general expression for this transfer coefficient.

(iv) When a *series of reactants* is oxidized (reduced) by *two* different reagents, the *ratio* of the two *rate constants* is the same for all members of the series in the region of chemical transfer coefficients equal to 0.5 (i.e., in the region where $|\Delta F^{\circ\prime}| < |\Delta F^*|_{\Delta F^{\circ\prime}=0}$ in each case).

(v) When the *series of reactants* in Eq. (8.34) is oxidized (reduced) electrochemically at a given metal/solution potential difference, the *ratio* of the *electrochemical* rate constant to the *chemical* rate constant in Eq. (8.35) is the *same* for all members Ox of the series, in the region where the chemical and (work-corrected) electrochemical transfer coefficient is 0.5."

NOTE: M. always refers to the case $\Delta F^{o\prime} = 0$ which is the case for electron exchange. One has then $\Delta F^{*}|_{\Delta F^{o\prime}=0} = \lambda/4$ (if w^r and w^p are negligible). The second case to consider is that in which $\Delta F^{o\prime} \neq 0$ but it is small. Here, for instance $|\Delta F^{o\prime}| < |\Delta F^{*}|_{\Delta F^{o\prime}=0}$.

"(vi) The rate constant of a (chemical) *EE* reaction, k_{ex}, is related to the *electrochemical* rate constant at *zero activation overpotential*, k_{el}, for this redox system, according to Eq. (8.42) when the work terms are negligible

$$(k_{ex}/Z_{soln})^{\frac{1}{2}} \cong k_{el}/Z_{el} \tag{8.42}$$

where Z_{soln} and Z_{el} are collision frequencies, namely about 10^{11} liter mole^{-1} sec^{-1} and 10^4 cm s^{-1}. (In Eq. (8.42) $\cong$ should be replaced by $\geq$ when the ion–electrode distance in k_{el} exceeds one-half the ion–ion distance in k_{ex}.)"

NOTE: The relation of the kinetics of EE reactions to that of the cross reactions is analogous to the relation between the electric potentials of single electrodes and those of the galvanic cells. And llike in the electrochemical case the data on standard potential can be used for reactions in solution or at the electrodes, so here chemical and electrochemical rate constants are related.

8.22. Tests of These Relations

"Experimental tests of these various deductions have been summarized in recent surveys [33, 36, 37]. On the whole, the agreement is encouraging; there are four examples [33, 36, 37], however (all but one involving cobalt complexes), where Eq. (8.40) is in error by factors of 10^3 to 10^6.[(22)] Again, in the case of aromatic molecules or ions, electrode-reaction rates computed on the basis of homogeneous rates using Eq. (8.42) appear to be too fast [38]. Recently, the variation in electrochemical transfer coefficient α has been measured over a wide potential range and found to be in reasonable agreement with Eqs. (8.36) and (8.42) [39].

Comparisons of the experimental data of the type outlined in deductions (i) to (vi) test the similarity of various effects in the reactions being compared (e.g., absence of *spin restrictions*, absence of highly specific effects), and test the *effectively quadratic nature* of the two surfaces in Fig. 1 (The vibrational potential energy was taken to be *effectively a quadratic function* of *displacements* in Fig. 1, and the *ion–solvent free energy* to be a *quadratic function of fluctuations* in *local* orientation polarization, according to the assumptions listed earlier).

The principal discrepancy is expected to arise from highly specific effects (e.g., influence of strong adsorption at an electrode), from nonadiabatic effects (e.g., spin restrictions), or from different operative mechanism (e.g., presence of excited electronic states in one reaction and not in another). In deduction (i) any breakdown of the quadratic behavior of ΔF^*, particularly where $\Delta F^{0\prime}$ is large, could lead to serious numerical error."

NOTE: The M. relations are for ET reactions what the ideal gas law is for the theory of gases: a set of fundamental limiting laws. Here is a list of effects causing discrepancies from M. laws.

(i) Nonadiabatic effects (e.g., spin restrictions).
(ii) Highly specific effects (e.g., influence of strong adsorption at an electrode).
(iii) Bridging effects.
(iv) Presence of excited electronic states.

Note moreover that the vibrational PESs in ET are *quadratic* functions of displacements and also *quadratic* is the *free energy* function of local orientation fluctuations.

And finally the general expressions (8.37) and (8.38) for the activation energies are *quadratic* functions of $\Delta F^{\circ\prime}$ and of the work terms difference.

8.23. Absolute Values of A

"Other deductions from Eqs. (8.36) to (8.38) concern the numerical magnitudes of the quantities appearing in Eq. (8.36). Usually, experimental rate constants can be expressed as a function of temperature by

$$k = Ae^{-E_a/RT}, \tag{8.43}$$

where E_a, the activation energy, has the *experimental definition*

$$E_a = \frac{-R\partial \ln k}{\partial(1/T)}." \tag{8.44}$$

NOTE that E_a can be obtained from Eq. (8.43) by the derivation in Eq. (8.44) only if E_a is independent of temperature.

"The experimental value of A can be a rather revealing quantity, both for electrode and solution reactions: If $\kappa\rho$ were about unity and *if ΔF^* had no temperature dependence*,"

NOTE: $\Longrightarrow \Delta S^* = 0$ because

$$\Delta S^* = -\partial \Delta F^*/\partial T. \tag{8.45}$$

Comparing the Eqs. (8.36) and (8.43)

$$k = Z\kappa\rho \exp(-\Delta F^*/RT)$$

and

$$k = Ae^{-E_a/RT},$$

remembering that $\Delta F^* = \Delta H^* + T\Delta S^*$ so that we can write Eq. (8.36) as

$$\begin{aligned} k &= Z\kappa\rho \exp(-\Delta U^*/RT + \Delta S^*/R) \\ &\Longrightarrow k = Z\kappa\rho \exp(\Delta S^*/R)\exp(-\Delta H^*/RT), \end{aligned}$$

writing then the last equation as $k = A \exp(-\Delta H^*/RT)$, that is, writing A as

$$A = Z\kappa\rho \exp(\Delta S^*/R). \tag{8.46}$$

"*A would equal Z*... if a minor temperature dependence of $Z\kappa\varrho$ is ignored." "Thus, deviations of A from a value of *ca* 10^{11} liter mole^{-1} sec^{-1} (solution reactions) or 10^4 cm sec^{-1} (electrode reactions) reflect either

(i) A temperature dependence of ΔF^* or
(ii) A large difference of $\kappa\rho$ from unity.

Nonadiabaticity can make κ much less than unity, and so tend to make A/Z small. ρ probably never deviates much from unity, though there are some special circumstances[1] where it could be as large as 10.

When coulombic repulsions or attractions become important, w^r and w^p can become quite temperature-dependent in a way well-known when the solvent can be treated as a dielectric continuum.[(23)] The resulting value of ΔS^* can be quite different from zero, and that of A/Z quite different from unity. Addition of sufficient added electrolyte tends to make w^r and w^p small if the electrolyte introduces no other effects such as bridging. Then, the *coulombic contribution* to ΔS^* is also small. In homogeneous reactions which are not of the EE type, $\Delta F^{\circ\prime}$ and $\Delta S^{\circ\prime}$ do not vanish. This $\Delta S^{\circ\prime}$ provides *another contribution* to ΔS^* which can also be quite large. Both contributions, coulombic and $\Delta S^{\circ\prime}$, are included in Eq. (8.37)."

NOTE: In order to understand how $\Delta S^{\circ\prime}$ can enter in ΔS^*, just think of Eqs. (8.45) and of (8.37) and (8.38).

"In electrode reactions, $E - E^{\circ\prime}$ is usually held fixed as the temperature is varied, and so ΔS^* arises mainly from the $\mathrm{d}w^r/\mathrm{d}T$ and $\mathrm{d}w^p/\mathrm{d}T$ terms.

[1]If the coulombic repulsion is so large, and the dependence of the dielectric contribution to λ on R_0 so small, that $\kappa \exp(-\Delta F^*/RT)$ varies but slowly with R_0, the ΔR_0 appearing in the theoretical expression for ρ might be appreciable.

In the case of electrode reactions studied at high electrolyte concentrations, the experimental A/Z is typically unity [40], to within a factor of 10, suggesting that $\kappa\rho$ is also. Few detailed studies of A are available for homogeneous reactions at high electrolyte concentration. For those studied at low concentration, $\mathrm{d}w^r/\mathrm{d}T$ and $\mathrm{d}w^p/\mathrm{d}T$ effects are very apparent. Typically, EE reactions between ions of *like sign* cause a large ordering of solvent molecules near the activated complex, because of the *large charge* on it, and cause $\Delta S^*/R$ to be quite negative, about -10 to -15 in some cases."

Compare Eq. (8.46). Notice that "ordering" corresponds to $\Delta S^* < 0$ and that the gas constant R has the same units as S.

"This order of magnitude is the same as that calculated from differentiation of Eq. (8.37) with respect to temperature and using a dielectric continuum expression for the w's.

8.24. Absolute Value of E_a

Since E_a is defined only by Eq. (8.44), a *theoretical* value of E_a can be obtained only by inserting the theoretical expression for $k(T)$ into Eq. (8.44), a fact overlooked in a recent work [34] on $Fe^{2+} - Fe^{3+}$ exchange. For this reason, any temperature-dependent theoretical quantities such as w^r, w^p, and $\Delta F^{\circ\prime}$ contribute to E_a not only in their own right but also through their temperature derivatives. Therefore, E_a does 'not' equal ΔF^* exactly.(24)

The numerical value of E_a is obtained by inserting Eq. (8.37) or (8.38) into (8.44). There are several contributions to E_a. In a reaction in which $\Delta F^{\circ\prime}$ vanishes (e.g., in an EE reaction) or in an electrode reaction in which $E - E^{\circ\prime}$ vanishes, the *intrinsic reorganization terms* λ and λ_{el} are the *principal contributors* to E_a. These λ's contain

(i) A contribution from the coordination shell of each reactant and
(ii) A contribution from reactant–medium interactions

λ increases with increasing difference in '*equilibrium*' *bond lengths or angles* in each coordination shell before and after reaction and with increasing difference in '*equilibrium*' *polarization* of the solvent at each point of the medium before and after reaction. λ depends, too, on the *bond-force constants*. These effects are illustrated in Fig. 1.1a.[2]"

"Increasing or decreasing the $\Delta F^{o\prime}$ at fixed λ corresponds to raising or lowering the P surface relative to the R one."

"Since $\lambda/4$ is *approximately* the barrier height in Fig. 1, the theoretical expression for *the solvent polarization contribution to* λ bears one further comment. The dielectric continuum form [22] of this expression is easy to use but is necessarily approximate. The statistical mechanical form [22] of this expression is simple in its appearance but requires for its evaluation a good, simple statistical mechanical theory of equilibrium solvent–ion interaction.

Some information is available on force constants and bond lengths in coordination compounds. The various numerical calculations which have been made [34, 35] are not too far from the observed E_a's, but the exact values of the pertinent force constants and bond lengths are often somewhat uncertain as yet. These changes in

(i) Equilibrium bond lengths.
(ii) (and, in part, solvent polarization)

are believed to account for the *major* observed differences in rates of EE reactions. Until recently, the extreme slowness of the

---

[2]The above contributions to λ are illustrated in Fig. 1 when $\Delta F^{o\prime}$ is zero: When the equilibrium bond length undergoes a large change as a result of reaction, the two curves in Fig. 1 are considerably displaced from each other horizontally. They then intersect only at a high potential energy and so with a high E_a. The larger the force constants, the higher the potential energy at the intersection (Fig. 1) and the higher the E_a. *Similarly*, large changes in local equilibrium solvent polarization cause the two curves in Fig. 1 to be appreciably displaced from each other horizontally and increase E_a thereby. This polarization effect is large when the ion size is small and when the orientation polarization is large (i.e., when the static and optical dielectric constants are quite different). All of these effects are evident from the equations available for λ and λ_{el}.

homogeneous $Co(NH_3)_6^{2+}$ – $Co(NH_3)_6^{3+}$ exchange reaction was attributed to this source. However, recent crystallographic measurements [41] have revealed that the changes in equilibrium bond lengths were similar to those of a number of other (2+, 3+) coordination compounds which undergo an EE reaction at a much higher rate. The slowness of the $Co(NH_3)_6^{2+}$ – $Co(NH_3)_6^{3+}$ reaction may thus be due to a small value of κ.

Another reaction that is relatively slow is the $Co(phen)_3^{2+}$ – $Co(phen)_3^{3+}$ exchange, [42] the reaction being much slower than the $Fe(phen)_3^{2+}$ – $Fe(phen)_3^{3+}$ exchange. It is not yet known whether the slowness is due to a change in bond-length effect or to a small value of κ. The former would cause E_a to be larger in the cobalt reaction while the latter would cause κ to be smaller in that reaction. Coulombic effects would be expected to *cancel* when *ratios* of the two rate constants are compared in this manner.[25] Thus far, however, the $Fe(phen)_3^{2+}$ – $Fe(phen)_3^{3+}$ has been too fast to study, and it may be necessary to resort to indirect studies, utilizing Eq. (8.40), to explore these effects."

NOTE: As in electrochemistry it is possible to know a standard potential *indirectly* from the knowledge of others, or as in thermochemistry a formation enthalpy may be known from the knowledge of other enthalpies, so here Eq. (8.40) allows the same with rate constants (and with the equilibrium constant appearing in it).

8.25. Electrode Material

"We have not commented thus far on the nature of electrode material [43]. It affects the rate in several ways

(i) Because of its surface charge and
(ii) Because of its adsorption
(iii) It influences the double layer and
(iv) Other contributions to w^r and w^p.

When the difference $E - E'_0$ can be specified and controlled, differences in inner potentials in metal electrodes are automatically compensated by studying the reaction at a given $E - E'_0$. However, in other cases (some semiconductors, for example), $E - E'_0$ is unknown and has to be replaced by the theoretical expression from which it arose, an expression involving differences of electrochemical potentials at the electrode/solution interface. In that case the nature of the electrode material appears explicitly.

8.26. Other Applications and Concluding Remarks

There are a number of other applications that can be made of Eqs. (8.36) to (8.38) to various types of ET problems. For example, ET reactions of *excited states* are expected to obey Eq. (8.36) when the *appropriate* $\Delta F^{\circ\prime}$ and λ are introduced. The equations were used to formulate a theory of chemiluminescent reactions [44]. Again, the concepts were used to formulate a theory of solvated electron reactions, by allowing for the sensitivity of the charge cloud of the electron to solvent fluctuations [45].

We have considered some other applications elsewhere in the Elmau symposium. They include

(i) Thermal and photochemical ET.
(ii) The question of how much a standard free energy or energy deficit a reaction can tolerate and still occur.
(iii) Effect of vibrational readjustments on computed activation energies and on ratio of exchange currents at metals and degenerate semiconductors.
(iv) Atom-transfer reactions and possible modifications of the equations."

Errata: M. refers in points (i), (ii), (iii), and (iv) to papers of him which are nonexistent. They are papers numbered 14, 15, 16 in the references of the original paper.

"Although much is understood about the nature of ET processes, there are gaps in our knowledge. The uncertainties will be removed with

(i) Increased knowledge of bond lengths and force constants
(ii) Increased experimental knowledge of nonadiabatic effects, of
(iii) Specific interactions such as bridging and adsorption, and of
(iv) Solvent–ion–electrode interactions.

Our main tools in obtaining this knowledge may prove to be

(i) Comparative studies of rate constants, such as those listed earlier.
(ii) Measurements of the Arrhenius preexponential factors A under conditions where coulombic effects are either negligible or well-understood.
(iii) Crystallographic measurements of bond lengths and angles, and
(iv) Vibrational spectroscopic measurements of force constants.
(v) The study of photochemically induced ET reactions could also add to this knowledge (if free radicals are not formed), by providing direct information on the role of excited states."

M77. General Introduction to Faraday Society Discussions on Electrode Reactions of Organic Compounds, *Disc, Faraday Soc.* **45**, 7, (1968).

8.26.1. *Introduction and Survey*

"The last Faraday Society Discussion on Electrode Processes were held in 1947 and marked the onset of an extensive fruitful postwar development of fast reaction techniques in electrochemistry. Considerable information has since been obtained on rate constants of simple inorganic ETs, paralleling a similar development in homogeneous reactions, and on ET theory. A theoretically based relation exists between the two areas.

It has often been noted that the Faraday Society Discussions have marked a turning point in a field. The area of organic reactions at electrodes has been somewhat on the periphery of the field of organic chemistry, its relation to the subject of metal-catalyzed organic syntheses notwithstanding. The papers of the present Discussions illustrate, perhaps for the first time as a body, the rich variety of studies in the present and a role for current organic concepts.

The papers may be loosely classified into the following overlapping categories.

Primarily Theoretical—Hoytink, on energetics of reactions of aromatic ions leading to excited states and applications of ET theory; Mehl and Hale, on reactions at the interface of an anthracene electrode and an electrolyte and applications of the same ET theory; Hush and Segal, on the stability of CH_3F^- relative to CH_3F and possible implications for ET with bond fission; and Parsons, on the role of adsorption on reaction schemes and current-potential plots. Other papers employ several current theoretical concepts of physical organic chemistry.

Adsorption—Parsons, cited earlier; Brummer and Cahill, on interaction of H atoms with adsorbed species. Breiter, on anodic oxidation of CH_3OH; Conway et al., on reaction order, i–V plots and kinetic isotope effect for acetophenone reduction; Brown and Lister, on reaction order in acetone reduction; and Hickling and Wilkins, on inhibition of O_2 evolution by adsorbed species for Kolbe and other reactions.

Carbonyl Reduction—Conway et al., steady state studies, just cited; Brown and Lister, just cited, and hydrocoupling with acrylonitrile, study of intermediates via potential switching; and Zuman et al. on polarography of unsaturated ketones and aliphatic aldehydes.

Other Reductions—Lund, on polarographic reduction of heterocyclics; Gygax and Jordan, on polarographic and controlled potential electrolysis (c.p.e.) of hemoprotein models; and Littlehailes and Woodhall, on study of electrochemically and chemically prepared quaternary ammonium-sodium amalgams used in reductions.

Stereochemistry in Electrochemistry—Grabowski et al., on least structural change rule and electrode reactions, including inversion or retention of configuration in RCl reductions; Dietz and Peover, on effect of planarity on reduction rates of cis and trans substituted stilbenes and on related systems; and Bard et al. on comparison of polarographic reduction potentials, spin resonance data and m.o. calculations for sterically hindered substituted anthracenes.

Anodic Oxidation—Kuwana and Strojek, on spectral study (with optically transparent electrodes), cyclic voltammetry and c.p.e. for the oxidation of o-toluidine; Eberson and Nilsson, on anodic cyanation and similarities to data in homogeneous electrophilic substitutions; Tsutsumi and Koyama, on differences in anodic cyanation and anodic methylation, and relation to homogeneous relative reaction rates; Fleischman and Goodridge, on effect of square wave pulses in Kolbe and Hofer–Moest syntheses; Hickling and Wilkins, on square wave pulse and potential-switching studies in Brown–Walker and Kolbe syntheses.

8.27. Theoretical Review. Weak-Overlap Electron Transfer

In the organic reactions considered in this discussion there are those for which existing weak-overlap ET theory might suffice, in some cases with due allowance for the role of adsorption, and those (usually atom or PTs) in which the electronic interaction of the reactants is large. More specifically, the mechanism in the former case can be described in terms of the properties of two PESs, one for the reactants and one for the products [30]. The *weak interaction* causes a *small splitting* in the *intersection region* (Fig. 1). In the second case, the interaction is so large that the potential energy profile now resembles one of the alternatives in Fig. 2. It is pertinent, therefore, to review several aspects of these problems. In passing, one notes that the adsorption studies include those where the interest is centered on the adsorption phenomenon itself and those designed to find simple conditions for detailed kinetic investigation of reaction mechanisms.

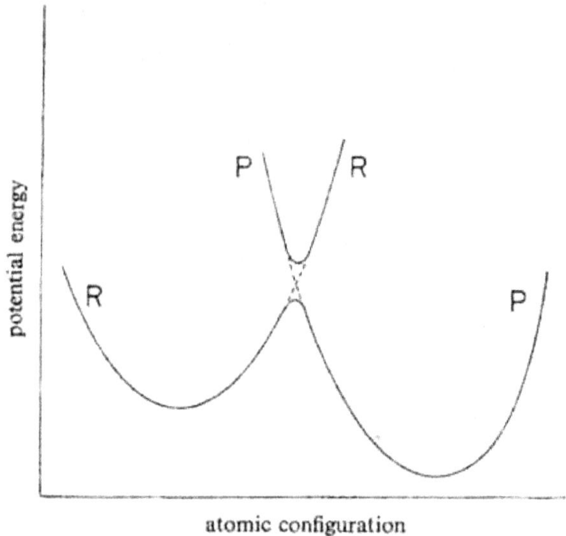

Fig. 1. Profile of potential energy surface of reactants (R) and that of products (P) plotted against configuration of all atoms in the system, for weak-overlap reactions; ---, surface for zero electronic interaction of the reacting species; —, adiabatic surface. The surfaces are given for some specified separation distance of the two reacting species. The plot for electrode reactions is similar but contains many surfaces (Fig. 2, Ref. [3]).

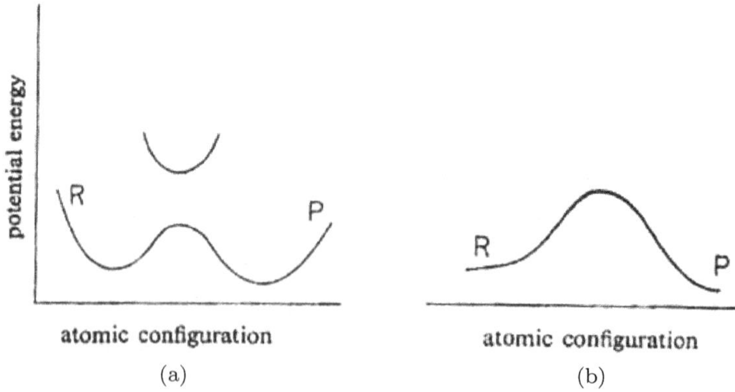

Fig. 2. Profile of potential energy surface for strong-overlap reactions, including atom and proton transfers. In system (a) the transfer occurs primarily near some given A_1A_2 separation distance; in (b) there is an appreciable change of A_1A_2 distance during the transfer (Appendix II, Ref. [10]).

We first recall that for weak-overlap ET reactions the following theoretical relation has been derived [46] and that there are a number of experimentally interesting consequences [7, 30].

$$k = \kappa\rho Z \exp(-\Delta F^*/RT) \tag{8.47}$$

$$\Delta F^* = w^r + (\lambda/4)(1 + \Delta/\lambda)^2 \tag{8.48}$$

where ρ is of the order of unity; Z is the collision number ($\sim 10^{14}\,\text{cm}^3\,\text{mole}^{-1}\,\text{sec}^{-1}$ for homogeneous reactions" 10^{14} was later changed to 10^{15}, see **M170** "and $\sim 10^4\,\text{cm}\,\text{sec}^{-1}$ for electrode reactions); κ is a transition probability, being unity for quantum mechanically adiabatic reactions; w^r is the work required to transport the reactants (or reactant plus electrode) together; $\lambda/4$ is an 'intrinsic barrier,' related to the barriers of certain isotopic exchange reactions";

NOTE: In which $\Delta F^{0\prime} = 0\ldots$

"Δ is $\Delta = \Delta F^{\circ\prime} + w^p - w^r$ for homogeneous reactions and is $\Delta = n_e F(E - E^{\circ\prime}) + w^p - w^r$ for electrode reactions, where n_e is the number of electrons transferred."

NOTE: In Fig. 2, M. represents the "Profile of PES for strong-overlap reactions, including ATs and PTs. In system (a) the transfer occurs primarily *near some given* A_1A_2 separation distance; in (b) there is *an appreciable change* of A_1A_2 distance during the transfer (Appendix II, Ref. [10])." In this Figure, M. shows how one passes from the typical PE profile for weak overlap ET of Fig. 1 to strong-overlap reactions in Fig. 2a and, in Fig. 2b, to the standard chemical kinetics representation of activation barrier.

"One set of corollaries of the equation arises because of an additivity property of λ and permits one to deduce relations such as Eqs. (8.49) to (8.51) and capable of experimental test [30, 47]

$$k_{12} \cong (k_{11}k_{22}K_{12}f_{12})^{\frac{1}{2}}, \tag{8.49}$$

$$(k_{11}/Z)^{\frac{1}{2}} \cong k_{\text{el}}/Z_{\text{el}}, \tag{8.50}$$

$$k_{12}/k_{\text{el},E} \cong \text{constant}, \tag{8.51}$$

Typo corrected in Eq. (8.49)

where k_{12} is the homogeneous reaction rate constant for CRs,

$$A_1^{ox} + A_2^{red} \xrightarrow{k_{12}} A_1^{red} + A_2^{ox}. \quad (8.52)$$

K_{12} is its equilibrium constant in the given medium, k_{ii} is the rate constant for the isotopic exchange reaction

$$A_i^{ox} + A_i^{red} \xrightarrow{k_{ii}} A_i^{red} + A_i^{red} \quad (8.53)$$

f_{12} is a known function of k_{11}, k_{22}, K_{12} and arises from the $\Delta^2/4\lambda$ term in the expansion of (8.48), k_{el} and $k_{el,E}$ are the rate constants of Eq. (8.54) at $E = E^{0\prime}$ and at any fixed E, respectively,

$$A^{ox} + M(e) \rightarrow A^{red}. \quad (8.54)$$

Conditions for deriving Eqs. (8.49) to (8.51) from (8.48) are noted in Ref. [8]."

NOTE: Expanding Eq. (8.48) we get

$$\Delta F^* = w^r + \left(\frac{\lambda}{4}\right)\left(1 + \frac{2\Delta}{\lambda} + \frac{\Delta^2}{\lambda^2}\right) = \frac{\lambda}{4} + \frac{\Delta}{2} + \frac{\Delta^2}{4\lambda}.$$

"From Eq. (8.48) one can derive a relation for $\partial\Delta F^*/\partial\Delta F^{o\prime}$, the Brønsted slope of homogeneous reactions.

$$\alpha = \frac{1}{2}(1 + \Delta/\lambda) \cong \frac{1}{2}(1 + \Delta F^{o\prime}/4\Delta F_0^*) \quad (8.55)$$

where ΔF_0^* is the intercept at $\Delta F^{o\prime} = 0$ of a plot of $-RT \ln k$ against $\Delta F^{o\prime}$. A related expression is obtained for electrode reactions.

Equation (8.48), which has a variety of consequences, has been derived with varying degrees of generality.[26] However, underlying each derivation are *two principles*, which might be termed '*conservation*' and '*quadratification*.' The former is revealed in various ways. For example, the function $\boldsymbol{P}_u(\boldsymbol{r})$ related to the orientation polarization at each point $\boldsymbol{r}$ in the medium when the system is an activated complex, was shown to be given by [48].

$$\boldsymbol{P}_u = \alpha_u[(1 - n)\boldsymbol{E}^r + n\boldsymbol{E}^p], \tag{8.56}$$

where n proves to be

$$n = \frac{\partial \Delta F^*}{\partial \Delta F^{\circ\prime}} \text{ in the homogeneous system and}$$

$$n = \frac{\partial \Delta F^*}{\partial [n_e F(E - E^{\circ\prime})]} \text{ in the electrode reaction one.}$$

($\mathbf{E}^r$ and $\mathbf{E}^p$ are electrical fields; α_u is a susceptibility.)

Again, a more general (statistical mechanical rather than dielectric continuum) treatment of the environment yields Eq. (8.57) for the effective potential energy governing the Boltzmann-weighted *configurational distribution* of the activated complexes in many-dimensional configuration space, that is, of systems at the intersection of the two surfaces in Fig. 1 [22, 30]

$$U^{\text{eff}} = (1 - n)U^r + nU^p. \tag{8.57}$$

(U^r and U^p are the potential energies of a system possessing the charge distribution of reactants and that of products, respectively.) Thus, there is a 'conservation,' in the sense of a 'give and take' between the properties of U^r and those of U^p, or between $\boldsymbol{E}^r$ and $\boldsymbol{E}^p$.

The 'quadratification' arises from two of the approximations made in deriving Eq. (8.48). One is that the vibrational motion within the coordination shell of the reactants is harmonic (or, over the change of distances involved, can be treated as that of an equivalent harmonic oscillator). Thereby, the vibrational potential energy is a quadratic function of *fluctuations in vibrational coordinates.* The second assumption is that the surrounding medium can be treated as dielectrically unsaturated, that is, that *the free energy* of the system is a quadratic function of *fluctuations* in dielectric polarization of the medium.[27]

We note further that for weak-overlap ET reactions the theoretical interpretation for the Brønsted slope, $\partial\Delta F^*/\partial\Delta F^{o\prime}$, or for its electrochemical counterpart $\partial\Delta F^*/\partial[n_e(E - E^{o\prime})]$, is [22]

$$\alpha = \langle\partial U^r/\partial q\rangle/(\langle\partial U^r/\partial q\rangle - \langle\partial U^p/\partial q\rangle), \qquad (8.58)$$

where q is the reaction coordinate and the averages $\langle\ \rangle$ are over the activated complexes in the many-dimensional configuration space, distributed in a Boltzmann population [49].

8.28. Theoretical Review. Atom and Proton Transfers

A somewhat different situation confronts us for ATs, PTs, or very-strong-overlap ETs. Now, the usual splitting of reactants' and products' PESs is *so large* in the *activated complex region* that a discussion of mechanisms based on intersecting surfaces and on a calculation of their properties is no longer meaningful. Thus, quantities such as *slopes* $\partial U^r/\partial q$ and $\partial U^p/\partial q$ are no longer well-defined[(28)] and one must seek elsewhere for a derivation of a theoretical expression for ΔF^* and for a theoretical interpretation of α. The counterpart of reaction (8.52)

$$A_1^{ox} + A_2^{red} \xrightarrow{k_{12}} A_1^{red} + A_2^{ox} \quad \text{ET}$$

is now

$$A_1B + A_2 \rightarrow A_1 + BA_2 \quad \text{AT} \qquad (8.59)$$

where B is the atom or proton being transferred. In an electrode reaction, the reaction may involve bond ruptures such as the following:

$$AB + M(e) \rightarrow A + B^- + M \quad \text{bond breaking (BB)} + \text{ET} \qquad (8.60)$$

$$C + AB + M(e) \rightarrow CA + B^- + M \quad \text{AT} + \text{ET} \qquad (8.61)$$

$$AB + M(e) \rightarrow A + BM(e) \rightarrow \text{etc.''} \quad \text{BB} + \text{adsorption (ADS)} \qquad (8.62)$$

NOTE: I have written the name of the class of each reaction to the side of the above symbols.

"As noted earlier, the theoretical constructs associated with two intersecting PESs should now be replaced by others. At the present time, resort must be made to approximate treatments. (Even the three-electron system $H + H_2 \rightarrow H_2 + H$, is only now beginning to yield to more exact approaches.) One simple model is the bond-energy bond-order model of H. S. Johnston [12, 50]. He assumed that the *total bond order* in Eq. (8.59) was *constant* along the reaction path, utilized bond-order bond-energy relations, found the potential energy maximum along the path, and calculated activation energies using parameters obtained from nonkinetic data. He obtained reasonable agreement with the data.

This model offers a simple method for exploring some of the topics cited earlier. The model obeys, incidentally, the *conservation* theme described earlier, the 'conservation' now being one of a bond order. Recently, we showed [51] that calculations of potential energy barriers based on the Johnston model are well-approximated by Eq. (8.48) even though the potential energy profile along the reaction coordinate now has a vastly different shape (Fig. 2b). Correspondingly, for ATs, when there are only minor solvation corrections and when the entropy changes cancel or are minor, Eq. (8.48) and its consequences such as Eqs. (8.49) and (8.55) once again obtain [52, 53].[(29)] (**Typo** corrected in Eq. (8.49). However, theoretical interpretation of the Brønsted slope α is now different. It was shown to be [1

$$\alpha = n^{\neq}_{BA_2}, \tag{8.63}$$

where $n^{\neq}_{BA_2}$ is the bond order of the BA_2 bond *being formed* in the activated complex of reaction (8.59). That is, the interpretation of α no longer depends on quantities such as $\partial U^r/\partial q$, which are not meaningful for the present profiles, but is expressed instead in terms of properties of the new model, *bond orders* in the present instance.

When large solvation changes occur, as in PTs, there can be *two contributions* to the barrier. One involves *desolvation* of one A_1 and *solvation* of the other in reaction (8.59). (Now, B is H^+ there.) A second involves the barrier, if any, for the gas-phase PT. The treatment of these reactions in solution is currently being explored by us using a combination of a method used for gas-phase PTs and one related to that employed for treating solvation in ETs. In the meanwhile, Eq. (8.48) has been used as a *conjecture* to analyze data on PTs, Brønsted slopes, and kinetic isotope effects [53]. While the possible applicability of related theoretical considerations to the electrode reactions (8.60) to (8.62) has not yet been explored, we intend to investigate this problem. Experimental kinetic data for various electrode reaction *series* on these reactions and on analogous homogeneous reactions would provide a useful guide in analyzing this somewhat complex theoretical situations.

I shall conclude this section with an example relating homogeneous and electrochemical results on reduction of acetophenones. The polarographic wave for reduction of substituted acetophenones in high acid media is reversible and is believed to be associated with the reaction

$$\begin{aligned} &\mathrm{X - C_6H_4 - CO - CH_3 + H^+ + M(e)} \\ &\rightleftharpoons \mathrm{X - C_6H_4 - \dot{C}OH - CH_3 + M.} \end{aligned} \tag{8.64}$$

Data are available on the effect of X on the half-wave potential, $E_{\frac{1}{2}}$ [54]. The equilibrium results may be compared with kinetic data [55] for addition of radicals R· (obtained from di-t-butyl peroxide) to substituted acetophenones.

$$\mathrm{X - C_6H_4 - CO - CH_3 + R\cdot} \xrightarrow{k} \mathrm{X - C_6H_4 - \dot{C}R - CH_3} \tag{8.65}$$

The slope of kinetic plots, $\log k_x$ against σ_x, where σ_x is Hammett's substituent constant for substituent X, is designated by ρ_c. The slope of the $E^x_{\frac{1}{2}}$ against σ_x plot for reaction (8.64) can be written analogously as $(2.3RT/n_eF)\rho_{\text{el}}$. (This ρ_{el}, which has the same units as ρ_{c}, differs from that defined in Ref. [54] by a factor of $2.3RT/n_e F$.) If the principal effect of the X in reaction (8.65) were to alter the $\Delta F^{o\prime}$ for that reaction, changes in $\Delta F^{o\prime}$ would be equal to changes in $E_{\frac{1}{2}}$ in Eq. (8.64), assuming the difference of chemical potentials of the radicals in the right hand side of Eqs. (8.64) and (8.65) to be independent of X. Thus, the Brønsted slope for reaction (8.62) can now be written as

$$\alpha = \frac{\partial \Delta F_{\text{X}}^{*}}{\partial \Delta F_{\text{X}}} = \frac{2.3RT}{nF}\,\frac{\partial \log k_{\text{X}}/\partial \sigma_{\text{X}}}{\partial E_{\frac{1}{2}}^{\text{X}}/\partial \sigma_{\text{X}}} = \frac{\rho_{\text{c}}}{\rho_{\text{el}}}.$$

ρ_c was found to be 1.59 [55] and ρ_{el} ranges from 3.2 to 4.5 for various media and investigators [54]. The corresponding values of α thus lie in the range 0.50 to 0.35. It would be desirable, however, to make the comparison of this equilibrium electrochemical and kinetic homogeneous data in a similar solvent medium, and to extend the comparison to other systems.

8.29. Conclusions

The papers of the present discussion illustrate a rich variety of mechanistically interesting reactions in the field of organic reactions at electrodes. Some are presumably weak overlap ETs, for which existing theory may suffice, with detailed *analysis of adsorption* when needed. Some are ATs involving *concerted* or *dissociative* bond rupture (i.e., $S_{\text{N}2}$ or $S_{\text{N}1}^{o}$, respectively). For such reactions a different theoretical framework is being constructed and being tested. It has several features in common with the first type of theory, but also has some certain differences arising from the *different shape* of the potential energy profiles in the two cases. A theoretical interpretation of the Brønsted slope, for example, is in one system given by Eq. (8.58) and in the other by Eq. (8.63). Theory and experiment have much to profit by continuing for these reactions the useful partnership they have enjoyed in the past decade for weak-overlap ETs."

NOTES

1. The cross relation of Nernst is exact while Marcus' cross relations, the one about the λ's and the one about the k's, are approximate.
2. Q: What does it mean normalization for function $g_i(n)$?

   M: One should use not use the word "normalization" but, rather, "scaling."
3. Q: Shouldn't it be expanded in powers of n around $n = \frac{1}{2}$?

   M: In powers of $n - \frac{1}{2}$.
4. **M**: This equation is going to be valid only for small $\Delta E^\circ/4E$. This is OK for ET. A correct relation is a hyperbolic function equation which can be reduced to the ET equation for small $\Delta E^\circ/4E$. If you plot the barrier for ET you have the quadratic barrier.
5. **M**: One should better speak of any "reaction variable."
6. Q: But the degree of reaction variable does not appear in the theory of M. from which Eq. (8.4) is derived. n appears for the first time in Eq. (8.11).

   M: It was the $-m$ ...
7. **M**: I symmetrized the expressions, you remember. In ET and in AT.
8. **M**: Look at the 1965 paper where I didn't assume that the force constants before and after ET were the same. We used the fact that they may be different. In the limit in which ΔE° goes to zero the result becomes exact. This is the compensation. I have a perturbation in the 1965 paper, the perturbation being that $(k_1 - k_2)/(k_1 + k_2)$. Later on people called it a "torsion parameter" known by τ, I believe.
9. **M**: Suppose that there was a large work term. That would cause a breakdown not having that work term in it. Not correcting for work terms then, if the work terms differ quite a bit for the

two self-exchange reactions then you have a breakdown. Only if you correct k_{11} for the work term, correct k_{22} for its work term, then the relationships would be correct. These work terms are independent of n. You should correct all the rate constants for the work terms.

10. **M**: That fraction, as an example, may occur in the formation of the precursor complex. Before the things get closer and before the reorganization really starts. "Any change in free energy before you start to reorganize, such as work terms."
11. **M**: Supposing you have an organic reacting with an inorganic reactant then for the self-exchange reaction you would have the work term for organic–organic and inorganic–inorganic and surely the meaning of those are not going to be that of organic–inorganic interaction.
12. "The Brønsted slope α, is expected to be 0.5, when $\Delta F^{\circ\prime} = 0$. However, deviations would occur

    (i) if $c \neq 0$ in Eq. (8.20): $k_{12} \cong (k_{11}k_{22}K_{12}^{1+c})^{1/2}$, that is, if an appreciable fraction of the free energy change were independent of the reaction parameter n, or
    (ii) if $g_i' \neq 0$ in Eq. (8.13): $x^{\neq} = 2[-\Delta E^{\circ} - \frac{1}{2}(E_{11}g_1' - E_{22}g_2')]/(E_{11}g_1'' + E_{22}g_2'')$ and, at the same time, the asymmetry $\varepsilon = \frac{E_{11}-E_{22}}{E_{11}+E_{22}} = \frac{E_{11}-E_{22}}{2E}$ is large."

    Q: In (i) does it all boil down to work terms or is there something else?

    M: It depends on what you include in work terms, for instance steric effects. Should we include them? If you include them, then you would have a work working "sterically."
13. Q: Where was this demonstrated for the first time, referring explicitly to parabolas?

    M: You find it in some tutorial, for example, the one in Peter Rock's book.

14. Q: what does *pseudo-vibrational* mean?

    M: It means that at the crossing of the parabolas, where the splitting is, the parabolas are not exactly parabolas anymore and there, at the crossing, the movement is not going anymore over a pure harmonic vibration.
15. **M**: That is the equation that says: don't use the parabolas beyond a certain ΔE° because there is no inverted effect. You cannot use a parabola up to that point. You get a deviation. So, for the forward reaction there is only the forward work term and for the backward reaction there is only the reverse work term.
16. **M**: It gives a symmetry property. If you look at the right side of the equation and you make a transformation: $r \to p$, $p \to r$, $n \to 1 - n$, then you get the equation for ΔF_R^{*p}. Written in this way, a symmetry property comes out.
17. **M**: This is ΔF_R^* at the value of n. n at the transition state is derived by minimization. You write ΔF_R^* as a function of n. Only at the transition state $\frac{\partial \Delta F_R^*}{\partial n}$ becomes zero. ΔF_R^* is gone through a maximum.
18. Q: Apparently one can do with distribution functions what one does with wave functions of systems made up by two widely separated parts.

    M: OK.
19. **M**: There was, supposedly, a Lake Eyring which dried up....
20. **M**: When they are separated it would be actual but when they are together they are not really independent but there is an approximate cancellation.
21. Q: Can you give examples of effects that are *canceled* and others that are *revealed*?

    M: The harmonic approximation occurs in both the isotopes and in the other thing, so it approximately cancels, the dielectric continuum approximation which I used initially occurs in all the effects and it cancels out. In the final formula that involves a cross relation no dielectric constant, no properties of the vibrations of the molecules appear there, no properties of the solvent appear there, they all cancel out in that approximation, so many properties that appeared initially cancel, so many of them do. What if you have different force constants? Then to the extent that the symmetrization is OK, in the 1965 paper I expanded in symmetric and asymmetric powers in order to get that simple relation.

22. Q: Please tell something about the orders of magnitude of differences in constants of rates that are acceptable or unacceptable.

    M: A factor of 10 is quite acceptable if you deal with rates considering that there are 14 orders of magnitude in the rate constants.
23. Q: Can you give a reference for the dependence of work terms on temperature?
    M: If you look at the expression for the work terms in the original paper, then you see the dielectric constant appearing there, so you just differentiate that appropriately with respect to temperature. There is quite an entropic contribution, not just an energetic contribution to the ion dipole interaction for the work terms whereas for the reorganization it is $\frac{1}{\varepsilon_r} - \frac{1}{\varepsilon_s}$ so ε_s plays the role because is normally so much bigger than ε_r, see the dependence of the work term on the dielectric constant and on how it depends on the temperature. There is an entropic contribution, not just an energetic contribution to the ion dipole interaction for the work terms. If your $\Delta G°$ is accompanied by a large $\Delta S°$, and Sutin and I speak about this in a paper and also in a 1985 review, then that can have consequently a temperature effect, you can even get a negative activation. Is a negative activation barrier a nonsense? Only when you think of a barrier as an energy barrier, but when it is a free energy barrier it is no longer nonsense. There are many reactions actually, gas-phase reactions where when you get to very low temperatures the rate increases when the temperature decreases. *The transition state is changing with temperature.*

24. **M**: ΔF^* has a temperature dependence, E_a is an empirical quantity, E_a does not depend from temperature, by its definition. It is a two parameters expression and you suppose a temperature independent E_a.
25. Q: Why does the ratio cancel the coulomb effects? Maybe because the ratio of constants corresponds to a difference in free energies and in the free energies there are the coulombic effects which are now the same because of the common (+2, +3) charges of the two couples of ions?

    M: OK.
26. Q: Do you refer to the two derivations with dielectric continuum theory and with statistical mechanics or to something else?

    M: That' right and in the statistical mechanics approach it involves a symmetrization. In the 1960 paper, before I did the symmetrization I assumed that the two force constants were the same. Later, in the 1965 paper, I would do more generally, I considered two different force constants and introduced symmetrical and antisymmetrical functions of the force constants and then neglecting terms involving the antisymmetrical ones. Varying degrees of generality.
27. Q: What about the relation of the different degrees of dielectric saturation (saturation, partial saturation, unsaturation) to your theory?

    M: There simply is a curve giving $\boldsymbol{P}(r)$ as a function of $\boldsymbol{E}(r)$. This curve goes first up almost linearly, it is proportional, then there is a plateau. The beginning of the curve describes the unsaturation, the plateau describes the saturation. In between there is a range in which there is a partial saturation.
28. Q: In which sense are they no longer well-defined? Geometrically they are defined at every point of the potential energy profile.

    M: In the case of AT's you no longer have parabolas, you see. [Here M. refers to the case when you have two parabolas.] In the

AT case you have distortions, you don't anymore have the U^r and U^p curves. It is the U^r and U^p that are not there anymore.

29. **M**: See, what it may be is that the AT expression would be derived without talking about the solvent. So, I was not prepared to tell what would happen with a solvent. In principle maybe in a gas-phase reaction the products might have more flexibility, internal flexibility, than the reactants. So that is not taken into account, it just focuses on the energetics. So, it was restricted, probably more restricted than it had to be, because it might well apply to reactions in solution, you see, if you have strong solvation effects, then you introduce the hyperbolic functions. Why? When you have the strongly interacting situation, then when the ΔG° gets more negative it just goes downhill.

References

1. R. A. Marcus, *J. Phys. Chem.* **72**, 891, (1968).
2. R. A. Marcus, *Electrochim. Acta.* **13**, 995, (1968).
3. R. A. Marcus, *Discuss. Faraday Soc.* **45**, 7, (1968).
4. (a) R. A. Marcus, *J. Chem. Phys.*, **24**, 966 (1956); (b) R. A. Marcus, *Discussions Faraday Soc.*, **29**, 21 (1960); (c) R. A. Marcus, *J. Phys. Chem.*, **67**, 883, 2889 (1868); (d) R. A. Marcus, *Ann. Rev. Phys. Chem.*, **15**, 155 (1964); (e) R. A. Marcus, *J. Chem. Phys.*, **43**, 679 (1965); (f) a review of this work and of that of other investigators (e.g. Levich and Dogonadze, Hush) is given in Ref. 4(d). A factor of $\kappa\rho$ omitted in Eq. (1), is about unity fora n adiabatic reaction and is not relevant for the present purposes.
5. When the reaction is partially or completely diffusion controlled, k is only one contribution to the observed rate constant. For example, when the diffusion step is followed by an irreversible reaction step, $k_{\text{obsd}}^{-1} = k^{-1} + k_{\text{diff}}^{-1}$. Compare R. A. Marcus, *Proceeding of the Symposium on Exchange Reactions, Upton, New York,* **1**, 281, (1965) and Ref. [7].
6. The cross-reaction is $\text{A}_1^{\text{ox}} + \text{A}_2^{\text{red}} \rightarrow \text{A}_1^{\text{red}} + \text{A}_2^{\text{ox}}$, where A_1^{ox} and A_1^{red} differ only in their redox state. The exchange reactions are $\text{A}_1^{\text{ox}} + \text{A}_1^{\text{red}} \rightarrow \text{A}_1^{\text{red}} + \text{A}_1^{\text{ox}}$ and $\text{A}_2^{\text{ox}} + \text{A}_2^{\text{red}} \rightarrow \text{A}_2^{\text{red}} + \text{A}_2^{\text{ox}}$.
7. R. A. Marcus, *J. Phys. Chem.* **67**, 853, 2889, (1963).
8. The work terms are omitted in Eq. (8.4). When they are included and $f_{12} \cong 1$, the right side of Eq. (8.4) has an additional factor $\exp[-(w_{12}^r + w_{12}^p - w_{11}^r - w_{22}^r)/2RT]$. (Note that $w_{ii}^r = w_{ii}^p$.) When f_{12} is not close to unity, the appropriate correction of Eq. (8.4) is made using Eq. (8.2).

9. N. Sutin, *Ann. Rev. Phys. Chem.* **17**, 119, (1966).
10. (a) N. Sutin, *Ann. Rev. Phys. Chem.* **17**, 154, (1966); (b) N. Sutin, *Proc. Exchange Reactions Symp.*, Upton, N.Y., **7** (1965); (c) H. Haim, N. Sutin, *J. Am. Chem. Soc.* **88**, 434, (1966).
11. A. O. Cohen, R. A. Marcus, unpublished data.
12. C. Parr, H. S. Johnston, *J. Am. Chem. Soc.* **85**, 2544, (1963).
13. The mean-square deviation for these calculations by Mrs. Audrey Cohen was 1.5 kcal mole^{-1}.
14. (a) E.g., J. E. Leffler, E. Grunwald, *Rates and Equilibria of Organic Reactions*, p. 157, John Wiley and Sons, Inc., New York (1963); (b) see also R. P. Bell, *The Proton in Chemistry*, Chapter 10, Cornell University Press, Ithaca, NY (1959); J. O. Edwards, *Inorganic Reaction Mechanisms*, Chapter 3, W. A. Benjamin, Inc., New York (1964).
15. We note that Eqs. (8.6) and (8.2) form a continuous function for ΔF^* for all values of $\Delta F_R^{\circ\prime}$. Also, in each case in Eq. (8.6) the rate constant or that of the reverse reaction is now essentially diffusion controlled in solution, and one calculates k_{obsd} accordingly, Ref. [5].
16. In this reaction, there is a relatively *abrupt charge transfer*, namely near a value of the reaction coordinate defining the intersection of reactants' and products' PESs. In the case of proton or strong-overlap ETs, the *change of charge distribution* is expected to occur *more gradually*, that is, *over a wider interval along the reaction coordinate.*
17. For simplicity, the method discussed is the one originally used in H. S. Johnston, *Adv. Chem. Phys.* **3**, 131, (1960).
18. While $(p_i - 1)\ln 2$ varies widely from bond to bond, it averages around 0.05 to 0.1, so that the energy barrier (Eq. 8.9b) of an atom-exchange reaction might be about 5% to 10% of the bond energy, a reasonable figure. In a few cases $p_i - 1$ was negative, so the maximum occurred at the end points and one sets $E_{ii} = 0$ instead of Eq. (8.9b).
19. In this notation, the n and $n - 1$ in the last two terms of Eq. (8.11) are the arguments of the functions g_i rather than multiplying factors.
20. (a) Equation (8.8a) follows from Eq. (8.11) by setting $g_i(n) = 4n(1 - n)$ in the interval $(0 < n < 1)$ and so outside of the infinitesimal regions around $n = 0$ and $n = 1$. In those regions, one could choose the g_i to approach zero extremely rapidly and, thereby, approximate Eq. (8.8b) as closely as desired. The details of this choice of g_i are unimportant for our purposes, but a study of the g_i in footnote 20b is revealing; (b) to obtain Eq. (8.10) one sets $n = n_2$, $g_1(n) = g_2(n) = -2(1 - n)\ln(1 - n)/\ln 2$.
21. E.g., one finds $(\Delta E^\circ \ln 2)/E = \ln[n/(1 - n)]$. Thus, $n = \exp(2y)/[1 + \exp(2y)] = \exp(y)/2\cosh y$ and $1 - n = \exp(-y)/2\cosh y$. Substitution in Eq. (8.14) yields Eq. (8.15).
22. R. A. Marcus, *J. Chem. Phys.* **43**, 679–701, (1965).

23. R. A. Marcus, *J. Chem. Phys.* **38**, 1858, (1963).
24. R. A. Marcus, *J. Chem. Phys.* **39**, 1734, (1963).
25. S^r can be expressed in terms of a ratio of a vibrational partition function of the activated complex to the rotational-vibrational one of the reactants. s^r can be expressed several ways, one being intuitive and all of which give a similar answer. Compare Ref. [14a], p. 133. See also R. A. Marcus, *J. Chem. Phys.* **43**, 1601, (1965), Eq. (17); E. W. Schlag, G. L. Haller, *J. Chem. Phys.* **42**, 584, (1965); D. M. Bishop, K. J. Laidler, *J. Chem. Phys.* **42**, 1688, (1965).
26. Of the six translational degrees of freedom of the two reactants (masses m_1, m_2), five coordinates become three translations and two *principal* rotations of the activated complex and a sixth may *mix with other coordinates to yield the reaction coordinate and others.*
    The partition function ratio of the *five* coordinates *of the activated complex* to the *six of the reactants* is

$$\frac{[2\pi(m_1+m_2)kT]^{\frac{3}{2}}(8\pi^2 IkT)}{\sigma(2\pi m_1 kT 2\pi m_2 kT)^{\frac{3}{2}}(h^5/h^6)}$$

    where I is the relevant principal moment of inertia, written suggestively as μR^2, σ is a symmetry number ($\sigma = 1$ or 2 according as the reactants are unlike or like), and $\mu = m_1 m_2/(m_1+m_2)$. This ratio equals $\frac{hZ}{kT}$, where Z is defined as $\frac{(8\pi kT/\mu)^{\frac{1}{2}}R^2}{\sigma}$ and equals about 10^{11} liter mole^{-1} sec^{-1}.
27. In writing Eq. (8.26) $\frac{1}{2}(w^r + w^p)$ was neglected relative to ΔF_0^*, since it is normally much smaller. To avoid this approximation, the ΔF_0^* in Eq. (8.26) can be replaced by $\Delta F_0^* - \frac{1}{2}(w^r + w^p)$.
28. E. S. Lewis, C. H. Funderburk, *J. Am. Chem. Soc.* **89**, 2322, (1967); Eq. (8.25) shows that the diffusion-control aspect is incidental, rather than necessary.
29. All zero-point effects have been included in the λ's.
30. R. A. Marcus, *Discuss. Faraday Soc.* **29**, 21–31, (1960).
31. E.g., I. Amdur, G. G. Hammes, *Chemical Kinetics: Principles and Selected Topics*, p. 47, McGraw-Hill Book Co., New York (1966).
32. We have replaced $-m$, in Ref. [4] by n, to conform with present notation.
33. R. A. Marcus, *Ann. Rev. Phys. Chem.* **15**, 155, (1964).
34. V. G. Levich, in *Advances in Electrochemistry and Electrochemical Engineering*, Vol. 4, p. 249, P. Delahay, C. W. Tobias, Editors, Interscience, New York (1966).
35. N. S. Hush, *Trans. Faraday Soc.* **57**, 557, (1961).†

† Hush's calculations can also be treated as a means for calculating the force constants and equilibrium bond lengths. The latter can be introduced into Eqs. (8.20.4) and (8.20.5).

36. N. Sutin, *Annu. Rev. Phys. Chem.* **17**, 119–142, (1966).
37. W. L. Reynolds, R. W. Lumry, *Mechanisms of Electron Transfer*, Ronald Press, New York (1966).
38. P. A. Malachesky, T. A. Miller, T. Layloff, R. N. Adams, *Symposium on Exchange Reactions*, p. 157. International Atomic Energy Agency Brookhaven National Laboratory (1965).
39. R. Parsons, E. Passeron, *J. Electroanal. Chem.* **12**, 524, (1966).
40. J. E. B. Randles, K. W. Somerton, *Trans. Faraday Soc.* **48**, 937, (1952); M. I. Tempkin, *Zh. fiz. Khim.* **22**, 1081, (1948).
41. M. T. Barnet, B. M. Craven, H. C. Freeman, N. E. Lime, J. A. Ibers, *Chem. Commun.* **24**, 307, (1966).
42. B. R. Baker, F. Basolo, H. M. Neuman, *J. Phys. Chem.* **63**, 371, (1959).
43. R. Parsons, *Surf. Sci.* **2**, 418, (1964).
44. R. A. Marcus, *J. Chem. Phys.* **43**, 2654, (1965).
45. R. A. Marcus, *J. Chem. Phys.* **43**, 3477, (1965); *Adv. Chem. Ser.* **50**, 138, (1965).
46. A theoretical review is given in Ref. [33], where comparison is made with the interesting results of Levich and Dogonadze, Hush, Gerischer, and other investigators.
47. For a recent survey of experimental data, see Ref. [9].
48. Reference [56]. We have changed the notation somewhat, for example, n replaces $1 + m$.
49. It follows from Eq. (8.58) that when the P surface in Fig. 1 intersects the R surface near the latter's minimum, $\langle \partial U^r/\partial q \rangle$ vanishes and α is zero. When the intersection occurs instead near the minimum of the P surface, $\langle \partial U^p/\partial q \rangle$ vanishes and α is unity. When instead the magnitude of the average slopes in Fig. 1 are equal at the intersection, that is, when $\langle \partial U^p/\partial q \rangle = -\langle \partial U^r/\partial q \rangle$, α equals 0.5. In these three cases, one sees that the activated complex, which occurs at the intersection of the two surfaces, resembles the reactants, the products, and a complete compromise between the favorable configurations of reactants and products, respectively.
50. H. S. Johnston, *Adv. Chem. Phys.* **3**, 131, (1960).
51. Reference [1]. See Eq. (8.6) for the necessary modification of Eq. (8.48) when $|\Delta/\lambda| \geq 1$.
52. A pioneering application of Eq. (8.48) to atom transfers in solution was made by N. Sutin in Ref. [9] p. 154 and in Ref. [10a] p. 7 and by A. Haim, N. Sutin, *J. Am. Chem. Soc.* **88**, 434, (1966).
53. A. O. Cohen, R. A. Marcus, *J. Phys. Chem.* **72**, 4249, (1968).
54. P. Zuman, *Substituent Effects in Organic Polarography*, Plenum Press, New York (1967).
55. E. S. Huyser, D. C. Neckers, *J. Am. Chem. Soc.* **85**, 3641, (1963).
56. R. A. Marcus, *J. Chem. Phys.* **24**, 966–978, (1956).

CHAPTER 9

Activated-Complex Theory, Application to Electron Transfer Reactions. Marcus' Notes on Electron Transfer, on Proton Transfer and on Electrode Kinetics

In this Chapter, **M49**, **M50**, **M74**, **M105**, **M118**, **M119**, and **M134** are considered.

In **M49** and **M50**, M. generalizes the activated-complex theory by extending it to curvilinear reaction coordinates. Excerpts from the papers are reproduced here that are of interest for electron transfer (ET) theory. Some concepts, formulas, and symbols appearing here have been used in the fundamental **M53**. There is an abridged description of **M74** with Marcus' Notes at the end. The Sections of **M105** here reproduced are the ones important for ET reactions, Marcus' Notes are at the end. Oral interviews with Marcus about **M118**, **M119**, and **M134** are finally presented, each one preceded by a brief presentation of the paper.

M49. Generalization of Activated-Complex Theory of Reaction Rates. I. Quantum Mechanical Treatment

"In activated-complex theory the reaction coordinate has been assumed to be Cartesian and, in quantum mechanical treatment at least, to be dynamically separable from the other coordinates. The effect of any rotational constants of the motion on the internal motion of the activated complex has normally been neglected, and equilibrium between reactants and activated complex has been assumed.

In the present paper and in a companion one (**M50**) on the classical mechanical formulation, this activated-complex theory is generalized by extending it to curvilinear reaction coordinates ... The assumption of separability is made in the quantum formulation... Separability is *not* assumed in the classical-mechanical formulation however."

"In curvilinear coordinates $q^1, \ldots, q^n$ the Schrödinger equation has the form [Pauli...]

$$H\psi = \alpha_1\psi, \tag{9.1}$$

where H is the Hamiltonian operator, α_1 the total energy, and ψ the wavefunction of the entire system

$$H = \frac{-\hbar^2}{2}\sum_{s,t=1}^{n}\frac{1}{g^{\frac{1}{2}}}\frac{\partial}{\partial q^s}g^{\frac{1}{2}}g^{st}\frac{\partial}{\partial q^t} + U. \tag{9.2}$$

The q^s are generalized coordinates, and U is the potential energy; g^{st} is a contravariant tensor conjugate to the metric tensor g_{st} appearing in the line element ds in mass-weighted space; g is the determinant of the g_{st}

$$g^{st} = \sum_{i=1}^{n}\frac{1}{m^i}\frac{\partial q^s}{\partial x^i}\frac{\partial q^t}{\partial x^i}; \quad g_{st} = \sum_{i=1}^{n}m^i\frac{\partial x^i}{\partial q^s}\frac{\partial x^i}{\partial q^t}, \tag{9.3}$$

$$ds^2 = \sum_{s,t=1}^{n} g_{st}dq^s dq^t; \quad \sum_{r=1}^{n} g_{sr}g^{rt} = \delta_s^t \tag{9.4}$$

where x^i is a Cartesian coordinate of an atom of mass m^i; the coordinates of the kth atom are given by $i = 3k, 3k+1, 3k+2$. Both g_{st} and g^{st} are symmetric tensors.

Under certain conditions on U and on g^{st}, Eq. (9.1) can be separated into m individual equations, each depending on its own set of variables. The wave function ψ then becomes

$$\psi = \prod_{\mu=1}^{m} \psi_{\mu}$$

where ψ_{μ} is the wavefunction for the μth set. As a particular case of this separation one could select one set of variables to consist of a single variable, the reaction coordinate, q^r, and select a second set to consist of all remaining coordinates. That is, $m = 2$ then. Under the assumed conditions g^{st} vanishes when s and t belong to different sets."

" ...it is supposed that it has been possible to describe a coordinates system such that one of the coordinates is approximately dynamically separable from all remaining coordinates in the activated-complex region and tends to lead in this vicinity from the reactants 'region' to the products' one. This coordinate is then the '*reaction coordinate*.' The potential energy surface (PES) is thereby *approximated* by one which permits separation of variables. The reaction coordinate forms one of the sets μ mentioned earlier. It is described by setting μ equal to r. Since only the nonseparable PES in the vicinity of the activated-complex region is being approximated by a separable surface, the properties in the separated system for the degrees of freedom other than q^r are those of the activated complex rather than of the reactants."

"Since the wavelength (more precisely, the reciprocal of the component of the wave vector along q^r) is large in the activated-complex region because of the low kinetic energy there, and since the potential energy changes rapidly there, most of the scattering may in fact occur in that region. Indeed, the phase integral expression for tunneling points up to this local scattering characteristic."

"The wave function of all degrees of freedom but q^r is denoted by ψ':

$$\psi' = \prod_{\mu \neq r} \psi_{\mu}(q^{\mu_1}, \ldots, q^{\mu_{h_{\mu}}}).$$

For any given value of q^r and of the conjugate momentum p_r the state of the remaining degrees of freedom can be regarded as describable by a (discrete) quantum number λ: In the case of an isolated gaseous activated-complex molecule confined in a volume, even the translational state can be regarded as quantized. In the case of an *activated complex in solution, a macroscopic subsystem can be regarded as the complex.* It can be placed in a box and the n 1 degrees of freedom regarded as giving rise to discrete eigenvalues, characterized by the quantum number λ . . ."

"The probability of the system being in a quantum state described by λ and of being in any small element $\Delta q^r \Delta p_r$ is denoted by $P(\lambda, p_r, q^r, T)\Delta q^r \Delta p_r$. We suppose that the reactants are in statistical equilibrium with this system. Since the probabilities of the system being in $\Delta q^r \Delta p_r$ and of having any given value of λ are independent and since the number of quantum states in $\Delta q^r \Delta p_r$ is $\Delta q^r \Delta p_r / h$, one obtains

$$P(\lambda, p_r, q^r, T) = \exp\frac{\left[\frac{-\alpha_1(\lambda, p_r)}{kT}\right]}{hQ_1} \tag{9.5}$$

where the given values of p_r, q^r, and λ automatically fix the *total energy* α_1 and where $\exp\left(\frac{-\alpha_1}{kT}\right)$ is the Boltzmann factor. Q_1 is the partition function of the reactants."

M50. Generalization of Activated-Complex Theory of Reaction Rates. II. Classical Mechanical Treatment

9.1. Introduction

"A number of derivations of the activated-complex equation for chemical reaction rates have been published.[1]

[1]For example; H. Eyring, *J. Chem. Phys.* **3**, 107, (1935); H. Eyring, J. Walter, G. Kimball, *Quantum Chemistry*, Chapter 16, John.Wiley & Sons, Inc., New York (1944); E. Wigner, *Trans. Faraday Soc.* **34**, 29, (1938); *Z. Physik. Chem.* **B19**, 203, (1932); G. H. Vineyard, *J. Phys. Chem. Solids* **3**, 121, (1957).

Several assumptions normally made in the classical mechanical form of the theory are following.

(i) For the reaction to occur some $n - 1$-dimensional hypersurface in the n-dimensional configuration space must be crossed. (The hypothetical system constrained to exist on this surface is the 'activated complex.' The surface is called the 'reaction hypersurface.')
(ii) The probability of finding the system in any part of the $2n$-dimensional phase space on the reactants' side of the above surface is that calculated from equilibrium statistical mechanics.
(iii) A system striking the above hypersurface has unit probability of crossing it and recrossing can be neglected. Thereby, the transmission coefficient is unity.
(iv) The kinetic energy along the reaction coordinate has a very simple form $\frac{p^2}{2\mu}$, where p is the momentum conjugate to this coordinate and μ is a constant, and there are no cross terms with p in the total kinetic energy expression.[2]

In addition, the Born–Oppenheimer approximation is normally employed. Sometimes this approximation breaks down, the reaction becomes quantum-mechanically nonadiabatic. The rate is then occasionally calculated with the aid of the Landau–Zener equation, and some approximations are contained therein."

[2]Normally, the potential energy is expanded about a saddle point, retaining only quadratic terms, a normal coordinate analysis is made, and rotation–vibration interaction is neglected. The reaction hypersurface then becomes a hyperplane, the reaction coordinate thereby becomes Cartesian, and Assumption iv is then valid. However, when some of the internal coordinates are not vibrational, this simple expansion is not necessarily appropriate (e.g., for motions of solvent molecules in some solution reactions).

9.2. Derivation of the Rate Equation for a Reaction Hypersurface Dependent on Coordinates Alone

"When the 'reaction hypersurface' depends on the coordinates alone, it is independent, thereby, of any constants of the motion. Otherwise, the latter would appear as parameters in the equation of the hypersurface. The equation of this hypersurface S may be written as $f(q^1, \ldots, q^n) = 0$ (on S).

A choice of coordinates can be made so that S is a coordinate hypersurface for one of them, q^r. Thus, q^r is constant on S, and can be taken as zero on it. This surface will be a q^r-coordinate hypersurface both in mass-weighted and in ordinary configuration space.

The reaction rate is *the net rate at which systems cross S.* It can be computed under the equilibrium assumption for the reactants as follows: The probability that a system in equilibrium with the reactants will lie in a volume element of phase space

$$\prod_{i=1}^{n} dq^i dp_i$$

is denoted by

$$\rho \prod_{i=1}^{n} dq^i dp_i,$$

where ρ is the equilibrium phase space density:

$$\rho = \frac{e^{-\frac{H(p,q)}{kT}}}{\int e^{-\frac{H(p,q)}{kT}} \prod_{i=1}^{n} dq^i dp_i}, \text{''} \tag{9.6}$$

where $H = \frac{1}{2}\sum_{i,j=1}^{n} g^{ij} p_i p_j + U(q^1, \ldots, q^n)$

"On dividing the above probability by dq^r and multiplying by $\dot{q}^r$, the probability that the reacting system will cross the element

$\prod_{i \neq r} dq^i$ of the hypersurface S in unit time is found to be

$$\left(\int \rho \dot{q}^r \prod_{i=1}^{n} dp_i \right) \prod_{i \neq r} dq^i ,$$

where the integration is over all p_i such that only passages from the reactants' side of S to the products' side are counted. The rate constant is then obtained by integrating over the coordinates

$$k_{\text{rate}} = \int \rho \dot{q}^r \prod_{i=1}^{n} dp_i \prod_{i \neq r} dq^i . \tag{9.7}$$

By definition of a rate constant of a homogeneous reaction (it has units of moles per volume and time), the q-integration in Eq. (9.7) is such that three translational coordinates of the activated complex are integrated over a unit volume. For a heterogeneous reaction the integration in Eq. (9.7) is such that the two translational coordinates of the activated complex parallel to the interface of the two phases are integrated over a unit area of the interface. In the denominator of Eq. (9.6), the integration over the translational coordinates of each reactant is over unit volume.

On one side of S (the reactants') q^r is negative, and on the other side it is positive. Accordingly, in order to count only passages from one first side of S to the other, the integration in Eq. (9.7) is such that $\dot{q}^r$ is confined to the interval $(0, +\infty)$.

According to Footnote 10, $\dot{q}^r$ is given by

$$\dot{q}^r = \sum_{j=1}^{n} g^{rj} p_j . \tag{9.8}$$

For any given value of $\dot{q}^r$, Eq. (9.8) represents the equation of a hyperplane in momentum space. Integration in Eq. (9.7) may therefore be performed as follows: For any given value of $(q^1, \ldots, q^n)$ the

p_i's are integrated over the infinite half-space in momentum space, corresponding to all variations in p_i subject to

$$\sum_{j=1}^{n} g^{rj} p_j$$

lying between zero and infinity. By integrating p_r from $-\sum_{j \neq r} g^{rj} p_j / g^{rr}$ to ∞ and by integrating the remaining p_j from $-\infty$ to ∞, this integration can be performed. During a subsequent integration over all q^i other than $i = r$, q^r is kept at the value zero. Since $\dot{q}^r$ is $\frac{\partial H}{\partial p_r}$, $\dot{q}^r \exp\left(\frac{-H}{kT}\right)$ equals $-kT\left(\frac{\partial e^{-\frac{H}{kT}}}{\partial p_r}\right)$. The p_r integration yields

$$k_{\text{rate}} = \left(\frac{kT}{h}\right) \exp\left[-\frac{(F^{\ddagger} - F)}{kT}\right], \tag{9.9}$$

where

$$e^{-F^{\ddagger}/kT} = \int e^{-H^{\ddagger}(p,q)/kT} \prod_{i \neq r}^{n} \frac{dq^i dp_i}{h^{n-1}}, \tag{9.10}$$

$$e^{-F/kT} = \int e^{-H(p,q)/kT} \prod_{i=1}^{n} \frac{dq^i dp_i}{h^n}, \tag{9.11}$$

and $H^{\ddagger}$ is given by Eq. (9.12). *It is the value of H when* $q^r = \dot{q}^r = 0$. Thus, it is the value of H for a system constrained to exist on the hypersurface of S.

$$H^{\ddagger} = T^{\ddagger} + U^{\ddagger}, \tag{9.12}$$

where

$$U^{\ddagger} = U(q^1, \ldots, q^n) \text{ at } q^r = 0, \tag{9.13}$$

. . . .

$F^{\ddagger}$ is the free energy of the *constrained system* and F is the free energy of the *unconstrained reacting system.* In both free energies and in all subsequent free energy expressions the usual *product of factorials*, which *corrects* for *indistinguishability of like particles*, is omitted for brevity. These factors cancel in computing Eq. (9.9)....

Integration over the momenta in Eqs. (9.10) and (9.11) can readily be performed. One obtains

$$e^{-\frac{F^{\ddagger}}{kT}} = (2\pi kT)^{\frac{(n-1)}{2}} \int \frac{e^{-\frac{U^{\ddagger}}{kT}} d\sigma}{h^{n-1}}, \tag{9.14}$$

$$e^{-\frac{F}{kT}} = (2\pi kT)^{\frac{n}{2}} \int \frac{e^{-\frac{U}{kT}} d\tau}{h^{n}}, \tag{9.15}$$

where $d\sigma$ is .... The *area* element of a coordinate hypersurface on which q^N is constant for mass-weighted configuration space .... This area element is the *volume* element in an $n - 1$ dimensional space in which dq^N is zero."

$d\tau$ is given by (13) in the original paper.

"On introducing an effective mass $m^{\ddagger}$ defined in the next section the expression for the rate constant becomes

$$k_{\text{rate}} = \left(\frac{kT}{2\pi}\right)^{\frac{1}{2}} \frac{\int e^{-\frac{U^{\ddagger}}{kT}} \left(m^{\ddagger}\right)^{-\frac{1}{2}} dS}{\int e^{-\frac{U}{kT}} dV} \tag{9.16}$$

9.3. Effective Mass

An effective mass $m^{\ddagger}$ for motion normal to S in ordinary n-dimensional configuration space may be defined in several ways. A definition suited to our purpose is the following: When the momentum $\mathbf{p}$ is normal to S in this ordinary configuration space the proportionality factor of $p^2/2$ in the kinetic energy is designated by $1/m^{\ddagger}$. To evaluate $m^{\ddagger}$, one may proceed thus.

The covariant components of a vector $\mathbf{v}$ of *unit* length (magnitude) normal to the q^r-coordinate hypersurface S in this space are equal to $v_i = \delta_i^r (a^{rr})^{-\frac{1}{2}}$. The covariant components of $\mathbf{p}$, p_i are therefore equal to $\delta_i^r p (a^{rr})^{-\frac{1}{2}}$, where p is the magnitude of $\mathbf{p}$. On noting that the kinetic energy is given by the first term in Eq. (9.17)

$$H = \frac{1}{2} \sum_{i,j=1}^{n} g^{i,j} p_i p_j + U(q^1, \ldots, q^n) \tag{9.17}$$

and on introducing the above values for the p^i's, the kinetic energy is found equal to $\frac{g^{rr} p^2}{2a^{rr}}$. Hence, we have

$$m^{\ddagger} = \frac{a^{rr}}{g^{rr}}. \tag{9.18}$$

9.4. Integration Over External Coordinates

In a dilute gas, an activated complex may be regarded as an isolated particle. In a liquid or dense gas, its motions may be strongly coupled to those of the of the surrounding molecules. In the latter case it will be useful to consider as the activated complex a *macroscopic* subsystem, near the center of which is the actual reactant or pair of reactants and on the boundary of which the correlation of the motion of the solvent molecules with those of the reactants is negligible. This subsystem is regarded as imbedded in the remainder of the infinite (or practically infinite) system. For homogeneous reactions rigid translations or rotations will later be performed on the subsystem, and the solvent molecules of the remaining part will be permitted to continuously adjust themselves. For heterogeneous systems rigid translations of the macroscopic activated complex parallel to the interface will be performed with similar adaptations of the remaining molecules occurring.

The activated complex of a homogeneous reaction in a gas or liquid, defined above, has as coordinates three translations (x, y, z), two rotations of an axis *fixed in the complex* (θ, ϕ), and $n - 5$ other coordinates, which will be called the internal coordinates of the complex, though one of them (rotation about the body-fixed axis) has a property analogous to the five 'external' ones: *The potential energy of the entire system is invariant to changes in the five external coordinates.*

In the case of a heterogeneous reaction on a uniform interface the potential energy function for the activated complex is *invariant* to the *two* Cartesian coordinates x and y parallel to the interface of the two phases. Presumably, such a case occurs in electrochemical ETs to a good approximation when the reactant is not adsorbed. In reactions involving localized adsorption on perfect crystals the potential energy is a periodic function of x and y. For any heterogeneous reaction the remaining $n - 2$ are the internal ones of the activated complex, though in the particular case of a nonuniform surface, $U^{\ddagger}$ and q^r below depend on *all* n coordinates.

The integral appearing in the denominator of Eq. (9.16) is evaluated for a system where the reactants are far apart, when there is more than one of them, or far from the interface in the heterogeneous reaction. The function U in this integral is independent of the three translations of the center of mass of each reactant, which are called external coordinates for the denominator of Eq. (9.16). (However, U is also independent of some of the other coordinates, of course.)

Since the properties of the reaction hypersurface depend only on the *internal coordinates*, they can be selected so that the coordinate q^r *is one of them.*

The reduced mass $m^{\ddagger}$ is shown in Appendix I to be independent of the values of the external coordinates. It is normally a function of the internal coordinates, though it is a constant in special cases, as discussed later. The area element dS is shown in Appendix II to be a product of a function of the external coordinates alone and of a function of the internal coordinates alone, the latter denoted by $R^2 dS_{\text{int}}$ for bimolecular reactions and by dS_{int} for homogeneous unimolecular reactions or for heterogeneous reactions, as discussed in the Appendix.

$$\begin{pmatrix}\text{homogeneous}\\ \text{bimolecular}\end{pmatrix} dS = \sin\theta d\theta d\phi dx dy dz R^2 dS_{\text{int}}, \quad (9.19)$$

$$\begin{pmatrix}\text{homogeneous}\\ \text{unimolecular}\end{pmatrix} dS = \sin\theta d\theta d\phi dx dy dz dS_{\text{int}}, \tag{9.20}$$

$$(\text{heterogeneous})\, dS = dx dy dS_{\text{int}}, \tag{9.21}$$

where θ and ϕ *define* a body-fixed axis of the complex, R is the distance of two atoms or any two points of the complex *on this axis*, and x, y, z have been defined earlier. In the computation of dS_{int} in Eq. (9.19) the two atoms or points are constrained so that one is fixed on the cited body-fixed axis and the other can move only along that axis. The two points can be the centers of mass of each reactant, for example (Appendix II). In the computation of the dS_{int} of Eq. (9.21) one point of the complex is constrained to move along any fixed line normal to the xy plane parallel to the solid–liquid interface. This point can be the center of mass of the reactant.

Similarly, the volume element dV in Eq. (9.16) can be shown to be the product of volume elements $\prod_a dx_a dy_a dz_a$ for the external coordinates of all reactants a and of dV_{int}, the volume element of all remaining coordinates. (There is only one term in Π_a when the reaction is unimolecular, of course.) These remaining coordinates are coordinates in a space where the center of mass of each reactant is fixed and where the reactants are far apart.

Integration may now be performed over the external coordinates in the numerator and denominator of Eq. (9.16). One obtains

$$\begin{pmatrix}\text{bimolecular}\\ \text{homogeneous}\end{pmatrix} k_{\text{rate}} = (8\pi kT)^{\frac{1}{2}} \int R^2 (m^{\ddagger})^{-\frac{1}{2}} e^{-U^{\ddagger}/kT} \frac{dS_{\text{int}}}{Q}, \tag{9.22}$$

$$\begin{pmatrix}\text{unimolecular homogeneous}\\ \text{or uniform heterogeneous}\end{pmatrix} k_{\text{rate}} = \left(\frac{kT}{2\pi}\right)^{\frac{1}{2}} \int (m^{\ddagger})^{-\frac{1}{2}} e^{-U^{\ddagger}/kT} \frac{dS_{\text{int}}}{Q} \tag{9.23}$$

where constraints on the numerator integration have just been described and where Q is given by Eq. (9.24)

$$Q = \int e^{-\frac{U}{kT}} dV_{\text{int}} \tag{9.24}$$

It is the *configurational integral of the reactants* when they are far apart."

....

"It will be convenient to choose the internal coordinates in such a way that one of the coordinates q^r is constant on the reaction hypersurface." (p. 2629)

M74. Electron Transfer Reactions (Chemische Elementarprozesse...)

In this review, the first after the publication in 1965 of Part VI of the series of papers on the theory of ET reactions, it is very well and systematically described how the various Marcus' results depend on the *assumptions* and the *approximations* of the theory.

9.5. Introduction

"ET processes of chemical interest are of various types:

(i) Certain homogeneous redox reactions
(ii) Electron exchange reactions
(iii) Electrode processes
(iv) Solvated electron reactions
(v) Certain chemiluminescent processes
(vi) Intramolecular ET and
(vii) ETs between ions in solution and semiconductor electrodes.

The distinction between redox reactions of the atom transfer type and those of ET type should be noted.

This contribution will discuss theoretical and experimental results in this field and will be based on several recent papers of the author (*Ann. Rev. Phys. Chem.* **15**, 155, (1964), *J. Phys. Chem.* **67**, 853, 2889, (1963), *J. Chem. Phys.* **43**, 679, (1965))."

9.6. Potential Energy Surfaces and Mechanisms of Electron Transfer

"We consider an ET between two 'reacting' species. These species consist of two dissolved ions (molecules) in the homogeneous case and one such ion and electrode in the electrochemical case."

Of course also of neutral molecules and electrode in the electrochemical case . . . .

"Some insight into the mechanism of ET is provided by examination of the behavior of the entire system on its PES. The surface for the system of *reactants and surrounding medium* is a function of all the translational, rotational, and vibrational coordinates in this system. Similarly, the surface for a system consisting of the *products and the medium* is a function of these coordinates also.

In the absence of *electronic coupling* between the orbitals of the two reacting species, the two PESs intersect. It is then not possible for the system to go from one surface to the other, that is, to undergo ET (Fig. 1). A suitable coupling (Fig. 2) removes the degeneracy at this intersection, in the usual quantum mechanical manner, and the system may now go from one original PES to the other by a *fluctuation* which permits it to pass through the 'intersection' region of coordinate space. If the coupling is strong enough the product is thereby formed *adiabatically* from the reactant.

Otherwise the formation occurs nonadiabatically.

For an adiabatic process in reaction kinetics there is *continuous quantum mechanical equilibrium* between electrons and nuclei and no abrupt electronic rearrangement. Such processes are primarily of interest here. In summary, ET can occur if there is a suitable *electronic coupling* between the reacting species and if there is a *suitable fluctuation* of coordinates (e.g., bond distances, orientation of solvent molecules, position of ions in atmosphere, including those in

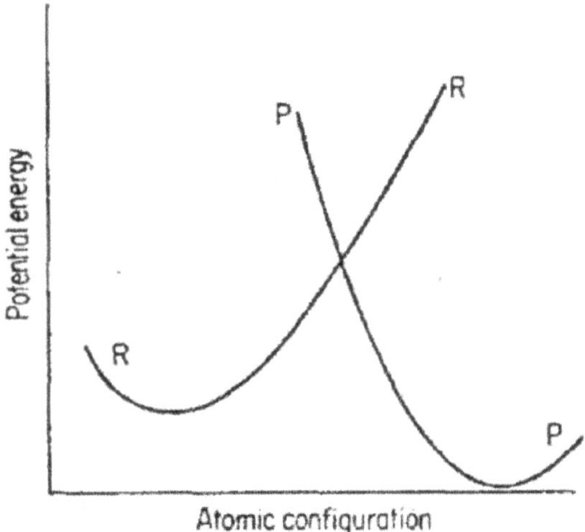

Fig. 1. R-reactants, P-products, no coupling between the states Ψ_R and Ψ_P.

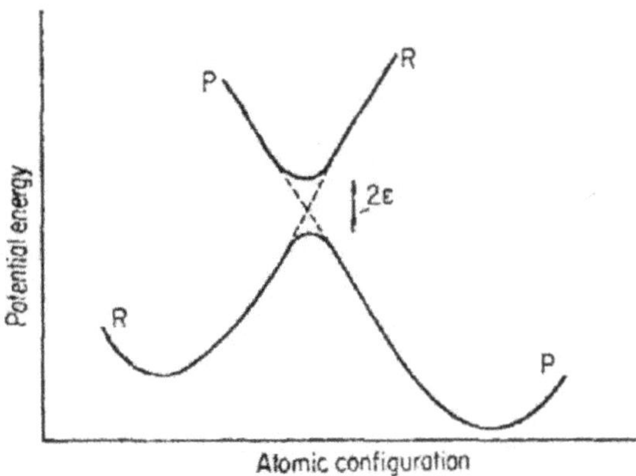

Fig. 2. R-reactants, P-products, coupling between the states Ψ_R and Ψ_P, ε-exchange energy.

the double layer) leading from values appropriate to the reactants to ones appropriate to the products.

NOTE: In Marcus' theory of ET the rearrangement is abrupt in the sense that it happens in a small intersection region.

9.7. Assumptions

To calculate the rate of the reaction, it is convenient to introduce several *assumptions*

(i) The rotational, translational, and *relevant* vibrational coordinates of both reacting species and of the molecules and ions in the medium are treated classically.
(ii) The probability of the system being in the intersection region and having any specified molecular velocities is that computed from equilibrium statistical mechanics.
(iii) Each reacting species in the activated complex has an inner coordination shell not shared by the other reactant. Consequently, any 'bridging' of the reactants is *of the outer sphere type*.
(iv) The potential energy of the entire system is the sum of *intra* and *inter particles* terms, the second term being a quadratic function of the permanent charge distributions on the particles ....
(v) The splitting of the doubly degenerate energy level at the intersection surface is sufficiently small that the potential energy of the lowest electronic state of the entire system differs but slightly from that at the original intersection, for each point of the *intersection region.*
(vi) Reaction occurs primarily by the system passing over the barrier rather than by a nuclear tunneling motion through it.
(vii) The reaction coordinate *in the region near the intersection surface* does not involve the rupture of a chemical bond.

Assumption (i) is reasonable under the usual experimental conditions, since the highest pertinent molecular frequency is usually a metal–ligand one in the inner coordination shell of an ionic reactant, and available data indicate this frequency to be typically 300 to 500 cm^{-1}. Assumption (ii) is the standard one in activated-complex theory. Assumption (iii) could be removed by extending the theory.

When assumption (v) is not fulfilled, a condition which could conceivably happen sometimes when *inner type bridging* occurs (i.e., when assumption (iii) fails), the theory could again be extended by calculating independently the splitting of the two energy levels, as a function of coordinates of the intersection surface. Assumption (vi) will undoubtedly fail at sufficient low temperatures. Assumption (vii) can be removed by calculating the adiabatic PES for the case of bond ruptures. . . .

Supplementing these assumptions, it is convenient to introduce several approximations which convert the equations for the reaction rate into a rather simple functional form, a form particularly useful for theoretical predictions of correlations among the experimental data

(i) The partial dielectric (or, more precisely, electric) saturation approximation for the entire medium outside the inner coordination shell is valid.
(ii) The effective potential energy function for coordinates in the inner coordination shell of each reactant is a quadratic function of coordinate displacements.
(iii) In any given species, the force constant of any bond s may differ when that species is a reactant in general if the given species is a reactant, k_s^r, as compared with when it is a product, k_s^p. These two force constants are expressed in terms of symmetric and antisymmetric functions of them, namely of $\frac{2k_s^r k_s^p}{(k_s^r + k_s^p)}$ and of $\frac{(k_s^r - k_s^p)}{(k_s^r + k_s^p)}$, respectively, and it is shown in **M53** that the second of these can be neglected. This approximation simplifies the equations of **M53** considerably.

Using these assumptions and approximations, and using the method of equivalent equilibrium distributions described in in **M30** and further refined in **M53**, one eventually obtains Eq. (9.27) for the rate constant of the electrochemical reaction (9.25) or of the homogeneous reaction (9.26) where Eqs. (9.25) and (9.26) denote different soluble redox couples.

$$Ox + ne \rightarrow Red \tag{9.25}$$

$$Ox_1 + Red_2 \rightarrow Red_1 + Ox_2 \tag{9.26}$$

$$k_r = Z\gamma\rho \exp\left(\frac{-\Delta F^*}{R_g T}\right) \tag{9.27}$$

where electrochemical:

$$\Delta F^* = \frac{w^r + w^p}{2} + \frac{\lambda_{el}}{4} - \frac{nF(E - E_0^{'})}{2} + \frac{[-nF(E - E_0^{'}) + w^p - w^r]^2}{4\lambda_{el}} \tag{9.28}$$

Homogeneous:

$$\Delta F^* = \frac{w^r + w^p}{2} + \frac{\lambda}{4} + \frac{\Delta F^{0'}}{2N_a} + \frac{[\Delta F^{0'} + w^p - w^r]^2}{4\lambda} \tag{9.29}$$

$R_g = kN_a$, where k is Boltzmann's constant and N_a is Avogadro's number.
w^r = work required to bring reactants together until they are at the most probable of the separation distances contributing effectively to reaction.
w^p = corresponding quantity for the products, for the same separation distance.
λ, λ_{el} = reorganization factor described in **M41** J. Phys. Chem. **67**, 853 and, with more generality in **M53**, J. Chem. Phys. **43**, 679 (1965). They are independent of E, $E_0^{'}$, and $\Delta F^{0'}$. λ as well as λ_{el} is a sum of the contribution from the inner coordination shell (λ_i) and of the contribution from the medium outside the coordination shell (λ_0). λ_i is dependent on the changes in bond distances and bond angles and on the force constants of all the vibrational coordinates of each reacting species in its reacting state and in its product state. λ_o is given for a dielectric continuum treatment (D_{op}—square of the refractive index,

D_s—static dielectric constant) by

$$\lambda_o = \left(\frac{1}{2a_1} + \frac{1}{2a_2} - \frac{1}{r}\right)\left(\frac{1}{D_{op}} - \frac{1}{D_s}\right)(ne)^2;$$

a_1 and a_2 are the radii of the spherical particles undergoing reaction including the coordination shell and $r = a_1 + a_2$ is taken as the mean distance between the centers of the reactants in the activated complex. In the electrochemical case $a_1 = a_2$, and r is twice the distance from the center of the reacting particle to the electrode surface. A formal statistical mechanical expression for λ_o is given in **M53**.

n = Number of electrons transferred
F = Faraday
E = Potential of half-cell (American convention, sign bivariant)
E_0' = Standard potential of half-cell
$\Delta F^{0'}$ = The apparent standard free energy of reaction,[1] equal to nF times the difference of standard potentials of the two redox systems, 1 and 2.
Z = Collision number in solution (10^{11} l mole^{-1} sec^{-1}) or, in the electrochemical case, the collision frequency with the electrode (10^4 cm sec^{-1}).
γ = An averaged probability of remaining on the lowest PES per passage across the intersection region (= 1 for an adiabatic reaction).
$\rho = \frac{\langle \Delta r \rangle}{\langle \Delta q \rangle}$, where $\langle \Delta r \rangle$ is the average range of separation distances *contributing effectively to reaction* and $\langle \Delta q \rangle$ is the average range of distances along the reaction coordinate[2] *effectively occupied by the equivalent equilibrium distribution.* Explicit formal expressions for ϱ are given in **M53**.

Typically, it is probably of the order of magnitude of unity[3]; an exception would occur if a low frequency orientation dielectric relaxation coordinate were the major component of the reaction coordinate, thereby making ρ less than one. We shall take $\rho\gamma$ to be unity ....

Sometimes reactions (9.25) and (9.26) are preceded or followed by other reactions, so that then neither *Ox* or *Red* denote the stable species present, merely those participating in the ETs (9.25) and (9.26). Then, k_r is the rate per unit concentration of these species rather than of the stable ones, and w^r, w^p, $\Delta F^{0'}$, nFE_0', λ, and λ_{el} refer to the free energy terms involving such species and not the stable ones.

9.8. Deductions

The characteristics of λ most important for predicting correlations are

(i) In a homogeneous reaction λ is essentially an additive property of the two redox couples, 1 and 2.

$$\lambda = \lambda_1 + \lambda_2 \tag{9.30a}$$

(ii) Let λ_{ex} be the value of λ for the homogeneous isotopic exchange reactions:

$$Ox_j^* + Red_j = Ox_j + Red_j^* \tag{9.32}$$

where the starred species denote isotopic labeling. If this reaction corresponds to an electrode reaction (9.25), then

$$\lambda_{el} = \frac{1}{2}\lambda_{ex}^{(4)} \tag{9.31b}$$

when the most probable distance of the reactant from the electrode in the *pre-electrode layer* equals twice that between the reactants in the homogeneous reaction.

As noted in **M41** *J. Phys. Chem.* **67**, 853, a very common condition experimentally is that the last term in Eqs. (9.28) and (9.29) can be neglected. From the resulting equation and the earlier properties of λ and λ_{el} one would then predict

(i) When the *effect of E on the work terms can be ignored*[5] the electrochemical transfer coefficient should be 1/2. The slope of the electrochemical plot of $\left(\frac{R_gT}{nF}\right) lnk_r$ versus electrode potential is called the electrochemical transfer coefficient.

(ii) The k_{el} deduced from the electrochemical exchange current at standard conditions should bear a simple relation to the rate constant of the homogeneous isotopic exchange reaction, k_{ex} for the case that the work terms are small in both experiments,

$$\frac{k_{el}}{Z} = \sqrt{\frac{k_{ex}}{Z_{ex}}}\text{''} \tag{9.33}$$

where Z_{ex} is the *collision number* in solution, 10^{11} l mole^{-1} sec^{-1}, and Z is the *collision frequency* with the electrode, 10^4 cm sec^{-1}.

NOTE: The number 10^{11} was later changed to 10^{12} in **M170**.

"(iii) In the comparison of the redox reaction of a series of reagents with a given chemical agent and with an electrode at a given E, the *ratio of the rate constants* should be the same for each member of the series, regardless of whether or not the reaction is irreversible.[(6)]

(iv) The "chemical transfer coefficient" [**M41**]—defined as the slope of the ΔF^*'s versus ΔF^0 plot in *a series* of homogeneous redox reactions—should be $\frac{1}{2}$, a result which provides some insight into the electrochemical coefficient when the work terms *cannot be ignored*.[(7)]

(v) The bimolecular rate constant k_{12} of the cross-reaction of two redox systems, 1 and 2

$$Ox_1 + Red_2 = Red_1 + Ox_2, \tag{9.34}$$

is related to the rate constants of the isotopic exchanges in the two systems, k_{11} and k_{22}, and to the equilibrium constant K_{12}, as in Eq. (9.35), when the work terms in $\Delta F^*_{12} - (\Delta F^*_{11} + \Delta F^*_{22})/2$ cancel or can be ignored.

$$k_{12} = \sqrt{k_{11}k_{22}K_{12}}^{(8)} \tag{9.35}$$

When these work terms cannot be ignored, the appropriately corrected form of Eq. (9.35) is given in Eq. (12) of **M41**. A more elaborate equation based on Eq. (9.30) itself, rather than on the approximate form of Eq. (9.30) is given in Eq. (13) of **M41**.

Any numerical calculation of the rate constant k_r itself requires a knowledge of equilibrium bond distances and force constants in the coordination shells of the reactant and the product, as well as of certain properties of the medium, for these contribute to λ. Some illuminating calculations based on uncertain values of these constants have been given and provide some insight into the problem of *a priori* calculations of reaction rates. In the present paper, we are particularly interested in *correlations*, however, for they are independent of the present uncertainties of *a priori* estimates of k_r's.

Equations (9.27) to (9.29) apply when each reacting species has a given inner coordination shell. However, sometimes a species may be present in several forms which differ in their inner coordination shell. Again, fluctuations in the number of strongly adsorbed ions on the electrode correspond to fluctuations in the "inner coordination shell" of the electrode. Subject to conditions summarized in footnote 9 of **M41**, the above five deductions still apply when these fluctuations in composition of the inner shells occur.

Comparisons of the five deductions with the experimental data has been given in **M41**. The agreement with the data may be considered encouraging. A considerable quantity of data supporting deduction 5 has since been obtained.

From the References:

"When one of the "reacting" species participating in an ET is an electrode, any strongly adsorbed solvent molecules and ions can be considered as the "inner coordination shell" of this "reactant."

"Occasionally, it is asserted in the literature that reaction rate should rigorously be expressed in terms of the activity of the reacting ion. This statement is not only incorrect but presupposes a knowledge of the single ion activity coefficient. Rather, the rate should be expressed operationally in terms of concentrations, and all *pertinent* salt effects should be incorporated into the theory of the proportionality constant, i.e., of the rate constant. It is readily verified that in careful transition state theory formulation of the latter constant no single ion activity coefficient appears."

NOTES

1. p. 352: "$\Delta F^{0'}$ = the apparent standard free energy of reaction..."

   M: The standard free energy is normally defined for a certain concentration of the surrounding electrolyte and this is related to an equilibrium concentration "in the prevailing medium," and so that $\Delta F^{0'}$ can vary with the concentration, you see, whereas the true $\Delta F^{0'}$ would be defined for some standard sol concentration.
2. p. 352: "$\varrho = \frac{\langle \Delta r \rangle}{\langle \Delta q \rangle}$... Typically, it is probably of the order of magnitude of unity; an exception would occur if a low frequency orientation dielectric relaxation coordinate were the major component of the reaction coordinate, thereby making ρ less than one."
3. Q: (1) Why is it probably of the order of magnitude of 1?
   (2) Please explain the exception...

   M: Those are sort of thermal fluctuations of the coordinates, one is a thermal fluctuation of a distance of two ions within a cage and Δq is fluctuation of some other coordinate, so, you know, these fluctuations, if there are of comparable frequencies of motion then at room temperature, they are comparable in size, probably,... this is just ballpark argument, the effective mass for the two coordinates might be different, so that would have to be taken into account if one wanted to be more precise...
4. Does this mean that it is twice easier to have the same ET reaction at the electrode than in homogeneous phase?

   M: Yes, in a free energy sense, if you meann easier in a rate sense no because the free energy occurs in an exponential, but easier in a free energy sense yes. In other words for a thermal process half the free energy is needed to reorganize one compared to the other. And the reason is that provided the image charge takes on its full value, in other words that the ion can come up and touch the electrode so that the distance from its image is twice the radius of the ion, provided that can happen, then the image charge produces the field around the ion and hence reduces the need for reorganization when the charge of the ion is changed.

5. p. 353: "When the effect of E on the work terms can be ignored..."

   Q: When can it be ignored? Maybe in the case of uncharged species?

   M: Yes, then you can, I mean except for nonelectrostatic effects, you can have changes in hydrogen bonding, those are all parts of bringing reactant and electrode close together, so they were part of the electrostatic work terms, they were parts of the work terms generally, they may be small ... but this says when you can ignore the effect of the E on the work terms. E will affect the local potential that the ion sees, the potential function depends on the distance from the electrode, because E affects the overall potential difference between the electrode and the solution, so it's going to affect the potential profile near the electrode, if there is any effect on that potential profile then it is going to affect the work term, because you bring the ion from long distance some place into that potential profile.
6. **M**: Well, if you measure the rate constant for the forward reaction, then what happens with the back reaction is irrelevant if you're really measuring the rate constant for the forward reaction... That may be when you're both in the kind of linear regime, when you're able to make the expansion and ignore the quadratic term, that may be an assumption there, I'm not sure.
7. **M**: First of all, that term should be $\frac{1}{2}$ only when you can ignore the quadratic term in ΔG^*, only when you're in the linear regime, so that should have been stated in there. If it isn't... if it turns out that the work terms in the electrochemical reaction are important, and I have no idea if they are, but if it turns out that they are

important in the potential that would cause a deviation from the simple value in the electrochemical case, then you might resort to seeing if the *chemical* transfer coefficient has a simple value, provided no work terms are important in the chemical case.

8. p. 354: "$k_{12} = \sqrt{k_{11}k_{22}K_{12}}$"

   Q8: Does the formula mean that the rate constant of the cross-reaction depends on the tendency to react of the single reagents and on how much they tend to react with each other, this last tendency being measured by their equilibrium rate constant?

   M: In a sense yes, because if K_{12} were equal to unity, in other words if equilibrium constant is one, no free energy term being considered, then k_{12} is a geometric mean of the self-exchange reactions ....

M105. Chapter II Activated-Complex Theory: Current Status, Extensions, and Applications

"The activated-complex theory of chemical reactions has proved to be very useful in interpreting rate data on a wide variety of chemical reactions... in conjunction with additional concepts, the theory has also found extensive use in... ET reactions both in solution and at electrodes."

9.9. Assumptions and Derivation

9.9.1. *Assumptions*

"In a system having N coordinates, say, one attempts to define a set of configurations occupying $N-1$ dimensions. This set, termed a *hypersurface*, separates two parts of the space and is chosen (as much as possible) so that once the system has crossed this hypersurface, only the products can result. *All the configurations belonging to the hypersurface* are called the *activated complex*.

To demonstrate what is involved in the definition of this hypersurface in a little more detail, we consider an example of a reaction

$$AB + C \rightarrow A + BC \tag{9.36}$$

in which all three particles lay along a straight line. The number of coordinates N needed to describe the relative distances is two. They can be chosen to be the AB distance and the distance between C and the center of mass of AB. The curves of constant potential energy are indicated on the usual skewed plot of Fig. 2.1, where the coordinates are mass-weighted values of the above two coordinates, as discussed later.

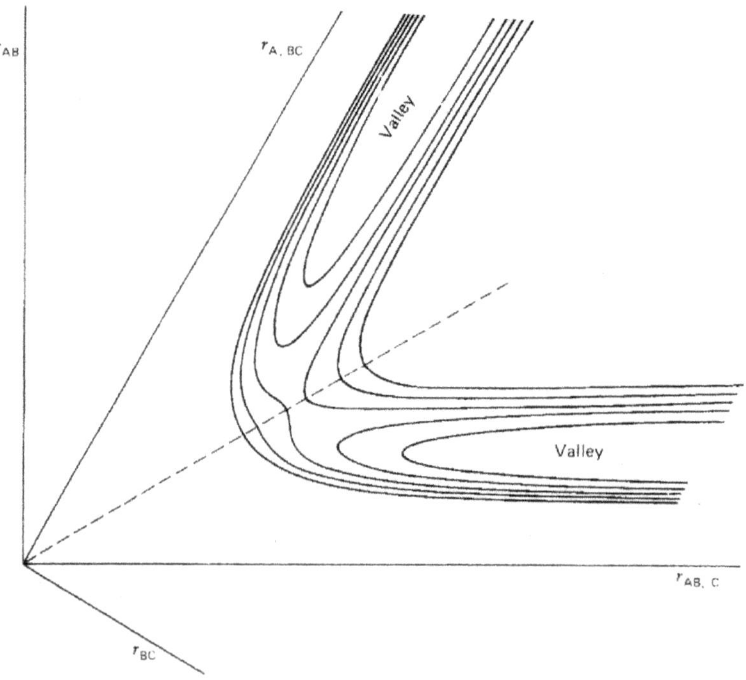

Fig. 2.1. Potential energy contours for reaction (9.36), plotted as a function of mass-weighted coordinates r_{AB} and $r_{AB,C}$ (cf. eq. 2.14). The dotted line denotes the activated complex.

A single point on this plot represents the values of the two coordinates, and the motion of that point represents the dynamical behavior of the system. Stable configurations of reactants are in the valley in the lower right-hand valley of the figure, and those of the products lie in the in the upper left-hand valley. The configurations that constitute the activated complex occupy $N - 1$ dimensions (i.e., one dimension on the figure) and are represented by the dotted line in the figure. The dotted line is seen to separate the reactants' region from the products' one, and the motion away from the dotted line is downhill, on either side of the line.

The assumptions of activated-complex theory are the following

(i) A set of configurations, the $(N - 1)$-dimensional hypersurface in an N-dimensional coordinate space, can be *defined* as having the property that a system that has crossed it has thereby reacted.

(ii) There is an *equilibrium* between reactants and systems crossing the surface in the *forward* direction. Such an equilibrium, which does *not* involve equilibrium between reactants and systems crossing in the reverse direction, has been termed a *quasi-equilibrium.*"[1]

(iii) In a quantum mechanical treatment it is meaningful to speak of energy levels $E_n^{\ddagger}$ of the activated complex.

(iv) When quantum mechanical effects are important in the rate of *crossing* of the surface (e.g., in tunneling), one *usually* assumes that the motion across the surface can be treated as a rectilinear motion and the relevant quantum mechanics for rectilinear tunneling can be applied.

9.9.2. *Derivation*

With these assumptions one can derive the customary activated-complex theory expression as follows: In the activated-complex theory one first calculates a *probability density* for finding the system in a given quantum state n and per unit length along the direction *perpendicular* to the $(N-1)$-dimensional hypersurface of configurations of the activated complex.[3]

One then multiplies the above by the appropriate *velocity* and sums over all quantum states n of the activated complex. We first derive an expression for the rate constant of a gas reaction in a given volume V and temperature T. The effect of tunneling is first neglected.

We denote by s the coordinate perpendicular to this $(N-1)$-dimensional hypersurface and denote by p_s the corresponding momentum. The value of s on the hypersurface is denoted by $s^{\ddagger}$. The probability of finding the system in a phase space volume element $dsdp_s$ and at the same time finding the system in a quantum state n of the activated complex is denoted by $\rho^{\ddagger}dsdp_s$."

NOTE: $\rho^{\ddagger}$ is the probability density...

"The latter is equal to

$$\rho^{\ddagger}dsdp_s = \frac{dsdp_s}{h}\frac{e^{-E/kT}}{Q}, \tag{9.37}$$

since $dsdp_s/h$ is the number of translational quantum states in $dsdp_s$. Here, Q is the partition function for the reactants at a given temperature and volume. E is the total energy and can be written near $s^{\ddagger}$ as

$$E = E_n(s) + \left(\frac{p_s^2}{2\mu}\right) + V_1(s), \tag{9.38}$$

where $E_n(s)$ is the energy of all degrees of freedom of the system on a hypersurface at any s (other than the s-coordinate), $\frac{p_s^2}{2\mu}$ is the kinetic energy along the reaction coordinate at that s, μ is the effective mass for motion along that coordinate, and $V_1(s)$ is the minimum potential energy on the hypersurface at this s. To discuss the crossing of the

[3]For a system with N degrees of freedom, the Nth of which is the reaction coordinate, the quantum number n denotes a quantum number for each degree of freedom; that is, it denotes $(n_1, \ldots n_{N-1})$.

hypersurface at $s = s^\ddagger$, the $E_n(s)$ and $V_1(s)$ will be replaced by their values $E_n^\ddagger$ and $V_1^\ddagger$ at $s^\ddagger$, and $p_s^{\ddagger 2}/2\mu$ will indicate the value of the s-kinetic energy at $s^\ddagger$.

When Eq. (9.37) is divided by ds, a probability per unit length is obtained, and when this result is multiplied by $\dot{s}$, the velocity along the reaction coordinate, the *contribution of the specified quantum states to the reaction rate* is calculated. Upon integration over all values of p_s in the positive direction—that is, from 0 to ∞—and summation over all quantum states n of the activated complex, an expression for the rate constant of the reaction k_r is then obtained:

$$k_r = \sum_n \int_0^\infty \frac{\dot{s}dp_s}{hQ} \exp\left[\frac{-\left(E_n^\ddagger\right) + (p_s^2/2\mu) + V_1^\ddagger}{kT}\right]. \tag{9.39}$$

Since $\dot{s}$ equals $\frac{p_s}{\mu}$, $\dot{s}dp_s$ equals $d\left(\frac{p_s^2}{2\mu}\right)$ and integration of Eq. (9.39) then leads to

$$k_r = \frac{kT}{h}\frac{Q^\ddagger}{Q}e^{-V_1^\ddagger/kT} \equiv \frac{kT}{h}e^{-\frac{\Delta A^\ddagger}{kT}}. \tag{9.40}$$

$\Delta A^\ddagger$ is the (Helmholtz) free energy of activation and $Q^\ddagger$ is the partition function of all degrees of freedom of the activated complex apart from that of the s-motion

$$Q^\ddagger = \sum_n \exp\left(\frac{-E_n}{kT}\right), \quad \Delta A^\ddagger = -kT\ln\frac{Q^\ddagger}{Q} + V_1^\ddagger. \tag{9.41}$$

The activation energy E_a of the reaction readily follows from Eq. (9.40) by differentiating lnk_r with respect to $\frac{1}{kT}$

$$E_a = \frac{\partial \ln k_r}{\partial\left(\frac{1}{kT}\right)}. \tag{9.42}$$

Insertion of Eqs. (9.40) into (9.42) yields terms $\partial \ln Q^\ddagger/\partial\left(\frac{1}{kT}\right)$ and $-\partial \ln Q/\partial\left(\frac{1}{kT}\right)$, denoted by $\langle E_n^\ddagger\rangle$ and $\langle E_n\rangle$, which are average energies of the activated complex, apart from that of the s-motion, and of the reactants, respectively. The kT term in the preexponential factor of (9.40) contributes another kT to the activation energy. Thus

$$E_a = kT + V_1^{\ddagger} + \langle E_n^{\ddagger} \rangle - \langle E_n \rangle \quad (\text{arbitrary } T). \qquad (9.43)$$

When the temperature is sufficiently low that all species are essentially in their lowest quantum state of energy E_0 (reactants) or $E_0^{\ddagger}+V_1^{\ddagger}$ (activated complex), $\langle E_n^{\ddagger} \rangle$ and $\langle E_n \rangle$ become the energies in the lowest states of the activated complex and reactants, respectively.

One then obtains

$$E_a = V_1^{\ddagger} + E_0^{\ddagger} - E_0 \quad (T = 0\ K). \qquad (9.44)$$

(**Typo** corrected).

9.9.3. *Extensions of Concepts*

In the previous section, we were mainly concerned with the crossing of a hypersurface at $s = s^{\ddagger}$, an $s^{\ddagger}$ that was independent of n, the quantum state of activated complex. To treat systems where $s^{\ddagger}$ depends on n or where quantum mechanical tunneling occurs through a wide barrier or where a free energy criterion is considered (see Section 9.11), it is desirable to extend some of the concepts of the previous section.

The two-dimensional system of Fig. 2.1, will serve for purposes of illustration. One may imagine that curvilinear reaction coordinate curves have been introduced (curves on which the *vibrational coordinate* is constant). One example of such curves is illustrated by a family of curves in Fig. 2.2, of which L, L', and L'' are members. Hypersurfaces (one-dimensional in the present figure) of constant s are illustrated by the dotted lines.

Curve L itself passes along the floor of the valley of the reactants in the lower right of Fig. 2.2, over the saddle point, and then down to the floor of the products in the upper left of Fig. 2.2 (L is thereby the curve of lowest potential energy V leading from reactants to

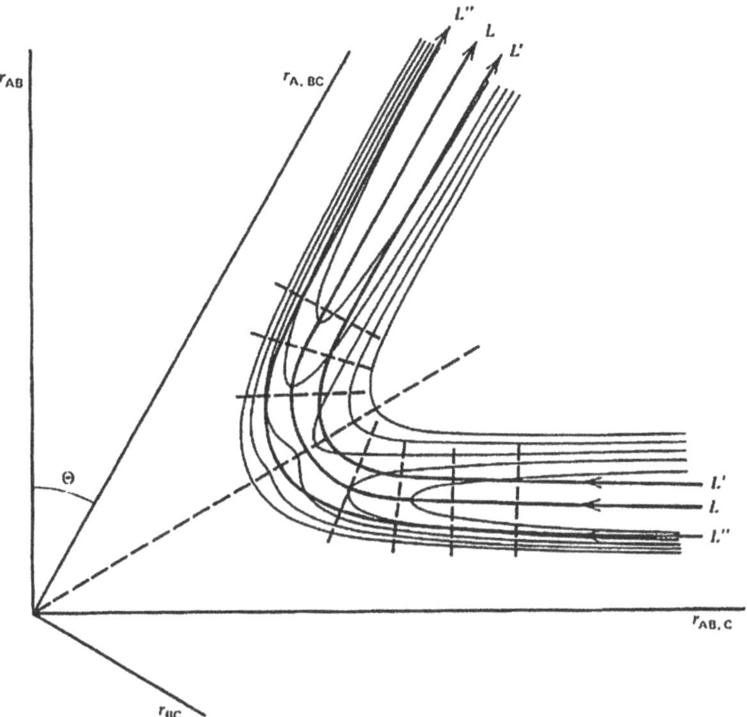

Fig. 2.2. Illustration of hypersurfaces at different values of s (the dotted lines normal to L-curves) and of reaction coordinate curves L, L', L'' differing in value of the vibrational coordinate (r in Fig. 2.6).

products). Along L, V has a value $V_1(s)$, the minimum value of V on any hypersurface at a given s [cf. Eq. (9.38)].

Classically the point representing the reacting system would oscillate in the reactants' valley, in which the reactants are far apart, over the dotted line occupying a saddle point region, and into a valley in which the products are far apart.

If we now use Eq. (9.38) for all values of s .... $E_n(s)$ is the energy, *kinetic plus potential*, the potential measured relative to that on L, for motion on the hypersurface. In this example of Figs. 2.1 and 2.2, $E_n(s)$ is the usual vibrational energy. A plot of the potential energy V along curve L, that is, $V_1(s)$—for a surface somewhat less

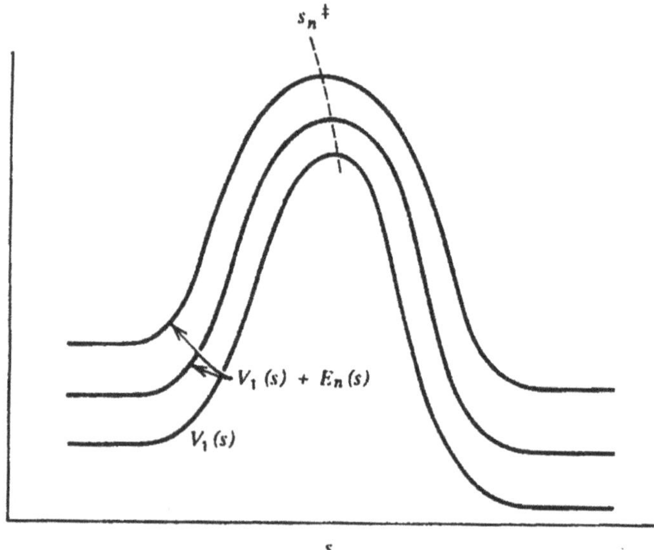

Fig. 2.3. Plot of effective potential energy $V_1(s) + E_n(s)$, for two states n, along the reaction coordinate, and plot of $V_1(s)$.

symmetrical than that given in Figs. 2.1 and 2.2—is given in Fig. 2.3, together with a plot of $E_n + V_1(s)$ for some n.

9.9.4. *Activated Complexes Whose Position is State-Dependent*

Apart from a quantum mechanical treatment of the s-motion, Eqs. (9.40) to (9.44) do not include the possibility, which is sometimes important, that $s^‡$, the position of the $(N - 1)$-dimensional hypersurface, may depend upon the quantum state n of the activated complex. Specifically, the maximum of $E_n + V_1(s)$ can depend on n as in Fig. 2.3. A simple example of this dependence occurs when part of the barrier to a bimolecular reaction arises from a centrifugal potential of the colliding molecules. The orbital angular momentum quantum number l then contributes to n.

For the case where $s^‡$ depends on n, and so is denoted by $s_n^‡$, Eq. (9.39) is still valid, except that $E_n^‡$ and $V_1^‡$ are the values at the maximum of $E_n(s)+V_1(s)$ (i.e., at $s = s_n^\ddagger$). For each n, the integration of Eq. (9.39) over p_s is performed at that $s_n^\ddagger$. As before, $\dot{s}dp_s$ equals $d\left(\frac{p_s^2}{2\mu}\right)$ and one ultimately obtains

$$k_r = \frac{\left(\frac{kT}{h}\right) Q_{\text{eff}}^\ddagger}{Q} \tag{9.45}$$

where

$$Q_{\text{eff}}^\ddagger = \sum_n \exp\left[\frac{-(E_n^\ddagger + V_1^\ddagger)}{kT}\right] \tag{9.46}$$

and

$$E_n^\ddagger + V_1^\ddagger \equiv E_n^\ddagger(s_n^\ddagger) + V_1^\ddagger(s_n^\ddagger).\text{''} \tag{9.47}$$

9.9.5. *Tunneling Contribution*

"In discussing the classical s-motion across a hypersurface in the interval $(s^\ddagger, s^\ddagger + ds)$, the s-motion can be treated as rectilinear. In a quantum mechanical treatment of the s-motion, a rectilinear coordinate can be used for sufficiently small *degrees of tunneling*, since only regions close to the usual saddle point in the PES in Fig. 2.1 would then contribute to k_r. With larger amounts of tunneling, the curvilinear nature of the s-motion should be considered, as one sees from Fig. 2.2."

"To consider tunneling... it will be assumed that the motion along a reaction coordinate s can, in the activated-complex region, be *separated* from the remaining motions and that a rectilinear expression for the kinetic energy of the reaction coordinate s can be used. That is, curvilinearity of the s-coordinate curves is neglected in the usual treatment. After this 'adiabatic separation of variables' of the internal motion at each s to obtain $E_n(s)$, one then solves the separated wave equation of the s-motion in the vicinity of the potential

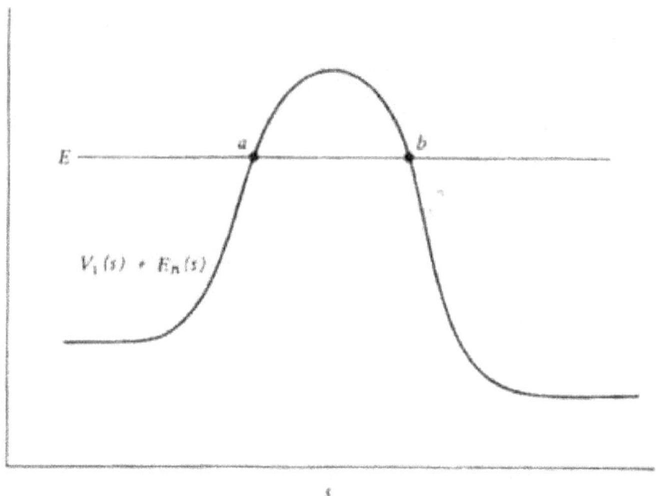

Fig. 2.4. Plot of effective potential energy $V_1(s) + E_n(s)$ versus reaction coordinate. Point a and b are "turning-points" (i.e., points where the classical momentum p_s vanishes).

energy maximum

$$\left[-\frac{\hbar^2}{2\mu}\frac{d^2}{ds^2} + V_1(s) + E_n(s)\right]\Psi_{En}(s) = E\Psi_{En}(s) \qquad (9.48)$$

The effective barrier for the tunneling, $V_1(s) + E_n(s)$, is depicted for some n in Fig. 2.4.

Since the tunneling is through a $V_1(s) + E_n(s)$ barrier, rather than through the $V_1(s)$ barrier along L, the system actually tunnels along the entire family of L curves in Fig. 2.2. The contribution from each L curve is weighted in a subtle way [by the contribution of the vibrational wave function to $E_n(s)$].

Solution of Eq. (9.48) yields the probability $\kappa(E, n)$ that the system crosses the activated-complex region. The energy for the remaining degrees of freedom has some value $E_n^{\ddagger}$ in the activated-complex region. The arguments leading from Eqs. (9.38) to (9.40) can again be used, in a somewhat modified manner, but now including tunneling.

The probability of finding the system in a phase space volume element $dsdp_s$ at some specified s near to, but to the left of, point a in Fig. 2.4, and at the same time of finding the system in a quantum state n for the remaining degrees of freedom, is $\rho dsdp_s$

$$\rho dsdp_s = \frac{dsdp_s}{h}\frac{\exp\left\{-\left[E_n(s)+V_1(s)+\left(\frac{p_s^2}{2\mu}\right)\right]/kT\right\}}{Q} \quad (s < a) \tag{9.49}$$

recalling Eq. (9.38). After dividing by ds, one obtains the probability per unit length at the given s. When this result is multiplied by $\dot{s}$, the velocity along the reaction coordinate at that s, the *probability flux* of system incident on the reaction barrier is obtained. Of this incident flux, only a fraction $k(E,n)$, passes into the products' region. Upon multiplying the above expression by $k(E,n)$, integrating over all possible values of p_s, and summing over all quantum states n of the activated complex, an expression is obtained for the rate constant k_r

$$k_r = \sum\int \kappa(E,n)\exp\left\{-\left[E_n(s)+V_1(s)+\left(\frac{p_s^2}{2\mu}\right)\right]\Big/kT\right\}\frac{\dot{s}dp_s}{hQ}. \tag{9.50}$$

Since $E = E_n(s) + V_1(s) + \left(\frac{p_s^2}{2\mu}\right)$ and $\dot{s}dp_s$ is dE, one obtains

$$k_r = \frac{kT}{h}\int_E \kappa(E,n)e^{-\frac{E}{kT}}\frac{dE/kT}{Q}. \tag{9.51}$$

When the dependence of $E_n(s)$ on s in the vicinity of $s^\ddagger$ is negligible, and written as $E_n^\ddagger$, then $\kappa(E,n)$ depends on E and n only via $E - E_n^\ddagger$. Denoting the latter by $\epsilon + V_1^\ddagger$, $\epsilon = E - E_n^\ddagger - V_1^\ddagger$ and so is the 'translational energy' at $s^\ddagger$. Equation (9.51) can be written as

$$k_r = \frac{kT}{h}\int \kappa(\epsilon)e^{-\frac{\epsilon}{kT}}\frac{d\epsilon}{kT}\frac{Q^\ddagger}{Q}\exp\left(\frac{-V_1^\ddagger}{kT}\right). \tag{9.52}$$

When there is no tunneling, $\kappa(\epsilon)$ in Eq. (9.52) is zero when ϵ is negative and equals unity when ϵ is positive. In this case, the integral over ϵ becomes unity and Eq. (9.52) reduces to Eq. (9.40). Similarly, Eq. (9.51) can be shown to reduce to Eq. (9.45) in the absence of tunneling.

A semiclassical value for $\kappa(\epsilon)$ in Eq. (9.52), when $\epsilon < 0$, is

$$\kappa(\epsilon) = \frac{1}{1+\alpha^2} \tag{9.53}$$

where

$$\alpha = \exp\left(\int_a^b |p_s| \frac{ds}{\hbar}\right). \tag{9.54}$$

and a and b are defined in Fig. 2.4 and

$$p_s = \pm\left\{2\mu\left[E - V_1(s) - E_n^{\ddagger}\right]\right\}^{\frac{1}{2}}. \tag{9.55}$$

A related expression can be written for $\kappa(\epsilon)$ for $\epsilon > 0$. As an example of Eqs. (9.53) to (9.55), one may consider the case where $V_1(s)$, the potential energy along curve L in Fig. 2.3, is a parabola, $V_1^{\ddagger} - \frac{1}{2}k(s - s^{\ddagger})^2$. The integral in Eq. (9.54) is given by

$$\int_a^b |p_s| \frac{ds}{\hbar} = \int_a^b \left|\left\{2\mu\left[-\epsilon - \frac{1}{2}k\left(s - s^{\ddagger}\right)^2\right]\right\}^{\frac{1}{2}}\right| \frac{ds}{\hbar} \tag{9.56}$$

The points a and b are those where the integrand vanishes. The integral is a standard one and equals $\frac{(-\pi\epsilon)}{h\nu}$, where ν is the 'frequency' of s-motion $\left(\frac{1}{2\pi}\right)\sqrt{\frac{k}{\mu}}$. ϵ is negative in the tunneling region.[(2)]"

9.10. Activated-Complex Theory for Solutions

"The derivation in Section 9.9 is referred to a system of a given volume and temperature, whereas in solution the reaction proceeds at a given pressure P and temperature T."

The rate constant is now

$$k_r = \frac{kT}{h}\exp\left(-\frac{\Delta G^{\ddagger}}{kT}\right) \tag{9.57}$$

where $\Delta G^{\ddagger} = G^{\ddagger} - G$.

9.11. Location of the Activated Complex and the Point of Maximum Free Energy

"The position $s_n^\ddagger$ of the activated complex for any n was seen earlier to be that for which $V_1(s) + E_n(s)$ has its maximum value. When there are many states n or when inadequate information is available about the s dependence of each, the procedure of locating each $s_n^\ddagger$ is too formidable and resort must be made to some more approximate argument. Just as in a simpler system where $s^\ddagger$ is frequently chosen to occur at the maximum of $V_1(s)$, so too is the $s^\ddagger$ in these more complicated systems usually defined as the one occurring at the maximum of some free energy $A(s)$, in a plot of $A(s)$ versus s. In the present section we shall consider a basis for such a choice of s using Eqs. (9.45) to (9.47).

We use $\epsilon_n(s)$ to denote the sum

$$\epsilon_n(s) = E_n(s) + V_1(s). \tag{9.58}$$

By definition of $s_n^\ddagger$, one has

$$\epsilon_n(s) \le \epsilon_n(s_n^\ddagger). \tag{9.59}$$

Thus, in Eqs. (9.45) to (9.47)

$$Q_{\text{eff}}^\ddagger = \sum_n \exp\left[\frac{-\epsilon_n\left(s_n^\ddagger\right)}{kT}\right] \le \sum_n \exp\left[\frac{-\epsilon_n(s)}{kT}\right], \tag{9.60}$$

The first sum in Eq. (9.60) is the correct one, and we wish, therefore, to make the last sum approach the first as closely as possible. To do so, one must choose an s, $s^\ddagger$, which makes the last sum as small as possible. Since the sum is related to the free energy $A(s)$ of a system moving on a hypersurface, at a given s

$$\exp\left[\frac{-A(s)}{kT}\right] = \sum_n \exp\left[\frac{-\epsilon_n(s)}{kT}\right], \tag{9.61}$$

this choice of $s = s^\ddagger$ also serves to locate the largest value of $A(s)$ in a plot of $A(s)$ versus s and so provides a basis for the approximate procedure."

9.12. Remarks on Applications

9.12.1. *Electron-Transfer Reactions*

"Like the unimolecular reactions, ET reactions involve *features additional* to those contained in activated-complex theory. In certain respects, electron transfer reactions are the simplest of all reactions: simple electron transfers involve no breaking of chemical bonds, no formation of new chemical bonds, but only *fluctuations of bond lengths and bond angles and fluctuations of molecular orientations* of the solvent in such a way as to facilitate ET. The first problem is to define the *nature of the activated complex* for this ET. Such a discussion leads to one of the PES and to the role of the Franck–Condon principle in the ET.

In any chemical reaction there is a change in configuration of the atoms and molecules from one that is appropriate to the reactants to one that is appropriate to the products. In reactions that involve only the transfer of an electron, there are the changes in configuration cited above: the bond lengths in a particular reactant will depend on the charge of that reactant and hence will differ before and after the electron transfer step. Sometimes this difference is very small and may be ignored, but at other times it constitutes the principal factor making the electron transfer rate constant small. Again, the solvent molecules in the vicinity of the reactant are more oriented, *on the average*, toward a more highly charged reactant. Thus, when the charge of that reactant is altered by ET, the *average degree of orientation* of the solvent molecules around it *eventually relaxes* from the one appropriate to reactants to one appropriate to products. Thus, there is a net change in the *distribution of orientations* of many solvent molecules.

When the reactants are far apart, the reaction coordinate is the distance between the two particles. Gradually, as they come closer, the reaction coordinate begins to include the other kinds of motion discussed above—those involving fluctuations of bond lengths and bond angles in the coordination shells of the reactants and fluctuations

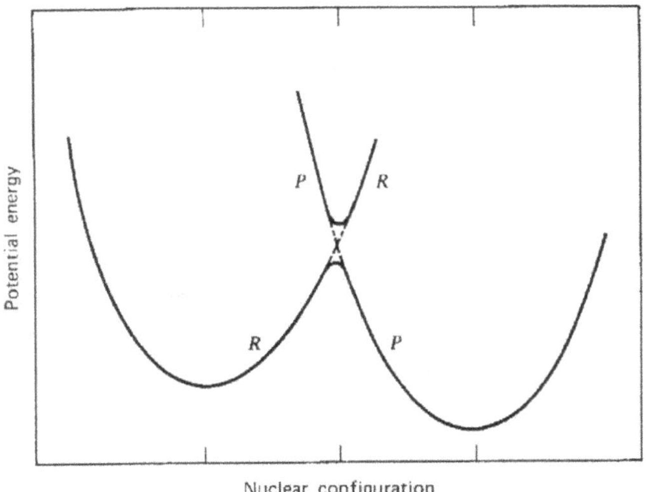

Fig. 2.7. Profile of potential-energy surfaces of reactants (R) and of products (P), plotted versus nuclear configuration of all the atoms in the system—= surface for zero electronic interaction of the reacting species—= adiabatic surface.

of solvent molecule reorientations. Along the reaction coordinate, such coordinates change from values appropriate to reactants to those appropriate to products. In this region, in fact, the reaction coordinate begins to resemble, instead of a translation, a vibrational-like motion as indicated schematically in Fig. 2.7.

Figure 2.7 describes the *profile*, labeled R, of the many dimensional surface of the potential energy of the two reactants and solvent, as a function of the various coordinates in the system mentioned above. It also contains a profile, labeled P, of the potential energy surface of the products and solvent, as a function of the same coordinates. These two curves do not have minima at the same configuration because the equilibrium bond lengths are different in any species when it is a reactant compared to when it is a product and because the typical favorable orientations of the solvent molecules are also different.

When the reactants are far apart, fluctuation in these coordinates can occur but without leading to an ET, that is, without causing the system to move from curve *R* to curve P in Fig. 2.7. When the reactants are close together, there is appreciable electronic coupling between them, causing a splitting of these potential energy surfaces, as indicated by the dotted lines in Fig. 2.7. Under such conditions a suitable fluctuation of coordinates from values appropriate to reactants toward values appropriate to products will lead in many cases to the system moving from the reactants curve to the products curve. For example, when the splitting of those surfaces is large enough, the system stays on the lowest split surface. That is, the system moves along the lower line in Fig. 2.7. In this case, the reaction is said to be adiabatic as far as the electronic motion is concerned. If, on the other hand, the splitting is very small, the chance of the system moving from the reactants' curve to the products' curve becomes correspondingly small, and on the occasions that it does so, the reaction is said to be electronically nonadiabatic.

For reaction to occur, we have seen that it is necessary for the system to reach the intersection region in Fig. 2.7. *This need is another way of stating the Franck–Condon principle for this electron transfer problem*: The Franck–Condon principle states that the process of going from the *R* surface to the *P* surface in Fig. 2.7 is a "'vertical'" one, to avoid changing atomic positions (and momenta). Thus, if the ET occurred to the left of the intersection region in Fig. 2.7, the transfer would require the absorption of light and thus be a photochemical reaction rather than a thermal reaction.

An important part of the problem of calculating the reaction rate lies in calculating the probability of reaching in this many dimensional coordinate system the intersection of the two PESs, *R* and *P*, in Fig. 2.7. If activated-complex theory is applied to this system, this intersection of the two surfaces constitutes, in effect, the activated complex for the reaction."

NOTE: The intersection surface is the transition state. Marcus' X and X^* represent the system "Immediately before and immediately after" the crossing.

"In applying activated-complex theory one circumstance that makes these reactions different from most other reactions involving ions in solution should be noted: In the electron transfer reactions of the type discussed here ("'outer sphere"' ETs), *there is a large separation distance between the charges of the reacting species before and after the transfer.* In most other reactions, the change of charge distribution does not occur as dramatically. This considerable change of charge distribution occurs in *a relatively small increment of reaction coordinate* in Fig. 2.7 as the system moves along the dotted line.

We consider the solvent polarization change accompanying the motion in Fig. 2.7. The actual many dimensional surface, whose profile is approximated in Fig. 2.7, actually has many local minima (since solvent orientations do not obey harmonic oscillator laws of motion). The many configurations in the general vicinity of these minima, suitably weighted by a Boltzmann factor, contribute to the solvent dielectric polarization.[(3)] The solvent polarization represented by any group of configurations on, or near, the intersection region in Fig. 2.7 cannot be in thermal equilibrium with the charges of the reactants (or with that of products). Thus, this polarization fluctuates from the equilibrium polarization appropriate to reactants as the system moves toward the intersection region in Fig. 2.7 and can be termed a *nonequilibrium polarization.* The electronic polarization of the system is typically, on the other hand, largely in equilibrium with the field created by the charges *and by the remaining polarization* (orientational, vibrational) of the solvent.

To calculate the free energy of a system containing this yet to be determined polarization function one has to find a reversible path for forming this nonequilibrium polarization system. Such a path was found by charging up the system to form the desired orientation-vibration polarization function of position and by holding that polarization fixed, changing the charges to the desired values (i.e., of reactants or of products), allowing the electronic polarization to adjust to these changes.

This expression for the free energy of the fluctuation involves an integral whose integrand is a quadratic function of the polarization (in the usual dielectric unsaturation approximation). One can minimize it to obtain the unknown polarization function, subject to the condition that the activated complex consists of configurations on the intersection of the R and P surfaces in Fig. 2.7. In this way the free energy of activation $\Delta G^{\ddagger}$ was calculated. The details have been described elsewhere. Here, we give the functional form of the equations . . . ."

"9.12.2. *Remarks on Other Reactions*

The treatment of ETs in the previous section raises the question of whether analogous predictions can be made for other reactions (viz., atom and proton transfers [PTs]). In the case of atom transfers the potential energy surfaces are typically not of the *small-splitting* (*weak-interaction*) type in Fig. 2.7. Instead, the splitting is very large, and a different method of calculation must be used."

NOTES

1. **M**: There is an equilibrium population not of all the states but of the states that would contribute in the forward direction, so there is a kind of equilibrium, quasi because it is really not an equilibrium.
2. Q: Why "frequency" in quotes? In the case of tunneling does one then consider the s-motion in the tunneling region a forth-and-back motion instead of the normal s-motion always in the forward direction?

   M: Normally, in the simple tunneling, the movement doesn't go back and forth, it goes through once, now there are certain special situations where you have to take into account that kind of movement back and forth, but by and large it just goes once. "Frequency" because tunneling is not really a physical frequency of motion, it has units of frequency but don't think of tunneling as a particle going back and forth under a barrier, I mean it's true that something like that happens in very subtle cases when you are near the top of the barrier,... but in typical tunneling regimes there is a certain probability ... there is sort of a one pass tunneling object.

Q: So, does this "frequency" refer only to the dimension somehow?

M: A frequency is associated with a force constant and some mass, so if the curve is an inverted parabola you still can speak of an *imaginary* frequency. The frequency is imaginary in that case, but it's not an actual going back and forth, it's an imaginary frequency. You know, like they speak in TS theory when you diagonalize in the vicinity of the TS, if everything goes well you have one imaginary frequency and that's a frequency along the reaction coordinate.

3. Q: Is the Boltzmann weighting what is responsible for the smoothing of the surface transforming it in the free energy parabolas?

   M: I don't think so, well... the Boltzmann weighting is always there, but I think what's responsible for that smoothing is more like the central limit theorem, you can have very complicated distributions where you have many many coordinates involved, then surfaces may look quadratic in some parameters,...

M118. Electron Transfer in Homogeneous and Heterogeneous Systems

Probably the first Marcus' review to be read by a student of his ET theory. In Equation (4.1) there is a **typo**: the symbol ΔG^* is to be used in it instead of $\Delta G^{\ddagger}$. A very important point is that on p. 481, where the universal importance of fluctuations in all chemical reactions is pointed out, see the very nice enlightening opening page, p. 3 of the "Introduction to Statistical Mechanics" by David Chandler. Here, Marcus writes on p. 481: "Typically, reactions occur as a result of a suitable *fluctuation of the coordinates* which permits the system the system to cross the *critical set of configurations (the* $N - 1$ *surface) known as the activated complex*. In reactions where chemical bonds are broken or formed these fluctuations in coordinates involve in a major way the increase in distance in the breaking bonds or decrease in distance in any bonds being formed. Sometimes bonds merely rearrange, and the fluctuations then involve *simultaneous changes in several bond distances and bond angles*. In each case, the fluctuations leading to reaction are those from a typical configuration of the reactants to a typical one of the products. We investigate in Section IV the type of fluctuations involved in simple ETs."

NOTES

1. p. 479: "Activated complexes are defined as a set of configurations of the system on this special N 1 dimensional surface."

   Q: A "set" or all of the configurations?

   M: All of the configurations but because they're Boltzmann-weighted some of the configurations on the surface will hardly ever be reached.
2. p. 480: "... when quantum effects are important for tunneling through the activated-complex region..."

   Q: How is the "region" defined?

   M: Imagine an inverted parabola going down in both directions in the vicinity of the saddle point, go kT below it, then that's sort of the activated-complex region, but of course the important thing is the hypersurface, so it is not a region, it is the whole $N - 1$ dimensional surface, but if you want to speak of being in that general region then you say you are in that general region when, if you look at the imaginary frequency, where $h\nu$ equals kT. I refer to that sort of a region, but actually one doesn't need even to discuss the region, really, because what's important is what's going on that particular $N - 1$ dimensional surface, then not so much on either side, except for tunneling purposes, then the imaginary region is what's on either side, that's what's the system *senses* because of the tunneling.

3. p. 481: "In some respects ET reactions are the simplest of all chemical reactions."

   M: Well, when the ET involves only the transfer of an electron, and not a simultaneous breakage of bonds, then there is no reaction simpler than that .... Most reactions involve breaking of some bonds and forming of the others, and that's more complex... you need then sophisticated structure methods, so when an ET reaction involves no breaking of bonds, and many of them involve no breaking of bonds, then you can't get simpler than that. But, of course, there are some ET reactions that might involve forming a complex to bind them together, and transfer and then that's more complicated, that's more than a simple ET, it's transfer of other atoms or groups of atoms too.
4. p. 482 "The abscissa in this figure is a 'generalized coordinate' involving changes in bond lengths of both reactants and reorientation of solvent molecules in the vicinity of the reactants, such that the system proceeds from a configuration appropriate to reactants... to a configuration appropriate to the products."

   M: The generalized reaction coordinate is an *energy difference*, and as you hold the energy difference constant between the R and P surfaces then you are at the same value of the generalized reaction coordinate, but you are at an $N-1$ coordinates hyperspace for that, so different processes are happening in different parts of that hyperspace, and many coordinates are changing from one hyperspace to the parallel hyperspace at a different energy difference. So, one has to be more specific, and what I used as generalized coordinate was really the energy difference. It was a contribution, although I didn't realize fully at the time of the 1960 paper, and then I brought it out very explicitly in the 1965 paper.
5. p. 485: "Frequently, dielectric continuum theory has been used to calculate this equilibrium position-dependent dielectric polarization and to calculate, approximately, its contribution to the free

energy of ions or of pairs of ions. Standard textbook electrostatics are used."

Q: Do you have a reference?

M: I think there was a book on electrolytes using continuum theory by Edward Amis, he had a book (Solvent Effects on Chemical Phenomena), that I think did this sort of calculations, and I think there was a reference by Gurney... and surely there are better sources now.

6. p. 490: "when specific solvent effects are absent each λ increases linearly..."

   Q: Can you list specific solvent effects?

   M: Hydrogen bonding would be a specific solvent effect, organic-type bonding and aromatic–aromatic...

7. p. 495: "when the solvent orientation contribution to λ can be approximated using dielectric continuum theory"

   Q: When is the approximation valid and when is it not?

   M: Well, in some situations where there can be very specific interactions that dielectric continuum theory doesn't take that into account, so that's one example, and then dielectric continuum theory is really something which is averaging over sort of a range of distances, but that's an approximation... Dielectric continuum theory is normally defined for uniform systems, it may describe the state around an ion as not uniform, it depends on some nonuniformity parameter where the field is not constant... dielectric continuum theory refers more to sort of situations where you don't have to worry about molecular dimensions, can average over those and in a sense molecular dimensions are comparable with the size of the ions, so is questionable. David Chandler and other people have done molecular dynamics calculations *of* λ.

M119. Energetic and Dynamical Aspects of Proton Transfer Reactions in Solution

(i) In the paper, there is a detailed discussion about the work terms, in terms of what they comprise.

(ii) When considering the hydrogen transfer reaction Cl + HBr → ClH + Br, Marcus introduces a new application of the Franck–Condon principle analogous to the one introduced in the case of ET. Now it is the H atom the light particle that moves while "the momentum of the 'slow' coordinate Cl-Br being substantially conserved in that region," p. 64.

(iii) Following: "Here the protonic motion is very nonadiabatic, and a significant increase of its vibrational action (and energy) occurs. Thus, in the reverse reaction vibrational energy should facilitate the PT, an effect which might be observable in a *suitably stabilized* (e.g., *intramolecularly* hydrogen-bonded) system using short laser pulses."

NOTES

1. p. 63 middle: "A diagram of the surface used is given in Fig. 1 in the usual skewed-axes form... the transition state is the line of steepest ascent from the saddle point, indicated in Fig. 1 by the dotted line."

   M: That is the line bisecting. The transition state by definition is a hypersurface that has one dimension less than the entire dimension of the system, so it is the whole line that divides the two dimensional space in two separate spaces. A hypersurface one dimension less. So here where you have a two-dimensional space, it has one dimension, if it would be an n-dimensional space, it would have $n-1$ dimensions.

2. p. 65 top "The work term can be composite of several terms... in large molecules... an appreciable steric restriction may occur,

and contribute a term w_r^{st}. Such steric factors might be reduced somewhat by favorable $\Delta G_R^{o\prime}$...

M: $\Delta G_R^{o\prime}$ is $\Delta G^{o\prime} + w_p - w_r$, in other words, $\Delta G_R^{o\prime}$ is $\Delta G^{o\prime}$ corrected for the difference of the work terms.

3. p. 65 middle: "Another contribution to the work term can also occur, when the intermediate product of the third step in the reaction

$$A_1 \ldots HA_2 \rightarrow A_1 + HA_2$$

is not the separated products but rather is a metastable intermediate which later ruptures, compare equation

$$AH + R_1R_2C = N^+ = N^- \rightarrow A^- + R_1R_2CH - N^+ \equiv N$$

Whenever this last step has an activation barrier w_{dec}^r which exceeds the barrier for the intermediate to reform the reactants, this w_{dec}^r should in effect be added to the previously computed free energy barrier. We then have

$$w^r = w_{\text{des}}^r + w_{st}^r + w_{\text{dec}}^r$$

Of these w_r contributions, only the first two contribute to the w_r in equation $\Delta G_R^{o\prime} = \Delta G^{o\prime} + w^p - w^r$."

Q: First: what do you mean by the index "dec"? Then, if I understand well, the state $A_1 + HA_2$ instead of being the last step is followed by the further step $A_2^- + HA^+$, say.

M: If you look at products there, that term on the right may not be the final product. The index "dec" is for "decompose." I think what you have there... there is an ET, that particular step is just a PT that could be followed by dissociation, maybe even by an ET, all is possible. This is a PT, and then that may be followed by some other reaction, sure. Then what are the reactions that are involved... I may imagine that that big compound might break up or may react in some other way, an ET maybe.

4. p. 66 top: "When attention is focused on the electrons of the reactants, and the electrons of the solvent are treated, for reasons of simplicity, as forming a polarizable dielectric continuum, one obtains a nonlinear Schrödinger equation."

   Q: Is there some paper of yours where this nonlinear equation is written down? Or is it to be found in the book of Pekar?

   M: Yes, where I treated the solvated electron, but I think a more fundamental treatment... dealing with subtleties more than I do, because there are additional subtleties, is in a paper due to Chandler, a paper that he wrote with James T. Hynes, Hynes overlooked some point and Chandler corrected it, and kindly brought Hynes into it... I recall it was great. The book of Pekar may have been translated in English now, it could have been, I read it in German... it was some nonequilibrium polarization,... an electron moving through and partially dragging an electronic cloud with it, the systems are not really being in equilibrium,... I read it and I benefited from it. But I ended up sort of deriving things my own way. This is in the solvated electron paper.
5. p. 66 top-middle "The Schrödinger equation for the wave function ψ of the electrons of the reactants, for any nuclear configuration r_n of the reactants and (positions) of solvent molecules is obtained by minimizing the following functional $\mathcal{F}(\psi)$ with respect to ψ at a given $\boldsymbol{P}$.

$$\mathcal{F}(\psi) = \frac{\hbar^2}{2\mu}\int |\nabla\psi|^2 dr_1 + \int V(r_1, r_n)\,|\psi|^2\,dr_1 + \int W_{\text{rev}}$$

   where... $V(r_1, r_n)$ includes the potential energy arising from interactions within the reactants and with the solvent molecules, apart from that included in the relatively long range polarization term W_{rev}."

   Q: What does the index "rev" mean? Can you explain something more about the last term?

M: It may be some electrostatic term of the electrons interacting with other electrons. There are the electrons interacting with nuclei, there is that kinetic energy term, and then solute electrons interacting with the external electrons . . . .

6. p. 66 mid-bottom: "In the case of ET reactions it was possible to introduce a simplifying approximation, writing ψ as a linear contribution of two terms with weak overlap between them, one term being the same as for the reactants and the other being the same as for the products,. . . To the extent that the electronic wave function for the transition state of the reaction in equation:

$$\mathrm{AH} + \mathrm{R_1R_2C} = \mathrm{N^+} = \mathrm{N^-} \rightarrow \mathrm{A^-} + \mathrm{R_1R_2CH} - \mathrm{N^+} \equiv \mathrm{N}$$

could be similarly approximated for this purpose, the previous results for ET could be adapted to that for PT,. . ."

Q: If I understand well, you mean that your treatment for ET can be directly extended to PT if the wave function for the transition state of the earlier reaction can be written, say, as

$$\psi = a\psi_R\,(\mathrm{AH}, \mathrm{R_1R_2C} = \mathrm{N^+} = \mathrm{N^-}) + b\psi_P(\mathrm{A^-}, \mathrm{R_1R_2CH} - \mathrm{N^+} \equiv \mathrm{N}).$$

Can we in this case speak of a "precursor complex" and a "successor complex" (remember yours X* and X) as in the case of ET?

M: Yes, I suppose you could, because precursor just means you brought the two reactants together and sort of consider them in a particular position, same for the products, that part is independent of the earlier statement.

M134. Theories of Electrode Kinetics (Levich Festschrift)

(i) M. refers to three important contributions of Levich: (1) the quantum aspects of ET; (2) the temperature dependence of ET at very low temperatures; (3) the behavior of highly exothermic reactions "where the process may occur in the so-called inverted region."

(ii) A romantic sentence: "theory and experiment in ET reactions have had a very fortunate marriage..."

(iii) The oxonium discharge reaction at a metal electrode is considered. The process is represented as happening in three successive steps. The first of them

$$H_3O^+ + M(e) \rightarrow H_2O + H - M$$

has been treated by Bockris, Conway, and coworkers applying TS theory, by Levich and coworkers assuming that "the individual or collective motion of the reactants and those of the products and the solvent medium can be treated as harmonic oscillators, and the use of first-order perturbation theory (Fermi's Golden Rule) or other methods to calculate the reaction rate,"

M. uses mass-weighted coordinates which yield the skewed axes potential energy contour plot in Fig. 1, and his classical profile of potential energy in Fig. 2, to describe and visualize the two approaches. In the potential energy contour, drawn at fixed values

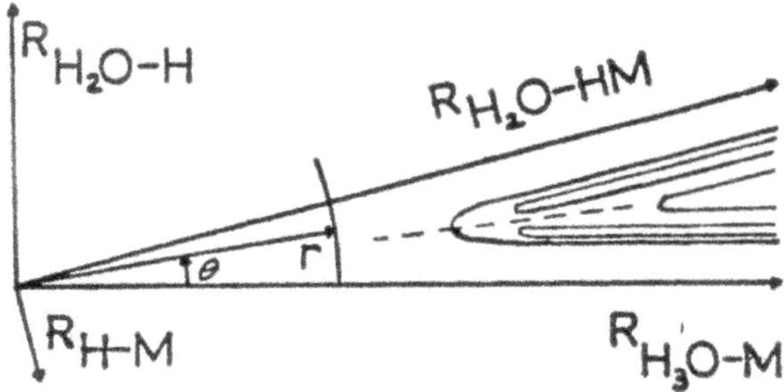

Fig. 1. Potential energy contour plot (schematic) for the three center reaction H_2O–H–M at a fixed value of the other coordinates q and at a given metal-solution potential difference. The mass of H_2O is taken as concentrated on the O, and the mass of M as infinite. The rotated axes are scaled H_2O–HM and H–M distances, again with the H_2O mass centered on the O. Polar coordinates r and θ are also indicated. The configurations along the dashed line form the conventional transition state.

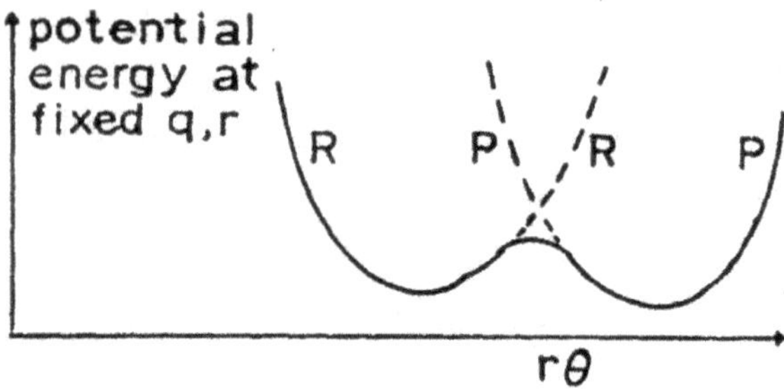

Fig. 2. Solid line: Profile of the potential energy V, in Fig. 1, at fixed r and q, versus the protonic distance coordinate rq. Dashed lines: a two electronic state ("weak overlap") model of this profile with one curve for reactants R and the other for products P.

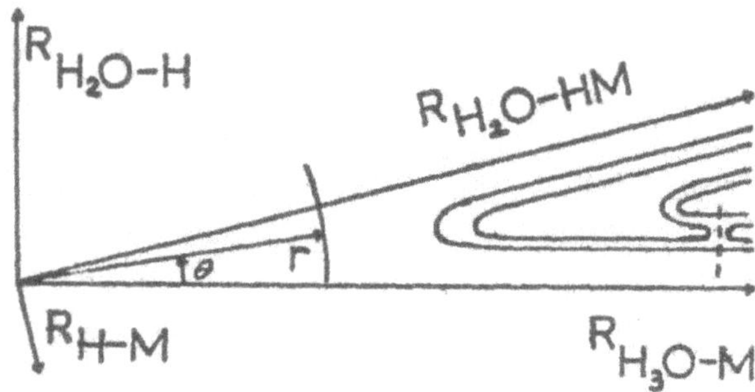

Fig. 3. Legend as in Fig. 1, but the reaction is now highly exothermic. The dashed line again denotes the transition state.

of the solvent coordinates q and at a given metal-solution potential difference, the two valleys of reactants and products and the saddle point separating them are schematically represented. Two polar coordinates, r and θ are used to describe the motion along the valleys and the motion bringing the system from one valley to the other. The TS state theory proper describes the ET process at small r's where the two valleys merge in the TS region. The Levich theory describes the process where r is larger and the two valleys are well separated. The difference between the regions where the two theories are applied is also seen in Fig. 2. There the lower solid curve describes the electronic PES for adiabatic PT, where because of the small proton mass the Franck–Condon principle may be applied to the rapid proton jump—similarly to what was done by M. in the ET case—and the Franck–Condon effects on solvation are omitted. The lower solid curve has first the character of the reactants and then that of the products passing through an intermediate region in the vicinity of the TS where it has both characters.

At larger r's the dashed lines represent, together with the adjoining parts of the solid curve, two separated R and P harmonic oscillator like potential energy curves. Each of the two localized electronic–protonic quantum states has a series of protonic vibrational states. "A *Franck–Condon reorganization* is needed when the proton jumps between well-separated channels." The reaction probability is calculated at any r "using Fermi's Golden Rule and perturbation theory. One takes into account the vibrational overlap integrals of the two localized protonic states... and also of the solvation states associated with each of those two protonic states."

We have seen how two different approaches are used to describe the PT at smaller and at larger r's. Problem: how to find the value of the r where most of the PT occurs? M. suggests the use of the semiclassical theory to treat this dynamical question.

In this paper, M. attempts an answer to the question: "Does the Franck–Condon principle play the same role in PT that it does in ET? When there is a jump between *two distinct wells*, the jump is so quick that the solvation appropriate to the first well in Fig. 2 does not have time to adjust to the proton jump into the second well. In that case, there is a *Franck–Condon reorganization of solvation* (the q-coordinates) .... If the PT occurs at an r where *the wells have merged* the Franck–Condon effect vanishes.

NOTES

1. p. 474 bottom: "one calculates or assumes some PES, *decides* on the transition state..."

Q: What does it mean "decides"? On p. 477 bottom: "an appropriate *method*, dynamical where necessary, for defining a 'transition state.'" And on p. 478 top: "a *method* for determining a typical r at which the PT occurs and for determining dynamically a 'transition state.'"

M: If you have a saddle point, the saddle point would be the center of the transition state. If you have PT, the proton would often tunnel before it reaches the saddle point, it would cut corners, in that case the saddle point isn't the transition state because you come short of it. There are some situations that are even more complicated, see the variational transition state theory. So, this is what I meant by "decides."

2. p. 476 top: "As r decreases, the barrier height separating the two wells *usually*decreases."

   M: There is a reason for that. As you go in, and r decreases there is a repulsion and you go up in energy, as you get closer to the saddle point the barrier decreases, finally at the saddle point there is no barrier at all.

3. p. 478 center: "The charge distributions of the two channels are quite different but tend *to approach* each other when the two channels merge, for example, in Fig. 1 at the r corresponding to the saddle point region."

   Q: Can we pictorially describe the initial and final channels and the transition state the following way:

$$\underset{(1)}{H_2OH^+ + M(e)} \rightarrow \underset{(2)}{H_2O\text{-}H^+\text{-}\text{-}\text{-}M(e)} \rightarrow \underset{(3)}{H_2O\text{-}\text{-}\text{-}H^+\text{-}\text{-}\text{-}M(e)} \rightarrow \underset{(4)}{H_2O\text{-}\text{-}\text{-}H^+\text{-}M(e)} \rightarrow \underset{(5)}{H_2O +^+ HM(e)}?$$

   where (1) and (2) are in the entrance channel for $r = \infty$ and for $r = r_{\text{saddle-point}}$ respectively, (4) and (5) are analogously in the exit channel, (3) represents the structure at the saddle point and (2) and (4) are two structures in which H^+ is shifted ≈ 0.25 Å to the left and to the right of the saddle point centered position just before and just after the jump, because "the jump is probably usually a very short one (e.g., $\approx$0.5 Å)?" Moreover: are there in the case of PT the analogous of the X and X* precursor and successor complexes of ET theory?

M: The precursor and successor complexes may not exist. Sutin uses the precursor and successor complex a lot. What I had in mind is probably different from what Norman had in mind. Really the precursor complex is better defined as Norman defined it, you bring the reactants together they are in a cage, so to speak, you have first reorganization, then transfer separation of the resultant complex, so the precursor complex is something that happens before the major reorganization starts. Also in the case of PT there may be a precursor complex, if the precursor complex is a reality or not, or if it is a mere convenience to break up the free energy barrier is another question, like for example part of the free energy involves the two reactants getting together, . . . as long as there is not some equilibrium between reactants and precursor complex,. . . it is just a convenience to break up the free energy to visualize it better.

4. Q: In the above theory you tacitly assume that after the proton has been transferred there is the ET process:

$$^{+}\text{HM(e)} \rightarrow \text{HM}$$

How does that happen?

M: Once the proton is transferred, the image charge that comes from somewhere in the interior unites with the proton charge, so that is really an H atom attached to the metal.

CHAPTER 10

Unusual Activation Parameters, Electron and Nuclear Tunneling, Mechanisms of Charge Separation, Spin Exchange and Electron Transfer in Photosynthesis; Electron Transfer versus Proton Transfer at Electrodes and in Solution, Hydrogen Evolution, Quantum Effects, Classical and Semiclassical Calculations of Electron Transfer Rates

Here are considered: **M110, M139, M140, M141, M145, M146, M154, M156, M163**

M110 Electron Transfer Reactions with Unusual Activation Parameters. A Treatment of Reactions Accompanied by Large Entropies Decreases

"The activation enthalpies for the oxidation of $Fe(H_2O)_6^{2+}$ by poly(pyridine)iron(III) and—ruthenium(III) complexes (reactions (10.1) and (10.2) for example)

$$Fe(H_2O)_6^{2+} + Fe(bipy)_3^{3+} \rightarrow Fe(H_2O)_6^{3+} + Fe(bipy)_3^{2+} \quad (10.1)$$

$$Fe(H_2O)_6^{2+} + Ru(bipy)_3^{3+} \rightarrow Fe(H_2O)_6^{3+} + Ru(bipy)_3^{2+} \quad (10.2)$$

are small or slightly negative. This has led to the suggestion that these reactions occur by a 'non-Marcus' path, which includes at least one distinctive feature not considered in a current outer-sphere electron transfer (ET) model. We wish to point out that small or even *negative activation enthalpies* are in themselves not necessarily inconsistent with this model and indeed are *predicted* by it for the system under consideration.

In terms of the model referred above, the rate constant for an outer-sphere ET reaction is given by

$$k = pZe^{-\Delta G^*/RT} \tag{10.3}$$

where p is the probability of ET *in the activated complex* and Z, the collision frequency between two *uncharged* particles in solution, is taken to be $10^{11}\ \mathrm{M}^{-1}\ \mathrm{sec}^{-1}$. The following relations between the kinetic parameters for a cross-reaction and the *component* self-exchange reactions can also be derived from this model.

$$k_{12} = (k_{11}k_{22}K_{12}f_{12})^{1/2} \tag{10.4}$$

$$\log f_{12} = (\log K_{12})^2/4\log(k_{11}k_{22}/Z^2) \tag{10.5}$$

$$\begin{aligned}\Delta G_{12}^* &= \frac{\Delta G_{11}^*}{2} + \frac{\Delta G_{22}^*}{2} + \frac{\Delta G_{12}^0}{2} + \frac{(\Delta G_{12}^0)^2}{8(\Delta G_{11}^* + \Delta G_{22}^*)} \\ &= \frac{\Delta G_{11}^*}{2} + \frac{\Delta G_{22}^*}{2} + \frac{\Delta G_{12}^0}{2}(1+\alpha)\end{aligned} \tag{10.6}$$

$$\alpha = \frac{\Delta G_{12}^0}{4\left(\Delta G_{11}^* + \Delta G_{22}^*\right)} \tag{10.7}$$

where k_{12}, K_{12}, ΔG_{12}^*, and ΔG_{12}^0 refer to the cross-reaction and k_{11}, k_{22}, ΔG_{11}^*, and ΔG_{22}^* refer to the exchange reactions. Usually α is small; it is negative in reactions of negative ΔG_{12}^0 and becomes increasingly so with decreasing ΔG_{12}^0.

The value of ΔS_{12}^* can be obtained by differentiating Eq. (10.6) with respect to temperature $[\Delta S = -\partial(\Delta G)/\partial T]$

$$\Delta S_{12}^* = \left[\frac{\Delta S_{11}^*}{2} + \frac{\Delta S_{22}^*}{2}\right]\left(1 - 4\alpha^2\right) + \frac{\Delta H_{12}^0}{2}(1+2\alpha) \tag{10.8}$$

The value of ΔH^*_{12} can similarly be obtained from Eq. (10.6) by using a Gibbs–Helmholtz equation $\left[\Delta H = \partial\left(\frac{\Delta G}{T}\right)/\partial\left(\frac{1}{T}\right)\right]$ for ΔH^*_{12}, ΔH^*_{11}, ΔH^*_{22}, and ΔH^0_{12}

$$\Delta H^*_{12} = \left[\frac{\Delta H^*_{11}}{2} + \frac{\Delta H^*_{22}}{2}\right](1 - 4\alpha^2) + \frac{\Delta H^0_{12}}{2}(1 + 2\alpha) \quad (10.9)$$

It is readily apparent from Eq. (10.9) that ΔH^*_{12} will decrease with decreasing ΔH^0_{12} and that ΔH^*_{12} will be small or even negative if ΔH^0_{12} is sufficiently negative.

The relationship between ΔG^*, ΔH^*, and ΔS^* and the usual *experimentally derived* quantities $\Delta G^{\neq}$, $\Delta H^{\neq}$, and $\Delta S^{\neq}$ is

$$\begin{aligned} \Delta G^{\neq} &= \Delta G^* - RT\ln(hZ/kT) \\ \Delta S^{\neq} &= \Delta S^* + R\ln(hZ/kT) - \frac{1}{2}R \\ \Delta H^{\neq} &= \Delta H^* - \frac{1}{2}RT \end{aligned} \quad (10.10)$$

noting that Z is proportional to $\sqrt{Z}$.

The following are among the more important assumptions made in deriving Eqs. (10.4) to (10.6).

(i) Differences in the *stabilities* of the *precursor and successor complexes* for the various reactions (including work terms for bringing the reactants together and separating the products[1]) have been neglected. Noncancellation of these stabilities can lead to difficulties since the derived relations are really between *Franck–Condon reorganization energies* rather than between *free energies of activation*."

The formulas for k_{12} and $\log f$ are then properly modified including the stability constants of precursor and successor complexes.

(ii) "It has been assumed that none of the reactions involved in the rate comparisons are *nonadiabatic*. If allowance is made for the possibility that one or more of the reactions could be nonadiabatic," then once more the formulas for k_{12} and log f properly modified including the nonadiabatic probabilities.

The usual Marcus' formulas assume that, if adiabatic, the interaction energies are not large enough to decrease significantly the free energies of activation. M's theory is adiabatic but not too much, the splitting is small and so the activation barrier is not much lowered with respect to the crossing.

(iii) "*Nuclear tunneling* contributions have been neglected in deriving Eq. (10.4). A nuclear tunneling contribution for the usual ET reactions is discussed elsewhere in terms of *Franck–Condon vibrational overlap factors* and is presumed to be small. However, in the so-called '*inverted*' region, where the intersections of potential energy surface (PESs) of the reactants and products occurs *on the left-hand side* of the reactants curve. . ., the ET can only occur *via* a nonadiabatic mechanism. Indeed, it has been shown that in this 'inverted' region, *nuclear tunneling effects may become important.* Because of nuclear tunneling, rates will remain constant at the diffusion-controlled limit rather than decreasing again as the standard free energy change for the reaction becomes very negative. In other words, the 'inverted' behavior predicted by the simple model may not be seen because of nuclear tunneling considerations.

(iv) In deriving Eq. (10.4) it has been assumed that no *very rapid preequilibrium changes* (e.g., a spin change) take place *prior to the ET step*. If such a change does occur in one of the reactants, then the system will still be described by Eq. (10.4)" but with f now properly modified.

10.1. Application to Reactions (10.1) and (10.2)

It is shown that the negative values of $\Delta H_{12}^{\ddagger}$ are predicted by the simple model but there is a discrepancy between observed and calculated free energies of activation. The interpretation of such discrepancies in the light of the earlier (i) to (iv) possible reasons for it are discussed.

10.2. A Molecular Interpretation of the Negative Activation Enthalpies

After having demonstrated that ΔH_{12}^{*} will decrease with decreasing ΔH_{12}° and that ΔH_{12}^{*} will be small or even negative if ΔH_{12}° is sufficiently negative, the reason for the negative $\Delta H_{12}^{\ddagger}$ is explored in molecular terms. The main reason for it is that the standard entropy changes for the reaction, that is ΔS_{12}^{0}, are very negative indicating that the quantum states of the products are more widely spaced than those of the reactants.[(2)]

"The standard entropy changes for the reactions ΔS_{12}^{0} are very negative (Table I) indicating that the quantum states of the products are *more widely spaced* than those of the reactants, as discussed in footnote 29.

(29) The quantum states in Fig. 1 can be regarded as quantum states of a pair of reactants (products) plus solvent medium, each quantum state being a vibrational–rotational–translational state of

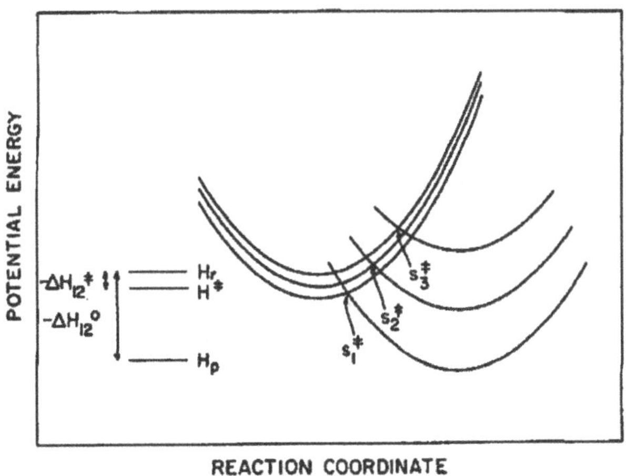

Fig. 1. Positions $s_i^{\ddagger}$ of activated complexes for a reaction of very negative ΔS_{12}^{0}. Potential energy curves for a few of the many quantum states of the system are given. For brevity, the usual adiabatic splittings at the intersections of corresponding curves for a given pair of reactants and products, to yield three adiabatic curves, have been

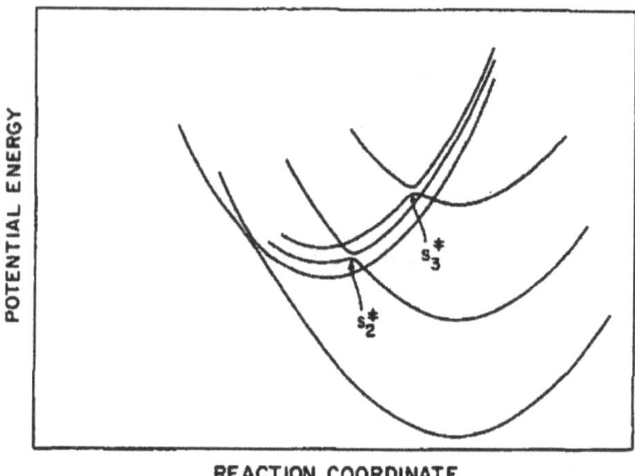

Fig. 2. This figure is the same as Fig. 1, except that two low potential energy curves "intersect" in an "inverted" configuration and the high ones intersect in a normal one." Only one example of the first and two of the second are shown. In the first case, a dynamical calculation involves corrections for nonadiabaticity (eg, as

this system. The (many-particle) translational states would be those appropriate to the given volume of the system of reacting pair plus surrounding medium. The quantum states, being so many in number are very closely spaced on the average, the spacing being much less than kT. It is possible for a reacting system to proceed to any of the quantum states of the products in Figs. 1 and 2, though with varying degrees of probability. The splitting at the intersection of the curves for reactants with those of products occurs, thereby, with all the curves but is shown in Fig. 2 only for some of the main intersections, for illustrative purposes. Since the entropy of a system can be written as $S = -k \sum_i W_i \ln W_i$, where W_i is the probability of finding the system in some quantum state i, it is clear that S^0 is given by a similar expression, using values of W_i appropriate to the standard state of the system. When, as in Figs. 1 and 2, the states are much more closely spaced for the reactants than they are for the products, in the standard thermodynamic state for each, the W_i's for the reactants are on the average much less than they are for the products (remembering that

$\sum_i W_i = 1$), and so S^0 for the reactants is much larger than that for the products. I.e., ΔS_{12}^0 is very negative for the systems under consideration. Conversely, a very negative ΔS_{12}^0 implies an average closer spacing of the states of reactants than of those of the products.

Figure 1 gives a schematic description of this behavior using the usual pair of intersecting potential energy curves (PECs) for reactants and products. Here, however, we have sketched the curves for several of the quantum states of the reactants and for several of the corresponding quantum states of the products. (The case where curves for two low-lying quantum states of the system 'intersect' in an 'inverted' configuration, while the high ones intersect in a normal one, is shown in Fig. 2.) For a few quantum states labeled 1, 2, 3, ..., the intersection of corresponding curves occurs at values of the reaction coordinate given by $s_1^\ddagger, s_2^\ddagger, s_3^\ddagger, \ldots$. The mean energy of the reactants is obtained from the energy of each quantum state of the system for a given pair of reactants, multiplied by the probability of finding the system (reactant pair plus surrounding solution) in that state and summed over all states. (The probability involves the usual Boltzmann weighting factor.) In solution, the average energy E of a specified system (reactant pair plus solution) is essentially equal to the thermodynamic heath content H of that system, since PV is negligible. This average for reactants is indicated by a horizontal line H_r in Fig. 1, while the corresponding quantity for the products is indicated by the line H_p.

The positions of the activated complexes, indicated by the $s_i^\ddagger$'s in the figures, are seen to be quantum state dependent. The internal energy of any quantum state i of the activated complex is the value of this energy at $s_i^\ddagger$. The mean energy for the activated complexes is a Boltzmann-weighted average and is indicated by the horizontal line labelled $H^\ddagger$ in Fig. 1. It is defined more precisely for quantum state dependent activated complexes in a recent article (**M105**). The method described there shows how a free energy maximum criterion can be used to define an *effective activated complex* at some mean $s_i^\ddagger$

for a reaction whose activated complexes each occur at positions which are quantum state dependent.

In Fig. 1, where ΔH_{12}^{0} is very negative, $\Delta H_{12}^{\ddagger}$ can be negative also; the principal reason, in the case of the system under consideration, is that the more energetic states of the activated complex are less reactive than the less energetic ones, a result directly due to the wider spacing of the product quantum states (compared to reactant states spacing) and to the lack of appreciable activation energy for the lowest state (for this diagram). Thus, only the lower energy states of the reactants react readily, and the mean energy of the species which do react is less, therefore, than the mean thermal energy of the reactants. By Tolman's theorem, the activation energy is therefore negative.

With the approximation (**M105**) embodied in the free energy maximum criterion, one then uses the usual ET theory (**M16**, **M30**, **M41**, **M48**, **M53**) to obtain Eqs. (10.3) to (10.10)."

NOTES

1. Q: Note reads: "If the precursor complex is very short-lived, then the rate constant should be expressed either in terms of the work terms or in terms of the stability constants of the precursor complexes per unit length of the reaction coordinate."

   M: In such a case, the work term is not an independent variable. Moreover, imagine of having a precursor complex C formed from A and B with rate constant k_1 which either decomposes back to A and B with rate constant k_2 or goes on to form the successor complex with rate constant k_3. We then have the relation $[C] = \frac{k_1}{k_2+k_3}[A][B]$ and $\frac{k_1}{k_2+k_3}$ is the stability constant of the precursor complex.
2. **M**: One should consider that the *activation barrier* is a *free energy barrier* and, of course $\Delta G^{\neq} = \Delta H^{\neq} - \mathrm{T}\Delta \mathrm{S}^{\neq}$. The barrier $\Delta G^{\neq}$ must always be positive but if $\Delta S^{\ddagger}$ is negative and big in absolute value, then it may also be possible for $\Delta H^{\neq}$ to be negative. The microscopic interpretation is the following: If $\Delta S^{\ddagger} < 0$, then the

density of states of the products is smaller than the density of states of the reagents. In terms of the PECs, there are crossings between the *R* and *P* curves largely spaced and rapidly going up in energy. If the population of these crossings is of a Boltzmann kind, then the higher crossings contribute very little to $H^{\neq}$ while there are many low-lying *R* states contributing to the average energy of the reagents. It may then happen that $H^{\neq} < H_r$.

M139 (=M146) Electron and Nuclear Tunneling in Chemical and Biological Systems

NOTE: Marcus' **NOTES** to this and the following papers appear in oral interviews reported at the end of the papers.

In this paper, the characteristics of nuclear and electron tunneling are dealt with and distinguished in relation with different possible shapes of PESs. The results are applied to photosynthetic systems. The relation of nuclear tunneling to the energy gap law in radiationless transitions is also described.

The macroscopic system considered by Marcus is made up of the two reactants plus surrounding environment which may consist of solvent molecules only or also of ions participating to the reaction as ionic atmosphere. The number of nuclear coordinates of the system is of the order of 10^{23}. M. represents the PESs with the usual R and P profiles where the abscissa represents "*some path* in the many-dimensional nuclear coordinate space leading from atomic configurations appropriate to the stability of the reactants to those appropriate to the stability of the products."

NOTE that the abscissa does not represents the *reaction coordinate* for ET.

The different types of coordinates involved in ET are described. The abscissa in Fig. 1 involves "*some concerted combinations* of all of these different coordinates." The different shapes of the profiles are described as functions of the distance between the reactants, the *R* and *P* profiles cross when the reactants are at great distance (Fig. 1)

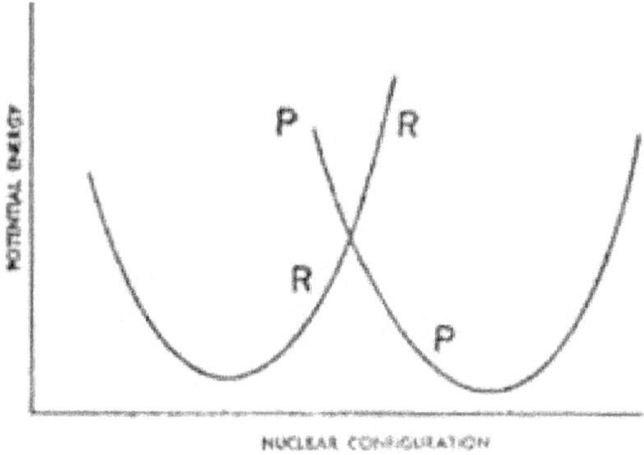

Fig. 1. Profile of a plot (see Refs. 1–5) of the potential energy of the system of reactants plus environment (R) and products plus environment (P) versus the configuration of the nuclei of the entire system. The plot is made for an approximately thermoneutral reaction, and for the case of no electronic interaction between the reactants.

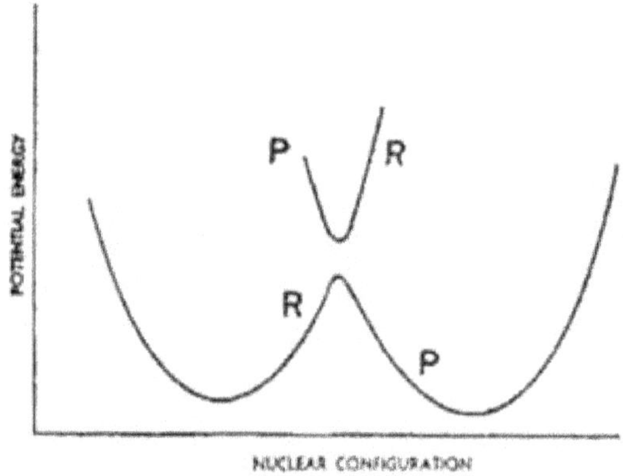

Fig. 2. Same as in Fig. 1, but for a finite electronic. coupling.

but a splitting appears at closer distances (Fig. 2), the Landau–Zener transition probability $\kappa(E)$ is briefly described.

"The *electron tunneling* rate is reflected in the extent of the splitting ΔE of the R and P surfaces in Figs. 2 and 3, being of the order of $\Delta E/h\,\text{sec}^{-1}$."

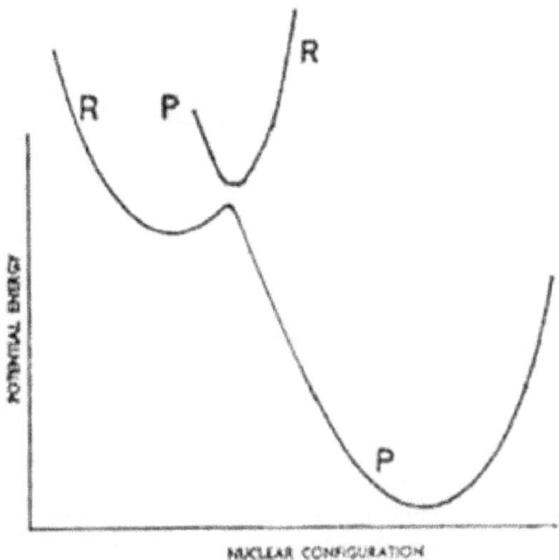

Fig. 3. Same as in Fig. 2, but for a very exothermic reaction.

NOTE that "the effective barrier to this tunneling is not depicted... nor do we need it to calculate the transition probability κ, the appropriately weighted value of $\kappa(E)$: the properties of the surfaces at the intersection region, with the aid of a quantum-mechanical expression such as that of Landau and Zener, permit the calculation of κ."

NOTE: M. never introduces a barrier such that to be found, for instance, on p. 196 of Kauzmann's Quantum Chemistry. This point has been discussed on Chapter 1.

"When the splitting is very small, κ is also small, and the reaction is said to be electronically nonadiabatic. When the splitting is large, κ is close to unity and the reaction is termed adiabatic and *the concept of electron 'tunneling' is not useful to describe* κ. A key question is whether reactions are electronically adiabatic or nonadiabatic and we shall consider a criterion for establishing this aspect of the ET.

Nuclear tunneling occurs in Figs. 2 and 3 when the energy is below the potential energy maximum of the lower surface. In this instance at low enough temperatures *the reaction ultimately becomes independent of temperature*: all systems react from their lowest vibrational state or states, and at low enough temperatures the probability of finding the system in that state is essentially temperature-independent.

We shall need in our discussion of some biological systems still another figure. When the reaction is '*downhill*,' that is, *exothermic*, the P surface in Fig. 2 is lowered, as in Fig. 3. The barrier is then less and the reaction is therefore faster... (Figs. 1–3 effectively apply to electrochemical systems also). When the reaction is highly exothermic, and *when at the same time the difference in equilibrium nuclear configurations (horizontal distance between minima) is small*, e.g., the difference in each bond length in a reactant and the same bond length in a product is small, as in Fig. 4, either the R and P surfaces may no longer cross or they will cross only at a very high energy.

A situation where nuclear tunneling can occur is illustrated in Fig. 4. In this case, *even at room temperature* such a system *reacts by*

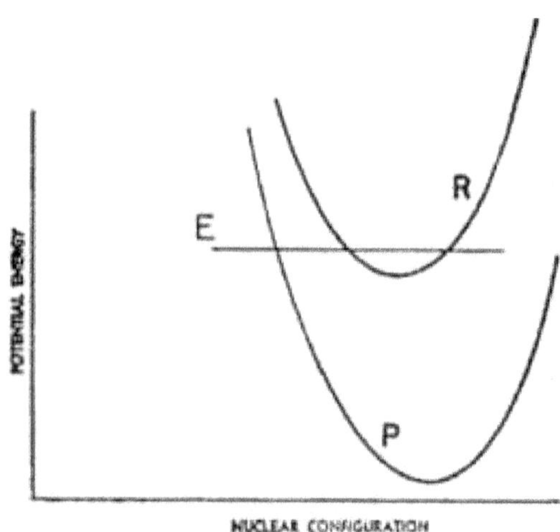

Fig. 4. Same as in Fig. 2, but for a very exothermic reaction that, at the same time, has a small difference in stable configurations of reactants and products.

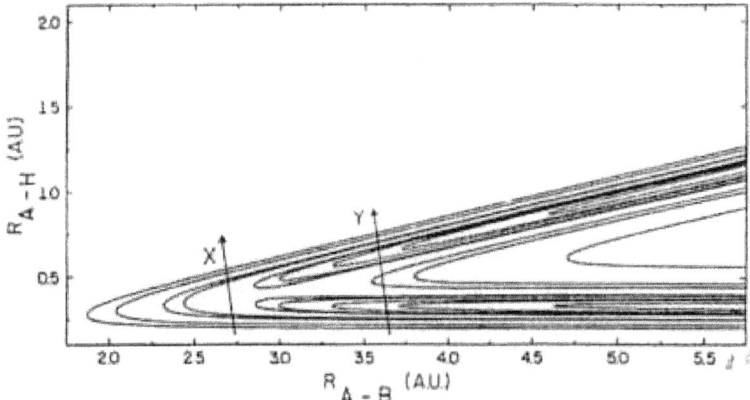

Fig. 5. Skewed-axis plot (52) for a proton transfer reaction $A - H^+ + B \rightarrow A + B - H^+$, R_{A-H} denotes a scaled $A - H^+$ distance, and R_{A-B} denotes a scaled distance between B and the center of mass of $A - H^+$. The lines give the potential energy contours, and the system moves from one valley to the other, either via the saddle-point (cf. line X) or before then (cf. line Y).

nuclear tunneling, either because there is no crossing of the R and P curves or because the 'intersection' point is energetically relatively inaccessible. In this case, the electronic coupling could still be either strong or weak, and its strength would affect the ET rate."

"From Figs. 2 and 3 one sees that initially the potential energy barrier (potential energy at the 'crossing point' minus that at the minimum of the R curve *decreases* with increasing reaction exothermicity. However, as one sees from Fig. 4, the effective barrier *increases* with increasing exothermicity (**M30, M57**), because (a) the 'intersection' occurs only at high energies, and (b) when tunneling occurs, the tunneling distance at any energy E increases as the P is lowered still further, i.e., when the exothermicity is increased."

"When tunneling occurs, i.e., when (b) occurs, the decrease of rate with increasing exothermicity at sufficiently large exothermicities is more gradual than when reaction occurs via (a), i.e., the intersection case (**M30, M57**); the rate varies exponentially with the *first* power of the exothermicity instead with the *second* power. The *first* *power behavior* is well-known in the field of radiationless transitions, where the phenomenon is known as *energy gap law.*

We shall later invoke this behavior of decreasing rate with increasing exothermicity in highly exothermic systems to explain a result in biological ETs. For the moment, however, we note from Fig. 4 that the physical origin of the behavior in (b) above is clear: the greater the vertical difference in the minima of the R and P curves, the greater will be the *nuclear tunneling distance* from the R to the P curve at any fixed energy E, and hence the smaller will be the reaction rate; similarly, in (a) *the higher will be the 'intersection point,'* if any, and so the higher will be the energy barrier.

This behavior can also be understood in terms of *Franck–Condon vibrational energy overlap integrals*: the greater this vertical distance, the greater will be the oscillations of the vibrational wave function of the P system at the energy E and the smaller will be the overlap integral with the vibrational wave function of the R system at the same total energy E. A semiclassical evaluation of this overlap integral leads to the descriptions (a) and (b) given earlier.

We turn next to *the criteria for recognizing electron and/or nuclear tunneling.*"

10.3. Electron and Nuclear Tunneling Recognition

10.3.1. *Electron Tunneling*

The rate constant k_r of an ET reaction can be written as

$$k_r = A\exp(E_a/kT) \tag{10.11}$$

where E_a is the activation energy. A is the pre-exponential factor" of the form $Z\kappa\exp\left(\frac{\Delta S^*}{k}\right)$ where Z is "of the order of 10^{13} sec^{-1} for first-order reactions, 10^{11} $M^{-1}\,sec^{-1}$ for second-order reactions, and 10^4 $cm\,sec^{-1}$ for reactions at electrodes, multiplied in each case by $\kappa\exp\left(\frac{\Delta S^*}{k}\right)$" (see Refs. **M30, M48, M53, M71**). "Here κ *is a*

Landau–Zener type transition probability for going from curve R to curve P in the 'intersection region,' suitably weighted over various translational and vibrational states. ΔS^* *is the entropy of activation due to any loss or gain in rotational–vibrational freedom of the system.* Thus,

$$A = 10^{13}\kappa \exp\left(\frac{\Delta S^*}{k}\right) \text{sec}^{-1} \quad \text{(first-order)} \tag{10.12}$$

$$A = 10^{11}\kappa \exp\left(\frac{\Delta S^*}{k}\right) \text{M}^{-1}\,\text{sec}^{-1} \quad \text{(second-order)} \tag{10.13}$$

$$A = 10^{4}\kappa \exp\left(\frac{\Delta S^*}{k}\right) \text{cm sec}^{-1} \quad \text{(electrode reaction)} \tag{10.14}$$

When the electron tunneling rate is very small and when nuclear tunneling is absent, κ is small and so when $\Delta S^* \cong 0$ we shall have $A \ll 10^{13}\,\text{sec}^{-1}$ and $A \ll 10^{11}\,\text{M}^{-1}\,\text{sec}^{-1}$ in the first- and second-order cases, respectively. *Because of these assumptions of no nuclear tunneling and* $\Delta S^* \cong 0$, some care must be exercised in concluding from a small value of A the presence of a slow electron tunneling rate. Accordingly we turn next to one of these questions, nuclear tunneling.

10.3.2. *Nuclear Tunneling*

A *typical manifestation of nuclear tunneling* is seen from *the temperature behavior of* k_r: when the reaction rate constant k_r displays an activation energy in one temperature range and tends to became temperature-independent at lower temperatures, one has evidence of nuclear tunneling through the barrier, as in Fig. 2. We have also already noted that in Fig. 4 one has nuclear tunneling even at room temperature. Unless the intersection region (if any) is reasonably accessible energetically, it is possible that the k_r for this system will be largely temperature-independent at conveniently accessible temperatures.

10.4. Examples of Electron and Nuclear Tunneling

We consider examples of the A factor in Eq. (10.15).

$$Ox_1 + Red_2 \overset{k_{12}}{\rightarrow} Red_1 + Ox_2 \tag{10.15}$$

In the $MnO_4^- - MnO_4^{2-}$ exchange reaction, when ΔS^* is calculated from a dielectric continuum expression, $A\exp(-\Delta S^*/k)$ is of the order of $10^{11}\,M^{-1}\,sec^{-1}$ give or take a factor of 10. So here the κ in Eq. (10.7) is of the order of unity or perhaps 0.1. For reactions such as $Fe^{2+} - Fe^{3+}$, *the product of the ionic charges is some three times larger, and ΔS^* is also.* The calculation of κ from $A\exp(-\Delta S^*/k)$ is therefore somewhat more uncertain, since the dielectric continuum theory is probably somewhat more uncertain for systems of large ionic charges. Nevertheless, results by Dr. Sutin in this symposium indicate a rather small κ in this way. A's for various ETs at electrodes measured *in high salt concentration where the ΔS^* due to electrostatic effects is presumably small*, were about 10^3 to $10^5\,cm\,sec^{-1}$, close to the theoretical value of $10^4\,cm\,sec^{-1}$ for such systems (see Eq. 10.8). Thus for them κ is of the order of 1, again give or take a factor of 10. The self-exchange reaction between an aromatic molecule and its anion is very rapid, close to diffusion control. *The value is reasonably well accounted for by the solvent reorganization energy, reflecting the reorientation of solvent molecules occurring along the abscissa in Fig. 2.* Thus, in this case κ is probably close to unity also. . . ." Other examples follow.

"Thus, in such cases, apart from the $Fe^{2+} - Fe^{3+}$ system, κ appears to be of the order of unity, to within a factor of 10 or so, and so these reactions would therefore be more or less adiabatic (within this factor) and *the electron tunneling concept is less useful, i.e., electron tunneling does not contribute more than this factor to the slowness of the reaction in these cases.* When the system reaches the intersection region in Fig. 2 it tends to remain on the lowest potential surface there and so react ≥ 0.1 of the time for those systems.

We consider next nuclear tunneling. Reactions are usually not studied at the extremely low temperatures that" a series of authors "used in their respective experiments (This Symposium). Their observation of *a temperature-dependent rate constant at higher temperatures and a temperature-independent one at low temperature reflects the occurrence of nuclear tunneling at the low temperatures. More typical in chemistry are the studies at not-so-low temperatures, and here a deviation from linearity of the Arrhenius plot of* $\ln k_r$ *versus 1/T is taken as evidence. More commonly, the existence of large isotope effects using H, D, and T in the case of proton transfer (PT) reactions is the usual evidence for nuclear tunneling, since otherwise it is difficult to account for the size of the large isotope effects.*

Another example of tunneling occurs in radiationless transitions, in the case of the energy gap law even at room temperatures..."

NOTE: There are then four kinds of nuclear tunneling, one at low temperatures and three at ambient temperature. In the case at ambient temperature there is one in the case of the isotopic effect and another in the case of the energy gap law. *Marcus has then discovered the possibility of nuclear tunneling in the case of the inverted free energy effect.*

10.5. Electron Transfer in Biological Systems

"The *sequence of ET reactions* which we shall consider is in the *bacterial photosynthetic system*, wherein an *electronic excitation* of a *bacteriochlorophyll dimer* $BChl_2$ is presumed to occur after *light absorption* and *excitation transfer* in the system. This excitation is followed by an ET to bacteriopheophytin BPh, and then to an *iron–ubiquinone acceptor, X*, with the indicated half-lives:

$$\mathrm{BChl}_2^* + \mathrm{BPh} \rightarrow \mathrm{BChl}_2^+ + \mathrm{BPh}^- \quad \tau \leq 10\,\mathrm{psec} \tag{10.16}$$

$$\mathrm{BPh}^- + X \rightarrow \mathrm{BPh} + X^- \quad \tau \cong 150\,\mathrm{psec} \tag{10.17}$$

$$X^- \rightarrow \text{etc. (Latin for et cetera).} \tag{10.18}$$

When the X is already reduced before the formation of $BChl_2^*$, one has instead of Eq. (10.17) some back reactions to form *a triplet and the ground state singlet* $BChl_2$

$$BPh^- + BChl_2^+ \rightarrow Bph + BChl_2^{*T} \quad \tau \cong 10\,\text{ns} \qquad (10.19)$$

$$BPh^- + BChl_2^+ \rightarrow BPh + BChl_2 \quad \tau \cong 10\,\text{ns} \qquad (10.20)$$

Following reactions Eqs. (10.17) and (10.18), the chlorophyll dimer cation, $BChl_2^+$, is neutralized by an electron from cytochrome c

$$\text{cyt } c^{\text{II}} + BChl_2^+ \rightarrow \text{cyt } c^{\text{III}} + BChl_2 \quad \tau \cong 1\,\mu\text{sec} \qquad (10.21)$$

where the value of $\tau \cong 1\,\mu$sec is cited for the reaction in *Chromatium* with a low potential cytochrome c ....

We shall consider the question of nuclear and electron tunneling and nuclear rearrangement for these reactions.

Reaction (10.16) is seen to be extremely fast. Were it a factor of $\leq$100 faster it would have a $k_r \cong 10^{13}$ sec^{-1}, the theoretical maximum. From the studies of ET rates between an aromatic molecule and its anion and between $Fe(phen)_3^{2+}$ and $Fe(phen)_3^{3+}$ referred to earlier... one would expect *very little rearrangement of bond lengths and very little barrier to reaction* in these highly aromatic systems, and thus very little activation energy. The barrier in References, such as it is, is probably due to *reorientation of the polar solvent molecules, but it is small because of the large size of the aromatic systems. In the membrane there are presumably less polar groups in the vicinity of the reactants in (10.16), and so there would be even less reorientation of polar groups than in the above aromatic systems. So, in the absence of other effects, reaction (10.16) should be very rapid.*

If the R and P curves for reaction (10.16) are, for the above reasons (smallness of nuclear rearrangement), *situated almost over each other*, one would need nuclear tunneling in order to reach curve P from curve R, giving rise to a $\kappa < 1$ in Eq. (10.12). Such a behavior would have little temperature dependence, as compared with that for the case where the R and P curves 'intersect' some distance above the minimum of the R curve.

If the reactants in Eq. (10.16) were situated quite some distance apart, the electron tunneling rate would be small, again giving rise to a $\kappa \ll 1$ and a largely temperature-independent rate. At this time the experiments haven't distinguished between these *three possibilities*: (a) a smallness in $\kappa(\leq 0.01)$ due to a slowness in nuclear tunneling, (b) a smallness in $\kappa(\leq 0.01)$ due to a slowness in electron tunneling,"

NOTE: κ is then a measure of both.

"and (c) a little rearrangement leading to a slight activation energy. *A measurement of the temperature dependence would tend to distinguish (c) from (a) and (b).*

Reaction (10.17) is about fifteen (**Typo** corrected in the original) times slower, but remarks analogous to the above apply. Reactions (10.16) and (10.20) provide an interesting comparison with each other. If *the geometry*, e.g., *separation distance and mutual orientation of the reactants*, is the same for both, then the fact that Eq. (10.20) is some 1000-fold slower than Eq. (10.16) is at first glance perhaps puzzling until one looks at *the exothermicities*. The midpoint potential of the $BChl_2^+ - BChl_2^-$ couple is about 0.45 volts.... The absorption and fluorescence spectra have maxima at 863 and 902 nm, respectively, corresponding to 1.44 and 1.38 volts. Thus, *the minima of the* $BChl_2$ *and* $BChl_2^*$ *curves are separated vertically by about 1.41 volts*, which yields a midpoint potential of the $BChl_2^+ - BChl_2^*$ couple of $-(1.41 - 0.45)$, i.e., -0.96 volts. The midpoint potential of the $BPh - BPh^-$ couple is about -0.4 volts. Accordingly, reaction (10.16) is 0.56 volts downhill. Thus, *the slowness of (10.20) relative to (10.16) may reflect the energy gap law—and the added resulting nuclear tunneling needed to go from the R to the P curve—discussed in Section II.*

Reaction (10.17) is also quite downhill, about 0.35 volts, since the midpoint potential of $X - X^-$ is about -0.05 volts. Its slowness relative to Reaction (10.16) presumably reflects other factors.

Reaction (10.19) is probably slow for reasons different from the exothermicity-induced slowness of (10.20), since (10.19) would not be very exothermic. This reaction and the influence of applied magnetic fields have been discussed by several researchers... and references... The radicals BPh^- and $BChl_2^+$ are formed in a singlet state, and the triplet state can be formed in (10.19) as a result of *hyperfine interactions with the nuclear spins.* A characteristic frequency or this interaction is of the order of the rate corresponding to 10 ns (R. Haberkorn and M. E. Michel-Beyerle, pers. comm.).

Reaction (10.21) is still slower. The *activation energy* of this reaction in the room temperature range has been measured ($E_a \sim$ 5 Kcal mole^{-1}), and its presence reflects an appreciable *reorganization* which accompanies the reaction. The nonzero activation energy also indicates that the minima of the R and P curves for this reaction are appreciably *displaced from each other along the abscissa, so that the curves can intersect even though the reaction is quite exothermic.* The slowness of Reaction (10.21) (**typo** corrected) also probably reflects *the slowness of the cytochrome c self-exchange reaction in solution...* The value of A in Eq. (10.11) for reaction (10.21) at room temperature is of the order of 5×10^9 sec^{-1}, well below the 10^{13} sec^{-1} value... Presuming nuclear tunneling at room temperature to be minor, the small A value could arise (a) because of the electron tunneling rate may be slow, particularly since there appears to be a very large separation distance between the reactants in Reaction (10.21)..., and (b) because of possible entropic effects, as indicated by the ΔS^* term. A knowledge of ΔS^0 for this reaction, if not already available would be desirable.

10.6. Proton Transfer Reactions

We conclude with some brief remarks on PT reactions, reactions which occur later in the *photosynthetic chain*."

NOTE: Marcus summarizes here the results of a preceding paper on this topic.

NOTES

1. p. 110 middle: "The abscissa represents *some path* in the many-dimensional nuclear coordinate space leading from atomic configurations appropriate to the stability of the reactants to those appropriate to the stability of the products."

   p. 111 Note: "In the symbol $\kappa(E)\ldots$ the E is really that part of E along the reaction coordinate, i.e., along the abscissa in Figures...

   Q: One may have the impression that any path represented by the abscissa is a reaction coordinate.

   M: Let me be more explicit about the reaction coordinate: in the 1960 paper and in much more detail in the 1965 paper I used N-dimensional spaces and in that N-dimensional space the reaction coordinate at a point was an energy difference between products and reactants at that point, *the vertical difference in energy of two potential energy surfaces served as the reaction coordinate.* So, for any value of that reaction coordinate there is *an* $(N-1)$*-dimensional hypersurface where all points have the same value of the reaction coordinate.* So in the path along that reaction coordinate, you go from one of those $(N-1)$-dimensional surfaces to the next one, to the next one, to the next one... On the $(N-1)$-dimensional surface you have a distribution which is a Boltzmann distribution, so there can be parts on that surface that are of lower potential energy and parts that are high, so you have a canonical distribution on that surface, so the paths are those within that canonical distribution, which means they are almost any but in Boltzmann-like distributed fashion, in a canonically distributed fashion. When you speak of the transition state (TS) theory you don't write down the actual path, you write down what you think the TS is, you talk of the reaction coordinate *in the vicinity of the TS*, and of course you can cross that TS at any place along but weighted in a Boltzmann fashion. So, all of those different places that you're crossing with different energies, different potential energies, they correspond to different paths, but you say nothing about the past history of the path, you're only talking about the crossing of the TS. The expression in the paper is loose because if you use TS theory, you're only talking about a distribution at the TS, you're not really talking about the details of the path in going from the reactants to that, so that was a loose speaking. If you have some distribution of points in the N-dimensional space, you have a distribution on that $(N - 1)$-hypersurface which is said to be the TS. The reaction coordinate is actually defined as an energy difference in that case, the main point is that you can go along many paths to reach that intersection surface, and not just one path. That is sort of loose terminology.

2. p. 110 top: "In treating ET reactions and recognizing the nuclear rearrangements which *facilitate* them"

   Q: Are "facilitate" and "induce" synonymous or is there some subtle difference between the two expressions?

   M: They are really the same. You have a fluctuation, by the fluctuation you reach the intersection region and then ET is possible.
3. p. 110 in note: "W. F. Libby made the stimulating suggestion that the Franck–Condon principle should be applicable to ET reactions. The description in terms of potential energy surfaces provides a way of visualizing this principle and its implications."

   Q: In the case of spectroscopic transitions we have the vertical transitions between potential energy surfaces of different electronic states. But here the transition is horizontal at constant energy and (as in the spectroscopic case) at fixed nuclear coordinates and between different degenerate electronic states. How does then the typical M potential energy profile describe the FC principle in the ET case? I believe in this way: the M profile is partly an R profile and partly a P profile. The R and P parts have a point in common at the TS. There the precursor and successor complexes X^* and X are represented by this common point. There, at a common nuclear configuration, we have the ET process $X^* \to X$.

M: The transition happens not only at constant energy but also at constant position and at constant momentum. The transition is energetically horizontal, that's right. You are going from one parabola to another at the intersection. Everything happens on top of the adiabatic surface.

Typo: p. 111 bottom: the $w(E)$ in note 2 is just a misprint for $\kappa(E)$.

4. p. 111 bottom: "The electron tunneling rate is *reflected* in the extent of the splitting ΔE."

Q: Why are splitting and electron tunneling probability related? In adiabatic ET everything happens on the lower surface of M's classical profile. It looks then at first sight surprising that the amount of the splitting between lower and higher profiles should be involved at all in the ET process. I believe (please correct if I am wrong) that ε_{12} is involved somehow *indirectly*, because the higher the splitting, the lower the probability of a nonadiabatic process and consequently the higher the probability of an adiabatic one because the two probabilities sum up to 1. I understand that there is a barrier for electron tunneling but why is this barrier for electron tunneling related to the splitting?

I believe it is so because, as you write on p. 184 of **M48** (the great review in Annual Reviews of Physical Chemistry) "$\varepsilon_{12}/\pi\hbar$ equals $p\nu_e$," where p is the tunneling probability and "ν_e is number of times the electron in its orbit 'strikes' the above barrier per second (ν_e is the classical frequency of electronic motion...). It is then so that the higher the splitting, the higher the probability of an adiabatic ET, that is, the higher the probability of passage of the electron through the potential energy barrier and so the *splitting* and the *width and height* of the barrier for electron tunneling are somehow related. I mean, the higher the splitting then the higher the probability of the adiabatic processes and this can be considered somehow equivalent to a lower or thinner barrier for electron tunneling probability? Is there a theoretical formal quantitative relation between the two?

M: That is like in the case of PT, the proton tunnels... In the adiabatic transitions everything happens on the lower surface, but *in the transition region between adiabatic and diabatic processes things happen on both surfaces. Both surfaces are needed in the description.* The barrier for electron tunneling is just a convenient approximation... there is a symmetric and antisymmetric combination for the electron to be on the two sides... if the splitting is very small the electron would readily tunnel, it would be both sides if the splitting went to zero... as soon as there is coupling you permit to the electron to go from one to the other... what you can express in Landau–Zener fashion.

5. p. 115 middle: When nuclear tunneling occurs... the rate varies exponentially with the first power of the exothermicity instead of with the second power."

Q: When writing of the second power dependence you refer, I believe, to the activation free energy dependence from the standard free energy of reaction in your formula:

$$\Delta G^* = w^r + \frac{\lambda}{4} + \frac{\Delta G^{\circ\prime}(R)}{2} + \frac{\Delta G^{\circ\prime}(R)^2}{4\lambda}$$

where $\Delta G^{\circ\prime}(R)$ appears at first *and second* power. Is it so? Whereas the energy gap law gives a rate constant dependence of the kind (see Turro's book) $k_{IC} \cong 10^{13}\exp(-\alpha\Delta \mathrm{E})$ (with $\Delta E = \Delta G^{\circ}$)?

M: The first power dependence is the quantum treatment, in other words... in the inverted region it goes not down parabolically, it does it linearly. The earlier formula is the classical version.

If you consider nuclear tunneling, if you do as Jortner did, after you reach the maximum then you have a linear dependence.

Errata, **Typos** p. 116 middle: "A is a pre-exponential factor of the order of 10^{13} sec^{-1} ..." I believe it should be corrected this way: "A is a pre-exponential factor where a collision number Z of the order of 10^{13} sec^{-1} appears...". Then in formulas (10.12), (10.13), and (10.14) below one should write A = instead of just A. Moreover, comparing formulas (10.12), (10.13), and (10.14) with what you write in **M118**, p. 478 bottom, in (10.14) you use centimeter but in (10.13) you tacitly use liters instead of cubic centimeter and so instead of a maximum for k_{rate} of 10^{14} cc $\cdot$ mole^{-1} $\cdot$ sec^{-1} one has 10^{11} liters $\cdot$ mole^{-1} $\cdot$ sec^{-1}, that is, 10^{11} M^{-1} $\cdot$ sec^{-1}.
One should probably signal this shift from one volume unity to another in passing from Eqs. (10.13) to (10.14).

6. p. 116 bottom: "When the electron tunneling rate is very small and when nuclear tunneling is absent, κ is small...".

Q: Which is the relation between κ, the Landau–Zener type transition probability, and electron tunneling? The LZ transition probability gives the probability of an adiabatic transition with attendant change of character of the electronic wave function, that is, of ET, when the molecules collide at velocity v, that is, it gives the probability of a *dynamical* process in which the nuclei move. When one considers instead the electron tunneling, at least in its usual elementary formulation, the molecules are *fixed* at a certain distance and one considers the electron to be transferred hitting the PE wall a number of times every second, see earlier **3**. In this case, you write that $\varepsilon_{12}/\pi\hbar = p\nu_e$. In this expression, $\varepsilon_{12}/\hbar$ appears while in the LZ formula we have $\varepsilon_{12}^2/\hbar$. May this be a clue, an indication that the LZ formula takes care of the tunneling?

M: *The Landau–Zener and the electron tunneling are equivalent kinds of description.* The electronic wave function is actually imaginary, it has an exponential decay component in the region between the two sides, you don't have to use tunneling that is just a convenient description. When one uses the semiclassical wave function one can express everything in terms of a tunneling integral, but you don't have to do that, you may use the actual wave functions. It is true that when you do tunneling calculations there is no splitting, but the treatment would be based on splitting. *LZ discusses the coupling of nuclear and electronic motion, and the probability of a transition from one curve to the other on passage through the intersection region that's LZ.* Now the tunneling calculation is one way of getting approximately the splitting that enters into the LZ formula, the tunneling is an approximate way of getting at that, so what binds the two approximations is the ε_{12}, there is a semiclassical approximation for ε_{12} that involves tunneling. In the case of the splitting between curves in the region of inverted free energy effect, if you consider the diabatic curves, the diabatic curves in the inverted region would cross and there is *a splitting* which causes them *to repel* each other, they are prevented from crossing by the diagonalization you have when you go from diabats to adiabats. *It really has a vertical splitting*, in other words if you do a straight line approximation of each curve and you intersect the two curves and you get into the off-diagonal element, you get a picture just the way it looks. Take the two adiabatic curves in the inverted region, OK, now join the diabatic curves, just do by hand, you have a point of intersection, now look at that point of intersection in relation to the lower adiabatic and upper adiabatic curves, that difference is ε_{12}. In fact if you draw the adiabatic curves in the correct way and you join them to form the diabatic curves, *then at the point of intersection go up and go down, the amount you go up to reach an adiabatic is* ε_{12} *and the amount you go down is* ε_{12} *also.* To take the two adiabatic curves and *to join with a segment parallel to the abscissa is not the good way to look at it* because one should look at the diabatic curves and at the particular point where their energy are equal, now put in the off-diagonal element, that is an equation that you can solve, the energy is H_{11}, which is identical to H_{22} at that point, + or − H_{12}. Look at the Erratum in **M57**.

7. p. 117 middle: "Unless the intersection region (if any) is reasonably accessible energetically, it is possible that the k_r for this system will be largely temperature-independent at conveniently accessible temperatures."

   Q: What do you mean by "conveniently accessible temperatures"?

   M: For instance in the case of the experiment of Miller, Calcaterra, and Closs Anyway, in that famous experiment the temperature is room temperature, that was high enough because some of the frequencies involved were surely high frequencies, so that means you could react from your zero point level or you would... just for a gain of a little tunneling, excite to higher levels...
   Errata p. 117 middle bottom: "... $A \cdot \exp(-\Delta S^*/k)$ is of the order of 'should be corrected':.... A is of the order of." Lower down: "The calculation of κ from $A \cdot \exp(-\Delta S^*/k)$" should be changed in "The calculation of κ from $A = Z\kappa \exp(-\Delta S^*/k)$."

8. p. 120, middle: "The barrier... is probably due to reorientation of the polar solvent molecules, but it is small because of the large size of the aromatic systems. In the membrane there are presumably less polar groups in the vicinity of the reactants in $BChl_2^* + BPh \rightarrow BChl_2^+ + BPh^-$, and so there would be even less reorientation of polar groups than in the above aromatic systems."

   Q: Can we say that in general the larger the size of the molecule, the lesser the intensity of the electric field experienced by the solvation molecules and so the less the reorganization energy and that the less solvated the molecule is, the lesser the reorganization energy?

   M: *The change* of the electric field decides you are comparing reactants with products.

9. p. 121 top: "At this time the experiments haven't distinguished between these three possibilities: (a) a smallness in $\kappa (\leq 0.01)$ due

to a slowness in nuclear tunneling, (b) a smallness in $\kappa(\leq 0.01)$ due to a slowness in electron tunneling, and (c) a little rearrangement leading to a slight activation energy."

M: If you have a small nuclear rearrangement you are not going to have to excite the nuclear motion much... and that would not contribute to an activation energy.

10. p. 121 middle: "The midpoint potential of the $BChl_2^+ - BChl_2^-$ couple is about 0.45 volts ....the minima of the $BChl_2$ and $BChl_2^*$ curves are separated vertically by about 1.41 volts, which yields a midpoint potential of the $BChl_2^+ - BChl_2^*$ couple of $-(1.41–0.45)$, that is, -0.96 volts.

    Typo: The minus sign on $BChl_2$ is a misprint, the right couple is just $BChl_2^+ - BChl_2$. The earlier numbers agree then with the following scheme.

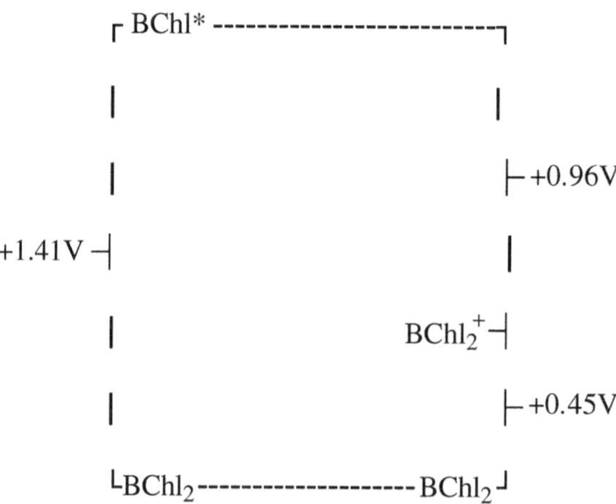

11. p. 123 bottom: "The surface in Fig. 5... has a saddle point which occurs near a symmetrical configuration... (Line X in Fig. 5 crosses the saddle point). At such a saddle point the two valleys in the surface (corresponding to *stable* $A - H^+ + B$ and $A + B - H^+$ systems, respectively) have merged. If the PT occurs near *this* saddle point, namely near line X in Fig. 5, there is no applicability of the weak-overlap (perturbation theory)

Franck–Condon concepts of ET transfer, since the two channels have *merged*."

M: I can have two rivers and they come together, where they join is a merger. Imagine you have two valleys sort to parallel until they intersect each other, they merge together and they are connected by a saddle point. The system may not want to get there, it may have not enough energy to reach that point, so it tunnels before it reaches that point. Between the two barriers if the distance is large there is a very large barrier... instead of having a saddle point you may have something like a basin.

12. p. 124 top-middle: "There will be a classically nonallowed transition if the vibrational energy of $A - H^{+}$ for some path Y *exceeds* the potential energy barrier at that Y."

    Q18: Shouldn't it be *allowed* instead of *nonallowed*?

    M: It should be less, not exceed. Either change the first part (of the sentence) or the second.

M140. Electron Transfer and Tunneling in Chemical and Biological Systems

In Fig. 1, the usual Marcus' potential energy profile is shown for "normal" ET reactions, whereas in Figs. 2 and 3 the profiles are shown for nuclear tunneling in the case of normal and "abnormal" ET reactions, that is, for reactions showing inverted free energy effect.

10.7. Nuclear Tunneling

"At low enough temperatures the thermal fluctuations in the system are too small (reflected in a small Boltzmann factor) to permit the system to go over the top of the barrier in Fig. 1. Instead the system in one of its lower vibrational states tunnels through the potential energy barrier created by the *R* and *P* surfaces, as in Fig. 3.

In Fig. 2, when the two surfaces do not intersect at a thermally readily accessible region, or at all, nuclear tunneling could occur even at room temperature.

10.8. Temperature Behavior

The temperature behavior of the rate constant k_r is interesting. In the case illustrated in Fig. 1, a plot of $\ln k_r$ versus $1/T$ would have a finite slope at the higher temperatures and, because of nuclear tunneling, essentially zero slope at very low temperatures. At high temperatures there is no nuclear tunneling in the case of Fig. 1. At low temperatures the reaction proceeds exclusively by nuclear tunneling from the lowest vibrational state (s) of the reacting pair (**typo** corrected), and so is then temperature independent. At high temperatures the rate constant is of the form

$$k_r = A \exp\left(\frac{-E_a}{kT}\right) \tag{10.22}$$

and the value of A can be of particular significance for electron tunneling...

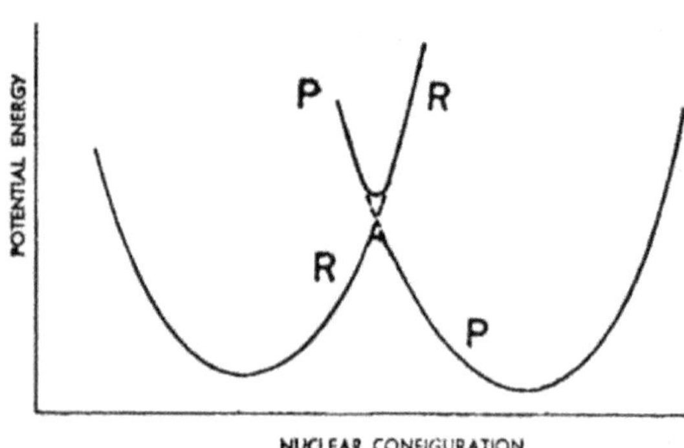

Fig. 1. "Normal" electron transfer reaction. Profile of a plot of the potential energy of the system of reactants plus environment (R) and products plus environment (P) versus the configuration of the nuclei of the entire system. The plot is made for an approximately thermoneutral reaction, and (dotted lines) is made for the case of no electronic interaction between the reactants and (solid lines at intersection) for the case of finite electronic interaction.

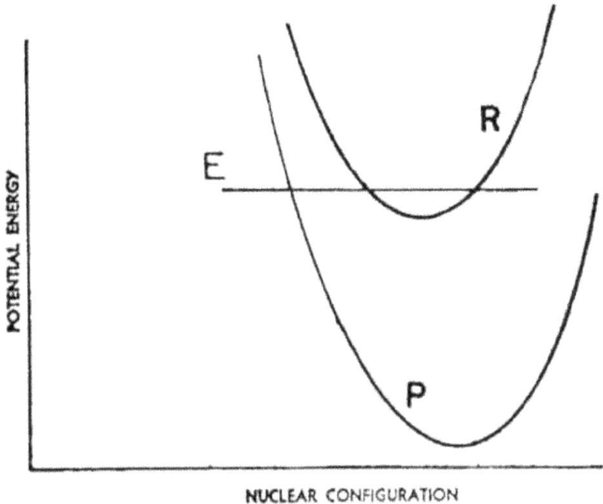

Fig. 2. "Abnormal" electron transfer reaction. Same as in Fig. 1, but for a reaction where the R and P curves do not intersect at readily accessible thermal energies, i.e., for a very exothermic reaction.

Instead of the temperature behavior in the above example, ln k_r may be independent of $1/T$, a result that could be due either to the intersection of the P curve in Fig. 1 with the R curve *near the latter's minimum*, or due instead to *nonintersection* as in Fig. 2. In the case shown in Fig. 2, an increase of the mean thermal energy decreases the *nuclear tunneling path length* between the two curves and results in a higher nuclear tunneling rate. However, such a temperature effect on nuclear tunneling is usually assumed to be small. Thus, a situation in which ln k_r is essentially *independent of temperature at all temperatures* does not in itself imply nuclear tunneling—such behavior could be the result of nuclear tunneling in the case of Fig. 2, but not in the Fig. 1 case, *where the P curve intersects the R curve at its minimum*."

Marcus relates then the possible different shapes of his potential energy profiles to the various magnitudes determining the rate constants, that is to ΔG^0, ΔS^0, κ, and λ.

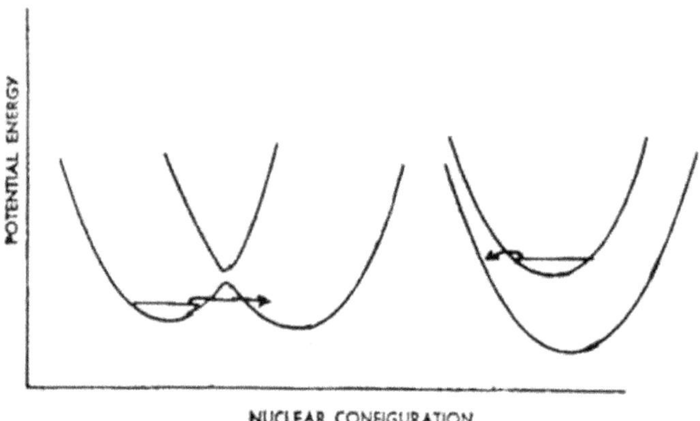

Fig. 3. Nuclear tunneling through the barrier.

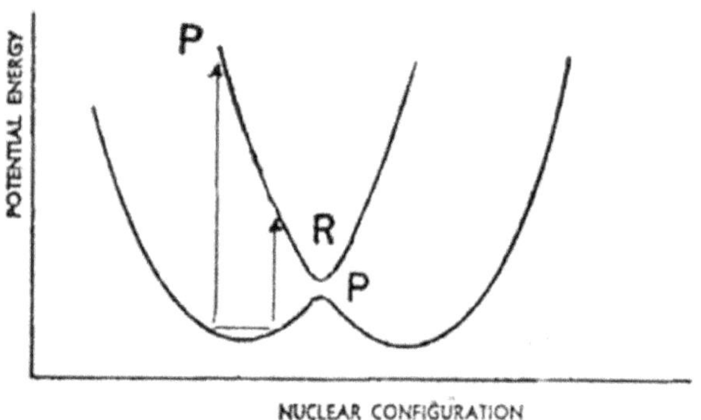

Fig. 4. An electron transfer caused by light absorption in a single step. The vertical lines give a measure of the range of light frequencies absorbed.

M. describes then his classical formula for the kinetics of ET reactions, that is

$$k_r = B \exp\left[-\frac{\lambda}{4kT}\left(1 + \frac{\Delta}{\lambda}\right)^2\right] \tag{10.23}$$

$$\text{where } B = \kappa Z \exp\left(\frac{-w^R}{kT}\right) \quad \text{(bimolecular)} \tag{10.24}$$

$$\text{and } B = \kappa \nu \exp\left(\frac{-w^R}{kT}\right) \quad \text{(unimolecular)} \tag{10.25}$$

A very good definition of w^R: "the free energy (electrostatic, nonpolar, steric, conformational) to bring the reactants to the most favorable distance r for ET."

"Δ denotes the standard free energy of reaction at the distance r for electron transfer , ΔG_r^0:

$$\Delta = \Delta G_r^0 \text{''} \tag{10.26}$$

"Both w^R and λ are 'reorganizational' terms, but one of them *precedes* the step controlled by the free energy change Δ. Only the free energy barrier created by the other (λ) can be affected by a favorable free energy change Δ."

Note that in ET reactions by the term unimolecular it is meant a reaction between two reactants bound together in the same molecule.

The Landau–Zener transition probability, when velocity weighted, yields for $\kappa\nu$ in Eq. (10.25), in the limit of low κ

$$\kappa\nu = \left(\frac{\pi}{\lambda kT}\right)^{\frac{1}{2}}\left(\frac{\epsilon^2}{\hbar}\right) \quad \text{(small } \kappa\text{; unimolecular)} \tag{10.27}$$

The earlier Marcus' equations represent the classical limit of Levich's quantum-mechanical rate expression (vide infra). In Levich's theory, "the system is treated as a collection of harmonic oscillators, with frequencies for the products and environment assumed to be the same as those of the reactants and environment . For such reactions ΔS^0 was assumed to be zero. Quantum-mechanical first-order perturbation theory was used ("Golden Rule"), and Franck–Condon overlap integrals involving the vibrational wave functions of the reactants and products were calculated. In the high temperature limit Eq. (10.23) was obtained with B given by Eq. (10.27) and Δ given by Eq. (10.28) instead of Eq. (10.26)

$$\Delta = \Delta E_r^0 \tag{10.28}$$

"With Δ given by Eq. (10.28) instead of Eq. (10.26) and Eq. (10.23) of course no longer obeys microscopic reversibility, unless $\Delta S^0 = 0$."

"The quantum-mechanical theory provides not only the 'high temperature' (classical) limiting behavior embodied in Eq. (10.23), but also the behavior at temperatures where nuclear tunneling occurs and so is of interest in describing the entire temperature range."

"Equation (10.23) is based on the systems' reaching the intersection of the R and P curves in Figs. 1 or 2 and ceases to be appropriate when nuclear tunneling becomes important (unless one incorporates into κ a nuclear tunneling factor). In the case of Fig. 2, nuclear tunneling could be very important, and in the case of Fig. 1 it could be important at low temperatures. Instead of Eq. (10.23) one can obtain (when $-\Delta \gg \lambda$) an expression whose exponent varies much more slowly with Δ. It can depend linearly instead of quadratically on Δ, yielding the so-called 'energy gap law'") for the rates of radiationless transitions."

M. describes then how to apply the earlier equations to reactants in excited states and to electrochemical reactions.

10.9. Theoretical Value of *A*

The value of the pre-exponential factor A in Eq. (10.22) is given by

$$A = \kappa Z e^{\frac{\Delta S^*}{kT}} \quad \text{(bimolecular)} \tag{10.29}$$

$$A = \kappa \nu e^{\frac{\Delta S^*}{kT}} \quad \text{(unimolecular)} \tag{10.30}$$

Where ΔS^* is a temperature derivative

$$\Delta S^* = -\left(\frac{\partial}{\partial T}\right)\left[w^r + \left(\frac{\lambda}{4}\right)\left(1 + \frac{\Delta}{\lambda}\right)^2\right] \tag{10.31}$$

Thus, when ΔS^* is either negligible or can be calculated, the measurement of A provides information on κ in the bimolecular and in the unimolecular case.

10.10. Effect of Electric Fields and of Standard Free Energy of Reaction on Reaction Rate

"We consider here the effects of varying the standard free energy and of varying an applied electric field on the reaction rate. The field can arise from other charges near the reactants (e.g., in or near the membrane when the reactants are in a membrane) as well as any externally applied field..."

"By adding different substituents to one of the reactants the standard free energy term ΔG_r^0 can be varied when the substituent affects the free energy of the oxidized form and the reduced form differently... Under certain conditions the λ term is unaffected, namely, *when the substituent does not affect the difference of equilibrium bond lengths of the two forms or the proximity of polar groups near the ionic charge.* Under these conditions, introduction of a substituent *merely raises or lowers the P curve in Fig. 1 relative to the R curve vertically without displacing it laterally or changing the curvature of either curve (surface). When the variation in a substituent also leads to a change in λ, either or both of these additional changes in the curves in Fig. 1 would also occur.*"

NOTE: $= \lambda_i + \lambda_0, \lambda_i = \frac{1}{2}\sum_j k_j(Q_j^r - Q_j^p)^2$

and $\lambda_o = (\Delta e)^2 \left(\frac{1}{2a_1} + \frac{1}{2a_2} - \frac{1}{R}\right)\left(\frac{1}{D_{op}} - \frac{1}{D_s}\right)$.

A substituent can then affect λ_i if it affects the difference in equilibrium bond lengths of R and P or if affects the distance R between the reactants.

"For the case of an applied electric field for a reaction between oxidized A and reduced B at a separation distance r

$$A_{\text{red}} + B_{\text{ox}} = A_{\text{ox}} + B_{\text{red}} \quad (\text{distance } r) \tag{10.32}$$

We denote the charges on the reactants by e_A^R and e_B^R and those of the products by e_A^P and e_B^P. The applied field produces a *potential* ψ_A and ψ_B at the centers of charge of A and of B. The $\Delta G_r^0(\psi)$ equals the value at $\psi = 0$, $\Delta G_r^0(\psi = 0)$, plus an added contribution

$$\Delta G_r^0(\psi) = \Delta G_r^0(\psi = 0) + \Delta e_A(\psi_A - \psi_B) \qquad (10.33)$$

where Δe_A denotes $e_A^P - e_A^R$ and hence denotes $-(e_B^P - e_B^R)$."

"The region of Δ which defines the regime appropriate to Fig. 1, is $-\lambda < \Delta < \lambda$."

NOTE: From Marcus' Nobel lecture: $\Delta G^* = \frac{\lambda}{4}\left(1 + \frac{\Delta G^0}{\lambda}\right)^2$. If $\Delta = -\lambda$, $\Delta G^* = 0$, that is there is no activation energy and in Marcus' profile the P curve cuts the R curve at its minimum. If $\Delta = \lambda$, $\Delta G^* = \lambda$ and the R curve cuts the P curve at its minimum.

"In this regime the rate constant varies with $\Delta = \Delta G_0^R$ according to Eq. (10.23) as

$$kT \ln k_r = \text{constant} + \frac{\lambda}{4} + \frac{\Delta}{2} + \frac{\Delta^2}{4\lambda}. \qquad (10.34)$$

The slope of a plot of $kT \ln k_r$ versus Δ is

$$\frac{dkT \ln k_r}{d\Delta} = \frac{1}{2}\left(1 + \frac{\Delta}{\lambda}\right). \qquad (10.35)$$

Thus, this slope varies smoothly from a value of 0 at $\Delta = -\lambda$, through a value of 1/2 at $\Delta = 0$ and finally to a value of 1 at $\Delta = \lambda$, in the domain $-\lambda < \Delta < \lambda$ appropriate to Fig. 1. Thus the applied field will have negligible effects on k_r in the vicinity of $\Delta = -\lambda$, an effect about equal to $1/2\ \Delta e_A(\psi_A - \psi_B)$ in the vicinity of $\Delta = 0$ and an effect of about $\Delta e_A(\psi_A - \psi_B)$ in the case of a reaction that is very 'uphill' ($\Delta \cong \lambda$)."

NOTE: The curve to which Marcus refers can be seen as Fig. 6 in **M270** and as Fig. 7 in **M278** (Nobel Lecture).

"Outside the domain ($-\Delta < \lambda < \Delta$) appropriate to Fig. 1, namely when $\lambda < |\Delta|$, caution is needed in applying Eq. (10.23). Calculations by the following Eq. (10.37) show that at high vibration frequencies $kT \ln k_r$ is much less sensitive to changes in Δ than implied by Eq. (10.34) when the latter is applied in the range $\lambda < |\Delta|$. Thus, electric fields would then have relatively small effects in this region, much as they do in the case of Fig. 1 when ($\Delta \cong -\lambda$).

The effect of electric fields on rates have been extensively studied in the case of electrode reactions at metal electrodes... For example, A in Eq. (10.33) is the reactant and B is the metal electrode. In that case $\psi_B - \psi_A$ is the potential difference between the electrode and the solution at the distance where A is undergoing ET. The arguments have been extended to semiconductor electrodes.

10.11. Charge Transfer Spectra

ET can be assisted by light not only by producing electronically excited species, which then are reactive, as in Eq. (10.36)

$$A^*_{\text{red}} + B_{\text{ox}} \rightarrow A_{\text{ox}} + B_{\text{red}} \tag{10.36}$$

but also more directly: Absorption of light in the presence of sufficient electronic coupling between the reactants yields the vertical transition depicted in Fig. 4. This is a *charge transfer transition.* These charge transfer spectra, or *intervalence spectra*, as they are sometimes called, have been extensively studied... The *intensity* of the absorption band depends on factors such as the *extent of electronic coupling* of the electronic orbitals of the two reactants, and the intensity of the spectra has been used to estimate the splitting 2ϵ in Fig. 1. If the splitting ϵ is too small, the intensity of the band can, of course, be too small to allow an adequate signal-to-noise ratio. Under favorable circumstances it would be possible to learn also *about* λ from charge transfer spectra in several ways: when the optical absorption of a metal ion reactant yields a charge transfer *between the metal and the ligands* in that reactant, the oxidation state of the metal is thereby changed. A comparison of the absorption and fluorescence spectra could then provide information on λ. If the fluorescence is from the electronic state formed in absorption, a displacement (Stokes shift) between the long wave absorption maximum and short wave fluorescence maximum can be expressed (for small ε) in terms of the $0 \rightarrow 0$ energy and the contribution of that reactant to λ. At least it can be so expressed in the absence of complications or when those complications are corrected for. Several cases involving tris-bipyridyl complexes of metal ions have been discussed recently and compared with λ contribution to the ET rates.

Again, if at low temperatures the vibrational structure is resolved, information about λ can be further obtained. For example, assuming the *dipole matrix element* to be independent of the vibrational coordinates *over the range of coordinates involved in the absorption* (Condon approximation), then when the even transitions ($0 \rightarrow 0, 0 \rightarrow 2, 0 \rightarrow 4 \ldots$) of a symmetric vibration occurs, but not the odd ones ($0 \rightarrow 1, 0 \rightarrow 3, \ldots$) in the optical change of electronic state, the surfaces, plotted versus that vibrational coordinate, are not horizontally displaced from each other. (Any antisymmetrical mode q has its minimum at $q = 0$ and so when plotted versus q the surfaces are not horizontally displaced.) The occurrence of ($0 \rightarrow 0, 0 \rightarrow 2, 0 \rightarrow 4 \ldots$) transitions of a symmetric mode under such conditions arises from difference of vibrational frequencies in the initial and final states, and so the intensities can provide a measure of those differences. Other spectral results for various aromatic molecules describe the *displacement of the potential energy minima* for the vibrational modes... (this example itself is not a charge transfer one, of course).

Large Stokes shifts occur when charge transfer occurs in a *polar solvent*... because of interactions with the polar environment, and so they should not be used when only the vibrational contribution is needed for application to a system in a nonpolar environment.

10.12. Equations. Inclusion of Nuclear Tunneling

Equation (10.23) does not include quantum effects associated with nuclear tunneling. While such neglect appears to be appropriate for most ET reactions, Eq. (10.23) clearly does not apply when tunneling becomes important.

A quantum expression for the rate constant of ETs was given by Levich and Dogonadze many years ago for the case of (a) very small electronic interaction (nonadiabatic ET), (b) treatment of the polar solvent medium via a Fourier series as a collection of harmonic oscillators, all of frequency, and (c) frequencies unchanged by the reaction. Although condition (b) limits its application to biological systems, exactly the same equations arise, regardless of whether the oscillators are oscillators of the solvent or of the reactants.

For an exothermic ($\Delta < 0$) first-order reaction one has

$$k_r = \kappa' \nu I_p \left(\frac{a}{\sinh \gamma}\right) \exp\left(-\frac{\Delta}{2kT} - a \coth \gamma\right). \quad \text{(Levich)} \tag{10.37}$$

(The rate for the reverse endothermic reaction is Eq. (10.37) multiplied by $\exp(\frac{\Delta}{kT})$.) In Eq. (10.37) I_p is the modified Bessel function and

$$p = -\frac{\Delta}{h\nu}, \quad a = \frac{\lambda}{h\nu}, \quad \gamma = \frac{h\nu}{2kT}, \quad \kappa' = (\varepsilon/\hbar\nu)^2. \tag{10.38}$$

The factor in Eq. (10.37) multiplying $\kappa'\nu$ arises from a sum of Franck–Condon integrals for overlap of the reactants' with the products' vibrational wave functions, a sum over all final and all (Boltzmann-weighted) initial states. Jortner has also obtained Eq. (10.37)... κ' defined by Eq. (10.38) can be termed the 'nonadiabaticity factor,' prompted by the result in the following Eq. (10.41), at 0 K (**typo** corrected).

Equation (10.37) has been extended to systems with more than one frequency and/or to oscillators which are both displaced and distorted (i.e., frequency differences between reactants and products)...."

Compare **M140** for References on the above and for relevant Franck–Condon factors.

"Instead of Eq. (10.37a) semiclassical equation has been used by Hopfield, written for brevity in the form of a single frequency

$$k_r = B' \exp\left[\frac{-(\Delta+\lambda)^2}{2\lambda h\nu}\coth\left(\frac{h\nu}{2kT}\right)\right] \quad \text{(Hopfield)} \qquad (10.39)$$

where

$$B' = \nu\left(\frac{\varepsilon}{\hbar\nu}\right)^2\left[\frac{1}{2\pi a}\coth\left(\frac{h\nu}{2kT}\right)\right]^{\frac{1}{2}}. \qquad (10.40)$$

In the classical limit ($\hbar \longrightarrow 0$), Eqs. (10.37) and (10.39) both yield Eq. (10.23), with B equal to the $\kappa\nu$ in Eq. (10.27) and Δ being given by Eq. (10.28).

Jortner has shown that Eqs. (10.37) and (10.39) can differ numerically by several orders of magnitude at low temperatures, for the case $p = 0$. The limiting behavior at low temperatures is therefore of particular interest. At low temperatures one has

$$k_r = \nu\left(\frac{\varepsilon}{\hbar\nu}\right)^2 e^{-a}a^p/p! \quad \text{(Levich, low } T) \qquad (10.41)$$

The coefficient of $\nu\left(\frac{\varepsilon}{\hbar\nu}\right)^2$ in Eq. (10.41) is the well-known Franck–Condon factor for the overlap of the lowest vibrational wave functions of one electronic state with the p'th vibrational wave function of another. By comparison Eq. (10.39) yields

$$k_r = \nu\left(\frac{\varepsilon}{\hbar\nu}\right)^2\left(\frac{1}{2\pi a}\right)^{\frac{1}{2}} e^{-\frac{(p-a)^2}{2a}}. \quad \text{(Hopfield low } T) \qquad (10.42)$$

The coefficients of $\nu\left(\frac{\varepsilon}{\hbar\nu}\right)^2$ in Eqs. (10.41) and (10.42) differ by being normalized Poisson and Gaussian functions, respectively, in a (hypothetical) distribution of p (the distribution is hypothetical since there is only one p, namely, $-\Delta/h\nu$.). The two will be approximately equal numerically, therefore, when both p, (i.e., $-\Delta/h\nu$.) and the 'mean' value of p in the 'distribution,' namely, $\underline{a}$ (i.e., $\lambda/h\nu$) are both large. *Large $\lambda/h\nu$ is the so-called "'strong coupling'" regime.*

Equations (10.37) and (10.38) can differ greatly when ν is large: A plot of $\ln k_r$ versus Δ gives a parabolic curve in the case of Eqs. (10.23) (Marcus) and (10.39) (Hopfield). For large ν the curve is strongly asynunetric in the case of Eq. (10.37) (Levich), being ultimately only linearly dependent on Δ when $-\Delta \gg \lambda$ (energy gap law).

In the case of very exothermic reactions, care is needed in using all of these equations, particularly in the case of Fig. 2. Significant errors due to anharmonic effects can arise when $|\Delta \gg \lambda|$ and can considerably influence the nuclear tunneling, or in Fig. 1 when the contribution of one reactant to λ is negligible relative to that of the other (preventing compensation effects). In radiationless transitions, for example, anharmonic effects can lead to errors of orders of magnitude....

Finally, in cases where ε is large enough for the perturbation theory to break down, Eqs. (10.37) and (10.39) need modification of course.

10.13. Biological Systems

The sequence of reactions which we shall briefly consider in the case of bacterial photosynthesis are the following (**M139**), beginning with the reaction of the electronically excited bacteriochlorophyll dimer $BChl_2^*$

$$BChl_2^* + BPh \rightarrow BChl_2^+ + PBh^- \tag{10.43}$$

$$BChl_2^+ + BPh^- \rightarrow BChl_2^{*T} + BPh \tag{10.44}$$

$$\text{and/or} \rightarrow \text{BChl}_2 + \text{BPh} \tag{10.45}$$

$$\text{BPh}^- + \text{QFe} \rightarrow \text{BPh} + \text{Q}^-\text{Fe} \tag{10.46}$$

$$\text{cyt } c^{\text{II}} + \text{BChl}_2^+ \rightarrow \text{cyt } c^{\text{III}} + \text{BChl}_2 \tag{10.47}$$

where BPh and Q denote a bacteriopheophytin and a quinone molecule, and T denotes a triplet state.

In the case of electron transfer between aromatic species... as well as between iron phenanthroline species..., the fast rates of homogeneous electron transfers show that the λ is about 0.5 eV and about 0.4 eV" in the two cases. "These studies, made in polar solvents, also suggest that the major contribution to λ is actually from the reorientation of polar solvent molecules, with negligible contribution from the vibration of the reactants.

Correspondingly, one would expect the vibrational λ contribution from BPh to be very small in the membrane, BPh being also an aromatic (porphyrin) system. The $BChl_2$ may have a somewhat *larger vibrational contribution to* λ *due to some looseness in the binding* of one BChl to the Mg^{+2} of the other, and the possibilities thereby of *some geometrical reorganization on loss of an electron.* Interestingly enough, the Stokes shifts for $BChl_2$ is appreciable: the absorption and fluorescence spectra have maxima at 863 and 902 nm, respectively..., whereas for the monomer BChl... (and hence presumably for BPh) (**typo** corrected) the Stokes shift is only a few nm. This appreciable difference presumably reflects the *greater sensitivity of the dimer* to some geometrical change upon change of electronic configuration (here, excitation).

In the case of quinones, electron transfer rates in a polar solvent again reveal a relatively small λ, about 0.7 volts. It again appears to be due largely to reorganization of the polar solvent, with presumably relatively little vibrational contribution.... The presence in reaction (10.46) of a Fe quinone linkage may contribute to the latter's vibrational λ, because of the possibilities it offers for geometrical changes (Fe – O displacements) on charge transfer.

If the dielectric constant of the membrane is about 5 one expects the environmental contribution to be roughly one-half that in highly polar solvents...; m there equals −0.5 for a self-exchange reaction). On this basis one expects the environmental contribution to λ to be about 0.2 to 0.3 volts in the membrane. The fast hopping mechanism for conduction in aromatic crystals is an indication of a relatively small vibrational λ for aromatic systems.

The λ for the cytochrome c self-exchange is somewhat uncertain. It is not yet definitely known whether the activation energy in the self-exchange... is due to a λ for this system or to a *conformational* w^R. The looseness of the binding in the fifth and sixth positions of the heme could contribute to λ...

Thermodynamic data, such as the Δ for individual steps in reactions Eqs. (10.43) to (10.47), are becoming available. The *midpoint potential* of the $BChl_2^+ - BChl_2$ couple for Rps. spheroids is about 0.45 volts.... From the average of the absorption and fluorescence maxima for $BChl_2 \rightarrow BChl_2^*$ (1.44 and 1.38 volts with an average of 1.41 volts), the midpoint potential for $BChl_2^+ - BChl_2^*$ is about −(1.41 − 0.45), i.e., −0.96 volts. The midpoint potential of the BPh–BPh^- couple is about −0.4 volts (in vivo value for Rps. Viridis..., a contested value however. It may be about −0.6 volts....) With −0.4 volts, reaction (10.43) would be downhill by about 0.56 volts, and reaction (10.45) would be 0.45 − (−0.4) or 0. 84 volts downhill. The midpoint potential of the QFe-Q^-Fe couple is about −0.18 volts, so that reaction (10.46) is (0.4–0.18) or 0.22 volts downhill. The midpoint of the low potential cyt c in reaction (10.47) is about 0 volts, e.g., ..., so that reaction (10.45) is about 0.45 volts downhill.

Reaction (10.43) has a half-time of ∼10 psec... and perhaps even as small as 3 psec; reaction (10.46) has a half-time of the order of 150 psec, and both reactions are temperature independent. The half-time for reactions (10.44) and (10.45) is of the order of 10 ns, the former being in the *hyperfine coupling mechanism range*.... Reaction

(10.47) is the slowest of all, about 1 μsec half-time in Chromatium for reaction with a low potential cytochrome....

The speed of reactions (10.43) and (10.46) is due to their small λ. If we assume an environmental contribution to λ of about 0.2 to 0.3 volt, for the reason discussed earlier, and a small effective vibrational contribution of say 0.2 volts for reaction (10.46) , the net λ is 0.5 volts. The exponential factor in Eq. (10.23) then yields results 'consistent' with the rates of reactions (10.43), (10.45), and (10.46) (κ' not yet known). If one assumes an environmental ν of the order of $300\,\text{cm}^{-1}$, $h\nu/2kT$ at room temperature is 0.725, $(h\nu/2kT)\coth(h\nu/kT)$ is 1.17, and so Eq. (10.39) gives essentially the same results as Eq. (10.23) for the environmental contribution to the barrier for these reactions. Since $\lambda/h\nu$ is large (about 13) and $-\Delta/h\nu$ for these reactions is large, Eq. (10.37) also gives essentially the same result as Eqs. (10.23) and (10.39). Only when a large ν is used does Eq. (10.23) deviate from Eqs. (10.37) or (10.39) at room temperature; when $\lambda/h\nu$ is no longer large Eq. (10.39) deviates from Eq. (10.37). (Indeed many theoretical calculations are concerned precisely with the domain where $\lambda/h\nu$ is no longer large.... as revealed in their nonparabolic dependence of $\ln k_r$ on Δ in the exothermic region $-\Delta > \lambda$.) A more accurate discussion would include the appropriate ν's for *both the environmental and the vibrational* λ's, and such calculations can readily be made, extending Eqs. (10.37) and (10.39). Reaction (10.47), the slowest of all, is the best candidate for a small κ, because of a large separation distance.... The latest experimental A in Eq. (10.22) for that reaction... is about $5 \times 10^9\,\text{sec}^{-1}$. Thus, unless there is a ΔS^*, κ is of the order of 10^{-3}, an appreciable nonadiabaticity.

10.14. Appendix A: Footnotes

1. In the case of Fig. 2, if the R and P surfaces intersect, one can show that when the energy in this plotted coordinate exceeds the potential energy of the intersection, the probability of transition

from the R to the P surface, per double passage past that point, passes through a maximum as a function of ε, *being zero for small and large* ε. In the case of Fig. 1, the probability of transition at the intersection increases with increasing ε.

2. One uses Eq. (10.44) of **M48** (Nonadiabatic: $\kappa \cong 2\pi\varepsilon_{12}^2/\hbar\bar{v}|s_1 - s_2|$), evaluates the slopes of the parabolas at the intersection and notes that there are two passages of the intersection point when κ is small, per 'excursion' in the vicinity of that point, so that Eq. (10.44) is to be multiplied by 2."

10.15. Appendix B: Encounter Complexes

"If encounter complexes are formed, the overall rate constant k_r is given by a somewhat more complicated expression than Eq. (10.23). The ET reaction of A_{red} with B_{ox}, written now in several steps, is

$$A_{red} + B_{ox} \underset{k_{-1}}{\overset{k_1}{\rightleftharpoons}} (A_{red}, B_{ox}) \underset{k_{-2}}{\overset{k_2}{\rightleftharpoons}} (A_{ox}, B_{red}) \overset{k_3}{\rightarrow} A_{ox} + B_{red}. \quad (B.1)$$

The two encounter complexes are those in parentheses, with an electron transfer step with rate constants k_2 and k_{-2}. Writing expressions for $(d/dt)(A_{red}, B_{ox})$ and $(d/dt)(A_{ox}, B_{red})$, setting them equal to zero (steady state approximation), and noting that the overall reaction rate is $k_3(A_{ox}, B_{red})$, one obtains Eq. (B.2) for the overall rate constant k_r instead of Eq. (10.23).

$$\frac{1}{k_r} = \frac{1}{k_{enc}} + \frac{1}{k_{act}\gamma}, \quad (B.2)$$

where k_{enc} is k_1, the rate constant for the formation of the encounter complex (A_{red}, B_{ox}), γ is the probability that the fate of (A_{ox}, B_{red}) is to undergo ET, $[\gamma = k_3/(k_{-2} + k_3)]$, and k_{act} is the activation-controlled rate constant (k_{act} equals k_2 times an equilibrium constant k_1/k_{-1} for the first step in Eq. (B.1)). k_{act} is given by Eq. (10.23) of the text or, in the case of nuclear tunneling, by the later modifications in the text. Eq. (B.2) reduces to $k_r = k_{enc}$ or $k_r = k_{act}\gamma$, according as $k_{enc} \ll k_{act}\gamma$ or $k_{enc} \gg k_{act}\gamma$, respectively.

When the rate of formation of $(A_{\mathrm{red}}, A_{\mathrm{ox}})$ is diffusion-controlled, k_{enc} equals the well-known rate constant for a diffusion-controlled reaction (p. 129 of **M30**)

$$k_{\mathrm{enc}} = k_{\mathrm{diff}}. \tag{B.3}$$

When the rate of formation of the encounter complex $(A_{\mathrm{red}}, B_{\mathrm{ox}})$ is activation-controlled instead of diffusion-controlled, due, for example, to a substantial reorganization of the solvent structure to form an encounter complex, k_{enc} can be

$$k_{\mathrm{enc}} = Z \exp(-w_1^R/kT) \tag{B.4}$$

Instead of Eq. (B.3); w_1^R is the free energy barrier for forming this encounter complex from the reactants and equals w^R when the complex has no stability.

The particular case that $\gamma = 1$ and $k_{\mathrm{enc}} = k_{\mathrm{diff}}$ was discussed earlier (Eq. 6.1 of **M118**).

NOTES

1. p. 20 top: "$\Delta S°$ describes the shape of energy surface P of the products in Figs. 1 and 2 relative to that of the reactants' R. For example, when the shape of the surface of the products near its minimum (minima) has a relatively small curvature... while that of the reactants has a high curvature... the products (and environment) have much more freedom of movement than do reactants (and environment), and $\Delta S°$ is quite positive."

   M: When the entropy is concerned, one should really be thinking of the whole potential energy surface, imagine to go from a 10^{23} dimensional surface to another one then there will be a large entropy change if the topography, the topology, of the potential energy surfaces changes. Thinking about entropies I would forget about the parabolas and their curvatures, I would look at the many-dimensional aspect. So, the earlier statement is not fundamental.

2. p. 21 top-middle: "The value of B in $k_r = B\exp\left[-\frac{\lambda}{4kT}\left(1+\frac{\Delta}{\lambda}\right)^2\right]$ is given by $B = \kappa Z\exp(-w^R/kT)\ldots$ where κ is an electron tunneling factor." On p. 16 bottom you define κ as "a velocity-weighted Landau–Zener transition probability."

   Q: Are the two definitions equivalent?

   M: It is more complicated than that, but roughly that is the case. If you have a nonadiabatic reaction then the rate is proportional to the Landau–Zener factor, which is also proportional to the velocity of the particles and is also proportional to the velocity distribution. So, that velocity that always comes in when you consider collisions... So that's what I meant by velocity weighting. If you have a Landau–Zener factor, you have to average over velocities. The main point is that they are both transition probabilities, if you are going to use Landau–Zener then you have the velocity weighted, in some way or another. In the tunneling expression when the electron jumps there are subtleties in there, there is the nuclear motion and I did something about that in my first paper on ET in 1956 because we can't just say that the electron is tunneling, but we have to do something about the nuclear motion. In the paper of the other R. Marcus (R. J. Marcus...) and Eyring, they made a serious mistake. You have to specify which is the frequency of the electron going back and forth, before it tunnels it comes in. The Landau–Zener transition probability is really a tunneling factor, but it depends upon velocity, it is dynamic, so it got the electron tunneling in there but in a subtle way. See the book of Walter Kauzmann Quantum Chemistry, I refer to it in some articles because there you find a sort of physical interpretation of the Landau–Zener expression, a nice physical interpretation, that would be a good book to look at.
3. p. 22 top: "w^R and w^P denote prior conformational terms in the unimolecular case."

   Q: w^R in the bimolecular reactions is the work to bring the system from infinity to R. Here the equivalent of R is, I believe, the right conformation prior to ET. But what about the equivalent of infinity? Is it maybe the average conformation in solution?

M: Suppose you have two reactants, they are sitting in one place, the reaction is unimolecular, and maybe they change their conformation or they reorient, you know, that is like a conformational change, and then in order to begin to react there is sort of a barrier to reaction, that's a part of w^R, this part is independent of $\Delta G°$, a sort of free energy of reorganization that comes in, it is the free energy associated to the selection from the average conformation of the one conformation that allows the two reactants to react. So, the average conformation in solution is the equivalent of the infinity in the bimolecular reactions.

4. p. 22 bottom: "With Δ given by $\Delta = \Delta E_r^0$ instead of $\Delta = \Delta G_r^0$, Equation $k_r = B\exp\left[-\frac{\lambda}{4kT}\left(1+\frac{\Delta}{\lambda}\right)^2\right]$ no longer obeys microscopic reversibility, unless $\Delta S° = 0$.

   M: Somebody used ΔE instead of ΔG but that is wrong. In order to see it, look at the expression for k. Now use the same expression for k for the back reaction, changing $+\Delta$ to $-\Delta$, now take the ratio of the two, which would be the equilibrium constant and an equilibrium constant is related to a free energy. If Δ is ΔE, and that is what Levich and Dogonadze originally did, they had it wrong. There are some reactions with a huge entropy of reaction.
5. p. 23 top-middle:" Equation $k_r = B\exp\left[-\frac{\lambda}{4kT}\left(1+\frac{\Delta}{\lambda}\right)^2\right]$ is based on the system reaching the intersection of the R and P curves in Figs. 1 or 2 and ceases to be appropriate when nuclear tunneling becomes important (*unless one incorporates into* κ *a nuclear tunneling factor*)."

   Q: Can we say at this point that κ is the correction to the ordinary collision formula taking care in general of both the electronic and nuclear tunneling?

M: If you want to do a reaction that has some nuclear tunneling in it, then for electron transfer you use equations different from the earlier equation, you use some modified equations like the Jortner's equation. κ usually refers to electron tunneling. It depends on how you treat it, for example, when people treat proton transfer with coupled electron transfer in proteins or molecules then, depending on what the situation is, there are very limiting situations where the protonic wave functions overlap and also the electronic wave functions have to overlap. So, in that case you have some sort of a κ that is the product of these two overlaps, they are related to the squares of those overlaps.

6. p. 32 middle. "The presence in reaction $BPH^- + QFe \rightarrow BPh + Q^-Fe$ of a Fe quinone linkage may contribute to the latter's vibrational λ, because of the possibilities it offers for geometrical changes (Fe-O displacements) on charge transfer."

Q: Can you please define when to use charge transfer and when electron transfer? I believe that charge transfer is more general than electron transfer, in the case of proton transfer for instance one has a concomitant charge transfer. Still more in general one could speak of charge density change or transfer or of change in charge distribution... I believe one should speak of electron transfer when an electron hops from a localized state in the reactants to a localized place in the products. Please comment on that...

M: Well, in this case, in that particular case, one could have used electron transfer as well. Now, charge transfer is a more general term, if you transfer a proton, or if you are involved in semiconductors and you transfer a hole you have a charge transfer, charge transfer is more general term than electron transfer. With charge *density* transfer it starts getting into complications, because if it is less than a full charge transfer then it is just a little bit of transfer. In this particular case in this paper, it would have been better to use electron transfer.

M141 Mechanisms of Charge Separation and Subsequent Processes Group Report

10.16. Theoretical Considerations

"Basically, three theories approach the general problem of rate reactions and are potentially applicable, among other processes, to electron, proton, or H-atom transfer, as long as these are 'one step' processes. The best known and the most widely referred to in the literature (because it is best understood by the experimentalist) is the one elaborated by Marcus. This theory is considered classical because the nuclear motions are treated on a classical basis, quantum mechanics being applied to the electron motion.

A quantum theory of charge separation was developed by Levich and Dogonadze and extended later by researchers to treat more than one vibrational frequency. Recently this problem was approached on essentially the same basis by Jortner. A related route, referred to below as semiclassical, was described by Hopfield, who in effect evaluated the Franck–Condon vibrational factors in an approximate way...."

"The theoretical predictions yield the relations schematically represented in Figs. 1 and 2."

In Fig. 1, a typical variation of the rate constant, $\ln k$, as a function of the temperature, $1/T$, is represented. It is the Arrhenius plot. The vertical dashed line separates the high temperature limit in which the rate is classically controlled (A) from the low temperature limit (B) where the rate is quantum controlled.

In Fig. 2 $\ln k$ is plotted against $-\Delta G$ and the three behaviors, classical, semiclassical, and quantum are shown. The classical plot (C) is a parabola, as is the semiclassical (SC) with a somewhat smaller curvature, while the quantum curve is asymmetric.

"Fluorescence quenching experiments have been interpreted by Rehm and Weller... as ET's between aromatic compounds in which one compound acts as donor, the other as acceptor. The results are

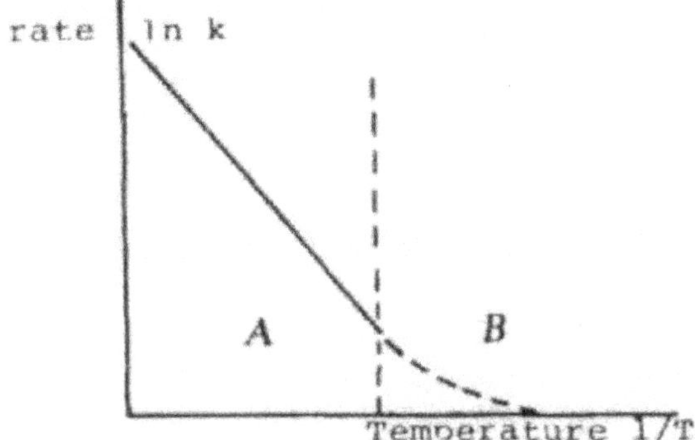

Fig. 1. Typical variation of the rate constant k of a reaction as a function of temperature. The vertical dashed line separates the high temperature limit in which the rate is classically controlled (A) from the low temperature limit (B) where the rate is quantum controlled.

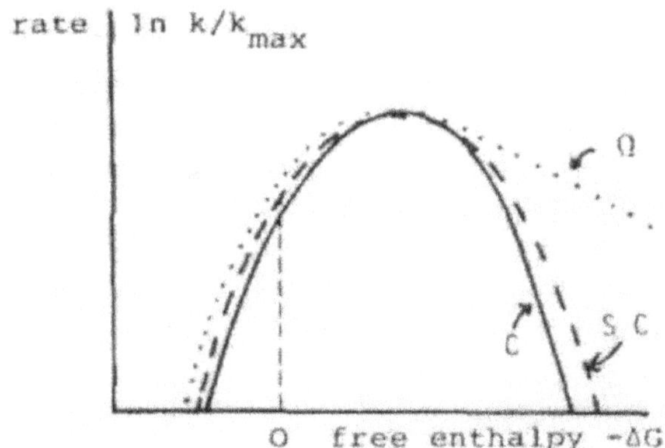

Fig. 2. Calculated reduced rate $k/k_{\max}$ as a function of the one step reaction. The ratio $k/k_{\max}$ is used to normalize the curves for the three treatments: C, classical; SC, semiclassical; and Q, quanturm.

plotted in Fig. 3. The data fit the parabolic form of the equation if ΔG is not too negative, but obviously fail at very negative values of ΔG where a diffusion-controlled limiting rate appears. The expected falloff in rate is not observed. A possible explanation is that the electron transfer that quenches the fluorescent state does not occur in a

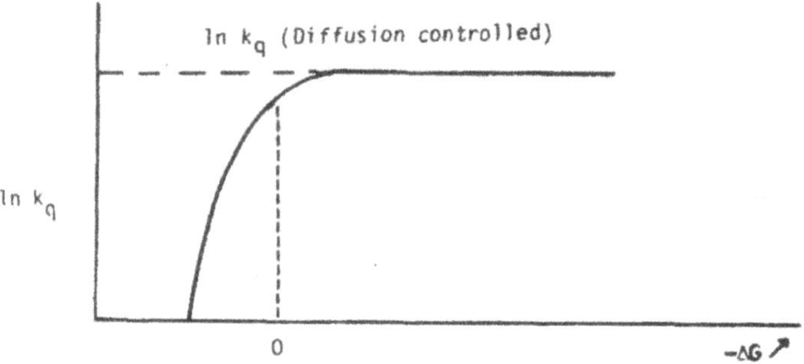

Fig. 3. Plot of $\ln k_q$ as a function of increasing negative value of ΔG.

single step. It is known that exciplex formation may mediate the ET and Weller suggested that in processes where ΔG is large exciplex formation may provide an intermediate step that has to be taken into account in formulating the mechanism.

New systems should be looked for in which exciplex formation is inhibited, as *in molecules bearing the electron acceptor at one end and the donor at the other end of a rigid structure*, such as those studied by Davidson et al.... or where exciplex formation is unlikely, as in some coordination compounds."

NOTE: This is the first time in which this hypothesis appears in Marcus' papers. Later there will be the famous experiment of Miller, Calcaterra, and Closs.

Slow or fast electron tunneling rates can be inferred from experimental data. "Two sources of information are the Arrhenius pre-exponential factor (when there are no other contributions to the usual $10^{13}\,\mathrm{s}^{-1}$ or $10^{11}\,\mathrm{l\,mole}^{-1}\,\mathrm{s}^{-1}$ factors for unimolecular or bimolecular reactions) and charge transfer spectra. An example is the reaction

$$\mathrm{Cyt}\ c^{\mathrm{II}} + \mathrm{BChl}_2^{+} \rightarrow \mathrm{Cyt}\ c^{\mathrm{III}} + \mathrm{BChl}_2$$

for which the pre-exponential factor is of the order of 10^9 to $10^{10}\,\mathrm{s}^{-1}$, indicating slow electron tunneling unless there are other contributing factors."

10.17. Efficiency of Light-Induced Separation

For the content of this section we refer the reader to Marcus' paper.

10.18. Definition and Determination of Redox Potentials of Excited Molecules

The definition and determination of redox potentials of excited molecules is discussed.

"Knowledge of redox potential of molecules in an excited state is obviously of crucial importance in light-induced processes... The argument will be illustrated for the oxidation potential of an excited donor molecule D, which captures an electron according to the reaction

$$D^+ + e^- \rightleftharpoons D \tag{10.48}$$

"The ground state oxidation potential for this reaction is given by: constant $-E_D^{\text{ox}}$, where the constant refers to the reference electrode. The excitation of the donor involves the energy change:

$$D \overset{h\nu}{\rightleftharpoons} D^* \quad \Delta E_{0,0}(D) \tag{10.49}$$

which corresponds to the zero–zero transition that can be obtained from spectroscopic data (for instance, from the frequency of the crossing point of the absorption and fluorescence spectra (see Fig. 4) Addition of Eqs. (10.48) and (10.49) yields the desired half reaction

$$D^+ + e^- \rightleftharpoons D^*$$

for which the oxidation potential is: constant $-$ $E_{D^*}^{\text{ox}}$ (10.50)

with:

$$\text{const} - E_{D^*}^{\text{ox}} = \text{const} - E_D^{\text{ox}} + \Delta E_{0,0} \tag{10.51}$$

so that one can write:

$$E_{D^*}^{\text{ox}} = E_D^{\text{ox}} - \Delta E_{0,0}(D) \tag{10.52}$$

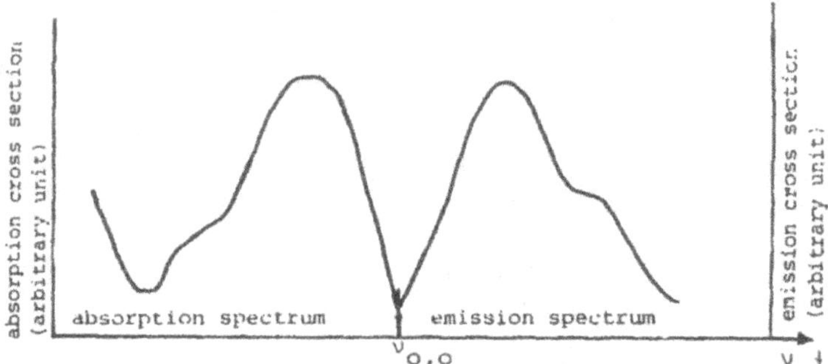

Fig. 4. Typical absorption and emission spectra of an organic compound allowing the determination of the 0, 0 band and thus $\Delta E_{0,0}$.

Analogous considerations for the reduction potential, E_A^{red}, of acceptor molecule A lead to

$$E_{A*}^{\text{red}} = E_A^{\text{red}} + \Delta E_{0,0}(A) \tag{10.53}$$

These procedures imply that only negligible entropy changes are involved in electronic excitation, so that the spectroscopically determined values of $\Delta E_{0,0}$ can be considered as free enthalpy changes, ΔG. It has been found that this requirement is fulfilled with ± 0.03 eV for most compounds studied....

The free enthalpy change ΔG_{et}^* connected with an excited state ET reaction can now be easily obtained from Eqs. (10.52) or (10.53)

$$\Delta G_{\text{et}}^* = E_D^{\text{ox}} - E_A^{\text{red}} - \Delta E_{0,0} \tag{10.54}$$

The question of whether ET reactions are likely to be accompanied (or followed) by proton transfer can be handled in a manner quite similar to that used above for ET alone. The reactions (and their free enthalpy changes) which have to be included are the proton dissociation reaction of $DH\cdot^+$, the radical cation of the electron donating molecule, and of $AH\cdot$, the protonated radical of the electron acceptor

$$DH\cdot^+ \rightleftharpoons D\cdot + H^+ \ (2.303 \text{ RT } pK_{DH\cdot}^+) \tag{10.55}$$

$$AH\cdot \rightleftharpoons A\cdot^- + H^+ \ (2.303 \text{ RT } pK_{AH\cdot}) \tag{10.56}$$

The use of appropriate thermodynamic cycles leading to the overall reactions (in the excited state):

$$DH^* + A \rightleftharpoons D\cdot + HA\cdot \tag{10.57}$$

or

$$DH + A^* \rightleftharpoons D\cdot + HA\cdot \tag{10.58}$$

gives for the free enthalpy changes $\Delta G^{*\prime\prime}_{\text{ht}}$ (ht = hydrogen transfer) 'involved in excited state H-atom transfer reactions (10.57) and (10.58):

$$\Delta G^*_{\text{ht}} = E^{\text{ox}}_{DH} - E^{\text{red}}_A - \Delta E_{0,0} + 2.303\ \text{RT}\ (pK_{DH\cdot^+} - pK_{AH\cdot}), \tag{10.59}$$

where $\Delta E_{0,0}$ is the zero–zero transition free energy required for the excitation of DH or A, respectively. Typical values for aromatic phenols (DH = ArOH) and aromatic nitrogen heterocyclic compounds (A = Ar(N)) are $pK_{DH\cdot^+} \approx 0$ and $pK_{AH\cdot} \approx 15$, respectively so that

$$2.303\ \text{RT}\ (pK_{DH\cdot^+} - pK_{AH\cdot}) \approx 0.9\,\text{eV},$$

which means that additional proton transfer can considerably facilitate an ET reaction between suitable electron (and proton) donor molecules and electron (and proton) acceptor molecules. As reactions (10.57) and (10.58) do not involve a net charge separation, their free enthalpy changes are virtually independent of the polarity of the medium. This makes H-atom transfer a likely redox mechanism in biological systems with low polarity environments (lipids, interior of proteins).'

'The well-known relations between the oxidation potential, E^{ox}_D, and the (adiabatic) ionization potential, IP_D

$$E^{\text{ox}}_D = \text{IP}_D - \Delta G^{\text{solv}}_{D+} - \text{constant}, \tag{10.60}$$

and between the reduction potential, E^{red}_A, and the (adiabatic) electron affinity, EA_A:

$$E^{\text{red}}_A = \text{EA}_A + \Delta G^{\text{solv}}_{A^-} - \text{constant}, \tag{10.61}$$

in which $\Delta G^{\text{solv}}_{D^+}$ and $\Delta G^{\text{solv}}_{A^-}$ are the solvation free enthalpies connected with the reactions

$$D_g + D_s^+ \rightleftharpoons D_s + D_g^+ \tag{10.62}$$

$$A_g + A_s^- \rightleftharpoons A_s + A_g^- \tag{10.63}$$

(where the subscripts g and s refer to the gas phase and the solution), can be combined to yield the free enthalpy change involved in excited state electron transfer reactions according to Eq. (10.54)

$$\Delta G^*_{\text{et}} = E_D^{\text{ox}} - E_A^{\text{red}} - \Delta E_{0,0} = \text{IP}_D - EA_A - \Delta E_{0,0} - (\Delta G^{\text{solv}}_{D^+} + \Delta G^{\text{solv}}_{A^-}) \tag{10.64}$$

Equation (10.64) shows that the difference $E^{\text{ox}} - E^{\text{red}}$ of the same molecule, rather than being closely connected (as is often said) only with the longest wavelength zero–zero transition, strongly depends on solvent polarity through the solvation free enthalpy term $(\Delta G^{\text{solv}}_{D^+} + \Delta G^{\text{solv}}_{A^-})$. For a medium-sized molecule $A = D$ in highly polar solvents the magnitude of this term is of the order of 4 eV and decreases to about 2 eV as the dielectric constant approaches 2. Thus Eq. (10.64) also indicates that ET reactions between A^* and D to yield ionic products A^- and D^+ are expected to be strongly solvent dependent and to decrease in rate with decreasing solvent polarity....

In the case of molecules adsorbed on a solid-state surface... the free enthalpy change connected with the excited state ET can also be discussed on the basis of Eq. (10.64). It should, however, now be modified so as to contain only the solvation free enthalpy connected with the adsorbed molecule, since the solvation of the counter charge residing in the solid-state material does not depend on the surrounding medium and can be neglected. The solvent effect expected for such heterogeneous systems should, therefore, be smaller than in homogeneous systems.'"

Among the other topics discussed in the paper there are

(i) The effect of electric fields on charge separation, the interplay of chemical and electrochemical potentials, of adsorption and of diffusion;
(ii) The using of magnetic fields in understanding problems related to charge separation;
(iii) The formation of a charge carrier pair from a triplet pair TT $\rightarrow$ $+-$ and the modulation of the charge yield by a magnetic field;
(iv) The significance of the triplet-doublet (TD) process in charge separation;
(v) The study of spin precession in a cation (or hole)—anion (or electron) pair and of how the reaction rates depend on the spin state of the complex in which spin motion is occurring;
(vi) The magnetic field effect on the rate for irreversible separation of the electron–hole pair that yields an estimate of the ratio of the rate constants for recombination in the singlet and triplet channels;
(vii) The magnetic field effect observed in bacterial photosynthesis.

NOTES

1. p. 130 bottom: 'In Fig. 2 the classical plot (C) is a parabola, as in the semiclassical (SC) with a somewhat smaller curvature, while the quantum curve (Q) is asymmetric. Depending on the reduced values of the various frequencies of motion, $h\nu_i/kT$, the curves for the three treatments are similar to or different from each other.'

   M: Those are comparisons of quantum and classical curves. In the classical case you have a nice inverted parabola, that's classical. In the quantum case you get tunneling. On the *inverted side* when you look at the barrier by drawing the curves, you see it is *a very narrow barrier*, very thin, whereas it is very thick on *the other side*. So that means you're going to get more tunneling at very high negative ΔG^0's compared with the other side of the parabola, so then you get that asymmetry due to that quantum-mechanical tunneling and when ν is high, you would get that asymmetry. But when ν is low it is the same as classical. I've defined reduced frequency of motion the frequency divided by kT, and so when ν is small then the quantum curve becomes identical with the classical.

M145. On spin exchange and electron transfer rates in bacterial photosynthesis.

10.19. Introduction

'A mechanism currently assumed in bacterial photosynthesis... involves ET between an electronically excited bacteriochlorophyll special pair $(\mathrm{Bchl})_2^*$ and a bacteriopheophytin (Bph), followed by ET to an iron–ubiquinone complex FeQ,

$$(\mathrm{Bchl})_2^* + \mathrm{Bph} \rightarrow (\mathrm{Bchl})_2^+ + \mathrm{Bph}^- \tag{10.65}$$

$$\mathrm{Bph}^- + \mathrm{FeQ} \rightarrow \mathrm{Bph} + \mathrm{FeQ}^- \tag{10.66}$$

and by subsequent steps. Studies of magnetic field effects on the back reactions (10.67) and (10.68)

$$(\mathrm{Bchl})_2^+ + \mathrm{Bph}^- \rightarrow (\mathrm{Bchl})_2^{*T} + \mathrm{Bph} \tag{10.67}$$

$$(\mathrm{Bchl})_2^+ + \mathrm{Bph}^- \rightarrow (\mathrm{Bchl})_2 + \mathrm{Bph} \tag{10.68}$$

in which T denotes a triplet state and $(\mathrm{Bchl})_2$ is the ground state singlet, have led to an estimate of the spin exchange integral for the radical pair in (10.65) and, in turn, to the electron transfer integral for (10.65). A discrepancy then arises in the calculated compared to the experimental ET rate.'

'The spin exchange integral J is the splitting of the triplet and singlet states of the free radical pair $(\mathrm{Bchl})_2^+\mathrm{Bph}^-$ and is estimated from magnetic field effects to be, at most, 10^{-3} cm.' There are several mechanisms that contribute to J .... Probably the dominant one is due to a *virtual transfer*[(1)] of the electron from Bph^- to the empty excited orbital of $(Bchl)_2^+$, where it undergoes a strong intramolecular exchange interaction with the hole"[(2)]

On the basis of this value for J the rate constant k for ET reaction between two fixed sites calculated using the Fermi Golden Rule or other models such as those of Levich, Jortner, or Hopfield, is less than or of the order of 10^{10} sec^{-1}, as compared with the experimental value equal to or exceeding 2×10^{11} sec^{-1}.

"One possible explanation of the discrepancy between these calculated and observed values of the k of Eq. (10.65) is that the electron in Bph^- or the hole in $(Bchl)_2^+$ resides on several sites, such that there is one pair of adjacent sites that has a large exchange integral for the Bph^-–$(Bchl)_2^+$ interaction and more remote sites that have a very small exchange integral for Bph^-–$(Bchl)_2^+$."

The two states explanation, of the "close state" and of the "distant state," is shown to be able to yield a maximum rate constant of Eq. (10.65) of 10^{13} sec^{-1} instead of 10^{10} sec^{-1}.

NOTES

1. Q: What is a *virtual* electron transfer? How is it defined?

   M: There was not still the structure, so we operated in the dark and we guessed there was an intermediate, it turned out not to be correct. The virtual ET, that's the superexchange, but I have to think about it. There is a kind of transition that comes in there, it is called virtual because it is not an actual transition, just like in superexchange intermediate state the electron really doesn't reside there but the wave function (*of the electron*) appears.
2. Q: Is the electron–hole exchange interaction the same as the electron–electron exchange interaction except that the wave function of the second electron is substituted by a hole wave function?

   M: Yes, that's right.
3. p. 4185 2nd column top-middle: "The Fermi Golden Rule expression, applied to the rate constant k for ET reactions between two

fixed sites, is

$$k = \frac{2\pi\varepsilon^2}{\hbar} \sum_{i,f} |\langle\psi_f/\psi_i\rangle|^2 p_i \delta(E_f - E_i) \qquad (10.69a)$$

...Eq. (10.69a) assumes a continuous distribution of final vibrational states f; p_i is the probability of finding the reactants in an initial vibrational state i.

Equation (10.69a) can also be written

$$k = \frac{2\pi\varepsilon^2}{\hbar} \frac{1}{\delta E} \sum_{i,j} |\psi_f|\psi_i\rangle|^2 p_i \qquad (10.69b)$$

In which the sum over f is over all states f whose energy E_f lies in the interval $(E_f, E_f + \delta E)$."

Q: **Typo** I believe a bra symbol is missing in Eq. (10.69b). Can you please explain the difference between the two formulas, in particular in relation to the δ functions in the two cases?

M: It certainly is missing. In the limit in which δE goes to zero, then I guess it would be the same. δE has the dimension of $1/E$. Maybe one way of seeing it is this: this delta function is a very sharply peaked thing, so if it doesn't have a continuous distribution, you get zero, typically... so it really applies to the case where you have a continuous distribution of X. Now, if you have a continuous distribution and you have a certain density of states, one operation that sometimes is done is replacing f by a typical density of states, of final states, you see that quite often. So, you get rid of the sum over f, it is now ρ_f. Now, if you look at the other expression, you have the sum over the final states divided by the energy range you are considering, that's really a ρ_f. The number of states there in the region δE is a density of states. You are summing over final states of a range δE, so this is just a way of rewriting Eq. (10.69a) and of approximating a delta function effectively by a double step function, which goes up, stays constant for a while, then comes down. So, essentially it is an approximation to a delta function. It is a double step function. As you know you can approximate a delta function in many ways, in many limiting ways.

4. p. 4186 2nd column top-middle: "In the former case the mechanism of Eq. (10.65)

$$(\mathrm{Bchl})_2^* + \mathrm{Bph} \rightarrow (\mathrm{Bchl})_2^+ + \mathrm{Bph}^- \tag{10.65}$$

now is

$$(\mathrm{Bchl})_2^* + \mathrm{Bph} \rightarrow [(\mathrm{Bchl})_2^+ + \mathrm{Bph}^-]_\mathrm{C} \tag{10.70}$$

$$[(\mathrm{Bchl})_2^+ + \mathrm{Bph}^-]_\mathrm{C} \rightleftharpoons [(\mathrm{Bchl})_2^+ + \mathrm{Bph}^-]_D \tag{10.71}$$

M: In the chemical equilibrium in the last line the systems are in locked positions.

5. p. 4187 2nd column top: "... a spin hopping (rate constant k) between two sites differing in energy by Δ. In the limit of a slow hopping rate ($k \ll \Delta$) between the two sites..."

Q: $k \ll \Delta$ means that the physical dimension of k is energy. Is it possible?

M: They don't have the same units. You can only write that if they have the same units. If you have a certain splitting and associate with it a rate constant, then express that splitting in terms of a rate constant. It is not possible for a rate constant to have the dimensions of an energy.

M154. Similarities and Differences between Electron and Proton Transfers at Electrodes and in Solution. Theory of Hydrogen Evolution Reaction

Abstract

"Depending on the initial energies, a PT may proceed either via a saddle point in a PES or by crossing from the reactants' to the products' valley before the saddle point is reached. In the second path, the analogy to weak-overlap ETs is pointed out. The present study is intended to unify previously divergent viewpoints, by showing how they are *special cases of a more general picture.* Expressions are obtained from the reaction rate in terms of the properties of the PES and of other properties of the system, using a hydrogen evolution reaction as an example.

10.20. Introduction

In a PT, for example, in the hydrogen evolution (10.72) reaction at an electrode M,

$$H_2O^+ + M(e) \rightarrow H_2O + H\text{-}M \tag{10.72}$$

some dynamical effects result from the lightness of the H particle and two questions arise

(i) Can a conventional TS theory be used to calculate the reaction rate, with a TS near some saddle point region of the PES?
(ii) Does the PT occur so quickly that, as in weak-overlap ET, a 'nonequilibrium' solvent dielectric polarization arises?"

Marcus presents here a unified treatment of previous ones, a treatment already outlined in **M134**.

10.21. Potential Energy Surface and Reaction Paths

"We first consider the PES for reaction (10.72) as a function of *two* of the coordinates, the H_2O-H and the H-M distances, using the usual mass-weighted skewed axes coordinates (Fig. 1). There are also the bending motions of the O – H – M, the stretching motion of the other O – H stretches, and the coordinates of the solvent environment. The actual potential energy is a function of all of these coordinates Path α in Fig. 1 is a path through the saddle point and path β is a path at any fixed O-M distance.

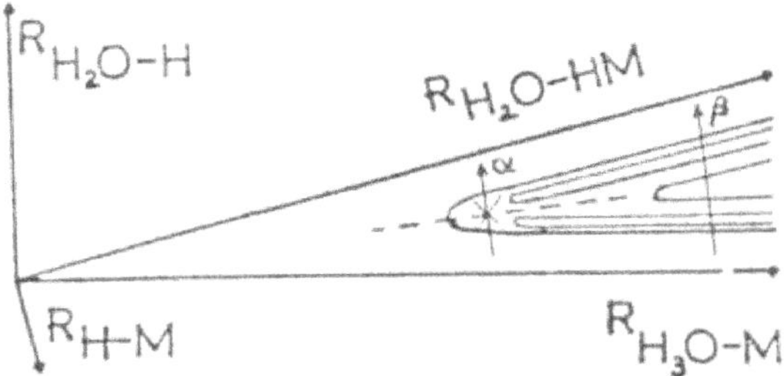

Fig. 1. Potential energy contour plot (schematic) for the three center reaction H_2O–H–M at a fixed value of the other coordinates and at a given metal-solution potential difference, for the case where the reaction is almost thermoneutral (symmetric). The rotated axes are scaled H_2O–HM and H–M distances (between O and the center of mass of HM and between H and M).[9] The configurations along the dashed line form the conventional transition state, and X denotes the saddle point on the potential energy surface. Reaction paths α and β are described in the text.

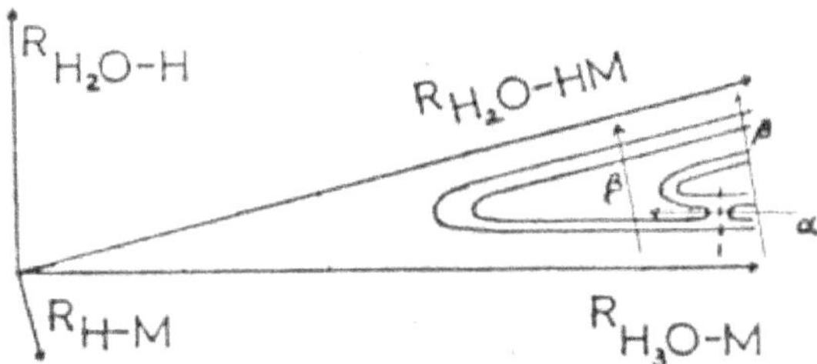

Fig. 2. Legend as in Fig. 1, but the reaction is now highly exothermic. The dashed line denotes the conventional transition state, passing through the saddle point of the potential energy surface.

In the highly exothermic case, caused by a very favorable overpotential, a schematic diagram of the surface can resemble that in Fig. 2. Paths α and β are again drawn.

One task will be to decide whether the system ever reaches the saddle point region as in path α in Figs. 1 or 2, or whether, either because the O-H vibrational energy is sufficiently high or because tunneling of the H from the O to the M may be large, the system moves from the reactants' valley to the products' valley *before* the saddle point region is reached, as in path β in Fig. 1 and as in the β-path at a large O-M distance in Fig. 2."

Paths α and β can be *competitive*, conceivably one (α) prevailing in very exothermic or very endothermic conditions and the other (β) prevailing under more thermoneutral conditions. A prescription for estimating the relative importance of the two paths is described in the present paper. Path β is assumed in Ref. (6)" (by Dogonadze, Kuznetsov, and Levich). "Path α is used in the usual TS theory."

10.22. Energetics for Paths α and β

In order to illustrate the different reaction paths Marcus uses a PS which treats changes in the O-H-M potential energy due to changes in O-H and H-M distances by a bond energy-bond order (BEBO) method, and which treats the remaining motions largely in terms of their dielectric polarization behavior. In an approximation to BEBO, compare **M69**, the BEBO surface was described by

$$V_e = n_2 \Delta V^0 + \Delta V'[n_1 \ln n_1 + n_2 \ln n_2]/\ln 2 \qquad (10.73)$$

"where n_1 and n_2 are the bond orders of the H_2O-M and H-M bonds, respectively. At any point along the minimum potential energy path it is assumed that the sum $n_1 + n_2$ remains constant, namely unity." In Fig. 3 any pair of points P and P' lying on the minimum potential energy path, in the reactants' and products' valleys respectively, have been joined by a β path.

"The difference of bond orders of the H-M bond at any pair of points P and P' in Fig. 3 will be denoted by Δn,

$$\Delta n = n_2(P') - n_2(P). \qquad (10.73a)$$

Along each PP' path the potential energy in Figs. 1–3 is given by Eq. (10.73) (with $n_1 + n_2 = 1$ at the points P and P') and, along PP', by Eq. (10.73) using bond length-bond order relations.

The interactions of the solvent with the H_2O-H^+-M subsystem along the various paths is also to be included and, for brevity

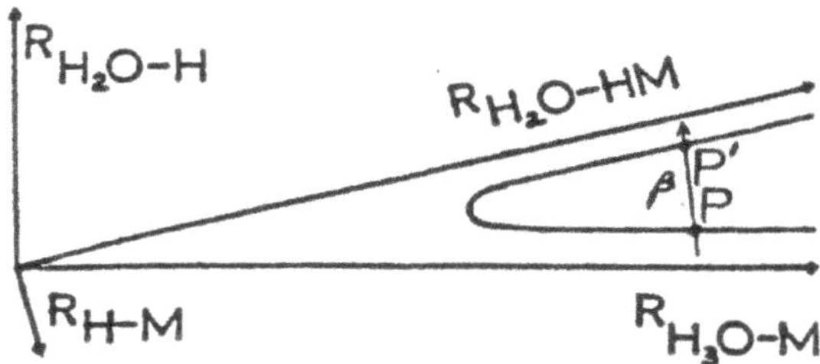

Fig. 3. Legend as in Figs. 1 and 2 . The curve denotes the minimum potential energy path, which proceeds along the reactants' valley, over a saddle point, and along the products' valley. The saddle point, may be situated as in Fig. 1 or as in Fig. 2 or, in the case of a very endothermic reaction, in the products' valley. Points P and P' are corresponding pairs points which lie at the intersection of a β-path with the minimum potential energy path.

and simplicity, we shall do so classically (various high-frequency modes, treated structurally, could also be included and treated quantum mechanically)." Marcus considers then a particular value of the orientation-vibration polarization of the remaining coordinates $\underline{P}_u(\underline{r})$ at each point $\underline{r}$ of the solvent medium.

"An expression for the *free energy of solvation* of the reactants G^r_{sol} for any given $\underline{P}_{\text{u}}(\underline{r})$ function is

$$G^r_{\text{sol}} = -(1 - D_{\text{op}}^{-1})\frac{1}{8\pi}\int \underline{D}^r \cdot \underline{D}^r \text{dr} - \int \underline{P} \cdot \underline{D}^r \text{dr} + 2\pi c \int \underline{P} \cdot \underline{P} \text{dr} \qquad (10.74)$$

(cf. Eq. (5.2) in **M119** and Appendix 1 in **M60**) where

$$\underline{P} \equiv \underline{P}_u / D_{\text{op}}, \quad c^{-1} = \frac{1}{D_{\text{op}}} - \frac{1}{D_s}$$

D_{op} and D_s are the optical and static dielectric constants, and $\underline{D}^r$ is the electric field directly due to the charges. G^r_{sol} varies as the point P moves along the minimum PEC in Fig. 3 since $\underline{D}^r$ varies with position along that line. The first term in Eq. (10.74) is the solvation term when there is no orientation-vibration polarization $\underline{P}_u$, the second term is a dipolar interaction of $\underline{P}_u$ with the charges in a medium of dielectric constant D_{op}, and the last term is the orientation-vibration polarization energy stored up in the polarized dielectric; it vanishes when D_s equals D_{op}, as does $\underline{P}_u$.

When the system in Fig. 3 is at the point P' the solvation free energy is that of the products G^p_{sol} for the H_2O-H-M configuration $P' \ldots . \underline{P}_u$ is given by the same expression as Eq. (10.74) but with r superscripts replaced by the p's.

In the vicinity of P in Fig. 3 we let the system have a local H_2O-H-M *protonic vibrational state* of energy E^r_v, a solvation free energy G^r_{sol}, and an electronic energy V^r_e. The free energy of the system near P, $G^r(P)$, is then given by

$$G^r(P) = G^r_{\text{sol}} + E^r_v + V^r_e. \tag{10.74a}$$

G^r_{sol} includes the electrode potential term since the $\underline{D}$ in Eq. (10.74) includes fields due to charges on the electrode and in the ion atmosphere (double layer).

10.23. Energetics along a β-Path (Proton Jump)

When the system proceeds along a path β in Figs. 1–3, from point P to point P' (Fig. 3), it does so by a *protonic motion* so rapid that $\underline{P}_u$ is *constant along the PP' path.* With *energy conserved during a protonic jump* from the P to P' valleys, and with *the entropy* associated with the orientation-vibration polarization $\underline{P}_u$ also unchanged during the transition *at fixed* $\underline{P}_u$ we have

$$G^r(P) = G^p(P'). \tag{10.75}$$

Thus far, the polarization $\underline{P}_u$ in Eqs. (10.74) to (10.75) is arbitrary. As in ET reactions we choose it so that G^r_{sol} is a *minimum for a system at the point* P, subject to the constraint imposed by (10.75). Thereby one finds

$$\delta G^r = 0 = -\int (\underline{D}^r - 4\pi c\underline{P}) \cdot \delta \underline{P}_u \mathrm{d}\underline{r} \tag{10.76}$$

$$\delta G^r - \delta G^p = 0 = -\int (\underline{D}^r - \underline{D}^p) \cdot \delta \underline{P}_u \mathrm{d}\underline{r}. \tag{10.77}$$

Multiplying the second equation by a Lagrange multiplier m, adding and setting the coefficients of $\delta \underline{P}_u$ equal to zero, as the most general solution of Eqs. (10.76) and (10.77), one finds at each point $\underline{r}$ in the medium that

$$4\pi c\underline{P}_u / D_{\mathrm{op}} = (1+m)\underline{D}^r - m\underline{D}^p, \tag{10.78}$$

where D^r and D^p denote the electric fields for systems at P and at P' in Fig. 3, directly due to the charges. The similarity of the procedure embodied in Eqs. (10.74), (10.76) to (10.78) to that used for the transfer of another light particle, the electron, compare **M16**, may be noted.

Introduction of this $\underline{P}_u$ into (10.74) and into the corresponding expression for G^p_{sol} yields (10.79) and (10.82)

$$G^r_{\mathrm{sol}} = G^r_{\mathrm{sol}}(\mathrm{eq}) + m^2\lambda \tag{10.79}$$

where the first term is the *equilibrium* solvation for a system at point P,

$$G^r_{\mathrm{sol}}(\mathrm{eq}) = -(1 - D_s^{-1}) \int \underline{D}^r \cdot \underline{D}^r \mathrm{d}r / 8\pi \tag{10.80}$$

and $m^2\lambda$ is the *fluctuation term*," compare Eq. (69) in **M53**, "due to $\underline{P}_u$'s being different from its equilibrium value in $G^r_{\mathrm{sol}}(\mathrm{eq})$. λ is given by Eq. (10.81).

$$\lambda = \int (\underline{D}^p - \underline{D}^r) \cdot (\underline{D}^p - \underline{D}^r) \mathrm{d}r / 8\pi c. \tag{10.81}$$

The difference of charge distribution on the right hand side of Eq. (10.81) is expected to be roughly proportional to the Δn in Eq. (10.73), and so to depend on the point P. Similarly, the solvation for a system at point P' in Fig. 3 is

$$G^{p}_{\text{sol}} = G^{p}_{\text{sol}}(\text{eq}) + (m+1)^2\lambda. \tag{10.82}$$

The value of m is determined from Eqs. (10.75), (10.79), and (10.82)

$$-(2m+1)\lambda = G^{p}_{\text{sol}}(\text{eq}) - G^{r}_{\text{sol}}(\text{eq}) + E^{p}_{v} - E^{r}_{v'} + V^{p}_{e} - V^{r}_{e} \tag{10.83}$$

when the system near point P in a given vibrational state v of energy $E^{r}_{v'}$ is transformed by the proton jump into a system near P' in a proton vibrational state v' of energy $E^{p}_{v'}$.

10.24. Rate Expression for the β-Path

We denote by $\kappa_{vv'}(E^0_t)$ the probability of a reactive $v \to v'$ *protonic transition*, when the initial translational energy along the line of centers in Figs. 1–3 is E^0_t and the initial protonic vibrational energy is E^0_v. The TS expression for the reaction rate is given by (cf. Eq. (14), **M56**)

$$k_{\text{rate}} = \sum_{v,v'} \int_{E^0_t=0}^{\infty} \kappa_{vv'}(E^0_t) e^{-(E^0_v+\text{E}^0_t)/kT} \text{dE}^0_t\, f/Q_v kT \tag{10.84}$$

where

$$\begin{aligned} f &= \frac{kT}{h} e^{-(\Delta G^{r}_{\text{sol}}+m^2\lambda)/kT} \frac{(2\pi\mu kT)}{h^2} \frac{1}{(2\pi\mu kT)^{3/2}/h^3} \\ &= \text{Z}e^{-(\Delta G^{r}_{\text{sol}}+m^2\lambda)/kT}. \end{aligned} \tag{10.85}$$

$\Delta G^{r}_{\text{sol}}$ is the increment in equilibrium solvation free energy on going from ∞ to P. In Eq. (10.85), the translational partition function for the three translational degrees of freedom of the reactant, and for the delocalized reactant on the surface were introduced.... Q_v is the protonic vibrational partition function for the reactants in Figs. 1–3, and Z is the collision frequency $(kT/2\pi\mu)^{1/2}$ for collisions with unit area of the electrode."

$$\left(\frac{kT}{h}\frac{(2\pi\mu kT)}{h^2}\frac{1}{(2\pi\mu kT)^{3/2}/h^3} = (kT/2\pi\mu)^{1/2} = Z\ldots\right).$$

"The calculation of $\kappa_{vv'}$ proceeds as follows: One first locates the point P of deepest O-M penetration, namely where the translational energy along the O-M direction vanishes, that is, where, in the reactants' valley we have, in this β-path mechanism

$$E_v^0 + E_t^0 = E_v^r + \Delta V_e^r \tag{10.86}$$

E_v^r and ΔV_e^r refer to the protonic vibrational energy and the increment in potential energy (from the H_3O^+ moving from ∞) at point P in Fig. 3. In calculating E_v^r the vibrational quantum number v is taken as constant (v) within the reactants' valley (adiabatic treatment for the H-vibration in the valley). E_v differs from E_v^0 only because of changes in cross sectional profile (e.g., in vibration frequency) during motion along that valley of the PES.

If the value of E_v^r at P is sufficiently large to overcome any V_e barrier from reactants to products along the β-path at that P, the corresponding $\kappa_{vv'}$ is set equal to unity. Otherwise, a tunneling calculation for $\kappa_{vv'}$ is used....

The barrier along this β-path is modified somewhat by the presence of the G_{sol} term for the cited $\underline{P}_u$. Thus, changes in $V_e + G_{\text{sol}}$ along this β-path serve as the *effective barrier* to proton motion along that path (Appendix).

Thus far, the electrode-solution potential difference ϕ has not been specifically introduced. It is implicitly present in each G_{sol} terms. By evaluating these terms one obtains φ. As in ET theory, (cf. **M129**, p. 180), a 'standard potential' φ^0 for reaction (10.72) in the prevailing medium can also be introduced by setting the free energy activation terms ΔG^r and ΔG^p equal at that φ. One can then express the rate of the forward reaction in terms of the difference $\varphi - \varphi^0$.

10.25. Decision as to Paths α and β

If for any point P for any given E_v^0 and E_t^0 in Eq. (10.84) the point P is closer to the origin than the saddle point, then path α will dominate rather than path β *for that* (E_v^0, E_t^0) *pair*. If point P, in the case of Fig. 1, is quite close to the saddle point, the Δn defined by Eq. (10.73a) becomes very small and so the term $\underline{D}^r - \underline{D}^p$ arising from a difference in charge distributions at P and P' also becomes small, and so does, thereby, the λ in Eq. (10.81). Thus, with this approximate vanishing of the $m^2\lambda$ in (10.85) one has again retrieved the usual TS theory result. On the other hand, when the barrier along a β-path can more easily be overcome, either by excess vibrational energy in E_v^r or by a sufficiently large value of $\kappa v v'$, one obtains a β-path mechanism rather than one proceeding via the saddle point in the PES.

The above remarks also serve to point out the similarities and the differences between PTs and weak-overlap ETs. *The weak-overlap ET proceeds via the counterpart of a β-path and has the $m^2\lambda$ terms* of the previous section. The PT can proceed via an α-path, wherein the $m^2\lambda$ is absent, or via a β-path, depending on the initial conditions. The remarks made earlier on the hydrogen evolution reaction are also intended to apply to other (e.g., homogeneous PTs).

10.26. Appendix: Variation in $\underline{\mathbf{D}}$ along a β-Path

$\underline{D}$ varies along a β-path, since the electronic structure of the system varies along that path. $\underline{D}(\underline{r})$ is equal to (cf. Eq. (5.4) in **M119**)

$$\underline{D}(\underline{r}) = -\nabla_r \int |\psi(\underline{r}_j|^2 \sum_i \frac{e_i}{|\underline{r} - \underline{r}_i|} \prod_j d\underline{r}_j + C \qquad \text{(A.1)}$$

where ψ is the electronic wavefunction for any nuclear configuration, and is a function of the coordinates of all electrons j of the reactants, and the sum is over all i, i.e., over all electronic and nuclear charges e_i in positions $\underline{r}_i$. Thus, this $\underline{D}(r)$ can be calculated from an *electronic structure calculation.* The C in Eq. (A.1) is the contribution to $\underline{D}$ arising from other charges on the electrode and from the ion atmosphere (including double layer)."

NOTES

1. p. 5 top: "Fig. 3. Legend as in Figs. 1 and 2. The curve denotes the minimum potential energy path, which proceeds along the reactant's valley, over a saddle point..."

   M: That β is not going over the saddle point. The β path is a tunneling path.
2. p. 7 top: "With energy conserved during a protonic jump from the P to P' valley, ... we have

$$G^r(P) = G^p(P). \tag{10.75}$$

   Thus far, the polarization $\underline{P}_u$ in the above Eq. (10.75) is arbitrary. As in ET reactions we choose it so that G^r_{sol} is a minimum for a system at the point P, subject to the constraint imposed by Eq. (10.75). Thereby one finds

$$\delta G^r = 0 = -\int (\underline{D}^r - 4\pi c\underline{P}) \cdot \delta \underline{P}_u d\underline{r} \tag{10.76}$$

$$\delta G^r - \delta G^p = 0 = -\int (\underline{D}^r - \underline{D}^p) \cdot \delta \underline{P}_u d\underline{r} \tag{10.77}$$

   Q: It is $G^r(P) = G^r_{\text{sol}} + E^r_v + V^r_e$ Eq. (10.77a), and so if, at P, G^r_{sol} is at a minimum, E^r_v is a constant vibrational energy at P and V^r_e is a constant electronic energy, the differential of G^r at P is zero. The same is true of the differential of G^p at P'. Which kind of extra information is then added by condition (10.77)?

   M: The curve in Fig. 3 is meant to be one of the potential energy contours, P is not at the minimum, no, not at the minimum of the potential energy curve. You see, imagine that you have a free energy surface, and now this free energy surface is a function of these coordinates here in this subspace, and now at the point P you are in a vibrational subspace, I mixed potential energy and free energy for sort of economy but that may well lead to confusion. The point P is where the H motion has reached a turning point, in case of reaction, then the point P' is where that H motion, sort of like a motion along the radius, has reached a turning point in the case of the products. Now, if the system jumps from P to P' and it jumps so very quickly, so quickly that the other nuclei do not have time to change their positions... in other words it is like for an electron... so that the Franck Condon Principle would be obeyed, then going from P to P' would not change the kinetic energy nor the potential energy or any distribution of the all rest of the system and that means that the free energy of the P system is equal to the free energy of the P' system. Applying this sort of version of the Franck Condon... $\delta G^r = 0$ now... in other words if you look at P there is an arbitrary polarization, and there you have fluctuations in it, then with respect to those fluctuations G is a minimum, subject to all constraints that are present. Subject, for example, to the given specified distance at that point. $\delta G^r = 0$ *refers to fluctuations.* Why do we need Eq. (10.77)? You see, there is a constraint, that δG^r *is equal to* δG^p *for every arbitrary fluctuation.* In other words, we want *fluctuations subject to the condition the Franck–Condon principle be satisfied.* So, if you take that δG on both sides, δG^r and δG^p, the you take δ of the difference to be zero, you have to satisfy by calculus of variations both equations simultaneously so you bring into a Lagrangian. It is exactly what I did in 1956. The point is that $\delta G^r = 0$ and $\delta G^p = 0$ must be satisfied *both simultaneously*, and the only way you can do it is by a Lagrangian multiplier. *You have to consider the difference in order to satisfy both equations simultaneously.* That's q uite a feat, what you are you doing when you are using Lagrangian multipliers turns out you are making partial derivatives equal and so on. Supposing you have two curves and one of much bigger radius than the other, and you want to find the condition that the two curves just touch each other at a point of contact, so that they don't cross, that involves a Lagrangian, that problem is to minimize subject to a constraint.

M156. On Quantum, Classical, and Semiclassical Calculations of Electron Transfer Rates

Abstract

"Similarities and differences between quantum, semiclassical, and classical treatments of ET reactions are described. All of these methods can be described as '*vibrationally assisted electron-tunneling*'....

The nonadiabatic classical expression provides a useful lower bound, because it has no nuclear tunneling, to the nonadiabatic quantum result. The quantum correction is calculated for some actual systems. The 'semiclassical' approximation is sometimes larger and sometimes smaller (in the 'normal' and 'inverted' regions, respectively) than the quantum result. This behavior is readily understood in terms of the WKB theory.

The quantum correction to the classical cross relation of rate constants was calculated for a case of interest and found to be minor...."

10.27. Introduction

During an ET reaction there can be a readjustment of bond lengths and angles of the reactants and of reorientational and other structural changes in the surrounding polar environment, both before and after the *electron transfer act*. In many compounds, such as $Ru(NH_3)_6^{+2,+3}$, $Ru(bpy)_3^{+2,+3}$ (bpy = bipyridyl), and $MnO_4^{-1,-2}$, the *change in bond lengths* appears to be *very small*, *the change* then being largely in the *orientations* of molecules in the *surrounding environment*. Typically, one expects a similar situation for other metal–ligand complexes, having phenanthroline and bipyridyl ligands. In contrast, in ions such as $Fe(H_2O)_6^{+2,+3}$, as well as other aquo ions, *appreciable changes in bond lengths* can occur (~0.14 Å in the case cited), and so *both of the types of contributions* (intramolecular vibrational and environmental) then contribute."

10.28. Classical Expression for Rate Constant

Marcus' theory of ET treats the motion of the nuclei in a classical manner. "The reaction may be 'adiabatic' or 'nonadiabatic' and this aspect is treated via the Landau–Zener theory of curve crossing. For a qualitative description of the phenomenon, use can be made of the profiles of potential energy plots in many-dimensional space, for the *reactants plus environment* (R) and for the *products plus environment* (P), such as those in Fig. 1 and, in the case of sufficiently highly exothermic reactions, those in Fig. 2.

In Fig. 1 reaction occurs when the system initially on the lower R curve ends up on the lower P curve, either *by surmounting the energy barrier* in Fig. 1 or by *nuclear tunneling through it.*

In Fig. 1, when *the splitting* of the R and P surfaces is *large enough* at the intersection of these surfaces, a system initially on the R surface tends to end up, in a motion from left to right, on the P surface, and the reaction is *adiabatic*. (The κ in Eq. (24) of **M48** is

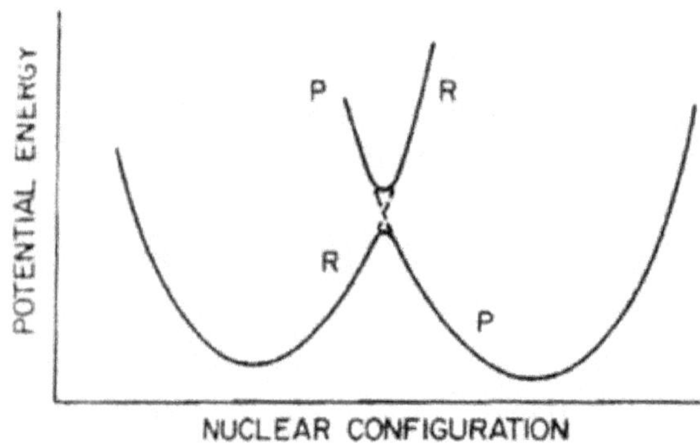

Fig. 1. Plot of profiles of potential energy surface of reactants (R) and products (P) in many-dimensional coordinate space.[2b,c]

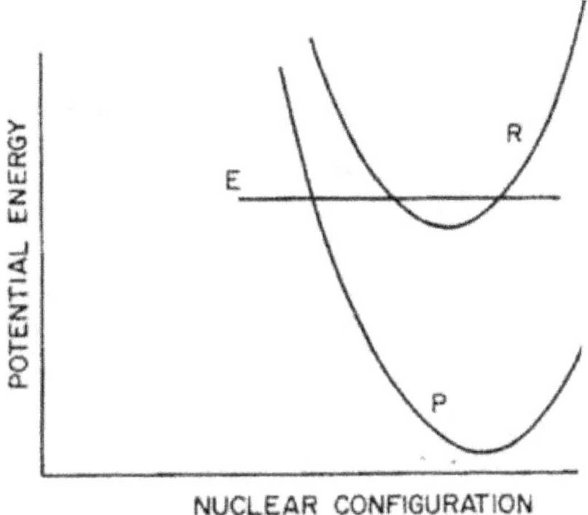

Fig. 2. Legend in Fig. 1 but for the "inverted" region.

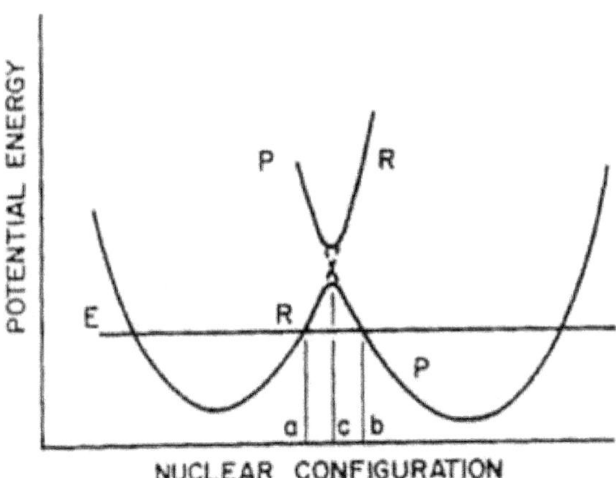

Fig. 3. Projection of Fig. 1 onto the V versus vibrational coordinate plane. Two of the classical turning-points of motion at a given vibrational energy E are labeled a and b.

then unity.] In reactions where the splitting is *small*, a reacting system passing through the intersection region of Fig. 1 has little probability of ending up as products, κ is thereby small and the reaction is nonadiabatic (cf. Eq. (25) of **M48**; footnote 3, **M141**).

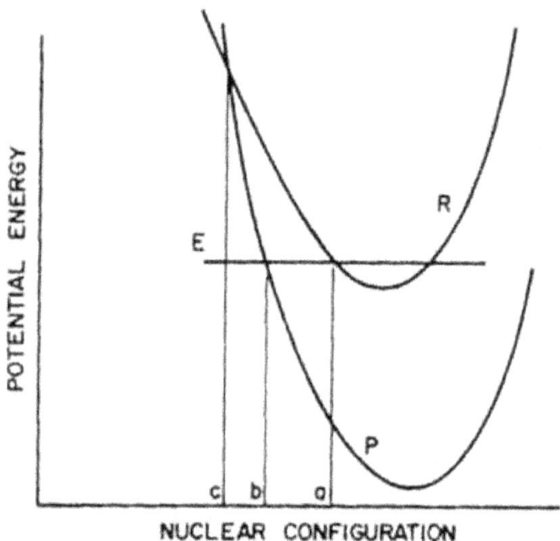

Fig. 4. Legend as in Fig. 3, but for the "inverted region."

In highly exothermic reactions such as those in Fig. 2, an *increase in exothermicity*, i.e., *a vertical lowering* of the P curve relative to the R one results in an intersection at a higher potential energy than before, and *the reaction is then expected to be slower*. This region was termed 'inverted' region and this phenomenon of *decreased rate* with *increased exothermicity*, at sufficiently high exothermicities, has its analog in radiationless transitions in the form of the *energy gap law*. It may have application to *the back reaction in bacterial photosynthesis*.

The form found for a *first-order* constant k_r (*reactants at fixed sites*) at temperature T, is

$$k_r = \kappa\nu \exp\left[\frac{-(\Delta G^0 + \lambda)^2}{4\lambda kT}\right], \tag{10.87}$$

where ΔG^0 is the standard free energy of reaction *of that step*, λ is a *nuclear reorganization parameter* and ν is a *typical frequency for motion in this reaction across the TS* (i.e., across the intersection in Figs. 1 or 2; $\sim 10^{13}$ sec^{-1}). For a bimolecular reaction between two species free to move in a solvent, the rate constant is given instead by

$$k_r = \kappa Z \exp\left(\frac{-w^r}{kT}\right) \exp\left[\frac{-(\Delta G_R^0 + \lambda)^2}{4\lambda kT}\right] \tag{10.88}$$

where Z is the collision frequency in solution ($\sim 10^{11}$ liter mole^{-1} sec^{-1})." A value later corrected in **M170**. "w^r is the work (*both coulombic and noncoulombic*) to bring the reactants from ∞ to a separation distance r, ΔG_R^0 is $\Delta G^0 + w^p - w^r$, w^p being the work required to bring the products from ∞ to r. λ is the sum of two terms, one due to the *intramolecular changes* (λ_i) and the other to *changes in environment* (λ_o)

$$\lambda = \lambda_i + \lambda_o \tag{10.89}$$

Contributions to λ_i arise from changes in *geometry* and *force constants*. When λ_i is mainly due to changes in bond lengths Δq in the reacting species with a force constant k, λ_i is essentially

$$\lambda_i = \frac{1}{2}\sum_j k_j(\Delta q_j)^2 \tag{10.90}$$

where the sum is over the vibrations of the reactants and where k_j is related to the k_j^r (reactants) and k_j^p (products) by being equal to $2\frac{k_j^r k_j^p}{(k_j^r + k_j^p)}$ (**M53**). Using a general expression forλ_o (**typo** corrected) (**M53**) one finds Eq. (10.91) for the case where the reactants are, in *their polar interactions with the environment*, treated as two spheres (radii a_1 and a_2) separated by a distance r in a medium of static dielectric constant ε_s and optical dielectric constant ε_{op}. The *reorganization* described by ë$_o$ is that in the *dielectric polarization of the environment*.

$$\lambda_o = (\Delta e)^2\left(\frac{1}{2a_1} + \frac{1}{2a_2} - \frac{1}{r}\right)\left(\frac{1}{\varepsilon_{op}} - \frac{1}{\varepsilon_s}\right) \tag{10.91}$$

where Δe is the charge transferred from one reactant to the other.... A significant contrast between Eqs. (10.90) and (10.91) may be noted.

Equation (10.90) has two sums of terms, one set from one reactant and one set from the other. The $1/r$ term prevents such characterization in Eq. (10.91). Instead of the reorganization being the sum of that of the reactants, it is the reorganization of the entire environment.

In the limit of a nonadiabatic reaction, use of the Landau–Zener formula for κ converts Eq. (10.87) to (**M140**)

$$k_r = \frac{2\pi}{\hbar}\varepsilon^2 \frac{1}{(4\pi\lambda kT)^{\frac{1}{2}}} \exp\left[\frac{(\Delta G^0 + \lambda)^2}{4\lambda kT}\right] \tag{10.92}$$

where ε is the ET transfer integral" (the splitting). "Similarly, the κ in Eq. (10.88) is found to be replaced by $\frac{\left(\frac{2\pi\varepsilon^2}{\hbar v}\right)}{(4\pi\lambda kT)^{\frac{1}{2}}}$.

The dependence of rate on separation distance r occurs via

(i) The $1/r$ term in Eq. (10.91).
(ii) The w^r and w^p terms in Eq. (10.88) and in the nonadiabatic case.
(iii) The ε term in Eq. (10.92). The decay of ε with r is exponential-like.

In the case of reactions at a metal–solution interface Eq. (10.88) is again obtained for the case where the reactant is not specifically bound to the interface, with Z now being the collision frequency of the reactant with *unit area* of the electrode, the $1/2a_2$ terms now being absent from Eq. (10.91), r being the distance of the reactant *from its dielectric image* (i.e., twice the reactant-interface separation distance), and ΔG^0 now being replaced by the $(E - E_0)$. E is the electrode-solution potential difference and E_0 is the 'standard' potential for this half-cell (in the prevailing environment)....

10.29. Quantum Aspects of the Nuclear Motion

10.29.1. *Multiphonon Expression*

The quantum theory for nuclear rearrangement in ET reactions was introduced by Levich and Dogonadze, who assumed the reaction to be a *nonadiabatic* one. Such an assumption permitted them to use Fermi's Golden Rule for a *first-order* rate constant for reactants in *fixed sites*

$$k_r = \frac{2\pi\varepsilon^2}{\hbar}\sum_{i,f}|\langle\psi_f \mid \psi_i\rangle|^2\delta(E_f - E_i)\rho_i, \qquad (10.93)$$

where ρ_i is the Boltzmann population of systems in an initial *vibrational* state i. The ψ_i and ψ_f are the *vibrational* wave functions of the *entire* system in its reactants' and products' form, respectively. δ is the Dirac delta function, which ensures *energy conservation* in the ET reaction. E_i and E_f are initial and final energies.

Levich and Dogonadze treated the *vibrations of the polar solvent* as harmonic of a given frequency, *omitting* any damping and the intramolecular vibrations, and then evaluated the double sum. A similar sum occurs in *optical transitions in solids* (R. Kubo and Y. Toyozawa). The result for the *first-order* rate constant was

$$k_r = \frac{2\pi\varepsilon^2}{\hbar}F(-\Delta G^0), \qquad (10.94)$$

where $F(-\Delta G^0)$ is the Franck–Condon factor; it depends on $-\Delta G^0$, on the vibration frequency ν and on the temperature T:

$$F(-\Delta G^0) = \frac{1}{h\nu}I_p(x)\exp\left[-x\cosh\gamma - \left(\frac{\Delta G^0}{2kT}\right)\right] \qquad (10.95)$$

$$x = \frac{\left(\frac{\lambda}{h\nu}\right)}{\sinh\gamma}, \quad \gamma = \frac{h\nu}{2kT}, \quad p = \frac{-\Delta G^0}{h\nu} \qquad (10.96)$$

$I_p(x)$ is the modified Bessel function of order p and argument x. λ is given by Eq. (10.91). When the asymptotic expression for $I_p(x)$, $\left(\frac{1}{2\pi x}\right)^{\frac{1}{2}}\exp\left[x - \left(\frac{p^2}{2x}\right)\right]$, is introduced and when one lets $\frac{h\nu}{2kT} \to 0$, Eq. (10.94) reduces to its classical limit given by Eq. (10.92). To obtain Eq. (10.94) from Eq. (10.93), E_i was written as a vibrational energy $\left(n+\frac{1}{2}\right)h\nu$, and E_f was written as a vibrational energy $\left(m+\frac{1}{2}\right)h\nu$ plus ΔU^0, the potential energy difference between the minima of the R and P curves in Figs. 1 or 2. The resulting expression would *violate microscopic reversibility* when ΔU^0 is different from ΔG^0, and so we replaced the potential energy change of Levich and Dogonadze by ΔG^0. In this way, the correct classical limit, given by Eq. (10.92), is obtained.

This *defect* of multiphonon theory occurs when $\Delta U^0 \neq \Delta G^0$. Is arises since the initial and final states of the environment were treated quantum-mechanically in a simplified way which ignores any ΔS^0 of reorientational effect of the solvent molecules. ΔS^0 is large, for example, in the reaction Fe^{+2} – $Ru(bpy_3^{+3})$.

Equations (10.97–) to (10.96) assume there is *only one frequency*, and so apply *either* when there is a Δq or a change in dielectric polarization (subject to the caveat noted earlier) but *not both*. The treatment leading to Eqs. (10.94) to (10.96) was extended... to include both intramolecular vibrations and environmental changes...."

10.29.2. *Semiclassical Expression*

In the semiclassical expression for k_r, the Franck–Condon term in Eq. (10.94) is semiclassical

$$k_r = \frac{2\pi\varepsilon^2}{\hbar} F_{sc}(-\Delta G^0) \tag{10.97}$$

"where the semiclassical Franck–Condon term $F_{sc}(-\Delta G^0)$ is

$$F_{sc}(-\Delta G^0) = \frac{1}{(2\pi\lambda h\nu \coth\gamma)^{\frac{1}{2}}} \exp - \left[\frac{(\Delta G^0+\lambda)^2}{2\lambda h\nu \coth\gamma}\right] \tag{10.98}$$

Here, λ is given by Eq. (10.90) for intramolecular changes or by Eq. (10.91) for changes in dielectric polarization. Eq. (10.98) was obtained by Hopfield for intramolecular changes (with ΔG^0 replaced by ΔU^0)....

When compared with the exact result, we found the semiclassical results to be sometimes too high (low exothermicities) and sometimes too low (quite high exothermicities). The results and the origin of these discrepancies are discussed later.

10.29.3. *A Simplified Multiphonon Expression*

We have already noted that when ΔS^0 is quite different from zero, it is incorrect to use the multiphonon (or the semiclassical) theory to treat the relevant reorganizational changes. When the ΔS^0 arises largely from the environmental changes, as it frequently does, one can use, instead, the classical expression derived earlier for those changes and use a quantum treatment for the intramolecular vibrations." On the basis of this observation Marcus derives a simplified multiphonon expression.

10.30. Summary of Numerical Results

"A comparison of the various approximations with the exact result for the values of the parameters below is given in Table I for two actual systems. . . .

The breakdown of the semiclassical approximation, Eq. (10.98), has been pointed out previously (Jortner) for $P = 0$, i.e., for $-\Delta G^0 = 0$, a breakdown also seen in in Tables I and II. The discrepancy is less at small enough $\frac{\lambda_i}{h\nu_i}$ . . . . The semiclassical approximation (10.98) first improves with increasing P, that is, with increasingly negative $-\Delta G^0$ . . . . However, when P becomes quite large the semiclassical approximation again breaks down. . . . The reasons for this behavior are considered next. In the 'normal case,' represented by Fig. 1, the square of the overlap of the vibrational wave functions ψ_i and ψ_f for an energy below that of the intersection of the R and P curves, is represented in WKB theory for the wave functions approximately by a *nuclear tunneling factor*

$$T_{i\to f} \cong e^{-2\int_a^c \frac{|p_i|dq}{\hbar}} e^{-2\int_c^b \frac{|p_f|dq}{\hbar}} \qquad (10.99)$$

Here, a and b are two classical turning points of the motions on the R and P curves at this energy (Fig. 3) and the intersection occurs at $q = c$; p_i and p_f are the *imaginary* momenta, calculated as a function of the coordinate q for the given total energy and for the vibrational potential energies V_i and V_f. Instead of Eq. (10.99), one calculates in the semiclassical method $|\psi_i|^2$ at $q = c$, and so instead of Eq. (10.99) one has

$$T_{i\to f}^{SC} \cong e^{-2\int_a^c \frac{|p_i|dq}{\hbar}}. \tag{10.100}$$

Clearly Eq. (10.100) can be much larger than Eq. (10.99), because the second factor in Eq. (10.99) is now missing, and so results in a considerable error when c and b are far apart. That is, here the 'semiclassical' approximation given by Eq. (10.100) and hence by Eq. (10.98) yields too high a result—too much nuclear tunneling. This effect is the greater the greater the relevant barrier for tunneling, $\frac{\lambda_i}{4}$ for $P = 0$, in Fig. 3.

In the case of Fig. 4, replacing the vibrational wave functions by their WKB counterparts and evaluating the overlap integral by contour integration yields

$$T_{i\to f} \cong e^{\frac{-2}{\hbar}\left(\int_c^a |p_i|dq - \int_c^b |p_f|dq\right)} \tag{10.101}$$

Instead, in the 'semiclassical' approach in Eq. (10.98) one evaluates $|\psi_i|^2$ at the intersection $q = c$. Thereby, the semiclassical value of $T_{i\to f}$ is, instead of Eq. (10.101)

$$T_{i\to f}^{SC} \cong e^{\frac{-2}{\hbar}\int_c^a |p_i|dq} \tag{10.102}$$

This resulting tunneling rate predicted by Eq. (10.102) and hence by Eq. (10.98) is now too small because of the absence of the second term in the exponent of Eq. (10.101). If the P curve were almost vertical near the intersection, the integral over $|p_f|$ in Eqs. (10.99) and (10.101) from c to b would be negligible and so the two results would then again agree.

In summary, when nuclear tunneling occurs in the exact results, the semiclassical Eq. (10.98) can only be relied upon when the P curve in Figs. 3 and 4 intersects the R curve sufficiently steeply. Otherwise, it predicts too much tunneling at low $|\Delta G^0|$'s and too little tunneling at moderate to high $|\Delta G^0|$'s. When nuclear tunneling is unimportant, the classical expression can be used instead....

By evaluating the κ — as in Marcus' theory — in a way which includes nuclear tunneling, one can treat both the nonadiabatic and adiabatic limits....

One final topic which will be considered here is the question of ln k_r versus ΔG^0 plots. These plots are highly symmetrical (parabolic) in the classical or semiclassical approximation but are highly skewed when quantum effects become important. However, the result can be replotted in a different way which once again restores their symmetry."

10.31. Discussion

"SOLOMON: What are the important normal modes which must be considered for the biological electron transfer proteins in the configurational coordination diagram you present. As the different possibilities might have very different vibrational frequencies, could a reasonable assignment be reached through an experiment sensitive to, for example, a temperature dependence?

MARCUS: The most important modes depend upon the *specific* ET reaction. In some reactions, such as those between metal-phenanthrolines involving bipyridyl complexes, the important modes appear to be those of the dielectric polarization in the environment outside the complex ions. In other reactions such as those involving iron acquo ions, both metal ligand modes and environmental modes contribute roughly equal amounts. In the case of the cytochrome c–cytochrome a reaction, the answer is not yet known but may involve environmental modes and, perhaps, a metal-axial ligand stretching mode depending upon how much the metal-axial ligand bond length changes upon change of redox state. When one can measure the rate constant k_r at sufficiently low temperatures T so that the ln k_r versus $1/T$ plot becomes curved (and eventually flat at low enough T), *the temperature where the curvature sets in is related to the characteristic frequency of the predominant vibrational reaction mode.* At least it does so when the curvature is not due to some new mechanism occurring, of course.

....

MARCUS: On the *nonadiabatic theory* for ET rates, *an overlap integral for the vibrational wave function occurs (the Franck–Condon factor)*, as does *an electronic matrix element* ε. When WKB theory is used for the vibrational wave functions,*the nuclear overlap integral becomes equivalent to a nuclear tunneling formula*, as in my paper, when the vibrational energy is below the energy of the intersection (Figs. 3 and 4). While it is best to evaluate *the electronic matrix element* ε with the full electronic wave functions and Hamiltonian. . . *one can obtain a rough estimate for it* by approximating the Hamiltonian, yielding *an overlap integral of electronic wave functions. Use of WKB theory for the latter then yields an electron-tunneling expression* analogous to Hopfield's and serves to relate ε to the earlier calculation by Chance and DeVault."

NOTES

1. In p. 5: "The form found for a first-order rate constant k_r (reactants at fixed sites) at temperature T, is

$$k_r = \kappa\nu \exp\left[\frac{-(\Delta G^0 + \lambda)^2}{4\lambda kT}\right]\text{''} \tag{10.103}$$

Q: Compare now this equation with the earlier:

$$k_r = \kappa\nu \exp\left(-\frac{w^R}{kT}\right)\left[-\frac{\lambda}{4kT}\left(1+\frac{\Delta}{\lambda}\right)^2\right]$$

Here a work term appears which is missing in Eq. (10.103). Are there cases when it should be considered and cases when it shouldn't?

M: In the first case, the system is fixed in position, at the electrode, so there is no work term that comes in, bringing the reactant from a distance. The reaction is a unimolecular reaction in which the reactant is fixed at the electrode. w^R is the work of bringing the reactants together from infinity, there may be a coulombic repulsion, it may be something that has to happen before the reorganization term starts setting in, maybe the reactant has to reorient itself, and the reorganization doesn't handle that, it handles the situation once you fix the reactants, then there is the reorganization. So there are various events that come in that are not part of the reorganization and either are not affected by ΔG^0. I tried to divide the exponential into two parts, one that is not affected by ΔG^0 and another that is affected by ΔG^0.

2. p. 29: Consider Eq. (10.104) (Levich...)

$$k_r = \kappa' \nu I_p \left(\frac{a}{\sinh \gamma}\right) \exp\left(-\frac{\Delta}{2kT} - a \coth \gamma\right) \quad (10.104)$$

where $a = \lambda / h\nu$. On p. 30 you write: "Large $\lambda / h\nu$ is the so-called 'strong coupling' regime".

M: I think the term "strong coupling regime" stems from polaron theory, that if you have an electron in a polar medium then it is strongly coupled to the medium. If one looks up for instance the book on polaron theory by Pekar one sees that an electron in polar medium interacts either weakly or strongly with that polar medium depending on how much polar the medium is. If the medium is almost non polar then one is in the weak coupling regime, when it's polar the electron really digs a deep hole in the medium. When the system is polar λ happens to be large in any ET reaction or related reactions or process in that medium and when $h\nu$ is small then the situation is fairly classical, if the frequency was very high you would hardly notice that polar $\frac{\lambda}{h\nu}$, and it may have been phrased in different terms in those days, is a measure of the extent of that coupling, of that electron to the surrounding medium.

3. p. 14: "In the 'normal case," represented by Fig. 1...

$$T_{i\to f} \cong e^{-2\int_a^c \frac{|p_i|dq}{\hbar}} e^{-2\int_c^b \frac{|p_f|dq}{\hbar}} \tag{10.105}$$

In the case of Fig. 4....

$$T_{i\to f} \cong e^{\frac{-2}{\hbar}\left(\int_c^a |p_i|dq - \int_c^b |p_f|dq\right)}," \tag{10.106}$$

Q: Please explain the limits of the integral in this last case...

M: The formulas come out in the sort of rigorous way of treating the tunneling problem, in other words when you set it up and you put in your wave functions and you do the tunneling in a proper way—and that was done way back early 1930s or so—if, in other words, one uses a semiclassical theory, the formulas are consequence of semiclassical theory. The tunneling in Fig. 4 happens through a smaller barrier, it is effectively the barrier from b to c, it's not the whole barrier. The main point is that the barrier you can see is thinner than the barrier from a to c, that result is a rigorous result coming from semiclassical theory.*The difference there reflects how thin the barrier is, and the consequence of that is the asymmetry, whereas if you look at the first case you see you're tunneling through two barriers instead of a difference of barriers. That difference causes the asymmetry.* It is a rigorous consequence of semiclassical theory, but note that semiclassical theory is an approximation.

4. p. 17: "When one can measure the rate constant k_r at sufficiently low temperatures T so that ln k_r versus $1/T$ plot becomes curved (and eventually flat at low enough T), the temperature where the curvature sets in is related to the characteristic frequency of the predominant vibrational reaction mode."

M: Often the difference between classical and quantum in a vibrational problem is given by the ratio of $\frac{h\nu}{kT}$, so for example if you take a quantum partition function and you see that $\frac{h\nu}{kT}$ is small, you get the standard classical partition function.

M163. Quantum Effects in ET Reactions

Abstract

"Classical, semiclassical, and quantum theories of outer-sphere ET reactions in polar media are discussed. For each, the Franck–Condon overlap factors for the hexaamminecobalt, hexaaquoiron, and hexaammineruthenium self-exchange rates and for the $Fe^{2+}-Ru(bpy)_3^{3+}$ cross-reaction are evaluated and compared. The quantum effect on the rates is relatively unimportant in the 'normal' ΔG^0 region. Direct and saddle point evaluations of the Franck–Condon factors are made and compared.

10.32. Introduction

An outer-sphere ET reaction in a polar solvent is characterized by changes in the force constants and bond lengths and bond angles of the reactants and by fluctuations in the surrounding solvent. In many systems the inner-sphere changes are very small, so that the reaction is controlled by fluctuations in the solvent polarization (e.g., $Ru(NH_3)_6^{3+/2+}$ and $Cr(2, 2' - \text{bipyridyl})_3^{3+/2+}$). On the other hand, some redox systems involve substantial internal reorganization (e.g., $Fe(H_2O)_6^{3+/2+}$ and the $Co(NH_3)_6^{3+/2+}$). In such systems inner-sphere effects are important.

In this paper, we briefly describe classical, semiclassical, and quantum theories of ET. It has been suggested that reactions in which inner-sphere reorganization is important are not adequately described by classical theory, bur require a quantum mechanical treatment. A quantum mechanical treatment is available for *nonadiabatic* ETs, and was developed at first for the solvent modes and later for the bond vibrations."

Nature and magnitude of quantum effects are discussed for the above reactions.

"It is expected that if nuclear tunneling is to be important, it will be so for systems in which a *high-frequency mode* undergoes a significant displacement." Examples are given.

"Nuclear tunneling will, other things being equal, be more important for high rather than for low-frequency modes as one can see from the nature of harmonic oscillator eigenstates."

Marcus considers for illustration purposes the one-dimensional model surface sketched in Fig. 1.

"Nuclear tunneling depends on the overlap of reactant and product wave functions in the classically nonallowed region, and therefore is directly related to the amplitude of the reactants' wave function in the region $q > b$. This wave function extends further into the classically forbidden region, for any given energy, *the higher the vibration frequency.* It follows that tunneling from a state of given energy is more probable for a high-frequency mode than for a low-frequency mode, at a given energy.

In the present paper, it is found that for the reaction rate constant a reasonable *order of magnitude estimate* for the contribution of configurational changes of *high-frequency quantum modes* in the

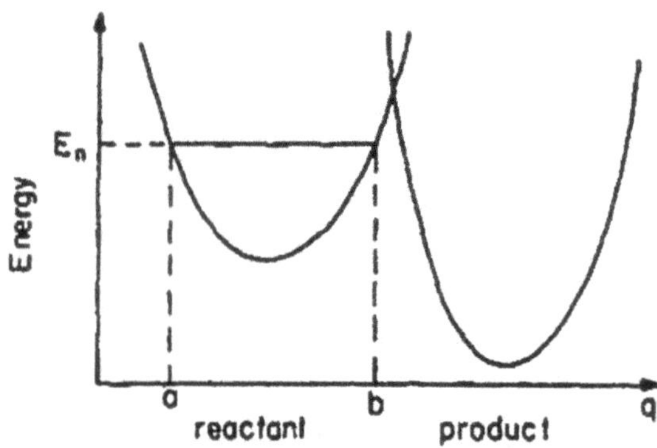

Fig. 1. Model harmonic potentials for electron transfer vs. a generalized configuration coordinate q[9].

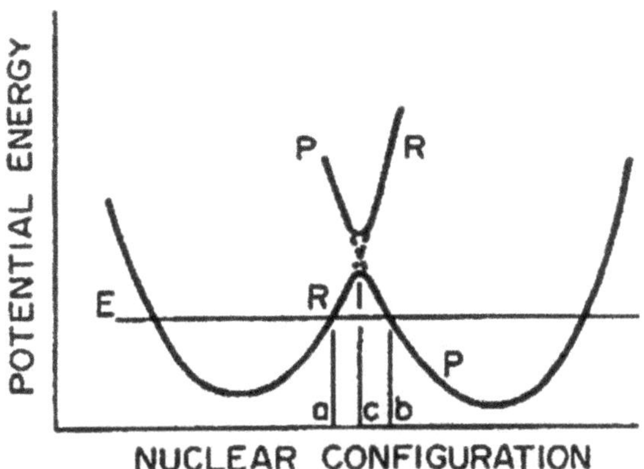

Fig. 2. Curve similar to Figure 1, but for a nearly thermoneutral reaction ($\Delta E \simeq 0$). Points a and b here are classical "turning points" of motion on the reactants' and products' potential energy curves, for the given energy E. Point c is at the intersection of the two potential energy surfaces. The actual nuclear tunneling distance is ab (Cf. Ref. 26).

first coordination layer, for typical metal–ligand frequencies, can be provided *by a classical expression* (**M30**, **M48**).

10.33. Quantum Treatment

Franck–Condon Factor. An approximate quantum-mechanical rate expression based on the golden-rule transition probability is applicable to ET systems in the nonadiabatic limit. Within the Condon approximation the transition probability involves the product of the square of an *electron exchange integral*, and a *thermally weighted sum G of Franck–Condon factors*:

$$G = \frac{1}{Q} \sum_n \sum_m e^{-\frac{E_n^{\mathrm{vib}}}{kT}} |\langle n|m\rangle|^2 \delta(E_n - E_m) \qquad (10.107)$$

where Q is the reactants' vibrational partition function, and n and m designate initial and final *vibronic states*, respectively. E_n and E_m are initial- and final-state energies, E_n^{vib} is the initial-state vibrational energy, and $|n\rangle$ and $|m\rangle$ are treated as *harmonic oscillator eigenfunctions*, equal to a product over the system's degrees of freedom of single-mode harmonic oscillator functions."

The single-mode harmonic oscillator overlap integrals required for evaluating G directly are known. The expressions used by Marcus for these integrals are "presented in the Appendix... in terms of $f = \frac{\omega'}{\omega}$, ω' and ω being the frequencies associated with $|m\rangle$ and $|n\rangle$ respectively, and in terms of the dimensionless change $X^{\frac{1}{2}}$ in equilibrium coordinate value from $|m\rangle$ to $|n\rangle$. For a normal mode $X = F(\Delta Q)^2/2\hbar\omega$, where ΔQ is the change in the normal coordinate, $\omega/2\pi$ is the vibration frequency and F is the force constant for the mode ($\omega^2 = F$).

In the case of $X \neq 0$ but $\omega' = \omega$, one obtains the well-known limiting form for $n > m$

$$\langle n|m\rangle = X^{\frac{(n-m)}{2}} \left(\frac{m!}{n!}\right)^{\frac{1}{2}} e^{-X/2} L_m^{n-m} \qquad (10.108)$$

where L is an associated Laguerre polynomial.

An approximate simple formula for the multimode case has also been derived in **M156**.... This relation was applied to the hexaacquoiron self-exchange reaction and to the $Fe^{2+} - Ru(bpy)_3^{3+}$ cross-reaction and shown to give good agreement with the exact quantum values.

Quantum Treatment of the Solvent. The interaction of the solvent with the reactant ions is implicitly included in Eq. (10.107) as a set of *one or more* harmonic modes. Usually only a single frequency, $\hbar\omega_1 = 1\ \text{cm}^{-1}$, is used in calculations. However, in view of the significant decrease in the real part of the dielectric constant of water at $170\,\text{cm}^{-1}$ (and the corresponding peak in the imaginary part) we have chosen to use a two-frequency quantum description of the solvent interaction: $\hbar\omega_1 = 1\ \text{cm}^{-1}$ and $\hbar\omega_2 = 170\,\text{cm}^{-1}$. A dielectric *dispersion* in the solvent was first treated for ET by Ovchinnikov and Ovchinnikova.

As a first approximation to this *two-frequency description* we divide the outer-sphere reorganization energy into two parts, writing λ_{out} which is four times the solvent reorganization energy (**M69**), as

$$\lambda_{\text{out}} = \lambda_1 + \lambda_2 \tag{10.109}$$

where

$$\begin{aligned}\lambda_1 &= \lambda_{\text{out}}\left(\frac{1}{\varepsilon_s} - \frac{1}{\varepsilon_{ir}}\right) \Big/ \left(\frac{1}{\varepsilon_s} - \frac{1}{\varepsilon_{op}}\right)\\ \lambda_2 &= \lambda_{\text{out}}\left(\frac{1}{\varepsilon_{ir}} - \frac{1}{\varepsilon_{op}}\right) \Big/ \left(\frac{1}{\varepsilon_s} - \frac{1}{\varepsilon_{op}}\right)\end{aligned} \tag{10.110}$$

$\varepsilon_{ir} = 5.0 =$ real part of the dielectric constant on the 'plateau' between $1\,\text{cm}^{-1}$ and $170\,\text{cm}^{-1}$;
$\varepsilon_s = 78.3 =$ static dielectric constant;
$\varepsilon_{op} = 1.78 = n_D^2$.

Thus, the quantum treatment of the solvent interaction (the solvent is taken to be aqueous in this paper) involves *two* harmonic modes included in the degrees of freedom of the system. In performing the quantum-mechanical calculations *for the solvent* Eq. (10.108) was again used but X was obtained in the following manner. It is first recalled that for an internal normal mode i of the reactants X_i which equals $F_i(\Delta Q_i)^2/2\hbar\omega_i$, can be rewritten as $\frac{\lambda_i}{\hbar\omega_i}$, since $\lambda_i = F_i(\Delta q_i)^2/2$. By analogy, we use for X for the solvent $\frac{\lambda_1}{\hbar\omega_1}$ and $\lambda_2/\hbar\omega_2$ where λ_1 and λ_2 have been defined in Eqs. (10.109) and (10.110). The numerical values employed for $\lambda_{1,2}$ are given later in the paper, while $\hbar\omega_{1,2}$ are given above.

Saddle Point Method. For a system having several vibrational normal modes of different frequencies, the direct evaluation of Eq. (10.107) can require considerable computing time. However, G can easily be evaluated approximately by replacing the δ function in Eq. (10.107) by its Fourier integral representation, and then using the saddle point method. After some manipulations one obtains

$$G = (2\pi Q)^{-1} \int_{-\infty}^{\infty} e^{-i\Delta Et + f(t)} dt \tag{10.111}$$

and, after using the saddle point method to approximate the integral, one obtains Eq. (10.112)

$$G \sim |2\pi f''(t_0)|^{-\frac{1}{2}} Q^{-1} e^{-i\Delta E t_0 + f(t_0)} \tag{10.112}$$

where ΔE is the energy (endoergicity) of the transition; t_0 is the stationary phase value of t in the integrand in Eq. (10.111), and f, f'' and t_0 are given in the Appendix.

In the case of a self-exchange reaction, product modes in the oxidized species are equivalent to reactant modes in the reduced species so that the formulae simplify considerably. In a thermoneutral self-exchange reaction $t_0 = -i/2kT$. For other cases Eq (A.6) in the Appendix may be solved numerically, for example, by iterating from the approximate root

$$t_0 \sim -i\frac{(\Delta E + \lambda)}{2kT\lambda} \tag{10.113}$$

where $\lambda = \sum_{j=1}^{N} \lambda_j$, and each $\lambda_j = \frac{1}{2} F_j (\Delta Q_j)^2$. Equation (10.113) gives the exact saddle point in the high temperature limit, when frequency changes are neglected and provides a reasonable starting point for iteration in other cases.

Classical Treatment. When all the degrees of freedom of the system are treated in the classical limit $\frac{\hbar\omega}{kT} \to 0$, and when frequency changes are neglected, Eq. (10.111) reduces to Eq. (10.114).

$$G = (4\pi kT\lambda)^{-\frac{1}{2}} \exp\left[\frac{-(\Delta E + \lambda)^2}{4kT\lambda}\right] \tag{10.114}$$

This equation is similar in form to the classical expression for G, but contains energies rather than free energies. This difference arises because Eq. (10.111) tacitly assumes zero entropy of reaction, and indeed the initial equation (10.107), with its assumption of harmonic oscillators, does not contain any important ΔS^0 term, whereas the actual ΔS^0 can be quite large. The classically derived expression (cf. **M156**) is more general in this respect, since it doesn't assume harmonic oscillations for all motions. As defined earlier, $\lambda_j = \frac{1}{2}F_j(\Delta Q_j)^2$ and $\lambda = \sum_{j=1}^{N}\lambda_j$. It has been shown (**M53**) that frequency changes may be included in an approximate manner by using an average force constant to calculate λ_j, rather than using the initial force constant. F_j above is an averaged force constant

$$F_{av} = 2FF'/(F + F') \tag{10.115}$$

where F and F' are the force constants in the reactant and product states, respectively. The classical value of the Franck–Condon sum (Eq. 10.114) is computed by using λ's calculated with average force constants given by Eq. (10.115).

'Semiclassical' Treatment. Consider first a one-dimensional case with a coordinate Q. The $\delta(E_m - E_n)$ of Eq. (10.107) can be introduced into $|\langle n|m\rangle|^2$. When the commutator of the initial and final Hamiltonians, $\mathscr{H}_n$ and $\mathscr{H}_m$, is neglected, $\delta(E_n - E_m)$ in the integral becomes $\delta(\mathscr{H}_n - \mathscr{H}_m)$, which in turn is $\delta(V_n - V_m)$ since the kinetic energy terms in $\mathscr{H}_n$ and $\mathscr{H}_m$ cancel; V_n and V_m are the potential energies of the reactants and products, respectively. By using the identity $\sum_m |m\rangle\langle m| = 1$, we may reduce the thermally weighted double sum of squared overlap integrals in Eq. (10.107) to a single sum over n of $\langle n|\delta(V_n - V_m)|n\rangle \ldots$. These integrals are readily evaluated, yielding a sum of factors proportional to $|\chi_n(Q)|^2$, where Q is that value of the coordinate for which the reactant and product potential energies are equal, and χ_n is the wavefunction of the reactants. The remaining sum over n in Eq. (10.107) is then readily evaluated to yield

$$G = (2\pi\lambda\hbar\omega \coth\gamma)^{-\frac{1}{2}} \exp\left[\frac{-(\Delta E + \lambda)^2}{(2\lambda\hbar\omega\coth\gamma)}\right] \tag{10.116}$$

where $\gamma = \frac{\hbar\omega}{kT}$, and E and λ are defined as in Eq. (10.113), but λ is for the single mode being considered. . . .

This method of obtaining G's which originated in the theory of optical spectra of solids, is sometimes termed 'semiclassical' because of *neglect of commutators* of $\mathscr{H}_n$ and $\mathscr{H}_m$, although the term 'semi-classical' has a variety of other meanings (corresponding to other approximations) in the literature."

10.34. Calculations and Discussions

Detailed calculations and discussions follow for the four reactions considered.

Cross-Reactions. "Quantum effects on the classical cross relation (cf. **M41**) are found to be relatively small, in the 'normal' ΔG^0 regime. In this relation, the rate constant k_{12} of the cross relation is related to those (k_{11}, k_{22}) of the self-exchange reactions." when the work terms are either small or nearly cancel, via Eq. (10.117)

$$k_{12} \cong (k_{11}k_{22}K_{12}f_{12})^{\frac{1}{2}} \tag{10.117}$$

where K_{12} is the equilibrium constant of the cross-reaction and f_{12} is given by Eq. (10.118)

$$\ln f_{12} = (\ln K_{12})^2 \Big/ \left[4 \ln\left(\frac{k_{11}k_{22}}{Z^2}\right)\right] \tag{10.118}$$

where Z is the collision frequency in solution. Expressed in terms of the classical G's, this expression can be rewritten as Eq. (10.119), where

$$\bar{G}_{12} = (\bar{G}_{11}\bar{G}_{22}K_{12}\bar{f}_{12})^{\frac{1}{2}} \tag{10.119}$$

$$\ln \bar{f}_{12} = (\ln K_{12})^2/[4\ln(\bar{G}_{11}\bar{G}_{22})] \tag{10.120}$$

$$\bar{G}_{ij} = (4\pi\lambda_{ij}kT)^{\frac{1}{2}}G_{ij} \tag{10.121}$$

The classical results in Tables III and IV are those for a classical adiabatic result

$$k_{12} = Z\bar{G}_{12} \qquad (10.122)$$

where Z is defined earlier (and is taken to be $10^{11}\,\mathrm{M}^{-1}\,\mathrm{s}^{-1}$. Equation (10.122) is valid when work terms for formation of the precursor and successor complexes are neglected and when nonadiabaticity is negligible. To assess a quantum correction, we obtained the "quantum results" in Tables III and IV by using Eqs. (10.121) and (10.122) but with the G_{ij} in Eq. (10.121) replaced by its quantum value. The "semiclassical" values in Table III were calculated by introducing the semiclassical value of G_{ij} into Eqs. (10.121) and (10.122)....

Conclusion. We have shown that the Franck–Condon contributions to the rates of the hexaamminecobalt, hexaammineruthenium, and hexaacquoairon self-exchange reactions at 300 K can be reasonably well approximated by the classical expression (factors of 4.3, 1.2, and 3.5, respectively). These corrections are relatively minor, in view of the uncertainties in various quantities involved in the rate expression. A nonadiabatic model was assumed, but analogous results would be expected for an adiabatic model....

The quantum effect on the cross-reaction relation (Eq. 10.117) for hexaaquiron(II) with tris-bipyridylruthenium(III) is negligible (a factor of 0.94), since some *cancellation of quantum effects* occurs in the calculation of cross-reaction rates."

We conclude that a reasonable *order of magnitude* estimate for the contribution of configurational changes of high-frequency quantum modes in the first coordination layer, for typical metal–ligand frequencies, to the rate constant can be provided by a classical expression. Preexponential factors and activation energies are expected to be more sensitive to use of the classical approximation (they are to other approximations also)...."

NOTES

1. p. 745 1st column low: "**Hexaaquoiron(II/III) Self-Exchange Reaction**... It has been suggested that in a system like this one, in which a high-frequency mode undergoes a significant bond length change, quantum effects should be large. But calculation of the sum over Franck–Condon factors yield a quantum value of about 3.5 times the classical value."

   M: Some people think that quantum effects are always important. Actually, the quantum effects are *high in the inverted region but in the normal region are not that high.*

CHAPTER 11

Mixed Reaction-Diffusion Rates, Quantum Effects in the Inverted Region, Frequency Factor in Electron Transfer: Its Role in the Highly Exothermic Regime, Highly Exothermic Electron Transfer, Electron and Proton Transfer, Nonadiabatic Processes, Quantum-like and Classical-like Coordinates, Applications to Nonadiabatic Electron Transfers

M30a. Discussion Comment on Mixed Reaction-Diffusion Controlled Rates

"We deduce here an expression applicable to systems where *intermolecular forces influence the diffusion.* The reaction scheme normally consists of a diffusion step (11.1) to form some $A \ldots B$ pair, which in turn can either diffuse apart, reverse step of in (11.1) or react:

$$A + B \underset{k_{-1}}{\overset{k_1}{\rightleftharpoons}} A \ldots B \tag{11.1}$$

$$A \cdots B \xrightarrow{k_2} \text{Products} \tag{11.2}$$

NOTE: k_1 is the diffusion rate constant, k_2 the activation rate constant.

Let $c(r)$ denote the *concentration of B* with respect to some A at an A-B distance r. The *net flux of diffusion* of B toward an A, $J(R)$, at $r = R$ must equal the reaction rate

$$J(R) = k_2 c(R), \tag{11.3}$$

NOTE: the activation reaction rate is equal to the flux due to diffusion.

where $c(R)$ is the concentration of $A \ldots B$'s (NOTE: equal to the concentration of B's surrounding a certain A at distance R) and R is a *typical* $A \ldots B$ distance.[1]

The diffusion equation to be solved is

$$J(r) = 4\pi r^2 \left[D(dc/dr) + vc\right], \tag{11.4}$$

NOTE: on the spherical surface of area $4\pi r^2$ are the B's diffusing toward A.

where ν, the relative velocity caused by the forces, equals a velocity *per unit force*, u, multiplied by the force, dw/dr, $w(r)$ being the work required to bring A and B from infinity to r. D is the *sum* of the *diffusion constants* of A and B. The term u is readily shown to be D/kT (at equilibrium, $J = 0$ and $c = c_0 \exp(-w(r)/kT)$, c_0 being the value of c at $r = \infty$).[2]

One has therefore

$$J(r) = 4\pi r^2 \left(D\frac{dc}{dr} + \frac{Dc}{kT}\frac{dw}{dr}\right) \tag{11.5}$$

In the steady state, $J(r)$ is independent of r. Solving Eq. (11.5) subject to boundary condition Eq. (11.3) (NOTE: i.e., the flux at R is equal to reaction rate at R, i.e., all of the B's reaching A at the distance R forming an $A \ldots B$ pair do react) and setting $c = c_0$ at $r = \infty$, we find for the *overall* rate constant, k, which is the *flux per unit* c_0, J/c_0:

$$\frac{1}{k} = \frac{\exp w(R)/kT}{k_2} + \frac{1}{4\pi D / \int_{r=R}^{\infty} \exp(w/kT)\, dr/r^2} \tag{11.6}$$

An alternative method of deducing Eq. (11.6) which is perhaps somewhat more illustrative of physical processes involved and which shows more clearly the relationship with Debye's equation for a purely diffusion-controlled reaction rate, is the following.

In Eqs. (11.1) and (11.2), we can set $dc_{A\ldots B}/dt = 0$ in the steady state and obtain for the overall rate constant

$$k = k_1[k_2/(k_2 + k_{-1})]. \quad (11.7)$$

k_1 is the rate constant computed by Debye by integrating Eq. (11.5) subject to the Smoluchowski boundary condition, $c = 0$ at $r = R$. (This boundary condition prevents back diffusion.) k_1 is the flux per unit c_0 and was found in this way to be

$$k_1 = 4\pi D \bigg/ \int_R^\infty \exp(w/kT)dr/r^2. \quad (11.8)$$

Similarly, k_{-1} is the flux when $c(R) = 1$ and $c(\infty) = 0$. NOTE: Recall: the velocity of a reaction equals the rate constant when the concentrations are = 1...). We find, upon integrating Eq. (11.5)

$$k_{-1} = 4\pi D\exp[w(R)/kT] \bigg/ \int_R^\infty \exp(w/kT)\,dr/r^2. \quad (11.9)$$

Combining Eqs. (11.7) to (11.9), (11.6) is once again obtained.

In Eq. (11.6), incidentally, it is convenient to define a rate constant k_e which would be the value of k if $c(R)$ always had its equilibrium (Boltzmann) value, which it does when $k_2 \ll k_{-1}$.This value of k_e, it is seen from Eqs. (11.6) or (11.7), equals $k_2\exp[-w(R)/kT]$. When k_e is very small relative to the diffusion term, we see from Eq. (11.6) that $k = k_e$ and the reaction rate is "activation controlled." When the converse is true, k equals Debye's expression $4\pi D/\int_R^\infty \exp(w/kT)dr/r^2$ and the reaction is diffusion controlled.[(3)]

NOTE: From Eq. (11.6), the first term in the denominator to the right is $k_2\exp[-w(R)/kT]$, that is, this constant k_e.

NOTES

1. **M**: Just contact distance.
2. Q: Is there a concentrations distribution as in the ionic atmosphere and/or a hard spheres distribution?

   M: I guess you have both simultaneously. You know, the ionic atmosphere, the usual Debye concept, doesn't say much about hard spheres... a hard spheres distribution takes into account a kind of discretization of the solvent... take the hard spheres distribution, add Coulombic forces to it, shielding, and so on, then one would get you something more sophisticated than the usual ionic atmosphere, they are different sides of the same coin, hard spheres distribution says nothing about the ionic charges aspect...

   Q: So, this is some kind of general expression where you can describe hard spheres and ionic atmosphere, and so on, depending on the $w(r)$.

   M: Yes, I suppose so, with the hard spheres distribution you have kind of a damped oscillation, first coordination layer, second coordination layer... and I suppose you could always formally write it that way, in other words you use a $w(r)$ that matches some hard spheres distribution, or the hard spheres distribution where the spheres are charged.
3. **M**: k_e is the real rate constant, the actual rate constant, which includes the work term. It is split up into two parts, a part is associated with the chemical reaction and a part associated with the approach, so the two parts are here separated the way I separated them in electron transfer (ET)... where I had a $w(r)$ and the reorganization, so it it's just a way of writing a rate constant, separating it into two parts, k_2 would be the effective rate constant when the system is already at R.

M164. Quantum Effects for ET Reactions in the Inverted Region

"**Abstract**. Quantum effects in outer-sphere ET reactions in the *inverted region* are considered. The results of quantum, 'semiclassical,' and classical calculations on model systems are presented.

A series of highly exothermic reactions of tris-bipyridyl complexes involving electronically excited reactants is discussed with regard to the possible importance of quantum effects and of *alternate reaction pathways* in understanding the failure of the series of reactions to exhibit pronounced 'inverted' behavior. *Electronically excited products* or alternate *atom-transfer (AT) mechanisms* provide possible explanations for the large discrepancy.

11.1. Introduction

In the *usual range* of standard free energies of reaction ΔG^0, outer-sphere homogeneous ET reactions have rates which increase with increasingly negative ΔG^0. However, when $-\Delta G^0$ is very large *both classical* (**M30**, **M57**) *and quantum theories* predict that the ET rate will ultimately decrease with increasingly negative ΔG^0 (inverted region), namely when $-\Delta G^0$ is greater than λ, four times the total reorganization energy of the reaction. Experimental studies have shown little or no decrease of the rate constant in this 'inverted' region. There have been suggestions that

(i) *Quantum effects* are responsible, suggestions that
(ii) *Electronically excited products* may be responsible (they correspond to reactions with a smaller $-\Delta G^0$), and suggestions that
(iii) Where the rate of ET is *inferred from* and, in fact, *equated to* the rate of fluorescence quenching, the fluorescence quenching in the inverted region may be due instead to a faster alternate non-ET initial step, *exciplex formation.*

In this paper, we consider the importance of *nuclear tunneling* first for a model system and then for an actual system using realistic vibration frequencies and bond length changes for the data of Creutz and Sutin. The discrepancy is found to remain very large, some quantum effects notwithstanding. An alternate pathway of forming *electronically excited product* is explored; it reduces the discrepancy considerably.

Another possible alternate pathway is an *AT*. Still another possibility (*longer range ET*) is also considered.

11.2. Theory

Quantum Treatment. An approximate quantum-mechanical rate expression based on the golden-rule transition probability is applicable to ET systems in the nonadiabatic limit.[(1)]

Within the Condon approximation the *transition probability* involves the product of the square of an *electron exchange integral* and a thermally weighted sum, G, over Franck–Condon factors

$$G = \frac{1}{Q} \sum_{n} \sum_{m} \mathrm{e}^{-E_n^{\mathrm{vib}}/kT} \, |\langle \chi_n | \chi_m \rangle|^2 \, \delta(E_n - E_m) \quad (11.10)$$

where Q is the reactants' vibrational partition function and n and m designate initial and final vibronic states, respectively. E_n and E_m are initial- and final-state energies. E_n^{vib} is the initial-state vibrational energy, and $|\chi\rangle$ is treated as a harmonic oscillator eigenfunction assumed equal to a product over the system's degrees of freedom of single-mode harmonic oscillator functions.

The overlap integrals required for evaluating G directly by the sum of Eq. (11.10) are well-known (e.g., Ref. **M163**). The solvent interaction is included in Eq. (11.10) via two harmonic modes that have frequencies $\hbar\omega_1 = 1\,\mathrm{cm}^{-1}$ and $\hbar\omega_2 = 170\,\mathrm{cm}^{-1}$ . . . .

Classical Treatment. When all the degrees of freedom of the system are treated in the classical limit $\hbar\omega/2kT \rightarrow 0$, and when frequency changes are neglected Eq. (11.10) *reduces* to

$$G = (4\pi kT\lambda)^{-\frac{1}{2}} \exp\left[- (\Delta E + \lambda)^2 / 4kT\lambda\right] \quad (11.11)$$

where $\lambda = \sum_{j=1}^{N} \lambda_j$, ΔE is the energy of reaction, and λ_j is four times the reorganization energy for the jth mode. For a vibrational normal coordinate, $\lambda_j = \frac{1}{2} F_j (\Delta Q_j)^2$, where F_j is the force constant and ΔQ_j is the equilibrium displacement from reactant state to product state, of the jth normal coordinate. Equation (11.11) is similar in form to a classical expression (**M30, M57**) which allowed for large entropies of reaction when they occurred. However, unlike this classical expression, it contains energies rather than free energies, since Eq. (11.10) does not contain any large entropy terms. The other classical expression (**M30, M57**) is more general in this respect (**M156**).

It has been shown (**M53**) that frequency changes may be included in an approximate manner by using *average* force constants to calculate λ rather than using the actual force constants. The average force constant is

$$F_{\text{av}} = 2FF'/(F + F') \tag{11.12}$$

where F and F' are the force constants in the reactant and product states, respectively. We use F_{av} when evaluating the classical value of the Franck–Condon sum (Eq. 11.11)....

Semiclassical Treatment. A 'semiclassical' treatment of ET has been given and discussed elsewhere (**M156, M163**) The semiclassical expression for the thermally weighted Franck–Condon sum is

$$G = (2\pi\lambda\hbar\omega\coth\gamma)^{-1/2}\exp[-(\Delta E + \lambda)^2/(2\lambda\hbar\omega\coth\gamma)] \tag{11.13}$$

The variables of Eq. (11.13) are defined as for Eq. (11.11) and $\lambda\hbar\omega\ \coth\gamma$ is an abbreviation for $\sum_{j=1}^{N}\lambda_j\hbar\omega_j\coth\gamma_j$, where γ_j is $\hbar\omega_j/kT$. 'Semiclassical' has come to denote a variety of different methods in the dynamics literature, one of which yields Eq. (11.13).

Comparison of the Three Treatments. Fig. 1 is a plot of G, the Franck–Condon sum, calculated classically and quantum mechanically, versus ΔG^0, the standard free energy of reaction for a *model system*. ΔG^0 is the same as ΔE in Eqs. (11.10 and 11.11), since Eq. (11.10) tacitly assumes zero for $\Delta S^{0(2)}$ when $F_i = F_i'$. The model system represents metal-bipyridyl systems (e.g., $\text{Ru (bpy)}_3^{2+} + \text{Os (bpy)}_3^{3+}$). The internal reorganization in such systems is negligible ($\lambda_{\text{inner}} = 0$) and the outer-sphere reorganization energy $\frac{1}{4}\lambda_{\text{out}}$

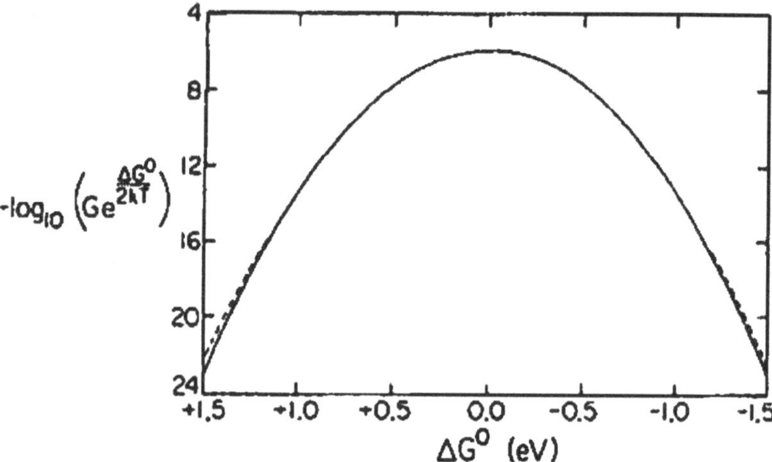

Fig. 1. Model $M(bpy)_3^{3+/2+}$: —, classical Franck–Condon sum; --- quantum Franck–Condon sum ($\lambda_{out} = 54$ kJ/mol; $\lambda_{inner} = 0$; temp = 300 K).

is $\approx 13.4\frac{\text{kJ}}{\text{mol}}$. The ordinate is a plot of $\log_{10}\left(Ge^{\Delta G^0/2kT}\right)$ versus ΔG^0. As shown in **M163**, both the classical and quantum values of the ordinate is symmetric in ΔG^0, when plotted in this manner.

Fig. 2 is a plot similar to Fig. 1.**(3)** The λ's and frequencies used are for the *hypothetical system* described in Table I. This system differs from that in Fig. 1 by including *two high frequency internal modes* and having both a larger inner-sphere and a larger outer-sphere reorganization energy.... In Fig. 2, the ordinate is a log plot of the Franck–Condon sum, G versus ΔG^0, and so Fig. 2 unlike Fig. 1 is not symmetrical about $\Delta G^0 = 0$.**(4)**

The two plots are qualitatively alike. The classical value for the ordinate in each plot is generally less than the quantum value, as expected since the classical theory does not include *vibrational tunneling.* In the *normal region* (i.e., $-\Delta G^0 < \lambda$) the classical and quantum values agree very well. But as *the free energy decreases into the inverted region*, the quantum value decays *less rapidly* than the classical. Because of the *high frequency* internal modes included in the second system, the discrepancy between the classical and

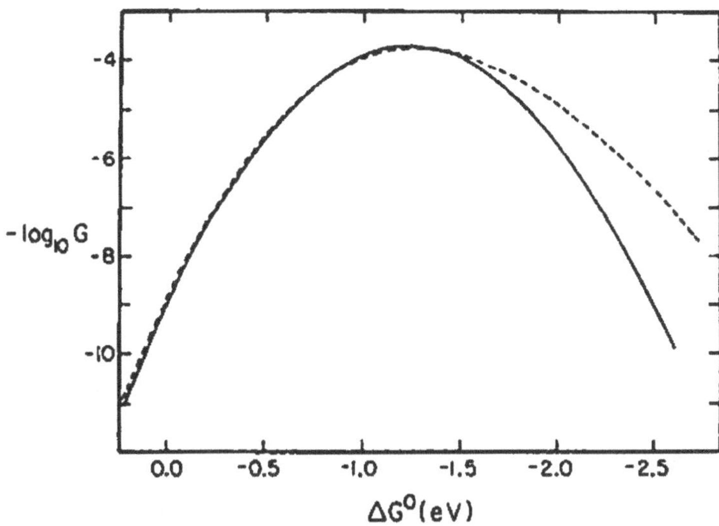

Fig. 2. Hypothetical systems (λ's and frequencies in Table I); —, classical Franck–Condon sum (assuming also eq 3); - - -, quantum Franck–Condon sum.

quantum values only becomes appreciable in Fig. 2. Similar results were observed earlier by Jortner *et al.* using other model systems.

The 'semiclassical' values are compared with the quantum for selected values of ΔG^0 in Table II. They are smaller when the system is in the inverted region and high otherwise. This effect is due to an approximation to vibrational tunneling inherent in the 'semiclassical' method which, as discussed in recent papers (**M156, M163**), is valid only when the slope of the products' potential energy curve (PEC) is extremely steep near its intersection with the reactants' PEC. (Only then is the semiclassical nuclear tunneling distance *ac* in Figs. 3 and 4 of Ref. **M156** or of Fig. 2 of Ref. **M163** equal to the effective nuclear tunneling distance *ab* there.)

The quantum values plotted in Fig. 1 and 2 were calculated both by direct evaluation of Eq. (1.10) and by the saddle point method described in **M63**. . . ."

"**Reactions Having Large Negative Free Energies**. Both the classical and the quantum theories described earlier predict that the ET rate will ultimately decrease when ΔG^0 becomes increasingly negative, i.e., when $-\Delta G^0$ *exceeds the total* λ *for the system*. The classical theory predicts quadratic dependence in the very negative ΔG^0 region (as seen in Figs. 2 and 3). But experimental studies of highly exothermic reactions have shown little or no decrease of the rate constant in the inverted region, due to a variety of possible reasons discussed earlier.

We first explore the kinetic effect of formation of products in their lowest electronic state, for reaction of *excited* $Ru(bpy)_3^{2+}$ with tris(bipyridyl) ruthenium, -osmium, and -chromium quenchers studied experimentally by Creutz and Sutin. The reactions are listed in Table III. Given are the standard free energies of reaction calculated from the known reduction potentials in Table IV. The reactions consist of *ET quenching* of the *lowest luminescent excited state* of $Ru(bpy)_3^{2+}$ or $Ru(Mebpy)_3^{2+}$ where bpy = 2,2'-bipyridyl and Mebpy = 4,4'-dimethyl-2,2'-bipyridyl.

The nature of the ruthenium (II) *complex excitation — metal to ligand charge transfer* — contributes a significant internal reorganization energy to the ET reaction. From vibrational progressions in the low-temperature luminescence and absorption spectra of $Ru(bpy)_3^{2+}$, it appears that a high frequency mode, $\hbar\omega = 1300\,\mathrm{cm}^{-1}$, is excited in the luminescing state.

NOTE: This is the lowest single-mode harmonic oscillator state of the electronically excited state.

We have found the associated λ_{inner} to be $1300 \pm 100\,\mathrm{cm}^{-1}$($15.5 \pm 1\mathrm{kJ/mole}$) (NOTE: for formation of electronic ground state products), by fitting the following *line-shape function* to the emission spectrum

$$\text{Intensity } \propto e^{-x}\frac{x^n}{n!} \tag{11.14}$$

NOTE: the function has a maximum being the product of a decreasing exponential function and of an increasing function...

where n is the vibrational quantum number in the ground electronic state, and $x = \lambda/\hbar\omega$; $\hbar\omega$ is the frequency of the vibrational mode ($\hbar\omega = 1300\,\text{cm}^{-1}$ in the present case). Equation (11.14) gives the square overlap of the lowest single-mode harmonic oscillator state of the electronically *excited* state with the nth vibrational state of the *lowest* electronic state of the ruthenium (II) complex,[(5)] when both states have the same frequency but *the equilibrium position of the nth state is displaced* relative to that of the zeroth state. Because the vibrational quantum is so large relative to kT ($kT = 208.5\,\text{cm}^{-1} = 2.494\,\text{kJ/mol}$ at 300 K) transitions from vibrational states higher than the zeroth need not be considered in the *emission equation* (Eq. 11.14).

In the appendix, it is shown that when the emission and absorption line shapes are due to a high frequency vibration, ($\hbar\omega \gg kT$), the Stokes shift is approximately twice λ_{inner} for the transition from electronic ground state to electronic excited state. Using the average of the singlet–triplet absorption maxima at 77 K... and the average of the emission maxima at 298 K... one obtains

$$\lambda_{\text{inner}} = \frac{1}{2}(18300 - 16200)\,\text{cm}^{-1} = 1050\,\text{cm}^{-1} = 12.5\,\text{kJ/mole}$$

for the ruthenium charge-transfer transition.

(NOTE: 18300–16200 is the Stokes shift. . . .)

This estimate for λ_{inner} is in fair agreement with the value $\lambda_{\text{inner}} = 15.5\,\text{kJ/mole}$ obtained above by fitting Eq. (11.14) to emission spectra. $\lambda_{\text{inner}} = 15.5\,\text{kJ/mole}$, inferred for the metal-to-ligand charge transfer, will be assumed for the contribution of the $^*\text{Ru(bpy)}_3^{2+}$-Ru $(\text{bpy})_3^{3+}$ subsystem to the ET reactions in Table III. . . .

Using $\lambda_{\text{inner}} = 15.5\,\text{kJ/mole}$, $\lambda_{\text{out}} = 54\,\text{kJ/mole}$, and the ΔG^0's in Table III, we calculated rate constants for the reaction to form *ground state* products. In the *adiabatic limit* the classical rate constant is given by Eq. (11.15) (**M53**) when work terms are negligible

$$k_{et} = Z(4\pi kT\lambda)^{\frac{1}{2}}\,G \qquad (11.15)$$

where G is the *classical* Franck–Condon sum given by Eq. (11.11), with ΔE replaced by ΔG^0, and $\lambda = \sum_j \lambda_j$ is the sum over inner- and outer-sphere λ's. Z is the collision frequency in solution $\sim 10^{11}$ M^{-1} sec^{-1} (**M57**, **M53**, **M69**). For simplicity, the quantum rate constant was assumed to be given by the same expression (Eq. 11.15) but with the quantum Franck–Condon sum (Eq. 11.10) used for G. In this way, the quantum expression reduces to the classical in the limit $\hbar \to 0$. Strictly speaking Eqs. (11.10) and (11.11) for the G's (classical and quantum) were derived for nonadiabatic ETs.

NOTE: From Eqs. (11.15) and (11.11) we have: $k_{et} = Z\exp\left[-\frac{(\Delta E+\lambda)^2}{4kT\lambda}\right]$. Changing ΔE with ΔG^0 we see that this rate constant is the classical

$$k_{bi} = \kappa\rho Z_{bi}\exp(-w^r/kT)\exp[-\Delta F^*(R)/kT]$$

Of **M53** when $\kappa = \rho = 1$ and $w^r = 0$.

The classical and quantum rates and the observed rates are plotted in Fig. 3 (solid line for classical, dashed line for quantum).**[(6)]**

The plotted values are not the ET rate constants themselves, but rather the rate constants corrected for diffusion (**Erratum**: **M69** wrong citation) k_{obs},

$$k_{\text{obs}} = \left(\frac{1}{k_{et}} + \frac{1}{k_d}\right)^{-1} \tag{11.16}$$

Where k_d is the diffusion limit: $\sim 3.5 \times 10^9$ M^{-1}s^{-1}.

The difference between the quantum and the classical calculations in the very negative ΔG^0 region is again not negligible, because of the high frequency internal mode involved in the present reactions, $\hbar\omega = 1300\,\text{cm}^{-1}$, and the fact that its contribution to λ_{inner} is not negligible. Still, the classical and quantum calculations are in qualitative agreement and *neither* explains the observed rates in the inverted

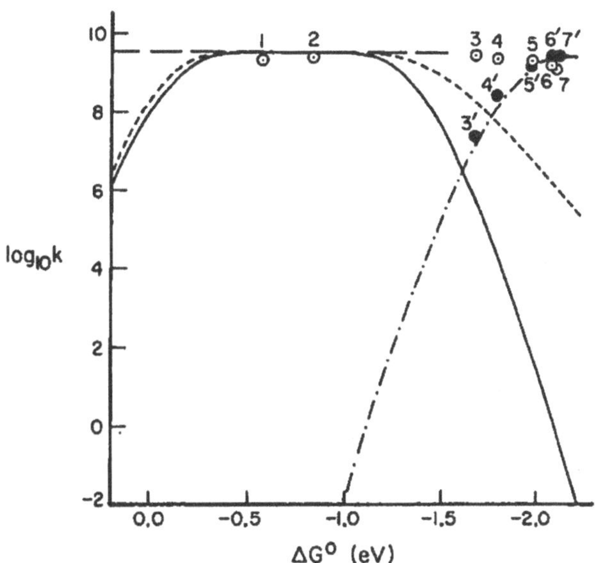

Fig. 3. k (calculated and experimental) for bipyridyl systems ($k = (k_{et}^{-1} + k_d^{-1})^{-1}$ with $k_d \simeq 3.5 \times 10^9 \mathrm{M}^{-1}\,\mathrm{s}^{-1}$; ΔG° is for formation of ground-state products); —, classical to ground-state products; - - -, quantum to ground-state products; - - -, calculated classical rate to *Ru(III) products; •, calculated quantum rate to *Ru(III) products; ⊙, experimental rate constant. The numbers correspond to the numbers in Table III. Primes indicate calculated rates to excited-state products $\lambda_{\text{out}} = 54$ kJ/mol; $\lambda_{\text{inner}} = 15.5$ kJ/mol; temp = 300 K; Ru(III) excitation energy = 1.76 eV).

region, as Fig. 3 demonstrates. The discrepancy would be even greater if a nonadiabaticity factor κ (**M57**) were introduced.

In order to assess the possibility of the ET *products* being formed in *excited electronic states* we have calculated quantum mechanically the rates of ET to *excited product* states. The calculation requires a λ_{inner} for formation of these products.... The effect on λ_{inner} of forming electronically excited ruthenium (III) in the reactions of Table III is not known, so λ_{inner} for reactions to form excited-state ruthenium (III) products is taken to be the same as the λ_{inner} for formation of electronic-ground-state products, $\lambda_{\text{inner}} = 15.5\,\text{kJ/mole}$.

The excitation energies in Table V were used, together with the reduction potentials of Table IV, to yield the ΔG^0's (Table III) for formation of electronically excited ruthenium (III) products. The three reactions involving quenching of excited ruthenium (II) by ruthenium (III) appear to proceed more favorably to an excited ruthenium (III) product than to the ground state. The quantum mechanically calculated rates to excited ruthenium (III) are indicated by solid circles in Fig. 3, and are in good agreement with experiment (open circles) for the three reactions involving ruthenium (III) quenchers. . . ."

The calculated versus experimental values of rates are also discussed for reactions of ET quenching of excited ruthenium (II) by the chromium (III) and osmium (III) complexes. The quenching by the osmium complex may also proceed by the mechanism of H-atom transfer followed by proton exchange with the solvent. A third possibility is described later in this section.

"To allow comparison, we have also calculated classically the rate constants for ET to form electronically excited ruthenium (III) products. The same λ's and $\Delta G^{0\prime}s$ were used as for the quantum calculations discussed earlier. The classical rates to excited products are shown in Fig. 3 by the 'dash-dot' line and agree well with the quantum values (solid circles). We note that *excited-state formation corresponds to the normal free energy region, while ground state product formation lies in the inverted region.*

NOTE: It is so because $\Delta G^{0*\prime}s > \Delta G^{0\prime}s$, compare Table III.

There is a third possible explanation for the large rate constants observed for reactions (11.12) and (11.13) in Fig. 3 (reaction of two electronically excited ruthenium (II) complexes with the osmium (III) complex). The distance between the centers of the reactants in the activated complex, r, may in this case of an electronically excited reactant, be greater than the distance of the closest approach. The distance of closest approach equals $a_1 + a_2$ where a_1 and a_2

are the radii of the two reactants. The value of the outer-sphere reorganization energy used in the rate constant calculations above ($\lambda\frac{1}{4}_{\text{out}} = 13.4\,\text{kJ/mole}$) was calculated by using the classical expression (**M30**) for λ_{out} (Eq. 11.17) and assuming $r = a_1 + a_2$.

$$\lambda_{\text{out}} = (\Delta e)^2 \left(\frac{1}{\varepsilon_{op}} - \frac{1}{\varepsilon_s}\right)\left(\frac{1}{2a_1} + \frac{1}{2a_2} - \frac{1}{r}\right) \tag{11.17}$$

In Eq. (11.17), Δe is the *change in charge of the reactant*, ε_{op} is the optical dielectric constant, and ε_s is the static dielectric constant of the solvent. If r were greater than $a_1 + a_2$, then the outer-sphere reorganization energy would be calculated to be greater than 13.4 kJ/mole, as may be seen from Eq. (11.17): the reactions in which Os (byp)_3^{3+} quenches electronically excited ruthenium (II) complexes to form a *ground* electronic-state ruthenium (III), (reactions 3 and 4 of Table III), have large negative free energies and they lie in the 'inverted region.' In this case, increasing r and hence increasing λ_{out} has, as is seen from Eq.(11.11), the effect of increasing the calculated ET rate.[7] At least, it has this effect of rate enhancement if the reactions do not become too nonadiabatic at the larger r.

Indeed, if $r = 1.3\,(a_1 + a_2)$ and $a_1 \cong a_2$, then the quantum-mechanically calculated rate constants (corrected for diffusion according to Eq. (11.16)) for the ET reaction between $^*\text{Ru(bpy)}_3^{2+}$ and $^*\text{Ru(Mebpy)}_3^{2+}$ and Os(bpy)_3^{3+} (reactions 3 and 4 of Table III), are $k_3 = 8 \times 10^8\,\text{M}^{-1}\text{s}^{-1}$ and $k_4 = 3 \times 10^8\,\text{M}^{-1}\text{s}^{-1}$, respectively. These values are within 1 order of magnitude of the experimental values obtained by Creutz and Sutin, $k_3 \cong 3.2 \times 10^9\text{M}^{-1}\text{sec}^{-1}$, and $k_4 \cong 2.6 \times 10^9\,\text{M}^{-1}\text{sec}^{-1}$. If $r = 2(a_1 + a_2)$ and $a_1 \cong a_2$, the quantum mechanically calculated values of the rate constants are $k_3 = 2 \times 10^9\,\text{M}^{-1}\text{sec}^{-1}$ and $k_4 = 1 \times 10^9\,\text{M}^{-1}\text{sec}^{-1}$, essentially in agreement with the experimental values. . . .

However, ET at too large an r makes the reaction increasingly nonadiabatic and then reduces the reaction rate.*The appropriate r is the one which achieves a maximum rate.*

Work terms were neglected. . . .

At least in the Creutz and Sutin systems, it appears that the lack of significant inverted behavior is indicative either (a) of the third possibility above or (b) of alternate reaction pathways becoming competitive at large negative ΔG^0's rather than (c) of nuclear tunneling. Nuclear tunneling due to the very high frequency modes involved in transitions from the electronically excited reactants *is* a significant effect at very large negative ΔG^0's, but *doesn't explain the lack of inverted behavior*, as one sees from the dashed line in Fig. 3.[(8)]

Conclusions. Rate calculations for a hypothetical system and for the bipyridyl systems studied by Creutz and Sutin suggest that the quantum effects are expected to be small in the normal region (i.e., for small to moderate ΔG^0's) even for systems having fairly large internal frequencies. At large negative ΔG^0's quantum effects may frequently be significant. For most of the reactions considered in the 'inverted' region, the calculated and experimental results agree within 1 order of magnitude, *provided that electronically excited products are formed.* An alternate AT pathway may occur in reactions where the calculated rate constant for an ET is appreciably less than the experimental one in this 'inverted' region. A third possibility of ET at a larger distance is also considered.

11.3. Appendix. Relation between λ and the Stokes Shift

We consider the case where excitation of a single harmonic vibrational mode is responsible for the emission and absorption line shapes. We define $X \equiv \lambda/h\nu$, where $^1/_4\lambda$ is the *inner-sphere* reorganization energy for transition from the electronic ground state to the luminescing state, and ν is the frequency of the mode.[(9)] (ν is assumed to be the same in both electronic states.)

NOTE: $\lambda = \lambda_i + \lambda_o$, M. supposes that $\lambda_i = \frac{1}{4}\lambda$, that is, equal to the ΔG^* of an exchange reaction.

We assume for brevity $h\nu \gg kT$, and then the luminescence will occur from essentially only the lowest vibrational state in the electronically excited state. Equation (11.14) gives the emission line shape as

$$I_e(l) \propto e^{-X} X^l / l! \tag{A1}$$

where l is the quantum number of the vibrational level in the ground electronic state to which luminescence occurs. The energy of the corresponding quantum emitted is $E_e = E_{0\leftarrow 0} - lh\nu$, where $E_{0\leftarrow 0}$ is the electronic-excitation-energy of the luminescing state relative to the ground state. The energy E_e of this quantum at the *emission maximum* is

$$E_e = E_{0\leftarrow 0} - l^* h\nu \tag{A2}$$

where l^* is the value of l which maximizes Eq. (A1), $l^* = X$.

Similarly, since $h\nu \gg kT$, absorption occurs essentially only from the lowest vibrational level in the electronic ground state, so the absorption intensity is

$$I_a(m) \propto e^{-X} X^m / m! \tag{A3}$$

where m is the vibrational quantum number of an electronically excited vibronic level to which absorption occurs. $I_a(m)$ is maximized with respect to m, and the energy E_a of the *absorption maximum* is

$$E_a = E_{0\leftarrow 0} + m^* h\nu \tag{A4}$$

where m^* is found by maximization of Eq. (A3) to equal X.

The Stokes shift is $E_s = E_a - E_e$. From Eqs. (A2) and (A4) we have

$$E_s = \left(m^* + l^*\right) h\nu = 2Xh\nu \tag{A5}$$

but $X = \lambda / h\nu$, so

$$E_s = 2\lambda \tag{A6}$$

Equation (A6) is a well-known approximate formula."

NOTE: $X = (m^* + l^*)/2$, appears as the average vibrational excitation of the ground and excited electronic states.

NOTES

1. p. 748 2nd column top: "in the nonadiabatic limit."

   M: Not in the adiabatic limit because the Golden Rule is a perturbation expression.

2. p. 749 1st column middle: Following: "Figure 1... The ordinate is a plot of $\log(Ge^{\Delta G^0/kT})$ versus ΔG^0... both the classical and quantum values of the ordinate are symmetric in ΔG^0, when plotted in this manner."

   p. 749 1st column bottom: "In Figure 2 the ordinate is a log plot of the Franck–Condon sum, G, versus ΔG^0, and so Fig. 2 unlike Fig. 1 is not symmetrical about $\Delta G^0 = 0$."

   Q: Please explain the symmetry and asymmetry in the two different plots. Do you get the symmetry because on multiplying G by $e^{\Delta G^0/kT}$ you cancel the $\mathrm{e}^{-E_n^{\mathrm{vib}}/kT}$ in $G = \frac{1}{Q}\sum_n \sum_m \mathrm{e}^{-E_n^{\mathrm{vib}}/kT} |\langle \chi_n \mid \chi_m \rangle|^2 \, \delta(E_n - E_m)$?

   M: Supposing you have a curve that is symmetrical about $\Delta G^0 = -\lambda$, you can get it symmetrical about a new ΔG^0, have it symmetrical about, say, $\Delta G^0 = 0$, that is different by a constant, namely by λ, from the former origin.

3. p. 749 1st column middle: "ΔG^0 is the same as ΔE in Eqs. (11.10) and (11.11), since Eq. (11.10) tacitly assumes zero for ΔS^0 when $F_i = F_i'$."

   M: Levich and Dogonadze in their work didn't have any entropy changes. If you have a bath of harmonic oscillators and in the same bath you have reactants and products, then the entropy associated with the harmonic oscillators is not going to be large, typically it's going to be somewhat similar for reactants and products, the big entropy changes are there when you have a lot more sort of reorientation or ordering in one than in the other, and that happens when you have *hindered rotations*. In other words, if you want a large entropic effect you have to bring in rotation and then hinder it. If you just have frequencies and you modify them a little bit, you are not going to change entropy very much. If you look at the partition function of the system made up by the collection of the bath's harmonic oscillators, from that partition function you get free energy and entropy. See, for instance, Moelwyn-Hughes' Physical Chemistry book, Chapter VIII on partition functions. The point is that the molecules are hindered in rotation, there are not just vibrations, and so in a model such as what Levich, Dogonadze, and other people used for other purposes before them there are purely harmonic oscillators, it's a highly idealized model, it's better than nothing, in fact can be quite useful, but the system is not a collection of harmonic oscillators. The harmonic oscillator model has been a very useful model in physics for a long time, I mean Levich and Dogonadze gave a good example of an application, but there are applications before that and after that, so it's a very useful model, in fact now it's called, I think, the spin-boson model. Anthony Leggett and somebody else wrote a two parts review article, and that's a widely cited harmonic oscillator model for the solvent or solid, ... it's widely used but as model for solutions, it has its uses and has strong limitations, particularly it can't capture systems that involve hindered rotations and so it can't capture reactions which have large entropies of reaction, if you have just harmonic oscillators before and after, then your entropy of reaction is close to zero. In solution the fields can become strong, and when you go from reactants to products and you have big changes of charges, you have enormous effects, so all of that would come into free energy of systems with hindered rotations going over to vibrations, depending on the charge and so on...

4. p. 749 1st column middle: "Fig. 2 is a plot similar to Fig. 1."

   M: The quantum-mechanical correction makes Fig. 2 not symmetric. When the quantum effect is strong, instead of going down so sharply the curve in the inverted region will go down more slowly.

There is a smaller effect on the normal side because the effective barrier is thicker, you have two parabolas with tunneling from one to the other, the quantum-mechanical correction introduces a kind of asymmetry.

When you have high frequencies the quantum-mechanical effects are very strong. You can see the asymmetry, for example, in the famous plot of Closs, Miller, and Calcaterra, where you had a few high frequencies there.

5. p. 750, 1st column bottom: "Eq. (11.14), intensity $\propto e^{-x}\frac{x^n}{n!}$, gives the square overlap of the lowest single-mode harmonic oscillator state of the electronically *excited* state with the nth vibrational state of the *lowest* electronic state of the ruthenium (II) complex..."

   Q: Is this intensity, this emission equation, the same thing as the FC factor appearing in $G = \frac{1}{Q}\sum_n \sum_m \mathrm{e}^{-E_n^{\mathrm{vib}}/kT} |\langle \chi_n | \chi_m \rangle|^2 \delta(E_n - E_m)$?

   M: Yes, providing you are starting only from the lower state. Only from the lower state, otherwise it becomes some Bessel function.

6. p. 751 1st column middle: "The classical and quantum rates and the observed rates are plotted in Fig. 3."

   Q: In Fig. 3, there appears a flattening on top of the diagram, that is, there is, corresponding to a certain ΔG^0 interval, an independence of the rate constant on ΔG^0, which is the mark of the reaction not being activation controlled, that is, of being diffusion controlled.

   M: In fact, where the diagram levels off is roughly the diffusion constant that controls the rate. The diffusion constant is of the order of 10^{10} depending on whether there are coulombic forces.

7. p. 751 2nd column bottom: "... the reactions in which $Os(bpy)_3^{+3}$ quenches electronically excited ruthenium(II) complexes to form a *ground* electronic-state(III) (reactions 3 and 4 of Table III) have large negative free energies, and they lie in the 'inverted region.' In this case increasing r and hence increasing λ_{out} has, as is seen

from Eq. (11.11): $G = (4\pi kT\lambda)^{-\frac{1}{2}} \exp[-(\Delta E + \lambda)^2/4kT\lambda]$, the effect of *increasing* the calculated ET rate."

M. $\lambda = k\frac{a^2}{2}$, where a is a displacement of the minimum. So, if at constant ΔG^0 I make a larger, the P curve intersects the R curve at lower free energies and the free energy barrier to reaction is reduced. This happens in the inverted region, contrary to what happens in the normal region because *in the inverted region there is the difference in sign between* ΔG^0 *and* λ, ΔG^0 is negative in the inverted region while λ is positive.Note that λ is directly related to the free energy barrier for the reaction only in the case of self-exchange reaction.

8. p. 752 1st column middle: "Nuclear tunneling due to the very high frequency modes involved in transitions from the electronically excited reactants is a significant effect at very large negative ΔG^0's, but *doesn't explain the lack of inverted behavior*, as one sees from the dashed line in Fig. 3."

Q: Even considering nuclear tunneling there is an inverted effect with higher rates due to the nuclear tunneling. Are the high frequency modes more important than the low frequency modes in favoring nuclear tunneling just because corresponding to them there is a smaller distance between the R and P PECs?

M: The main point is that the tunnel effect is associated with high frequencies, always high frequency modes favor nuclear tunneling. Quantum effects are always larger with high frequency modes, always you compare $h\nu$ with kT, and when $h\nu$ is small compared with kT quantum effects are hardly observable. When $h\nu/kT$ is larger you just have a few quanta... that's true in general, not just for nuclear tunneling. When you compare $h\nu$ with kT, it's also true that if you have higher $h\nu$ the wave function would penetrate more into the forbidden region. The fact that the distances between the R and P curves are smaller is a consequence of the fact that when you put your energy levels higher, then of course the distance from one PES to the other is smaller, but the big point is that making the energy levels higher, brings out more quantum effects.

9. p. 752, 2nd column top "We define $X \equiv \lambda/h\nu$, where $\frac{1}{4}\lambda$ is the *inner-sphere* reorganization energy for the transition from the electronic ground state to the luminescing state, and ν is the frequency of the mode."

   Q: (i) We know that $\lambda = \lambda_i + \lambda_o$. Why do you choose $\lambda_i = \frac{1}{4}\lambda$, which is the total λ when $\Delta G^0 = 0$?

   (ii) Once I have found the vibrational energy associated with the charge-transfer transition, why should this excitation be the same as that for the ET reaction?

   M (i): When the two parabolas have the same bottom, the same horizontal bottom, namely $\Delta G^0 = 0$, when it is just a matter of quadratics, then the vertical line from the bottom of the R curve to the P curve is four times as high as the vertical line from the common bottom of the R and P curves to their intersection point.

   Consider two parabolas in which the free energy G is plotted against the generalized coordinate q. Let the equations for the two parabolas be

$$G_r(q) = \frac{k}{2}(q - q^r)^2 \quad \text{and} \quad G_p(q) = \frac{k}{2}(q - q^p)^2$$

   for reactants and products respectively. In the most simple case the parabolas have, as here, the same constant k, the vertices (the bottoms) are at $q = q^r$ and at $q = q^p$ respectively, and $\Delta G^0 = 0$, that is, the vertices have the same value of the ordinate G. The intersection point of the parabolas, at $q = q^\ddagger$ and $G = G^\ddagger$, has the abscissa $q^\ddagger = \frac{1}{2}(q^r + q^p)$. Let's define the magnitude $\lambda \equiv G_p(q^r)$. We see then that

$$\lambda = \frac{k}{2}(q^r - q^p)^2$$

   and $\Delta G^\ddagger = G_r(q^\ddagger) = \frac{\lambda}{4}$

In the general case in which $\Delta G^0 \neq 0$, the above equations are generalized by:

$$q^{\ddagger} = \frac{1}{2}(q^r + q^p) + \frac{\Delta G^0}{k(q^p - q^r)} \quad \text{and}$$

$$\Delta G^{\ddagger} = G_r(q^{\ddagger}) = \frac{\lambda}{4}\left(1 + \frac{\Delta G^0}{\lambda}\right)^2$$

There isn't really a λ_i and a λ_o in the case where the two curves don't have the same curvature, in the 1965 paper I introduced symmetrical and antisymmetrical functions of the force constants and then neglected terms involving the antisymmetric ones, in practice there is not a λ for reactants and a λ_o, there is a more complicated equation but not separated λ's, the λ only comes in for the symmetrizing.

I shouldn't have used 1/4 there. I should have used just the word λ. That was a mistake.

(ii) A CT transition is an ET reaction. You have several contributions, one of which is going from the bottom of the *R* PES to the bottom of the *P* PES. You have then in addition some λ contribution. Going from bottom to bottom you have the ΔG^0, it is part of the vertical energy change, the radiative part, that's the vertical transition, you know, that comes off as light. If you draw two parabolas one of which is low down and the other one representing the excited electronic state, so you have a CT transition, and the vertical distance you can express in terms of λ and then ΔG^0. It is $\lambda + \Delta G^0$. The ΔG^0 would be entirely associated with excitation, then of course there is λ, the reorganization.

M170. On the Frequency Factor in ET Reactions and Its Role in the Highly Exothermic Regime

"Abstract

Consideration of current information on the dependence of the ET rate on the *radial separation distance* and on the reactants' *radial distribution function* suggests for adiabatic transfers a frequency factor closer to $10^{12}\,M^{-1}\,sec^{-1}$ than to $10^{11}\,M^{-1}\,sec^{-1}$. One effect is to raise the λ values estimated from self-exchange rate constants, and to extend thereby the range of ΔG^0's in which the " inverted region" is *masked* by a diffusion-controlled reaction rate.

11.4. Introduction

The bimolecular rate constant of an ET reaction in solution may be written as

$$k_{\text{act}} = \kappa C e^{-\Delta G^*/kT} \tag{11.18}$$

where C is a constant, the constant C is the ρZ in **M48** and **M53**, which is frequently taken to be $\sim 10^{11}\,\text{Mole}^{-1}\,\text{sec}^{-1}$, the collision frequency at unit molar concentration in the gas phase. κ is a factor which is unity for adiabatic ETs and less than unity for nonadiabatic ones. ΔG^*describes the free energy barrier to reaction. It contains work terms for bringing the reactants together and contains vibrational and solvent orientational terms, arising from *a reorganization which permits the ET to occur.*

In the present paper, we examine the value of C more closely, taking into account some estimates of the radial distribution function and of the dependence of the ET rate on the intervening *separation distance r*. We shall be more concerned with obtaining an order of magnitude estimate of C rather than with attempting to evaluate precisely each of the factors that contribute to the bimolecular rate constant.

11.5. Theory

A detailed calculation of the ET rate involves, among other things, a knowledge of the motion along the *reaction coordinate* regardless of whether the reaction is *electronically* adiabatic or nonadiabatic. *Contribution to this motion* can come from

(i) Intramolecular vibrations in the reactants, including any in their *coordinations shells*

NOTE: Henry Taube writes in the Foreword to the book "Concepts of Inorganic Photochemistry" by Adamson and Fleischauer: "The term *coordination chemistry* has come to include much of inorganic chemistry so that now, in general usage, it can be said to embrace the chemistry of inorganic particles in compact environments. It is not restricted to orthodox coordination complexes but is concerned also with systems in which the atoms or ions are in direct contact with solvent molecules — the interaction can range from the very selective to the very labile and nonspecific — as well as to the particles in either nonmolecular or molecular solids."

(ii) Electrically polarized vibrations of the surrounding solvent molecules outside the coordination shells
(iii) Orientational motions of those solvent molecules, and
(iv) The translational motion of the approach of the two reactants (cf. **M48**).

When motion (iv) does not contribute significantly to the reaction coordinate,[1] one can calculate the bimolecular constant in *two steps*: one first calculates the reaction rate constant at *fixed* separation distance r, a first-order rate constant $k(r)$, and then suitable averages over all r.

NOTE: Recall the radial distribution of the electron around the proton in the hydrogen atom: $P(r)4\pi r^2 dr$ where $P = \psi^2 \ldots$

---

[1]That is, if q denotes the reaction coordinate, q then has *contributions* mainly from the *coordinates* in (i)–(iii). Only then is it *meaningful,* one can show, to speak of a first-order rate constant $k(r)$ for each separation distance r. In hard spheres collision theory only (iv) contributes to the reaction coordinate. The averaging is particularly straightforward *when the rate of diffusion is fast*. Then the probability that the pair of reactants has a separation distance between $(r, r + dr)$ at any time equals essentially the equilibrium (no reaction) probability $g(r)4\pi r^2 dr$, $g(r)$ is the equilibrium radial distribution function for the pair.

One can then write the bimolecular reaction rate constant k_{act} as

$$k_{\text{act}} = \int_0^\infty g(r) 4\pi r^2 k(r) dr \tag{11.19}$$

We have used the subscript "act" to denote "activation-controlled rate constant." Later we return to the case where, instead, diffusion is a slow step.

NOTE: k_{act} is now calculated over all distances, using $g(r)$ and under the hypothesis of a very fast diffusion, so that the $k_{\text{diff}} \gg k_{\text{act}}$ doesn't take part in the k_{obs}.

It is convenient to write $k(r)$ in terms of an electronic transition probability factor $\kappa(r)$ (e.g., a velocity averaged Landau–Zener factor), a free energy barrier $\Delta F^*(R)$ in **M53** and *a mean frequency factor* ν *associated with motions* (i)–(iii).

$\Delta G'(r)$ depends on r mainly via contributions (ii) and (iii) of the vibrational and orientational polarization of the solvent molecules (**M48**, **M53**). Contribution (i) to $\Delta G'(r)$ is independent of r (**M48**, **M53**) when the dependence of intramolecular vibrations of each reactant on r is negligible. We now have

$$k(r) = \kappa(r) \nu e^{-\Delta G'(r)/kT} \tag{11.20}$$

We proceed to estimate C from Eqs. (11.19) and (11.20) by considering individually $g(r)$, $\kappa(r)$, ν, and $\Delta G'(r)$. $g(r)$ is a damped oscillating function of r. For purposes of making an estimate of the terms in Eq. (11.19) we write $g(r)$ in a suggestive form

$$g(r) = g_0(r) e^{-w(r)/kT} \tag{11.21}$$

Where $g_0(r)$ is largely a geometric factor, being similar to the value it would have for an uncharged and nonspecifically interacting pair. *All of the specific-interaction contributions to* $g(r)$ *are contained in what might be called the "work term"* $w(r)$.$g_0(r)$ in a liquid rises sharply from a value of about zero in the short range *repulsive* potential region to a maximum, characteristic of the behavior of nearest neighbors, then falls to a value somewhat below unity, and then, after more oscillation, reaches a limiting value of unity at larger r.[(1)]

We write the sum of $w(r)$ and $\Delta G'(r)$ as $\Delta G^*(r)$:

$$\Delta G^*(r) = w(r) + \Delta G'(r) \tag{11.22}$$

The factor ν in Eq. (11.20) is in units of s^{-1} and is a *mean of the frequencies* of each of the motions (i)–(iii), each weighted by its contribution (one can show) to $\Delta G'$. For example, if a coordinate does not contribute to $\Delta G'$, it does not contribute to ν. Typical frequencies for (i), (ii), and (iii) are $\sim$400 cm^{-1} (for metal ligand bonds), $\sim$200 cm^{-1} (for the solvent water), and (iii) $\sim$1 cm^{-1} (again for water) (**M163**). The contributions of (ii) and (iii) to ΔG^* are estimated to be roughly equal (**M163**), so that when weighted (iii) contributes negligibly to ν in the case of water as the solvent. The relative contributions of (i) and (ii) will vary from system to system, and we shall for concreteness select a mean ν of 300 cm^{-1}, which is about 10^{13} sec^{-1}.

Several rather rough estimates have been made for the dependence of κ on r, which in each case has an exponential dependence on r. We can then write

$$\kappa(r) = \kappa_0 e^{-\alpha(r-\sigma)} \tag{11.23}$$

where κ_0 is the value of $\kappa(r)$ at $r = \sigma$, and σ *is the value of r where* $g_0(r)$ *has its maximum.*

NOTE: The most probable distance at which the reaction happens is now defined as the distance at which $g_0(r)$ has its maximum.

If κ_0 is about unity, the reaction is usually termed adiabatic. When there are no intervening molecules between the two reactants, α has been estimated to be 2.6 $Å^{-1}$. When there are intervening molecules, α has been estimated to be about 1.4 $Å^{-1}$ (elsewhere, a more precise formula than (11.23) will be used, one which joins a nonadiabatic equation for $k(r)$ with an adiabatic equation at the r where both are equal).

The function $g_0(r)$ falls from its maximum (at $r = \sigma$) of between 2 and 3 to a value of about unity in a distance of about one half a molecular diameter or perhaps somewhat less.[2]

If that diameter is about 6 Å, then $\exp(\alpha - r)$ is a more rapidly varying function of r in the vicinity of $r = \sigma$ (it falls to $1/e$ of its value in 0.4 Å in the case of no intervening material). Taking $\Delta G^*(r)$ to be relatively slowly varying also, the integral in Eq. (11.19) can be evaluated approximately by taking $g_0(r) = 0$ for $r < \sigma$, replacing r^2, $g(r)$, and $\Delta G^*(r)$ by their values at $r = \sigma$, and integrating from $r = \sigma$ to ∞.

NOTE: $\kappa(r)$ only remains to be integrated ...

One obtains:

$$k_{\text{act}} \cong g_0(\sigma) e^{-\Delta G^*(\sigma)/kT} 4\pi\sigma^2 \nu\kappa_0/\alpha \tag{11.24}$$

Using the previous estimate of $\nu \sim 10^{13}\text{sec}^{-1}$, $g_0(\sigma)$ of between 2 and 3, and $\alpha \sim 2.6\,\text{Å}^{-1}$, one obtains for a pair of reactants, each of radius 3 Å

$$k_{\text{act}} \sim 10^{12} k_0 e^{-\Delta G^*(\sigma)/kT} \text{M}^{-1}\text{sec}^{-1} \quad (\sigma = 6\,\text{Å}) \tag{11.25}$$

NOTE: This is finally Eq. (11.18) with the new Z.

For larger reactants the numerical factor in Eq. (11.25) is of course larger than 10^{12}, since it is proportional to σ^2. For the case of an electronically adiabatic reaction $\kappa_0 \sim 1$.

When there is intervening material, the most probable r is somewhat greater than σ, the mean $g_0(r)$ is less than $g_0(\sigma)$, $\alpha \sim 1.4\,\text{Å}^{-1}$, and so the constant multiplying $\exp(-\Delta G^*/kT)$ in Eq. (11.25) is still of the same order of magnitude. Thus a value of $10^{12}\text{M}^{-1}\text{sec}^{-1}$ appears to be a reasonable choice....

It is useful to compare this value of $10^{12}\text{M}^{-1}\text{sec}^{-1}$ for C with the gas phase hard sphere collision frequency $Z = (8\pi kT/\mu)^{1/2}\sigma^2$, for molecules of reduced mass $\mu \sim 50$ and radius 3 Å. The latter equals

---

[2]Strictly speaking, we should use instead a $g_0(r)$ for a pair of solute molecules. When the solute molecules are much larger than the solvent's $g_0(\sigma) \cong 1$.

$2 \times 10^{11}\text{M}^{-1}\text{sec}^{-1}$ at 300 K. Physically, the reason why the value of C is higher than Z (some five times higher in the present estimate) is twofold. Collisions of hard spheres occur only on contact, while ET can occur at various r which are larger than the "contact distance" σ.

NOTE: see for instance: F. F. Di Giacomo, A Semiempirical Potential Energy Surface for Chemiluminescent Electron Transfer Reactions, *Z. Physik. Chem. N. F.* **122**, 1–13, (1980).

Second, the $g_0(r)$ is larger than unity,[3] the "dilute-gas" hard spheres value, although it can be close to unity if the solute molecules are appreciably larger than those of the solvent.

Thus far we have assumed diffusion to be rapid. If we use, in the first approximation, the estimate in Eq. (11.24) for k_{act}, the observed rate constant is given in the steady state by Eq. (11.26) for the case of no back reaction is really (**M30, M118**)

$$\frac{1}{k_{\text{obs}}} = \frac{1}{k_{\text{act}}} + \frac{1}{k_{\text{diff}}} \tag{11.26}$$

where k_{diff} is the diffusion-controlled rate constant. Its value depends on the charges of the reactants, being of the order of $10^{10}\text{M}^{-1}\text{sec}^{-1}$ in water for uncharged species. When $k_{\text{diff}} \gg k_{\text{act}}$, k_{obs} equals k_{act}, and equals k_{diff} when $k_{\text{act}} \gg k_{\text{diff}}$.

In the so-called normal region, where the bimolecular rate constant increases with increasingly negative standard free energy of reaction ΔG^0 (**M30, M53**), *the reorganizational term* $\lambda(r)$ *in* $\Delta G^*(r)$ *favors small r.*

NOTE: See $1/R$ in the λ_o in $\lambda_o = (ne)^2\left(\frac{1}{2a_1} + \frac{1}{2a_2} - \frac{1}{R}\right)\left(\frac{1}{D_{op}} - \frac{1}{D_s}\right)$.

(The work term $w(r)$ in $\Delta\text{G}^*(r)$ favors small or large r, depending upon its sign.) The closer the reactants are together, the less the distant solvent molecules experience a change of potential energy, via a change in the local electric field, when the ETs, and so the less

[3]See Note 2.

the need for solvent reorganization to occur, and so the smaller is $\Delta G^*(r)$.

In the so-called inverted region (**M30**), which should occur at very negative ΔG^0, the bimolecular reaction rate constant decreases, instead, with increasingly negative ΔG^0due to the difficulty in this region of intersection of the PEC of the reactants with that of the products.[(2)] A larger r produces a larger solvent reorganizational term parameter λ_o (in the notation of **M118**), causes thereby an increased lateral displacement of the two surfaces, and permits them to intersect more easily (NOTE: i.e., at lower energies), in this highly exothermic region. Thereby*in the "inverted" region an increased r actually causes a decrease in* ΔG^*. The "inversion" will be less evident when this decrease in ΔG^* is large enough.

NOTE: λ_o is intrinsically positive. When $\Delta G^0$is also positive, both of them contribute in the same direction to ΔG^* and a smaller λ_o favors a higher reaction rate. A smaller λ_o is favored, in its turn, by a smaller r in it. When ΔG^0 is negative and $|\Delta G^0| > \lambda_o$, that is, in the inverted region, the more negative the ΔG^0 the higher the ΔG^* barrier. In this case, a higher λ_o *lowers* the activation barrier and we have a higher λ_o when r is larger.

Some further words about the inversion region are in order. In a quantum-mechanical treatment of the *reorganizational motion*, e.g., via the evaluation of Franck–Condon factors, only the *intra*molecular (**typo** corrected in the paper) vibrations[(3)] need to be treated quantum mechanically, since the frequency of the other coordinates is relatively low, compare **M163**. Moreover, using typical metal–ligand frequencies the quantum correction to the classically calculated reaction rate was found to be typically small in the normal region (from 20% to a factor of 4 in a recent calculation (**M163**). The nuclear tunneling was more noticeable in the inverted region (**M163**), because the tunneling tended to *decrease* the tendency of the rate to decrease with increasing $-\Delta G^0$ at very negative ΔG^0. This nuclear tunneling does not, however, cause the effect to disappear, see **M164**. In the field of radiationless transitions this effect of decreased rate with increased exothermicity is again very well-known and is termed there the "energy gap" law.

Some experimental evidence for the "inverted effect" has been offered. Indeed, the preferential formation in other highly exothermic reactions of electronically excited states of products, instead of the more exothermically formed ground state products, would be a further manifestation of this effect.

The calculated inverted effect is *less* when one takes into account

(i) The higher estimate of the frequency factor (10^{12} vs. 10^{11} $M^{-1}sec^{-1}$)[(4)] and
(ii) The role discussed earlier of the dependence on r of the solvent reorganizational contribution to the λ in $\Delta G^*(r)$.

The inverted effect is masked *to some extent* by the presence of k_{diff} in Eq. (11.26).[(5)]

For many reactions which have been studied no evidence of the inversion has been obtained. In some cases alternate pathways may exist, hydrogen AT, for example, or formation of electronically excited products. In others, where the rate has been inferred only from the quenching of fluorescence of one of the reactants, quenching may occur via an exciplex formation, rather than via ET, when exciplex formation is possible. In others the calculated inverted effect may be less due to the *two factors* cited above. In still others, anharmonic effects can contribute to decreasing (or increasing) the rate and high frequency modes which play less of a role in the normal region may also contribute importantly to accepting the excess energy in the "inverted region." The latter would reduce the tendency for k_{obs} to *decrease* with increasingly negative ΔG^0 in the very negative ΔG^0 region.[(6)]

There has been some tendency to use an equation which was derived originally for *ATs* (**M69**). This equation displays no inversion. However, except in the case of ATs, the application of such an equation must be regarded as purely empirical at present....

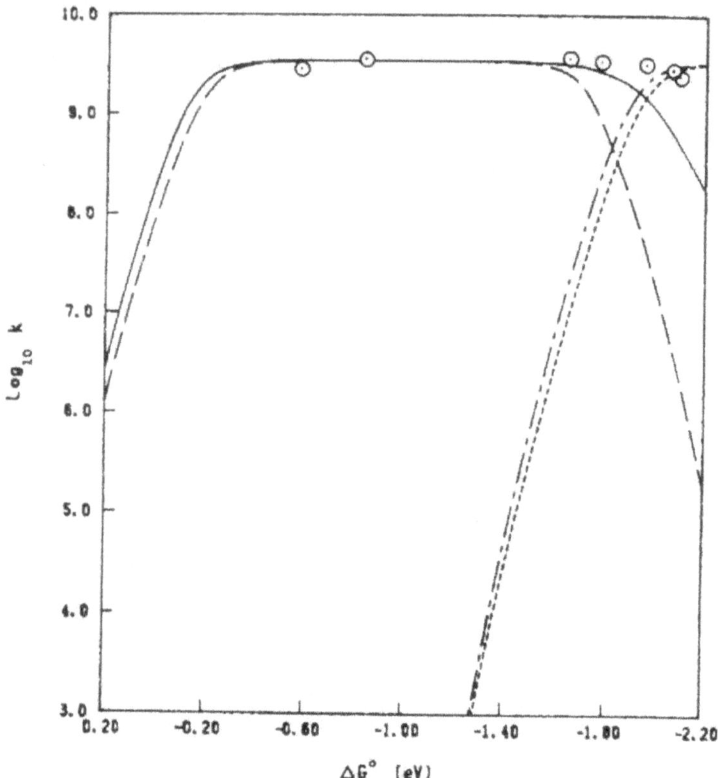

Fig. 1. Experimental (open circles) and calculated rate constants for reaction (9) versus ΔG°. Dashed line—classical result; solid line—quantum mechanical result; solid circles—quantum mechanical calculation to form an electronically excited product; dashed-dotted line—classical result to form an electronically excited product. Numerical constants used are given in [13], together with an identification of the individual reactants in the plot. The units of k are $M^{-1}\,s^{-1}$.

NOTE: reaction (9) in the legend of the original paper is here reported as Eq. (11.26) and Ref [13] is **M164**.

As an example of the effect of using a value for C in Eq. (1.18) of $10^{12}M^{-1}sec^{-1}$ instead of $10^{11}M^{-1}sec^{-1}$, we consider the series of highly exothermic reactions

$$^{*}RuL_3^{2+} + ML_3'^{3+} \rightarrow RuL_3^{3+} + ML_3'^{2+} \qquad (11.27)$$

The asterisk denotes an electrically excited reactant, L and L′ are various permutations of the ligands bpy and mebpy, and M denotes Cr, Os, and Ru.

We use a C of 10^{12} instead of the 10^{11} used in **M164**, and use a λ_o which is greater than that used in **M164** (there denoted by λ_{out})

$$e^{-\Delta\lambda_o/kT} = \frac{1}{10} \tag{11.28}$$

so as to give the same self-exchange rate constants. Fig. 1 is obtained instead of Fig. 3 in **M164**. One sees, on comparison with the Fig. 3 in **M164** that the results are now closer to the observed values in the highly exothermic region.

The effects described herein may also be present in electrode reactions. . . ."

NOTES

1. p. 866 bottom: "$g(r)$ is a damped oscillating function of r. For purposes of making an estimate of the terms in Eq. (11.11) we write $g(r)$ in the form

$$g(r) = g_0(r)e^{-w(r)/kT} \tag{11.29}$$

where $g_0(r)$ is a largely geometric factor, being similar to the value it would have for an uncharged and nonspecifically interacting pair. . . $g_0(r)$ in a liquid rises sharply from a value of about zero in the short range *repulsive* potential region to a maximum, characteristic of the behavior of nearest neighbors, then falls to a value somewhat below unity, and then, after more oscillations, reaches a limiting value of unity at larger r."

Q: These functions are similar to the classical Arrhenius formula. Does the temperature there favor the reaction because it favors more energetic collisions?

M: It depends on what one means by "more energetic collisions,". . . because the energy can be in translational energy, can be in internal energy... so when you say more energetic collisions it can be even slower collisions with higher energy particles. When one uses the distribution that is appropriate to the Boltzmann distribution to reach the TS, and it is any motion associated with the reaction coordinate, you know that the temperature is whatever it is in an equilibrium distribution.

2. p. 869 middle: "In the so-called inverted region, which should occur at very negative $\Delta G°$, the bimolecular reaction rate constant decreases, instead, with increasing negative $\Delta G°$ due to the difficulty in this region of intersection of the PECs of the reactants with that of the products."

   M: The difficulty refers to the difficulty of having big values of $\Delta G°$. One should read "difficulty in energetically reaching this region of intersection."

3. p. 869 bottom: "In a quantum-mechanical treatment of the reorganizational motion... only the intermolecular vibrations need to be treated quantum mechanically"

   Q: **Typo**? Probably you meant "intramolecular."

   M: It is intramolecular.

4. p. 870 top-middle: "The calculated inverted effect is less when one takes into account the higher estimate of the frequency factor (10^{12} versus $10^{11}\,M^{-1}sec^{-1}$) ..."

   M: The reason is the following: consider the equation: $k \approx A\exp(-\lambda/4kT)$. For a *given* k to a greater A (which contains Z) there must correspond a larger λ and a larger λ means a smaller inverted effect.

5. Following: "The inverted effect is masked *to some extent* by the presence of k_{diff} in $\frac{1}{k_{\text{obs}}} = \frac{1}{k_{\text{act}}} + \frac{1}{k_{\text{diff}}}$"

   Q: If $k_{\text{act}} \ll k_{\text{diff}}$, then there is no masking. From the earlier equation we have $k_{\text{obs}} = \frac{k_{\text{act}} \cdot k_{\text{diff}}}{k_{\text{act}} + k_{\text{diff}}}$. It not immediately evident how big should k_{diff} be in order to mask a smaller k_{act}.

M: If $k_{act} \ll k_{diff}$ you can neglect k_{diff}, therefore there is no masking. k_{diff} is typically 10^{10}, a factor of 10 slower than the collision frequency. That's typical, it depends on whether the reactants are charged or not. It's about 10^{10} instead of 10^{11} $\mathrm{L\,M^{-1}\ sec^{-1}}$, so if you get into the inverted region, if it's below 10^{10} then you would see it. If one is on the inverted region and above 10^{10} then it would be masked. It's not a big region but it extends over quite a ΔG^0 region. We have plots, and Sutin has plots, I think, of the masking, you see that there, wherever there is a flat top it's masked.

6. p. 870 middle: "anharmonic effects can contribute to decreasing (or increasing) the rate, and high frequency modes which play less of a role in the normal region may also contribute importantly to accepting the excess energy in the 'inverted region.' The latter would reduce the tendency for k_{obs} to decrease with increasingly negative ΔG^0 in the very negative ΔG^0 region."

M: If you think of the intersection of two parabolas, and if you make one of them anharmonic, then the intersection is going to occur in a different region and, depending on whether the anharmonicity is positive or negative, it's going to change to the extent to which you are in the inverted region. In other words, if you look at the intersection of the two curves, then, depending on what the anharmonicity is, one way or the other, you can see that either it is going to intersect the upper curve in the inverted region higher up or lower down, depending on the anharmonicity. If you have a curve that's way down and going high up and the other curve is sort of near its bottom, the anharmonicity will only be important in the curve that's way down, the curve that's near its bottom is largely in the nearly harmonic region.

If you look for instance at Fig. 2 in **M164,** at the classical curve compared with the quantum, you'll see that in the normal region there is no much difference and in the inverted region the quantum curve is higher than the classical, there is less slowing in the rate.

Q: So, it refers also to the free energy.

M: Yes, that's right, in fact we should always be talking about free energy, because you have tremendous anharmonicities in the individual motions themselves, but the free energy can still be roughly quadratic. In other words in the models that people have used in computers where they use an actual model of solvent that it's pretty anharmonic, nevertheless when they calculate the free energies like Chandler, or Tom Miller, the free energies come out to be roughly quadratic, so there is a big difference, one has to be very careful.

Q: You very often speak of the cumulants expansions and of the central limit theorem, are there papers of yours where you specifically use them?

M: I didn't, I did not originate work in this area, the terms were first used either by Mukamel or by Warshel, I think. I didn't realize I was doing a cumulant expansion, until one of them in their own paper pointed out that if one uses the cumulant expansion one gets the quadratic expression in the free energy. I was just thinking of quadratic expression in the free energy but that turned out to be equivalent to cumulant expansion for the partition function.

Q: And what about the central limit theorem?

M: Well, somehow that's connected... because the central limit theorem can come about as a result of treating a collection of more or less independent objects and going just up to the second order cumulant expansion.

M173. Further Developments in ET

"The inverted region in ET reactions is studied for the reaction of electronically excited ruthenium(II) tris-bipyridyl ions with various metal(III) tris-bipyridyl complexes. Numerical calculations for the *diffusion-reaction equation* are summarized for *the case where ET occurs over a range of distances*. Comparison is made with the experimental data and with a simple approximation. The analysis reveals some of the factors which can cause a *flattening* of the $\ln k_{obs}$ versus ΔG^0 curve in the inverted region. Ways of improving the chance of observing the effect are discussed.

Some time ago it was predicted (**M30**, **M53**) that, in a series of weak-overlap ET reactions, the rate would first increase when ΔG^0 was made more negative, and then, when ΔG^0 became very negative, eventually decrease. Evidence for such an "inverted effect" has been given in a number of papers, but in many other studies the reaction rate reaches a limiting value, rather than a decreasing value, when $-\Delta G^0$ becomes large. Possible explanations for the latter result have been suggested

(i) Alternate pathways for the reaction when ΔG^0 is very negative (such as H-transfer (**M164**, **M170**), electronically excited product states (**M164**), or, when the reaction was observed via quenching of fluorescence, exciplex formation)
(ii) Quantum-mechanical nuclear tunneling (**M110**, **M164**)
(iii) Masking by diffusion, and
(iv) Reduction of the inverted effect (by ET over a distance (**M170**))

Quantum-mechanical tunneling *reduces the magnitude* of the predicted effect but does not eliminate it in the weak-overlap systems,... (**M164**). Moreover, there is a 1:1 correspondence between the quantum mechanically calculated charge-transfer spectrum (emission and absorption vs. $h\nu$) for a weak-overlap redox system and the plot (Eq. 11.37 and 11.38 given later) of k_{act} versus the energy of reaction, ΔE, and hence in a series of reactions of *given* ΔS^0, versus $-\Delta G^0$. Here, k_{act} is the activation-controlled quantum mechanically calculated rate constant. Thus, the well-known existence of a *maximum in the charge transfer versus wavelength spectrum implies* that there will be a *maximum in the* $\ln k_{act}$ versus $-\Delta G^0$ *plot* when the ET is a weak-overlap reaction.(1) This correspondence removes any question that nuclear tunneling would eliminate the inversion, since that tunneling occurs to the same extent in both the charge-transfer spectrum and the $\ln k_{\text{act}}$ versus $-\Delta G^0$ plots,[(2)] and the former has a well-known maximum. It also removes any argument that large anharmonicities in practice eliminate the effect: the correspondence applies regardless of whether the vibrations are harmonic or anharmonic, as long as the ET is a weak-overlap one. (The effects of having a very strong-overlap ET remain to be investigated.)

In a recent paper, an approximate calculation was made of effects (ii) to (iv) above (**M170**), using an approximate analytical solution for the diffusion problem, for the case where the reaction occurs readily over a short range of separation distances of the reactants. In the present report, we summarize the results of our recent calculations on a numerical solution of the same problem. A more complete description is given elsewhere (**M174**). One additional modification made here to **M170** is to ensure that the current available rate constant data at $\Delta G^0 = 0$ (Appendix) are satisfied.

11.6. Theory

The *diffusion-reaction equation* for *the pair distribution function* $g(r,t)$ of the reactants, which react with a rate constant which at any r is $k(r)$, is given by

$$\frac{\partial g(r,t)}{\partial t} = \frac{1}{r^2}\frac{\partial\left(r^2 J_r\right)}{\partial r} - k(r)\,g(r,t) \tag{11.30}$$

where J_r is the *inward radial flux density* (per unit concentration) due to *diffusion* and to any *forced motion arising from an interaction potential energy*, $U(r)$,[(3)] assumed to depend only on the separation distance r. The magnitude of J_r is given by

$$J_r = D\frac{\partial g}{\partial r} + \frac{Dg}{k_B T}\frac{dU}{dr} \equiv De^{-U/k_BT}\frac{\partial}{\partial r}\left(ge^{U/k_BT}\right)^{(4)} \tag{11.31}$$

where D is the sum of the diffusion constants of the two reactants.

NOTE: $\frac{\partial}{\partial r}(ge^{U/k_BT}) = \frac{\partial g}{\partial r}e^{U/kT} + ge^{U/kT}\frac{dU}{dr}\frac{1}{k_BT}$. On multiplying it by De^{-U/k_BT} we get $D\frac{\partial g}{\partial r} + \frac{Dg}{k_BT}\frac{dU}{dr}$.

NOTE: In **M30a** there is the formula $J(r) = 4\pi r^2 \left(D\frac{dc}{dr} + \frac{Dc}{kT}\frac{dw}{dr}\right)$ with w in place of U and c in place of g.

The observed rate constant, k_{obs}, at time t is then given by

$$k_{\text{obs}} = \int_0^\infty k(r)g(r,t)4\pi r^2 dr \tag{11.32}$$

The steady state solution to Eq. (11.30) satisfies $\partial g/\partial t = 0$, that is, it satisfies

$$(1/r^2)d(r^2 J_r)/dr = k(r)g(r) \tag{11.33}$$

For the experimental conditions investigated thus far, the steady state solution is an excellent approximation to the solution of Eq. (11.30) and we consider this case. However, in proposing some experiment in the picosecond regime to enhance the chance of observing the inverted effect, we consider the time-dependent Eq. (11.30).

The rate constant $k(r)$ is typically assumed *to depend exponentially on r,* varying as $\exp(-\alpha r)$. Theoretical estimates have been made for α of 1.44 Å^{-1} when there is intervening material between the reactants, and 2.6 Å^{-1}when there is not...

These values of α are sufficiently large that $k(r)$ falls off rapidly with r. When this "reaction distance" is small relative to the distance over which the function $h(r) = g\exp(U/k_BT)$ (NOTE: cf. Eq. 11.31) changes significantly, that is, over which $(h(r) - h(\sigma))/(h(\infty) - h(\sigma))$ becomes appreciable, one can introduce an approximate analytic solution to Eq. (11.33) (cf. **M174, M30a**)

$$\frac{1}{k_{\text{obs}}} = \frac{1}{k_{\text{act}}} + \frac{1}{k_{\text{diff}}}^{\textbf{(5)}} \tag{11.34}$$

where, in the present case, we have (from Eq. (11.32) with $g(r) \equiv 0$ for $r < \sigma$)

$$k_{\text{act}} = \int_\sigma^\infty k(r)e^{-U/k_BT}4\pi r^2 dr \tag{11.35}$$

and where (Debye)

$$k_{\text{diff}} = 4\pi D \Big/ \int_{\sigma}^{\infty} \frac{e^{U/k_B T}}{r^2} dr \tag{11.36}$$

Equation (11.34) was actually derived for the case where reaction occurs *at some contact distance* $r = \sigma$. A derivation of Eq. (11.34) for the present case of a *volume distributed* rate constant $k(r)$ is approximate and is given elsewhere (**M174**).

For $k(r)$ we shall assume at first, as in **M170**, that *the reaction is adiabatic at the distance of closest approach* $r = \sigma$, and that is *joined* there to the nonadiabatic solution which varies as $\exp(-\alpha r)$. The adiabatic and nonadiabatic solutions can be joined smoothly .... For simplicity... we will join the adiabatic and nonadiabatic expressions at $r = \sigma$. We subsequently consider another approximation in which the reaction is treated as being *nonadiabatic even at* $r = \sigma$.

The well-known perturbation theory expression for the nonadiabatic rate constant is given by

$$k(r) \cong \frac{2\pi}{\hbar} |V(r)|^2 \,(\text{F.C.}) \tag{11.37}$$

where (F.C.) is the Franck–Condon factor and $V(r)$ is the electronic matrix element for the ET. (F.C.) is given by

$$(\text{F.C.}) = \frac{1}{Q} \sum_{i,f} e^{-E_i/k_B T} \, |\langle i \mid f \rangle|^2 \, \delta\left(E_f - E_i + \Delta E\right) \tag{11.38}$$

where i and f denote initial and final (reactants' and products') *nuclear configuration states*, including those of the solvent; ΔE is the energy of reaction; and Q is $\sum_i \exp(-E_i/k_B T)$. The solvent will be treated classically (**M30**), to avoid the quantum harmonic oscillator treatment of the polar solvent which is sometimes used. (The latter yields a large error for ΔS^0 when ΔS^0 is large (**M163**). The *contribution of the polar solvent* to the Franck–Condon factor is (cf. Kestner, Logan, Jortner (1974), **M30**)

$$(\text{F.C.})_{\text{solvent}} = (4\pi\lambda_{\text{out}}k_B T)^{-1/2} \times \exp[-(\Delta G^{0\prime\prime} + \lambda_{\text{out}})^2/4\lambda_{\text{out}}k_B T]^{(6)} \tag{11.39}$$

where $\Delta G^{0\prime\prime} = \Delta G^0 + E_f^v - E_i^v$ and the superscript v denotes inner shell vibrational energy.

NOTE: An expression similar to the above appears in Eq. (11.40) of **M164**:

$$G = (4\pi kT\lambda)^{-\frac{1}{2}}\exp\left[-\frac{(\Delta E + \lambda)^2}{4kT\lambda}\right] \tag{11.40}$$

and is there originally defined for $\lambda = \sum_{j=1}^{N} \lambda_j$. Then it is used in Eq. (11.41) there

$$k_{et} = Z(4\pi kT\lambda)^{\frac{1}{2}}\, G, \tag{11.41}$$

but with λ as the sum over the inner and outer-sphere λ's and with ΔE replaced by ΔG^0.

The *matching* of the adiabatic and nonadiabatic expressions for $k(r)$ at $r = \sigma$ yields a value for $V(\sigma)$ given by (**M174**)

$$\frac{2\pi}{\hbar}\, |V(\sigma)|^2\, (4\pi\lambda k_B T)^{-1/2} \sim 10^{13} s^{-1} \tag{11.42}$$

and, for a reorganization parameter λ of about 70 kJ/mole, yields $|V(\sigma)|^2 \sim 0.023$ eV. This value and

$$|V(r)|^2 = |V(\sigma)|^2 \exp[-\alpha(r - \sigma)] \tag{11.43}$$

were introduced into Eq. (11.37) as our first approximation to $V(r)$.

NOTE: This formula gives the variation of the splitting with r.

The series of ET reactions for which we calculated rate constants involve quenching of the lowest excited electronic state of Ru $(bpy)_3^{2+}$. This *Ru(II) state is a *metal-to-ligand charge-transfer* *state* in which an excess electron appears to be localized on one of the bipyridyl ligands, and this electron may be transferred to a metal-centered orbital on the oxidant, at least when an unexcited oxidant is formed. A calculation of the distance dependence of $V(r)$ for this particular transfer would be desirable but, lacking that, the simple exponential form indicated in Eq. (11.43) has been used instead.

The actual numerical integration of Eqs. (11.31) and (11.33) was performed by converting Eq. (11.33) to a pair of ordinary differential equations, then using a standard integration routine for integrating the latter, integrating outward from $r = \sigma$ to large r until $g(r)$ had its correct functional value at large r, $g(r) \sim 1 - c/r$ where c is a constant. (This functional form is the solution of Eqs. (11.31) and (11.33) at r large enough that $k(r) = U(r) = 0$ and for U vanishing more rapidly than $1/r$). Because $g(\sigma)$ was unknown to a multiplicative constant initially, we actually performed the integration for a function $G(r) = g(r)c_1$, with c_1 unknown and with a preassigned value for $G(r)$ at $r = \sigma$.

The terms c_1 and c could be determined from the numerical values of G at large r, and then $g(r) = G(r)/c_1$. The value of k_{obs} was calculated from the total flux at $r = \infty$:

$$k_{obs} = 4\pi D \lim_{r\to\infty} \left(r^2 \frac{dg}{dr} \right) = 4\pi Dc \tag{11.44}$$

11.7. Results

Calculations were performed for the system studied by Creutz and Sutin

$$^{*}\text{Ru(II)bpy}_3\text{+M (III) bpy}_3 \rightarrow \text{Ru (III) bpy}_3\text{+M(II)bpy}_3 \tag{11.45}$$

where the bpy's are various bipyridyls, M is one of several metals, and the asterisk denotes an electronically excited molecule. The questions we address is how, for a model which has the "experimental" rate constant at $\Delta G^0 = 0$ ($k_{obs} \sim 4 \times 10^8 \text{M}^{-1}\text{sec}^{-1}$)

(Appendix) and the observed diffusion-limited rate constant ($k_{\text{diff}} \sim 3.5 \times 10^9 \text{M}^{-1}\text{sec}^{-1}$), do the values predicted for k_{obs} at quite negative ΔG^0's *compare* with those calculated from Eq. (11.46) and with the experimental results? Is the effect of ET *over a range of distances* sufficiently large to explain the observed results (i.e., very little falloff of rate constant with increasing $-\Delta G^0$'s?)

We use a λ_{in} of 15.5 kJ/mole associated with a frequency of 1300 cm^{-1} (**M164**), and λ_{out} of 54 kJ/mole at $r = \sigma$. All calculations were performed with $T = 298\,\text{K}$. The dependence of λ_{out} on r (**M53**) is incorporated in the calculations. An equilibrium Debye–Hückel expression for the ion-atmosphere-shielded Coulombic repulsion of the reactants is assumed given by

$$U(r) = \frac{z_1 z_2 e_o^2}{\varepsilon r} \frac{e^{\kappa \underline{a}}}{(1 + \kappa \underline{a})} e^{-Kr} \tag{11.46}$$

for the case where the two reactants have the same radius. Here, κ is the reciprocal of the Debye–Hückel *screening length*, ε is the static dielectric constant, the $z_i e_o$ values are the ionic charges of the reactants, and $\underline{a}$ is the distance of closest approach of the ions in the ion atmosphere to a reactant ion. The distance $\underline{a}$ is $r_i + r_a$, where r_i is the radius of a reactant ion and r_a is the radius of the principal ion of opposite sign in the ion atmosphere. When $r_i \geq r_a$, $\underline{a}$ lies between $2r_i$ and r_i, being $2r_i$ when $r_i = r_a$ and being r_i when $r_a = 0$. Using the current approximate radii we shall, for concreteness, take $\underline{a} = 3\sigma/4$. (In Eq. (11.46), the reactants are assumed to have the same radius. A more general expression than Eq. (11.46) is cited in **M174**. At the prevailing ionic strength of about 0.52 M, κ^{-1} is about 4.2 Å. Because of this large ionic strength, $U(r)$ is quite small, even at $r = \sigma$.

Using $\alpha = 1.5\,\text{Å}^{-1}$ and, at first, $V(\sigma) = 0.023\,\text{eV}$, k_{act} at $\Delta G^0 = 0$ is found to be $1.2 \times 10^{10}\text{M}^{-1}\text{sec}^{-1}$ which is substantially higher than the current experimental value (Appendix) of $\underline{\text{ca}}\ 4 \times 10^8\text{M}^{-1}\text{sec}^{-1}$. Assuming the validity of the latter, *either* $V(\sigma)$

is less than 0.023 eV, i.e., the reaction is not adiabatic at the *contact distance* $r = \sigma$, *or* λ is higher than estimated, *or* Eq. (11.46) underestimates $U(r)$. We consider first using a different $V(\sigma)$, namely, 0.0045 eV, which yields the current "experimental" rate constant at $\Delta G^0 = 0$. (The same final results for the $\ln k_{\rm obs}$ versus ΔG^0 plot would be obtained, essentially, if one used instead a different $U(\sigma)$, as long as there is agreement of $k_{\rm act}$ at $\Delta G^0 = 0$.)

The numerical solution of Eq. (11.33) and the rate constant data of Fig. 1 agree at the data's maximum ($\sim 3.5 \times 10^9 \mathrm{M}^{-1}\mathrm{sec}^{-1}$) when one chooses $3.0 \times 10^{-6} \mathrm{cm}^2 \mathrm{sec}^{-1}$ for the sum of the D's of the two reactants. This D is somewhat near those estimated rather indirectly (electrochemically) for the individual D's of ferric and ferrous phenanthroline complexes ($\sim 1.9 \times 10^{-6}$ and $3.7 \times 10^{-6} \mathrm{cm}^2 \mathrm{sec}^{-1}$, respectively).

Since reaction may also yield electronically excited products when ΔG^0 is sufficiently negative, we include this reaction, as we did in **M164**. The mean excitation energy used for the formation of the electronically excited Ru(III) product is 1.76 eV (**M164**)...

We first compare the present numerical results for the solution of the steady state Eqs. (11.32) and (11.33) with the approximate solutions given by Eqs. (11.34) and (11.35) and the experimental value for $k_{\rm diff}$. The results agreed to about three percent when ΔG^0 was varied from +0.6 to −3.0 eV. The experimental value for $k_{\rm diff}$ and Eq. (11.36) imply a value of $D = 3.5 \times 10^{-6} \mathrm{cm}^2 \mathrm{sec}^{-1}$, compared with the $3.0 \times 10^{-6} \mathrm{cm}^2 \mathrm{sec}^{-1}$ found when Eqs. (11.32) and (11.33) were solved. Had the same D been used for both the exact (Eqs. 11.32 and 11.33) and the approximate (Eq. 11.34) solutions, their agreement for the rate constants would have been about 10% instead of 3%, which is still very close.

The results of solving Eqs. (11.32) and (11.33) are next compared with the experimental data in Fig. 1, using $V(\sigma) = 0.0045$ eV. The solid line refers to the formation of ground state products, and the dotted line to the formation of an electronically excited Ru(III)

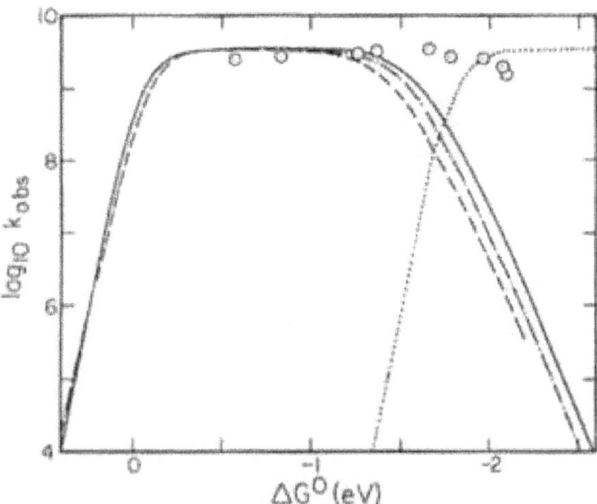

Fig. 1. Calculated and experimental rate constants for Reaction 14 vs. $\Delta G°$. Key: —, r-dependent λ_{out}: — . —, fixed λ_{out}; – – –, from Ref. 1 in which reaction occurred only at $r = \alpha$; and ··· current result (r-dependent λ_{out}) for formation of an electronically excited product.

NOTE: Reaction 14 in the legend of the original paper is here reported as Eq. (11.45) anf Ref. 1 is **M30**.

product. For further comparison with the solid line, a calculation was made with λ_{out} held fixed (54 kJ/mole, the value at $r = \sigma$) and is given by the dash-dot line. In order to obtain agreement with the solid line at $\Delta G^0 = 0$, $V(\sigma)$ was reduced to 0.0039 eV in calculating the dash-dot line. The dashed line is the result of a calculation (**M164**) in which reaction was treated as occurring adiabatically, but only at some contact distance σ, and in which Eq. (11.34) was used, together with the experimental value for k_{diff}. The λ_{out} value used for this last curve was again 54 kJ/mole, the present $\lambda_{out}(\sigma)$.

In Fig. 2, we give a comparison of the solid line of Fig. 1 with that obtained using $V(\sigma) = 0.023$ eV and a larger λ ($\lambda_{out}(\sigma) = 83$ kJ/mole). A slightly smaller D ($2.7 \times 10^{-6}\text{cm}^2\text{sec}^{-1}$) was required to make the latter calculation yield the experimental value of the maximum observed rate constant, $3.5\times10^9\text{M}^{-1}\text{sec}^{-1}$. Both curves have the same k_{obs} at $\Delta G^0 = 0$.

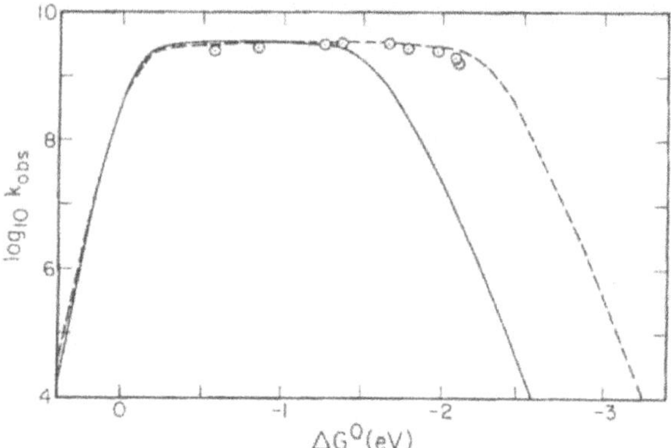

Fig. 2. Calculated rate constants for Reaction 14 vs. $\Delta G°$. Key: —, take from solid line in Figure 1, $V(\sigma) = 0.0045$ eV, $\lambda_{out}(\sigma) = 54$ kJ/mol; and – – $V(\sigma) = 0.023$ eV and $\lambda_{out}(\sigma) = 83$ kJ/mol.

NOTE: Reaction 14 in the legend of the original paper is here reported as Eq. (11.45).

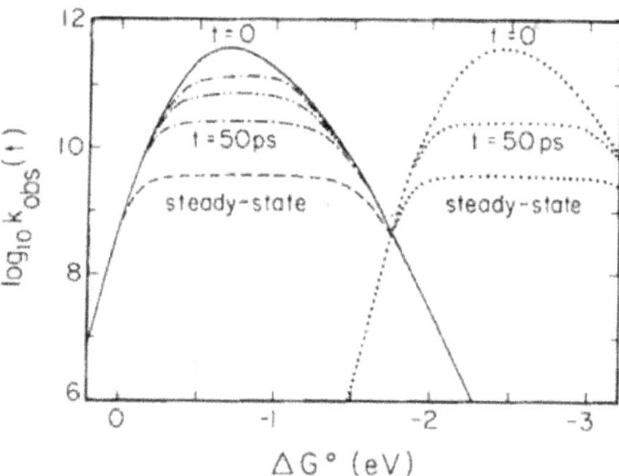

Fig. 3. Time-dependent calculations of $k_{obs}(t)$ vs. $\Delta G°$ for various observation times. Key: — · —, 1 ps; —. .—, 5 *ps*; and ···, $k_{obs}(t)$ for formation of an excited state Ru(III).

11.8. Discussion

The results comparing *the exact Eqs. (11.32) and (11.33)* with *the approximate Eqs. (11.34) and (11.35)* show that the latter provide a good approximation for the present conditions, at least. The results in Fig. 1 show that, to account for the experimental results at very negative ΔG^0's using the present value of λ_{out} (54 kJ/mole), it is necessary to postulate the formation of electronically excited products. This was also the case in an earlier result (**M164**). The sum of the two rate constants in Fig. 1 yields agreement with the data in Fig. 1 to a factor of about 2. If, as for the dashed line in Fig. 2, the value of λ were actually appreciably larger, the formation of ground state products alone would suffice to obtain agreement. (Classically, the maximum in the k_{act} versus ΔG^0 curve occurs at $\Delta G^0 = -\lambda$ and so is shifted to more negative ΔG^0's when λ_{out} is increased.)

Returning to Fig. 1, one sees that holding λ_{out} fixed at its value at $r = \sigma$ (dash-dot line) does not cause a large deviation from the more correct result (r-dependent λ_{out}, solid line) in the inverted region. A similar approximation was used, of course, for the dashed line, where a $k(\sigma)$ was used instead of a $k(r)$.

We also have explored the solution of the time-dependent Eq. (11.30) to study the plot corresponding to Fig. 1 *when the observation of fluorescence quenching in reaction (11.45) is made at short times*. In these short time calculations we have assumed, for simplicity, that reaction occurs only at $r = \sigma$ .... Results for $k_{obs}(t)$ are given for several times in Fig. 3, and curves are also given for the formation of electronically excited products. The value of $k_{obs}(t)$ is obtained as the slope at time t of a plot of $[M(III)bpy_3^{+2}]^{-1} \ln[{}^*Ru(II)bpy_3]$ versus t.[7] The results show the *enhancement of the predicted inversion effect at small times*, and an experimental study of this or related systems at such times would be desirable, and may, in fact, distinguish between possibilities cited earlier that $V(\sigma) < 0.023$ eV or that $\lambda > (15.5 + 54)$ kJ/mole; at short times there would be a *double maximum* in the total rate constant versus ΔG^0 plot in the first case and a *single maximum* in the second.

The details of these short time calculations, made for the case that $U(r) \cong 0$, are given in **M174**. Searching for the inverted effect in unimolecular systems (reactants linked to each other) would also be very desirable since their rates would not be diffusion limited.

11.9. Appendix. "Experimental" Rate Constant at $\Delta G^0 = 0$

The self-exchange rate constant for reaction (11.45), when M is Ru and when an *excited* Ru(II) product is formed, has been estimated to be about $10^8\,M^{-1}\,s^{-1}$. The self-exchange rate constant for reaction (11.45), when M is Ru and when the products and reactants are in their ground electronic states, has been estimated to be $1.2 \times 10^9\,M^{-1}s^{-1}$, which is the observed rate constant for the oxidation of $Ru(bpy)_3^{2+}$ by $Ru(phen)_3^{3+}$, for which $\Delta G^0 \sim 0.01$ eV. Corrected for diffusion (Eq. 11.34), the k_{act} for the latter is $2 \times 10^9 M^{-1}sec^{-1}$. Sutin has noted that the cross-relation (**M30**, **M53**) should be applicable to a nonadiabatic ET *if* the electronic matrix element, $V(r)$, for the cross-reaction is equal to the geometric mean of the matrix elements for the self-exchange reactions. Assuming that that condition is approximately satisfied, the exchange rate constant for reaction (11.45) when $\Delta G^0 = 0$ is estimated to be the geometric mean, $(20 \times 1)^{1/2} \times 10^8 M^{-1}sec^{-1}$, i.e., $4.5 \times 10^8 M^{-1}sec^{-1}$. Corrected for diffusion using Eq. (11.34), this becomes $4 \times 10^8 M^{-1}sec^{-1}$, the value given in the text.

Interestingly enough, the rate constant at $\Delta G^0 = 0$ for reaction (11.45) when *Ru(II) and M(III) are replaced by *Cr(III) and Ru(II), respectively, is $\sim 2 \times 10^8\,M^{-1}sec^{-1}$ in 1 M H_2SO_4..

NOTES

1. p. 236 top-middle: "the well-known existence of a maximum in the charge transfer versus wavelength spectrum implies that there will be a maximum in the $\ln k_{act}$ versus $-\Delta G^0$ plot when the ET is a weak-overlap reaction (Fig. 1)."

   M: If you draw a Gaussian curve and you ask yourself how you can generate a Gaussian absorption curve from two parabolas,

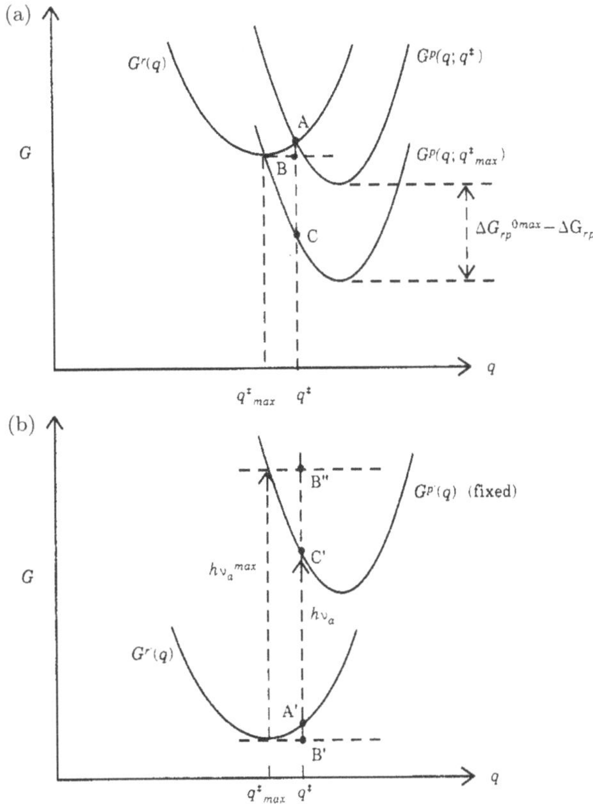

Fig. 1. Plot of free energies G for the r and p electronic states vs the coordinate q defined in the text. (a) Thermal electron transfer reaction, with plots for two different values of $\Delta G^0_{\rm rp}$ indicated; (b) optical transition, with two different absorption frequencies indicated. Reproduced from **M245**.

then you will see that one part of the spectrum corresponds to the normal region and the other part to the inverted region. In other words, *draw two parabolas so dispiaced that you get an absorption spectrum, and look at the part that deals with the red side and to the part that deals with the blue side*, blue side meaning higher energies of the absorption spectrum and red side meaning low energies, frequencies below the maximum. Forget about the maximum, think of what is on either side of the maximum. Now lower the upper parabola down so that you have a curve crossing and you can get a thermal ET reaction without occurring of light. If you do an absorptin spectrum you can't have a thermal ET reaction unless you lower the upper parabola down. If you now lower the upper parabola so fa down that you are in the inverted region, you will see that one part of the spectrum corresponds to the normal region and the other part to the inverted region.

If you take the two theoretical expressions for an ET rate and for an absorption spectrum, and you will do simple FC quantum mechanical overlaps, you see that the two expressions look identical, the only difference being the delta of the energy difference. In the delta function argument, in one case you see $c_f - c_i$ and in the other case $c_f - c_i - h\nu$, otherwise the parts of the overlap functions are identical.

Q: Different ET reactions compared with a single CT reaction.

M: Yes, in other words supposing now that you have a CT reaction, meaning that the product electronic state is not one of those in Fig. 1(a) but is something in Fig. 1(b), in other words for absorption the product state has to be much higher, not comparable, so the analogue with the same property so to speak for an absorption spectrum of Fig. 1(a) is given in Fig. 1(b), in other words so the Fig. 1(b) is an absorption spectrum analogue to what's happening in thermal ET in Fig. 1(a).

Q: But for different reactions.

M: Yes, it would be a different process because now the product state is much higher instead of just being about the same energy.

Q: Whereas in CT you consider a single process.

M: Yes, but you can have *a series* of CT reactions just by raising and lowering the upper curve. Yes, you can well imagine that just like you can have *a series* of ET reactions where the barrier changes with substituents, you can have a series of absorption spectra where the absorption frequency maximum changes with the substituent.

Q: In the paper one has the impression that you consider for the ET a series, a family of reactions, and for CT you reason on one single reaction.

M: Yes, that's right, the point is that *if you have an absorption spectrum, if you want to imitate it by ET reactions you have to use a family of ET reactions*.

Q: So, in the absorption spectrum it is intrinsically considered a whole family of ET transitions.

M: Yes, I guess so, even though it is a single curve, *an absorption spectrum corresponds not to one energy, it corresponds to a whole distribution of energies, but a reaction corresponds to one standard free energy, so you have to consider a family of reactions to correspond to one absorption spectrum.*

Q: So, it is not necessary to consider a family of CT reactions.

M: No, it isn't. Now, at the time when I wrote the article I didn't realize, but later saw some indication of it, that Jortner may have pointed out this analogy between CT spectrum and ET rates some time before I did.

Q: But you were anyway the first to discover the inverted free energy effect.

M: Yes.

2. p. 236 middle: "tunneling occurs to the same extent in both the charge-transfer spectrum and the ln k_{act} versus $-\Delta G^0$ plots."

M: In the CT spectra you don't just have a vertical transition, what happens is that *some of the absorption spectrum comes from the classically forbidden region* and in the ET case, that would be the region contributing to the tunneling. The transition is in this case due to overlapping vibrational wave functions whose tails extend beyond the classical turning points.

3. p. 236 bottom: "The *diffusion-reaction equation* for *the pair distribution function* $g(r, t)$ of the reactants, which at any r is $k(r)$, is given by

$$\frac{\partial g(r,t)}{\partial t} = \frac{1}{r^2}\frac{\partial\left(r^2 J_r\right)}{\partial r} - k(r)\, g(r,t) \tag{11.47}$$

where J_r is the *inward radial flux density* (per unit concentration) due to *diffusion* and to any *forced motion arising from an interaction potential energy* $U(r)$ ..."

M: Diffusion is always driven by a concentration gradient, if you don't have a concentration gradient you won't see diffusion. You may have reactants going back and forth... if you have reactants with the same concentration you won't see a diffusion process. I mean, they are diffusing but you won't see it. What's going in cancels what's going out.

If you are considering a reaction involving diffusion of reactants, then the best way of treating it is treated in a classic paper of F. Collins and G. E. Kimball of 1950, and they will certainly have a correct treatment on diffusing particles toward each other, because Kimball was a master. That's the Kimball of Eyring, Walter and Kimball. The way of treating the diffusion is similar to the way in a gas phase collision you treat the relative motion and the motion of the center of gravity. You do a similar thing for diffusion problems in liquids, you have the diffusion of the reactants moving to each other, there is the diffusion equation for one reactant and the diffusion equation for the other and then you make a transformation, so you get the equivalent of what one normally does in the gas phase, motion of center of mass and motion of the reactants with respect to each other,... it's just more complicated. There is a danger if you fix one reactant and thinking of the other as moving toward it, because you wouldn't think that way in the gas phase, would you? So you wouldn't think that way in the diffusion problem. I treated the diffusion that way since I have learned about it.

That treatment is interesting because what happens is the following: supposing you have a salt AB in water, if you're looking at the salt just diffusing then the effective diffusion coefficient, you know, when you're doing that kind of separation of variables, and you think of diffusion coefficient of the salt as a whole, it turns out it to be related like a reduced mass equation $\frac{1}{D_{\text{eff}}} = \frac{1}{D_A} + \frac{1}{D_B}$ to the diffusion coefficient of the ions in which it dissociates, so that the slower moving species determines the effective diffusion coefficient. On the other end, if you're looking at two ions diffusing toward each other in a reaction, in other words if you're looking now at a kind of center of position coordinate — the relevant coordinate — then for that the effective diffusion coefficient is the sum of the diffusion coefficients, and all of this comes out in a rigorous way.

4. p. 237 top, Eq. (11.31): $J_r = D\frac{\partial g}{\partial r} + \frac{Dg}{k_B T}\frac{dU}{dr}$.

Q: On p. 129 of **M30a** there is $J(r) = 4\pi r^2 \left(D\frac{dc}{dr} + \frac{Dc}{kT}\frac{dw}{dr}\right)$ instead of $J_r = D\frac{\partial g}{\partial r} + \frac{Dg}{k_B T}\frac{dU}{dr}$, where $g(r)4\pi r^2 dr$ is the probability that the pair of reactants has a separation distance between $(r, r + dr)$ (**M170**) and $c(r)$ in **M30a** is "the concentration of B with respect to some A at an A-B distance r." The formula $J(r) = 4\pi r^2 \left(D\frac{dc}{dr} + \frac{Dc}{kT}\frac{dw}{dr}\right)$ refers to a spherical shell of volume $4\pi r^2 dr$ with center in A. c is the concentration inside the spherical shell, at radius r in the interval between r and $r + dr$. We have then a concentration dependent on r. The dimensions of $c(r)$ are then $\frac{1}{V}$ where $V = 4\pi r^2 dr$. On multiplying by $4\pi r^2$ the dimension becomes $\frac{1}{L}$. We have then not the usual concentration per volume but a concentration per length. The g appearing in $J_r = D\frac{\partial g}{\partial r} + \frac{Dg}{k_B T}\frac{dU}{dr}$ has then also the dimension $\frac{1}{L}$, this time a probability per length. There is some analogy between the two formulas...

M: The first formula for the flux, that's a very common formula. That $g(r) = c(r)/c_0$ sounds right because $g(r)$ usually goes to unity at long distances. V is simply a volume, the dimension is 1 over volume. Those are different definitions, that $g(r) = c(r)/c_0$ is the correct statement, by the way. $J(r)$ and J_r are different, $J(r)$ is a flow per unit area and is a total flow through the area. But the relationship between g and c is $g(r) = c(r)/c_0$. My guess is that if you write $g(r) = c(r)/c_0$ where c_0 is concentration at infinity you we'll be able to relate one equation to the other.

5. p. 237 middle: "The rate constant $k(r)$ is typically assumed to depend exponentially on r, varying as $\exp(-\alpha r)$. Theoretical estimates have been made for α in different systems, varying between 1.1 Å^{-1} to 2.6 Å^{-1}.

   These values of α are sufficiently large that $k(r)$ falls off rapidly with r. When this 'reaction distance' is small relative to the distance over which the function $h(r) = g\exp(U/k_BT)$ (cf. Eq. 11.40) changes significantly, that is, over which $(h(r) - h(\sigma))/(h(\infty) - h(\sigma))$ becomes appreciable, one can introduce an approximate analytic solution to Eq. (11.33) (cf. **M174**, **M30a**)

$$\frac{1}{k_{\text{obs}}} = \frac{1}{k_{\text{act}}} + \frac{1}{k_{\text{diff}}}\text{"} \tag{11.48}$$

Q: On p. 867 top of **M170** you wrote:"

$$g(r) = g_0(r)e^{-w(r)/kT} \tag{11.49}$$

where $g_0(r)$ is largely a geometric factor, being similar to the value it would have for an uncharged and nonspecifically interacting pair. *All of the specific-interaction contributions to $g(r)$ are contained in what might be called the 'work term' $w(r)$*."

Now we have $h(r) = g\exp(U/k_BT)$. Please explain the meaning of $h(r)$ . . . Are the g's in **M170** and here the same? I mean, is $h(r) = g_0(r)e^{-w(r)/kT}e^{U/k_BT}$? Please explain the different signs at exponentials.

M: The value of 2.6 Å^{-1} probably refers to vacuum. U and w are two different symbols for the same thing. There is positive sign in front of U in $h(r) = g\exp(U/k_BT)$ because g would be related to e^{U/k_BT}.

6. p. 238: "The well-known perturbation theory expression for the nonadiabatic rate constant is given by

$$k(r) \cong \frac{2\pi}{\hbar} |V(r)|^2 \text{(F.C.)} \tag{11.50}$$

where (F.C.) is is the Franck–Condon factor and $V(r)$ is the electronic matrix element for the ET. (F.C.) is given by

$$\text{(F.C.)} = \frac{1}{Q} \sum_{i,f} e^{-E_i/k_B T} |\langle i \mid f \rangle|^2 \delta(E_f - E_i + \Delta E) \tag{11.51}$$

where i and f denote initial and final (reactants' and products') *nuclear configuration states*, including those of the solvent; ΔE is the energy of reaction; and Q is $\sum_i \exp(-E_i/k_B T)$. The solvent will be treated classically, compare **M30**, to avoid the quantum harmonic oscillator treatment of the polar solvent which is sometimes used. (The latter yields a large error for ΔS^0 when ΔS^0 is large (**M163**). The *contribution of the polar solvent* to the Franck–Condon factor is (cf. Kestner, Logan, Jortner (1974), **M30** and **M33**)

$$\begin{aligned}&\text{(F.C.)}_{\text{solvent}}\\ &\quad = (4\pi\lambda_{\text{out}} k_B T)^{-1/2} \exp[-(\Delta G^{0\prime\prime} + \lambda_{\text{out}})^2 / 4\lambda_{\text{out}} k_B T]\text{''}\end{aligned} \tag{11.52}$$

M: Levich was the original one to use it. F.C. as is written is a perturbation expression that comes from perturbation theory, and so if the reaction were adiabatic it wouldn't strictly conform to that, that's really for nonadiabatic cases, I used it approximately even in adiabatic cases. Whenever you use Eq. (11.51), that implies a weak overlap, it's a perturbation, and that's equivalent to nonadiabaticity. F.C. does not take care of the electron tunneling, the electron tunneling is in the matrix element, the electronic matrix element. If you would use the word electron tunneling, then when electron tunneling is difficult, the matrix element H_{AB} is small. Electron tunneling is sort of way of visualizing the electronic matrix element, it is not a rigorous approach.

7. p. 244 top: "Results for $k_{obs}(t)$ are given for several times in Fig. 3, and curves are also given for formation of electronically excited products. The value of $k_{obs}(t)$ is obtained as the slope at time t of a plot of $[\mathrm{M(III)bpy}_3^{+2}]^{-1}\ln[{}^*\mathrm{Ru(II)bpy}_3]$ versus t."

   Q: Please explain the above formula. The reaction to which it refers is:

$$^*\mathrm{Ru(II)bpy}_3 + \mathrm{M(III)bpy}_3 \rightarrow \mathrm{Ru(III)bpy}_3 + \mathrm{M(II)bpy}_3$$

   M. M(III) must be in excess, this M(III) is hardly changing with time, *Ru... is an excited state, so this is going to be changing, so, let me write down something, as M(III) is hardly changing with time, I am going to call that k_{obs}, now, if M(III) changes very slowly with time, M(III) will be approximately constant, and then we have the Ru , the formulas are: $-\frac{d\mathrm{Ru}^*}{dt} = k\mathrm{Ru}^*\mathrm{M}_3 \approx k_{obs}\mathrm{Ru}^* \Rightarrow \mathrm{Ru}^* = \mathrm{Ru}^{*o}\exp(-k_{obs}t) \Rightarrow -\ln\mathrm{Ru}^*$ versus t gives the slope k_{obs}, in detail: $\ln\mathrm{Ru}^* = \ln\mathrm{Ru}^{*0} - k\mathrm{M(III)}t \Rightarrow$ slope of $\ln\mathrm{Ru}^*$ versus t is equal to $k\mathrm{M(III)}t$, or else, $\frac{1}{\mathrm{M(III)}}\ln\mathrm{Ru}^* = \frac{1}{\mathrm{M(III)}}\ln\mathrm{Ru}^{*0} - kt$, so if you plot this term versus t, then the slope is k. I called it $k(t)$ because M(III) varies with time, this is the time rate, so this intercept varies with time, it depends on how much M(III) varies with time. Look at the experimental paper. The slope is varying with time because M(III) is varying with time.

M174. Theory of Highly Exothermic ET Reactions

11.10. Introduction

It has been predicted that the rate constant of a series of homogeneous ET reactions

$$\mathrm{ox}_1 + \mathrm{red}_2 \rightarrow \mathrm{red}_1 + \mathrm{ox}_2 \tag{11.53}$$

in which ox_1 or red_2 is varied (*at constant intrinsic reorganization energy* λ) should first increase with increasingly negative standard free energy of reaction ΔG^0 at small ΔG^0. It should then achieve a maximum at some value of ΔG^0 and thereafter *decline* as ΔG^0 continues to become still more negative. The *region of decline* was termed the 'inverted' region (**M30**). The existence of an inverted region was first predicted on the basis of a classical theory (**M30**, **M53**). The quantum-mechanical correction given by quantum-mechanical perturbation theories predicts a *smaller* but nevertheless finite inversion (**M156**, **M164**). The difference arises from nuclear tunneling. . . .

Again according to the theoretical expressions there is a 1:1 correspondence (Ulstrup and Jortner) between the optical line shape and the activation rate constant k_{act} versus the energy of reaction ΔE plot, (*for a weak-overlap system*). Thus, *for a given* ΔS^0, there should be a correspondence with a k_{act} versus ΔG^0 plot for an ET transfer reaction. We then argue in a concluding section that the existence of a well-known maximum in a CT absorption versus wavelength plot implies that there should be a maximum in the $\ln k_{\text{act}}$ versus ΔG^0 plot.

On the other hand, many studies of highly exothermic reactions have found a diffusion-limited rate constant which extends to quite negative ΔG^0's rather than the predicted declining rate constant. . . . These studies frequently involve measuring the *rate of quenching of fluorescence* by a series of reactants, *where the quenching was presumed or demonstrated to proceed by ET.* In most cases, the reason for the absence of decrease in the rate is unknown, although several possibilities have been suggested. They include

- (i) Competing mechanisms at large $-\Delta G^0$, such as
  - (a) H-atom transfer (**M164**, **M170**)
  - (b) Formation of products in excited electronic states (**M164**), or
  - (c) When reaction is observed by quenching of fluorescence, exciplex formation
- (ii) Quantum effects (**M110**, **M156**, **M164**) (nuclear tunneling)

(iii) A modifying effect of ET occurring over a range of distances r (**M170**), and
(iv) The increase of the reorganization parameter λ with r in case (iii), thereby reducing the extent of inversion.

In the present paper, we report calculations which incorporate effects (ii–iv) and, in part, (i) and compare with the experimental results of Creutz and Sutin and with a simple approximation (**M170**) to the problem. It is also proposed that experiments conducted at very short times following the onset of reaction will enhance the chances of observing inverted behavior that, in bimolecular systems, is masked by diffusion in *conventional steady state rate measurements. Unimolecular systems, in which the reactants are linked to each other*, should be even better in this respect, since they are unaffected by diffusion. A brief summary of the present study has been given elsewhere (**M173**).

11.11. Theory

Diffusion. In *extracting* the 'activation rate constant' from an observed rate constant that is near the diffusion limit, it can be shown that the observed rate equals the *harmonic mean* of the activation rate and the diffusion-limited rate, *when reaction occurs at some specific encounter distance* σ (**M30**)

$$k_{\mathrm{obsd}} = 1/(1/k_{\mathrm{act}} + 1/k_{\mathrm{diff}}) \tag{11.54}$$

where the diffusion rate constant k_{diff} is given by (**M30**, Noyes, Debye...)

$$k_{\mathrm{diff}} = 4\pi D \Big/ \int_{\sigma}^{\infty} \exp(U/k_B T) r^{-2} dr \tag{11.55}$$

In Eq. (11.55) D is the *sum* of the reactants' diffusion coefficients, $U(r)$ is the intermolecular potential of the reactants, and k_B is the Boltzmann constant.

NOTE: In general the observed rate constant is a combination of activation and diffusion rates, with, of course, two possible limiting cases.

ET's can occur over a range of reactant separation distance (NOTE: see for instance F. F. Di Giacomo, A Semiempirical Potential Energy Surface for Chemiluminescent Electron Transfer, *Z. Physik. Chem. N.F.* **122**, 1–13, (1980)), rather than only at a specified distance. In such cases, the observed bimolecular rate constant k_{obsd} is related to the unimolecular rate constant $k(r)$, the rate of reaction of pairs of reactants having *fixed* internuclear, center-to-center, *separation distance r*, via a pair distribution function $g(r)$:

$$k_{\text{obsd}} = 4\pi \int_0^\infty g(r)k(r)r^2 dr \tag{11.56}$$

NOTE: In **M170** the same equation is for k_{act}.

In Eq. (11.56) we have assumed that k and g are radially symmetric. When the system has a $k(r)$ instead of only a k at $r = \sigma$, k_{act} is defined by using Eq. (11.56) with $g(r)$ replaced by its equilibrium value, $\exp(-U(r)/k_B T)$ for $r > \sigma$ and, in the present model, by zero for $r < \sigma$,[1] since k_{act} would be *the observed rate constant if diffusion would be infinitely fast.* Thus

$$k_{\text{act}} = 4\pi \int_\sigma^\infty k(r)r^2 \exp(-U(r)/k_B T)dr \tag{11.57}$$

NOTE: In **M170** $g(r) = g_0(r)e^{-w(r)/kT}$. Here $g_0(r)$ is neglected and $U(r) = w(r)$.

We shall wish to compare Eq. (11.56) with the use of Eq. (11.54), (11.55), and (11.57), for reactions occurring over a range of separation distances. To this end we solve the following Eq. (11.58).

In the present case, the reactants are substantially larger than the solvent molecules and so we shall assume that the short range intermolecular contribution to g(r) can be neglected. Then *g(r) in Eq. (11.56) may be obtained as the solution to a diffusion equation,* which is given by Eq. (11.58) for the case of radial symmetry.

$$\frac{\partial}{\partial t}g(r,t) = \frac{D}{r^2}\frac{\partial}{\partial r}\left(r^2\frac{\partial g}{\partial r}\right) + \frac{D}{k_B T r^2}\frac{\partial}{\partial r}\left(r^2 g\frac{dU(r)}{dr}\right) - gk(r) \tag{11.58}$$

The first term on the right arises from the *diffusive* flux, the second term from the *conductive* flux due to the long-range intermolecular potential $U(r)$ between the reactants, and the third term from the *loss* of reactants due to reaction....

NOTE: In **M170** $g(r) = g_0(r)e^{-w(r)/kT}$ was just given as 'suggestive.' The exponential function was given in **M30a** as solution of a diffusion equation, which is what will be done here. Note also that from Eqs. (11.54) and (11.57) one obtains *approximate* calculated rate constants and from Eqs. (11.56) and (11.58) *exact* rate constants.

For two reactants having charges $z_1 e$ and $z_2 e$, e being the electronic charge, U in the Debye–Hückel approximation is given by Eq. (11.59)

$$U(r) = \frac{z_1 z_2 e^2}{2\epsilon r}\left[\frac{\exp\kappa a_1}{1+\kappa a_1} + \frac{\exp\kappa a_2}{1+\kappa a_2}\right]\exp(-\kappa r) \tag{11.59}$$

where a is the *distance of closest approach* and r is the *separation distance* of the two centers. In Eq. (11.59), ϵ is the static dielectric constant of the solvent, κ is the inverse of the Debye–Hückel screening length, and a_i is the radius of ion i, r_i, plus that of the principal ions of opposite sign in the ion atmosphere, r_i^a.

$$a_i = r_i + r_i^a \tag{11.60}$$

... In the present paper, the two reacting ions are of the same size and are both positively charged, and so $a_1 = a_2 \equiv a$, that is

$$U(r) = \frac{z_1 z_2 e^2}{\epsilon r}\frac{\exp\kappa a}{1+\kappa a}\exp(-\kappa r) \tag{11.61}$$

and a is the distance of closest approach between a reacting ion and the *principal ion of opposite sign* in the ion atmosphere.

At large internuclear separations, the concentration of reactants must equal the bulk (no reaction) concentration. Thus, one of the boundary conditions on Eq. (11.58) is $\lim g(r, t) = 1$ as $r \to \infty$. When a *volume distributed* rate constant $k(r)$ is used instead of the usual *surface* one $k(\sigma)$, the boundary condition at the *distance of closest approach* $r = \sigma$ is obtained by requiring *total* flux (diffusive plus conductive) across $r = \sigma$ to be zero. This inward-directed flux (per unit concentration) is given by $4\pi r^2 D$ times the left hand side of

$$\frac{\partial}{\partial r} g(r, t) + g \frac{dU}{dr} \Big/ (k_B T) = 0 \text{ at } r = \sigma (t \geq 0) \quad (11.62)$$

and so Eq. (11.62) provides the second boundary condition.

NOTE: Compare the earlier equation, $4\pi r^2 D \left(\frac{\partial g(r,t)}{\partial r} + \frac{g}{k_B T} \frac{dU}{dr} \right) = 0$ with the equation $4\pi r^2 D \left(\frac{dc}{dr} + \frac{c}{k_B T} \frac{dw}{dr} \right) = J(r)$ from **M30a**.

A derivation of Eqs. (11.54), (11.55), and (11.57) as an approximate solution to Eqs. (11.58) and (11.62) at steady state, is given in Appendix B.

Unimolecular Rate. The ET reaction may be adiabatic, nonadiabatic, or somewhere in between (**M16**, **M48**, **M53**). A first-order quantum perturbation treatment of nonadiabatic ET reactions yields the familiar result (**M156**)

$$k(r) = \frac{2\pi}{\hbar} |V(r)|^2 (FC) \quad (11.63)$$

In Eq. (11.63) $V(r)$ is the matrix element between the reactant and product electronic states of the perturbation that gives rise to ET. The quantity FC is the *thermally weighted sum* of Franck–Condon factors given by Eq. (11.64)

$$FC = \frac{1}{Q} \sum_{i,f} e^{-E_i/k_B T} |\langle i \mid f \rangle|^2 \delta(E_f - E_i + \Delta E) \quad (11.64)$$

and has dimensions of $(\text{energy})^{-1}$.[(2)] In Eq. (11.64) i and f designate initial and final (reactants' and products') *nuclear configuration states*. The *reactant state* includes the pair of reactant molecules and the solvent surrounding them. Q is the nuclear partition function of the initial state. The functions $|i\rangle$ and $|f\rangle$ will be treated, for simplicity, in the harmonic oscillator approximation in the case of the *intramolecular* vibrations.

In the classical limit $\hbar\omega/k_BT \rightarrow 0$, and when frequency changes in individual vibrational modes are neglected, the FC given in Eq. (11.64) reduces to the expression in Eq. (11.65) (Levich, Ulstrup Jortner...).

$$FC = (4\pi\lambda k_BT)^{-1/2}\exp[-(\Delta E + \lambda)^2/(4\lambda k_BT)] \qquad (11.65)$$

As has been discussed in **M163** the quantum nonadiabatic result Eqs. (11.63) and (11.64) plus a dynamical (harmonic oscillator) assumption *for the motion of the solvent* does not allow for any large entropies of reaction. Compare Note (52).

Note (52): The entropy of reaction that is treated by the nonadiabatic formula, Eqs. (11.63) and (11.64) arises, in the *harmonic oscillator approximation*, from any noncancelling changes in the vibration frequencies that typically occur in oxidation-reductions. However, the inner-sphere contribution to ΔS^0 is usually minor, and for the polar solvent no changes in frequency are used (**M163**). The actual ΔS^0 is in large part due to reorganization of the solvent molecules, and this effect, (e.g., reflected in electrostriction) is neglected in the harmonic oscillator treatment of the solvent (with fixed ionic radii). Because of the low frequencies involved, solvent reorganization can to a good approximation be treated classically at ordinary temperatures. The correct classical treatment of the solvent can be recovered simply by replacing ΔE with ΔG^0 in Eq. (11.64) *when summing over the solvent modes*. This is equivalent to the procedure used by Ulstrup and Jortner, who use the classical expression for the relevant Franck–Condon approximation.[(3)]

To avoid this difficulty one can use, instead, a more nearly correct treatment of the polar solvent, one which is classical but in which no harmonic oscillations of the solvent are assumed (**M48**, **M16**, **M53**).

In this case, the FC factor for the solvent is (Kestrner, Logan, Jortner)

$$(FC)_{\mathrm{solv}} = (4\pi\lambda_{\mathrm{out}}k_BT)^{-1/2}\exp[-(\Delta G^0 + E_f^{\mathrm{v}} - E_i^{\mathrm{v}} + \lambda_{\mathrm{out}})^2/(4\lambda_{\mathrm{out}}k_BT)] \quad (11.66)$$

where the v superscripts denote vibrational energy. Equation (11.66) may be compared with the quantum results that we obtained in **M163**, where a quantum treatment of the solvent water was used, described by two modes which have frequencies of 1 and $170\,\mathrm{cm}^{-1}$. The latter correspond to significant declines in the real part of the dielectric constant of water at those frequencies.[(4)] The 1-cm^{-1} mode was treated classically and the 170-cm^{-1} quantum mechanically (**M163**). The quantum $(\mathrm{FC})_{\mathrm{solv}}$ at room temperature was only 20% different from the classical value given by Eq. (11.66), and so in the present paper we shall use Eq. (11.66) for the solvent contribution.

We turn next to the estimate of $V(r)$. An *adiabatic* model *corresponding* to the *nonadiabatic* model of Eq. (11.65) yields

$$k_{ad} = \nu\exp[-(\Delta E + \lambda)^2/(4\lambda k_BT)] \quad (11.67)$$

(cf. **M48**, **M16**, **M53** with ΔG^0 replaced by ΔE). In Eq. (11.67) ν is a typical *frequency for nuclear rearrangement*, $\nu \sim 10^{13}\mathrm{sec}^{-1}$. If one assumes at first that at some distance, for example, at the *van der Waals' contact* ($r = \sigma$), the reaction is *adiabatic* and that *it becomes nonadiabatic* for larger r's (**M170**), one can then evaluate the preexponential factor in Eqs. (11.63) to (11.65) approximately by matching Eqs. (11.63) to (11.65) with Eq. (11.67) at $r = \sigma$. Thereby expression (11.68) is obtained when the joining is made at $r = \sigma$.

$$(2\pi/\hbar)\,|V(\sigma)|^2\,(4\pi\lambda k_BT)^{-1/2} \sim 10^{13}s^{-1} \quad (11.68)$$

NOTE: From substituting Eq. (11.65) in Eq. (11.63) and comparing with Eq. (11.67).

For a reaction for which the *nuclear reorganization energy term* λ is 70 kJ/mol, the $V(\sigma)$ calculated from expression (11.68) is about 0.023 eV. If instead of expression (11.68) the reaction is nonadiabatic at $r = \sigma$, the actual value of $V(\sigma)$ is less than this, and we explore this possibility. Also, a more elaborate calculation, a Landau–Zener-type theory for the adiabatic-nonadiabatic aspect, could have been included:

For an exponential dependence of the *matrix element* on r, $V(r)$ is given by

$$|V(r)|^2 = |V(\sigma)|^2 \exp[-\alpha(r - \sigma)] \tag{11.69}$$

where $r - \sigma$ is on the average (and for spherically symmetric reactants, exactly) the edge-to-edge distance between the reactants. The theoretically estimated or experimentally inferred values of α range from 2.6 to 1.1 Å^{-1}. The value of 2.6 refers to a theoretical calculation where the electron tunnels from one reactant to the other via a vacuum (Jortner). When medium is present, a value of 1.44 Å^{-1} was roughly estimated (Hopfield), using a calculation based on an electron tunneling through a square barrier of about 2 eV.[4]

NOTE: Recall that Marcus said that tunneling comes naturally from semiclassical mechanics.

... We shall use a value of 1.5 Å^{-1}....

Method of Calculation. The equilibrium (no-reaction) steady state solution to Eq. (11.58) is $g(r) = \exp[-U(r)/k_BT]$, when the two boundary conditions (i) $\lim g(r) = 1$ as $r \to \infty$ (cf. **M170**, p. 867 top) and (ii) Eq. (11.62) at $r = \sigma$ are employed. *Reaction*

---

[4]The familiar WKB formula (A. Messiah, *Quantum Mechanics*, Vol. 1, Chapter 6, Wiley, New York (1966)) for the probability of an electron of mass m tunneling from σ to r through a potential V when E is the total energy of the electron, is

$$T \propto \exp\left\{-(2\hbar)\int_\sigma^r [2m(V(\rho) - E)]^{1/2} d\rho\right\}.$$

If in the case of a nonadiabatic ET $E - V(\rho)$ is taken to be roughly the binding energy B of the transferred electron, then $T \propto \exp[-2(2m_eB)^{1/2}(r - \sigma)/\hbar]$. Comparisons with Eqs. (11.63) and (11.69) indicate $\alpha \cong 2(2m_eB)^{1/2}/\hbar$. For $B = 2\,eV$, this expression yields $\alpha = 1.5$ Å^{-1}.

will cause deviation from this solution. If we rewrite the diffusion equation in terms of $h(r) = g(r)\exp(U/k_BT)\ldots$[(5)]"

The numerical solution to Eq. (11.58) is described and an exact solution is constructed for $g(r)$ and k_{obsd}.

"11.12. Steady State Results

With $k(r)$ determined as described previously, we are in a position to examine numerically the effect on k_{obsd} of a reaction rate constant *contributed from a range of internuclear separation distances.* The steady state (long time) solutions of Eqs. (11.56) and (11.58) will be examined first, since they are more easily found and correspond to existing experimental measurements.

The detailed calculations presented in this section are for the quenching of the bipyridyl complexes of Ru(II) by various metal(III) bipyridyl complexes studied experimentally by Creutz and Sutin."

The values of σ's of $V(\sigma)$, of λ's, of α's, of D's and of $U(\sigma)$ are investigated and discussed that would give a theoretical value of k_{calcd} in agreement with k_{obsd}.

"Calculations were made for the formation of ground state products and of an electronically excited $\mathrm{Ru^{III}bpy_3}$ product. . . .

With the parameters discussed earlier and the $k(r)$ discussed in the preceding section, we have calculated *the reactant pair distribution function* $g(r)$ and the observed rate constant k_{obsd} as a function of ΔG^0. We first test the approximate Eqs. (11.54) and (11.55), using for k_{diff} the maximum value observed for k_{obsd} (which we will call the 'experimental' k_{diff} since $k_{\mathrm{act}}^{\max} \gg k_{\mathrm{diff}}$). . . .

The results for calculated and experimental k_{obsd} versus ΔG^0 are plotted in Fig. 1 . . . .

The closeness of the solid and dash-dot curves in Fig. 1 shows that the effect of having an r-dependent λ_{out} instead of a λ_{out} fixed at $r = \sigma$ is small. The approximation used in **M30** of treating the reaction as occurring at $r = \sigma$ and as being adiabatic there agrees well with the present results (ct. solid and dash-dot curves in Fig. 1),

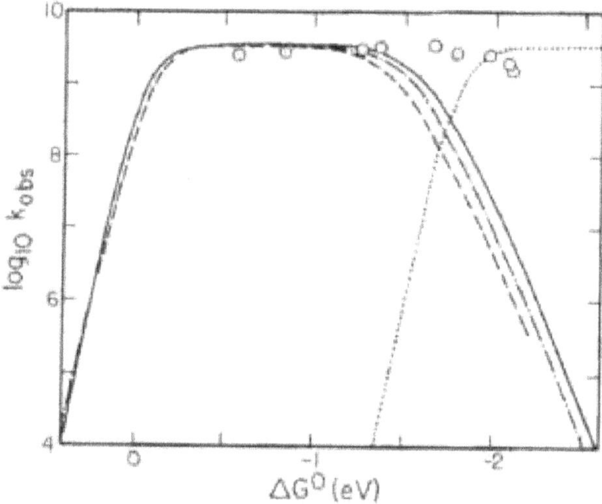

Fig. 1. Calculated and experimental rates of electron-transfer quenching of ruthenium(II) bipyridyls vs. $\Delta G°$. The experimental points (circles) are due to Creutz and Sutin.[14,15] The solid line and dotted curves are for formation of ground-state products and an electronically excited product, respectively, using an r-dependent λ_{out}, with $\lambda_{out}(\sigma) = 54$ kJ/mol and $V(\sigma) = 0.0045$ eV. The dash–dot curve is for formation of ground-state products with λ_{out} fixed at $\lambda_{out}(\sigma)$. The dashed curve is the calculation reported in Ref. [7] in which reaction occurred only at $r = \sigma$.

because of compensation. (The nonadiabaticity for the solid curve decreases the rate but the reaction over a distance causes an enhanced rate, compared with the rate for the dash-dot curve.)

To be consistent with the experimental data in Fig. 1, if one uses the above λ's, *it is necessary to introduce the formation of an electronically excited Ru(III) product*, namely, the dotted curve there."

The possibility of using other values of λ_{out} and of λ_{in} to get a better agreement of calculations and experiment is also investigated.

"In Fig. 2 calculations having a larger but still r-dependent λ_{out}... are given.... As is evident from the approximate Eq. (11.66), *the greater* λ_{out} *the less the tendency to inversion*, other things being equal. Indeed, one sees from Fig. 2 that, if λ_{out} equalled 83 kJ/mol, it would *not* be necessary to invoke the excited electronic state of Ru(III).

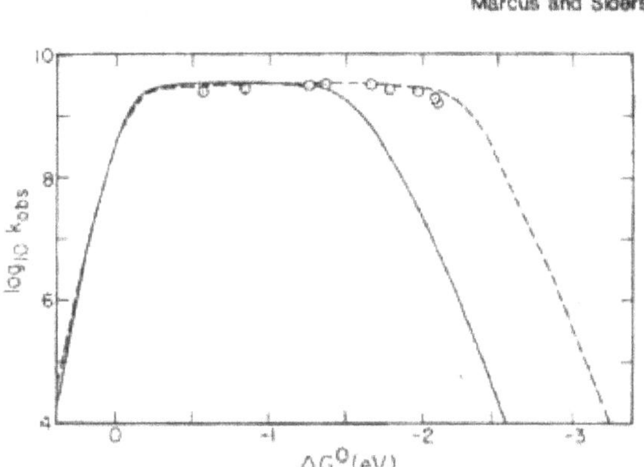

Fig. 2. Calculated and experimental rates of electron-transfer quenching of ruthenium(II) bipyridyls vs. ΔG°. The solid curve is taken from Figure 1. The dashed curve is with an r-dependent λ_{out}, $\lambda_{out}(\sigma) = 83\,\text{kJ/mol}$, and $V(\sigma) = 0.023\,\text{eV}$. The experimental points (circles) are those of Creutz and Sutin.[14,15]

11.13. Short Time Experiments

Reactions that are fast relative to diffusion are controlled by the rate of diffusion rather than by their 'activation' rates, and so *diffusion can mask interesting rate behavior*. In the case of reactions that can be induced *in a very short time*, for example, by a pulse of light, such as reaction (11.70), followed by reaction (11.71), this masking effect may be reduced. In a fast bimolecular reaction (11.71) in which the

$$\text{red}_2 + h\nu \rightarrow \text{red}_2^* \tag{11.70}$$

$$\text{ox}_1 + \text{red}_2^* \rightarrow \text{red}_1 + \text{ox}_2 \tag{11.71}$$

reactants ox_1 and red_2^* are initially randomly distributed, reaction causes the reactant pair distribution function, $g(r, t)$, to depart from its equilibrium value. Since the reactants closest together tend to react first, $g(r, t)$ becomes increasingly depleted near $r = \sigma$ as time increases. *At long time* $g(r, t)$ approaches the steady state distribution function discussed previously. However, at small t, the distribution of reactants is closer to the equilibrium one, even for quite fast reactions, and the observed rate constant is then nearer the value that it would have in the limit of infinitely rapid diffusion. That is, as $t \to 0$, k_{obsd} approaches the activation rate constant k_{act} given by Eq. (11.57). Thus, if the rates of fast reactions such as reaction (11.71) can be measured *at sufficiently short times*, the *masking effect of diffusion* can be circumvented."

NOTE: If the reaction is fast the k_{act} is big and the reaction's velocity is regulated by the smaller k_{diff} which masks then the k_{act}. But in a reaction that happens in a very short time the diffusion just hasn't time to set in and the only rate constant determining the velocity of the reaction is then the k_{act}. It is "as if" there was a big k_{diff} not determining the reaction rate.

"For simplicity of presentation, we shall consider first the time-dependent problem for the case that $U = 0$, a realistic case at the present high ionic strength. The following time-dependent solution to Eqs. (11.56) and (11.58) with $U = 0$ is well-known and will suffice to provide order-of-magnitude *estimates for the rate enhancement* to be expected at short times. When reaction occurs only at a fixed internuclear separation σ, with bimolecular rate constant k_{act}, and in the absence of long-range forces between the reactants, k_{obsd} is given by (Noyes)

$$k_{obsd}(t) = \frac{1}{\frac{1}{k_{act}} + \frac{1}{k_{diff}}}\left[1 + \frac{k_{act}}{k_{diff}} e^{x^2} \operatorname{erfc}(x)\right] \tag{11.72}$$

where $\operatorname{erfc}(x)$ is the complementary error function

$$\operatorname{erfc}(x) = (2/\sqrt{\pi}) \int_{x(t)}^{\infty} e^{-u^2} du \tag{11.73}$$

$$x = (Dt)^{1/2}(1 + k_{act}/k_{diff})/\sigma \tag{11.74}$$

NOTE: the t-dependence is in x.

and k_{act} is for reaction occurring at $r = \sigma$, but we shall use. . . .

$$k_{act} = 4\pi \int_{\sigma}^{\infty} k(r) r^2 dr \tag{11.75}$$

NOTE: This is Eq. (11.57) with $U = 0$.

D is again the sum of the reactants' diffusion coefficients. k_{diff} is the diffusion-limited rate constant and is the same as in Eq. (11.55) but with $U = 0$, i.e.

$$k_{\text{diff}} = 4\pi D\sigma \tag{11.76}$$

In obtaining Eq. (11.72) the usual boundary condition, Eq. (11.77), on the flux at $r = \sigma$, was satisfied.

$$4\pi D\sigma[\partial g(\sigma)/\partial\sigma] = k_{\text{act}} g(\sigma) \tag{11.77}$$

That is, NOTE: $k_{\text{diff}}[\partial g(\sigma)/\partial\sigma] = k_{\text{act}} g(\sigma)$

At *large t* the second term in the brackets in Eq. (11.72) vanishes, so that Eq. (11.72) reduces to the steady state expression, Eq. (11.54). As $t \to 0$, on the other hand, k_{obsd} as given by Eq. (11.72) approaches k_{act}.

NOTE: At $t = 0$, x in Eq. (11.74) equals zero, $e^{x^2} = 1$ and $\text{erfc}(x) = \sqrt{\pi}/2$ in Eq. (11.72), and for $k_{\text{act}} \ll k_{\text{diff}}$ Eq. (11.72) → Eq. (11.54).

The rate behavior for large values of k_{act} at sufficiently short times is, thus, not masked by diffusion.... At observation times of the order of 0.5 ps, which may be accessible by using present subpicosecond techniques, the rate constants are greatly enhanced, and there is a pronounced double maximum in the plot in Fig. 3[(6)] ...

It may, of course, be equally useful or more useful to look experimentally for inverted behavior in ET reactions between *redox centers that are linked chemically.* Having the reactants linked together would entirely circumvent the problem of slow diffusion. Also, if the chemical link were rigid, the reaction would be forced to occur at a single, well-defined reactant separation distance."

"11.14. Analogy between Charge-Transfer Spectrum and Plot of k_{act} versus ΔG^0

The probability of the optical dipole-induced transition from the ith vibrational level of electronic state $|a\rangle$ to the fth vibrational level of electronic state $|b\rangle$ is given by

$$\Gamma(\Delta E_1 \pm h\nu) = C \sum_{i,f} e^{-E_i/k_B T} \, |\langle i \mid f \rangle|^2 \, \delta[E_f - E_i + (\Delta E_1 \pm h\nu)] \tag{11.78}$$

using the Golden Rule and the Condon approximations. In Eq. (11.78), C is a proportionality constant

$$C = (2\pi |\langle a|\mu|b\rangle|^2 / Q\hbar),$$

ΔE_1 is the difference in energy of the zero-point vibrational levels of electronic states $|b\rangle$ and $|a\rangle$ for a particular system, and $h\nu$ is the energy of the radiation emitted (+) or absorbed (−). E_f and E_i are the vibrational energies associated with $|f\rangle$ and $|i\rangle$.

Comparing Eq. (11.78) with Eqs. (11.63) and (11.64), we see that Γ/C *is the same function of* $\Delta E_1 \pm h\nu$ *that* k_{act}/C' *is of* ΔE, where

$$C' = 2\pi |V(r)|^2 / Q\hbar.$$

Thus, since Γ, and hence Γ/C, has a maximum as a function of $\Delta E_1 \pm h\nu$ (where this argument is varied by varying $h\nu$) in the absorption plot, k_{act} *must have a maximum* as a function of ΔE. In the k_{act} versus ΔE plot, ΔE is varied by studying *a series of reactants*, by varying one of the reactants, in which (ideally) the vibration frequencies and bond lengths of this series of reactants are fixed, as are those of the corresponding products, and so the ψ_i's, ψ_f's, E_i's, and E_f's are the same for each member of the series. ΔE is the only variable in this series. Because of the constancy of the ψ_i's, etc., the ΔS^0 is also a constant, and so a plot of k_{act} versus ΔE is merely a displacement of the plot of k_{act} versus ΔG^0. In summary, the maximum in the absorption coefficient versus absorption frequency plot, well-known in charge-transfer (and other) absorption spectra, implies a maximum in the plot of k_{act} versus ΔG^0. The condition on the argument is that Eq. (11.78) provides a suitable description of the former and that Eqs. (11.63) and (11.64) adequately describe the latter."

"**Appendix B**. *Derivation of Eq. (11.54) for Reactions over a Range of r's.* We obtain Eq. (11.57) first: If diffusion is sufficiently fast, the steady state solution to Eq. (11.58) is given by the *equilibrium* expression.

$$g(r) = \exp[-U(r)/k_B T] \quad \text{(fast diffusion)} \tag{B1}$$

For $r \geq \sigma$, and $g(r \leq \sigma) = 0$. The activation (**Typo** corrected) bimolecular rate constant may be obtained by substituting this equilibrium g into Eq. (11.56), yielding Eq. (11.57).

NOTE: The *pair distribution function* $g(r)$ is the solution of Eq. (11.58). If the diffusion is infinitely fast we have Eq. (11.57) for k_{act}. We see then that diffusion is present even in k_{act} through the pair distribution function.

To obtain an approximate steady state solution (cf. **M170**) of Eq. (11.58) under other conditions (NOTE: if diffusion is, for instance, not sufficiently fast) the equation is first rewritten as

$$\frac{D}{r^2}\frac{\mathrm{d}}{\mathrm{d}r}\left[e^{-U/k_B T} r^2 \frac{\mathrm{d}}{\mathrm{d}r}\left(g e^{U/k_B T}\right)\right] = k(r)g(r) \tag{B2}$$

NOTE: in the steady state $g(r,t)$ doesn't depend on t and so the partial derivatives become ordinary derivatives...

Integration yields

$$De^{-U/k_B T} r^2 \frac{\mathrm{d}}{\mathrm{d}r}(g e^{U/k_B T})|_{r=\sigma}^{r=R} = \int_\sigma^R k(r)g(r)r^2 \mathrm{d}r \tag{B3}$$

The flux is given by $4\pi r^2 D$ times the left-hand side of Eq. (11.62), and so the left hand side of Eq. (B3) is $1/4\pi$ times the flux at $r = R$ minus that at $r = \sigma$. The condition of zero net flux across the $r = \sigma$ boundary (Eq. (11.62)) implies that in the left-hand side of Eq. (B3) the term at the lower limit $r = \sigma$ vanishes. The *unimolecular* rate constant $k(r)$ is, as discussed in the text, *a rapidly decreasing function* of r. For r greater than some distance σ', where $\sigma' - \sigma$ is still a small quantity, $k(r)$ is essentially zero. Therefore, for $R > \sigma'$ the right-hand side of Eq. (B3) may be approximately replaced by its limit at $R \to \infty$, and, because of the vanishing of the left-hand side of Eq. (B3) at its lower limit, we then have (writing ∞ instead of R) (**typo** corrected).

$$De^{-U/k_BT}r^2\frac{\mathrm{d}}{\mathrm{d}r}(ge^{U/k_BT}) = \int_{\sigma}^{\infty} k(r)g(r)r^2\mathrm{d}r\,(r > \sigma') \quad \text{(B4)}$$

Substituting Eq. (11.56) for the integral over r into Eq. (B4) allows one to rewrite the latter as

$$De^{-U/k_BT}r^2\frac{\mathrm{d}}{\mathrm{d}r}(ge^{U/k_BT}) = k_{\mathrm{obsd}}/4\pi\,(r > \sigma') \quad \text{(B5)}$$

NOTE: we have then $k_{\mathrm{obsd}} = 4\pi De^{-U/k_BT}r^2\frac{\mathrm{d}}{dr}(ge^{U/k_BT})$, where

$$k_{\mathrm{obsd}} = 4\pi\int_0^{\infty} g(r)k(r)r^2\mathrm{d}r.$$

Rearranging Eq. (B5) as $\frac{k_{\mathrm{obsd}}}{4\pi D}\frac{e^{U/k_BT}}{r^2} = \frac{\mathrm{d}}{dr}(ge^{U/k_BT})$ and integrating from σ' to ∞ yields

$$\frac{k_{\mathrm{obsd}}}{4\pi D}\int_{\sigma'}^{\infty} e^{U/k_BT}\frac{\mathrm{d}r}{r^2} = [g(r)e^{U/k_BT}]|_{r=\sigma'}^{r\to\infty} \quad \text{(B6)}$$

The potential $U(r)$ vanishes, by definition, as $r \to \infty$, and we require $\lim g(r) = 1$ as $r \to \infty$. Thus, we obtain

$$k_{\mathrm{obsd}}/k'_{\mathrm{diff}} = 1 - g(\sigma')\exp[U(\sigma')/k_BT] \quad \text{(B7)}$$

where

$$k'_{\mathrm{diff}} = 4\pi D\bigg/\int_{\sigma'}^{\infty} e^{U/k_BT}r^{-2}\mathrm{d}r \quad \text{(B8)}$$

We now proceed to evaluate the second term in the right-hand side of Eq. (B7) in terms of the activation-controlled rate constant k_{act}. If the product $\exp[U(r)/k_BT]g(r)$ varies only slowly for $\sigma < r < \sigma'$, then k_{obsd} is given (using Eq. 11.56) approximately by

$$k_{\mathrm{obsd}} \cong 4\pi g(\sigma')e^{U(\sigma')/k_BT}\int_{\sigma}^{\infty} k(r)r^2e^{-U(r)/k_BT}\mathrm{d}r \quad \text{(B9)}$$

NOTE: For the interval between 0 and σ the formula used for $g(r)$ is not the B1 formula for fast diffusion but the $h(r) = g(r)\exp(U/k_BT)$ which is used (vide infra) if reaction will cause deviation from the fast diffusion solution.

which, using Eq. (11.57), becomes

$$g(\sigma')\exp[U(\sigma')/k_BT] \cong k_{\text{obsd}}/k_{\text{act}} \tag{B10}$$

If we substitute Eq. (B10) into Eq. (B7), we obtain

$$k_{\text{obsd}}/k'_{\text{diff}} \cong 1 - k_{\text{obsd}}/k_{\text{act}} \tag{B11}$$

Because $\sigma' - \sigma$ is a small quantity, k'_{diff} is approximately equal to k_{diff}, where k_{diff} is defined as in Eq. (B8), but with σ in place of σ'. Substituting k_{diff} for k'_{diff} in Eq. (B11) and rearranging yields Eq. (11.54).

Finally, in Fig. 4, to illustrate how much or little $g(r)\exp[U(r)/k_BT]$ varies in the interval $\sigma' - \sigma$, we plot $k(r)$, $g(r)$, and $\exp[U(r)/k_BT]$ versus r, for $\Delta G^0 = -1.3\,\text{eV}$. The quantity σ' is indicated approximately, chosen so that $k(\sigma') = k(\sigma)/3$. The unimolecular rate constant $k(r)$ was calculated in the same way as for the solid line in Fig. 1. From the results in Fig. 4, the product $g(r)\exp[-U(r)/k_BT]$ varies by $\sim$20% over the interval $\sigma < r < \sigma'$. A similarly small change is observed with other values of ΔG^0. This observation suggests that it is adequate to treat $g(r)\exp[U(r)/k_BT]$ as constant for $\sigma < r < \sigma'$.

NOTES

1. p. 623 1st column middle-bottom: 'ET's can occur over a range of reactant separation distance, rather than only at a specified distance. In such cases the observed bimolecular rate constant k_{obsd} is related to the unimolecular rate constant $k(r)$, the rate of reaction of pairs of reactants having *fixed* internuclear, center-to-center, *separation distance* r, via a pair distribution

function $g(r)$:

$$k_{\text{obsd}} = 4\pi \int_0^{\infty} g(r)k(r)r^2 dr \tag{11.56}$$

In Eq. (11.56) we have assumed that k and g are radially symmetric. When the system has a $k(r)$ instead of only a k at $r = \sigma$, k_{act} is defined by using Eq. (11.56) with $g(r)$ replaced by its equilibrium value, $\exp(-U(r)/k_B T)$ for $r > \sigma$ and, in the present model, by zero for $r < \sigma$.'

Q: In **M170** $g(r)$ was defined as $g(r) = g_0(r)e^{-w(r)/kT}$ where: '$g_0(r)$ is largely a geometric factor, being similar to the value it would have for an uncharged and nonspecifically interacting pair.' Here $g_0(r)$ is missing.

M: $g(r)$ is the hard spheres pair distribution function, and then the $w(r)$ takes into account for the interaction that the hard spheres don't have, they just have an interaction at contact, otherwise there is no interaction. $w(r)$ is the work to bring the reactants together and, considering only $w(r)$, I don't take into account the oscillations that occur because of the shell-like nature of the pair reactants, the actual pair distribution function, which is sort of something like a damped oscillator. The $w(r)$ doesn't really take that into account, it might be an electrostatic effect, it doesn't take into account the structural aspect of $g(r)$.

2. p. 624 1st column middle: '$FC = \frac{1}{Q}\sum_{i,f} e^{-E_i/k_B T}|\langle i \mid f\rangle|^2\delta$ $(E_f - E_i + \Delta E)\ldots$ has dimensions of (energy)$^{-1}$.'

M: Remember the Fermi's Golden Rule: $k = \frac{2\pi|H_{12}|^2}{\hbar} FC$. The units of k should be $\sec^{-1}$. It is a first-order rate constant. The units of $|H_{12}|^2$ is energy square, then consider the units of the Planck constant, energy $\times$ seconds. A delta function has units of the reciprocal of whatever is the delta function of. For example, the integral of a delta function of energy times dE is unity, which means that δ has units of 1/energy. If the i and the f are normalized the $\langle i \mid f\rangle$ has no dimension.

3. p. 624 1st column, bottom." As has been discussed elsewhere, for example, **M163**, the quantum nonadiabatic result Eq. (11.63): $k(r) = \frac{2\pi}{\hbar}|V(r)|^2(FC)$ and Eq. (11.64): $FC = \frac{1}{Q}\sum_{i,f} e^{-E_i/k_BT}$ $|\langle i \mid f\rangle|^2\delta(E_f - E_i + \Delta E)$ plus a dynamical (harmonic oscillator) assumption for the motion of the solvent does not allow for any large entropies of reaction.[5] To avoid this difficulty one can use, instead, a more nearly correct treatment of the polar solvent, one which is classical but in which no harmonic oscillations for the solvent are assumed."

   Q: Two questions:

   (i) In—for instance—p. 484 of *Physical and Chemical Sciences Research Report 1*, **M118**, you write: "However, a major problem arises, since the interactions of the solvent molecules with each other and with the reactants are strongly anharmonic."
   How does this anharmonicity relate to the harmonic oscillations assumptions for the solvent?

   (ii) Whereas it is easy to visualize changes in entropy due to ordering or disordering of solvent molecules, it is less easy to understand how a change in vibrational frequency gives rise to a change in entropy. Is it so because of a redistribution of energy among different levels? As a reference where to look it up for it you give the book of Physical Chemistry by Moelwyn Hughes.

   M: (i) In practice we don't use harmonic oscillators for the solvent, the harmonic oscillators in practice are really just used for high frequency vibrations and solvent's high frequency vibrations aren't really taking part in most of the polarization. The answer is somewhat complex in a way, because some people actually do treat the solvent simply just like another harmonic oscillator but

[5] p. 624 bottom note: "(52) The entropy of reaction that is treated by the nonadiabatic formula, Eq. (11.63) and Eq. (11.64), arises, in the harmonic oscillator approximation, from any noncancelling changes in the vibration frequencies that typically occur in oxidation-reductions."

one should really talk of the free energy which is harmonic even when things are strongly anharmonic. Some people relate that free energy to an effective oscillator frequency. The detailed molecular frequencies are anharmonic in my view.

(ii) If you look at the expression of the free energy of a harmonic oscillator, take the derivative to get the entropy, there is some entropy associated with it due to the possibility of being in different states. Take any book in Statistical Mechanics that gives the partition function for the harmonic oscillator and look for entropy in terms of the properties of that partition function.

4. p. 624 1st column bottom: "As has been discussed elsewhere, for example, **M163**, the quantum nonadiabatic results Eqs. (11.63)

$$k(r) = \frac{2\pi}{\hbar}|V(r)|^2(FC) \tag{11.63}$$

and (11.64) plus a dynamical (harmonic oscillator) assumption for the motion of the solvent does not allow for large entropies for any large entropies of reaction. To avoid this difficulty one can use, instead, a more nearly correct treatment of the polar solvent, one which is classical but in which no harmonic oscillations for the solvent are assumed. In this case the FC factor for the solvent is (Kestner, Logan, Jortner)

$$(FC)_{\text{solv}} = (4\pi\lambda_{\text{out}}k_BT)^{-1/2} \times \exp[-(\Delta G^0 + E_f^v - E_i^v + \lambda_{\text{out}})^2/(4\lambda_{\text{out}}k_BT)] \tag{11.66}$$

where the v superscripts denote vibrational energy. Equation (11.66) may be compared with the quantum results that we obtained in **M163** where a quantum treatment of the solvent was used, described by two modes which have frequencies of 1 and 170 cm^{-1}. The latter correspond to significant declines in the real part of the dielectric constant of water at those frequencies..."

Q: (i) Please explain the relation of entropy and "the significant declines in the real part of the dielectric constant."

M: Water has a high dielectric constant because it responds by rotations, really by hindered rotations, so it changes in the electric field. The dielectric constant would be much less if only vibrations were considered, they don't have a lot of entropy associated with them. In other words, a lot of that dielectric constant of 78 is due to the hindered rotational part of the motion of the water molecules. And then there is a lot of entropy associated with them. Once you get to high frequencies, the rotations can't respond, they're too slow, and so what happens initially is that you have the vibrations responding. The vibrations don't have the huge entropy changes normally, but rotations have. A vibration is vibrating a little bit in space but rotation can be rotating over a large part of space, large part of angular space, and so there is a lot of entropy associated with it, not with vibrations. So, if you look up for example for any species the contribution to the absolute entropy of a molecule, you will see that rotations play a big role, the vibrations usually play a relatively little.
The decline for example of the dielectric constant from 78 down to about 4.5 is largely due to the rotations not responding, the part from 4.5 to about 1.8 is due to the bending vibration not responding well at high frequencies.
The imaginary part is associated with absorption, the real part is what appears in $\mathbf{D} = \varepsilon\mathbf{E}$, often you write the real part there, so that's the part that one usually means by dielectric constant, but of course the dielectric constant is treated more generally in terms of the relationship between the dielectric displacement, the electric field or the polarization, etc., has both real and imaginary parts, if you look at the frequency dependence. And the imaginary parts are associated with absorption. And imaginary parts are associated to the real parts by the Kramers–Kronig relation.

Q: (ii): Why two frequencies and how were they chosen? Are they related to the optical and static dielectric constants?

(iii): Please explain what it means to "correspond to significant declines, and so on."

M: The two frequencies are not related to the optical and static dielectric constants. If you look at a plot of real and imaginary parts of the dielectric constant versus frequency, then you find two frequencies, one is electronic, the other is vibrational, rotational, everything. If you plot the imaginary part you would see a peak showing that there is absorption, it may be the infrared vibrational polarization part, it's associated with a bending motion of H_2O, those are dipoles that adjust to a field, which is called vibrational polarization, when you polarize a solvent you can reorient the dipoles. I give this plot in one of my papers, see Fig. 1 in **M283** with Song. There are two contributions there.

Q: Please consider the plots, Fig. 1*, from Böttcher's "Theory of Electric Polarization." Is the upper curve the real part and the bottom curve the imaginary part? I have sent also a plot, Fig. 2*, from Kittel's "Elementary Solid State Physics."

M: I think so, because the imaginary part is an absorption and you see what you have around 14 that's an absorption. The bottom part is the imaginary part.

5. p. 625 1st column high: "*Method of Calculation.* The equilibrium (no-reaction) steady state solution to Eq. (11.58) is $g(r) = \exp[-U(r)/k_BT], \ldots$ If we rewrite the diffusion equation in terms of $h(r) = g(r)\exp(U/k_BT)\ldots$"

Q: Substituting for $g(r)$ in the last equation one has $h(r) = \exp[-U(r)/k_BT]\exp(U/k_BT)$.

M: The U is really $U(r)$. This is what you have when there is no reaction. When there is no reaction $h(r)$ is identically unity.

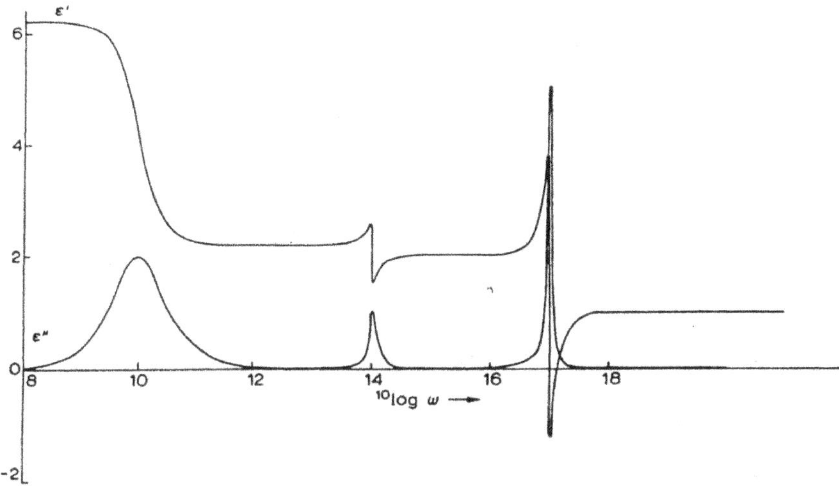

Fig. 1*. Dielectric dispersion and loss for a polar compound in the condensed phase. (Adapted from "Böttcher, Theory of Electric Polarization," Vol. II, Elsevier 1978).

6. p. 627 2nd column top: "Fig. 3 shows the behavior of k_{obsd} predicted by Eq. (11.72)

$$k_{\text{obsd}}(t) = \frac{1}{1/k_{\text{act}} + 1/k_{\text{diff}}} \left[1 + \frac{k_{\text{act}}}{k_{\text{diff}}} e^{x^2} \operatorname{erfc}(x(t)) \right] \quad (11.72)$$

at various times from $t = 0$ to $t = 1\ \mu$s. The time $t = 1\ \mu$s is sufficiently long that a steady state has been reached... At observation times of the order of 0.5 ps... the rate constants are greatly enhanced and there is a pronounced double maximum in the plot in Fig. 3..."

Q: **Erratum** Maybe it would be better to say that both curves in Fig. 3, one referring to a complex and the other to another complex, show a maximum. There are really two maxima, not a double maximum.

M: Yes, one shouldn't call it a double maximum. They are referring to two different systems.

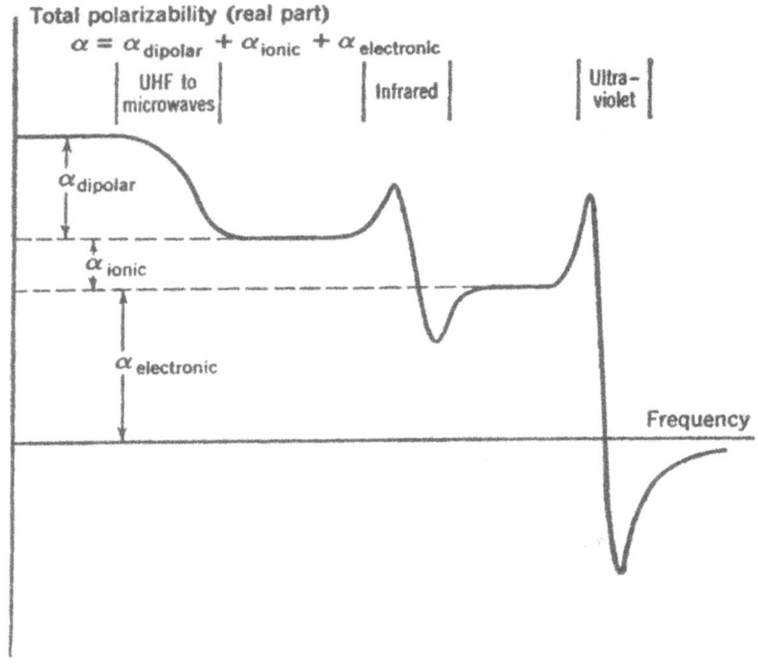

Fig. 2*. Frequency dependence of the several contributions to the polarizability (schematic). (Adapted from Kittel, "Elementary Solid State Physics," Wiley 1962).

M177. Electron and Proton Transfer

11.15. "Introduction

In the early days of ET one text which I found particularly helpful was Robinson and Stokes, *Electrolyte Solutions*. It provided an overview, as well as a detailed and current picture of electrolyte solutions, to one newly arrived in the field. It is a real pleasure to acknowledge in this Memorial Lecture my debt to Professor Robinson.

In the intervening years a number of Faraday Society Discussions related to the subject matter of the present Discussion have been held, including those on Oxidation-Reduction Reactions (1960), Proton Transfer Processes (1965), and Electrode Reactions (1968).

The present Discussion embraces all three and so emphasizes a trend whereby *a formalism has been developed which attempts to unify the three different fields and which has been extended to an increasingly broad class of reactions in chemistry.* We survey some of the developments in this area in the present lecture.

11.16. Weak-Overlap Electron Transfers

One of the virtues of studying simple weak-overlap ET's has been the absence of bond-breaking and bond-forming processes, with all their attendant uncertainties regarding the PES in the TS region. In this way it was possible, with approximate models for the coordination shell and for the solvent outside it, to allow f*or the reorganization prior to and following* the ET act. Libby, in his pioneering and stimulating suggestion on the application of the Franck–Condon principle to ET, thought of the coordination shell and other changes as arising from a *vertical* transition, as in spectroscopy, rather than of *a prior reorganization*. We introduced, instead, *a prior and a post reorganization....*

The apparent simplicity of the weak-overlap ET reaction, and certainly of the model permitted a detailed analysis (**M41**, **M48**, **M53**, **M71**, **M118**, **M129**) of topics such as the effect of driving force ΔG^0 on the reaction rate, effect of molecular parameters and of solvents (when not specifically interacting with the reactants) on the rate, relation between cross-reaction rates and those of isotopic exchange reactions, relation of homogeneous reactions rates to charge-transfer spectra (Hush), chemiluminescent ETs (**M57**), relation to electrochemical (EC) ETs, effect of driving force (activation overpotential) on the EC rate constant, and effect of amount of charge transferred. The interaction between experiment and theory in these fields has provided an exciting experience, a source of pleasure and occasionally of dismay. In the late 1950s and early 1960s, I visited Brookhaven National Laboratory and spoke often with Dick Dodson and Norman

Sutin, who were doing pioneering experiments in the field. Those visits were particularly stimulating. Taube was of course making giant strides, but at that time I did not have much contact with him.

Chemistry, of course, embraces much more than ET reactions, and it became natural to think about the relation of the formalism developed for weak-overlap ETs to other more complicated reactions (**M69**). (NOTE: another attempted unification...) The main ingredients of the formalism include work terms, w^r and $-w^p$, not necessarily coulombic, for bringing the reactants together and for separating the products, the *intrinsic barrier* $\lambda/4$ (additively related for the cross reaction (CR) to those of isotopic exchange reactions) and the standard free energy of reaction of the elementary ET step, ΔG^0, (Bixon, Jortner, **M41**, **M48**, **M53**, **M71**, **M118**, **M129**)

$$k \approx \kappa Z \exp(-\Delta G^{\ddagger}/kT) \tag{11.79}$$

where, in a classical treatment of the nuclear motion,

$$\Delta G^{\ddagger} = w^r + \frac{\lambda}{4}\left(1 + \frac{\Delta G^{0\prime}}{\lambda}\right)^2 \tag{11.80}$$

with $\Delta G^{0\prime} = \Delta G^0 + w^p - w^r$. κ *describes the nonadiabaticity* of the reaction ($\kappa \approx 1$ for an adiabatic reaction) and Z is the bimolecular collision frequency. Analogous equations are obtained for *unimolecular reactions*, for *EC reactions* and for *reactions at an interface*, in which each reactant is in a different phase. Z is replaced by the *appropriate analogue* in each case, and in the EC case ΔG^0 is replaced by the *activation overpotential.*

11.17. Other Classes of Reactions

Any *extension* of the concepts of ET to other classes of reactions must be aware of the differences.

NOTE: Marcus theory tends to unify and to extend.

In AT reactions, for example, *simultaneous* bond breaking and forming occur and *cannot be treated* by *a pair of intersecting harmonic-oscillator PE* or *quadratic free energy surfaces* (NOTE: this is the great advantage of the ET reactions). For this reason a rather different *simple model* was considered (**M69**), one which originated with Harold Johnston (BEBO). When further simplified (with PEs replaced in an intuitive way by free energies—forward and reverse rate constants obey microscopic reversibility **M69**)[(1)] and with work terms added, this yielded Eq. (11.81), with $\Delta G^{\ddagger}$ now given by

$$\Delta G^{\ddagger} = w^r + \frac{\lambda}{4} + \frac{\Delta G^{0\prime}}{2} + \frac{1}{2}\frac{\Delta G^{0\prime}}{y} \ln \cosh y, \quad y = 2\Delta G^{0\prime}(\ln 2)/\lambda \tag{11.81}$$

and with λ having the additivity property as before.

Provided that $|\Delta G^{0\prime}|/\lambda$ is less than *and not too close to unity*, this expression is well represented by the slightly simpler quadratic expression, Eq. (11.80) (**M69**). Equation (11.81) or, more usually Eq. (11.80), has now been applied to ATs, proton transfers (PTs), methyl-radical transfers, hydride-ion transfers, and *concerted* PTs. . .

In the electrochemical PT case two alternative opinions, discussed here by Krishtalik, have arisen and an effort at a unifying theory which included both as limiting cases was made (**M154**). Several central questions are those such as the following: *how far does the proton jump*, and hence *how much solvent rearrangement* has to occur? (In ETs, the center-to-center jump distance is always quite large, even when the reactants are in van der Waals' contact.)[(2)] When is the reaction coordinate in the TS region the *protonic coordinate*[(3)] and when, as expected for *sufficiently highly exothermic and thereby barrierless reactions*, is it the *intermolecular separation coordinate* for the two reactants?

11.18. Quantum Effects

Quantum corrections to Eqs. (11.79) and (11.80) are relatively minor at room temperature for typical reactions in the '*normal region*' ($\Delta G^{0\prime} < \lambda$), for example, see **M163**. The corrections become larger *in the inverted region* and *at low temperatures*. The *classical* Eqs. (11.79) and (11.80) have a simplicity which facilitates their application to and testing by experiment. In some cases, for example in the cross-relation between the rate constants of cross reactions and those of exchange reactions, it has been possible as a result to eliminate by cancellation the individual molecular properties, and so relate the rate constants to each other and to the equilibrium constant (the 'cross-relation').

NOTE: after the lack of bond breaking and making and the use of an approximate model, the above cancellation is a further reason for success of the simple classical theory.

To illustrate some of the features of the *quantum-mechanical* rate expression we consider for simplicity the case of a very highly exothermic nonadiabatic reaction. In the quantum theory of nonadiabatic ET reactions Levich and Dogonadze adapted to the problem, as they pointed out, an earlier result of Kubo and Toyozawa developed for other processes. The theory and its ensuing development employs what is now known as the *theory of radiationless transitions*.[4] The rate constant for ET between *fixed* sites is given by

$$k = \frac{2\pi|V^2|}{\hbar}(\text{F.C.}) \tag{11.82}$$

where we use the notation of Bixon and Jortner in this discussion. (F.C.) is the Franck–Condon factor and V the matrix element for the *ET transition*. One problem in ETs in polar media, as compared with radiationless transitions *involving only intramolecular vibrations*, is that the former can have *huge entropies of reaction*. The latter are not adequately modeled by quadratic potential energy functions, even though a suitable and applicable *free energy* function may be fairly quadratic as a function of some *charging parameter.*[5] As a result, an approach has been adopted in which the classical expression Eq. (11.80), which *does allow* for large possible entropy changes, is introduced into the *solvent contribution in (F.C.)*, to replace a quadratic potential energy expression there (Kestner, Logan, Jortner).

For the case in which the intramolecular vibration frequencies are high enough, and the reaction sufficiently exothermic, (F.C.) is given by

$$\text{(F.C.)} = \sum_{\nu=0}^{\infty} e^{-S} \frac{S^\nu}{\nu!} \frac{\exp[-(\Delta G^0 + \lambda_0 + \nu\hbar\omega)^2/4\lambda_0 kT]}{(4\pi\lambda_0 kT)^{1/2}} \tag{11.83}$$

where, for notational brevity, the oscillators have been taken to have a common angular frequency ω. S is the contribution λ_i of these vibrations to the λ of Section 11.16, in units of $\hbar\omega$, and λ_o is the contribution to λ of the solution outside the coordination shell.[6] One sees that the effect of the high frequency vibrations in this highly exothermic case is, *like* λ_o, to absorb large amounts $\nu\hbar\omega$ of the excess energy. In effect, it *reduces this exothermicity* and makes the reaction faster than would be the case if none were absorbed ($\nu = 0$).

Equation (11.83) is, of course, still a little cumbersome... There are approximations which we can introduce, which reduce the sum in Eq. (11.83) to a single term: We replace $\nu!$ by a continuous function, the gamma function, $\Gamma(\nu + 1)$, replace the sum by an integral over ν, and treat the integrand as a Gaussian about some maximum, which for convenience will be denoted by ν itself. If $\Gamma(\nu + 1)$ is then replaced by Stirling's formula, one obtains

$$\text{(F.C.)} \cong e^{-S} \frac{S^\nu \exp[-(\Delta G^0 + \lambda_o + \nu\hbar\omega)^2/4\lambda_o kT]}{\Gamma(\nu + 1)\hbar\omega} \tag{11.84}$$

Where ν is the solution of a transcendental equation, which when approximated by one iteration simplifies to

$$\nu = \frac{-\Delta G^0 - \lambda_o}{\hbar\omega} - \frac{2\lambda_o kT}{(\hbar\omega)^2} \ln \frac{(-\Delta G^0 - \lambda_o)}{S\hbar\omega}^{(7)}, \quad (11.85)$$

In Table 1, there is a comparison of "exact" and approximated FC factors.

"The *linear dependence* of $\ln k$ on ΔG^0 for these *highly exothermic reactions*, the well-known *energy-gap* law (Siebrand), is also seen from Eqs. (11.84) and (11.85) (cf. steepest descent derivation of that law): replacement of $\Delta G^0 + \nu\hbar\omega + \lambda_o$ in the exponent in Eq. (11.84) by the logarithmic term in Eq. (11.85) largely removes that ΔG^0,[(8)] while linear dependence on ν of the exponent of $S^\nu/\Gamma(\nu+1)$, that is, in $(S/\nu)^\nu$, and roughly of ν on ΔG^0 via Eq. (11.85) yields the gap law[(9)] ... Apart from an entropic term, $-(\Delta G^0 + \lambda_o)$ is the energy of the $0 \to 0$ transition, and so the equations relate the rate constant to the frequency of that transition. ...

11.19. The Inverted Region

Outside the range $|\Delta G^{0\prime}| < \lambda$ there is a considerable difference between Eqs. (11.80) and (11.81). Whereas Eq. (11.80) shows a decrease of rate with increasing driving force in the *inverted region* where $|\Delta G^0|/\lambda > 1$, Eq. (11.81) displays no such phenomenon. This difference is readily understood when one considers how the PES leading to Eq. (11.80) differs from that leading to Eq. (11.81).

While the quantum corrections to Eq. (11.79) in the inverted region significantly reduce the inverted effect, unless the relevant vibration frequencies are sufficiently low, *they do not eliminate it.* The inverted effect was predicted (**M30**) in 1960. Its analogue in radiationless transitions, the energy-gap law of Siebrand is well-known... The effect has also been invoked to explain the apparent slowness of the back reaction in bacterial photosynthesis. In this reaction, the bacterial chlorophyll dimer cation and bacterial pheophytin anion are *fixed*, presumably, rather than mobile, in the membrane. The inverted effect has also been invoked to explain the relatively larger rate constant estimated for forming a triplet state in this back reaction, compared with that estimated for the back reaction forming the ground state singlet, **M145**"

A series of other examples are reported...

"Effects which can thwart the observation of an inverted region in *bimolecular solution reactions* are several-fold

(i) Masking by diffusion control
(ii) The existence of alternative mechanisms, such as

  (a) Reactions *via* exciplexes
  (b) AT, or
  (c) Formation of electronically excited products, which reduces the magnitude of ΔG^0 for the elementary step.

Picosecond studies have been proposed to reduce the diffusion-control effect (**M174**). The AT alternative can be reduced in attractiveness by keeping the reactants physically separated

(i) In different phases or
(ii) In a frozen medium
(iii) By holding them apart by rigid chemical bonds or
(iv) By using a suitable choice of reactants with atoms which cannot undergo AT.

There is a predicted relationship between the inverted region and the high frequency tail of the related charge-transfer spectrum, when the weak-overlap and Condon approximations can be made for both (**M174**). When that high frequency tail is not obscured by a new absorption band it should be quite revealing of what to expect for the related *thermal* ET rate constant in the inverted region in this weak-overlap case.[10]

11.20. Effect of Separation Distance

One of the newer areas of interest has been the effect of separation distance r on the rate of ET. This effect, which provides a connection between geometry and rates, has been of considerable interest in biological ETs. Here reactants, more or less fixed in a membrane, may not have the *close contact* that they do in solution, and their ET rates may be dominated by this factor... when the rate constant at an edge-to-edge separation distance r, $k(r)$, behaves as

$$k(r) = k_0 e^{-\alpha r} \tag{11.86}$$

The maximum rate of ET is, at close contact, ca. $10^{13}\,\text{sec}^{-1}$[(11)] ... J. R. Miller (personal communication) estimated from experimental data that the maximum k_0 for reactions between molecules and/or ions was ca. $10^{14}\,\text{sec}^{-1}$ or somewhere between 10^{13} and $10^{14}\,\text{sec}^{-1}$ ... I have chosen a value of $10^{13}\,\text{sec}^{-1}$ so that Eq. (11.86) will yield at $r = 0$ *the maximum it can be*, which is *the frequency of nuclear motion*, $10^{13}\,\text{sec}^{-1}$. If, experimentally, k_0 proves to be $10^{14}\,\text{sec}^{-1}$, Eq. (11.86) can still be used but will break down for r values below that given by $\exp(-\alpha r) = 10^{13}/10^{14} = 0.1$, the reaction then becoming adiabatic at that r (ca. 2 Å)."

NOTES

1. p. 8 bottom: "...a rather different simple model was considered, one which originated with Harold Johnston (BEBO). When further simplified (with potential energies replaced in an intuitive way by free energies—forward and reverse rate constants obey microscopic reversibility)"

   M: The ratio of the forward and reverse rate constants is equal to an equilibrium constant, so they are related, and that relation is sometimes called microscopic reversibility. Now, I'm trying to remember, it may be that microscopic reversibility has a sharper definition, maybe when you're looking at individual contributions to the rate constants as one is related to the reciprocal of the other... a rate constant is a sum over many cross sections, etc., and the cross sections can be related by microscopic reversibility. What I don't remember is how microscopic reversibility is conventionally used, either is conventionally used with respect to the rate constants or with respect to the components of the rate constants, which are the rates at a much more detailed level than the averaging over rate constants.

If you want microscopic reversibility you can't use just energies because entropy comes in, you have to use free energy. The correct terminology for microscopic reversibility may be at a more detailed level than rate constants, I just don't remember. I know it is used at a more detailed level than rate constants but what I don't remember is if it also included the relationship between the rate constants themselves.

Q: Who did first discover the microscopic reversibility?

M: It must go way back... I wouldn't be surprised that there is something in Boltzmann... because he used an assumption which kind of destroyed the microscopic reversibility, the chaos assumption, I wouldn't be surprised if he knew it.

2. p. 9 top: "(In ETs, the center-to-center jump distance is always quite large, even when the reactants are in van der Waals contact.)"

   M: The reactants are in van der Waals contact when essentially you treat them as hard spheres, the spheres are touching each other, then under those conditions the center-to-center distance is very large, because the ions have typically the water molecules surrounding them that are strongly bound, so, you know, if the radius of an ion is, let's say, 1 Å, and you add 2.8, the diameter of a water molecule, so 2.8 and 1 is 3.8 so already you have a 7.6 Å center-to-center distance.

3. Following: "When is the reaction coordinate in the TS region the protonic coordinate..."

M: The *protonic coordinate* is a distance which locates the proton with respect to its initial and final positions. That would be what you want, you wouldn't want to know just how far the proton is from where it started, but you really would want to know the *proton reaction coordinate*, so to speak.

4. p. 9 middle: "In the quantum theory of nonadiabatic ET reactions Levich and Dogonadze adapted to the problem, as they pointed out, an earlier result of Kubo and Toyozawa developed for other processes. The theory and its ensuing developments employs what is now known as the theory of radiationless transitions."

   M: Siebrand and Jortner were involved in radiationless transitions work... Levich and Dogonadze did ET. The formalism for radiationless transitions and for ET is very similar, just the matrix element is different.

5. p. 9 middle-bottom: "One problem in ET in polar media, as compared with radiationless transitions involving only intramolecular vibrations, is that the former can have huge entropies of reaction. The latter are not adequately modeled by quadratic potential energy functions, even though a suitable and applicable free energy function may be fairly quadratic as a function of some charging parameter."

   Q: When is the free energy function suitable and applicable and when is it fairly quadratic?

   M: When you're transferring charge, if you go from, say, a plus and a minus to neutral, there is a tremendous sort of increase in the freedom of the solvent molecules surrounding the reactants, so you get a huge entropy change, so when you're transferring charge, unless is a *charge shift reaction*, you can get a huge change of entropy associated with the changing motion of the surrounding solvent.

   The quadratic energy function has a very little entropy contribution, it is all vibrations, there are no rotations in it. When you have a thing with a lot of rotations you can have something like the central limit theorem which still makes the free energy quadratic but the entropy change can be huge. The free energy function that I use had in it a lot of rotation that occurred, internal rotations, so it was suitable. It happens with that kind of reaction that the fluctuations from nonequilibrium—and I was dealing with fluctuations—are, as far as free energy is concerned, largely quadratic, as a result of the central limit there. Those fluctuations may be applicable for ET reactions and may not be applicable for reactions in which there are breaking bonds.

6. p. 9 bottom: "For the case in which the intramolecular vibration frequencies are high enough, and the reaction sufficiently exothermic (F.C.) is given by

$$(\text{F.C.}) = \sum_{\nu=0}^{\infty} e^{-S} \frac{S^{\nu}}{\nu!} \frac{\exp[-(\Delta G^0 + \lambda_0 + \nu\hbar\omega)^2/4\lambda_0 kT]}{(4\pi\lambda_0 kT)^{1/2}} \tag{11.83}$$

where, for notational brevity, the oscillators have been taken to have a common angular frequency ω. *S is the contribution λ_i of these vibrations to the* λ of Section 11.16, in units of $\hbar\omega$, and λ_o is the contribution to λ of the solution outside the coordination shell."

M: Just take two vibrational parabolas, think of the overlap integral using quantum mechanics, that's what $\frac{S^{\nu}}{\nu!}$ is. In other words, if you look at the a problem involving overlap of wave functions, FC factors, and you have two parabolas and you define an S which is $\lambda_i/h\nu$, then if you're starting from the ground state of the first one, going to various states of the other, you get that factor there, e^{-S}. If you are starting from a general state, the whole distribution of states, you get a Bessel function then.

7. Following: "Eq. (11.83) is, of course, still a little cumbersome... There are approximations which we can introduce, which reduce the sum in Eq. (11.83) to a single term: We replace $\nu!$ by a continuous function, the gamma function, $\Gamma(\nu + 1)$, replace the sum by an integral over ν, and treat the integrand as a Gaussian

about some maximum, *which for convenience will be denoted by* ν *itself.* If $\Gamma(\nu+1)$ is then replaced by Stirling's formula, one obtains

$$(\text{F.C.}) \cong e^{-S}\frac{S^{\nu}\exp[-(\Delta G^0+\lambda_o+\nu\hbar\omega)^2/4\lambda_o kT]}{\Gamma(\nu+1)\hbar\omega} \tag{11.84}$$

where ν is the solution of a transcendental equation, which when approximated by one iteration simplifies to

$$\nu = \frac{-\Delta G^0-\lambda_o}{\hbar\omega} - \frac{2\lambda_o kT}{(\hbar\omega)^2}\ln\frac{(-\Delta G^0-\lambda_o)}{S\hbar\omega}\text{"} \tag{11.85}$$

M: ν was a vibrational number, you can go to various vibrational states ν, and if you go to the maximum then you should write $\nu_{\max}$ but instead of having extra notation I callad it ν. To get Eq. (11.85) from Eq. (11.84) I just take a log of both sides, differentiate with respect to ν and find where the maximum is. I imagine that's what I did.

8. Following: "The *linear dependence of* $\ln k$ *on* ΔG^0 *for these highly exothermic reactions*, the well-known *energy-gap* law, is also seen from Eqs. (11.84) and (11.85) (cf. steepest descent derivation of that law): replacement of $\Delta G^0+\lambda_o+\nu\hbar\omega$ in the exponent in Eq. (11.84) by the logarithmic term in Eq. (11.85) largely removes that ΔG^0..."

   M: It was the use of the steepest descent that basically did it. You have an integral and if you write it as e^{-g} ... then you expand g in the quadratic essentially, so basically that's what I did.

   The replacement is not something which is arbitrary, it was something that came about as a result of the maximization process, and introduced symbols, that's all. The replacement is something that follows from a procedure.

9. Following: "linear dependence on ν of the exponent of $S^{\nu}/\Gamma(\nu+1)$, that is, in $(S/\nu)^{\nu}$, and roughly of ν on ΔG^0 via Eq. (11.85) yields the gap law."

M: The energy gap law is proportional to $e^{-\text{some coefficient} \times \text{some energy change}}$, and so not some energy change squared or cubed, just some energy change, in other words linear in the energy there, and that's where the linearity is.

10. p. 11 bottom: "There is a predicted relationship between the inverted region and the high frequency tail of the related CT spectrum, when the weak-overlap and Condon approximations can be made for both (**M174**). When that high frequency tail is not obscured by a new absorption band it should be quite revealing of what to expect for the related *thermal* ET rate constant in the inverted region in this weak-overlap case."

    M: Well, if you look at the absorption and look at the theory of the absorption, the part to the left of the maximum corresponds to the normal region in the corresponding ET rate problem and the part to the right to the maximum, which I called the high frequency tail, the part to the right to the maximum corresponds to the inverted region. If you write down the two expressions, there is a formal correspondence.

    The energy gap law was discovered about 1964 to 1965, much later than the inverted free energy effect.

11. p. 12, middle: "The maximum rate of ET is, at close contact, ca. $10^{-13}\,\text{sec}^{-1}$."

    Q: What if Donor and Acceptor are at a fixed distance with no vibrations? Why this time typical of the frequency of nuclear motion?

    M: You have to use some other coordinate to induce ET then. There may be another coordinate, for instance, if you have two species that are rattling around in a cage, it's a low frequency coordinate. It is hard to say, when there are various motions, what is carrying at most across the TS, that is one possible motion, but vibrational motion may do, some solvent motions might do it, a number of motions can be carrying across that $N-1$-dimensional hypersurface. The vibration would be the quickest one but there may be others just as important, in other words it is sort of a complicated mess there, many motions that are carrying it across, that's why in the 1965 paper I sort of threw everything into one basket. I didn't know what the main component was at the molecular level.

That 10^{-13} sets it very roughly at the vibrational frequency of nuclear motion. If you have pure vibrations and they are high frequency vibrations, and there are no translations that were permitted, no rotations permitted... vibrations are of different frequencies, for instance for a 1000 wave numbers then it would be 1000 times 3×10^{10}, it'd be 3×10^{13}, so it depends...

M194. Nonadiabatic Processes Involving Quantum-like and Classical-like Coordinates with Applications to Nonadiabatic Electron Transfers

"Nonadiabatic processes may involve both classical-like and quantum-like coordinates.[1] A semiclassical analysis is used to treat the contribution of the former to the Franck–Condon factor in the reaction rate expression, thereby avoiding the usual harmonic oscillator approximation. Microcanonical and canonical rate constants are calculated, yielding an expression which includes contributions from both types of coordinates. The results are applied to nonadiabatic ET reactions in solution and show how ΔG^0 enters the final rate expression, even though ΔE^0 is present in the initial Golden-Rule nonadiabatic formula. This result avoids an approximation which has arisen in the nonadiabatic ET literature.

11.21. Introduction

In nonadiabatic processes, some of the degrees of freedom may be treated as largely classical, and for them it is useful to *simplify* the usual Golden-Rule expression for the *transition rate* of the process. One such example occurs in nonadiabatic ET reactions in solution. In ET reactions, the many *orientational* coordinates of the solvent molecules are typically classical-like and play an important role in ET reactions (**M48**, **M16**, **M53**), considerably influencing their rate.

One purpose of the present paper is to provide insight and justification for a physically reasonable but somewhat *ad hoc* procedure which has sometimes been used for nonadiabatic ETs (**M174**). In that procedure, the *quantum FC overlap factors* for some coordinates are combined with a classical treatment (**M16**, **M53**) for the rearrangement of the remaining coordinates, typically the orientational coordinates of the solvent molecules.

In the final theoretical expression for the reaction rate, sometimes ΔG^0 but more frequently (Levich, Dogonadze, Jortner, Hopfield) ΔE^0 is written, where ΔG^0 is the standard free energy of reaction. ΔE^0 is the standard energy change of reaction, which occurs in the Golden-Rule expression for nonadiabatic processes. The appearance of ΔE^0, instead of ΔG^0, in the relevant final equations will result from using in such expressions a *harmonic oscillator* approximation for all the classical-like coordinates. For harmonic oscillators having the same frequencies for reactants as for products, ΔS^0 vanishes and so ΔE^0 and ΔG^0 become equal.[6]

For reactions in solution, however, the harmonic oscillator model is inadequate for treating the orientations of the solvent molecules. There can be very large entropic changes accompanying reaction, for example. This approximation is avoided in the present paper by the use of semiclassical theory and generalized (curvilinear) coordinates.

In simplifying the nonadiabatic expression we retain that part of the FC factor which relates to any highly quantized degrees of

[6](a) When $\Delta S^0 = 0$, $\Delta G^0 = \Delta H^0$. ΔH^0 is negligibly different from ΔE^0 for reactions in solution. (b) The entropy of activation (and of reaction) has also been obtained for the particular case that all modes of motion are treated as harmonic oscillators, but of different frequencies for reactants and products, by R.R. Dogonadze, A. M. Kuznetsov and M. A. Vorotyntsev... Such an entropy change is expected to be relatively small in comparison with major changes resulting in from changes in solvation and associated with restricting or freeing the orientations of solvent molecules.

freedom and only use semiclassical theory to *convert* the remaining part to a classical version. The final result also provides extension of an expression used for 'surface hopping' (Tully) between two electronic surfaces via classical trajectories, by allowing some coordinates to be treated purely quantum-like.

11.22. Theory

The Golden-Rule expression for the *unimolecular* rate constant $k_{\alpha\beta}$ for a nonadiabatic transition $\alpha i \to \beta f$ from electronic state α to electronic state β and from a quantum state i of the nuclear motion on surface α to a state f on surface β is

$$k_{\alpha\beta} = \frac{2\pi V^2}{\hbar} |\langle \varphi_f \mid \varphi_i \rangle|^2 \Delta(E_f - E_i + \Delta E^0), \tag{11.86a}$$

where V is the electronic matrix element for the *transition* (the Condon approximation is made), f denotes the set of quantum numbers ($f_1 \cdots f_M$ for the nuclear motion in electronic state β (M coordinates), i denotes those ($i_1 \cdots i_M$) for that *motion in electronic state* α, E_f and E_i are the corresponding energies, in excess of the zero-point energy of the β state and of the α state, respectively, and ΔE^0 is the standard energy of the reaction for process $\alpha \to \beta$ at 0 K. Equation (11.86a) has been extensively used in the nonadiabatic ET literature (**M174**, Levich, Dogonadze, Efrima and Bixon, Warshel, Kestner, Logan and Jortner, Ulstrup, Hopfield). In that case, α and β denote electronic states of the reactants and products, respectively.

NOTE: In Eq. (11.86a) it is clearly stated that we are considering the rate constant for transition between two *electronic states.* Note moreover that in Eq. (11.86a) the Condon approximation is used for the calculation of V and that the Franck–Condon factors $\langle i \mid f \rangle$ appear.

We shall suppose that of the M degrees of freedom the first N are classical-like and write

$$\varphi_i = \psi_i \chi_I, \quad \varphi_f = \psi_f \chi_F, \tag{11.87}$$

Where ψ_i (**typos** corrected) refers to $i = 1, \ldots, N$, χ_I to $I = N + 1, \ldots, M$, and ψ_f and χ_f to $f = 1, \ldots, N$ and $F = N + 1, \ldots, M$, respectively.

It is convenient to denote that part of the energies E_i and E_f in Eq. (11.86a) associated with the N classical-like coordinates by ϵ_i and ϵ_f and the part associated with the remaining coordinates by E_I and E_F respectively. We define ΔE^0_{IF} as the effective ΔE^0 for a transition $\alpha I \rightarrow \beta F$

$$\Delta F^0_{IF} = \Delta E^0 + E_F - E_I. \tag{11.88}$$

Equation (11.86a) now becomes

$$k_{\alpha\beta} = \frac{2\pi V^2}{\hbar} |\langle F \mid I \rangle|^2 |\langle f \mid i \rangle|^2 \Delta(\epsilon_f - \epsilon_i + \Delta^0_{IF}) \tag{11.89}$$

If *all* the modes of motion were *rotational*, rather than some being oscillatory, the 'primitive' semiclassical wave functions ψ_i would be a single term. When some or all of the modes of motion are oscillatory the phase of the semiclassical wave functions has more than one branch, each branch corresponding to a particular *set of signs* for the Nmomenta $(p_1 \cdots p_N)$. The normalized wave function ψ_i can be written as the sum in Eq. (11.90), using a Van Vleck determinant for each term

$$\psi_i = \sum a^{-1/2} \left| \frac{\partial^2 S_i}{\partial q \partial i} \right|^{1/2} \frac{e^{iS_i/\hbar}}{h^{N/2}} \tag{11.90}$$

where the sum is over the various branches of S_i and would contain 2^N terms if all modes of motion were oscillatory, S_i is the phase integral $\sum_k \int p_k dq_k$, taken from some turning point where all the p_k's vanish, plus various multiples of $\pi/4$, depending on the particular oscillatory branch, e.g. **M103, M100**, and a is a determinant of coefficients in the elementary distance expression (Appendix A). $\partial^2 S_i/\partial q \partial i$ denotes an $N \times N$ determinant having elements $\partial^2 S_i/\partial q_j \partial i_k$

$$|\partial^2 S_i/\partial q \partial i| \equiv \det \partial^2 S_i/\partial q_j \partial i_k.^{(2)} \tag{11.91}$$

We use this shorthand notation for determinants throughout, as well as an analogous one for differentials and for sets of variables:

$$du \equiv du_1 \cdots du_N, \quad u \equiv (u_1 \cdots u_N). \tag{11.92}$$

The normalization factor in Eq. (11.90) can be verified by first noting that in the product $\psi_i^* \psi_{i'}$ any highly oscillatory terms can be neglected. Only terms in which S_i and $S_{i'}$ are of the same branch then remain. We obtain

$$\begin{aligned}\langle i' \mid i \rangle &= \int \cdots \int \psi_i^* \psi_{i'} \sqrt{a} dq \\ &= \sum \int \exp\left[\sum_k (\partial S_i/\partial i_k)(i_{k'} - i_k)/\hbar\right] d(\partial S_i/\partial i)/h^N,\end{aligned} \tag{11.93}$$

where the sum $\sum$ is again over the branches of S_i and where we have used the shorthand notation (11.92). In Eq. (11.93) $S_{i'}$ in the exponent was expanded about S_i and $|\partial^2 S_i/\partial q \partial i|dq$ was replaced by its equivalent $d(\partial S_i/\partial i)$. Each $\partial S_i/\partial(i_k h)$ is the angle variable conjugate to the classical action $i_k h$ and its individual domain is the unit interval. However, different branches in the sum correspond, for oscillatory modes of motion, to different parts of this unit domain. (For any rotational motion the single branch covers the whole unit interval.) Upon removing the sum sign in Eq. (11.93) before the integral we recognize the fact that each angle variable ranges, thereby, over its full (0, 1) domain. The right-hand side of Eq. (11.93) then becomes, on integration, a product of Kronecker deltas $\delta_{i_1 i_{1'}} \cdots \delta_{i_k i_{k'}}$.[(3)] Thus, the *semiclassical* ψ_i given by Eq. (11.90) is properly normalized.

We consider next the FC factor $\langle i \mid f \rangle$ using Eq. (11.90) for ψ_i and an analogous expression for ψ_f.

NOTE: the idea is to evaluate FC factors using semiclassical functions.

The integral over the coordinates q becomes

$$\langle i \mid f \rangle = \sum \int \cdots \int \left| \frac{\partial^2 S_i}{\partial q \partial i} \right|^{1/2} \left| \frac{\partial^2 S_f}{\partial q \partial f} \right|^{1/2} e^{i(S_f - S_i)/\hbar} \frac{dq}{h^N} \tag{11.94}$$

and is evaluated, as is customary for *semiclassical* FC factors, by the *stationary phase method*. The only terms retained in the sum are those where S_f and S_i belong to the same branch, the remaining terms being highly oscillatory. The *stationary phase point* is determined by

$$\frac{\partial}{\partial q_k}(S_f - S_i) = 0,$$

i.e.,

$$p_k^f = p_k^i \quad (k = 1, \ldots, N) \tag{11.95}$$

and we have

$$\begin{aligned} S_f - S_i \cong\ & S_f^0 - S_i^0 \\ & + \frac{1}{2} \sum_{k,j} \frac{\partial^2}{\partial q_k \partial q_j} (S_f - S_j)(q_k - q_k^0)(q_j - q_j^0), \end{aligned} \tag{11.96}$$

where the various quantities on the right are evaluated at the stationary phase point $q = q^0$. Integration in Eq. (11.94) then yields

$$\langle i \mid f \rangle = \sum \left[\left| \frac{\partial^2 S_i}{\partial q \partial i} \right| \left| \frac{\partial^2 S_f}{\partial q \partial f} \right| \Big/ \left| \frac{\partial^2 (S_f - S_i)}{\partial q^2} \right| \right]^{1/2} \frac{e^{i(S_f^0 - S_i^0)/\hbar}}{h^{N/2}}. \quad (4) \tag{11.97}$$

The stationary phase points (11.95) all occur at the same q, but with different combinations of signs of the p_k's. If this q^0 is the only stationary phase point, the number of terms in the sum is the same as that in Eq. (11.90), one per branch of S_i. There may also be stationary phase points at other values of q, and if so they are then included in the sum.

Equation (11.97) can be simplified in a straightforward way to yield

$$\langle i \mid f \rangle = \sum \left| \frac{\partial p^i}{\partial i} \right|^{1/2} \left| \frac{\partial q}{\partial f} \right|^{1/2} \frac{e^{iW_{fi}/\hbar}}{h^{N/2}}, \tag{11.98}$$

where W_{fi} is the value of $S_f - S_i$ at the stationary phase point (11.95) and the sum is again over the branches of S_i.

From Eqs. (11.89) and (11.98) one obtains

$$k_{\alpha\beta} = \frac{2\pi V^2}{\hbar h^N} |\langle F \mid I \rangle|^2 \sum \left| \frac{\partial p^i}{\partial i} \right| \left| \frac{\partial q}{\partial f} \right| \Delta(\epsilon_f - \epsilon_i + \Delta E^0_{IF}).^{(5)} \tag{11.99}$$

The rate constant for a process $\alpha I \rightarrow \beta F$, when the initial *state of motion* of the *classical-like coordinates* is described by $(i_1 \cdots i_N)$, is found by integrating Eq. (11.99) over the f's. In doing this we first introduce a coordinate system to simplify the resulting expression: The intersection of the two potential energy surfaces $U_\alpha(q)$ and $U_\beta(q)$ occurs where $U_\alpha(q) = U_\beta(q) + \Delta E^0_{IF}$ and defines an $N-1$ dimensional hypersurface in the N-dimensional coordinate space. We define a coordinate q_N which is constant on this hypersurface, and which takes on different values for other values of $U_\beta(q) - U_\alpha(q)$.

Equation (11.99) is next integrated over the $(f_1, \ldots, f_N)$ variables, noting that $|\partial q/\partial f| df$ equals dq. Further, *each stationary phase point (11.95) occurs on the hypersurface* just cited, and at that point $\epsilon_f - \epsilon_i$ equals $U_\beta - U_\alpha$, since $p^i_k = p^f_k$ there for all p_k's. Thus, the $\Delta(\epsilon_f - \epsilon_i - \Delta E^0_{IF})$ in Eq. (11.99) can be replaced by $\Delta(U_\beta - U_\alpha - \Delta E^0_{IF})$. One can then write dq_N as $d(U_\beta - U_\alpha)/[\partial(U_\beta - U_\alpha)/\partial q_N]$ and define the absolute value S of this difference of slopes

$$S = |\partial(U_\beta - U_\alpha)/\partial q_N|. \tag{11.100}$$

In integrating over the f's in Eq. (11.99) one notes that $|\partial q/\partial f| df$ equals dq, then uses the $\delta(U_\beta - U_\alpha - \Delta E^0_{IF})$ in the integration over

$d(U_\beta - U_\alpha)$, and so obtains *the rate constant for a state* $(i_1, \ldots, i_N)$:

$$k(i_1 \cdots i_N) = \frac{2\pi V^2}{\hbar} |\langle F \mid I \rangle|^2 \times \sum \int \cdots \int \frac{\partial p^i}{\partial i} S^{-1} dq_1 \cdots dq_{N-1}/h^N. \quad (11.101)$$

The *microcanonical* rate constant $k_{IF}(\epsilon)$ for a given transition $\alpha I \to \beta F$ is obtained by integrating $k(i_1 \cdots i_N)$ over all i_k's such that ϵ_i lies in a narrow interval $(\epsilon, \epsilon + d\epsilon)$ and then dividing by the number $\int \cdots \int di_1 \cdots di_N$ of states in this interval,[6] that is, by $\rho_\alpha d\epsilon$, where ρ_α is the *density of states of the classical-like degrees of freedom* in electronic state α.

NOTE: It so appears that the microcanonical rate constant is the contribution of the average i_k state lying in the narrow energy interval.

One next notes that $|\partial p^i/\partial i| di$ equals dp^i, and that when the space of numbers $(i_1 \cdots i_N)$ is mapped onto the space of p_k^i's the various signs of p_k^i's, which created the various branches in the sum in Eq. (11.101), serve to map the i_k's onto all of the relevant p^i space, and so the sum in Eq. (11.101) disappears. We then have

$$k_{IF}(\epsilon) = \frac{2\pi V^2 |\langle F \mid I \rangle|^2}{\hbar \rho_\alpha(\epsilon) d\epsilon \hbar^N} \times \int \cdots \int S^{-1} dp_1 \cdots dp_N dq_1 \cdots dq_{N-1}.^{(7)} \quad (11.102)$$

The integration in Eq. (11.102) is restricted to the domain where $(q_1 \cdots q_{N-1}, p_1 \cdots p_N)$ lies *within the energy shell* $(\epsilon, \epsilon + d\epsilon)$, and so the integral is proportional to $d\epsilon$.

To obtain the *rate constant at a given temperature* $k_{IF}(T)$ for the $\alpha I \to \beta F$ transition, one multiplies Eq. (11.102) by $\rho_\alpha(\epsilon) \exp(-\epsilon/kT) d\epsilon/Q$, Q being the partition function for the N classical-like coordinates of the reactant, and extends the integration to all values of the $2N - 1$ variables

$$k_{IF}(T) = \frac{2\pi V^2}{\hbar Q}|\langle F \mid I\rangle|^2 \times \int \cdots \int e^{-\epsilon/kT} S^{-1} dp_1 \cdots dp_N \, dq_1 dq_{N-1}. \tag{11.103}$$

In passing we note that the stationary phase point for a transition $(i_1 \cdots i_N) \to (f_1 \cdots f_N)$ determines a point in the $2N$-dimensional phase space. When the additional condition on the energy ϵ_f is imposed, via the delta function in Eq. (11.99), the $i \to f$ transition corresponds to a point in $2N - 1$-dimensional phase space, a point in the domain in Eqs. (11.102) and (11.103). The integration over this space corresponds to the integration over all i's and f's, subject to the one constraint imposed by a delta function.

Equation (11.103) is first integrated over the momenta. The kinetic energy in ϵ is written as $\frac{1}{2}\sum_{i,j} g^{ij} p_i p_j$ and one obtains (Appendix A)

$$k_{IF}(T) = \frac{2\pi V^2}{\hbar}|\langle F \mid I\rangle|^2 \frac{\int \cdots \int e^{-U_\alpha/kT} S^{-1}\sqrt{a}\, dq_1 \cdots dq_{N-1}}{\int \cdots \int e^{-U_\alpha/kT}\sqrt{a}\, dq_1 \cdots dq_N}. \tag{11.104}$$

The *overall* rate constant $k(T)$ is obtained by multiplying Eq. (11.104) by the probability of finding the system in state I and summing over all I's and all F's

$$k(T) = \sum_{I,F} k_{IF}(T) e^{-E_I/kT}/Q_I, \tag{11.105}$$

where Q_I is the partition function $\sum_I \exp(-E_I/kT)$ *of the degrees of freedom associated with the quantum numbers I.*

NOTE: The probability of finding the system in state I is given by $e^{-E_I/kT}/Q_I$.

The $k(T)$ for the *reverse reaction* is obtained by interchanging (I, α) with (F, β). One can verify that microscopic reversibility is obeyed by these two rate constants, upon noting that U_α equals $U_\beta + \Delta E^0_{IF}$. The ratio of the two rate constants is found to equal, as it should, $\exp(-\Delta E^0/kT)$ times the ratio of the appropriate partition functions.

In the case of an adiabatic reaction one obtains Eq. (11.104), (**M50**) but with several differences [cf. Eqs. (33) and (34) of **M50**]: the S^{-1} is absent, the $N - 1$-dimensional surface element differs slightly from that element in Eq. (11.104), a consequence of the absence of a q_N term in Eq. (11.102), and a different factor precedes the integral.

To relate Eqs. (11.104) and (11.105) to an equation which has been used (**M174**) for treating nonadiabatic ETs when some of the degrees of freedom are treated quantum mechanically and some classically, two further steps will be used. We need to relate the integrals in Eq. (11.104) to a free energy of formation of a *nonequilibrium polarization state* (**M16**, **M53**) *appropriate to the TS of the reaction*, and require, thereby, the two integrals to have the *same* number of coordinates. This nonequilibrium state is the one which is 'centered' (**M53**, **M30**) on the intersection hypersurface.

To relate these integrals to this *free energy of formation*, we first replace S^{-1} to some average $\langle S^{-1}\rangle$ which we discuss later, and thus obtain an integral involving $\exp(-U_\alpha/kT)\sqrt{a}dq_1 \cdots dq_{N-1}$. We next replace this *surface distribution* by an '*equivalent equilibrium distribution*,' (**M53**, **M30**), by multiplying by $\exp(-U_N/kT)dq_N$, *integrating over* q_N, and then dividing by a one-dimensional configuration partition function Q_N for this motion of q_N.[7] Here, U_N is an *effective potential energy function* which vanishes on the

[7]NOTE 22: "As in Refs **M53** and **M30** we choose the quantity now labeled as U_N in Eq. (22) to be $m\left(U_\alpha - U_\beta - \Delta E^0_{IF}\right)$, [cf. Eq. (13) of **M53**: $U^* = U^r + m\left(U^r - U^p\right)$, where the potentials were instead defined relative to the same zero], m is a Lagrangian multiplier, which now depends on I and F; m is chosen using a procedure analogous to that in the above references, but for the given I and F,

intersection surface $U_\alpha = U_\beta + \Delta E^0_{IF}$, and which serves *to convert the surface distribution to a volume distribution centered on the intersection surface*, the 'e.e.d.'[(8)] of Refs. **M53** and **M30**.

We now have

$$k_{IF}(T) = \frac{2\pi V^2}{\hbar} |\langle F \mid I \rangle|^2 \frac{\langle S^{-1} \rangle}{Q_N} \frac{\int \cdots \int e^{-U^*_\alpha/kT} \sqrt{a}\, dq_1 \cdots dq_N}{\int \cdots \int e^{-U_\alpha/kT} \sqrt{a}\, dq_1 \cdots dq_N}, \tag{11.106}$$

where

$$U^*_\alpha = U_\alpha + U_N. \tag{11.107}$$

The possibility of having a *large entropy of activation* for the reaction is allowed in Eqs. (11.104) or (11.106): In the *TS region*, that is, *at the intersection* of the U_α and $U_\beta + \Delta E^0_{IF}$ PESs, the behavior of U_α can be very different from its behavior in the region *appropriate to the reactants.* (denominators of Eqs. (11.104) or (11.106)). For example, in an ET reaction the ionic charges of the reactants differ from those of the products, resulting frequently in *enormous* changes in *entropy of solvation.* These changes are reflected in a difference in the behavior of U_α in the '*reactants*' *region*' of the N-dimensional coordinate space compared with the behavior of U_β in the '*products*' *region.*" This difference is mirrored in the behavior of U_α on the *intersection hypersurface.*

An expression derived for the ratio of integrals in Eq. (11.106), using a 'dielectrically unsaturated' approximation for the *response of*

---

i.e., for the given ΔE^0_{IF} and hence for the given ΔG^0_{IF}. The Q_N in Eq. (21) is defined the following way: The surface integral prior to the introduction of U_N in Eqs. (21) and (22) was $\int \ldots \int \exp\left(-U_\alpha/kT\right) \sqrt{a}\, dq_1 \ldots dq_{N-1}$ and is written as $\exp\left[-I\,(q_N)/kT\right]$, *defining thereby a free energy* $I\,(q_N)$. One multiplies by dq_N, integrates over q_N, and expands $I(q_N)$ as a quadratic function of q_N about its value ($q_N = 0$) on the intersection surface (**M53**). Q_N is defined (**M53**) as the integral over this exponential and equals $[2\pi I''(0)kT]^{1/2}$, I'' being the value of d^2I/dq_N^2 at the intersection hypersurface.

the solvent outside the coordination shells of the reactants to changes in ionic charges, yielded[8]

$$\frac{\int \cdots \int e^{-U_\alpha^*/kT}\sqrt{a}dq_1 \cdots dq_N}{\int \cdots \int e^{-U_\alpha/kT}\sqrt{a}\, dq_1 \cdots dq_N} = \exp\left[\frac{-\lambda_o}{4kT}\left(1+\frac{\Delta G_{IF}^0}{\lambda_o}\right)^2\right], \tag{11.108}$$

where λ_o is a quantity in terms of the molecular properties. For example, if dielectric continuum theory is used, it depends on the change in charges of the reactants, the ionic radii, the separation distance between reactants, and the dielectric properties of the surrounding medium.[9]

NOTE: Here M. uses the symbol ΔF_o^* for ΔF_{sol}^*.

If any vibrations are included, in addition, in the classical-like coordinates $(q_1 \cdots q_N)$, λ_o includes a contribution from them also and then depends also on the changes in the corresponding equilibrium bond lengths and on vibration frequencies.[10]

ΔG_{IF}^0 in Eq. (11.108) is $\Delta G^0 + E_F - E_I$. That is, it is the *effective* ΔG^0 when the system goes from state I of the quantum degrees of freedom of reactants to state F for those of the products.**(9)**

We turn next to the remaining factor in Eq. (11.106), $\langle S^{-1}\rangle / Q_N$. We consider first a simple case, namely where U_α for each value of $(q_1 \cdots q_{N-1})$ has a term $\frac{1}{2}k'(q_N - a)^2$ and where U_β behaves as a displaced oscillator at that $(q_1 \cdots q_{N-1})$, and so has $\frac{1}{2}k'(q_N - b)^2 - \Delta E_{IF}^0$ there, k' is a force constant assumed independent of $(q_1 \cdots q_{N-1})$. Then $|\partial(U_\beta - U_\alpha)/\partial q_N|$ equals $k'\alpha$.**(10)** Further,

---

[8]One can see this by comparing Eqs. (1) and (13) of **M53** with Eqs. (31) and (81) there. The notation is somewhat different from the present notation.

[9]In **M53**, λ_o occurs in Eq (68), $\Delta F_o^* = m^2\lambda_o$, and by comparison with Eq. (89) there, $\Delta F_{\text{sol}}^* = m^2(ne)^2\left(\frac{1}{2a_1} + \frac{1}{2q_2} - \frac{1}{R}\right)\left(\frac{1}{D_{op}} - \frac{1}{D_s}\right)$, is the *coefficient of* m^2 in Eq. (89).

[10]This contribution to λ_o is labeled λ_i in **M53** and given by Eq. (84) there.

$Q_N = \int \exp[-k'q_N^2/2kT]dq_N$, where $q_N = 0$ is the value of q_N on the intersection hypersurface.

Note the sign of integral instead of the usual summation sign.

Thus, Q_N equals $\sqrt{2\pi kT/k'}$ and so $\langle S^{-1}\rangle/Q_N$ equals $1/\sqrt{2\pi kTk'a^2}$.

NOTE: k' in the denominator of the first radical cancels one k' of k'^2a^2.

Moreover, $\int_0^\infty e^{-a^2x^2}dx = \frac{1}{2a}\sqrt{\pi}\ldots$

One can express this quantity in terms of λ_o, by noting that $\lambda_o/4$ in Eq. (11.108) is the energy barrier when $\Delta G^0_{IF} = 0$. For the simple quadratic model for U_α and U_β one finds that $\lambda_o/4$ equals $k'a^2/8$. Thus, we have

$$\frac{\langle S^{-1}\rangle}{Q_N} = \frac{1}{\sqrt{4\pi\lambda_o kT}}. \tag{11.109}$$

In the actual many-dimensional case the values of U_α and U_β on any $N-1$ dimensional hypersurface (*fixed* q_N) when suitably averaged at that q_N, depend approximately quadratically on a suitably chosen q_N, when one makes a 'dielectrically unsaturated' approximation (**M53**). One can show that Eq. (11.109) is approximately valid for this many-dimensional system also.[11]

[11] We use the e.e.d. to calculate the *averages* of U_α and U_β and *describe these averages using* the parameter m.**(11)** This m entered earlier (Note 7) into U_N and, like q_N, has played a major role (**M30**) in defining *the position of the intersection hypersurface*: m is varied by varying ΔG^0_{IF} (**M30, M53**). We write d/dq_N as $(d/dm)(dm/dq_N)$. The difference $U_\beta - U_\alpha$ *averaged over an e.e.d. at a given* q_N (and hence at a given m) can be shown from Eq. (68) of **M53**, $\Delta F_o^* = m^2\lambda_o$, and from an analogous equation for the *free energy of formation of the e.e.d.*, at this same q_N, *on the β surface from the products* to be $(m+1)^2\lambda_o - m^2\lambda_o$, plus a term independent of m. That is, $\langle U_\beta - U_\alpha\rangle$ at this q_N is $(2m+1)\lambda_o$.**(12)** (**Erratum** corrected) plus an m-independent term. Hence, $\langle d(U_\beta - U_\alpha)/dq_N\rangle$ equals $2\lambda_o dm/dq_N$. Further, the q_N – partition function Q_N is approximately equal to that for the reactants, namely $\int \exp(-\Delta F_0^*/kT)dq_N$, where, Eq. (68) of **M53**, $\Delta F_o^* = m^2\lambda_o$, ΔF_o^* is $m^2\lambda_o$. Integration of the latter over q_N yields $\sqrt{\pi kT/\lambda_o}\, dq_N/dm$. Thus, the approximate value of $\langle S\rangle^{-1}/Q_N$ is $1/\sqrt{4\pi\lambda_o kT}$.

Note referring to footnote 11 $\frac{d[(2m+1)\lambda_o]}{dq_N} = \frac{d[(2m+1)\lambda_o]}{dm}\frac{dm}{dq_N} = 2\lambda_o dm/dq_N$.

Equations (11.106) to (11.109) yield

$$k(T) = \frac{2\pi V^2}{\hbar}\sum_{IF}\frac{|\langle F \mid I\rangle|^2 e^{-E_I/kT}}{Q_I(4\pi\lambda_o kT)^{1/2}}\exp\left[-\frac{\lambda_o}{4kT}\left(1+\frac{\Delta G^0_{IF}}{\lambda_o}\right)^2\right] \tag{11.110}$$

an equation used in several recent articles (**M174**, Efrima and Bixon, Warshel).

We have noted earlier that in the rate expression for nonadiabatic transfers, e.g., by Levich and Dogonadze, Ulstrup and Jortner, an exponential containing ΔE^0 instead of the ΔG^0 in Eq. (11.110) has been sometimes written. In derivations of the latter result it was *assumed* that all the classical coordinates were equivalent to harmonic oscillators, and, as already noted, for such systems, ΔS^0 vanishes and so ΔE^0 and ΔG^0 become equal for the present reaction.[12]

Equation (11.110) is the rate constant at a fixed separation distance R. If the reaction involves, instead, contributions from a distribution of R's one can proceed as in **M53**. One would obtain Eq. (11.110) but now instead of ΔG^0_{IF} one finds $\Delta G^{0\prime}_{IF} = \Delta G^0_{IF} + w^p - w^r$, where w^r and w^p denote the work required to bring reactants together and the products together, respectively, to some *average* distance R *in the TS*. One also finds that Eq. (11.110) is multiplied by a factor $4\pi R^2 \delta R \exp(-w^r/kT)$ where δR is the *range of separation distance* contributing significantly to the reaction rate. An elaboration, in which a somewhat slow diffusion may be involved, is described in **M174**.

In obtaining Eqs. (11.99) to (11.106), all stationary phase points (11.95) were taken to be real, since the aim in the paper was to treat the

While $\langle S\rangle^{-1}$ is not exactly equal to $\langle S^{-1}\rangle$ except when S is a constant, as it was in the simple harmonic oscillator model, we see that approximately $\langle S^{-1}\rangle/Q_N$ equals $1/\sqrt{4\pi\lambda_o kT}$, as in Eq. (11.109).

[12]See above.

q_k's classically. One quantum correction is that of *nuclear tunneling* along the q's. *This tunneling occurs when one or more of the p_k's in Eq. (11.95) is complex valued.* In that case a factor $\exp(-2|\mathrm{Im}\, W_{fi}|)$ appears in the integrand of Eq. (11.99) (and in later equations), Im W_{fi}[13] being the imaginary part of W_{fi} at this complex-valued stationary phase point. Additionally, some subtleties appear in the sum $\sum$ in Eq. (11.101) and hence in the subsequent equations.[14]

In concluding this section we comment on the relation between Eq. (11.103) and an interesting paper of Schmidt. He employed Yamamoto's expression for the rate constant in terms of the *reactive flux correlation function* and introduced a *classical approximation* by a different procedure. To see the connection between that paper and Eq. (11.103) we begin the present Eq. (11.89), the $\langle F \mid I\rangle$ being absent in Schmidt's paper since the intent there was to convert all coordinates to classical. After replacing the delta function in Eq. (11.89) by its Fourier integral representation and noting that $\epsilon_i|i\rangle$ and $\epsilon_f|f\rangle$ equal $H_i|i\rangle$ and $H_f|f\rangle$, the H's being Hamiltonians, one finds

$$k_{\alpha\beta} = \frac{V^2}{\hbar}|\langle F \mid I\rangle|^2 \langle i \mid f\rangle \int_{-\infty}^{\infty} \langle f|e^{iH_f t} e^{-iH_i t}|i\rangle e^{i\Delta E^0_{IF} t} dt. \tag{11.111}$$

[13]For example, in one dimension the exponential factor $\exp\left(-2|\mathrm{Im}|W_{if}/\hbar\right)$ becomes $\exp(-2J/\hbar)$ where, when the slopes $\partial U_\alpha/\partial q_N$ and $\partial U_\beta/\partial q_N$ are of opposite sign, J is the *sum* of the absolute values of the phase integrals $\int p_N dq_N$ from the classical turning point on curve α to the crossing point and from the crossing point to the classical turning point on curve β. When the slopes are of the *same* sign, J is the difference in absolute value.

[14]The preexponential factor in the one-dimensional case (or in the case where the q_N coordinate is separable from the others) is a factor of 2 smaller in this case than the factor in Eq. (B1) (also v_x is replaced by $|v_x|$) ... The origin of this difference is also evident in the semiclassical wave function in the classically allowed region, there being two exponential terms with imaginary exponents when a turning point occurs, while in the classically forbidden region there is only one term, a decaying exponential.

To obtain $k_{IF}(T)$ as in Scmidt's paper, one sums over all final states f, uses the completeness relation to remove the $\sum_f |f\rangle\langle f|$, multiplies by a Boltzmann factor $\exp(-\beta\epsilon_i)/Q$, and sums over i. A classical-like approximation is then introduced by *neglecting the commutators* of H_f and H_i in Eq. (11.111). One obtains

$$k_{if}(T) = \frac{V^2}{\hbar Q}|\langle F \mid I\rangle|^2 \sum_i \left\langle i \left| \int_{-\infty}^{\infty} e^{(H_f - H_i + \Delta E_{IF}^0)t} dt \right| i \right\rangle e^{-\beta\epsilon_i} \tag{11.112}$$

$H_f - H_i$ equals $U_\beta - U_\alpha$, and a second use of the integral representation for the delta function then yields an expression frequently used in the physics literature

$$k_{IF}(T) = \frac{2\pi V^2}{\hbar Q}|\langle F \mid I\rangle|^2 \sum_i \langle i|\delta(U_\beta - U_\alpha + \Delta E_{IF}^0)|i\rangle e^{-\beta\epsilon_i}. \tag{11.113}$$

Because of the neglect of the commutator this type of expression is sometimes called 'semiclassical' but it differs from the semiclassical approximation used in the present paper.

Schmidt then replaces the trace in Eq. (11.113), i.e., the sum over i, by an integral over phase space and so obtains the desired classical expression. His procedure, beginning with Eq. (11.89), is clearly very direct. Its one disadvantage is that at an intermediate stage, namely Eq. (11.113), it can lead to a significant error, seen as follows: Eq. (11.113) has been shown (**M163**), when all coordinates are treated as harmonic oscillators, to lead to another (Hopfield) expression for the reaction rate. That expression can be inaccurate (calculated in **M163**) to be a factor of ~20 different from the quantum value for the rate constants of thermoneutral reactions such as the $Fe^{2+/3+}$ aquoion exchange at room temperature). The physical nature of the approximation has been identified (**M163, M156**), and one finds that Eq. (11.113) overestimates the quantum effect.

11.23. Appendix B: Relation of Eq. (11.104) to the Landau–Zener Formula

It is useful to consider the connection between Eq. (11.104) and that obtained using the one-dimensional Landau–Zener formula for nonadiabatic transitions, since the differences between Eq. (11.104) and the *adiabatic* expressions are then understood in a simple way (see Warshel for closely related derivation).

For weak electronic interaction (small matrix element V) the Landau–Zener probability P of *transition* from electronic state α to state β is given by

$$P = 4\pi V^2/\hbar v_x S_x \tag{B1}$$

where v_x is the (Cartesian) velocity $\dot{x}$ at the crossing point and S_x is the difference of slopes of the two potential energy curves at that point, the slope being expressed in Cartesian coordinates $S_x = d(U_\beta - U_\alpha)/dx$. The product $v_x S_x$, i.e., $(dx/dt)[d(U_\beta - U_\alpha)/dx]$ can also be written as $d(U_\beta - U_\alpha)/dt$, and hence as $\dot{q}_N \partial(U_\beta - U_\alpha)/\partial q_N$, *since* $U_\beta - U_\alpha$ *depends on only* q_N *by definition of this coordinate* in the text. Systems initially on *surface* α cross the intersection of the two potential energy curves, and if they do not go to β, reach a classical turning point on α and recross that intersection surface. Equation (B1) includes this probability of a double crossing.

The reaction rate constant $k(T)$ is equal to P_α, the probability of finding the system in a *phase space volume element* at the *crossing point*, per unit length along the reaction coordinate, namely by

$$P_\alpha = e^{-\epsilon/kT} dq_1 \cdots dq_{N-1} dp_1 \cdots dp_N / Qh^N, \tag{B2}$$

multiplied by the velocity $\dot{q}_N$ and by the P given by Eq. (B1), then integrated over all coordinates $(q_1 \cdots q_{N-1})$ on the $N-1$-dimensional surface of intersection of U_α and $U_\beta + \Delta E^0_{IF}$, and integrated over all momenta, noting that the ϵ in Eq. (B2) equals $U_\alpha + \frac{1}{2}\sum_{ij} g^{ij} p_i p_j$. (Even p_N is allowed to be positive and negative since the system crosses the intersection hypersurface twice, in the low transition probability limit, for each crossing in the forward direction.) Using Eq. (A7) of Appendix A one obtains

$$k(T) = \frac{2\pi V^2}{\hbar} \frac{\int \cdots \int e^{-U_\alpha/kT} S^{-1} \sqrt{a} dq_1 \cdots dq_{N-1}}{\int \cdots \int e^{-U_\alpha/kT} \sqrt{a} dq_1 \cdots dq_N}. \quad \text{(B3)}$$

If multiplied by the probability $|\langle F \mid I \rangle|^2$ Eq. (B.3) yields Eq. (11.104) in the text for $k_{IF}(T)$. Of course, a main purpose of the present paper was to derive Eq. (11.110) systematically from the Golden-Rule expression (11.86a) and not to begin with the classical expression (B2).

NOTES

1. Abstract: 'Nonadiabatic processes may involve both classical-like and quantum-like coordinates.'

   Q: When would you consider a coordinate classical-like or quantum-like?

   M: I would be inclined to say... supposing that you look at the coordinate, that you determine the kind of a frequency associated with it, rotation for instance has a frequency, it may be a damped frequency but there is a frequency, so I'd be inclined to go by whether $kT/h\nu$ is large or small.
2. p. 4495 1st column top: '$|\partial^2 S_i/\partial q \partial i| \equiv \det \partial^2 S_i/\partial q_k \partial i_j$.'

   Q: Which kind of coordinate is the i?

   M: In semiclassical mechanics, that S is an $\int p dq$ ... the partial with respect to i could be the action variable associated with a quantum number.
3. Following:

$$\langle i' \mid i \rangle = \int \cdots \int \psi_i^* \psi_{i'} \sqrt{a} dq$$

$$= \sum \int \exp\left[\sum_k (\partial S_i/\partial i_k)(i_{k'} - i_k)/\hbar\right] d(\partial S_i/\partial i)/h^N, \quad \text{(11.93)}$$

Upon removing the sum sign in Eq. (11.93) before the integral we recognize the fact that each angle variable ranges, thereby, over its full (0, 1) domain. The right-hand side of Eq. (11.93) then becomes, on integration, a product of Kronecker deltas $\delta_{i_1 i_{1'}} \cdots \delta_{i_k i_{k'}}$."

Q: We know that $\delta(t-x) = \frac{1}{2\pi}\int_{-\infty}^{\infty} \exp(i\omega(t-x))d\omega$ for Dirac's delta, but the imaginary unity i is missing at the exponential in Eq. (11.93).

M: **Typo**: You certainly want an imaginary unity in there if you want to make a delta function, yes... it's missing, should be present. Because if you think of the semiclassical wave functions and you write as $e^{iS/h}$, there'd be an i in there. Or consider the $i\int pdq/\hbar$... so there is an i missing in the formula, and the trouble is that in the paper I'm using i for the quantum numbers, so you get confused... the i's and f's there maybe should be replaced by m's and n's... Look at Eq. (11.94): there is an i in there that should be present in the other.

4. p. 4495 1st column bottom:

$$\text{“}\langle i \mid f\rangle = \sum \int \cdots \int \left|\frac{\partial^2 S_i}{\partial q \partial i}\right|^{1/2} \left|\frac{\partial^2 S_f}{\partial q \partial f}\right|^{1/2} e^{i(S_f - S_i)/\hbar} \frac{dq}{h^N} \tag{11.94}$$

... we have

$$S_f - S_i \cong S_f^0 - S_i^0 + \frac{1}{2}\sum_{k,j} \frac{\partial^2}{\partial q_k \partial q_j}(S_f - S_i)(q_k - q_k^0)(q_j - q_j^0), \tag{11.96}$$

where the various quantities on the right are evaluated at the stationary phase points $q = q^0$. Integration in Eq. (11.94) then yields

$$\langle i \mid f \rangle = \sum \left[\left| \frac{\partial^2 S_i}{\partial q \partial i} \right| \left| \frac{\partial^2 S_f}{\partial q \partial f} \right| \Big/ \left| \frac{\partial^2 (S_f - S_i)}{\partial q^2} \right| \right]^{1/2}$$

$$\times e^{i(S_f^0 - S_i^0)} / h^{N/2}\text{''} \tag{11.97}$$

M: The Eq. (11.96) is simply a Taylor expansion, and now the first derivative with respect to i is missing... in other words it is zero because the momenta are equal and the derivative of S_f with respect to the q is a momentum, and the two momenta are equal in a transition, they're conserved, so I think what happens there is that to get Eq. (11.97) one does a Gaussian expansion and takes into account that the momentum is unchanged.

To go from Eqs. (11.96) to (11.97) you do a multidimensional quadratic integral. If you take $e^{-(E_{ij}x_i^2)}$, $x_i x_j$, the whole quadratic integral, you will get the answer in terms of a determinant. That's a multidimensional integration.

5. "The stationary phase points Eq. (11.95) all occur at the same q, but with different combinations of signs of the p_k's. If this q^0 is the only stationary phase point the number of terms in the sum is the same as that in Eq. (11.90), one per branch of S_i. There may also be stationary phase points at other values of q, and if so they are included in the sum.

   Equation (11.97) can be simplified to yield

$$\langle i \mid f \rangle = \sum \left| \frac{\partial p^i}{\partial i} \right|^{1/2} \left| \frac{\partial q}{\partial f} \right|^{1/2} \frac{e^{iW_{fi}/\hbar}}{h^{N/2}}, \tag{11.98}$$

where W_{fi} is the value of $S_f - S_i$ at the stationary phase point Eq. (11.95) and the sum is again over the branches of S_i.

From Eq. (11.89)

$$k_{\alpha\beta} = \frac{2\pi V^2}{\hbar} |\langle F \mid I \rangle|^2 |\langle f \mid i \rangle|^2 \delta(\epsilon_f - \epsilon_i + \delta_{IF}^0) \tag{11.89}$$

and Eq. (11.98) one obtains

$$k_{\alpha\beta} = \frac{2\pi V^2}{\hbar h^N} \langle F \mid I \rangle^2 \sum \left| \frac{\partial p^i}{\partial i} \right| \left| \frac{\partial q}{\partial f} \right| \delta(\epsilon_f - \epsilon_i + \Delta E^0_{IF})\text{."} \tag{11.99}$$

Q: What happened to the disappeared exponential in Eq. (11.98)?

M: The exponential may have gone over to a delta function.

6. p. 4496 1st column top: "...the rate constant for a state $(i_1, \ldots, i_N)$:

$$k(i_1 \cdots i_N) = \frac{2\pi V^2}{\hbar} |\langle F \mid I \rangle|^2 \times \sum \int \cdots \int \frac{\partial p^i}{\partial i} S^{-1} dq_1 \cdots dq_{N-1} / h^N. \tag{11.101}$$

The *microcanonical* rate constant $k_{IF}(\epsilon)$ for a given transition $\alpha I \rightarrow \beta F$ is obtained by integrating $k(i_1 \cdots i_N)$ over all i_k's such that ϵ_i lies in a narrow interval $(\epsilon, \epsilon + d\epsilon)$ dividing by the number $\int \cdots \int di_1 \cdots di_N$ of states in this interval..."

Q: Does this mean that the *microcanonical* rate constant k_{IF} is the contribution of the average i_k state lying in the narrow energy interval?

M: Yes I guess so. But there is a little problem there. When you speak of a given energy of course you have zero states at that energy, right? You really have a little energy above, below, and so on. Classically, it's true, you can have a fixed energy. So even if formally you consider this fixed energy often one really considers a range of energies and there is a little uncertainty associated probably with the range of energy to consider, how small an energy range should be. When you represent a big system by a microcanonical ensemble, that's kind of an idealization, isn't it? So, there is a little approximation that comes in there that makes a little bit of difference, I guess, when you calculate some observable, if you consider a microcanonical system.

7. Following: "One next notes that $|\partial p^i/\partial i|di$ equals dp^i, and that when the space of numbers $(i_1 \cdots i_N)$ is mapped onto *the space of* p_k^i's the various signs of p_k^i's, which created the various branches in the sum in Eq. (11.101) serve to map the i_k's into all of *the relevant* p^i *space*, and so the sum in Eq. (11.101) disappears. We then have

$$k_{IF}(\epsilon) = \frac{2\pi V^2 |\langle F \mid I \rangle|^2}{\hbar \rho_\alpha(\epsilon) d\epsilon \hbar^N} \times \int \cdots \int S^{-1} dp_1 \cdots dp_N dq_1 \cdots dq_{N-1}, \text{"} \tag{11.102}$$

Q: Please explain the difference between *the space of* p_k^i's and *the relevant* p^i space. Why does then the sum in Eq. (11.101) disappear?

M: Let's consider the case of angular momentum because that's the easiest case. With angular momentum you have positive and negative rotations, so a one-dimensional rotation can be rotating one way or can be rotating the other way and of course there will be different p_k^i, one way and one the other, different branches, so to speak, but now if you're integrating in a way which integrates over the p's through positive and negative values, you take that whole set and you put it into one integral and the integral is covering both positive and negative directions.

8. p. 4496 2nd column middle: "To relate Eqs. (11.104) and (11.105)

$$k_{IF}(T) = \frac{2\pi V^2}{\hbar} |\langle F \mid I \rangle|^2 \frac{\int \cdots \int e^{-U_\alpha/kT} S^{-1} \sqrt{a}\, dq_1 \cdots dq_{N-1}}{\int \cdots \int e^{-U_\alpha/kT} \sqrt{a}\, dq_1 \cdots dq_N}. \tag{11.104}$$

$$k(T) = \sum_{I,F} k_{IF}(T) e^{-E_I/kT}/Q_I, \tag{11.105}$$

to an equation which has been used (**M174**) for treating nonadiabatic ETs when some of the degrees of freedom are treated quantum mechanically and some classically, two further steps will be used. We need to relate the integrals in Eq. (11.104) to a free energy of formation of a *nonequilibrium polarization state* (**M16**, **M53**)appropriate to the TS of the reaction, and require, thereby, the two integrals to have the *same* number of coordinates. This nonequilibrium state is the one which is 'centered' (**M53**, **M30**) on the intersection hypersurface.

To relate these integrals to this *free energy of formation*, we first replace S^{-1} to some average $\langle S^{-1}\rangle$ which we discuss later, and thus obtain an integral involving $\exp(-U_\alpha/kT)$ $\sqrt{a}dq_1\cdots dq_{N-1}$. We next replace this *surface distribution by an 'equivalent equilibrium distribution'* (**M53**, **M30**), by multiplying by $\exp(-U_N/kT)dq_N$, integrating over q_N, and then dividing by a one-dimensional configuration partition function Q_N for this motion of q_N. Here, $U_N = m(U_\alpha - U_\beta - \Delta^0_{IF})$ is an *effective potential energy function* which vanishes on the intersection surface $U_\alpha = U_\beta + \Delta E^0_{IF}$, and which serves to convert the surface distribution to a volume distribution centered on the intersection surface, the 'e.e.d.' of Refs. **M53** and **M30**."

Q: (i) I believe the U_α in Eq. (11.104) is the potential energy at the intersection, that is, $U_\alpha = U_\beta + \Delta E^0_{IF}$.

(ii) There are then three functions to consider: (a) The surface distribution, (b) The "e.e.d.", (c) The effective potential energy function which serves to convert the surface distribution to a volume distribution centered on the intersection surface, the "e.e.d." Why is the potential energy function designated as "effective"?

M: (i) I imagine so.

(ii) Effective because it is not a real one. *It is a construct designed to go from a surface to a volume.* In other words, when you have a distribution, let's say of some hypothetical system, the e.e.d. is in *three dimensional space*, the most probable localization is on the intersection surface, but there is something that is keeping near that surface, and that's an effective potential function, is a hypothetical one.

Three dimensional is in $3N - 1$ dimensional space. $3N$ is 10^{23} or so. If there are N coordinates and if we forget about the center of mass, that's 1 compared to 10^{23}, if you have N atoms you have $3N$ coordinates and N is of the order of 10^{23} because you are dealing with a big system, and the e.e.d. then is a distribution function in that space.

9. p. 4497 1st column top: "An expression derived for the ratio of integrals in Eq. (11.107), using a 'dielectrically unsaturated' approximation for the *response of the solvent* outside the coordination shells of the reactants to changes in ionic charges, yielded[8]

$$\frac{\int \cdots \int e^{-U_\alpha^*/kT} \sqrt{a}\, dq_1 \cdots dq_N}{\int \cdots \int e^{-U_\alpha/kT} \sqrt{a}\, dq_1 \cdots dq_N} = \exp\left[\frac{-\lambda_o}{4kT}\left(1 + \frac{\Delta G_{IF}^0}{\lambda_o}\right)^2\right], \qquad (11.106)$$

where λ_o is a quantity in terms of the molecular properties. For example, if dielectric continuum theory is used, it depends on the change in charges of the reactants, the ionic radii, the separation distance between reactants, and the dielectric properties of the surrounding medium.[9] If any vibrations are included, in addition, in the classical-like coordinates $(q_1 \cdots q_N)$, λ_o includes a contribution from them also and then depends also on the changes in the corresponding equilibrium bond lengths and on vibration frequencies.[10] ΔG_{IF}^0 in Eq. (11.106) is $\Delta G^0 + E_F - E_I$. That is, it is the effective ΔG^0 when the system goes from state I of the quantum degrees of freedom of reactants to state F for those of the products."

Q: I believe one should use λ instead of λ_o in Eq. (11.106).

M: **Typo** Yes, one should have used the symbol λ, definitely.

10. Following: "We turn next to the remaining factor in Eq. (11.105)

$$k_{IF}(T) = \frac{2\pi V^2}{\hbar}|\langle F \mid I\rangle|^2 \times \frac{\langle S^{-1}\rangle}{Q_N} \frac{\int \cdots \int e^{-U_\alpha^*/kT}\sqrt{a}\, dq_1 \cdots dq_N}{\int \cdots \int e^{-U_\alpha/kT}\sqrt{a}\, dq_1 \cdots dq_N}, \quad (11.107)$$

$\langle S^{-1}\rangle / Q_N$. We consider first a simple case, namely where U_α for each value of $(q_1 \cdots q_{N-1})$ has a term $\frac{1}{2}k'(q_N - a)^2$ and where U_β behaves as a displaced oscillator at that $(q_1 \cdots q_{N-1})$, and so has $\frac{1}{2}k'(q_N - b)^2 - \Delta E_{IF}^0$ there, k' is a force constant assumed to be independent of $(q_1 \cdots q_{N-1})$. Then $|\partial(U_\beta - U_\alpha)/\partial q_N|$ equals $k'\alpha$."

Q: (i) Do you mean that extending out of the intersection surface and orthogonal to it for each point $(q_1 \cdots q_{N-1})$ there is a parabola along q_N? How are related to each other the parabolas extending out of different surface points? The parabolas are apparently all equal to each other but I find it difficult to visualize the set of parabolas.

(ii) I believe the here undefined α is $a - b$.

M: (i) I just mean that extending out of there, there is an imaginary parabola which sort of describes a hypothetical localization of the e.e.d. For the effective distribution what I had in mind was a hypothetical thing which sort of confined the system to be *near* the intersection surface. And that would be a single parabola centered on the intersection surface. You have a hypersurface and the parabola is in one dimension perpendicular to a whole hypersurface. Let's draw a line in two dimensions, to represent $10^{23} - 1$ dimensions. Now, perpendicular to that you plot a potential energy, the parabola all along that seem, roughly with the same force constant, perpendicular to that line, so you really have a parabola that's there corresponding to every point of that line, every time you move along that line you're seeing the same parabola, it's like a ditch, just think of a ditch that's sort of parabolic in shape. In a ditch you have that at every point along the length of the ditch, the ditch is going up on both sides. I hate to be in the position to put you on a ditch but anyway, that's the way... hopefully there is no water in the ditch.

(ii) It's probably a – b, definitely.

11. From footnote 11: "We use the e.e.d. to *calculate the averages* of U_α and U_β and *describe these averages using* the parameter m."

    Q: This is apparently the second way of using the e.e.d. after that of having a volume distribution centered on the intersection surface substituting the surface distribution.

    M: Yes, that hypothetical distribution is used to calculate an average.

12. Following: "The difference $U_\beta - U_\alpha$ *averaged over an e.e.d. at a given* q_N (and hence at a given m) can be shown from Eq. (68) of **M53**, $\Delta F_o^* = m^2\lambda_o$, and from an analogous equation for the *free energy of formation of the e.e.d.*, at this same q_N, *on the* β *surface from the products* to be $(m+1)^2\lambda_o - m^2\lambda_o$, plus a term independent of m. That is, $\langle U_\beta - U_\alpha \rangle$ at this q_N is $(2m+1)^2 - m^2\lambda^0$ plus an m-independent term.

    Q12: I believe here there are **two typos**: the above $(2m+1)^2 - m^2\lambda^0$ should be $(2m+1)\lambda_o$.

    M: **Typos** Yes, it should be $(2m+1)\lambda_o$.

CHAPTER 12

Entropic, Electrolyte, and Orientation Effects in Electron Transfer, Electron Transfer in Chemistry and Biology, Application of Electron Transfer to Systems of Biological Interest, Barriers for Thermal and Optical Electron Transfer in Solution, Semiclassical Model for Orientation Effects in Electron Transfer, Orientation Effects on Electron Transfer between Porphyrins

In this Chapter and in the following, oral interviews with Marcus are reported on **M126, M201, M204, M207, M209, M210**. The interviews are presented in the form of **NOTES** to Marcus' articles.

M126. A Study of the Entropic and Electrolyte Effects in Electron Transfer Reactions

1. p. 10:

$$\text{“}\exp(\Delta S^{\ddagger}/k) = A/(kTe/h) \qquad (1.4)$$

$\Delta S^{\ddagger}$ is the entropy of the transition state minus that of reactants, each in some standard state. The value of $\Delta S^{\ddagger}$ depends on the units of A (e.g., A can be in l mole^{-1} sec^{-1} or cm^3 mole^{-1} sec^{-1},

say) and thereby the standard state is implied in $\Delta S^{\ddagger}$ (e.g., 1 mole cm^{-3} or 1 mole l^{-1})."
Q: 1 mole cm^{-3}? Does it make sense?

M: That l is liters, liters mole^{-1} sec^{-1} or cm^{3} mole^{-1} sec^{-1}, not one. This is a bimolecular reaction, it can't have units of 1 mole^{-1} sec^{-1}. In the case of $\Delta S^{\ddagger}$ the 1 is a one.

Q: But 1 mole cm^{-3}? Does it make sense?

M: Ja, sure, that's concentration, your standard state you describe in terms of specific concentration, you decide on the units, that will decide on the standard state. Normally people don't use mole per cubic centimeter, they use mole per liter, but in principle you can do the other, in other words it makes sense it's just that is not commonly done. It is done whenever you use the units of cubic centimeter per mole per second for a rate constant, then implied in that as far as theory is concerned, it this 1 mole cm^{-3}. The units have to correspond in each case.

2. Following: "This large negative $\Delta S^{\ddagger}$ is quite common for electron exchange reactions between small hydrated ions"

Q: Please explain, why so?

M: That's when you bring small hydrated ions together if they are of similar charge, you get a huge field, that huge field further orients everything and produces a negative entropy.

Q: A huge reorientation.

M: Yes, provided the ions are the same sign.

3. p. 12: "$w^r = -kT \ln g_{12}(R)$"

Q: Please explain why and how the work term is expressed in terms of the pair distribution function.

M. In ordinary collision theory the g_{12} is unity, so if you want to bring in a work term, then that's deviation from unity.

4. p. 15 top "the ion atmosphere in the [1-0, op] system *does not respond to the charges on the central ion*"

   p. 19 bottom: "There is no atmosphere contribution to G^{op}_{1-0}, since by definition the ion atmosphere and orientational coordinates do not rearrange in the [1-0, op] system; only the electronic polarization responds to changes in charges on the two central ions."

   Q: Please explain . . . the second statement is OK but in the first doesn't the electronic polarization of the ionic atmosphere respond to charges of the central ions as stated in the second sentence?

   M: Optical, op, means that only the electrons are responding, that means that the nuclear charge distribution isn't responding.
5. p. 16 top:

$$\text{“}\Delta G^{\ddagger atm}_{R,o} = 0\ (DH)\text{”} \tag{4.17}$$

   M: I see that for any geometry other than $R = 2a$, the $\Delta G^{\ddagger atm}_{R,o}$ would not be zero. For this special case there are two opposing effects that just cancel, one related to the change in the ionic atmosphere interacting with the ions individually in forming the TS and the other related to the ionic atmosphere effect on the repulsion of the two ions in the TS, in the nonequilibrium situation that constitutes this transition state. One can see the exact cancellation for $R = 2a$ in the equations mathematically but to understand it physically one would have to understand in great and subtle detail each of the two effects described earlier that enter into the equations. I have not tried to do so. The smallest value that R can have is 2a. I happened to notice that the cancellation is exact only at that distance, but otherwise the two facts appear to be unrelated.
6. p. 17 top:

$$\text{“}h(r_{12}) = C(r_{12}) + \int C(r_{12})\rho C(r_{32})dd_3 + \int C(r_{13})\rho C(r_{34})\rho C(r_{42})dd_3dd_4 + \cdots\text{”} \tag{5.3}$$

Q: **Typo** I guess$\cdots$ it should be

$$h(r_{12}) = C(r_{12}) + \int C(r_{12})\rho C(r_{32})dr_3 + \int C(r_{13})\rho C(r_{34})\rho C(r_{42})dr_3dr_4 + \cdots$$

7. p. 20: "various contributions to $\Delta S^{\ddagger}\cdots$

$$\Delta S^{\ddagger}_{\text{trans},R} = k\ln(Zh/kT) - (k/2) \tag{5.26}$$

$$\Delta S^{\ddagger}_{el} = k\ln\bar{\kappa} \tag{5.27}$$

$$\Delta S^{\ddagger}_{W^r} = k\ln g^0_{12} - (e_1e_2/D_sTR)[(1+p)/q^2]\{\theta(1+p) - [\kappa R(\theta-1)(1+p+\kappa L)/pq]\} \tag{5.28}$$

$$\Delta S^{\ddagger}_{w^r}(DH) = -(e_1e_2/D_sTR)(2\theta+\theta kR+\kappa R)/2(1+\kappa R)^2 \tag{5.29}$$

$$\Delta S^{\ddagger\text{solv}}_{R,o} = m^2(\Delta e)^2\theta/D_sTR \tag{5.30}$$

$$\Delta S^{\ddagger\text{atm}}_{R,o} = -m^2(\Delta e)^2[\kappa R^2/q^3D_sTR]\{q\theta+(\theta-1)[1+p-(\kappa L/p)]\}\text{"} \tag{5.31}$$

Q: (i) Please compare Eq. (5.27) with $\Delta S_e = k\ln\Omega/\Omega^*$, Eq. (18) from **M16**, where "the product of the electronic degeneracies of each of the reactants is Ω^* and that of the product is Ω." Please explain also the physical meaning.
(ii) What is the physical meaning of the $\Delta S^{\ddagger}_{W^r}$?
(iii) What is the physical meaning of $\Delta S^{\ddagger\text{atm}}_{R,o}$?
(iv) Why is the $\Delta S^{\ddagger\text{atm}}_{R,o}(DH) = 0$?

M: (i) The ΔS_e is part of the ΔS, is just the part related to the electronic degeneracy and not dealing with anything else.

(ii) You know, there is a certain entropy change associated with Debye Hückel and the rearrangements and everything taken into account, it's related to the change in entropy that comes when you bring the two ions together that changes the atmosphere and also changes the orientation of the water molecules, so both changes would come into there. When the two ions are near each other you're affecting everything, and don't forget that's maybe including the sort of role of the Franck–Condon effect, you know, all thrown into it, big mix-mash, that's related to the change in the atmosphere when you bring the two ions together in the presence of an atmosphere, if you let $\kappa = 0$ then that presumably reduces to the ordinary expression that you expect when there is no ionic atmosphere around, the θ should be related to the temperature derivative of the dielectric constant.

(iii, iv) The $\Delta S_{R,o}^{\ddagger atm}$ is there for two reasons, one it would be there even if there was no Franck–Condon effect because you are changing entropy when you bring ions together... and in addition to that there are changes going on electronically to make things more complicated as far as only changes in entropy. But basically once you got the free energy associated with the changes if you differentiate with respect to T and put a negative sign you got the entropy change.

8. Please look at p. 21 Table 1. There you see the values and the signs of the different contributions, differing in values and in signs. Can you comment on that?

   M: Because you lose some translational degree of freedom, the ΔS is negative, the nonadiabatic is negative because its probability is less than 1, the entropic work is negative, I have to look into a reason for that, let's say the system is more ordered around the more highly charged species, which wouldn't be surprising, the atmosphere reorganization is negative and very small, now . . . that $\Delta S^{\ddagger}$ is that calculated by what? Here it says that atmosphere reorganization DH is zero, I wonder if that is zero at first order, maybe at second order is not zero, it would be surprising that it'd be zero, I mean that's very surprising.

M201. A Model for Orientation Effects in Electron Transfer Reactions

NOTES

1. p. 5614 first column bottom: "The potential V is zero outside the well and has a constant negative value inside. Actually, there will also be a Coulombic term when the molecule is charged, but it is assumed, for the present, that in a medium with some polarity this contribution is small relative to the values of V_0 used below."

   M: If the molecule is charged the potential outside is varying like $1/r$, it is not necessarily the solvent polarization.
2. p. 5815 second column, bottom: "Energy levels for several states are shown in Fig. 6 as a function of eccentricity at constant volume for the infinite potential case . . ." See Fig. 6.

   Q: The shape of the well determines "attractive" and "repulsive" branches of the curves. But in this case we do not have nuclear–nuclear repulsions or Pauli repulsions between closed shells of electrons. Which one is then the physical reason for the shape of these curves?

   M: Consider the energy of a particle in the box. A disk has a very thin amount in one dimension, and large amounts in the other dimensions, the thin amount is going to make the energy very high. So for the disk, when you make it thinner and thinner the energy sweeps up.

   Q: Compare with the H_2^+ system. There you have a PECs with a repulsive branch . . .

   M: Definitely, because approaching the protons the electron moves in a smaller and smaller region of space . . . that's definitely a fact that causes the curves to go up. This fact and the nuclear repulsion . . . both factors enter there.

3. p. 5620 second column, middle just under the formulae: "Spherical states . . . The variables $(r_A, \vartheta_A, \phi_A)$ depend implicitly on $(r_B, \vartheta_B, \phi_B)$." Then on p. 5621 1st column just below middle: "It can be seen in Fig. 6 that θ_A equals Θ and θ_B equals $\pi - \Theta$, where θ_A and θ_B are the spherical polar angles in each well . . ."

   Q: Please explain, "Fig. 6" is probably a misprint for "Fig. 8."

   M: Supposing that you take *A* and you specify the position of an electron, the coordinates of that, have a point there, a point expressed by those coordinates, then if you now look of the coordinates centered in *B* those coordinates aren't independent, they are all determined by the coordinates you specify for *A*. There is no free choice. If you fix a point anyplace, doesn't matter. If you refer to coordinates with respect of *B* those coordinates are specified by the coordinates with respect of *A* and the relative positions of the two ellipsoids. **Typo**: There is the misprint.
4. p. 5621 first column bottom: "To exhibit the angular dependence of Ψ_{mtu} at large r, Ψ_{mtu} in Eq. (27) was projected onto the associated Legendre polynomials P_n^m (cos θ). If the angular probability distribution at large r were insensitive to the nonzero eccentricity of the spheroidal well one would find $|\langle P_n^m | \Psi_{\text{mtu}} \rangle|^2$ equal to zero except for a single value of n."

   Q: If I understand well you use this projection to measure how much the spheroidal wave functions differ from the spherical ones in their θ dependence.

   M: That's probably true.

M204. Electron Transfers in Chemistry and Biology

NOTES

M: "Writing that review was a wonderful experience for both of us, Norman and I, and he is still alive . . . he is three or four years younger than I am"

1. p. 266 first column middle: "Quantitative predictions of the theory ... include ... nonspecific solvent effects on the rate constant ..."

   Q: How can one define "nonspecific solvent effects"? Is it the same as the "$1/D_{op} - 1/D_s$ solvent effect" (**M188** p. 173 middle) so that every other solvent effect is "specific"? Moreover, can you list specific solvent effects coming to your mind?

   M: The theory doesn't predict specific solvent effects even if they occur ... The $1/D_{op} - 1/D_s$, that would be *the continuum version of the nonspecific solvent effect.* By "nonspecific" I mean that the solvent doesn't interact *in a chemical way*, with either of the reactants. Very strong, very specific binding that you don't capture in any sort of an electrostatic theory. For example, suppose there was a strong hydrogen bond ... then we need to take that into account. $1/D_{op} - 1/D_s$ is a continuum formula. But if you had a solvent molecule which binds to the oxidized form of the reactant, and didn't bind to the reduced form of the reactant, that's a very specific effect. Or even if it binds specifically to both in a way which is not in any way handled by any kind of electrostatics. If Van der Waals or London forces do not cancel in the transition state, they might contribute, they are additional forces. Then you could absorb them in a constant ...
2. Note ¶ p. 267 first column: "The actual profile of the many-dimensional surface is more complicated than Fig. 1 when there are solvation contributions—there are many local potential energy minima, for example. In that case only the profile of the *free energy* (in effect, a Boltzmann-weighted profile) of a system consisting of the pair of reactants, at a given separation, and their environment, along the reaction coordinate could have a quadratic appearance. Such an approximation is much less drastic than assuming that the potential energy is a quadratic function of the coordinates."

M: If you have many many variables, even if the distribution of each is far from harmonic, far from Gaussian, the collection of the whole thing is going to be Gaussian, we have a Gaussian distribution that's like a distribution of a quadratic surface, that's one way of putting it. In another way of putting it, if you want to use statistical mechanics and the language of the *cumulants*, you make an *expansion to second order cumulants*.

It is quadratic because of the *central limit* there, it is a Gaussian distribution, so it is a distribution equivalent to that of a quadratic surface. It doesn't depend on the fact that λ_i depends quadratically on Δq_j because it is also true of the λ_0. I have a collective variable of many many coordinates, it is $(\Delta E)^2$ because of the many coordinates, it ends up like a Gaussian distribution.

3. Q: Always referring to your classical profile, you write in note § on p. 267, first column:
"... the transfer has to occur at nuclear configurations ... at or near the intersection region."

Q: If the profile is one-dimensional, the transition state must be zero-dimensional, that is, must be represented by the point on top of the profile for an adiabatic process or by the crossing point. So, why "near"? Does the transition state extend somewhat outside of its $N-1$ dimensions in its vicinity in a neighborhood of reactive configurations in the N dimensional space? Is it so that—in a way somehow similar to what happens in spectroscopy for a spectroscopic line—there is a "width" of the transition state? Is it so because of Heisenberg uncertainty or because of nuclear tunneling or are there other reasons? Or maybe the transition may begin with a nuclear configuration near the transition region but in order to effectively and definitely happen it is to pass always through a nuclear configuration of the transition state?

M: You see, you have an intersection of many dimensional potential surfaces, but you can't localize a point that well because of the uncertainty principle. The whole LZ theory isn't just dealing with a point but with *a region around the point.* The transition state doesn't have a width, the width business was an old Eyring's idea, in an early derivation, but that width no longer appears in the derivation of Eyring, Walter and Kimball, it was never part of Wigner's argument, it disappeared along with the Lake Eyring. Eyring was great, brought large joys, he was wonderful.

4. Q: The legend under Fig. 2 in p. 268: "The levels in the wells are vertical ionization energies: the filled and open circles denote, respectively, ionization of *the reduced state* at its equilibrium nuclear configuration and at the equilibrium configuration appropriate to its *oxidized state* . . ."

   Q: What did you mean here by reduced state?

   M: I meant the reduced state of the reactant that is reduced.

5. Q: Always referring to Fig. 2b: in the situation there depicted do the half-filled circles mean that the transferring electron is *delocalized on both reactants, that it is in both wells*? That is, if the reactants system is A + B and the products system is A^{+}+ B^{-} the wave function representing the electron distribution is $|A + B\rangle + |A^{+} + B^{-}\rangle$?

   M: In other words, there is a probability of 50% that the electron is in one well, 50% in the other, it can't be in both at the same time. This can happen only at the intersection, where you can go from one to the other without any change . . .

6. p. 268 2nd column, middle: "For a reaction in which there is substantial electronic coupling between the reactants ("adiabatic reactions"), κ is about unity.

   Q: How big should the splitting be—order of magnitude—for the reaction to be an adiabatic one?

   M: It is just a sort of ballpark (approximately correct) . . . of the order of 100 wavenumbers or so . . . maybe 200 wavenumbers, it is of the order of kT. Ballpark.

7. Q: In **M118**, "Electron Transfer in Homogeneous and Heterogeneous Systems" Physical and Chemical Research Report 1, we have the following formula and notations:

$$\Delta G^{0'}(R) = \Delta G^{0'} + w^p - w^r$$

But in this paper on p. 269 1st column top the same formula is written as:

$$\Delta G^{0'} = \Delta G^0 + w^p - w^r$$

In a note in the same page you explain that $\Delta G^{0'} = \Delta G^0 \cdots$

M: The prime denotes the standard free energy of reaction in the actual medium, of course remember that w^p and w^r are functions of R ... The ΔG° is really under *ideal* conditions. While $\Delta G^{\circ\prime}$ is under *effective* conditions.

8. p. 278 1st column, middle: "Noncancellation of the hydrophobic/hydrophilic interaction in the cross relation . . . has been considered as possible reason for the breakdown in these unsymmetrical systems."

M: Well, suppose you have an organic and an inorganic reactant, now if you look at, say, Van der Waals forces or maybe more specific forces, the organic–organic interaction and the inorganic–inorganic interaction is not going to give the organic–inorganic as a mean, they are different interactions, so unless the mean interaction of the organic–inorganic are roughly the mean of the other two, then, you know, there wouldn't be cancellation. In other words, consider an ET between a hydrophobic and a hydrophilic reactant. The self-exchange reaction would in each case have a favorable work term, whereas the cross relation would not. Accordingly, the simple cross relation would need to be modified by the noncancelling work terms.

9. p. 280 2nd column middle: "For systems where the spin–orbit coupling is small, for example in ET involving organic systems,

the spin restrictions can be more important and would reduce the value of κ."

Q: Where do you treat the influence of spin on ET? Could you tell something on this topic? Why a *small* spin–orbit coupling leads to a small and not to a big value of κ?

M: If you have some reaction that violates spin conservation . . . Supposing that the ET involves the spin change, the overall spin of the reactants differing from the overall spin of the products, and the only way spin can change is from spin–orbit interaction, then if the spin–orbit interaction is small then the probability of an ET overcoming the spin problem would be small, there'd be a small κ.

11. p. 282 2nd column, middle-low: "In the Creutz–Taube ion the two ruthenium centers are connected by a pyrazine group. This group provides strong electronic coupling of the metal centers instead of weak coupling. The properties of the pyrazine-bridged system are those expected for a *delocalized* rather than a *localized* ground state . . ."

    Q: Your theory (outer sphere thermal ET) applies to systems in which the electron is *localized* on the reductant when the system is in the bottom of the R well in Fig. 2b and hops to the oxidant in the transition state where it is *delocalized* on both reductant and oxidant, only when passing through the transition state, while in the case of the Creutz–Taube ion the electron is delocalized already in the bottom of the R PES.

    M: That's correct, when passing the transition state there are two electronic configurations weakly coupled, but there is an approximately fifty/fifty coefficient to each . . . You know, if you think of the double well, if you increase the interaction so you get a much bigger splitting, eventually instead of a maximum there is going to be a minimum in between, the double well becomes a single well, very much a delocalized system . . . it becomes a single well problem instead of a double well.

12. p. 285 1st column middle: "In the adiabatic regime ν" in the equation

$$k = \kappa(r)\nu \exp(-\Delta G_r^*/RT) \tag{31}$$

"is an *effective* frequency for nuclear motion along the reaction coordinate in Fig. 1 . . . In the nonadiabatic regime ($\kappa \ll 1$) $\kappa\nu$ is actually independent of the frequency of nuclear motion . . ."

Q: (1) Why "effective"? (2) Why in the nonadiabatic regime there is the independence on the frequency of nuclear motion?

M: (1) "Effective" because it does not apply to a single motion. In the 1965 paper, there are a couple of approximations, so it is not exactly equal to the frequency. There is a relation of some range of distances to same range of the region, I called it ρ there, in other words the preexponential factor is not just that frequency, it is a magnitude with the unit of frequency, probably something in the ballpark of frequency, but it is an effective frequency.
(2) In the preexponential factor in the Fermi golden rule for the nonadiabatic regime there is no frequency in there . . . There is probably some physical reason for the independence on the frequency of nuclear motion but I have to think about it.

13. p. 285 1st column, bottom: ". . . the vertical ionization potential *of the redox site*."

Q: You write "of the redox site" and not "of the reactant." Do you mean that the vertical ionization potential to consider is not the usual one of the reductant when it is at infinite distance from the oxidant but that of the reductant more or less close to the oxidant?

M: The influence would be very small. The surrounding environment . . . just a minor effect. I should have said "reductant" instead of "the redox site."

14. p. 286 1st column, middle: "... the 'entropy of activation' $\Delta S^{\ddagger}$, which includes both ΔS_r^* and an entropic contribution from $\kappa\nu$, namely $R \ln (A'/10^{13})$ ..."

    Q: Why "entropy of activation" in quotes? Then: how can an entropy contribution come from a term made up of a frequency and a transmission coefficient?

    M: All right, first: there is a $\Delta S^{\ddagger}$ which is the usual Eyring $\Delta S^{\ddagger}$ and then there is a ΔS^* which I introduced, they are not the same thing, go back to the papers to see it. Eyring's definition is very clear but I wanted a ΔS to contain, if I remember rightly, parts that were not part of the *kT*/*h* part of the preexponential, but had a certain physical context, rather than a formal context, so that was the reason why I introduced ΔS^*, it's close to but is not exactly the same as $\Delta S^{\ddagger}$. Second: there [are] all sorts of subtleties involved there, I'm not sure where to begin. One is that one can write down the TS theory expression, and that's fine, but one can go one step further and take the $\Delta S^{\ddagger}$ part and *break it up into various parts*, like the collisional part and the other parts and so on, and, when one does that, one gets a somewhat different expression, because now the $\Delta S^{\ddagger}$ disappeared, but new quantities come up in its place, namely ΔS^* and part of the [term] that's related to the particles getting together for the reaction occurring and that's where that frequency factor comes in, so that there is a way of going from the Eyring expression systematically over to the new expression but that's a long and involved thing that would be hard for me to describe here. So, an alternative way of seeing what is there is this: starting from TS theory, which is a very formal theory, by introducing appropriate assumptions one can go over to the other expression that had a $\kappa\nu$ in it and ΔS^*, and let me give you an example: if you take the Eyring expression, that $\frac{kT}{h}\exp(\Delta S^{\ddagger}/R)$, you can break up that $\Delta S^{\ddagger}$, in the case of a bimolecular reaction, in a part that is bringing the particles together there and there is an entropy change there. When you do that, when you get rid of that entropy change, that converts the $\frac{kT}{h}e^{\Delta S^{\ddagger}}$ into a collision frequency times an $e^{\Delta S^{*}/R}$. kT/h has just units of frequency, nothing more. 10^{13} is the value of $\frac{kT}{h}$ at room temperature. 10^{13} is at room temperature another word for $\frac{kT}{h}$ and people may not see the connection. In the case of a bimolecular reaction, one part of the entropy comes from bringing the particles together and the other from reorienting them, in the case of a unimolecular reaction one is getting rid of the vibration because the TS tells that there is one vibration less, and having a certain reaction probability when you are there in ET reaction, that reacting probability when you are at the top of the barrier is the equivalent of some kind of an entropy factor, in other words it's not a big temperature dependent factor, so there are subtleties then on how these things come about, you can derive some of this stuff without any TS theory but employing assumptions that contain TS theory, you see, but specialize it a little bit or you can go systematically, which I didn't do, except for the collision part. I go systematically from TS theory, bringing in all the parts that come in into a $\Delta S^{\ddagger}$, and then end up with a $\kappa\nu$ for example in the case of a unimolecular reaction, and let me tell you how the ν comes about. There is a partition function associated with one extra vibration of the reactants compared with the products. If you write products of partition functions instead of $\Delta S^{\ddagger}$, then in the denominator you are going to have a vibrational partition function for the reactant, that's not present in the TS, a vibration disappears, becomes a translation, that partition function is $\frac{kT}{h\nu}$, classically. If you take a $\frac{kT}{h}$ and you divide it by $\frac{kT}{h\nu}$ you get a ν, so that's how the ν comes in, so that's where the ν comes from and the κ is related of course to, well, even though you got to the transition probability there is not a unit probability in some cases of carrying on, and that's where the κ comes in, and then since it is a preexponential factor, it's sort of contribution to the entropy of activation which you introduce by introducing this physical idea. TS theory is just sort of very general but when you introduce several physical ideas you sort of evaluate some terms in it. The two contributions to the entropy of activation are: one is the loss of a vibration frequency, that's where the ν comes from, and the other is a probability of sort of doing something and entropy is related to probability. Yes, those are the two contributions.

15. p. 288 1st column middle-bottom: ". . . a distance of particular interest for ET is the distance between the closest atoms (of the two reactants) that are strongly coupled to their respective redox sites."

    Q: How and when is an atom to be considered coupled or strongly coupled to the redox site? Consider an atom bound to the redox site. Is it to be considered coupled from the point of view of ET just because of the chemical bond? Or maybe the bond is to be of a particular nature in order to favor ET? Is there a precise definition for that, such that one can decide if an atom is *strongly* coupled or, say, *weakly* coupled to the redox site?

    M: Maybe when the atom is connected by an aromatic network . . . that's a coupling . . . . so it must be a bond of a particular nature. In other words, it depends on the molecular orbital . . . it is really distributed throughout the aromatic system. If it is so, then the electron is not just localized to the metal atom, it is distributed and it is this distributed orbital that is undergoing electron transfer. Take the electronic orbital that you are interested in and calculate the probability of transfer.

16. p. 288 1st column bottom: "Eq. (33b) can be rewritten as

$$\kappa\nu = 1 \cdot 10^{13} \exp(-\beta d) \ \sec^{-1}$$

    where d is equal to the actual separation of the centers of the two closest (in the case of aromatics) carbon atoms of the two reactants minus some amount to allow for the extension of the

π-electronic orbitals beyond the carbon nuclei. For concreteness we will use a value of 0.3 nm for the latter amount"

M: The 0.3 nm comes from the spacing of the planes in the graphite.

17. p. 291 2nd column middle: "The benzimidazole- and the pyrine-pentaammineruthenium complexes provide a large, aromatic ligand for *probing* hydrophobic protein regions such as the heme crevice."

    Q: What do you mean by "probing"?

    M: If you have a hydrophobic region and you have some ligand that can penetrate into it, that would be an example of probing.

18. p. 292 1st column bottom: "The reaction of ferrocytochrome *c* with $Co(ox)_3^{3-}$ is reasonably well-behaved at the high ionic strength (0.5 M) used for the studies. This is not the case for the reaction of cytochrome *c* with ferro- and ferricyanide . . . although *rate saturation behavior* has been reported at high concentrations of the complex . . ."

    Q: What is it this "rate saturation behavior"? Does it mean that the rate constant beyond a certain concentration of a reactant becomes insensitive to a further increase of its concentration? If so, why does this happen? Is it a phenomenon depending on a limited velocity of reaction at the reaction center? Or maybe on diffusion velocities of reactants or products?

    M: I think that, I am not sure, but what I may have had in mind, because I forgot now this, is that when you change the sol concentration eventually you may produce some rate constant which becomes independent from the sol concentration, like a saturation, I think that' the case, I think that's what I had in mind, but I'm not sure, I'd have to look up because it mentions a specific reaction.

19. p. 294 1st column bottom: "Mixing the t_{2g}-orbitals of the Fe center in cytochrome *c* with the π^*-orbitals of the porphyrin ring

effectively *extends* the metal d-electron density to the porphyrin edge, so that facile *electron transport* to and from the iron center via the exposed heme edge should be possible."

Q: Do you mean electron transfer by the above "electron transport"?

M: Maybe the actual metal orbital is distributed, in other words imagine your actual orbital is that of an aromatic system, so one wouldn't think of the electron as sort of moving because that wouldn't be quantum mechanically correct, there is some coherent wave function but you don't have an actual electron transfer specified in the wave function, it would be wrong to consider a transport. Hopping would happen when the two reactants are weakly coupled. With metal centers weakly coupled then you have hopping. If the molecular orbitals are strongly coupled then there would be a kind of superexchange, molecular orbitals not of the same energy, so you have a superexchange, My guess is that there is no transport. I guess if the orbitals of that iron center are mixed with some orbitals of the ring . . . if they weren't mixed then you have to jump from those very tight orbitals, iron orbitals, you have to actually jump to the other reactant, but if there is a mixing of the two, then there is a substantial amount of electron density that's on the porphyrin ring instead of being confined to the t_{2g}-orbitals, so that makes the electron density at the periphery of the porphyrin ring larger than it would otherwise be if there were essentially negligible coupling of the t_{2g}-orbitals to the π^*-orbitals of the porphyrin ring. So that is going to help any transition rate, because the electron is now closer to the other reactant, but if there were no coupling they would be isolated on the iron, then you have to tunnel through, that's smaller probability.

20. p. 294 1st column bottom: "The 'exposed' edge of the heme group of cytochrome *c* is surrounded by folds in the polypeptide

chain; these folds may prevent the close (edge-to-edge) approach of the hemes *required* for $\kappa \cong 1$"

Q: The requirement appears to be a *necessary* condition for $\kappa \cong 1$. Is it also sufficient?

M: At $\kappa \cong 1$ the exposed edges do not actually have to touch. At two angstrom from each other, that wouldn't change much from unity. If the coupling is strong . . . metal ions . . . then it would be sufficient. It all depends on the strength of the electronic coupling of the edge to the metal atom . . .

21. p. 294 2nd column top and middle: "In these complexes the heme edge-to-edge (carbon–carbon) separation is 0.94 nm." "For purposes of calculation, we use the former value for the heme edge-to-edge (carbon-to-carbon) separation in the cytochrome *c* exchange, and use the 0.3 nm correction discussed in subsection IVB."

    Typo IVB is a misprint for IVC. The misprint is steadily repeated throughout the paper.
22. p. 294 2nd column bottom: "X-ray crystallographic studies (0.18 nm resolution)"

    Typo misprint for 0.018 nm.
23. p. 296 1st column top-middle: "The large separation was estimated above the introduce . . ."

    Typo misprint: "the" for "to."
24. p. 297 1st column top: "It is seen that the adjacent heme-edge pathway is approximately 10^5 times more favorable than the remote His-33 pathway, which involves electron transfer *through* the protein."

    Q: Electron transfer or electron transport? When would one use an expression or the other? When for instance we have the classical ET reaction between Fe^{2+} and Fe^{3+} ions in solution do we have electron transport through the solvent if the ions are far away from each other and electron transfer by a direct hop if the ions are close to each other?

M: In the transport through the protein you have a coherent wave function . . . every carbon atom is decreased by a certain factor, take the r being along the chain, so . . . rather than being an electron jump. That question comes up a lot, this electron transport along DNA decreases exponentially along the distance, and you have this coherent process, a superexchange, the electrons are hopping to some higher orbital and hopping along . . . so in the case of DNA where you have transport along distances you have this coherent process and the wave functions overlap and then you have a hopping mode along long distances.

Q: Hopping or not hopping? Is superexchange hopping to some higher orbital? And how about the hopping mode along long distance?

M: First of all, as far as ET . . . ET is fine, instead of saying electron transport. Maybe hopping isn't the right word, because the electron doesn't reside on that high orbital in superexchange, so hopping isn't a good word. Hopping is reserved for the electron physically be present on an orbital for a substantial amount of time instead of just making use of the energy level by a superexchange mechanism.

At very long distances . . . the experiments from the chap in Switzerland, Giese, really Giese and Michel-Beyerle . . . so in the case where you got those very long transfers then, at least in those experiments of Giese, it would be a tunneling process, it becomes very small, so it is easier for the electron to hop into a higher energy orbital and then just hop along. In this case it is not superexchange, when it goes to higher orbital and *physically resides* there, that's not superexchange.

25. p. 302 2nd column ¶ note: "A binding constant calculated from $S = 0.01$ and a K_A obtained from Eqs. (25) and (57) is only about 2 to 5 M^{-1} for these oppositely charged reactants. The contrast

between this value and the extrapolated experimental binding constant suggests that for oppositely charged reactants in contact the local electrostatic charges may dominate the interaction and require that the *delocalized-charge expression* Eq. (25) be replaced by a more complex one . . ."

Q: Equation (25) is the following:

$$w_{ij} = \frac{\frac{z_i z_j e^2}{2D_s r}\left(\frac{\exp(B\sigma_i\sqrt{\mu}}{1+B\sigma_i\sqrt{\mu}} + \frac{\exp B\sigma_j\sqrt{\mu}}{1+B\sigma_j\sqrt{\mu}}\right)}{\exp Br\sqrt{\mu}}$$

What does it mean that this is a "delocalized-charge" expression?

M: It must be from Sutin's work . . . in that expression you are using the total charge, so in this sense is delocalized.

26. p. 303 2nd column bottom: "Such copper centers are present in the *electron carriers* plastocyanin, azurin, and stellacyanin, which each contain one copper atom per molecule."

    Q: What is it meant by electron carrier? A reductant such as Fe^{2+} is to be considered an electron carrier?

    M: It is some part in the electron transfer chain, I suppose so. In other words, if the electron transfers along a chain electron carrier and reductant is the same, except that a carrier implies the chain.

27. p. 304 2nd column middle: "Anions tend . . . to be more readily bound at interfaces than do cations of comparable size, a known result in electrochemistry, and there is expected an interplay between binding and electrostatics."

    Q: What do you mean by interplay between binding and electrostatics?

    M: It could be that there is also electrostatics, in other words electrostatics is just part of the story, I imagine that is what I had in mind.

28. p. 304 2nd column bottom: "The $Cr(phen)_3^{3+}$ excited state produced is a very powerful electron acceptor and was found to oxidize the reduced plastocyanin with a concentration-independent rate constant of $3 \cdot 10^6$ s^{-1}."

    Q: The rates are concentration dependent but shouldn't the rate constant always be concentration independent? You are probably saying that the reaction is not a bimolecular reaction.

    M: It depends on how you define a rate constant. You can have a pseudo rate constant, define a pseudo rate constant, and that pseudo rate constant may depend on the concentrations, so it depends. Suppose you have a bimolecular reaction, and you absorb the concentration of one of the reactants into the rate constant, the rate constant becomes a pseudo rate constant, and is of first order.
29. p. 306 1st column top: "Oxidized and reduced azurin carry small net negative charges (Table X) and hence have negligible coulombic repulsion, while ferro- and ferricytochrome *c* have substantial positive charges (Table VII)."

    Typo Misprints: the charges of the proteins are reported in Table V.
30. p. 308 2nd column middle-bottom: "In the absence of reaction, the rate constant for diffusion k_D multiplied by the average number of repeated encounters γ will equal the *mean collision frequency* Z of the pair in solution. Since k_D for small uncharged species with no specific interactions is of the order of 1/10 to 1/100 the value of Z, γ is of the order of 10 to 100 for such systems."

    Q: You give here a definition of Z as $Z = k_D\gamma$. Is the average number of repeated encounters referred to a time interval of one second?

M: The reactants diffuse toward each other, get into a cage and, once there, they typically have repeated encounters, 10 to 100 hits per unit time. Z is just simply a statistical number. The frequency basically in a liquid doesn't say anything about the mechanism, when you bring in the diffusion coefficient you say something about the difficulty of reaching the cage, then there are the repeated encounters inside it.

31. p. 309 1st column bottom: "The *effective collision frequency* in solution is the value Z_{soln} in the absence of the work term w^r, multiplied by an exponential factor $\exp(-w^r/RT)$"

    Q: Is the *effective collision frequency* the same as the above mean collision frequency, that is, is it $Z = Z_{\text{soln}} \exp(-w^r/RT)$?

    M: Yes, it is the same, that Z_{soln} is 10^{11}.

32. p. 309 2nd column top: "Another model is to take the equilibrium constant K_{A} as the product of the effective volume occupied by the separation coordinate over which reaction occurs, $4\pi r^2 \delta r$... multiplied by the additional free energy factor $\exp(-w^r/RT)$ ...

$$K_A = 4\pi r^2 \delta r \exp(-w^r/RT) \tag{57}$$

    M: K_A here has the dimensions of a volume per mole.

33. p. 309 2nd column bottom: "In the case of an adiabatic reaction κ is unity and the frequency ν in Eqn. (58), $k_1 = \kappa\nu \exp(-\Delta G_r^*/RT)$, is given as a weighted sum of nuclear frequencies ν_i, namely:

$$\kappa\nu = \nu = \left(\frac{\sum_j \lambda_j \nu_j^2}{\sum_j \lambda_j}\right)^{1/2} \quad \text{(adiabatic case)} \tag{60}$$

    when there are several vibrational motions contributing to ν and to $\lambda(\lambda = \sum_j \lambda_j)$ (Dogonadze). In the case of a nonadiabatic reaction ($\kappa \ll 1$) the product $\kappa\nu$ is given instead by Eq. (32)

(misprint for 33a)

$$\kappa(r)\nu = 2\pi \frac{H_{AB}^2}{\hbar(4\pi\lambda RT)^{1/2}} \tag{33a}$$

and is seen to be independent of the frequencies ν_j of the nuclear motion!"

Q: Can you give a physical interpretation of a statement that deserves an exclamation mark?

M: Because the starting expression looks as though is dependent on the frequency, but there is a factor in the k which turns out to be inversely proportional to the frequency. Now, that's an interesting point, I'm not sure . . . you see, the $\kappa\nu$ is like a rate, right? If you did the Fermi golden rule, and if you put in the density of states, calculate the density of states for this system, you probably get the $1/\sqrt{4\pi\lambda RT}$. All right, so one might ask: why is that the in the Fermi golden rule you don't have the frequency? I mean one would have to think in those terms. In other words, to understand how you get Fermi golden rule, why is it independent of a frequency. That's basically it, so then I haven't explained why it's independent of the frequency, and in a way it may be that only superficially is independent of a frequency, because λ contains terms that are more composite, are related, but in a more complicated way, to the frequency. Yes . . . λ . . . if $\lambda = k\frac{a^2}{2}$, so k is a frequency, or really a force constant, that goes into a frequency, so it is just that some parts of the frequency are there, but sort of dissected and rearranged. It contains some properties which depend on the frequency. It doesn't have the frequency explicitly in there, but it has a vestige, a remnant of it. The λ has some aspects of a frequency in it, among other properties.

34. p. 308 2nd column top-middle: "The physical events involved *in a bimolecular collision* in solution between two molecules A and B involve a random diffusion of the molecules toward

and away from each other, becoming in some instances a *close-contact encounter complex A · ·B* and then diffusing apart. We may write this process as

$$\mathrm{A} + \mathrm{B} \underset{k_{-D}}{\overset{k_D}{\rightleftarrows}} \mathrm{A} \cdot \cdot \mathrm{B}\text{''} \tag{51}$$

p. 308 2nd column bottom: "If during the encounters, ET occurs *with a first-order rate constant* k_1, and if the *reverse* ET rate is negligible relative to the rate of separation of A^+ and B^-,

$$\mathrm{A} \cdot \cdot \mathrm{B} \overset{k_1}{\rightleftarrows} \mathrm{A}^+\mathrm{B}^- \rightarrow \mathrm{A}^+ + \mathrm{B}^-\text{''} \tag{52}$$

p. 310 2nd column top "We denote by A – B an AB pair in such a properly oriented oriented configuration; A –B is formed from the encounter complex A··B by suitable reorientation of A and/or B:

$$\mathrm{A} \cdot \cdot \mathrm{B} \rightleftarrows \mathrm{A} - \mathrm{B} \rightarrow \mathrm{A}^+ + \mathrm{B}^-\text{''} \tag{62}$$

Q: From the above the successive steps are apparently the following:

(i) Formation of an encounter complex A·· B is a bimolecular reaction;

(ii) From this follows ET with a first-order rate constant. Question: Should one first pass by a suitable fluctuation to a precursor complex, and from that—*with a first-order constant*?—to a successor complex, then to A^++ B^-? Are these precursor and successor complexes tacitly assumed?

3. Shouldn't one write the last reaction with a further step, that is, as:

$$\mathrm{A} \cdot \cdot \mathrm{B} \rightleftarrows \mathrm{A} - \mathrm{B} \rightarrow \mathrm{A}^+ - \mathrm{B}^- \rightarrow \mathrm{A}^+ + \mathrm{B}^-?$$

4. Can you say something about the relation between encounter complexes and precursor complexes? Does the encounter complex always and necessarily precede the precursor complex?

6. Using the symbols X* and X for the precursor and successor complexes could one detail the above process as:

$$A\cdot\cdot B \rightleftarrows A - B \rightarrow X^*(A - B) \rightarrow X(A^+ - B^-)$$
$$\rightarrow A^+ - B^- \rightarrow A^+ + B^-?$$

M: These complexes should really be not necessarily made up of reactants bound to each other but could be reactants in a cage. Norman (Sutin) especially writes using that formalism . . . the slow step usually happens after the cage, it is not necessary to introduce the successor complex, it is some intermediate along the way, unless there is diffusion, so it is unnecessary to introduce precursor and successor complexes, I did introduce X and X* in my very first paper but that reflects *the whole reorganization process*, another context, the precursor complex here doesn't bring any real reorganization. The X that I had wasn't a precursor complex, we mean two different concepts. Encounter complex and precursor complex is the same, I think so. Many are the necessary steps in the electron transfer process, there are many barriers . . .

35. p. 310 2nd column bottom: "A first-order rate constant k_1 for a properly docked pair (active sites in contact) . . ."

    Q: Does "active sites in contact" mean that there are no solvent molecules in between? I mean, suppose you have $Fe^{2+}(H_2O)_6$ and $Fe^{3+}(H_2O)_6$ complexes with water, are they in close contact if the coordinated molecules *touch* even if ET happens also when there are water solvent molecules in between?

    M: If the solvent molecules are strongly bound to an ion then they stay bound, the electron transfer could occur and in a paper with Paul Siders, we discussed that, I think when we discussed the inverted effect. Whenever the contact reduces the free energy barrier enough.

Q: Maybe one should write: "because an electron can transfer through the strongly bound solvent molecules."

M: Well, you have the solvent molecules that are part of the coordination shell so you mean those solvent molecules between the two coordination shells, is that what you mean?

Q: Yes.

M: Yes, when they are in contact I guess the solvent molecules between the coordination shell sort of have been squeezed out, and you just have the solvent molecules in each of the coordination shells and between them you may have no water molecules but you have water molecules all along that complex.

Q: So, you have this transfer through the bound solvent molecules.

M: Yes, presumably. Now, there are detailed calculations of that, Marshall Newton I think did calculations, I never saw what he did, but I think in his calculations the two solvent shells were in close contact with no additional solvent molecules in between. I think that's the case.

36. p. 311 1st column middle: "It should be noted . . . that the *binding constants for natural partners*, which generally have opposite charges, tend to be considerably higher than the values calculated from Eqs. (25) and (57)

$$w_{ij} = \frac{\frac{z_i z_j e^2}{2D_s r}\left(\frac{\exp(B\sigma_i\sqrt{\mu}}{1+B\sigma_i\sqrt{\mu}} + \frac{\exp B\sigma_j\sqrt{\mu}}{1+B\sigma_j\sqrt{\mu}}\right)}{\exp Br\sqrt{\mu}} \tag{25}$$

$$K_A = 4\pi\sigma^2\delta\sigma\exp(-w^r/RT)\text{''} \tag{57}$$

Q: What is it meant here by "binding constants" and by "natural partners"? Are the binding constants just the K_A's?

M: Natural partners are probably the reactants in the native systems . . . binding constants is . . . if you have two reactants touching each other forming a complex that would be the binding constant. It is just a K_A, I think so. They are equilibrium constants, that's right.

37. p. 311 1st column bottom: "When the rate constant k given by Eq. (63)

$$k = SK_A\kappa\nu \exp\left(-\Delta G_r^*\right) \tag{63}$$

becomes sufficiently high, the reaction may, as in the spherically symmetric case, . . . become diffusion-controlled. Whereas the problem in the latter case involved the solution of a differential equation for diffusion and reaction, leading in the steady state to Eq. (53)

$$\frac{1}{k_{\text{obsd}}} = \frac{1}{k_D} + \frac{1}{K_A k_1} \tag{53a}$$

$$k_D = 4\pi D / \int_{r=\sigma_{12}}^{\infty} \exp(w^r/RT) r^{-2} dr, \tag{53b}$$

the differential equation to be solved now also involves a *rotational-diffusive* process in the reactants."

Q: What is it meant by "rotational-diffusive" process?

M: If you have a particular site on a sphere that is active, then the spheres have to diffuse to each other, but the spheres have to be in contact with the right active sites. We didn't treat it but it could be treated.

38. p. 312 2nd column bottom: "For concreteness, we treat a solvent-mediated electron transfer and take $2\lambda_1 = \lambda = 1$ eV and $V_0^e = 5$ eV, the value for a vertical low energy ion-to-solvent charge-transfer transition." In p. 312 1st column V_0^{e} is defined as ". . . the vertical ionization potential for donation of an electron to an adjacent bridging unit is V_0^{e}."

Q: Is the ion-to-solvent charge-transfer transition an ion-to-solvent electron transfer?

M: That is probably connected with the Mc Connell formula, ET but without reorganization probably. There is something acting as an intermediate that doesn't have time to reorganize. It is not an ion to solvent ET, regardless if the solvent molecule is a bridging unit between the two reactants. The electron isn't residing in that intermediate water molecule. It is a superexchange.

M206. Application of Electron Transfer Theory to Several Systems of Biological Interest

NOTES

1. p. 228 bottom: "... an exponential dependence on the separation-distance r:

$$k(r) = \exp[-\beta(r - r_0)] \tag{5}$$

Values of β are given in Table II. A typical value is seen to be in the range 1.0 to 1.2 A^{-1}, and we shall use a value of 1.2 A^{-1} when, as in Table I, the *intervening material* between the two redox sites is 'saturated,' that is, nonconjugated. When a portion of the intervening material is conjugated (e.g., aromatic) the value of β for that part of the separation-distance is expected to be lower."

Q: Is the intervening material part of the reactant, say an aromatic reactant or can the intervening material be a solvent molecule? So that a solvent like water is to be considered nonconjugated but benzene, say, is to be considered conjugated? More in general, should the chemical nature of solvents be considered as favoring or hindering ET, for instance favoring some kind of electron transport through the solvent molecules? I mean is the nature of the solvent important in ET beyond the characters of the solvent described by the dielectric constants?

M: I think that now we typically use the value 1 for β. When I speak of intervening material I mean . . . between donor and acceptor, it could be the solvent, but usually when people do these experiments they are doing experiments with hydrocarbon networks in between, like Closs did.

The solvent, saturated or not saturated roughly, there is probably not too much difference between the two sigma bonds and so on, . . . Usually the reactants come and touch each other, so essentially there is no solvent in between.

M207. The Relation between the Barriers for Thermal and Optical ET Reactions in Solution

NOTES

1. p. 119 middle-bottom: ". . . the relation between the energy of the intervalence transition and the barrier to thermal ET in terms of free energies rather than of energies. This distinction, which has frequently been overlooked, is important when the ET is between oppositely charged reactants or between reactants that interact very differently with the solvent.

   Q: Why in the two above cases is the distinction between ΔG and ΔE particularly important?

   M: The question is whether there is entropy of reaction or not. That's a big question. If you have two charges going to no charge or vice versa there is a huge entropy of reaction, so it is important to use free energies instead of energies.
2. p. 124 middle:

$$\text{``}E_{op} = (\lambda - T\Delta S^{0'}) + \Delta H^{0'} \tag{4}$$

$$= \lambda + \Delta G^{0'} \tag{5}$$

Evidently for a series of *related reactants or compounds* (constant $T\Delta S^{0\prime}$) a linear dependence of E_{op} on both $\Delta H^{0\prime}$ and $\Delta G^{0\prime}$ is predicted."

Q: Is the condition of "constant $T\Delta S^{0\prime}$" a general definition of "related reactants or compounds" in ET theory or is it to be considered only in the restricted sense that only if this condition is fulfilled we do have the above linear dependence?

M: I should have said with constant $T\Delta S^{0\prime}$ *and* λ. I should have said "same λ."
In those equations you have a linear relation between those things, but of course only if λ is not varying, it's a constant, then only if $\lambda - T\Delta S^{0\prime}$ is a constant, only then you have a linear relation, otherwise if the λ had a complicated dependence on the reactant then there wouldn't be a linear relation, it is more complicated.

3. p. 128 middle: "The free energy of *reactants in an environment* G^r can be written as

$$G^r = G_0^r + m^2\lambda\text{''} \tag{A1}$$

Q: The free energy is an extensive thermodynamic property. Here "reactants and environment" is a bit vague as system's definition. Is it referred to, say, a unit mass or a unit volume of the solution made up of the A^+ and B^- ions (at unit activities) plus solvent among which the ET reaction $A^+ + B^- = A + B$ happens? Or to a mole of the reacting ions plus solvent? Or is the free energy referred to just the couple of $A^+ + B^-$ plus the solvent molecules interacting with it, I mean, is the free energy referred to the *molecular* systems? Is in this case G referred to a mole of "reactants and environment system?

M: It can be anyone of the above, if you want a mole in the system, fine, whatever you want. And then all is defined relative to that. The main point is that it refers to the entire system.

If you want a couple of ions, you take the couple of ions and then the surrounding solvent and call that the system. All of the solvent molecules beyond those surrounding the ions have no more influence.

M209. A Semiclassical Model for Orientation Effects in ET Reactions

NOTES

1. p. 3090, 2nd column bottom: ". . . the quantization conditions can now be satisfied only approximately at the well boundary:

$$C^i_{m+1}\Psi^i_{m,m+1} \cong C^0_{m+1}\Psi^0_{m,m+1} \quad (\xi = \xi_0) \tag{6a}$$

$$C^i_{m+1}\left(\frac{\partial\Psi^i_{m,m+1}}{\partial\xi}\right)_{\xi=\xi_0} \cong C^0_{m,m+1}\left(\frac{\partial\Psi^0_{m,m+1}}{\partial\xi}\right)_{\xi=\xi_0} \tag{6b}$$

To satisfy Eq. (6a) both sides were *squared* and then integrated over η and ϕ at $\xi = \xi_0$ thereby averaging over the boundary. *Taking the square root*, and following the same procedure for Eq. (6b) one obtains an equality of $C^i_{m+1}R^i_{m,m+1}(\xi_0)$ and of $C^0_{m,m+1}R^0_{m,m+1}(\xi_0)$ and also of their derivatives . . ."

Q: Instead of requiring the equality of the wave functions at every point (ξ_0, η, ϕ) one requires the equality of the overall probability at the surface of the ξ_0 oblate spheroid. Is this condition an approximation? Moreover, why do we integrate the squares of the wave functions and then take the square root instead of integrating directly the wave functions instead of their squares? Do you do so in order to take advantage of the normalization of functions Φ_m and S_{mn}?

M: This equality of the overall probability is an approximation, it is not a rigorous procedure, I don't think so.

2. p. 3091 2nd column middle: "For comparison, results using spherical wells of similar volume and energy are also given in Table 1. They are seen to be significantly less accurate than the present approximation to the spheroidal problem, particularly at $\Theta = 0°$"

   Erratum It is really so at $\Theta = 90°$.
3. p. 3093 2nd column middle: "The accuracy of the present method supports *the simple conceptual* previously introduced."

   Typo: The word "model" is missing. It should be "the simple conceptual model."
4. p. 3094 1st column middle ". . . we select the principal $R_{mn} S_{mn}$ term in the exact sum (largest coefficient) and compare (in Tables II to VI given later) its R_{mn} and S_{mn}, inside and outside the well, with those of the corresponding approximate functions used in the present single-term calculation of H_{BA}. They are compared *on a relative basis* to emphasize their similar shape.
   p. 3094 2nd column top-middle: "For comparison purposes, the functions are equated at the smallest ξ."

   Q: Is it in this last sense that the comparison is "on a relative basis"?

   M: You pick a point at which they are equal and all is relative to that. I do not know how they picked the point at which the wave functions are equal.
5. p. 3095 1st column bottom: "In obtaining a *uniform* semiclassical solution for S^0_{mn}, S^0_{mn} is converted to a function $\left(1 - \eta^2\right)^{\frac{1}{2}} S^0_{mn}$, whose differential equation contains no first derivatives. The comparison function chosen for making the uniform approximation is $\left(1 - \nu^2\right)^{\frac{1}{2}} P_l^m(\nu)$"

p. 3095 2nd column middle-bottom: "The function $R^0_{mn}(\xi)$, the radial function outside the potential well, satisfies Eq. (4c)

$$\frac{d}{d\xi}\left\{\left(1+\xi^2\right)\frac{dR^i_{mn}}{d\xi}\right\}+\left\{\frac{d^2}{4}\xi^2k_i^2+\frac{m^2}{1+\xi^2}-\lambda^i_{mn}\right\}R^i_{mn}=0, \tag{4c}$$

with the i superscripts and subscripts there replaced by o' s. In the present study the following *primitive* semiclassical approximation for $R^0_{mn}(\xi)$ sufficed because of the absence of turning points for the ξ motion

$$R^0_{mn}(\xi)\cong\left[\exp\left(-\int\limits_{\xi_0}^{\xi}|p_\xi|\,d\xi\right)\right]\Big/\left(\xi^2+1\right)^{\frac{1}{2}}|p_\xi|^{\frac{1}{2}} \tag{A5}$$

The inner radial function $R^i_{mn}(\xi)$ satisfies Eq. (4c). The tendency toward an absence of turning points, i.e., for the effective energy for the ξ motion to exceed the effective potential energy for all ξ, increases with increasing d, increasing k_i^2 and decreasing n. For the (m,π) states and choice of parameters appropriate to the modeling of large aromatic systems discussed here there are no turning points for the ξ motion in the region $\xi < \xi_0$ and so a *primitive* semiclassical approximation suffices for $R^i_{mn}(\xi)$."

Q: (i) which is the difference between a *uniform* and a *primitive* semiclassical approximation and why in the case of S^0_{mn} one uses the uniform and in the case of $R^i_{mn}(\xi)$ one uses the primitive?

(ii) What is it meant by a pure ξ motion? Is it the motion along a hyperboloid with an asymptote of constant $|\vartheta|$ going through different ξ?

M: To get rid of a singularity, because the primitive semiclassical approximation can have a singularity, you can go to infinity, but if you use the uniform approximation, which is more correct, then that gets rid of that singularity. See, you get a $1/\sqrt{\text{velocity}}$ in one case, but at the turning point, when the energy equals the potential energy, in other words at that point where the velocity is zero, you get infinity, but of course the actual wave function has no infinity, so what you use is a better solution of the equation in the region of that crossing, that region where the potential energy crosses as a total energy you use a more accurate treatment. This means primitive without a uniform semiclassical approximation. In the primitive semiclassical usually you just have exponentials. A positive exponential, maybe a negative exponential of a minus, they are the asymptotics of a more general function, an Airy function, it depends on the problem, which is the uniform approximation. Maybe some of the papers written with O'Connor on semiclassical theory may bring that out.

If you have something like e^{kx}, that is some kind of semiclassical wave function, then it goes down indefinitely, with no turning points, so that primitive expression is OK. But where you have turning points then you have an outgoing wave and an incoming wave, and if you want to represent by turning points then you need a uniform approximation at that point, in that region. The ξ motion means going along at constant ϑ, so ξ is changing.

M210. Mutual Orientation Effects on ET between Porphyrins

NOTES

1. p. 1436 2nd column top: "ET was modeled between two such nonpenetrating sites, and the effects of orientation on the *thermal* matrix element (the electronic matrix element appearing in theories of nonadiabatic electron transfer) . . ."

   Q: Why is the matrix element named "thermal"?

   M: Because there is an optical matrix element for charge transfers . . . they wrote "thermal" to distinguish from optical.

2. p. 1438 1st column top-middle: "Since the total Hamiltonian for the system, H_{tot} is

$$H_{\text{tot}} = -\frac{1}{2}\nabla^2 + V_{\text{A}} + V_{\text{B}} = H_{\text{A}} + V_{\text{B}} = H_{\text{B}} + V_{\text{A}} \qquad (6)$$

one has for the perturbation H'_A to H_A

$$H'_A = V_B \qquad (7)$$

Due to the definition of V_B, the expressions in Eq. (3)

$$H_{BA} = \langle \Psi_B | H'_A | \Psi_A \rangle, \quad H_{AA} = \langle \Psi_A | H'_A | \Psi_A \rangle \qquad (3)$$

then reduce to

$$H_{BA} = -V_0^B \langle \Psi_B | \Psi_A \rangle_B, \quad H_{AA} = -V_0^B \langle \Psi_A | \Psi_A \rangle_B \qquad (8)$$

Q: Please check if my interpretation of the above formalism is correct. The off-diagonal matrix element is proportional to the overlap between the wave functions Ψ_B and Ψ_A inside the B potential well and the potential energy of the electron described by the wave function Ψ_A when inside the potential well B is proportional to its probability of being inside B. Moreover, this potential energy lowers the single-site potential energy $V_A - V_0$, so that because of interaction the overall potential energy of the compound system becomes $V = -V_0^A - V_0^B \langle \Psi_A | \Psi_A \rangle$.

M: Interpretation correct.

3. p. 1442 1st column bottom: "Experimental results on such face-to-face porphyrins appear to indicate fast, light-driven forward electron transfer (< 6 ps) with *slow back transfer* to yield ground-state products (~ 1 ns). It has been suggested that the back transfer is *slow* due to a large *avoided crossing* of the electronic surfaces, which can occur in *the inverted region*, or because of the large driving force and *small* activation energy of the process, that is, the "inverted effect, or both."

M: When you are in the inverted region the activation energy is very small, in the inverted region you tunnel... And in fact when John Miller, Calcaterra, and Closs did their famous study they found that the inverted effect has a very weak, small activation energy. There is tunneling.

4. p. 1443 2nd column bottom: "In the present model, H_{BA}, the main contribution to the thermal matrix element for ET may be written as a volume integral over well B. That is

$$H_{BA} = -V_0^B \int_{\text{well } B} \Psi_A \Psi_B^* d\tau \qquad \text{(A1)}$$

where $d\tau$ signifies a three-dimensional integral. In well B, $-V_0^B \Psi_B^*$ equals $(E - T)\Psi_B^*$ where T is the one-electron kinetic energy operator... Then Eq. (A1) becomes

$$H_{BA} = \int_{\text{well } B} \Psi_A (E - T) \Psi_B^* d\tau \qquad \text{(A2)}$$

Since

$$\Psi_B^*(E - T)\Psi_A = 0 \qquad \text{(A3)}$$

anywhere outside well A, the left-hand side of Eq. (A3) can be integrated over well B and subtracted from the integral of Eq. (A2) without changing the value of H_{BA}. We thus obtain

$$H_{BA} = -\int_{\text{well } B} (\Psi_B^* T \Psi_A - \Psi_A T \Psi_B^*) d\tau \qquad \text{(A4)}$$

Q: Why is $\Psi_B^*(E - T)\Psi_A = 0$ anywhere outside well A?

M: Outside A that $E - T$ is zero, E equals T. $E - T$ is different from zero only inside well A. I got this from Bardeen ...

5. p. 1444 1st column middle-bottom: "We assume here that in electron transfers involving porphyrin excited states the first excited singlet state is adequately described by the four-orbital model. For concreteness, the x-polarized transitions are considered. In representing the present wave functions, we omit all doubly occupied molecular orbitals; they are assumed to be unaffected by the presence or absence of the transferable electron. Furthermore, *the ξ- and η-dependent parts of the wave functions need not be considered explicitly.*

   Q: Why so?

   M: I use only the Φ part of the wave function in order to get selection rules. If you are looking for selection rules, that changes with angles . . . then the other parts are kind of constants. That's about it.

CHAPTER 13

Dynamical Effects in Electron Transfer Reactions, Dielectric Relaxation and Intramolecular Electron Transfer, Solvent Dynamics, and Vibrational Effects in Electron Transfer

In this Chapter, oral interviews with Marcus are reported on **M211, M213, M214**. The interviews are presented in the form of **NOTES** to the publications.

M211. Dynamical Effects in ET Reactions

NOTES

1. p. 4894 1st column top-middle: "There is another approach to reactions that was begun by Kramers and includes some recent applications and extension of electron transfer (ET) reactions. It takes into account the *diffusive nature* of the dielectric relaxation of the solvent. In sufficiently viscous solvents the relaxation time can become long."

   Q: Is an *orientational* fluctuation the same as a *rotational* fluctuation?

   M: The inverse of relaxation is fluctuations. And fluctuations carry you to the transition state. In this case, there are mainly orientational diffusional fluctuations. Yes, the solvent orientational diffusional is the same as rotational diffusional.

2. Following: "and then the population of the *reacting systems in the transition state region* can become smaller than the equilibrium one."

   Q: In order to concretely understand the above, please tell me if the following example is correct: let us imagine the ET reaction is the following

$$A_s^+ - B_s^- \rightarrow A_s - B_s$$

   where the subscript s means that the ions are solvated. A system of two solvated ions A_s^+ and B_s^- at a fixed distance is *a reactive system in the transition state region* if it is in a condition (for instance of the two ions being at a certain relative distance, of having a certain reciprocal orientation, etc.) that a suitable vibrational or polarization fluctuation may then allow the reaction to happen. Of all possible couples of ions in our system only the above couples represent then reacting systems in the transition state region.

   M: That's a fair statement, yes.

3. p. 4913 2nd column bottom, in Note 13: "These authors consider a dipole and the effect of the *reaction field* on τ_L.

   Q: What is it meant by "reaction field"?

   M: That long preceded me, went back to Onsager... what the reaction field is... if you have a dipole that's oriented... surrounded by solvent molecules... the field that those molecules exert is the reaction field. It is the field of the surrounding molecules acting on the species at the center.

4. p. 4894 2nd column top-middle: "...for a brief time at the start of the reaction the reaction can proceed with the initial distribution in phase space of the reactants, but when the dielectric relaxation rate is slow relative to this *inherent reaction rate*"

   Q: What is it meant by "inherent" reaction rate?

M: I am not sure that that is a great term. If dielectric relaxation is slow... let me be specific about it in a crude model. This you can find on p. 129 of the Faraday discussions of 1960... on ET processes. On p. 129 I think I have a note there. The idea is this: write $\frac{1}{k_{\text{obs}}} = \frac{1}{k_{\text{diff}}} + \frac{1}{k_{\text{react}}}$, and so when that k_{diff} is small compared to k_{react} ... you are really comparing the k_{diff} and the k_{react}. Of course $k_{\text{react}} = k_{\text{activ}}$.

Q: Does "inherent reaction rate" mean just $k_{\text{react}} = k_{\text{activ}}$?

M: Yes, that's right.

5. p. 4894 2nd column middle: "Dielectric *relaxation* has been treated by a diffusive type of mechanism (Debye relaxation) because its rate shows a *thermally activated behavior*"

   M: The activation barrier for diffusion is bigger than that for reaction.

6. p. 4895 1st column top: "Kramers' model gives a rate constant inversely proportional to the relaxation time τ_L of the solvent orientational fluctuations, which characterizes *the kinetics within the system.* In this respect, it is essentially different from the traditional theories which assume thermal equilibrium among the reactants."

   Q: What do you mean by "*kinetics within the system*"? Is it maybe the local kinetics of the single reactant instead of the overall kinetics of the macroscopic reacting system?

   M: I am not sure that using the words "kinetics within the system" was a good idea because it is not very specific... It probably should mean the solvent kinetics within the system.

7. p. 4895 2nd column top-middle: "Typically, the orientational polarization should be described by a vector function of position and the free energy expressed in terms of its contribution and that

of intramolecular vibrations. However, within the Debye model in which the free energy for polarization fluctuations is quadratic in polarization components, we can single out for notational brevity a scalar variable X proportional to a certain component of the polarization vector which passes through the two minimum energy points associated with reactants and products in the polarization coordinate space."
p. 4896 1st column bottom: "The sum of the solvation free energy of the reactant, a free energy associated with coordinate X and represented by $\frac{1}{2}X^2$, and of the potential energy of the q motion, represented by $\frac{1}{2}aq^2$, is denoted by $V^r(q, X)$

$$V^r(q, X) = \frac{1}{2}aq^2 + V(X), \tag{3.1}$$

where

$$V(X) = \frac{1}{2}X^2. \tag{3.2}$$

The corresponding quantity for the product is denoted by $V^p(q, X)$, measured relative to the same energy zero:

$$V^p(q, X) = \frac{1}{2}a(q - q_0)^2 + \frac{1}{2}(X - X_0) + \Delta G^0, \tag{3.3}$$

where ΔG^0 is the standard free energy of reaction (the vibrational entropy of reaction in the above model is negligible or zero.) The equilibrium values of q and X *for the product* are denoted by q_0 and X_0. The quantities $\frac{1}{2}aq_0^2$ and $\frac{1}{2}X_0^2$ *represent* the λ_i and λ_0 in earlier publications (**M17, M53**)."

Q: Does the above mean that we can write the following "dictionary" relating the present and previous papers, in which the formulas to the left are simplified forms of the more complicated ones to the right?

$$\frac{1}{2}aq_0^2 \Leftrightarrow \lambda_i = \sum_j k_j(\Delta q_j)^2$$

$$\frac{1}{2}X_0^2 \Leftrightarrow \lambda_0 = (\mathrm{ne})^2\left(\frac{1}{2a_1}+\frac{1}{2a_2}-\frac{1}{R}\right)\left(\frac{1}{D_{op}}-\frac{1}{D_s}\right).$$

Do we have also the correspondence

$$\frac{1}{2}X^2 \Leftrightarrow F = \int\left\{\frac{\mathbf{E}_c^2}{8\pi}+\frac{\mathbf{P}_u^2}{2\alpha_u}-\frac{\mathbf{P}\cdot\mathbf{E}_c}{2}-\frac{\mathbf{P}_u\cdot\mathbf{E}}{2}\right\}dV$$

with the added specification that X is the component of the polarization vector along the line joining the minimum energy points associated with reactants and products in the polarization coordinate space? And what of the electric fields?

Moreover: q_0 in the present paper refers to the product but $\frac{1}{2}aq_0^2$ in your previous papers refers to both reactants and products because Δq_j is the difference of equilibrium bond lengths of the corresponding j'th bond in both sides of reaction:

$$A_1(\mathrm{ox}) + A_2(\mathrm{red}) \rightarrow A_1(\mathrm{red}) + A_2(\mathrm{ox}).(\mathbf{M118}, p.487).$$

Are then the two notations consistent?

M: If we had not done as above we would have had a more complicated space... and so this was just our way of cutting through... it isn't rigorous but it has the right spirit. The electric fields can be written in terms of the polarization... and the intrinsic charge distribution and geometry, it is more complicated. $\frac{1}{2}X^2$ has the above correspondence. Whether it will be that simple, if there will be that simple correspondence, I am not sure. Here we chose the q to be zero, zero positions for the reactants.

8. p. 4895 2nd column middle: "When *the reactant* can be represented by one potential energy surface in (X, q) space and the

product by another, and when the electronic coupling between these two is very small, the transition state lies at the intersection of these two surfaces"

Q: In Fig. 1, the free energy surface for reactant and for product is represented. If I understand well, a similar figure describes the states of every *single* reactant system, $\left|A_s^+ - B_s^-\right\rangle$ say, and every product system $|A_s - B_s\rangle$, say.

M: Yes, it refers to single reactant and single product system, that's right. It is a free energy, a single reactant but with all its solvent molecules, that's right. Because it is a free energy this means that with respect to the macroscopic system there has been a lot of averaging. You go from 10^{23} dimensions to two... When one speaks of Brownian motion one averages over a lot of coordinates... this is the Brownian motion of the polarization, of the vibration coordinate, so it is not the particle itself, that you could see as a point moving on that surface...

9. p. 4895 2nd column middle-bottom: "Thereby, one can define a rate constant $k(X)$, which involves *at each X a suitably averaging over the population in the q coordinate*"

   Q: Apparently this means that, using a more explicit notation $k(X)$ is really $k(X, \bar{q}(X))$?

   M: Yes, that's right. I guess it's right... I guess that $\bar{q}$ is a function of X ... Consider that along q, X changes. q is a fast coordinate...

10. p. 4895 2nd column bottom: "The reaction can be viewed as being described by the diffusion–reaction equation in which the coordinate X diffuses via a Brownian motion under the influence of a potential $V(X)$, while the reaction occurs at each X with a rate constant $k(X)$ during the diffusion."
    p. 4897 1st column top: "The coordinate X diffuses under the influence of a potential $V(X)$, which represents the free energy of the reactant's surface."

Q: Please check if the following *analogy* is correct: in the case of a gas diffusion, there is diffusion from a point of higher concentration to one of lower concentration under the influence of the chemical potential, $\left(\frac{\partial G}{\partial n_i}\right)_{n_j,P,T}$ say. The gas particles move by Brownian motion toward an equilibrium concentration. Under the influence of $\frac{dV}{dX}$, each system moves diffusing with a zigzagging Brownian motion on the free energy surface depicted in Fig. 1 toward the minimum O of the reactant surface with some probability of hitting the transition state curve C in Fig. 1 and then reacting and successively moving toward the minimum O' of the product. Moreover, can we say that the Brownian motion on the free energy surface describes how the collective polarization coordinate $X$of the solvent molecules surrounding the reactant "moves," that is, changes in time with the molecules' geometrical orientations and consequent polarization?

M: That zigzagging happens only along the X coordinate, you are averaging over the q coordinate.

Q: Can we say that the Brownian motion on the free energy surface describes how the collective polarization coordinate X of the solvent molecules surrounding the reactant "moves," that is, changes in time with the molecules' geometrical orientations and consequent polarization?

M: Yes.

11. p. 4896 1st column middle: "Two kinds of average survival times are introduced in Sec. VII, one being the customary *first mean passage time*"

    Q: Is it maybe *mean first passage time*?

    M: Correct.

12. p. 4897 1st column top: "…the distribution function $P(X; t)$ present for the coordinate X at time t satisfies the

diffusion–reaction equation

$$\frac{\partial P}{\partial t} = D\frac{\partial^2 P}{\partial X^2} + \frac{D}{k_B T}\frac{\partial}{\partial X}\left(P\frac{dV}{dX}\right) - k(X)P \text{''} \qquad (4.2)$$

Q: I believe $P(X;t)$ refers to all possible points $(X;\bar{q}(X))$ on the free energy surface. My question is: which is the physical meaning of the second

$$\frac{D}{k_B T}\frac{\partial}{\partial X}\left(P\frac{dV}{dX}\right)?$$

I submit a tentative physical interpretation: Let me first write Kramer's equation in the form:

$$\frac{\partial P}{\partial t} = D\frac{\partial}{\partial X}\left(\frac{\partial P}{\partial X}\right) + \frac{D}{k_B T}\frac{\partial}{\partial X}\left(P\frac{dV}{dX}\right) - k(X)P$$

The population P tends to equilibrium diffusing under the influence of the "population gradient" $\frac{\partial P}{\partial X}$ in the first term on the right. Now, the steeper the change of P with X, the more quickly we expect the population to change in time, and that's the story told by the first term on the right. The situation is similar to the conservation equations for the flame. But we should also expect the change of P in time to be influenced by how steep is the variation of V corresponding to a value of X. $\frac{dV}{dX}$ is like a "force" which in the second term on the right is "pushing" on P. It is then reasonable to have as a second term to the right not just $\frac{\partial}{\partial X}(P)$ or $\frac{d}{dX}\left(\frac{dV}{dX}\right)$ but $\frac{\partial}{\partial X}\left(P\frac{dV}{dX}\right)$. Does it all sound reasonable? Moreover: here we have already seen the diffusion of X under the driving force $\frac{dV}{dX}$. That diffusion happens for a single reactant system via a Brownian motion on the free energy surface. Here, we have the diffusion of $P(X;t)$ driven by $\frac{dV}{dX}$, P, and $\frac{\partial P}{\partial X}$. Does $P(X;t)$ too move with a Brownian motion?

M: The second term is a convective motion. The second term, the diffusion term, that's a convective motion, it is under the influence of a force really, the force being $\frac{dV}{dX}$. That second term on the right of the diffusion is a force. The first term on the right is the spontaneous diffusion. $\frac{dV}{dX}$ is like a "force" which in the second term on the right is "pushing" on P, you are right. You have a diffusion term and a forced diffusion term. And the two contribute to the flux. That diffusion happens for a single reactant system via a Brownian motion on the free energy surface. This is a probability distribution, if you multiply by the number of systems you get a population.

13. p. 4897 1st column middle: "Therefore, if the reaction does not occur, τ_L^{-1} determines the rate with which $P(X; t)$ approaches the thermal equilibrium distribution function determined by the potential $V(X)$. Because of the reaction, $P(X; t)$ at any time t can differ from the equilibrium distribution function. Frequently, the quantity of interest is not $P(X; t)$ but rather the survival probability $Q(t)$

$$Q(t) = \int P(X; t)dX \tag{4.4}$$

$$\text{with } Q(0) = 1 \quad \text{and} \quad Q(\infty) = 0.\text{"} \tag{4.5}$$

Q: Please check if the way I understand the above is right. The distribution function at time $t = 0$ is the initial distribution (equilibrium or nonequilibrium) and describes how the reacting systems are distributed among the different polarizations. One considers the initial distribution function to be normalized, that is Eq. (4.4) is valid. The number of reacting systems becomes smaller and smaller as the reaction progresses. At the end of reaction there are no more reacting systems left so that $Q(\infty) = 0$ because the reverse reaction has been neglected by hypothesis. Is it so?

M: It is really the initial distribution function, how the polarization is distributed, what the polarization function is at every point. X characterizes a polarization, the polarization is really a function, there is a kind of isomorphism going over to using a single coordinate instead of the whole distributed coordinates. The polarization distribution per unit reactant is unchanged, it's independent of the reactants decreasing, so it is normalized to the amount. Certainly, the reverse reactions have been neglected... otherwise we would have some replenishment. Often, you know, if you have a reaction that proceeds forward and back, you can treat often the forward reaction and treat the backward reaction, then put the two together. And often you're treating the reaction where the reverse reaction is improbable, unless you are going to some steady state kind of situation. I mean sometimes there is a reverse reaction but often there is little or no reverse reaction, you have sort of reaction that's sufficiently downhill.

You shouldn't neglect reverse reaction... but we did in this paper and several people noted that... one should consider the back reaction.

14. p. 4897 2nd column middle: "Thus, in this slow reaction limit one has the usual thermal equilibrium expression for the reaction rate constant. Whether the reorientational motion of the solvent molecules is diffusional *or not* is unimportant in this limit..."

Q: Please comment on the "or not." Do you mean that a certain orientation around an ion may be reached either by diffusional orientation or because of the *orienting electric field* of the ion?

M: The process can be diffusional but doesn't have effect on the rate... because the first step is independent on how slow the diffusion is... is the crossing of the barrier... it is $\frac{1}{k_{obs}} = \frac{1}{k_{diff}} + \frac{1}{k_{react}} \cdots$

Q: Do you maybe mean that in order for the reaction to happen you first need activation and this activation implies always a reorientation of the molecules either if this reorientation is diffusional or not?

M: What I mean is that if the whole diffusional part occurs rapidly on a time scale compared with the reaction part... it doesn't matter whether diffusion is fast or ultrafast....

15. p. 4898 1st column top-middle "*(4) The nondiffusing limit.* In this limit the reaction from the reactant to the product proceeds so rapidly *at the initial values of the slow coordinate X* that the distribution of X is not restored by diffusion in the course of the reaction."

 Q: I believe it should be: "The reaction... proceeds so rapidly at the initial values $P(X; 0)$ of the reactants population that the distribution of X ..."

 M: That's right, the equilibrium distribution is not restored... it should be the equilibrium distribution is not restored. You have an initial distribution of X, right? For each of those if it reacts very rapidly with the rate constant, there is no time for diffusion among the X's. The reaction for each of those X's is so fast....

16. p. 4898 2nd column top-middle: "A function $g(X)$ is first introduced...

$$g(X) = e^{-\frac{1}{2}V(X)/k_BT} \Big/ \left[\int e^{-V(X)/k_BT} dX\right]^{1/2} \quad \ldots.. \tag{5.1}$$

 The solution $P(X; t)$ to Eq. (4.2)

$$\frac{\partial P}{\partial t} = D\frac{\partial^2 P}{\partial X^2} + \frac{D}{k_BT}\frac{\partial}{\partial X}\left(P\frac{dV}{dX}\right) - k(X)P \tag{4.2}$$

 is then written as

$$P(X, t) = g(X)p(X; t), \tag{5.3}$$

 where, using Eq. (4.2), the equation for $p(X; t)$ is

$$\frac{\partial}{\partial t}p(X; t) = -[H + k(X)]p(X; t) \tag{5.4}$$

with

$$H = -D\frac{\partial^2}{\partial X^2} + \frac{D}{2k_BT}\left[\frac{1}{2k_BT}\left(\frac{dV}{dX}\right)^2 - \frac{d^2V}{dX^2}\right] \tag{5.5}$$

...We further note that $g(X)$ is the *ground state wave function* of H with the eigenvalue zero

$$Hg(X) = 0. \tag{5.6}$$

When the potential $V(X)$ is given in the form $V(X) = \frac{1}{2}X^2$, the operator H is essentially the same as the Hamiltonian of a harmonic oscillator."

Q: What do you mean when you consider $g(X)$ a wave function?

M: $p(X; t)$ describes the deviation from equilibrium. For example, in the solution of Boltzmann equation in gases... $p(X; t)$ is a correction factor. Diffusion is extremely rapid... No wave function here. If you work in a more general space the polarization is a function of space, here $g(X)$ represents an equilibrium distribution of that polarization... because at each point you have fluctuations in polarization, you should include fluctuation at every point of the system, it is very complicated. No wave function. It is written wave function just by analogy with the Schrödinger equation. Often you see in the literature that people have seen the analogy between the diffusion equation and the Schrödinger equation. It is a formal connection. So, the "ground state wave function" refers to the whole system, it suggests that you might give similar methods for solving the problem. So, there is nothing that really "waves," definitely not. There are just the equations that are similar, you interpret them differently.

17. Following: "In this case, it is easily shown that the nonvanishing eigenvalues of H are equal to n/τ_L, with $n = 1, 2, 3, \ldots,$ where

τ_L defined by Eq. (4.3), $\tau_L = k_B T/D$, represents the relaxation time of the X motion.

Q: The Xmotion appears here as quantized. Are there then polarization quanta?

M: No, you just have eigenvalues of the solution. In many fields of mathematics and physics you have eigenvalues, in some cases there is quantization, there are no polarization quanta.

18. Following: "We introduce the Laplace transform $\bar{p}(X; s)$ of $p(X; t)$ as

$$\bar{p}(X; s) = \int_0^\infty e^{-st} p(X; t) dt. \tag{5.10}$$

Since the Laplace transform of $\partial p(X; t)/\partial t$ is $s\bar{p}(X; s) - f(X)$, Eq. (5.4) yields

$$[s + H + k(X)]\, \bar{p}(X; s) = f(X). \tag{5.11}$$

It is convenient notationally to rewrite Eq. (5.11) in terms of *bra and ket vectors* such as $|\bar{p}(s)\rangle$, $|g\rangle$ and $|f\rangle$, whose coordinate representatives are $\bar{p}(X; s)$, $g(X)$, and $f(X)$. We thus have

$$(s + H + k)\, |\bar{p}(s)\rangle = |f\rangle \tag{5.11$'$}$$

Here, k denotes an operator diagonal in the coordinate representation $\langle X|k|X'\rangle = k(X)\delta(X - X')$, with the diagonal element $k(X)$, while H represents an operator given in the coordinate representation by Eq. (5.5)."

Q: What is the meaning of $\langle X|k|X'\rangle$ and what does the ket $|X\rangle$mean?

M: The symbol $\langle X|k|X'\rangle$ is kind of transition... k is diagonal in $|X\rangle$ and $|X'\rangle$. k is an operator, a diagonal operator. If you have an equation that looks like an eigenvalue equation... k should be an operator. $\langle X|k|X'\rangle$ is a definition of the operator k. There is no $k(X)$ on the left-hand side. You have the definition of an operator by its diagonal elements. If you have something which is like an eigenvalue equation, then you can define bras and kets, you can define vector solution, just as you write $H|\Psi\rangle = E|\Psi\rangle$. You can do something similar here. You write a system, $|X\rangle$ is a distribution corresponding to $|X\rangle$, $|X'\rangle$ is a distribution corresponding to $|X'\rangle$. They are not populations. They are just vectors used to solve an equation formally. The equation is (5.11′). You rewrite in a vector and an operator form. Compare Eqs. (5.11) and (5.11′). $f(X)$ you would write as $\langle X|f\rangle$. You have that Eq. (5.11′) gives rise to Eq. (5.11). $f(X)$ is the realization of the vector $|f\rangle$ in the $\langle X|$ representation.

19. Following: "The formal solution to Eq. (5.11′) is given by

$$|\bar{p}(s)\rangle = (s + H + k)^{-1}|f\rangle \tag{5.12}$$

where $(s + H + k)^{-1}$ represents the Laplace transform of the operator $\exp[-(H + k)t]$."
NOTE: Compare 18: We see that Eq. (5.11) is the representation in polarization space of Eq. (5.11′) written in terms of operators and kets.

20. Following: "...We introduce next an operator identity

$$\frac{1}{s + H + k} = \frac{1}{s + H} - \frac{1}{s + H}k\frac{1}{s + H + k} \tag{5.15}$$

If Eq. (5.15) is used iteratively an expansion of $(s + H + k)^{-1}$ with respect to k is obtained."

M. You see that formalism a lot... instead of k you may see a V. For example, the Born approximation in scattering amounts to just using the first term... in the second term of the Born approximation k would be missing in the denominator of the last term in Eq. (5.15). In the Born approximation, you make some truncation of some sort, and you do that by cutting out that last k that's in the denominator, perturbation k, and perturbation V is a Born approximation. So, in the Born approximation that last k in the denominator would be missing and now that's a truncation, you see, now you have the $\frac{1}{s+H}$, H being H_0, and the k you have in the upstairs, it's in the downstairs where there is the problem, that's the reason why I make the expansion. In books in scattering theory where here you see H, you see H_0 and where you see k you have a V. You see that sort of stuff quite a bit.

21. Following: "Since $(s + H)^{-1}$ describes the thermalization process of the distribution function $P(X; t)$ of X within the potential $V(X)$ *in the absence of reaction*, this expansion is advantageous in describing a case close to the slow reaction limit (1) in Sec. IV. On the other hand, that expansion is not suitable for describing the nondiffusing limit (4). In this limit, the reaction should occur with the rate $k(X)$ at each X before the thermalization within $V(X)$ occurs, and this situation can be described by the operator $(s + k)^{-1}$, that is, *the details of H play a smaller role in this case.*"

    Q: I believe I understand the physics behind the symbols. To the Laplace transform $(s + H + k)^{-1}$ corresponds the function $\exp[-(H + k)t]$, that is, the population described by the distribution function may decrease following the potential $V(X)$ or because of the chemical reaction. The operator $(s + H)^{-1}$ describes the decrease because of $V(X)$ while $(s+k)^{-1}$ describes the decrease due to chemical reaction. If this process is dominating the "details of H" play a smaller role and instead of H we have h, which is somehow a part of H. There appears also a difference $H - h$ in Eq. (5.16) (following).

    M: Very clever remark (that the details of H play a small role). Very clever. That remark is Sumi's. In anything that you see is very clever... just substitute the word Sumi.

Q: You should write once more the paper stressing the physics...

M: Why don't you do that? If you took the equation and... you take the Laplace transform you would get $s + H$, if you remove the chemical reaction. If in the equation you remove the diffusion term then you take the Laplace transform, you get the $s + k$.

22. p. 4899 1st column bottom: "...For this reason we introduce a second operator identity

$$\frac{1}{s+H+k} = \frac{1}{s+h+k} - \frac{1}{s+H+k}(H-h)\frac{1}{s+h+k} \tag{5.16}$$

where h is at present an arbitrary *operator* diagonal in the coordinate representation with the diagonal element $h(X)$... *We consider, therefore, that h has a magnitude (norm) of the order of H, that is, of the order of the relaxation rate constant* τ_L^{-1} for the reorientational fluctuation of the solvent molecules."

p. 4898 2nd column middle: "When the potential $V(X)$ is given in the form $V(X) = \frac{1}{2}X^2$, the operator H is essentially the same as the Hamiltonian of a harmonic oscillator. In this case, it is easily shown that the nonvanishing eigenvalues of H are equal to n/τ_L, with $n = 1, 2, 3, \ldots$, where $\tau_L = k_B T/D$, Eq. (4.3) represents the relaxation of the X motion."

Q: Can we express this term as $\langle X|\, h \,|X'\rangle = h(X)\delta(X - X')$? And should we interpret it in the sense that there are no "transitions" between different instantaneous states of the polarization function for the entire system?

M: That's the relaxation part, yes. When you just have the diffusion term, then the equation looks like that of a harmonic oscillator, the eigenvalues are n times some constant. $\langle X|\, h \,|X'\rangle = h(X)\delta(X - X')$ is certainly true. The polarization function is a localized term, and so I guess that in the X representation it would be diagonal. It is a diagonal operator, it depends only on X. One has the mapping of a complicated polarization onto a simple thing, you see. I would assume that $|X\rangle$ represent the whole state of polarizations, and so it doesn't interact with another state of polarization, it interacts with the charges, I assume. The different states can interact with charges but they can't interact with each other. I think so. When you talk with Sumi you can ask him. This is my first impression.

23. Following: "...Both the *situation* described by Eq. (5.15) and that described by Eq. (5.16) should be incorporated in the approximation we are seeking."

    Q: The left side members of Eqs. (5.15) and (5.16) are the same, that is, apparently the physical process described by both equations is the same but the right sides are different, that is, there are two different descriptions of the same physical process. Can you describe them?

    M: Well, I imagine that he describes the right-hand side of Eq. (5.15) in some approximate way so that Eq. (5.16) is the beginning of that, no approximations are there but it's the beginning of something.

24. Following: "We do so by introducing the right-hand side of Eq. (5.16) into $(s+H+k)^{-1}$ on the right-hand side of Eq. (5.15), and obtain an operator identity

$$\frac{1}{s+H+k} = \frac{1}{s+H}\frac{s+h}{s+h+k} + \frac{1}{s+H}k\frac{1}{s+H+k}(H-h)\frac{1}{s+h+k}. \tag{5.17}$$

    In this identity H is, in effect, replaced by h in the first term."

Q: This is true, I believe, if one brings out the common factor $\frac{1}{s+H}$, that is, if we write

$$\frac{1}{s+H+k} = \frac{1}{s+H}\left(\frac{s+h}{s+h+k} + k\frac{1}{s+H+k}(H-h)\frac{1}{s+h+k}\right)$$

so that the first term in parenthesis becomes $\frac{s+h}{s+h+k}$.

M: Yes. Right.

25. Following: "...Equations (5.17) and (5.18) given below provide the starting point of the present development. We introduce the following *decoupling* approximation:

$$\frac{1}{s+H}k\frac{1}{s+H+k} \to k_e^{-1}\frac{1}{s+H}k\,|g\rangle\,\langle g|\,k\frac{1}{s+H+k} \tag{5.18}$$

where k_e denotes the thermal equilibrium rate constant given by Eq. (4.6)

$$k_e = \int k(X)e^{-V(X)/k_BT}dX \bigg/ \int e^{-V(X)/k_BT}dX$$
$$\equiv \nu \exp\left(-\Delta G^* k_B T\right) \tag{4.6}$$

Q: Please check my derivation of Eq. (5.18). I first get from the operators product in the left side of Eq. (5.18) the following product of corresponding matrix elements:

$$\langle g|\,\frac{1}{s+H}\,|g\rangle\,\langle g|\,k\,|g\rangle\,\langle g|\,\frac{1}{s+H+k}\,|g\rangle.$$

Now I multiply numerator and denominator by $\langle g|\, k\, |g\rangle$:

$$\frac{\langle g|\, k\, |g\rangle\, \langle g|\, \frac{1}{s+H}\, |g\rangle\, \langle g|\, k\, |g\rangle\, \langle g|\, \frac{1}{s+H+k}\, |g\rangle}{\langle g|\, k\, |g\rangle}.$$

Considering now that $|g\rangle = \langle g|$ because $|g\rangle = e^{-\frac{1}{2}V(X)/k_BT} / \left[\int e^{-V(X)/k_BT} dX\right]^{1/2}$ and giving then the symbol $|g\rangle\, \langle g|$ the same integral meaning as to $\langle g|\, |g\rangle$, which is evidently $=1$, I put $|g\rangle\, \langle g| = 1$ in place of the first and second $|g\rangle\, \langle g|$ in the numerator of the above fraction. I consider, moreover, that $\langle g|\, k\, |g\rangle = k_e$ (Eq. 5.19) so that:

$$\frac{\langle g|\, k \frac{1}{s+H} k\, |g\rangle\, \langle g|\, \frac{1}{s+H+k}\, |g\rangle}{k_e}$$

Inverting now the order of the first k and of $\frac{1}{s+H}$ in the numerator of the above fraction and inverting moreover the second k with $|g\rangle\, \langle g|$ one finally gets

$$k_e^{-1}\, \langle g|\, \frac{1}{s+H} k\, |g\rangle\, \langle g|\, k \frac{1}{s+H+k}\, |g\rangle$$

which is the matrix element of the operator to the right-hand side of Eq. (5.18). It looks OK.

M: (5.17) is an identity$\cdots$ That $|g\rangle$ is a solution of the H equation. Before that $|g\rangle$ you should put open bracket $\langle X|$, i.e., one should consider $\langle X|g\rangle$, you should write $\langle X|g\rangle = e^{-\frac{1}{2}V(X)/k_BT} / \left[\int e^{-V(X)/k_BT} dX\right]^{1/2}$. When you put that $|g\rangle$ inside, that means that you are neglecting all excited states. Because you have to use sort of the completeness thing, maybe by neglecting that, that is the decoupling approximation. You insert $|g\rangle\, \langle g|$ to have an infinite sum so the decoupling approximation may amount to neglecting the other terms in the sum. You replace something by its lowest perturbation value. You replace with the most common contribution to the wave function. Very clever. I think it was a great idea.... Often you make an approximation and an all lot of consequences come out of it. That's fine, it defines the term "decoupling approximation." You decouple that state from all the other excited states... part of the overall perturbation. Clever! That was really great! That is smart! That k you have there, regard as an operator OK? $\langle g| k |g\rangle = k_e$, all right... you could do in two steps... in your book make sure that you write $\langle X|g\rangle = \langle g|X\rangle$. If the functions are real then $\langle X|g\rangle$ and the $\langle g|X\rangle$ turn out to be equal, so it is the X representatives that are equal. And the $|g\rangle \langle g| = 1$ is the approximation that was done in the decoupling in a sense... it is really the sum of all those terms... you neglect all those terms and replace one with $|g\rangle \langle g|$. In other terms you introduce the completeness identity and then you replace all terms with $|g\rangle \langle g|$. I have a gut feeling that there is a faster way of getting the right-hand side of Eq. (5.18). Put the sum before the k on the left-hand side. Before the k you now have a $|g\rangle \langle g|$. Right? Good. Now before the $|g\rangle \langle g|$ insert $k_e^{-1}k_e$. Now you have $k_e^{-1}k_e$ in front of the $|g\rangle \langle g|$. Now, bring outside the k_e^{-1}. Bring out to the left. Now you have a k_e times a $|g\rangle$, right? Now you can write $k_e |g\rangle$ because k_e is the eigenvalue of k for $|g\rangle$. That's it, in two simple steps.

26. p. 4900 1st column middle: "*(1) The slow reaction limit.* The typical *magnitude of s* can be estimated as being *the decay rate.*"

    Q: Is it so because

$$Q(t) = \int P(X; 0)e^{-k(T)t}dX \tag{4.16}$$

and

$$\bar{p}(X; s) = \int_0^\infty e^{-st} p(X; t)dt?$$

M: I think maybe it is this, but I have to think through that... if you are transforming something, then typically large values of s correspond to small times for the process. You see, in the transformation if you have large values of s you are going to have small times. So, if you have large values of s, which means a fast rate constant. I think that is what it is there. Often when you are transforming something when you have large values of the Laplace variable you look at small times.

27. Following: "...from Eq. (5.4) one obtains

$$Q(s) \cong (s + k_e)^{-1}_{\dots}$$

Typo: The bar on Q is missing.

28. p. 4900 2nd column top: "**VI. DETERMINATION OF *h*....** Therefore, we consider that h can be obtained by comparison of $\langle g| (s + H + k)^{-1} |g\rangle$ with its approximate value $\bar{Q}(s)$ of Eq. (5.24)

$$\bar{Q}(s) = \frac{1}{s + sa(s)} \left\{ 1 + \left[a(s) \langle g| - \langle g| k \frac{1}{s+H} \right] \times \frac{s+h}{s+h+k} |f\rangle \right\} \tag{5.24}$$

obtained with $|f\rangle$ replaced by $|g\rangle$.

We expand $\langle g| (s + H + k)^{-1} |g\rangle$ in powers of s^{-1} at $s \approx \infty$, as

$$\langle g| (s + H + k)^{-1} |g\rangle = s^{-1} - s^{-2} k_e + s^{-3} \langle g| k^2 |g\rangle - s^{-4} \times \left[\langle g| k^3 |g\rangle + \langle g| kHk |g\rangle \right] + \cdots \text{further terms} \cdots \tag{6.1}$$

where Eq. (5.14b), $H |g\rangle = 0$ has been extensively used.

It can be shown that in the expansion of $\bar{Q}(s)$ of Eq. (5.24) with $|f\rangle$ replaced by $|g\rangle$...."

Q: Expansion (6.1) is obtained from the binomial expansion of $(s + H + k)^{-1}$, that is, considering the expansion

$$(x + y)^n = x^n + nx^{n-1}y + \frac{n(n-1)}{2!}x^{n-2}y^2 + \cdots$$

Putting $s = x$, $H + k = y$, and $n = -1$. But how to expand the complicated expression (5.24)?

M: $a(s) = k_e^{-1} \langle g| k \frac{1}{s+H} k |g\rangle$, Eq. (5.22). Just substitute $a(s)$ in Eq. (5.24) with this expression and then expand term by term. I am amazed that Sumi can go to the fourth power, that's amazing.

Q: He must have written pages and pages of algebra.

M: Well, paper is cheap, you know... You must be careful with the bookkeeping... now there may be some short cuts to doing that. There may be some little tricks. When you have ratios there are subtle ways of doing that... it may not be as brutal as it looks. I took his word for that, I sure I did. It is great that you go through all this thing, it really is. Fearless, fearless.

29. Following: "...the terms proportional to s^{-1}, s^{-2} and s^{-3} do not contain h and just coincide with the corresponding terms in Eq. (6.1). Equating the corresponding terms proportional to s^{-4}, one obtains a requirement that

$$\langle g| k^2 |g\rangle \langle g| kh |g\rangle - \langle g| k |g\rangle \langle g| k^2 h |g\rangle = 0 \qquad (6.2)$$

This requirement can be satisfied when $h(X)$, the coordinate representative of h, is chosen to be a constant independent of X, irrespective of the magnitude of the constant. That is,

$$h(X) = h, \text{ a constant.} \qquad (6.3)$$

The constant h can be determined by comparing the terms proportional to s^{-5}. After some manipulations, one obtains

$$h = \frac{1}{2}\frac{\langle g|\, k\, |g\rangle\, \langle g|\, k^2 Hk\, |g\rangle - \langle g|\, k^2\, |g\rangle\, \langle g|\, kHk\, |g\rangle}{\langle g|\, k\, |g\rangle\, \langle g|\, k^3\, |g\rangle - (\langle g|\, k^2\, |g\rangle)^2} \tag{6.4}$$

Since the numerator is proportional to k^4 and to H, while the denominator is proportional to k^4, the h given by Eq. (6.4) has in fact *a magnitude of the order of H* or, to be more exact, of the order of the lowest nonvanishing *eigenvalue of H*."

Q: If I understand it well, the way to determine h is the following: One compares the expansions of the exact $\bar{Q}(s)$ operator and of its approximation. The first three terms are the same for the exact and the approximate expansion and h does not appear in them. In the fourth term, h appears and in order for the fourth terms of the exact and the approximate expansions to be equal a condition is to be imposed on h , that of its constancy. Imposing then the equality of the fifth terms in the expansions one finally has an explicit form for h.

M: If h is a constant, then the left-hand side of Eq. (6.2) is zero, OK... The eigenvalue is just a constant. He could have said at the beginning: suppose we introduce something which is a constant, and now go through all this stuff, then there is a value of that constant, he could have said that, I suppose, he didn't have to regard it as an eigenvalue. He could have said at the beginning: there is a constant which cuts off the series to some extent, and then say where is the value of that constant? That would be an alternative approach.

30. p. 4901 1st column bottom: "The quantity $[-dQ(t)/dt]dt$ equals the fraction of the reactant which is transformed into the product state during the time interval $(t, t + dt)$. Therefore, an average

survival time, the 'mean first passage time,' is usually defined by

$$\tau_a = \int_0^\infty t\,[-dQ(t)/dt]\,dt = \int_0^\infty Q(t)dt." \qquad (7.1)$$

Q: In Eq. (7.1) time appears as weighted by the time derivative of the survival probability. Intuitively, when time is small the time derivative of the survival probability is large and vice versa. Is it so? Or maybe you can give me a better physical meaning of the formula.

M: Yes, eventually that time derivative goes to zero. At the beginning your rate of change it is a maximum, like a first order reaction. Correct. The minus sign is there because you want a positive thing. dQ is decreasing all the time and so if you want a positive quantity...

Q: It is like elementary chemical kinetics.

M: Exactly.

31. p. 4901 2nd column top-middle: "The definition of τ_a given by Eq. (7.1) emphasizes the fast decaying part of $Q(t)$ .... Therefore we introduce an average survival time τ_b of the second kind by

$$\tau_b = \int_0^\infty tQ(t)dt \Big/ \int_0^\infty Q(t)dt," \qquad (7.2)$$

Q: I realize that the integral in Eq. (7.1) is sensitive to the rate of change of $Q(t)$ in time, that is, near $t = 0$ the rate of change is higher. Why do you divide by $\int_0^\infty Q(t)dt$ in Eq. (7.2)? Do you do so for τ_b to have dimension of time and not of time squared?

M: Yes, it is defined to be a time, so you have to define it that way. It is defined to be a time and not a time squared. The idea is that in this definition you are using the actual value of t as a weighting function to do the averaging. It is a common way to define another time-like variable. It is like... the first is a zero moment, the second one is a first moment... you define a second moment with $t^2$and so on... normally when you define moments they give you different aspects of a problem...

32. p. 4901 2nd column bottom: "Thus, the two kinds of average survival times can be expressed in terms of $\bar{Q}(0)$ and $\bar{Q}'(0)$ as

$$\tau_a = \bar{Q}(0), \text{ and } \tau_b = -\bar{Q}'(0)/\bar{Q}(0). \tag{7.7}$$

Expanding $\bar{Q}(s)$ given by Eq. (5.24), one can obtain

$$\bar{Q}(0) = k_e^{-1} + k_e^{-2} \langle g| kH^{-1} \times [k\,|g\rangle \langle g| - k_e] \frac{h}{h+k} |f\rangle, \ldots \tag{7.8}$$

We consider first the calculation of the $\langle g| kH^{-1} |j\rangle \ldots$ with $\langle g|j\rangle = 0 \cdots$
It is first noted that

$$\langle g| kH^{-1} |j\rangle = \int_0^\infty \langle g| ke^{-Ht} |j\rangle \, dt\text{''} \tag{7.10}$$

Q: The definition of $g(X)$ is given on p. 4898 2nd column top:

$$g(X) = e^{-\frac{1}{2}V(X)/k_BT} \Big/ \left[\int e^{-V(X)/k_BT} dX\right]^{1/2} \tag{5.1}$$

in the form of a normalized Boltzmann distribution. You then write that it is also the ground state wave function of operator $H$with zero eigenvalue. What is $|j\rangle$?

M: You see here kH^{-1} and he doesn't convert it to some eigenvalue. My guess is, without knowing, because I forgot in the meantime what $|j\rangle$ was, that $|j\rangle$ is not an eigenfunction of H,

I guess. If it were, everything would be simple, first consider the calculation of that... so and so... if $|j\rangle$ is an eigenfunction of kH^{-1}, you can write down immediately what it is...

Q: What is it then this $|j\rangle$?

M: I guess it is that big term that is in Eq. (7.8) after the cross sign. That's quite possibly what it is... he is going to say: supposing that there is some $|f\rangle$ not in the eigenspace of $|g\rangle$ but it is in sort of a space... it is a component along $|g\rangle$ ... it is a general object, and then... let's consider the óperation on $|f\rangle$, then he may pick something for that $|f\rangle$ ... is $|f\rangle$ the unknown that he wants to solve for? $f(X)$ is essentially the unknown. He may not say now and says it later on... If it is an initial distribution, it does not have to be an eigenvalue of any simple operator... maybe you want to express the results in terms of an initial distribution... he wants to find a way of doing that... the solution of the problem... that is what he may be asking... and this is the way of doing it... it sounds that way to me. The $f(X)$ is what you stipulate to be the initial distribution... he has probably got a very good reason for doing it but I don't know what it is... Does he later find some $f(X)$, does he later obtain that? Now,$|j\rangle$ is not in the space of $|g\rangle$ ... why does he want to restrict to $\langle g|j\rangle = 0$ for the moment I don't know... but he maybe says later on why he wants to do that.

33. Following: "When $V(X)$ is given by Eq. (3.2), $V(X) = \frac{1}{2}X^2$, the operator

$$H = -D\frac{\partial^2}{\partial X^2} + \frac{D}{2k_BT}\left[\frac{1}{2k_BT}\left(\frac{dV}{dX}\right)^2 - \frac{d^2V}{dX^2}\right] \qquad (5.5)$$

given by Eq. (5.5) with this $V(X)$ is essentially the same as the Hamiltonian of a harmonic oscillator. For the harmonic oscillator, the matrix element, $\langle X| e^{-Ht} |Y\rangle$, of e^{-Ht} in the coordinate representation is well known (Kubo and Toyozawa). In the present case, we have

$$\langle X| e^{-Ht} |Y\rangle = \left[\left(1 - e^{-2t/\tau_L}\right) 2\pi k_B T\right]^{-1/2} \times \exp\left\{-\frac{1}{8k_B T} \times \left[(X+Y)^2 \tanh\left(\frac{t}{2\tau_L}\right) + (X-Y)^2 \times \coth\left(\frac{t}{2\tau_L}\right)\right]\right\} \quad (7.12)$$

Q: If I represent the reacting system by $\left|A_s^+ - B_s^-\right\rangle$, say, and the product system by $|A_s - B_s\rangle$, does $|X\rangle$ represent the set of all possible solvent *polarizations* corresponding to the set of all possible *solvent configurations* for the systems $\left|A_s^+ - B_s^-\right\rangle$ and $|A_s - B_s\rangle$ so that $|X(t)\rangle$ represents polarization X corresponding to a *particular local solvent configuration* at time t?

M: $|X\rangle$ and $|Y\rangle$ are eigenfunctions of a polarization operator that he may not have written down. $|X\rangle$ is supposed to represent *a state of polarization*... $|X\rangle$ represents some complicated integral expression of polarization function all over the place... the corresponding operator is some abstract polarization operator. $\langle X|$ is a variable but is also a bra operator.

Q: Which one is the variable of $|X\rangle$?

M: That may not come up as something that is of interest. Just imagine that there is a polarization operator that has some eigenfunctions of it, just like a position operator. You may have the same solvent polarization for the systems $\left|A_s^+ - B_s^-\right\rangle$ and $|A_s - B_s\rangle$ when out of equilibrium. The $|X\rangle$ is supposed *to take care of the solvent coordinates too.* $|X\rangle$ is the result of some arrangement of the coordinates.

34. p. 4902 2nd column bottom: "In this case of single exponential decay, the distribution function $P(X; t)$ for *the solvent configuration X* at time t . . . ."

    Q: Is "the solvent configuration X" a shortened form for "the solvent configuration of polarization component X"?

    M: Polarization is a macroscopic quantity and you have many solvent configurations. It would be better to call them ensemble of solvent configurations. Imagine you have a description of some distribution of polarization all over the system. You have an infinite number of polarizations, you have an infinite number of X's. It is just a symbol.
35. p. 4903 2nd column top: ". . .in the large τ_L limit k_r does not include ν and, thereby, does not include any nonadiabatic factor. . ."

    Q: Please provide a physical explanation. . . .

    M: The slow component that comes in is related to a solvent relaxation. Suppose you had a simple situation, you had a kind of diffusion of reactants. . . in that case the reciprocal of the rate constant is the reciprocal of the activation rate constant plus the reciprocal of the diffusion rate constant. In that way then the slowest process determines the outcome and the other doesn't even enter, so that would be the case here. The slowest time is τ_L, so the other term is there but it just doesn't show up. The nonadiabatic factor is bound to the actual reaction step and not to the diffusion part. Now, if one wanted to complicate things, one would look what interplay is there in diffusion and nonadiabatic effects because in the crossing there is also recrossing and so on, so one could get a lot more complicated, so what I have described describes the simplest view.
36. p. 4904 1st column top: "The rate constant obtained above for the condition of $\nu\tau_L >> \left[\Delta G^*/(\pi k_B T)\right]^{1/2}$ just coincides with

the rate obtainable by Kramers' method for the situation that the potential curve for the reaction has a cusp (as in Fig. 1) at the transition state configuration $X = X_c$ but such that the slope of the surface there is much steeper on the product's side than on the reactant's side."

M: Remember that there is a difference of two slopes in the Landau–Zener expression, there are two slopes there, and now if one slope is infinitely steep there would seem to be a problem... When you have a slope infinitely steep you wouldn't use the Kramerian approach... you have to derive from the beginning, which I don't think would be a problem but that's what one has to do because in fact the slope is infinite. If the slope becomes zero then you would see how Kramers would come in with one slope... It may be that statement is not quite right and what was meant was that once you get there, there was no recrossing after that, and the way one would say that is that one sets up the Kramers' equation and to recross the region only put the Landau–Zener in, and then forget about that steep slope and then see what one gets... I'll do it... in fact I may do as part of the class I'm teaching. Just looking at that limit... which one is the rate constant obtained for that condition? Which one is that? I'll check it out.

When you have a slope infinitely steep we wouldn't use Kramer's approach, you have to derive from the beginning... but since I haven't looked at the expression, I don't know. Have you looked at the expression to let the final slope become infinite? I remember that he has a cusp. If you had an infinite slope there would be no sort of back and forth, no diffusion across the cusp, so one of his basic ideas wouldn't apply. See, in one of his papers he has a slow diffusion across that cusp, but then slow diffusion means it goes back and forth, but if you have a slope that is an infinite drop off, the system can't go back and forth.

37. p. 4904 2nd column middle: "Explicit calculations were made for the case that the initial distribution $g(X)f(X)$ of *the coordinate X of the orientation of the solvent molecules...*"

    Q: This is a third definition of X in the paper until now. On p. 4895 2nd column top-middle we read: "a scalar variable X proportional to a certain component of the polarization vector which passes through the two minimum energy points associated with reactants and products in the polarization coordinate space"; on p. 4902 2nd column bottom we have "the *solvent configuration X*" and now we have the above. Isn't there some confusion in the paper between *variable*—the solvent orientation coordinate—and the *function*, I mean the polarization component?

    M: X is describing *a state of orientation*. Coordinate X is describing an ensemble of orientations of the water molecules of the entire system. There should really be only one definition in the paper. The polarization, that's like a macroscopic function, even if it is local it requires an ensemble description.
38. p. 4906 1st column top: "The multiexponential behavior above can be understood by considering that when τ_L is very large, and when at the same time the *reaction window is* not too narrow, the decay of $Q(t)$ is more or less composed of two stages. Firstly, in the fast stage the initial distribution of the coordinate X of the solvent orientation should acquire *a hole* with a size the same as that of the window of the reaction described by $k(X)$ in Eq. (4.1)

$$k(X) = \nu_q \exp\left[-\Delta G^*(X)/k_B T\right] \tag{4.1}$$

    with a rate much faster than the relaxation rate τ_L^{-1} of the solvent fluctuation. Then, in the slow stage *the widening of the hole, in which the two walls of the hole collapse and disappear into the whole successively*, should occur with a rate constant of the order of τ_L^{-1}."

Q: Can you please explain these holes?

M: If you start with a system at a certain time, then, in this two-dimensional model, you have reaction coming from a certain distribution of solvent configurations X ... the idea is that solvent is so slow that the system doesn't go all over that intersection line along the solvent coordinate so it jumps up along the fast coordinate part of the way, if you wouldn't have that fast coordinate you would only have to go along the X axis, but that means reaching a higher barrier, so it reduces the barrier by making use of that fast coordinate jumping as fast as it can, so it takes time to evolve along that X coordinate, so the first portion that is going to go in... and jumps up—but of course that involves bigger activation energy—maybe it first jumps where there is a compromise, between having a higher activation energy and not having to diffuse so far along the solvent coordinate... so that's the first thing that happens, but then you exhaust your population there, so the ensemble systems are forced to go further and that takes longer and then jump up from there... a different relaxation time and so on. I have to think this thing more, Francesco.... That means that when you react from that you create a hole, a vacancy there, and at that vacancy you have a certain rate, that $k(X)$, at that X you have sort of a hole, and then gradually you are losing stuff from other values of X, in other words you're creating holes at other values of X, so you get another $k(X)$, in other words as the time goes on $k(X)$ is changing because X is slowly changing as you empty one X and go over to another. Hole means vacancy, that there is no longer a system there. I don't think you will find that discussed in the paper, but that will be the physical basis.

39. p. 4910 top: "In the coordinate representation, the (W, Y) matrix element of the left-hand side of Eq. (5.18)

$$\frac{1}{s+H} k \frac{1}{s+H+k} \rightarrow k_e^{-1} \frac{1}{s+H} k \left|g\right\rangle \left\langle g\right| k \frac{1}{s+H+k} \tag{5.18}$$

is $\int k(X)A(W,X)B(X,Y)dX$, where $A(W,X)$ and $B(X,Y)$ denote the (W,X) and (X,Y) matrix elements of the operator $(s+H)^{-1}$ and $(s+H+k)^{-1}$, respectively."

Q: Does this mean that the above integral is

$$\int k(X) \langle W| (s+H)^{-1} |X\rangle \langle X| (s+H+k)^{-1} |Y\rangle \, dX?$$

M: Do the following: Do you have a pencil handy? All right: At the left-hand side of Eq. (5.18) put a $\langle W|$ and $|Y\rangle$ on the right-hand side. Now, in between there, where the k is, put an $|X\rangle \langle X|$. And you integrate over dX. Then you have $\langle W| \frac{1}{s+H} |X\rangle \langle X| k \frac{1}{s+H+k} |Y\rangle$. That's really what you wrote down in that integral. That $k(X)$ is really $|X\rangle k \langle X|$. What you wrote down is OK.

40. Following: "We define an average of an arbitrary function $C(X)$, say $\langle C(X)\rangle_{av}$, by

$$\langle C(X)\rangle_{av} = \int g(X)^2 k(X) C(X) dX \Big/ \int g(X)^2 k(X) dX. \tag{A.1}$$

Then, the left-hand side of Eq. (5.18; see above) is the same as

$$k_e \langle [A(W,X)/g(X)] \cdot [B(X,Y)/g(X)] \rangle_{av} ."$$

M: Multiply the inside of the integral $\int k(X)A(W,X)B(X,Y)\, dX$ by $g(X)^2$ and divide by $g(X)^2$. And remember that $k_e = \langle g| k |g\rangle \ldots$

41. p. 4910 1st column middle-bottom: "On each side of Eq. (5.18)

$$\frac{1}{s+H} k \frac{1}{s+H+k} \rightarrow k_e^{-1} \frac{1}{s+H} k |g\rangle \langle g| k \frac{1}{s+H+k} \tag{5.18}$$

one can multiply $(s+H)^{-1}$ by the identity operator $\sum_n |u_n\rangle\langle u_n|$, where the $|u_n\rangle$'s are eigenvectors of H. In the limit of small s, $(s+H)^{-1}\sum_n |u_n\rangle\langle u_n|$ tends to $s^{-1}|g\rangle\langle g|$."

p. 4910 2nd column bottom: "Since $(s+H)^{-1}$and $a(s) = k_e^{-1}\langle g| k\frac{1}{s+H}k|g\rangle$ approach, respectively, $s^{-1}|g\rangle\langle g|$ and $k_e s^{-1}$ in the neighborhood of the origin of s"

p. 4914 1st column bottom (Note 30): "...At sufficiently small s, $(s+H)^{-1}$ can be written as $s^{-1}|g\rangle\langle g|$, since the other eigenvalues of H are nonzero."

Q: Why $s^{-1}|g\rangle\langle g|$ and not $H^{-1}|g\rangle\langle g|$?

M: Well, let's see. One of the eigenfunctions is $|g\rangle$, right? Now, H operating on $|g\rangle$ has a zero eigenvector. $s^{-1}|g\rangle\langle g|$ is the first term of $(s+H)^{-1}\sum_n |u_n\rangle\langle u_n|$. When s is small... Why it is so for the other eigenvalues I have to think about it. What it may be is that for small s you have large times and the other eigenvalues don't contribute much and decrease more rapidly exponentially. I have to think that through.

See how simple things are Francesco?...

42. p. 4910 1st column bottom: "It is apparent that the decoupling Eq. (5.18) is exact under the condition that $\langle g|$ operates from the left on both sides of the equation."

M: If you look at Eq. (5.18), before k insert $k^{-1}k$. Now insert in that a $|g\rangle\langle g|$ between the k^{-1} and the $\frac{1}{s+H}$ and let the k^{-1} operating on the $|g\rangle$ there... $k^{-1}k_e$. Now it's a number, now you can pull it out to the left. And now you can get rid of the $|g\rangle\langle g|$ part.

43. p. 4911 2nd column top:" Then one obtains $\langle g| k|g\rangle\langle g| k^3|g\rangle/(\langle g| k^2|g\rangle)^2 = \cdots$ Eq. (C1). The expression in Eq. (C1) appears in the denominator on the right-hand side of Eq. (6.4) for h."

Q: Eq. (6.4) reads:

$$h = \frac{1}{2} \frac{\langle g|\, k\, |g\rangle \langle g|\, k^2 Hk\, |g\rangle - \langle g|\, k^2\, |g\rangle \langle g|\, kHk\, |g\rangle}{\langle g|\, k\, |g\rangle \langle g|\, k^3\, |g\rangle - (\langle g|\, k^2\, |g\rangle)^2} \tag{6.4}$$

that is, the (−) in the denominator of Eq. (6.4) has been misprinted in the lhs of Eq. (C1) as (/). There are two more misprints of this kind in the following. On column 2, p. 4911 bottom the nominator in Eq. (6.4) is also written with a slash instead of a minus and a slash is also arbitrarily inserted in $\langle g|\, kHk\, |g\rangle / \langle g|\, k^2\, |g\rangle$ which is, without the slash, the second term in the nominator of Eq. (6.4).

M: Yes, sure, that certainly should be a minus... yes, absolutely... Let me see... If k were a constant operator... then the difference in the denominator reduces to 0, OK? It could be that if you divide numerator and denominator of Eq. (6.4) by $\left(\langle g|\, k^2\, |g\rangle\right)^2$ this quotient would appear. I'll tell you why I'm not sure. Supposing k were a constant operator, and I don't know if that is the case Sumi considered... then the denominator there would be zero. But I don't see how you could get zero on the right-hand side of Eq. (C1). If you divide nominator and denominator then you can focus on the ratios, and looking at this expression here one could get unity... I don't know if it is possible... Sumi could have also said: divide numerator and denominator by the second term in the denominator... I just don't see the zero coming in in there. I don't know if it is OK... without looking at it... I just don't know... I think the right thing to do should be to derive that formula... it may not be so ugly... Again in C4 there is a ratio there...

That looks intended to make something dimensionless... maybe Eq. (13.41) is easy to do... I suspect what he did, he did for dimensional reasons... to get something dimensionless.

Try to compute Eq. (C5). I think there is a real merit in taking ratios... there may be lot of cancellations.

44. p. 4912 2nd column middle-bottom: "Taking the limit of $\lambda_i/\lambda_0 \to 0$ in Eq. (B5) for $a(s)$, we obtain

$$a(s) = k_e \int_0^\infty e^{-st}(1 - e^{-2t/\tau_L})^{-\frac{1}{2}} \times \exp\left(\frac{2e^{-t/\tau_L}}{1 + e^{-t/\tau_L}} \frac{\Delta G^*}{k_B T}\right) dt. \tag{E4}$$

When $\Delta G^*/k_B T \ll 1$, Eq. (E4) simplifies considerably. Now, we need to expand only $\left(1 - e^{-2t/\tau_L}\right)^{-12}$ in a power series in e^{-2t/τ_L}, obtaining the following Eq. (E5) when Res > 0:

$$a(s) = k_e \sum_{n=0}^{\infty} \frac{(2n)!}{[(2n)!!]^2} \frac{1}{s + 2n/\tau_L} \quad \text{for } \Delta G^*/k_B T \ll 1\text{"}$$

Q: What does it mean to expand $\left(1 - e^{-2t/\tau_L}\right)^{-1/2}$ in a power series of e^{-2t/τ_L}? Does it mean that one first expands $\left(1 - e^{-2t/\tau_L}\right)^{-1/2}$ in a binomial series like $(1 - x)^{-1/2} = 1 + \frac{1}{2}x + \cdots$, then one expands the exponential $x = e^{-2t\tau_L}$ as the usual exponential expansion and finally one integrates the result as in Eq. (E4)?

M: Absolutely, you don't expand the exponential. Just expand in the power series in that exponential. Just expand in powers of x.

Q: Just the binomial series.

M: That's right. If you expand the exponential, you have other exponentials to expand. Instead you can introduce a change of variable $x = e^{-2t/\tau_L}$ and then carry it over to integration. No further expansions unless you want to be like the Titanic.

45. p. 4914 2nd column top (Note 34): "Considering that k can be scaled by k_e, and both H and h by τ_L^{-1}"

    Q: What does the above "scaling" mean mathematically and physically? The first idea of scaling that comes to mind is that of a linear transformation in geometry that enlarges or diminishes objects. Here do you change k by k/k_e or by kk_e or by $k - k_e$?

    M: Scaled in other words means k/k_e. Often when you scale you get rid of some units.

46. Following: "we see that $\bar{Q}(0)k_e$ is composed of $(\tau_L k_e^2)^{-1} \langle g| (k-k_e) H^{-1} k |g\rangle$, $\langle g| h/(h+k) |g\rangle$, and $(\tau_L k_e^2)^{-1} \langle g| (k-k_e) H^{-1} h/(h+k) |g\rangle$, where $\langle g| k$ in Eq. (7.8) was replaced by $\langle g| (k - k_e)$, anticipating the procedure shown in Eq. (7.14)."

    Q: On p. 4901 there is Eq. (7.8) for $\bar{Q}(0)$

$$\bar{Q}(0) = k_e^{-1} + k_e^{-2} \langle g| k H^{-1} \times [k |g\rangle \langle g| - k_e] \frac{h}{h+k} |f\rangle \tag{7.8}$$

    Substituting $|g\rangle$ for $|f\rangle$ on the right and replacing $\langle g| k$ with $\langle g| (k - k_e)$ one gets

$$k_e^{-1} + k_e^{-2} \langle g| (k - k_e) H^{-1} \times [k |g\rangle \langle g| - k_e] \frac{h}{h+k} |g\rangle$$

$$\Rightarrow k_e^{-1} + k_e^{-2} \langle g| (k - k_e) H^{-1} k |g\rangle \langle g| \frac{h}{h+k} |g\rangle - k_e^{-1}$$

$$\langle g| (k - k_e) H^{-1} \frac{h}{h+k} |g\rangle$$

    that is, one finds the terms mentioned earlier in the paper. But what about the k_e's and the τ_L's?

    M: Remember k/k_e.

47. Following: "These three quantities can be calculated by performing the t, X, and Yintegrations, appearing in Eqs. (7.10)

and (7.14)

$$\langle g|\, kH^{-1}\, |j\rangle = \int_0^\infty \langle g|\, ke^{-Ht}\, |j\rangle\, dt \tag{7.10}$$

$$\langle g|\, ke^{-Ht}\, |j\rangle = \iint g(X)\,[k(X) - k_e]\,\langle X|\, e^{-Ht}\, |Y\rangle\, j(Y)dXdY \tag{7.14}$$

For these integrations, where we use Eq. (3.2) for $V(X)$, Eqs. (3.6) and (4.1) for $k(X)$, and Eq. (5.1) for $g(X)$,

$$V(X) = \frac{1}{2}X^2 \tag{3.2}$$

$$\Delta G^*(X) = \frac{1}{2}(X - X_c)^2(\lambda_0/\lambda_i) \tag{3.6}$$

$$k(X) = \nu_q \exp\left[-\Delta G^*(X)/k_BT\right] \tag{4.1}$$

$$g(X) = e^{-\frac{1}{2}V(X)/k_BT} \Big/ \left[\int e^{-V(X)/k_BT}dX\right]^{\frac{1}{2}} \tag{5.1}$$

we can change the t, X, and Y variables, respectively, to scaled ones t/τ_L, $X/(k_BT)^{\frac{1}{2}}$, and $Y/(k_BT)^{\frac{1}{2}}$ all of which are dimensionless. Then, the first of the three quantities mentioned above depends only on $X_c/(k_BT)^{\frac{1}{2}}$ and λ_i/λ_0 *which characterize the shift of the origin and the curvature of* $k(X)$ *relative to* $V(X)$"

M: Does $k(X)$ depend on $V(X)$? You see, if we scale X, then you scale X_c, and I don't know if that X_c is just the origin of X . . . I just don't know what he (Sumi) means as far as the origin. As far as the curvature: if you look at that $\Delta G^*(X)$ . . . the λ_i/λ_0 is like a force constant in a sense, you know. . . it is something like that. That force constant is what he means by curvature

M213. Dielectric Relaxation and Intramolecular Electron Transfer

NOTES

1. p. 4273 2nd column top: "In the present treatment we will assume, as in the previous two-state case, that the vibrational motion within the solute molecule adjusts itself so rapidly to the instantaneous orientations of the surrounding solvent molecules that *an equilibrium for the intramolecular vibrational modes* is always maintained during the process"

   M: Just think of thermal equilibrium. The intramolecular vibrational modes have a Boltzmann distribution. If by any chance the orientation polarization of the solvent affects those vibrations then they will be in equilibrium with that given orientation polarization of the solvent. It is an *equilibrium with the bath* surrounding at a certain temperature, *only the polarization is not in equilibrium.* Inside the single molecule we have all the possible vibrational modes according to the Boltzmann distribution.
2. Following: "In the two-state type of system the intramolecular vibrational motions can trigger ET"

   M: Think of that line in that space in Fig. 1 that constitutes the transition state, of the transition state and all directions in that space system. All right, when there is a component of that direction along the vibration, then that vibration is helping you cross the transition state. And typically there is a component. It's only when the motion across the transition state is parallel to the X axis that there isn't. *If one looks at Fig. 1 one sees that one can approach the transition state line at different directions. The vibrational coordinates contribute is proportional to the q component along that direction.*
3. Following: "This process of orientational relaxation of the polar solvent occurring after photoexcitation of the solute molecule can be regarded as diffusion of the polarization vector under the influence of a potential."

Q: To the molecules diffusing in real space is there a polarization vector associated that "diffuses," so to say, in polarization coordinates space?

M: That's right, that is exemplified by that equation (*Kramers's*) involving $X$where X is symbolic in a sense. It is symbolic of a nonequilibrium polarization in that equation, much more complicated. *What diffuses is the orientation of the molecules, it is an orientational diffusion, rather than a translational diffusion.*

4. Following: "Within the Debye model, in which the free energy for polarization fluctuations is quadratic in polarization components, we can select, for notational brevity, a scalar variable X proportional to a certain component of the polarization vector which passes through the two minimum energy points associated with the two free energy surfaces (the ground state and the excited state) in the polarization coordinate space. Only this component X is relevant to the process while *along the other two ones perpendicular to X the two free energy surfaces are 'parallel' only with a vertical shift.*"

   Q: The component of the polarization vector which passes through the two minimum energy points associated with the two free energy surfaces was already mentioned in **M211** p. 4896, see Fig. 1. There contours are shown referring to a three dimensional—and therefore graphically representable—free energy surface in which the free energy depends on the single polarization coordinate X and on the single vibrational coordinate q. The free energy surface you mention here is four dimensional at least with the free energy depending on three polarization coordinates. The surface is then more difficult to imagine. You mention two components of polarization perpendicular to X. Now, first, how do you chose two vectors perpendicular two each other and perpendicular to X? There is an ∞^1 number of ways of effecting this choice. Moreover, what does it mean that *along the other two ones perpendicular to X the two* *free energy surfaces are "parallel" only with a vertical shift*? Please explain... By the way: can the scalar variable X be considered a linear functional defined in polarization space?

M: I think it may be that it can be represented that way. Now let me treat your first question. Consider a different situation, consider a situation where you have a group of normal modes pretending that everything is harmonic for the reactants and for the products. Now it turns out that if the same force constants are used for the reactants and for the products you can select *a coordinate, the reaction coordinate, that is perpendicular to the hyperplane that constitutes the transition state.* You can treat one coordinate, a linear combination of those, I have done it, although I may have not written about that. And the Russians did it too. So, making a transformation with that coordinate that carries you along, that is perpendicular to that hyperplane, the reaction coordinate, and taking the average over all other coordinates, you don't ignore them, but you just take the average of them, so that means then that if you took your entire polarization of the entire system and Fourier decompose it, in some way or another, into components, one of the components was perpendicular to that hyperplane, somehow you put normal modes in for all of the components and you have just one component to carry you across. What it means is that if you take the two free energy surfaces and subtract them actually the difference of polarization would be parallel to that component, when you take a nonequilibrium expression. In other words... in this many dimensional space you have an infinite number of components. Because you are dealing with a nonequilibrium system where you are having three polarizations at every point. So,*imagine you discretize the polarization, you express it in terms of normal modes of polarization, of which there can be an infinite number, really. But you discretize, so you work with a finite number. In* *fact the Russians have done that in their polarization treatment, they use harmonic oscillators, they Fourier decompose.* So, if you think of these polarization components, discretized... think of normal modes, then *there will be one of those normal modes that is perpendicular to the intersection surface*, you can do that if you pretend everything is harmonic oscillators. So, you can pick one coordinate in that many-dimensional space which is the reaction coordinate, and *integrate all over the rest in a thermal fashion, and they don't affect the rate, they don't affect to carry you over. The X is that component of this big polarization field that is perpendicular to the intersection in that big polarization field. The only thing important is the longitudinal component, not the transverse. The transverse don't come in when you look at the details of the dielectric displacement interacting with the polarization. It turns out that it is the longitudinal component. This point of the two perpendicular components of Sumi doesn't come in at all. The perpendicular components don't interact with the dielectric displacement,* remember. *If you discretize, you can handle that infinity.* That (*problem of the two components*) is not in polarization space. Only the components along the line of centers contribute. Oh, I see what you mean... let me see... you can have polarizations parallel to the hyperplane. *He (Sumi) is not talking of longitudinal versus transverse. I don't know why he said two... because it is really many-dimensional. If he had said one perpendicular to X in that two-dimensional space... I don't see why he would say two. I would say $N-1$.* If you think of the space that we have there and you look at the surfaces, then... if you draw the surfaces, then along the line of centers those two surfaces will touch, along that line of centers, draw them fully, and so... *one would be vertically shifted from the other, and I don't know what parallel surfaces mean... I don't know why he uses the word "two."* I believe he should have used touching along that line... you can draw surfaces touching. In fact, if you think of it, again I don't know if I would have used that... he may have said: if you have the longitudinal one that counts, and if you come back now to transverse, the transverse components do not interact with the dielectric displacement. So, they are there and they wouldn't affect the intersection of one surface with the other, so that he may mean what he means by parallel. The point is that it would come into both equally, because they are not different, they are not interacting with the dielectric displacement, so they are not different in the two states, so I guess he really means transverse and they are not interacting with the dielectric displacement. They would have no effect on the energies... both together, but not separately. They don't contribute to the interaction, therefore they don't affect the ET, the only effect they have is to raise both surfaces together and they don't affect the intersection, that's maybe what he means.

5. Following: "The diffusion equation for the distribution function $P(X;t)$ present at coordinate X at time t is given by

$$\frac{\partial P}{\partial t} + D\frac{\partial^2 P}{\partial X^2} + \frac{D}{k_B T}\frac{\partial}{\partial X}\left[P\frac{dV_e(X)}{dX}\right], \qquad (5)$$

... $V_e(X)$ represents the excited-state free energy potential for polarization fluctuations. The corresponding free energy function for the ground state is written as $V_g(X)$. Within the Debye model, both $V_g(X)$ and $V_e(X)$ are quadratic functions of X

$$V_g(X) = \frac{1}{2}X^2 \qquad (6)$$
$$V_e(X) = \frac{1}{2}(X - X_e)^2 + \Delta G^0,$$

Where X_e represents *the relaxed value* of the polarization component X appropriate for the excited state of the solute molecule,"

Q: Is the "relaxed value" the minimum of the excited state free energy surface?

M: Probably. Relaxed value is minimum of the excited state.

6. following: "The emission spectrum $K(E;t)$ from the excited state at time t is given by

$$K(E;t) = \int P(X;t)\delta\left[E - V_e(X) + V_g(X)\right]dX \qquad (7)$$

upon using the Franck–Condon principle."

Q: Does $P(X;t)$have the dimension of polarization^{-1}? I believe so because the product of $P(X;t)$ and of *dX* is a probability.

M: Equation (7) is a statement of energy conservation. That at that point X you have a difference E of energy, that would be the Franck–Condon principle. At a certain X there is a certain frequency given by that delta function. And at that X there is a certain probability, a survival probability at time t. So, if you want to calculate the emission spectrum then you find out which is the probability that at time t there is that X. That presuming that the emission all along there is independent of everything else. For example, that there is no transition dipole which is changing along X. $P(X;t)$ is a pure number divided by X. The dimension is the reciprocal of X whatever the dimensions of $X$are. Yes you are right.

7. Following: "When the emission spectrum does not have a large asymmetry, its peak energy can be approximated by its average energy as

$$E_p(t) = \int EK(E;t)dE\text{"} \qquad (8)$$

Q: Here in order for the dimensions on both sides to be equal the dimension of $K(E;t)$ should be that of energy^{-1} ...

M: Yes, that's correct and you can see that from Eq. (7). The units of a delta function are 1/energy. Because the delta function times *dE* is going to be unity.

Q: I am reading you papers in a rather careful way.

M: Unique.

8. Following: "Then, introducing Eqs. (6) and (7), interchanging the order of integration, and using the normalization $\int P(X;t)dX = 1$, one obtains

$$E_p(t) = E_p(\infty) - X_e\bar{X}(t) \tag{9}$$

with

$$\bar{X}(t) = \int (X - X_e)P(X;t)dX, \tag{10}$$

where $E_p(\infty) = \Delta G^0 - \frac{1}{2}X_e^2$."

Q: The peak of the spectrum is found in correspondence of emission from the excited state with X_e polarization because the population of systems with such a polarization is the highest. In the excited state, the polarization X relaxes in time toward the value X_e thereby increasing the population of systems with this polarization. That is why the maximal value of $E_p(t)$ is reached for $t \to \infty$. This explains Eq. (9). I have made a figure that I include to graphically explain $E_p(\infty) = \Delta G^0 - \frac{1}{2}X_e^2$.

M: That would be at long time but at shorter it is the emission spectrum from X. The peak of the spectrum is for X. The peak is changing with time. *In time X relaxes to X_e and correspondingly the frequency relaxes from that corresponding to X to that corresponding to X_e* The emission energy will relax in time. The maximum is shifting as you go along. It is the same wave packet just shifting going down and so the intensity may be the same, it depends on how much radiation is lost in going down. Equation (9) shows how the peak energy changes with time. It doesn't show how the amount of the emission changes with time, just the energy. That's a frequency, it doesn't say anything about the intensity at that frequency. Instead of E it would have been clearer if we had written ν. Imagine you are on the excited curve just above the minimum on the excited curve, all right? The energy difference decreases as you go toward the minimum of the excited state. So, the energy is decreasing. What you got there in the figure is exactly right.

9. p. 4275 1st column top-middle: "In the solute–solvent cluster model it will be supposed that each neighboring solvent molecule near the solute molecule is in a *polymeric cluster*"

   Q: What is it a polymeric cluster?

   M: I think he just means a collection of molecules. He just meant a group of molecules.

10. p. 4275 1st column bottom: "Again, attention is focused in the present paper on the slowest relaxation. At very short times there will be smaller spectral shifts in the luminescence, with other time constants, due to the *other relaxation times.*"

    Q: For instance? Can you explain and expand somewhat on that?

    M: If you look at something like water, there is not just one relaxation time... in the dielectric dispersion, it is not just simply a Debye system, there is an absorption peak where the dielectric constant is about 12.5, due to a vibrational motion, probably a bending motion, so you have different types of motions and they have different relaxation times associated with them, and each of them is part of the polarization.

11. p. 4275 2nd column middle: "There is also the question of the significance of agreement or disagreement of *factors of* 2 in the rate constant when continuum models are used"

    M: I was wondering just whether he was thinking of something... I don't really know what he means... I do know that when you use the Kramers' approach... when diffusion is limited you get a factor of 2. But the correct treatment, if you didn't use that method, would be just let it diffuse over the barrier and then look up at what the probability is to end up to the infinite. When you do that, you get an answer that is different by a factor of 2 from the case when you say: OK, at the top of the barrier and then we have to reach a decision whether it is reacting or not... it doesn't permit any recrossing of the surface, and that error is an error of a factor of 2.

Q: This book would have never been possible without your help.

M: I hope it will be possible with my help. You ask a lot of very good questions. Michel-Beyerle remembers that what you wrote about reorganization was very well written and very clear, she enjoyed reading that.

M214. Solvent Dynamics and Vibrational Effects in Electron Transfer Reactions

NOTES

1. p. 60 middle: "In treating the solvent dynamics it is convenient to introduce a *scaled* progress variable (coordinate) X describing these fluctuations along the most probable path in some 'polarization space'.... One can well imagine that a Lagrangian multiplier m which appeared in an earlier paper [**M17**] would also serve as this X."

   Q: (1) Why and how scaled? Moreover, (2) how about using m for X?

   M: (i) The scaling means not having something constant in front of X, like in $\frac{1}{2}X^2$. Instead in $\frac{kX^2}{2}$ the old X would be $\sqrt{k}X$. So, X is scaled, you see.
   (ii) When you write the free energy expressed in terms of m, when m =0 you have the reactants, when $m = -1$ you have the products, so *you may regard m as a progress variable,* with m determined by the equation involving the free energy of reaction. *You can regard either X or m as a progress variable.*
2. p. 61 middle: "Equation (1)

$$\frac{\partial P}{\partial t} = D\frac{\partial^2 P}{\partial X^2} + \frac{D\partial}{k_B T\partial X}\left(P\frac{dV}{dX}\right) - k(X)P \qquad (1)$$

has been written for the case of an intramolecular ET, or for the case of a bimolecular ET in which, to simplify the calculation, the reactants are regarded during the polarization fluctuation process as being held in some solvent cage. In the latter case one also considers the translational diffusion of the two reactants toward and away from each other, resulting in their entering and leaving the cage. The D in Eq. (1) refers, instead, to the solvent orientational relaxation 'diffusion constant."'

Q: How is the translational diffusion taken care of?

M: The reactants are at a fixed distance. The translational diffusion is not taken care in that Eq. (1), you have some equation preceding that, the reactants diffuse toward being near each other and diffuse away from being near each other, that's a separate equation. *This is a solvent kind of relaxation diffusion.*

3. p. 62 middle-bottom: "(2) the 'wide reaction window' limit ($\lambda_i/\lambda_0 >> 1$), where the 'width' of the 'reaction window' in X-space is larger than the thermal width of the distribution of X"

   Q: Please check if I understand it well. The thermal width is the one that there is if and when $\lambda_i = 0$. If $\lambda_i \neq 0$ ΔG^* lowers and the window widens.

   M: That capital X refers to the solvent coordinate. If you call the axes q and X, the q axis is the vibrational coordinate and the X is the solvent coordinate. If q is the fast coordinate, you start with a distribution along the Xcoordinate and you don't move along that coordinate, you only move vibrationally, namely along fluctuations of the vibrational coordinate, so for example in the diagram that you showed you have some distribution of X, you would not diffuse along X, you would have fluctuations along the perpendicular coordinate and that's the way you would cross the TS line. When you have a lot of motion on a lot of q you have a wide reaction window in terms of q, you can reach the transition state line at quite different values of q.

4. Following: "(4) the nondiffusing limit, where τ_L approaches a value so large that $Q(t)$ becomes independent of τ_L. In effect, reaction occurs without solvent orientational dipolar fluctuations."

   M: You get out of this solvent distribution, you have a continuous set of distributions... and you react for each of those. What it means is that because you have such a heterogeneous set of fluctuations, you can react from anyone of those. You have a reaction from this solvent polarization, from that solvent polarization, slightly different... from different places. You have to cross that line, you have a whole set of distributions each one would have its particular polarization, that's fixed because you are not diffusing along that coordinate, and you react from that one.

CPSIA information can be obtained
at www.ICGtesting.com
Printed in the USA
LVHW082013070420
652563LV00003B/3